Geological, Geoenvironmental, and Geotechnical Engineering 2018

Transportation Research Record: Journal of the Transportation Research Board Volume 2672, Issue 52

ISBN: 978-1-5108-8257-7

Printed from e-media with permission by:

Curran Associates, Inc.
57 Morehouse Lane
Red Hook, NY 12571

Some format issues inherent in the e-media version may also appear in this print version.

Published by Sage Publications USA

ISBN (Print) 978-1-5108-8257-7
ISBN 2018 Printed Set (All Issues) 978-1-5108-7735-1

Printed by Curran Associates, Inc. (2019)

Additional copies of this publication are available from:

Curran Associates, Inc.
57 Morehouse Lane
Red Hook, NY 12571 USA
Phone: 845-758-0400
Fax: 845-758-2633
Email: curran@proceedings.com
Web: www.proceedings.com

Volume 2672 Number 52 December 2018

TRR
JOURNAL OF THE TRANSPORTATION RESEARCH BOARD

Contents

Geological, Geoenvironmental, and Geotechnical Engineering

Articles

TRANSPORTATION RESEARCH BOARD 2019 EXECUTIVE COMMITTEE*

TRB

TRANSPORTATION RESEARCH BOARD

CORE SPONSORS

Transportation Departments of the 50 States and the District of Columbia

Federal Government

U.S. Department of Transportation
- Federal Aviation Administration (FAA)
- Federal Highway Administration
- Federal Motor Carrier Safety Administration (FMCSA)
- Federal Railroad Administration (FRA)
- Federal Transit Administration
- National Highway Traffic Safety Administration (NHTSA)
- Office of the Assistant Secretary of Transportation for Research and Technology (OST-R)

U.S. Air Force Civil Engineer Center (AFCEC)
U.S. Coast Guard (USCG)
U.S. Department of Energy
U.S. Department of the Interior (DOI)
U.S. Environmental Protection Agency (EPA)

Nongovernmental Organizations

American Public Transportation Association (APTA)
Association of American Railroads (AAR)
California Air Resources Board

The National Academies of

SCIENCES · ENGINEERING · MEDICINE

The **National Academy of Sciences** was established in 1863 by an Act of Congress, signed by President Lincoln, as a private, nongovernmental institution to advise the nation on issues related to science and technology. Members are elected by their peers for outstanding contributions to research. Dr. Marcia McNutt is president.

The **National Academy of Engineering** was established in 1964 under the charter of the National Academy of Sciences to bring the practices of engineering to advising the nation. Members are elected by their peers for extraordinary contributions to engineering. Dr. C. D. Mote, Jr., is president.

The **National Academy of Medicine** (formerly the Institute of Medicine) was established in 1970 under the charter of the National Academy of Sciences to advise the nation on medical and health issues. Members are elected by their peers for distinguished contributions to medicine and health. Dr. Victor J. Dzau is president.

The three Academies work together as the National Academies of Sciences, Engineering, and Medicine to provide independent, objective analysis and advice to the nation and conduct other activities to solve complex problems and inform public policy decisions. The Academies also encourage education and research, recognize outstanding contributions to knowledge, and increase public understanding in matters of science, engineering, and medicine.

Learn more about the National Academies of Sciences, Engineering, and Medicine at **www.national-academies.org.**

The **Transportation Research Board** is one of seven major programs of the National Academies of Sciences, Engineering, and Medicine. The mission of the Transportation Research Board is to increase the benefits that transportation contributes to society by providing leadership in transportation innovation and progress through research and information exchange, conducted within a setting that is objective, interdisciplinary, and multimodal. The Board's varied activities annually engage about 7,000 engineers, scientists, and other transportation researchers and practitioners from the public and private sectors and academia, all of whom contribute their expertise in the public interest. The program is supported by state transportation departments, federal agencies including the component administrations of the U.S. Department of Transportation, and other organizations and individuals interested in the development of transportation.

Learn more about the Transportation Research Board at **www.TRB.org**.

TRR
JOURNAL OF THE TRANSPORTATION RESEARCH BOARD

Transportation Research Record
2018, Vol. 2672(52) 1–11
© National Academy of Sciences:
Transportation Research Board 2018
Article reuse guidelines:
sagepub.com/journals-permissions
DOI: 10.1177/0361198118755712
journals.sagepub.com/home/trr
SAGE

Use of a Laser Diffractometer to Obtain the Particle Size Distribution of Fine-Grained Soils

Jessica A. Galacgac[1] and Phillip S. K. Ooi[2]

Abstract

Obtaining the particle size distribution (PSD) of fine-grained soils in hydrometer testing has several shortcomings, particularly long test duration, whereas a laser diffractometer can provide a soil's PSD relatively expediently. PSDs of the finer fraction of several Hawaiian soils using laser diffraction were compared with those from hydrometer testing. The objective was to investigate some key test variables affecting laser diffractometer PSDs including circulation duration, soil pre-treatment with a deflocculant, and the values of test parameters. The following conclusions are drawn. PSD becomes finer with increasing circulation duration. Circulation duration can be excessively large when attempting to match the laser diffractometer and hydrometer PSDs for the untreated soil. Soil should be pre-treated with sodium hexametaphosphate to decrease the circulation or test duration. The larger the real portion of the refractive index (RI), the coarser the soil. RI values must be calibrated for different laser diffractometer models and for different mineralogical compositions as the RI is a function of not only particle size and mineral composition but also the particle shape and light wavelength and different laser diffractometer models utilize different light wavelengths. The smaller the imaginary component of the RI, the more pronounced the finer fraction. The imaginary component also affects the shape of the PSD, especially in the finer spectrum. The optimum real and imaginary values of RI are 1.4 and -0.1i for the laser diffractometer used and for the fines tested in this study, respectively.

Geotechnical engineers classify soils based on their particle size distribution (PSD) for transportation-related and other infrastructure projects. Whereas a sieve analysis is used to obtain the distribution of the coarser fraction, geotechnical engineers more commonly use sedimentation techniques, such as the hydrometer test and, to a lesser extent, the pipette method (*1,2*), to obtain the PSD of the finer fraction. Although the hydrometer test is fairly well established, it suffers from the following drawbacks:

1. It is time-consuming to conduct;
2. It requires a relatively large amount of soil sample (typically 50 g);
3. It assumes that Stokes' law (*3*) (which is valid for rigid, spherical, and smooth particles) is valid, when in fact soil fines are not rigid, spherical, nor smooth, and non-spherical particles take longer to settle than their equivalent spheres; therefore, hydrometer tests usually overestimate the clay fraction;
4. It assumes that all soil particles have the same specific gravity;
5. It assumes that the individual particles are effectively separated with the use of a deflocculant and that no interaction occurs during sedimentation;
6. Soil particles can settle on top of the hydrometer bulb during the first 2 min of the test when it remains in the test cylinder;
7. Periodical insertions and removals of the hydrometer may agitate the soil–liquid suspension following the 2-min reading;
8. The sides of the measuring cylinder may slow down the particles' descent; and
9. It assumes that flow is laminar when in fact Brownian motion can govern when the particle size is 0.001 mm or less (*4,5*).

The need to find a more expedient form of analysis has spurred interest in several newer methods of grain size analyses, e.g., electro-resistance particle counting, laser diffraction and optical-based image analysis (*6*). The focus here is on the use of the laser diffractometer for PSD measurement (*7–28*).

[1]Pacific Geotechnical Engineers, Inc., Waipahu, HI
[2]Department of Civil and Environmental Engineering, University of Hawaii at Manoa, Honolulu, HI

Corresponding Author:
Address correspondence to Phillip S. K. Ooi: ooi@hawaii.edu

A laser diffractometer sends a parallel beam of monochromatic light that bombards a soil–water suspension resulting in a scatter of the light source after hitting the soil particles. The pattern of scattered light can be correlated with the distribution of "optical diameters", or "equivalent sphere diameters." The equivalent sphere diameter is the diameter of a sphere having the same volume as the measured particle (*27*). Larger particles will produce a smaller angle of scatter, while smaller particles will produce the opposite. The scattered light includes those that are reflected (abrupt change in direction of light at an interface between two dissimilar media) and refracted (bending of light due to a change in speed) but not those that are absorbed (transfer of energy when light waves are re-radiated as heat, etc.) (*22*). Conversion of the scattered light pattern to optical diameters requires the use of light diffraction theories, a review of which will be included later.

Laser diffraction testing to derive a soil's PSD has the following advantages:

1. Test results can be obtained very quickly (within minutes);
2. Unlike the hydrometer, the PSD is independent of the density of the individual particles because the calculated size distribution is geometry-based; and
3. It provides a wide range of particle sizes including points in the range where Brownian motion can dominate and that can adversely affect hydrometer test results.

However, there has been a general reluctance for widespread adoption of laser diffraction because of:

1. A lack of comparisons available between hydrometer and laser diffraction test results;
2. The fact that two common theories/relationships exist to interpret the laser diffraction test results with the preferred theory relying on assumed test parameters; and
3. High equipment cost.

The objectives of this study are to investigate some of the key test variables that can affect the laser diffractometer PSDs, including circulation duration, the necessity to pretreat a soil with a deflocculant, and what test parameters to use. Side-by-side hydrometer testing was also conducted for comparison purposes. The motivation for this work is to advance understanding of laser diffraction testing so that more widespread acceptance can be enjoyed by the geotechnical engineering profession when designing transportation-related and other infrastructure.

Literature Review

One of the earliest publications on laser diffraction analysis was by Weiss and Frock (*7*). Since then, many studies have been published on this subject. Table 1 contains a list of some publications on laser diffraction testing, theories used for interpreting the PSDs, comparison test methods, materials tested, and test parameters (refractive indices or RI) used.

From Table 1, it can be seen that two diffraction theories (Fraunhofer and Mie) were used to convert the scattered light pattern to PSDs. The following section details these two theories.

Fraunhofer and Mie Diffraction Theories

Both Fraunhofer and Mie diffraction theories assume that the soil particles are spherical. Fraunhofer theory requires no material properties in its interpretation. It is based on the approximation that the laser beam is parallel, the detector is far away compared to the size of the diffracting soil particles, and the entire light source is diffracted (i.e., none is absorbed or refracted, which is obviously not true). Then, the intensity of the diffracted beam at any angle can be related to the projected area of a specific particle size. Fraunhofer theory is only accurate if the particles are large compared to the wavelength of the light source ($>10\lambda$ where λ = light wavelength) and therefore is less suitable for determining clay size fractions accurately (*10, 14, 17, 18, 23*) or if the ratio of the RI of the soil particles and the dispersing medium (typically water) deviates from unity (*10*). Being a limiting case of Mie's theory, Fraunhofer's theory is more useful with measurements of solids suspended in a gas (say air) where the RI of the two phases are more markedly different than if the solids were suspended in water (RI = 1.00029 and 1.33 for air and water, respectively).

When the suspended soil particles are small compared to the wavelength of light and/or if the RIs of the solid and dispersing medium are similar, Mie theory is more appropriate. Mie theory is a solution of Maxwell's equations describing the propagation of the electromagnetic wave of light in space (*20*). It considers the transmission of light through the particle and therefore, unlike Fraunhofer's theory, it requires the soil's RI as an input parameter. RI is a function of: (*i*) particle size; (*ii*) mineral composition; and (*iii*) the light wavelength. For light of a given wavelength, choosing a representative RI can be quite challenging since soil particles are multi-sized non-spheres and polymineralic in nature.

Refractive Indices

The RI of the soil mineral is usually expressed as a complex number consisting of a real portion (n) and an imaginary portion (ik) as follows:

Table 1. Summary of Some Previous Research on Laser Diffraction

Reference	Laser Diffractometer Model	Theory Used for Interpretation	Comparison Test Method	Test Material	Refractive Index	
					Real	Imaginary
7	Non-commercial	Fraunhofer	Sieve	Taconite ore	N/A	N/A
8	Microtrac 7991-0	Not specified	Sieve-pipette	Sands to clays	N/A	N/A
9	Malvern 3600E	Fraunhofer	Pipette	Silts and upper clay size ranges	N/A	N/A
10	Malvern 2600	Fraunhofer and Mie	None	Spherical polystyrene latex in ethanol and glycerine	1.58	Not specified
11	Malvern 2600	Not specified	None	Mostly kaolinitic and illitic South African soils	N/A	
12	Malvern	Mie	Sieve-pipette	Loam, silty clay, clay loam, and sandy loam	1.55	−0.1i
13	Coulter LS-230	Mie	Sieve-pipette	Building sand and Brazilian oxisols	1.56	−0.2i
14	Fritsch Laser Particle Sizer A22	Fraunhofer	Sieve-pipette	Fluvial and lacustrine fines and aeolian sands	N/A	N/A
15	Coulter LS-230	Fraunhofer and Mie	Sieve-pipette	3 weathered Brazilian oxisols	1.56	−0.15i to −0.16i
16	Malvern 2600	Not specified	None	Kaolinite	N/A	Not specified
17	Coulter LS-100	Fraunhofer	Sieve-pipette	Milled quartz and loess-derived luvisols	N/A	N/A
18	Coulter LS-100	Fraunhofer	Sieve-hydrometer	Clay from completely weathered granite and ash tuff	N/A	N/A
19	Malvern Mastersizer 2000	Mie	None	Aeolian derived accretionary mantle, with granitic sands	1.544	−0.008i
20	Beckman-Coulter LS-230	Mie	Sieve-pipette	Clay, silty clay, silty clay loam, slay loam, sandy clay loam, silt loam, loam, sandy loam, sand	1.5	−0.2i
21	Coulter LS-230	Fraunhofer	Sieve-pipette	Clay-with-flints	N/A	N/A
22	Malvern Mastersizer MS20	Mie	Sieve-pipette	Clay, silt, and sand fractions	1.53	−0.1i
23	Beckman-Coulter LS-230	Fraunhofer and Mie	None	Standardized latex particles	1.35 to 1.8	0 to −3i
24	Horiba LA-930	Mie	Sieve-hydrometer	Sand, loamy sand, sandy loam, loam, silt loam, clay loam, and silty clay	1.44 to 1.5	−0.1i – −0.2i
25	Fritsch Laser Particle Sizer A22	Fraunhofer and Mie	Sieve-hydrometer	Clay, silty clay, silty clay loam, clay loam, silt loam, loam, sandy clay loam, sandy loam, loamy sand	1.5 to 1.6	−0.1i
26	Malvern Mastersizer 2000	Mie	None	Eutric cambisols, orthic luvisols, haplic phaeozems, eutric fluvisol, calcaric cambisol, mollic gleysol	1.43 to 3.22	0 – −1i
27	Horiba LA-920	Not specified	None	Silty clay to sand, and commercially available silicone oxide particles	Not specified	Not specified
28	Malvern Mastersizer 2000	Mie	Optical microscope	Sands	1.52	−0.1i

$$\mathrm{RI} = n - \mathrm{i}k \tag{1}$$

where n is the ratio of velocity of light in a vacuum to velocity of light in the tested mineral, $i = \sqrt{-1}$, k is the extinction coefficient of the mineral, which represents the rate of reduction of optical light caused by absorption and scattering = $\alpha\lambda/4\pi$, α is the absorption coefficient, and λ is the light wavelength.

Table 2. Real Portion of Refractive Index, *n*, and Specific Gravity for Some Common Clay and Soil Minerals

Mineral	Refractive Index, *n*	Specific Gravity	Reference
Common Clay Minerals			
Chlorite	1.61–1.62	3.0	*29*
Gibbsite	1.57–1.59	2.42	*29*
Halloysite	1.55–1.57	2.55	*30, 31*
Illite	1.56–1.59	2.8	*29*
Kaolinite	1.549–1.565	2.65	*29*
Mica or Muscovite	1.53–1.7	2.7–3.3	*32*
Montmorillonite	1.55–1.57	2.5	*29*
Vermiculite	1.542–1.556	2.3	*29*
Other Common Soil Minerals			
Amorphous or fused SiO_2 (655 nm wavelength)	1.4564	2.2	*33*
Amorphous or fused SiO_2 (405 nm wavelength)	1.4696	2.2	*33*
Calcite	1.486–1.658	2.71	*29*
Feldspars	1.525	2.5–2.8	*32*
Quartz	1.544–1.553	2.65	*29*
Quartz at 655 nm wavelength	1.5419	2.65	*34*
Quartz at 405 nm wavelength	1.5571	2.65	*34*

The real portion, *n*, can be calculated using Snell's law and can be found in published tables for many common clay and soil minerals [Table 2 (*29–34*)]. As stated previously, *n* also varies with the wavelength of light. Values of *n* for amorphous SiO_2 (*33*) and quartz (*34*) associated with wavelengths of 655 nanometers (nm) or red light and 405 nm or blue light are also shown in Table 2 (the laser diffractometer used in this study manufactured by Horiba Instruments, Inc. utilizes two light sources having these two wavelengths). According to Horiba Instruments, Inc. (*35*), the light source with the smaller wavelength (i.e., 405 nm) is more appropriate for measurement of smaller particles. From Table 2, it can be seen that values of n for these common clay and soil minerals fall within a narrow range of 1.46 to 1.66.

As indicated previously, *k* is a measure of the attenuation when the electromagnetic wave propagates through the soil mineral. For transparent materials, the imaginary constant of RI, $k = 0$. The absolute value of *k* increases with increasing opaqueness as non-transparent materials will absorb more light. For a given mineral with a given absorption coefficient, *k* is proportional to λ, which varies from 390 nm for violet to 700 nm for red light in the visible range. As shown in Table 1, the most common adopted value of $k = -0.1$.

Pretreating Soil Before Laser Diffraction Testing

Most of the publications in Table 1 compared the laser diffraction PSD results with those from sieve and sedimentation tests (hydrometer or pipette), which require pretreatment of the soil with a deflocculant before testing. Deflocculation can be accomplished not just chemically but also using ultrasound or sonication. Chappell (*16*) found that pretreating soil with a deflocculant helped with disaggregation before laser diffraction analysis but he also found that application of ultrasound was comparable to pretreatment with a deflocculating agent. The principle behind ultrasonic disaggregation is the transmission of vibrating sound waves to the soil in suspension causing cavitation whereby microscopic bubbles initially form and subsequently collapse. Collapse of these bubbles generates pressure waves of sufficient energy to cause rupture or disaggregation of the soil aggregates. Chappell (*16*), however, did warn that continuous ultrasonic action for a prolonged period may break individual particles. Ryżak and Bieganowski (*26*) cautioned that both sonication and chemical dispersion should not be combined because aggregation of small particles into bigger ones can occur.

Quantity of Soil for Laser Diffraction Testing

Chappell (*16*) and Ma et al. (*36*) recommended about 1 g of soil as an optimum amount to use for laser diffraction testing. Use of too much soil can lead to multiple light scattering. On the other hand, use of insufficient material may mean testing a soil that is not representative of the sample, leading to erroneous PSDs. Ryżak and Bieganowski (*26*) and Kowalenko and Babuin (*27*) used machine-specific light obscuration ranges whereby if the light transmittance reading is lower than the recommended range, then the number of particles is too small for quality testing. Conversely, if the transmittance is higher than the recommended range, there is a risk of secondary scatter occurring. This state-of-the-art technique was employed in this study.

Apportionment of Soil for Laser Diffraction Testing

When comparing laser diffraction analysis with sedimentation methods, laser diffraction had a tendency to overestimate silt fraction and underestimate clay fraction (*17–19*).

Beuselinck et al. (*17*) explained that this is because of the platy shape of clay particles; average optical diameter becomes larger than their equivalent spherical diameter because of the tendency for the laser diffractometer to detect the larger dimension. To mitigate this, Eshel et al. (*20*) proposed using Mie theory with a generally lower value for the real portion of RI.

Keck and Müller (*23*) found that coarser sized particles were difficult to detect in a mainly finer sized particle population; the opposite was true for a soil sample with mainly coarser particles and little fines. Wen et al. (*18*) and Segal et al. (*24*) found that by testing the sand and fine fractions separately, their results showed a good correlation between sedimentation and laser diffraction testing.

Laser Diffractometer

The laser diffractometer used in this study is the Horiba Laser Scattering Particle Size Distribution Analyzer LA-950V2 (Figure 1). It is controlled using the software LA-950 Wet. The dispersion medium used was distilled water. A circulation speed setting of 5 was recommended by the manufacturer for use on soils and sediments. The dispersion medium was first de-bubbled to remove trapped air and a blank measurement was taken. This meant that the scattering data (background data for the dispersion medium) was zeroed or "blanked" to form a baseline value before adding the soil. Then, approximately 1 g of soil was added to the sample bath. The circulation speed was increased as needed to keep the soil particles suspended and flowing in the sample bath and to prevent particle clumping.

Dilution of the soil solution can be performed to change the light transmittance values so that they fall within pre-set acceptable ranges for each light source (blue and red). Activating the "Dilute" setting in the software drained a few milliliters of dispersant medium plus soil sample and added a few milliliters of only distilled water to increase the light transmittance values.

The soil sample can be sonicated at varying power levels and durations to aid in particle dispersion. The time of sonication at the manufacturer's recommended setting never exceeded 2 min to avoid particle breakage. PSDs with negligible changes from successive sonications indicate that the sample is "fully dispersed."

The PSDs were all derived using the more reasonable Mie theory. Once a measurement was taken, the soil and dispersion medium were drained and the laser diffractometer rinsed thoroughly. Two cycles of draining and rinsing were adequate to prepare the laser diffractometer for the next test.

Soils Tested

The soils tested were brought in by University of Hawaii civil engineering students from various parts of the island of Oahu, HI. The soils were first oven-dried and sieved. Only the portion passing the No. 200 sieve was used for laser diffraction and hydrometer testing, as recommended by Keck and Müller (*23*), Wen et al. (*18*) and Segal et al. (*24*). Fines for each soil were separately mixed, quartered, sub-sampled from the same quarter for testing, and finally soaked in a 4% solution of sodium hexametaphosphate (SHMP) and distilled water for about 8 h in accordance with ASTM D422 (hydrometer test) before laser diffraction and hydrometer testing, unless otherwise stated. To interpret the hydrometer results, a specific gravity of 2.8 was most appropriate for the tested soils, which were all basalt-based and of volcanic origin.

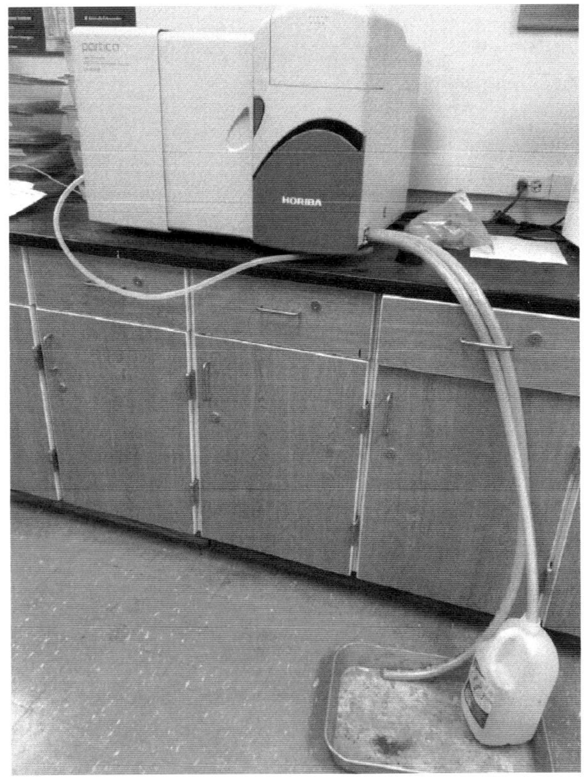

Figure 1. Horiba LA-950V2 laser diffractometer.

Test Results

This section presents the results of an investigation into the effects of circulation duration, the necessity to pre-treat a soil with a deflocculant, and the effects of RI on the PSDs from laser diffraction testing. For comparison purposes, PSDs were also obtained from hydrometer tests on companion samples.

Effects of Circulation Duration

It was theorized that the longer the circulation duration, the more the soil particles will disperse, thereby shifting the PSD to the right. To verify this, PSDs were obtained every 10 min for one sample over a total test duration of 1 h after the first

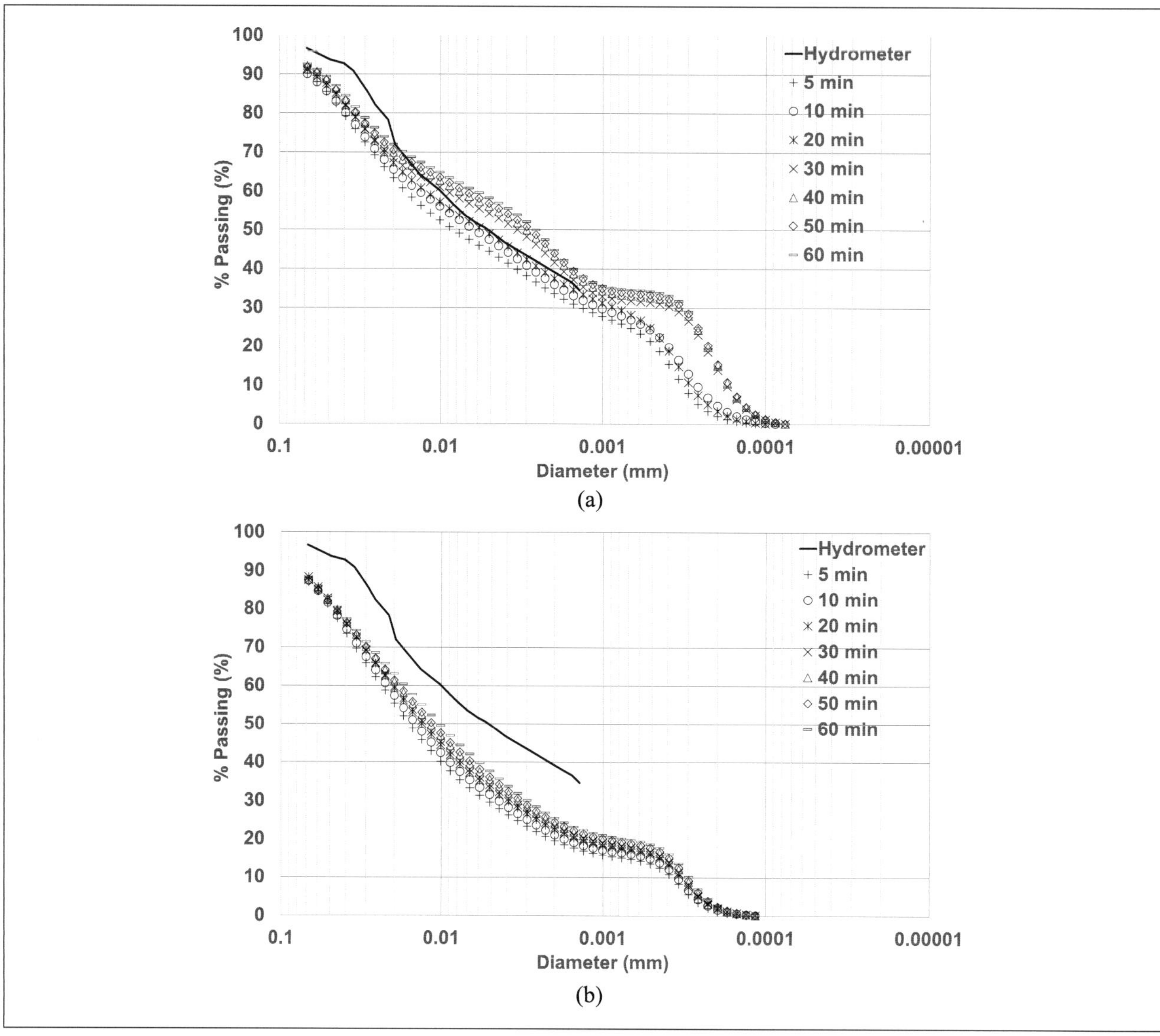

Figure 2. Effects of circulation time on the grain size curve: (*a*) RI = 1.44 – 0.1i; and (*b*) RI = 1.557 – 0.1i.

reading, which is labeled as 5 min. Five minutes roughly corresponds to the earliest time a PSD can be obtained because that is how long it takes to adjust the transmittance readings to be within the acceptable ranges, to sonicate and to take a reading. Figure 2, *a* and *b* present the PSDs assuming RI = 1.44 – 0.1i and RI = 1.557 – 0.1i, respectively. Throughout this paper, the laser diffraction data are represented using symbols while the hydrometer test data are depicted using a solid line. Using the software, the values of RI can be easily changed after each test to perform a sensitivity analysis of RI on the PSDs.

In Figure 2*a*, the 20-min measurement matched best with the hydrometer data assuming RI = 1.44 – 0.1i. When RI = 1.557 (Figure 2*b*), the circulation duration should be extended beyond 1 h to obtain a match. This also implies that the ideal circulation time is also a function of the value of the real portion of the RI used. Furthermore, it can be seen as the sample is circulated for longer, the shape of the grain size curve changes, especially with RI = 1.44; the change in shape is less pronounced when RI = 1.557.

Effects of Using a Deflocculant

A soil fine that was not pre-treated with SHMP was first tested in a dispersion medium consisting of just plain distilled water to obtain a PSD. Then, knowing the volume of distilled water in the sample bath, SHMP was added to create a 0.5% SHMP water solution consistent with the SHMP concentration in the hydrometer test. After adding SHMP, PSD measurements were taken at 2-h increments for a total duration of 8 h; 8 h was chosen because this corresponded to the minimum soaking period of soil in SHMP in accordance

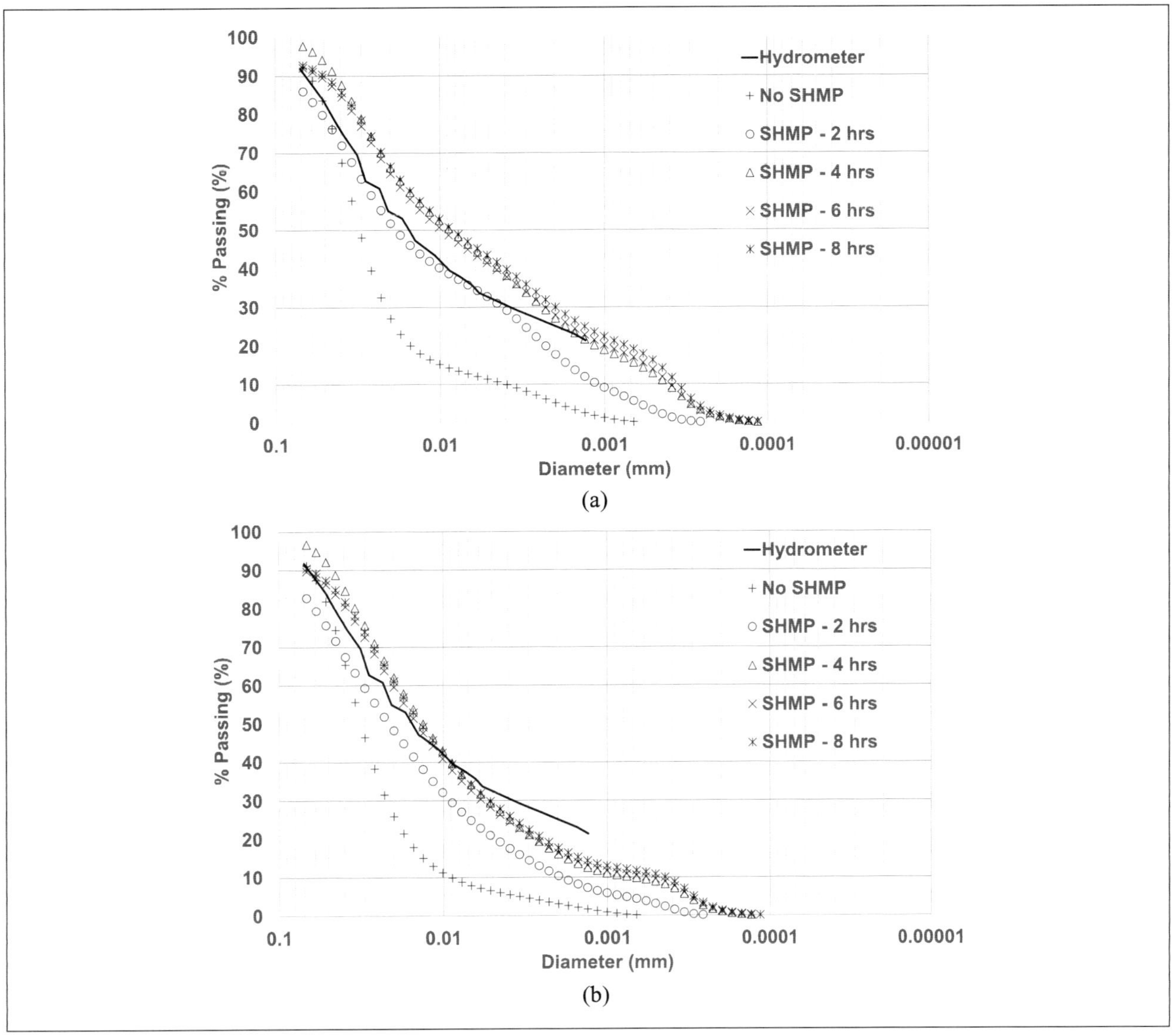

Figure 3. Effects of SHMP treatment of soil on the grain size curve: (a) RI = 1.44 – 0.1i; and (b) RI = 1.557 – 0.1i.

with ASTM D422 for the hydrometer test. Circulation of the soil in the dispersion medium was maintained throughout these 8 h. Figure 3, *a* and *b* show the PSDs from hydrometer and laser diffraction testing without and with SHMP treatment for several circulation durations for RI = 1.44 – 0.1i and RI = 1.557 – 0.1i, respectively.

In Figure 3, it can be seen that the PSDs without SHMP treatment are significantly coarser than those with SHMP treatment. Also, when the soil was not pre-treated with SHMP a priori but the SHMP was added during laser diffraction testing, it requires a circulation duration of 4 h to fully disperse the soil since the PSDs did not change much beyond 4 h. On the other hand, if the soil was pre-treated with SHMP (Figure 2), the PSDs did not change much from the initial reading. This suggests the importance of SHMP treatment before laser diffraction testing.

For an RI = 1.44 in Figure 3*a*, the 2-h measurement matched best with the hydrometer PSD. For an RI = 1.557, the 6-h reading showed a closer match to the hydrometer PSD. These circulation times seem excessive for a test that is supposed to be expedient, thus indicating that the value of the real portion of RI should be optimized by taking advantage of the fact that the PSDs did not change much for a soil pre-treated with SHMP, i.e., selecting the real portion of RI that corresponds to a minimal circulation duration.

Optimization of the Real Portion of the Refractive Index

By changing the real portion of the RI via a trial and error process, the optimum value can be found based on the following criteria:

1. Circulation duration should be about 5 min for soil pre-treated with SHMP; and
2. There should be relatively good agreement between the laser diffraction and hydrometer PSDs.

Based on these two criteria, an RI = 1.4 – 0.1i was deemed most optimum. Figure 4, *a–j* show the laser diffractometer and hydrometer PSDs for the 10 soil samples tested.

From Table 2, the real portion of the RI for common clay and soil minerals varies from 1.46 to 1.66. Yet, an optimum value of 1.4, which is below this range, was found to be most ideal. This can be attributed to two reasons: (*i*) Because of the platy shape of clay particles, average optical diameter becomes larger than their equivalent spherical diameter because of the tendency for the laser diffractometer to detect the larger dimension (*17*); and (*ii*) The real portion of the RI is a function of the light wavelength used in the laser diffractometer. From values for amorphous SiO_2 and quartz in Table 2, it can be seen that the longer the wavelength the lower the real portion of the RI. This suggests that the appropriate value of the RI not only depends on the soil mineral makeup but also on the light wavelength used in the laser diffractometer. Therefore, it may be necessary to calibrate the RI values for different laser diffractometer models and for different mineralogical compositions.

Sensitivity of Imaginary Portion of the Refractive Index

The imaginary portion of the RI represents the transparency and absorptivity of the soil mineral. The sensitivity of the PSDs to the imaginary portion of the RI was investigated by changing its value while keeping the real portion of the RI the same. Figure 5 presents the PSDs for k varying from −0.01 to −1. It can be observed that:

1. The smaller the value of the imaginary portion, the more pronounced the finer fraction becomes;
2. The imaginary component does affect the shape of the PSD, especially at the finer end; and
3. The most agreeable value of the imaginary portion of RI is −0.1i. This agrees with most of the publications in Table 1.

Rationale for Calibrating Against the Hydrometer Test

Although not perfect, the hydrometer test was selected as the basis for calibrating the RI because it is the "standard" procedure adopted by geotechnical engineers to obtain the PSDs of fine-grained soils. Most of the previously listed drawbacks of the hydrometer test are overcome by the laser diffractometer as follows:

1. Although the sample preparation time is the same, the test duration is only 5 min in laser diffractometer testing versus 48 h in hydrometer testing.
2. Only 1 g of soil is needed for laser diffractometer testing as opposed to 50 g with the hydrometer test.
3. Reliance on Stokes' law is avoided in laser diffractometer testing, although the soil particles are still assumed to be spherical in laser diffraction.
4. Because the laser diffractometer's calculated PSD is geometry-based, it is independent of the specific gravity of the individual particles in the soil makeup.
5. Use of a deflocculant is needed in both tests. Therefore, this drawback is common to both methods.

Drawbacks 6, 7, and 8 are peculiar to and only applicable to the hydrometer test. Regarding drawback 9, the laser diffractometer provides a wide range of particle sizes including points in the range where Brownian motion can dominate in sedimentation tests because the soil particles are not required to settle in laser diffractometer testing.

Summary, Conclusions, and Suggestions

A Horiba Laser Scattering Particle Size Distribution Analyzer LA-950V2 laser diffractometer was used to derive the PSDs of the finer fraction of several soils from the island of Oahu, Hawaii. These PSDs were compared to those from the hydrometer test to study the effects of circulation duration, the necessity to pre-treat a soil with a deflocculant, and the effects of the RI on the PSDs from laser diffraction testing. The following conclusions are drawn from this investigation:

1. A longer circulation duration will shift the PSD to the right; that is, make the soil appear finer.
2. Treating the soil with SHMP allows for a shorter circulation duration making the test more practical. Without SHMP pretreatment, an unreasonably long circulation duration is necessary, which defeats the purpose of using such a device. Therefore, it is recommended that soils be pre-treated with SHMP in the same manner as hydrometer test samples in order to decrease the circulation or test duration to a reasonable value.
3. The larger the real portion of the RI, the coarser the soil. Since the real portion of the RI is a function of not only particle size and mineral composition but also the particle shape and light wavelength, and since different laser diffractometer models utilize different light wavelengths, it may therefore be necessary to calibrate the RI values for different laser diffractometer models and for different mineralogical compositions.

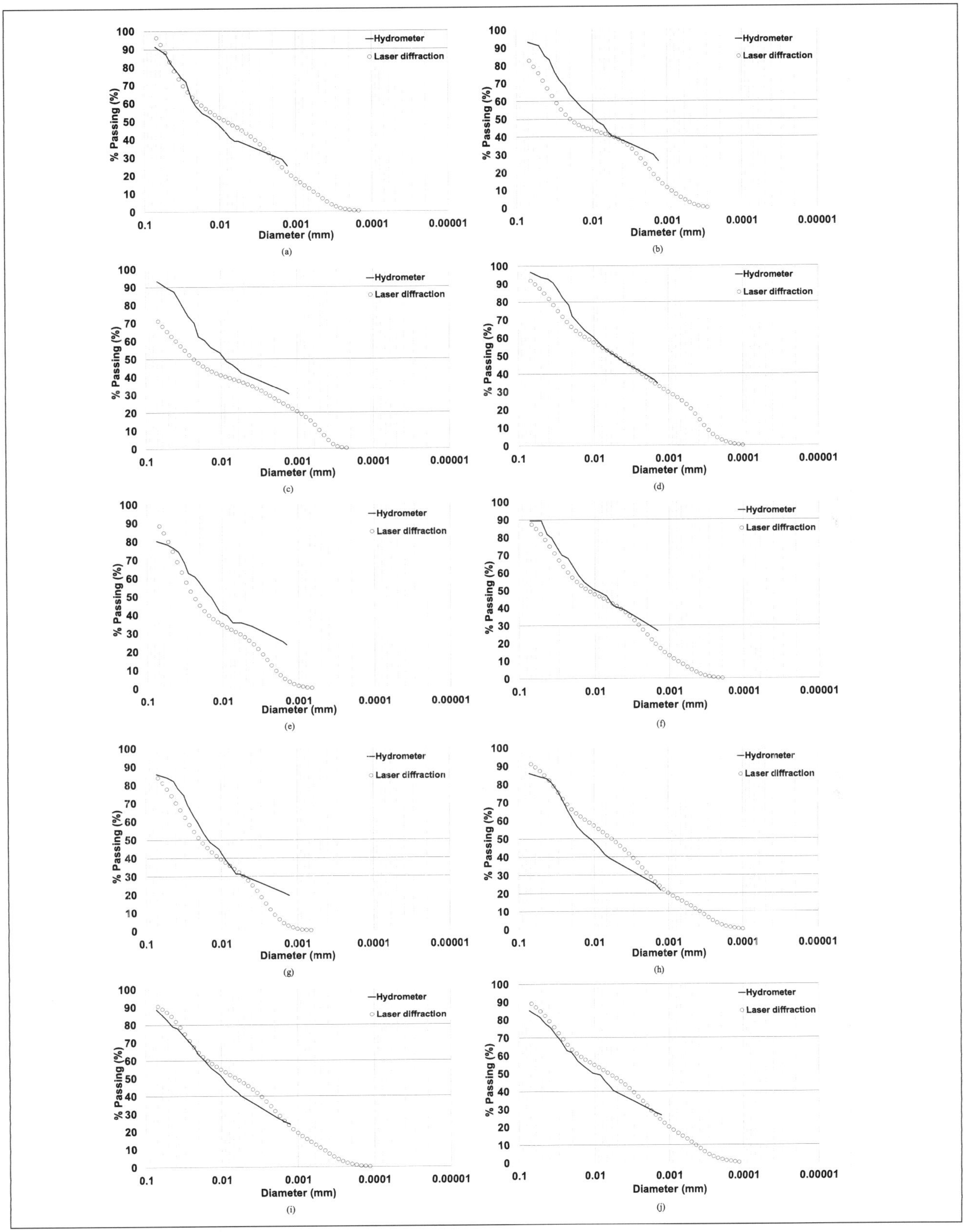

Figure 4. Comparison of laser diffractometer and hydrometer PSDs for 10 fine-grained soils with RI = 1.4 – 0.1i.

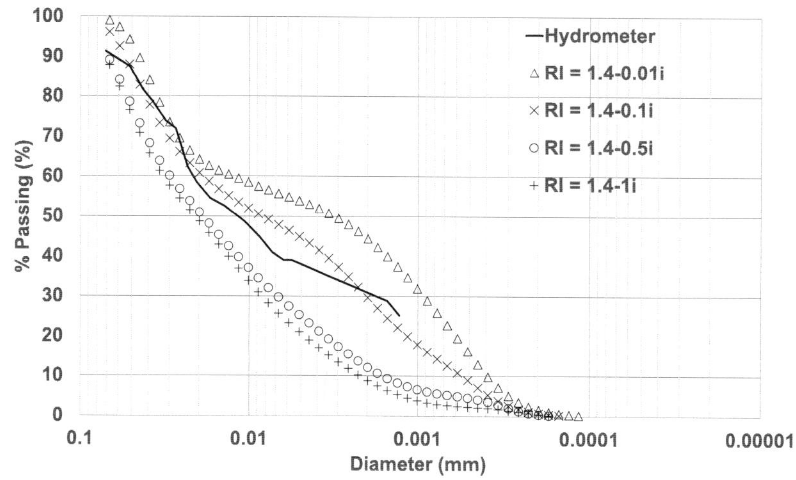

Figure 5. Effects of the imaginary portion of the RI on the PSD.

4. The smaller the imaginary component of the RI, the more pronounced the finer fraction. The imaginary component also affects the shape of the PSD, especially at the finer end.
5. Comparisons of the laser diffraction and hydrometer PSDs suggest that the real portion of the RI = 1.4 and the imaginary component of the RI = −0.1i provide good fits for 10 PSDs of fines assuming a 5-min-circulation duration.

Suggestions for Standardization

In light of this study, the following guidelines can be considered in an effort to strive for standardization of laser diffractometer testing:

1. Laser diffractometer testing should be performed on the fine and granular fraction separately.
2. From a practicality standpoint, the circulation/test duration should be limited to 5 min.
3. The manufacturer's recommended circulation speed for soils should be adopted.
4. Fines should be pre-treated with SHMP to minimize the circulation duration.
5. Sonication should be used to check whether the soil is fully dispersed. Sonication in excess of 2 min should be avoided especially when testing fines pre-treated with SHMP so as to avoid aggregation/flocculation.
6. Wherever possible, light obscuration ranges should be used to determine the right quantity of soil utilized for laser diffraction testing. If the light transmittance reading is lower than the recommended range, then the number of particles is too small for quality testing. Conversely, a transmittance reading higher than the recommended range runs the risk of secondary scatter.
7. The more rational Mie theory should be adopted when interpreting test results. Values of n and k should be calibrated for local geological minerals and laser diffractometers.

References

1. Coventry, R. J., and D. E. R. Fett. *A pipette and sieve method of particle-size analysis and some observations on its efficiency*. Div. Soils Divisional Report No. 38. Queensland, Australia. CSIRO. 1979.
2. Gee, G. W., and J. W. Bauder. Particle size analysis. In *Methods of Soil Analysis Part 1*, 2nd ed. (A. Klute, ed.), American Society Agronomy, Madison, W.I., pp. 383–411.
3. Stokes, G. G. On the effects of the internal friction of fluids on the motion of pendulums. *Transactions of the Cambridge Philosophical Society*, Vol. 9, No. 3, 1856, pp. 1–86.
4. Allen, T. *Particle size measurement*, 3rd ed. Chapman and Hall, New York, 1981.
5. Loveland, P. J., and W. R. Whalley. Particle size analysis. In *Soil and Environmental Analysis, Physical Methods* (Smith, K.A. and Mullins, C.E. ed.), Marcel Dekker, Inc., New York, 2001, pp. 281–314.
6. Goossens, D. Techniques to measure grain-size distributions of loamy sediments: a comparative study of ten instruments for wet analysis. *Sedimentology*, Vol. 55, 2008, pp. 65–96.
7. Weiss, E. L., and H. N. Frock. Rapid analysis of particle size distributions by laser light scattering. *Powder Technology*, Vol. 14, 1976, pp. 287–293.
8. Cooper, L. R., R. L. Haverland, D. M. Vendricks, and W. Knisel. Microtac particle-size analyzer: An alternative particle-size determination method for sediment and soils. *Soil Science*, Vol. 138, 1984, pp. 138–146.
9. McCave, I. N., R. J. Bryant, H. F. Cool, and C. A. Coughanowr. Evaluation of a laser-diffraction size analyzer for use with

natural sediments. *Journal of Sediments Petrology*, Vol. 56, No. 4, 1986, pp. 561–564.

10. de Boer, G.B., C. de Weerd, D. Thoenes, and H. W. Goossens. Laser diffraction spectrometry: Fraunhofer vs Mie scattering. *Particle Characteristics*, Vol. 4, 1987, pp. 14–19.
11. Levy, G. J., M. Agassi, H. J. C. Smith, and R. Stern. Microaggregate stability of kaolinitic and illitic soils determined by ultrasonic energy. *Soil Science Society of America Journal*, Vol. 57, 1993, pp. 803–808.
12. Wu, Q., M. Borkovec, and H. Sticher. On particle-size distributions in soils. *Soil Science Society of America Journal*, Vol. 57, No. 4, 1993, pp. 883–890.
13. Buurman, P., T. Pape, and C.C. Muggler. Laser grain-size determination in soil genetic studies. 1. Practical Problems. *Soil Science*, Vol. 162, 1997, pp. 211–218.
14. Konert, M., and J. Vandenberghe. Comparison of laser grain size analysis with pipette and sieve analysis: A solution for underestimation of clay fraction. *Sedimentology*, Vol. 44, No. 3, 1997, pp. 523–535.
15. Muggler, C. C., T. Pape, and P. Buurman. Laser grain-size determinationin soil genetic studies. 2. Clay content, clay formation and aggregation in some Brazilian soils. *Soil Science*, Vol. 162, 1997, pp. 219–228.
16. Chappell, A. Dispersing sandy soil for the measurement of particle size distributions using optical laser diffraction. *Catena*, Vol. 31, No. 4, 1998, pp. 271–281.
17. Beuselinck, L., G. Govers, J. Poesen, G. Degraer, and L. Froyen. Grain-size analysis by laser diffractometry: comparison with the sieve-pipette method. *Catena*, 1998, 32(3–4), pp. 193–208.
18. Wen, B., A. Aydin, and N. S. Duzgoren-Aydin. A comparative study of particle size analyses by sieve-hydrometer and laser diffraction methods. *Geotechnical Testing Journal*, Vol. 25, No. 4, 2002, pp. 434–442.
19. Campbell, J. R. Limitations in the laser particle sizing of soils. Advances in Regolith, *Proceedings of the Cooperative Research Centre for Landscape Environments and Mineral Exploration (CRC LEME) Regional Regolith Symposia*, Bentley: Cooperative Research Centre for Landscape Environments and Mineral Exploration (CRC LEME), 2003, pp. 38–42.
20. Eshel, G., G. J. Levy, U. Mingelgrin, and M. J. Singer. Critical evaluation of the use of laser diffraction. *Soil Science Society of America Journal*, Vol. 68. 2004, pp. 736–743.
21. Scott-Jackson, J., and H. Walkington. Methodological issues raised by laser particle size analysis of deposits mapped as clay-with-flints from the Palaeolithic site of Dickett's Field, Yarnhams Farm, Hampshire, UK. *Journal of Archaeological Science*, Vol. 32, No. 7, 2005, pp. 969–980.
22. Pieri, L., M. Bittelli, and P. Rosi Pissa. Laser diffraction, transmission electron microscopy and image analysis to evaluate bimodal Gaussian model for particle size distribution in soils. *Geoderma*, Vol. 135, 2006, pp. 118–132.
23. Keck, C. M., and R. H. Müller. Size analysis of submicron particles by laser diffractometry—90% of the published measurements are false. *International Journal of Pharmaceutics*, Vol. 355, No. 1–2. 2008. pp. 150–163.
24. Segal, E., P. J. Shouse, S. A. Bradford, T. H. Skaggs, and D. L. Corwin. Measuring particle size distribution using laser diffraction: implications for predicting soil hydraulic properties. *Soil Science*, Vol. 174, No. 12. 2009. pp. 639–645.
25. Stefano, C. D., V. Ferro, and S. Mirabile. Comparison between grain-size analyses using laser diffraction and sedimentation methods. *Biosystems Engineering*, Vol. 106, No. 2, 2010. pp. 205–215.
26. Ryżak, M., and A. Bieganowski. Methodological aspects of determining soil particle-size distribution using the laser diffraction method. *Journal of Plant Nutrition and Soil Science*, 2011. Vol. 174, No. 4, pp. 624–633.
27. Kowalenko, C., and D. Babuin. Inherent factors limiting the use of laser diffraction for determining particle size distributions of soil and related samples. *Geoderma*, Vol. 193–194. 2013. pp. 22–28.
28. Polakowski, C., A. Sochan, A. Bieganowski, M. Ryzak, R. Földényi, and J. Tóth. Influence of the sand particle shape on particle size distribution measured by laser diffraction method. *International Agrophysics*, Vol. 28, No. 2. 2014. pp. 195–200.
29. *CRC handbook of chemistry and physics*, 94th ed. CRC Press, Taylor & Francis Group, Boca Raton, F.L., 2013.
30. Deer, W.A., R.A. Howie, and J. Zussman. *An introduction to the rock-forming minerals*, 2nd ed. Longman Scientific and Technical, Harlow, Essex, England, 1992.
31. Carmichael, R.S. *Practical handbook of physical properties of rocks and minerals*. CRC Press, Inc., Boca Raton, F.L., 1989.
32. Beckman Coulter, Inc. *Some physical constants used in particle characterization*. Beckman Coulter, Inc., Miami, F.L., 1999.
33. Malitson, I. H. Interspecimen Comparison of the Refractive Index of Fused Silica. *Journal of the Optical Society of America*, Vol. 55, No. 10, 1965, pp. 1205–1209.
34. Ghosh, G. Dispersion-equation coefficients for the refractive index and birefringence of calcite and quartz crystals. *Optics Communications*, Vol. 163, No. 1–3, 1999, pp. 95–102.
35. Horiba Instruments, Inc. The LA-950 Laser Diffraction Technique Technical Note 159, 2009. http://www.horiba.com/fileadmin/uploads/Scientific/Documents/PSA/TN159. Accessed [April 14, 2016].
36. Ma, Z., H. G. Merkus, J. G. Smet, C. Heffels, and B. Scarlett. New developments in particle characterization by laser diffraction: size and shape. *Powder Technology*, Vol. 111, No. 1–2. 2000. pp. 66–78.

The Standing Committee on Soil and Rock Properties (AFP30) peer-reviewed this paper (18-00359).

Article

TRR
JOURNAL OF THE TRANSPORTATION RESEARCH BOARD

Transportation Research Record
2018, Vol. 2672(52) 12–22

DOI: 10.1177/0361198118758051
journals.sagepub.com/home/trr
SAGE

Laboratory Characterization of Clogging Potential of Porous Asphalt Mixtures

L. Chu[1] and T. F. Fwa[1]

Abstract

It is of practical significance to be able to evaluate the clogging potential of a porous asphalt mix in the laboratory during the mix design phase. This paper presents a laboratory evaluation procedure to assess the clogging potential of porous asphalt materials, and identify the mix properties that influence the clogging potential of a porous asphalt mixture. Eleven different porous mix designs were studied, with their top aggregate sizes varying from 9.5 to 19 mm, covering a porosity range of 10–28%. The procedure consists of a clogging treatment to introduce a selected clogging material into a test specimen in stages, with permeability measurements made after each stage of clogging. From the test results, two clogging potential indicators were defined: a clogging potential factor β, and the number of clogging cycles to reach the final state of fully clogged condition N_c. The effects of the following six mix properties on clogging potential of the mixes were studied: initial porosity P_0, initial permeability coefficient k_0, top aggregate size D_T, coefficient of uniformity C_u, coefficient of curvature C_c, and coefficient of infiltration C_i. Statistical analyses were performed and it was found that the two clogging potential indicators were significantly influenced by k_0, P_0, and D_T. Statistically significant predictive models for β and N_c were developed with these three mix properties as the predictor variables. The predictive models can serve to provide useful guidance for pavement engineers in the design of porous asphalt mixtures.

Porous pavements have been widely accepted as an effective form of pavement design to improve wet-weather driving safety (*1–3*) and reduce tire–pavement noise (*4–6*). Their porous structure allows surface runoff to flow within the structure, thereby reducing the thickness of surface water, giving rise to the wet-weather driving benefits of improved skid resistance, decreased hydroplaning risk, reduced headlight reflection, splash and spray, and better visibility of pavement markings on highways. The ability of porous pavements to reduce tire–pavement noise arises from their surface and internal voids that are able to dissipate sound energy, and weaken those noise-generation mechanisms such as air sucking, air pumping, air resonant and horn effect (*5, 7*).

Clogging is a common distress form that could severely reduce the effectiveness of porous pavements. Clogging is caused by dust, soils, debris, and fragments of road materials that are bound to be present on any road. Declogging or replacement of the porous surface course is required when its intended functional benefits are significantly jeopardized. To extend the useful service life of a porous pavement and to reduce the frequency of declogging maintenance, it is important for pavement engineers to select a mix design that has low clogging potential. Unfortunately, currently there is no sound engineering procedure to evaluate the clogging potential of a porous pavement material and guide pavement engineers in its mix design.

In this paper, a laboratory evaluation procedure to assess the clogging potential of porous asphalt materials is demonstrated. The objective was to identify the mix properties that influence the clogging potential of a porous asphalt mixture. It is hoped that such information can offer some useful guidance to pavement engineers in the selection of porous asphalt mix designs that have sufficiently low clogging potential. Eleven different porous mix designs were studied, with their top aggregate sizes varying from 9.5 to 19 mm, covering a porosity range of 10–28%. The top aggregate size of a gradation refers to the smallest sieve size that allows 100% of the aggregates to pass through it. The laboratory tests evaluate the clogging behaviors of the porous mixtures in terms of the rate of clogging, and the clogging effort needed to fully clog the mixtures. Statistical analyses

[1]School of Highway, Chang'an University, South Erhuan Middle Section, Xi'an, China

Corresponding Author:
Address correspondence to T. F. Fwa: ceefwatf@nus.edu.sg

Table 1. Gradation of Residual Soil as Clogging Material

Sieve size (mm)	1.18	0.600	0.425	0.300	0.150	0.075
% passing	100	94.3	90.0	84.8	72.4	58.2

were performed to examine how the clogging behaviors were influenced by the properties of porous asphalt mixtures such as porosity, permeability, and aggregate gradation characteristics.

Laboratory Evaluation Program

Scope of Experimental Test Program

The laboratory experimental program consists of the following two main components for each test specimen:

(i) Clogging treatment of test specimens to introduce clogging material into the specimens; and
(ii) Permeability measurements at all stages of clogging treatment to record the deterioration of specimen permeability in the process of clogging.

The detailed test procedures of these two components and the rationale of the methods adopted are presented in the following sub-sections.

Clogging Treatment. A key aspect of clogging treatment is the selection of clogging material. In Singapore, the major source of clogging materials on porous pavements has been the soils deposited from dirty wheels of vehicles, and from vehicles carrying earth or construction materials. Past studies by the authors (*8,9*) have established that the portion of residual soil with sizes smaller than 1.18 mm was mostly responsible for creating clogging in porous asphalt materials. Therefore, this component of the residual soil was employed as the clogging material in the clogging treatment. The gradation of this clogging material is given in Table 1.

The clogging experimental setup and procedure developed by Fwa et al. (*8,10*), as shown schematically in Figure 1, was adopted to introduce clogging in test specimens in the present study. The test apparatus consisted of a 150 mm diameter open-ended vertical cylinder. A 150 mm diameter cylindrical specimen was first secured in the apparatus before the clogging material was applied in multiple cycles of the clogging treatment. At the beginning of each cycle, an amount of 5.3 g clogging material was spread uniformly on the top face of the specimen, followed by filling the cylinder to a height of 41.5 cm to create a constant-head water flow to bring the clogging material into the specimen. The constant-head flow was maintained for 5 min before the water was released, and the permeability of the specimen was measured (see next sub-section for method of permeability measurement). This marked the end of a clogging treatment cycle. The next cycle of clogging treatment followed immediately with the application of another 5.3 g of clogging material. The clogging treatment was continued until there were negligible changes in the measured permeability in successive cycles of the clogging treatment.

Permeability Measurements. The experimental setup shown in Figure 1 is equipped with a high-precision pressure sensor and a high-speed data acquisition unit that provide accurate readings of changing water head for the determination of permeability coefficient by means of the falling-head method. A detailed description of the experimental setup and measuring procedure of permeability coefficient is found in earlier works by the authors (*11,12*). It is of significance to note that the experimental setup allowed permeability measurements to be made at the end of each stage of clogging treatment, without having to remove the specimen from the test cylinder.

Researchers in the field of porous media flow have found that water flows in porous coarse-grained materials are non-laminar, and that the traditional Darcy's equation is invalid (*13,14*). Several versions of modified Darcy's equation have been proposed by researchers. The following expression has been widely used and is adopted in the present study for the determination of permeability coefficient from the test results of falling-head tests (*11,12*):

$$k = \frac{v}{i^n} \tag{1}$$

where k is the permeability coefficient, v the flow velocity, i the hydraulic gradient, and n a numerical constant. Studies by Fwa et al. (*11,12*) have shown that n could be set as a constant of 0.70 without introducing any errors of practical significance.

Measuring specimen permeability as the clogging treatment progresses offers a convenient and practical way to define the degree of clogging. Fwa et al. (*10,15*) have considered the following two measures of percent clogging achieved in a test specimen,

$$\begin{array}{l}\text{\% clogged}\\ \text{based on porosity}\end{array} = \left[1 - \frac{\text{Porosity after cloggingg}}{\text{Porosity before clogging}}\right] \times 100\% \tag{2}$$

$$\begin{array}{l}\text{\% clogged based}\\ \text{on permeability}\end{array} = \left[1 - \frac{\text{Permeability coefficient after clogging}}{\text{Permeability coefficient before cloggging}}\right] \times 100\% \tag{3}$$

The definition by Equation 3 is adopted for the present study for three main reasons. First, it is cumbersome to remove a test specimen from the test cylinder at each intermediate clogging stage to measure porosity and re-install it

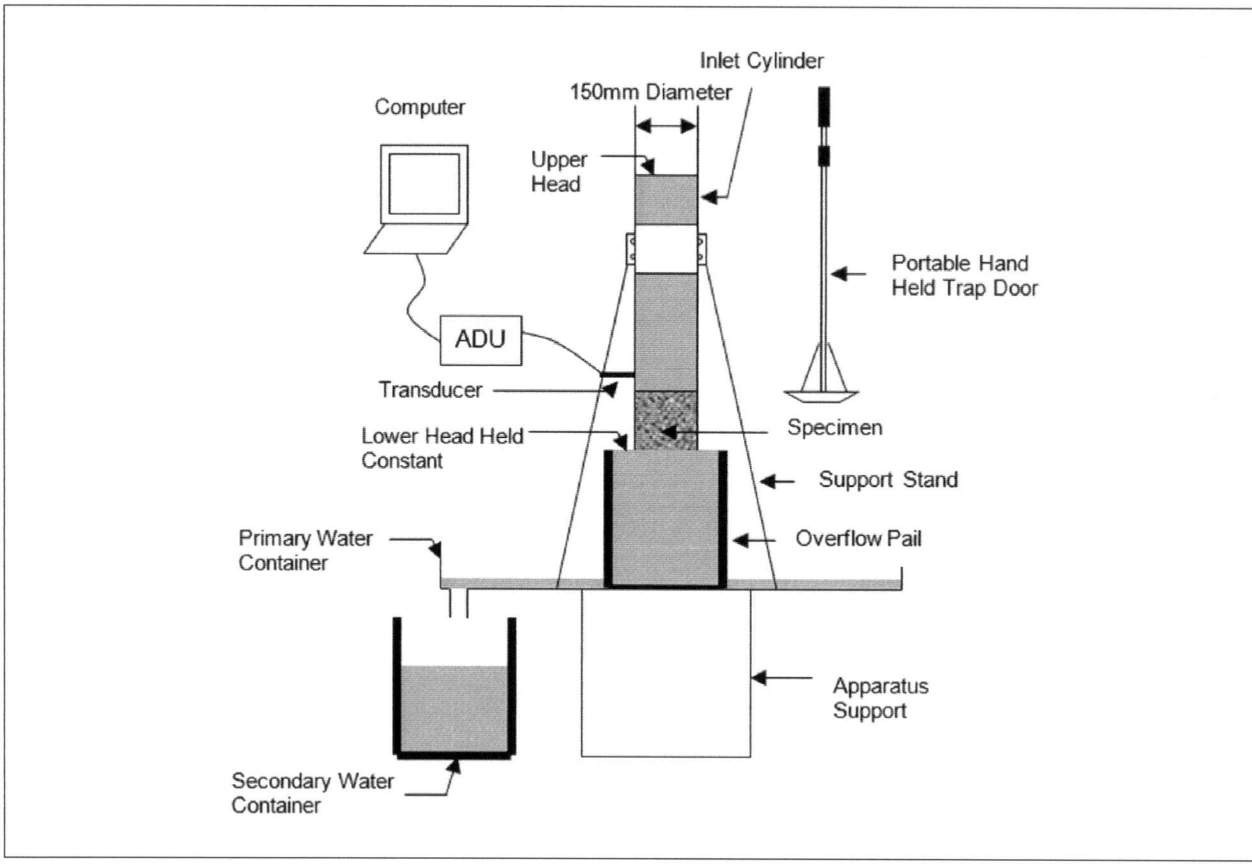

Figure 1. Schematic diagram of setup for laboratory clogging tests.

after measurement, but it is relatively straightforward to make permeability measurements without having to remove and disturb the test specimen. Second, considering an in-service pavement, it is practically impossible to measure changes in its porosity in the field in a non-destructive manner, but this can be done much more easily for permeability measurement, by means of a field device such as the permeameter developed by Tan et al. (*12*). Third, permeability can be considered as an indirect measure of the connectivity of pores of a specimen, which is an important property that can have an effect on tire–pavement noise generation and absorption, and on its drainage properties which affect wet-weather skid resistance. It is also noted when a specimen is fully clogged, most porous asphalt materials would become 100% clogged based Equation 3, but would have different values below 100% if Equation 2 is used. Therefore, Equation 3 provides a definition that is more convenient to use and easier to interpret.

Porous Asphalt Mixtures Studied

A total of eleven different porous mix designs were studied. Figure 2 shows the aggregate gradations of the mix designs, and Table 2 presents for each mix its mix proportions, top aggregate size, target porosity. The mix designs P1, P2, P3, P8, P9 and P11 are porous asphalt mixes that have been used in Singapore (*16,17*), mix P4 is a mix used in the UK (*18*), mixes P6 and P7 are the top layer porous asphalt designs used in the Netherlands (*19*), mix P5 is a mix used in the USA (*20*), and mix P10 is a mix adopted in India (*21*).

The asphalt binder used in all the mixes was polymer modified binder of grade PG76-22, and the aggregate used was granite. It can be seen from Table 2 that the mix designs studied cover the following ranges for the two key material properties of interest to the present study: the top aggregate size from 9.5 to 19 mm, and the target porosity from 10 to 30%. Three specimens were prepared for each mix design, and the actual porosity values of the three specimens fabricated for each mix are also recorded in Table 2. The actual range of porosity of the specimens was from 8.8 to 28.6%.

Clogging and Permeability Test Results

As there were three specimens for each mix design, altogether 33 sets of test data were obtained. For each test

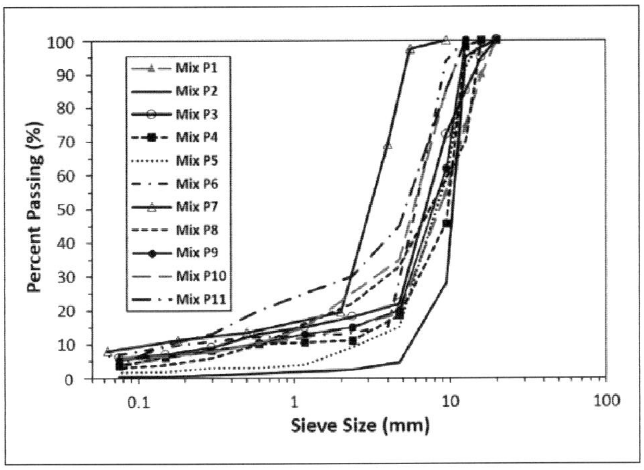

Figure 2. Aggregate gradations of porous asphalt mixes studied.

specimen, the measured results of the clogging and permeability tests included the following:

- Initial porosity, P_0;
- Initial permeability coefficient, k_0;
- Permeability coefficient after each stage of clogging treatment; and
- Number of clogging cycles, N_c required to fully clog a specimen.

The measured values of P_0, k_0 and N_c are summarized in Table 3. Also listed in the table are three mix gradation characteristics the meaning of which will be explained in the next sub-section. The measured permeability coefficient of a test specimen decreased as the clogging treatment progressed, until it was fully clogged. Figure 3 presents the permeability deterioration measurements of three of the eleven mix designs tested. These results are consistent with past studies by the authors that the permeability deterioration trend of a porous mixture caused by clogging can be very closely represented by a decreasing exponential function of clogging cycles with high coefficient of determination R^2 values of around 0.90 (*10,22*). Table 4 shows the results of the exponential functions for all the 33 specimens tested.

Characterizing Clogging Potential of Porous Asphalt Mixtures

Rate of Clogging and Clogging Cycles to Reach Fully Clogged State

The clogging potential of a porous asphalt mixture can be assessed by means of the following two clogging characteristics: (i) the rate of clogging, and (i) the clogging effort needed to fully clog the mixture. In this study, applying the definition of percent clogging based on permeability as given by Equation 3, the two clogging characteristics can be evaluated using the measured results presented in the preceding section. The clogging effort to fully clog a porous asphalt mixture can be represented by N_c, the number of clogging cycles at which the mixture becomes fully clogged (see Table 3).

Next, based on the definition of percent clogging given by Equation 3, the rate of clogging can be represented by the rate of change in the measured permeability coefficient. From the results of the clogging tests presented earlier in Table 4, it is seen that the permeability deterioration in the process of clogging can be described by the following expression:

$$k_N = k_0 e^{-\beta N} \tag{4}$$

where β is a positive regression constant, N the number of clogging cycles applied, k_N the permeability coefficient after N clogging cycles, and k_0 the initial permeability coefficient before clogging treatment.

Equation 4 describes the permeability deterioration curve of a porous asphalt mixture in the process of clogging. Before clogging treatment, $N = 0$ and the permeability coefficient is equal to k_0. k_0 is equal to the intercept of the deterioration curve with the vertical axis. The term $-\beta N$ represents the rate of deterioration. The lowest possible value of β is zero. $\beta = 0$ represents the case where a porous asphalt mixture is not affected by clogging materials at all, and its permeability coefficient value stays constant as k_0. The larger the non-negative β value, the higher is the rate of permeability deterioration, and the higher is the clogging potential of the porous asphalt mixture concerned. Therefore, β can be appropriately termed as a clogging potential factor that provides a meaningful measure of the rate of clogging.

Modeling of Clogging Performance and Clogging Potential

Based on the clogging performance of a porous asphalt mixture represented by its permeability deterioration curve such as those shown in Figure 3, the preceding section has shown that the clogging potential of a porous asphalt mixture can be characterized by two indicators, namely the clogging potential factor β, and the number of clogging cycles N_c to reach a fully clogged state. This section performs further analyses on the test results with the aims to (i) identify factors that influence the clogging behavior and clogging potential of porous asphalt mixtures, and (ii) develop predictive regression models for assessing the clogging potential indicators of a porous asphalt mixture.

Possible Factors Affecting Clogging Potential. A total of six mix properties are considered to be factors that could have an effect on the clogging performance and clogging potential of a porous asphalt mixture. Figures 4 and 5 show the plots of β and N_c, respectively, against each of the six factors. The six

Table 2. Properties of Mix Designs Studied

Mix code	Top aggregate size (mm)	Binder content (%)	Target porosity (%)	Actual specimen porosity (%)
P1	19	4.5	25	24.6, 23.8, 26.6
P2	19	5	25	26.5, 28.0, 28.6
P3	19	5	20	18.5, 18.6, 21.6
P4	16	5	20	18.2, 20.5, 19.1
P5	16	6.2	20	20.6, 21.3, 22.0
P6	12.7	6	20	18.9, 20.2, 22.6
P7	9.5	6.5	20	17.3, 20.1, 20.9
P8	16	5	15	14.0, 16.2, 16.4
P9	12.7	5	15	17.7, 18.6, 16.2
P10	12.7	4	15	13.7, 14.4, 14.7
P11	12.7	5	10	8.8, 11.4, 10.8

Table 3. Results of Clogging and Permeability Tests

Specimen code	Mix type	Initial porosity, P_0 (%)	Initial permeability coefficient, k_0 (mm/s)	Number of clogging cycles to reach fully clogged state, N_c
1	P1	26.6	19.2	17
2	P1	24.6	14.3	15
3	P1	23.8	11.1	14
4	P2	28.6	9.6	13
5	P2	28.0	8.4	11
6	P2	26.5	7.1	10
7	P3	21.6	11.7	14
8	P3	18.6	10.5	12
9	P3	18.5	7.7	10
10	P4	20.5	5.7	9
11	P4	19.1	5.0	8
12	P4	18.2	4.2	7
13	P5	22.0	9.8	12
14	P5	21.3	9.4	11
15	P5	20.6	9.2	10
16	P6	23.6	8.9	10
17	P6	20.2	8.4	9
18	P6	18.9	7.2	8
19	P7	20.9	7.5	9
20	P7	20.1	5.8	8
21	P7	17.3	5.2	8
22	P8	16.4	4.7	8
23	P8	16.2	3.4	7
24	P8	14.0	3.2	6
25	P9	18.6	4.2	8
26	P9	17.7	3.8	7
27	P9	16.2	3.6	6
28	P10	14.7	5.2	8
29	P10	14.4	4.8	7
30	P10	13.7	4.2	6
31	P11	11.4	2.6	7
32	P11	10.8	2.4	7
33	P11	8.8	2.1	5

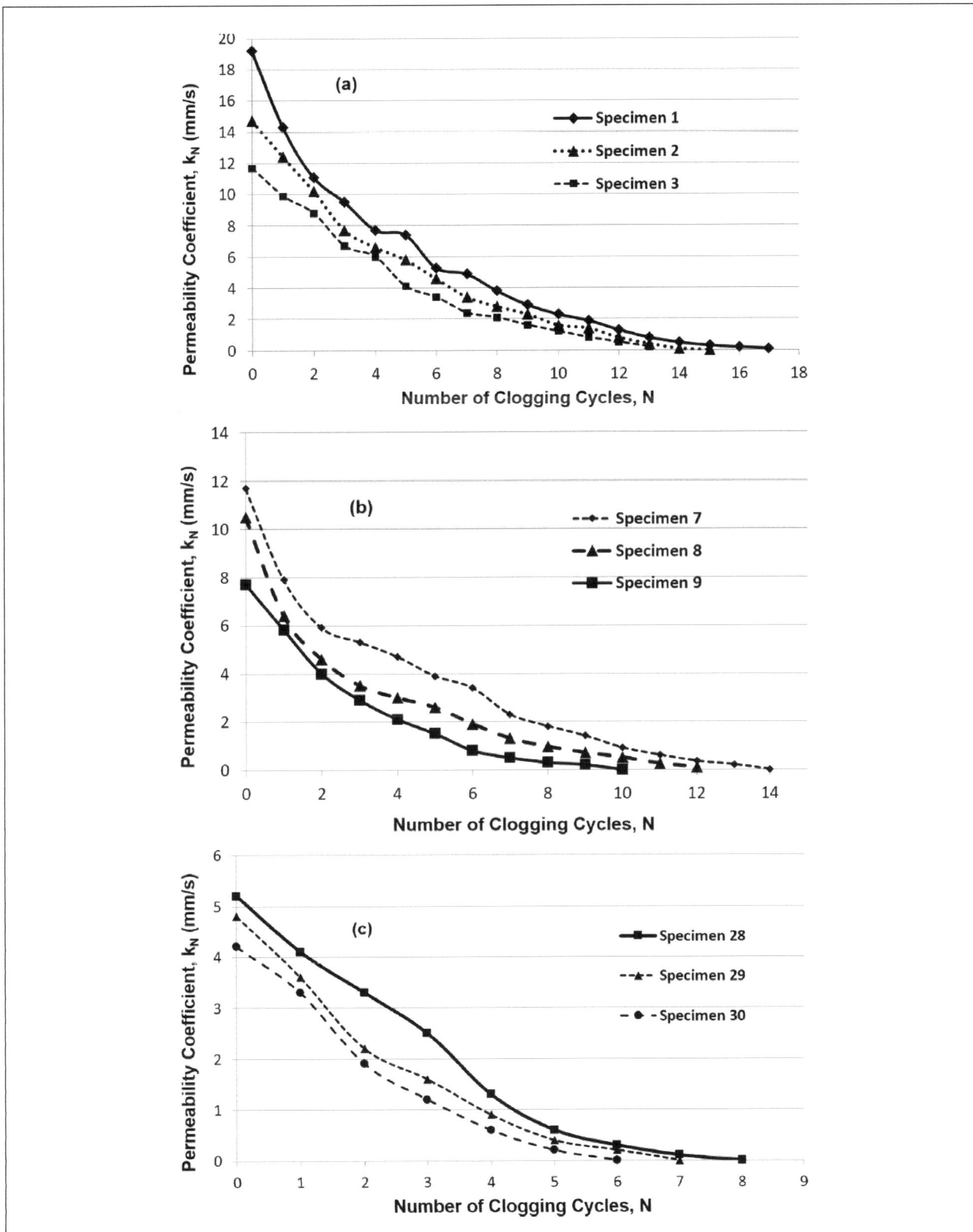

Figure 3. Examples of clogging test results. (*a*) mix P1; (*b*) mix P3; (*c*) mix P10.

Table 4. Regression Models of Clogging Induced Permeability Deterioration

Specimen code	Decreasing exponential model of permeability deterioration	Coefficient of determination, R^2
1	$k_N = 19.2\ e^{-0.255N}$	0.939
2	$k_N = 14.3\ e^{-0.251N}$	0.875
3	$k_N = 11.1\ e^{-0.243N}$	0.928
4	$k_N = 9.6\ e^{-0.248N}$	0.877
5	$k_N = 8.4\ e^{-0.310N}$	0.882
6	$k_N = 7.1\ e^{-0.323N}$	0.903
7	$k_N = 11.7\ e^{-0.268N}$	0.952
8	$k_N = 10.5\ e^{-0.328N}$	0.956
9	$k_N = 7.7\ e^{-0.384N}$	0.982
10	$k_N = 5.7\ e^{-0.365N}$	0.907
11	$k_N = 5.0\ e^{-0.414N}$	0.971
12	$k_N = 4.2\ e^{-0.512N}$	0.897
13	$k_N = 9.8\ e^{-0.301N}$	0.923
14	$k_N = 9.4\ e^{-0.318N}$	0.924
15	$k_N = 9.2\ e^{-0.342N}$	0.890
16	$k_N = 8.9\ e^{-0.346N}$	0.909
17	$k_N = 8.4\ e^{-0.389N}$	0.934
18	$k_N = 7.2\ e^{-0.399N}$	0.919
19	$k_N = 7.5\ e^{-0.433N}$	0.941
20	$k_N = 5.8\ e^{-0.409N}$	0.914
21	$k_N = 5.2\ e^{-0.459N}$	0.902
22	$k_N = 4.7\ e^{-0.396N}$	0.948
23	$k_N = 3.4\ e^{-0.417N}$	0.952
24	$k_N = 3.2\ e^{-0.415N}$	0.956
25	$k_N = 4.2\ e^{-0.504N}$	0.926
26	$k_N = 3.8\ e^{-0.469N}$	0.988
27	$k_N = 3.6\ e^{-0.588N}$	0.958
28	$k_N = 5.2\ e^{-0.460N}$	0.895
29	$k_N = 4.8\ e^{-0.476N}$	0.958
30	$k_N = 4.2\ e^{-0.520N}$	0.931
31	$k_N = 2.6\ e^{-0.437N}$	0.980
32	$k_N = 2.4\ e^{-0.430N}$	0.986
33	$k_N = 2.1\ e^{-0.423N}$	0.963

Note: N = number of clogging cycles applied; k_N = permeability coefficient in mm/s after N clogging cycles.

factors considered are: Initial porosity P_0, initial permeability coefficient k_0, top aggregate size D_T, coefficient of uniformity C_u, coefficient of curvature C_c, and coefficient of infiltration C_i. Table 5 gives the definition and tabulates the values of β, N_c, D_T, C_u, C_c, and C_i. The values of P_0 and k_0 are found in Table 3. The meaning and significance of the six factors are explained below:

- Initial porosity P_0: A mix with a higher porosity allows more fine clogging materials to infiltrate into its mass, it may also allow the flow to bring the fine materials out of its mass. The net effect on clogging will also depend on other related factors such as the size and tortuosity of pores. For the range of porosity values considered in the present study, Figures 4*b* and 5*b* show that higher porosity led to lower β and higher N_c, respectively. These results suggest that, within the porosity range of 10–28%, increasing the porosity of a porous asphalt mixture will help to reduce its clogging potential (i.e. reduce β and increase N_c values).
- Initial permeability coefficient k_0: A mix with a high porosity, large pore sizes, good connectivity of pores, and low tortuosity of connected pores will have high permeability. The effect of permeability on clogging will be dependent on the relative contributions of all these factors. For the range of permeability coefficient values considered in the present study, Figures 4*a* and 5*a* show that the effect is similar to that of porosity. That is, within the permeability range of 2.1–19.2 mm/s, increasing the permeability of a porous asphalt mixture has the benefit of reducing its clogging potential (i.e. lowering β and increasing N_c values).
- Top aggregate size D_T: For mixtures with similar shapes of aggregate gradation curve, those with larger top aggregate sizes tend to have larger pore sizes.

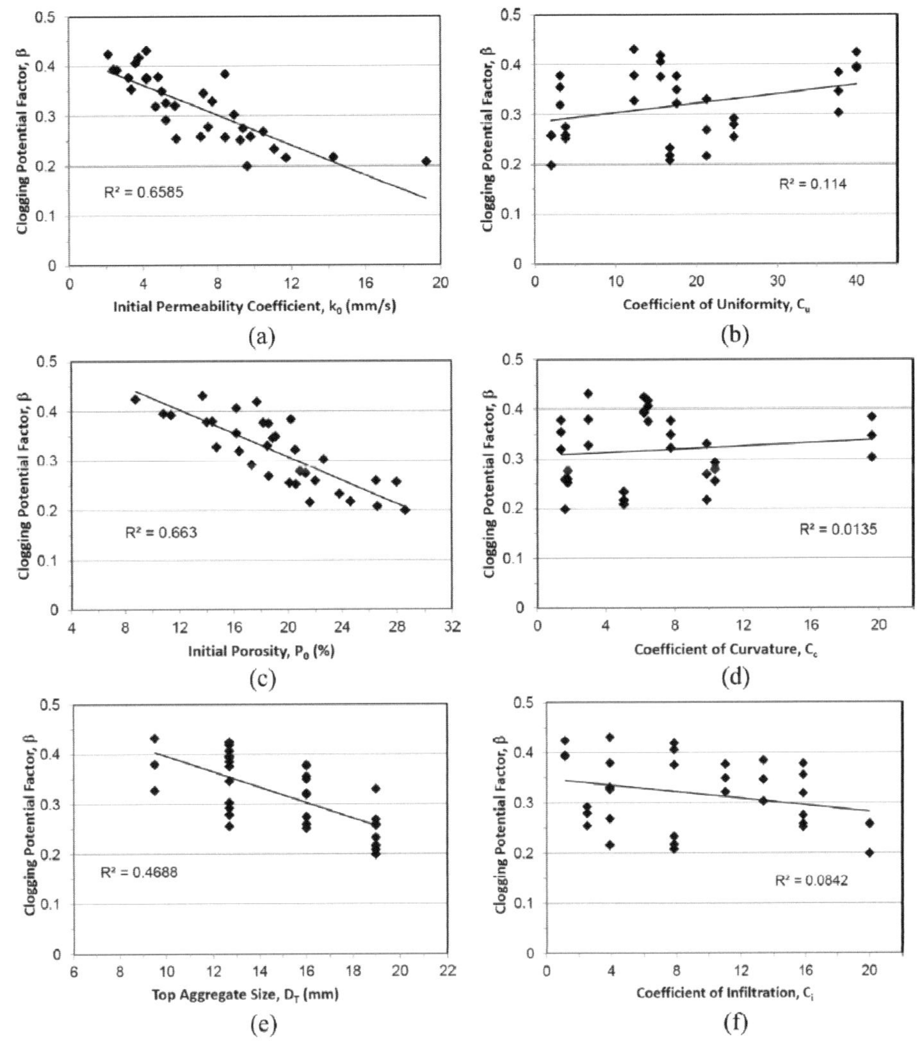

Figure 4. Relationships between clogging potential factor k_0 and six predictor variables. (*a*) β versus k_0 plot (*b*) β versus P_0 plot (*c*) β versus D_T plot (*d*) β versus C_u plot (e) β versus C_c plot (*f*) β versus C_i plot.

Such differences in pore sizes would influence the clogging potential of the mixtures. Figures 4*c* and 5*c* indicate that, for the range of top aggregate sizes from 9.5 to 19 mm, those mixtures with larger top aggregate sizes tend to display lower clogging potential (i.e. lower β and higher N_c values).

- Coefficient of uniformity C_u: C_u is defined as the ratio (D60/D10), where D60 is the aggregate diameter at 60% passing, and D10 is the aggregate diameter at 10% passing. A larger C_u value means a more well-graded gradation, that is, a gradation with a wider range of different grain sizes, leading to lower porosity and pore sizes (*23*). Figures 4*d* and 5*d* suggest that although mixtures with a lower C_u value tend to produce lower clogging potential (i.e. lower β and higher N_c values), the relationships are statistically weak.
- Coefficient of curvature C_c: C_c is defined as the ratio $(D30)^2/(D60 \times D10)$, where D30 is the aggregate diameter at 30% passing. The material is considered to be well graded if $1 < C_c < 3$ (*24*). Table 5 reveals that the C_c of the eleven mix types varies from 1.62 to 19.62, covering both the well graded and poorly graded range of C_c values. Based on the results of Figures 4*e* and 5*e*, C_c is found to have only a very weak effect on the clogging potential characteristics of porous asphalt mixes studied.
- Coefficient of infiltration C_i: C_i is defined as the ratio (D15/d85), where D15 is the aggregate diameter at 15% passing of the mix gradation, and d85 is the grain

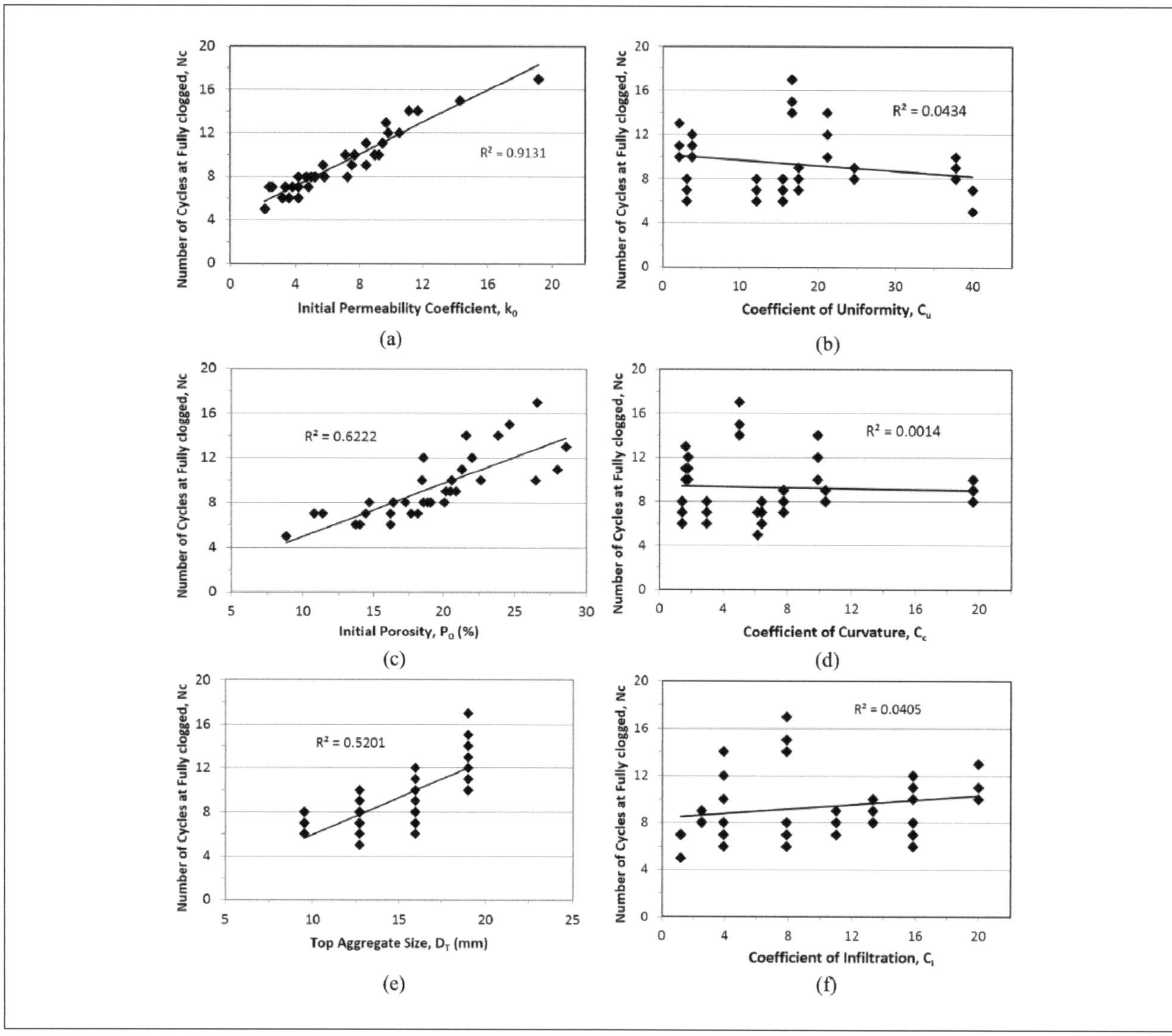

Figure 5. Relationships between number of clogging cycles when fully clogged N_c and six predictor variables. (*a*) N_C versus k_O plot (*b*) N_C versus P_O plot (*c*) N_C versus D_T plot (*d*) N_C versus C_u plot (*e*) N_C versus C_c plot (*f*) N_C versus C_i plot.

diameter at 85% passing of the clogging material. A value of 4 or larger means that the clogging material will infiltrate into the mixture tested (*22*). However, according to Figures 4*f* and 5*f*, this factor is also found to have very weak influence on the clogging potential characteristics of porous asphalt mixes studied.

Predictive Models of Clogging Potential of Porous Asphalt Mixture. Predictive models are useful for evaluating the likely clogging performance and clogging potential of a proposed porous asphalt mixture. This section develops the predictive models for the two clogging potential indicators established in the study, namely the clogging potential factor, β and the number of clogging cycles to reach the fully clogged state, N_c.

A stepwise linear multiple regression was performed to develop the two predictive models, each considering the six predictor variables identified in the preceding sub-section. The significance level for a variable to enter a model was set as 0.05, and that for removal is taken as 0.1. The predictive regression models obtained are:

$$\beta = 0.718 - 1.17\times10^{-2}k_0 - 2.44\times10^{-3}P_0 - 1.40\times10^{-2}D_T \quad R^2 = 0.716 \tag{5}$$

$$N_c = 1.571 + 0.608\,k_0 + 0.037\,P_0 + 0.186\,D_T \quad R^2 = 0.923 \tag{6}$$

Table 5. Parameter Values for Modeling Clogging Characteristics

Specimen code	Mix type	Clogging potential factor, β	Clogging cycles to fully clogged state, N_c	Top aggregate size D_T (mm)	Coefficient of uniformity, C_u	Coefficient of curvature, C_c	Coefficient of infiltration, C_i
1	P1	0.255	17	19	16.67	5.04	6.36
2	P1	0.251	15	19	16.67	5.04	6.36
3	P1	0.243	14	19	16.67	5.04	6.36
4	P2	0.248	13	19	2.04	1.62	2.00
5	P2	0.31	11	19	2.04	1.62	2.00
6	P2	0.323	10	19	2.04	1.62	2.00
7	P3	0.268	14	19	21.25	9.89	10.76
8	P3	0.328	12	19	21.25	9.89	10.76
9	P3	0.384	10	19	21.25	9.89	10.76
10	P4	0.365	9	16	17.50	7.78	3.55
11	P4	0.414	8	16	17.50	7.78	3.55
12	P4	0.512	7	16	17.50	7.78	3.55
13	P5	0.301	12	16	3.80	1.78	2.53
14	P5	0.318	11	16	3.80	1.78	2.53
15	P5	0.342	10	16	3.80	1.78	2.53
16	P6	0.346	10	12.7	37.78	19.62	2.13
17	P6	0.389	9	12.7	37.78	19.62	2.13
18	P6	0.399	8	12.7	37.78	19.62	2.13
19	P7	0.433	9	12.7	24.67	10.38	6.67
20	P7	0.409	8	12.7	24.67	10.38	6.67
21	P7	0.459	8	12.7	24.67	10.38	6.67
22	P8	0.396	8	16	3.17	1.39	2.53
23	P8	0.417	7	16	3.17	1.39	2.53
24	P8	0.415	6	16	3.17	1.39	2.53
25	P9	0.504	8	12.7	15.50	6.45	4.87
26	P9	0.469	7	12.7	15.50	6.45	4.87
27	P9	0.588	6	12.7	15.50	6.45	4.87
28	P10	0.460	8	9.5	12.24	2.97	8.05
29	P10	0.476	7	9.5	12.24	2.97	8.05
30	P10	0.520	6	9.5	12.24	2.97	8.05
31	P11	0.437	7	12.7	40.00	6.19	27.14
32	P11	0.430	7	12.7	40.00	6.19	27.14
33	P11	0.423	5	12.7	40.00	6.19	27.14

Note: C_u = (D60/D10); C_c = $(D30)^2$/(D60 × D10); C_i = (D15/d85), where D10, D15, D30, and D60 are aggregate diameter at 10, 15, 30, and 60 passing, respectively, of the porous asphalt mixture concerned, and d85 is the grain diameter at 85% passing of the clogging material.

For both the β and N_c models, k_0, P_0, and D_T were found to be the significant predictor variables, and the three gradation curve descriptors C_u, C_c, and C_i were not found to have significant effects on β and N_c. The two models indicate that high values of k_0, P_0, and D_T will give low β values (i.e. low rate of permeability loss in clogging treatment) and high N_c values (i.e. more clogging cycles to reach fully clogged). As an example, consider a practice in some countries that adopts a minimum design initial porosity of 20%; Figures 4 and 5 indicate that the average values of β and N_c for this design are about 0.3 and 10 cycles, respectively. Taking these performance criteria as a benchmark of porous asphalt mix design with sufficiently low clogging potential, the test results suggest that the desirable initial permeability k_0 and top aggregate size D_T would be 8 mm/s or more, and 16mm or larger, respectively.

Conclusion

A laboratory procedure has been developed to study the clogging performance of eleven porous asphalt mix types. The procedure consists of a clogging treatment to introduce a selected clogging material into test specimens in stages, and permeability measurements after each stage of clogging. Test results obtained allow two indicators, namely the clogging potential factor β and the number of clogging cycles to reach fully clogged state N_c, to be defined to describe the clogging performance and clogging potential of the mixtures studied. To understand the clogging behavior of the mixes tested, six mix properties were identified and their effects studied statistically. The analysis found that the two clogging potential indicators were significantly influenced by the initial porosity, initial permeability and the top aggregate size of the mix concerned. Statistically significant predictive models for β and N_c were developed with these three factors as the predictor variables. The predictive models can serve to provide useful guidance for pavement engineers in the design of porous asphalt mixtures. For new mixes that are not covered in the present study, this paper has provided a useful laboratory procedure for establishing their clogging potentials.

References

1. NCHRP. *Open-Graded Friction Courses for Highways. NCHRP Synthesis of Highway Practice 49*. Transportation Research Board, National Research Council, Washington D.C., 1978.
2. Isenring, T., H. Koster, and I. Scazziga. Experiences with Porous Asphalt in Switzerland. *Transportation Research Record: Journal of the Transportation Research Board*, 1990. 1265: 41–53.
3. McGovern, C., P. Rusch, and D. A. Noyce. *State Practices to Reduce Wet Weather Skidding Crashes*. Publication FHWA-SA-11-21. Federal Highway Administration, U.S. Department of Transportation, Washington D.C., 2011.
4. Bendtsen, H. *Porous Asphalt Pavement and Noise Reduction over a Long Period*. EURO- NOISE, Munich, Germany, 1998.
5. Raaberg, J., B. Schmidt, and H. Bentsen. *Technical Performance and Long-Term Noise Reduction of Porous Asphalt Pavement*. Report 112, Road Directorate, Danish Road 461 Institute, 2001.
6. Skarabis, J., and U. Stockert. Noise emission of concrete pavement surfaces produced by diamond grinding. *Journal of Traffic and Transportation Engineering (English Edition)*, Vol. 2, No. 2, 2015, pp. 81–92.
7. Sandberg, U., and J. A. Ejsmont. Tyre/Road Noise Reference Book, INFORMEX, Kisa, 466 Sweden, 2002.
8. Fwa, T. F., S. A. Tan, and Y. K. Guwe. Laboratory Evaluation of Clogging Potential of Porous Asphalt Mixtures. *Transportation Research Record: Journal of the Transportation Research Board*, 2002. 1681: 43–49.
9. Tan, S. A., T. F. Fwa, and Y. K. Guwe. Laboratory Measurements and Analysis of Clogging Mechanism of Porous Asphalt Mixes. *Journal of Testing and Evaluation*, Vol. 28, No. 3, 2000, pp. 207–216.
10. Fwa, T. F, E. Lim, and K. H. Tan. Comparison of Permeability and Clogging Characteristics of Porous Asphalt and Pervious Concrete Pavement Materials. *Transportation Research Record: Journal of the Transportation Research Board*, 2015. 2511: 72–80.
11. Fwa, T. F., S. A. Tan, and C. T. Chuai. Permeability Measurement of Base Materials Using Falling-Head Test Apparatus. *Transportation Research Record: Journal of the Transportation Research Board*, 1998. 1615: 94–99.
12. Tan, S.A., T. F. Fwa, and C. T. Chuai. Automatic Field Permeameter for Drainage Properties of Porous Asphalt Mixes. *Journal of Testing and Evaluation*, Vol. 27, No. 1, 1999, pp. 57–62.
13. Scheidegger, A. E. *The Physics of Flow through Porous Media*. University of Toronto Press, Toronto, Canada, 1963.
14. Bear, J. *Dynamics of Fluids in Porous Media*. Elsevier, New York, 1972.
15. Chu, L., T. F. Fwa, and K. H. Tan. Evaluation of Wearing Course Mix Designs on Sound Absorption Improvement of Porous Asphalt Pavement. *Construction and Building Materials*, Vol. 141, 2017, pp. 402–409.
16. Guwe, Y. K. *Draiange Analysis and Design of Porous Asphalt Pavement*. PhD thesis. Department of Civil Engineering, National University of Singapore, 2002.
17. Fwa, T F, C. C. Yoong, T. T. Nyunt, and S. L. See. Development of Environmentally Sustainable Pavement Mix. *International Journal of Pavement Research and Technology*, Vol. 6, No. 4, 2013, pp. 440–446.
18. Nicholls, J. C. *Asphalt Surfacings*. E & FN Spon, London, UK.
19. Hamzah, M. O., and Hardiman. Characterization of the Clogging Behaviour of Double Layer Porous Asphalt. *Journal of the Eastern Asia Society for Transportation Studies*, Vol. 6, 2005, pp. 968–980.
20. Kandhal, P. S., and R. B. Mallick. *Transportation Research Circular E-C005: Open Graded Friction Course: State of the Practice*. HRB, National Research Council, Washington, D.C., 1998.
21. Suresha, S. N., G. Varghese, and A. U. R. Shankar. Laboratory and Theoretical Evaluation of Clogging Behaviour of Porous Friction Course Mixes. *International Journal of Pavement Engineering*, Vol. 11, No. 1, 2010, pp. 61–70. doi:10.1080/10298430902730554.
22. Tan, S. A., T. F. Fwa, and C. T. Han. Clogging Evaluation of Permeable Bases. *Journal of Transportation Engineering-ASCE*, Vol. 129, No. 3, 2003, pp. 309–315.
23. Holtz, R., and W. Kovacs. *An Introduction to Geotechnical Engineering*. Prentice-Hall, Inc., Upper Saddle River, 1981.
24. ASTM D2487-11. *Standard Practice for classification of soils for Engineering Purposes (Unified Soil Classification System)*. ASTM International, West Conshohocken, PA, 2011.

The Standing Committee on Subsurface Drainage (AFS60) peer-reviewed this paper (18-00413).

Article

TRR
JOURNAL OF THE TRANSPORTATION RESEARCH BOARD

Transportation Research Record
2018, Vol. 2672(52) 23–33

DOI: 10.1177/0361198118756881
journals.sagepub.com/home/trr
SAGE

Use of an Artificial Neural Network Approach for the Prediction of Resilient Modulus for Unbound Granular Material

Sajib Saha[1], Fan Gu[4], Xue Luo[2], and Robert L. Lytton[3]

Abstract

The resilient modulus (M_R) is a fundamental material property that has a direct effect on the design and analysis of pavement structures. Many regression models have been developed previously to predict the coefficients of the M_R model from physical properties of base materials. However, the predicted model coefficients are confined to either a limited number of base materials or result in poor accuracy. To overcome this issue, a moisture- and stress-dependent model is adopted in this study to precisely estimate M_R of unbound base materials in unsaturated conditions, and a set of artificial neural network (ANN) models is developed to predict the coefficients of this model from base physical properties. The developed ANN models consist of seven input variables, ten hidden neurons, and one output variable. A large unbound base dataset was collected from the Long Term Pavement Performance (LTPP) database and used to train and generalize the network. Soil physical properties such as gradation (percent passing No. 3/8 sieve, percent passing No. 200 sieve), gradation shape parameter and scale parameter, index properties (i.e., plastic limit and plasticity index), maximum dry density, optimum moisture content, and test moisture content were selected as inputs for the ANN model. The M_R values estimated using the predicted coefficients were compared with the experimental data collected from LTPP and showed an R^2 value above 0.9, which is much higher than the M_R values computed using regression models. Finally, the M_R test results from different sources were used to validate the developed ANN models.

Unbound base materials beneath the pavement surface are characterized by the resilient modulus (M_R) in the mechanistic–empirical pavement design (*1*), which has a direct effect on the design and analysis of flexible pavement structures. The advantage of M_R over other material properties is that it represents the response to dynamic loading coming from vehicle movement. It is defined as the ratio of repeated cyclic loading (σ_{cyc}) divided by recoverable strain (ε_r):

$$M_R = \frac{\sigma_{cyc}}{\varepsilon_r} \qquad (1)$$

The resilient modulus can be estimated using three different approaches: (1) conduct lab tests with compacted soil specimen [e.g., repeated load triaxial test (RLT)]; (2) follow back calculation method from an in situ device [e.g., falling weight deflectometer (FWD)]; and (3) predict M_R model coefficients from soil physical properties. The first and second approaches are expensive, time consuming, and require expert labor. As a result, the third approach based M_R model is used in this study to characterize the unbound material under different conditions.

Various models have been developed to predict the resilient modulus of unbound base and subgrade materials. The most widely used generalized model developed in the National Cooperative Highway Research Program (NCHRP) project 1-28A is (*2*)

$$M_R = k_1 P_a \left(\frac{I_1}{P_a}\right)^{k_2} \left(\frac{\tau_{oct}}{P_a} + 1\right)^{k_3} \qquad (2)$$

where

I_1 = first invariant of the stress tensor;
τ_{oct} = octahedral shear stress;
P_a = atmospheric pressure; and
k_1, k_2, and k_3 = regression coefficients.

However, many studies have shown that the resilient modulus of unbound granular material is not only stress dependent but also moisture dependent (*3*). The AASHTO employed a moisture-dependent M_R model that was later

[1]Texas A&M Transportation Institute, Texas A&M University System, College Station, TX
[2]Center for Infrastructure Renewal, Texas A&M University System, College Station, TX
[3]Zachry Department of Civil Engineering, Texas A&M University System, College Station, TX
[4]National Center for Asphalt Technology, Auburn University, AL

Corresponding Author:
Address correspondence to Sajib Saha: sajibsaha@tamu.edu

adopted by the mechanistic–empirical pavement design guide (MEPDG) (*1*):

$$\log\frac{M_R}{M_{Ropt}} = a + \frac{b-a}{1+\exp[\ln\frac{-b}{a} + k_m(S-S_{opt})]} \quad (3)$$

where

M_R = resilient modulus at a given degree of saturation;

M_{Ropt} = resilient modulus at a reference condition;

a and b = minimum and maximum of $\log(M_R/M_{Ropt})$, respectively;

k_m = regression coefficient; and

$(S-S_{opt})$ = variation of degree of saturation, expressed as a decimal.

Other researchers (*4–6*) also proposed different moisture-dependent resilient modulus models for both unbound aggregates and subgrade. However, previous studies also demonstrated the importance of the matric suction in the resilient modulus model. The Pavement ME Design incorporated the suction in the resilient modulus model through NCHRP project 9-23A (*7*).

$$M_R = k_1 P_a (\frac{\theta + wc * \text{matric\_suction}}{P_a})^{k_2} (\frac{\tau_{oct}}{P_a} + 1)^{k_3} \quad (4)$$

where

θ = bulk stress; and

wc = water content.

There have been many other studies recommending the use of soil suction in modeling the resilient modulus (*8–12*). It has been shown that the matric suction can be different for the same variation of degree of saturation, that is ($S - S_{opt}$), from one material to another (*13*). Therefore, the degree of saturation and soil suction alone cannot accurately estimate the change of resilient modulus due to moisture. To consider the moisture and suction dependency of resilient modulus characteristics, a new model was proposed by Lytton (*14*):

$$E_y = k_1 P_a (\frac{I_1 - 3\theta f h_m}{P_a})^{k_2} (\frac{\tau_{oct}}{P_a})^{k_3} \quad (5)$$

where

I_1 = first invariant of the stress tensor;

P_a = atmospheric pressure;

θ = volumetric water content;

h_m = matric suction in the base matrix;

f = saturation factor, $1 < f < 1/\theta$;

τ_{oct} = octahedral shear stress; and

k_1, k_2, and k_3 = regression coefficients.

In this model, I_1 and τ_{oct} vary with the stress state, and h_m is related to the moisture content of unbound aggregates. Three steps are followed to determine the coefficients k_1, k_2, and k_3 using Equation 5.

1. Matric suction h_m in Equation 5 is estimated first at tested specimen water content. A separate artificial neural network (ANN) model is used to predict the coefficients of the soil–water characteristic curve (SWCC) parameters in the Fredlund–Xing equation described in later sections.
2. The volumetric water content, θ, and saturation factor, f, are calculated for each base material from the collected moisture content, maximum dry density (MDD), and specific gravity data. The saturation factor is an indicator of degree of saturation (*14*). It is multiplied by θ and h_m to express the stress on the soil mineral skeleton in the transition zone. The transition zone occurs between the air entry point suction value and the unsaturation point suction value. The saturation corresponding to the air entry point and the unsaturation point are considered as 100% and 85%, respectively, in this study. The following equation is used to calculate the saturation factor f in the transition zone:

$$f = 1 + \frac{S-85}{15}(\frac{1}{\theta} - 1) \quad (6)$$

where

S = degree of saturation (unit: %);

θ = volumetric water content (unit: decimal); it is considered to be 2 pF and 3.5 pF at the air entry point and the unsaturation point, respectively.

3. Coefficients k_1, k_2, and k_3 are calculated from collected M_R data at different combinations of confining pressure and deviator stresses. A Matlab code was generated to fit the curve for each base material.

Researchers have developed many correlation models to predict k_1, k_2, and k_3 from soil gradation, index, and strength properties (*15–18*). However, most of these models are either confined to a limited number of datasets or have a poor prediction accuracy. As a result, an ANN model is developed in this study to predict the coefficients from soil properties. The main advantage of a neural network model over nonlinear regression models is that it can capture the complex nonlinear scattered relationship between input and output parameters and train the model based on the evaluation of the error function. The application of the ANN approach for predicting the resilient modulus of subgrade soil has been studied by a few researchers (*19, 20*). Soil index properties [i.e., material percent passing No. 200 sieve, plasticity index (PI), optimum moisture content (OMC), MDD, variation in compaction water content] and stress variables (i.e., confining pressure, bulk stress, deviator stress) were used as input variables to predict the resilient modulus (*19*). A different neural network analysis was performed with nine different soil sources in Georgia (*20*). A similar study has been performed for $\ln(k_1)$, k_2, and k_3 coefficients by Nazzal and Tatari (*21*). Input variables for the correlation models developed by Nazzal and Tatari consisted of material percent passing No. 200 sieve, liquid limit (LL), PI, MDD, test dry density (TDD), OMC, test moisture content (TMC), and percentage of clay (% clay). However, previous ANN approaches had some limitations. The resilient modulus values were predicted either as model output for a single stress combination (*19, 20*) or using only the locally available soil data as input (*21*). To overcome these limitations, an ANN model is developed in this study using a large dataset of unbound base materials collected from the Long Term Pavement Performance (LTPP) database. Moreover, the resilient modulus equation coefficients (k_1, k_2, and k_3) were predicted as outputs to cover a wide range of stress combinations.

The paper is organized as follows. The next section describes the procedure to estimate suction at specific test specimen water content. Subsequently, ANN models are developed to predict the coefficients of the M_R model using the estimated suction. Next, the M_R values using the predicted coefficients are compared against other existing regression models. Unbound base course data collected by the Texas A & M Transportation Institute (TTI) and University of Illinois at Urbana-Champaign (UIUC) were used to validate the developed ANN models and are described next. The final section summarizes the major findings of this study.

Prediction of Suction at Test Specimen Water Content

Matric suction of unbound granular base material was required to accurately estimate the resilient modulus in unsaturated condition as shown in Equation 5. However, the quantity of available SWCC data of unbound base materials from the available literature was not sufficient to train an ANN model. Therefore, the Soil Survey Geographic Database (SSURGO) was used for data collection from the United States Department of Agriculture Natural Resources Conservation Service (USDA-NRCS). The database included soil physical and climatic properties as well as water content at specific suction pressure. Collected suction–saturation data points for each soil were fitted in the Fredlund–Xing equation as shown in the following equations (*22*):

$$S = C(h)\times[\frac{1}{\{\ln[e+(\frac{h}{a_f})^{b_f}]\}^{c_f}}] \tag{7}$$

$$C(h) = 1-\frac{\ln(1+\frac{h}{h_r})}{\ln[1+(\frac{1.45\times10^5}{h_r})]} \tag{8}$$

where

S = the degree of saturation (unit: decimal);

h = the soil matric suction (unit: psi); and

a_f, b_f, c_f, are h_r = soil fitting parameters.

A Matlab code was generated to fit the Fredlund–Xing equation with measured data points and calculate the SWCC fitting parameters. Out of a total 34,237 data points on soils collected from the NRCS, soils that satisfy gravel characteristics according to Unified Soil Classification System (more than 50% material retained on the No. 200 sieve is defined as coarse-grained material and more than 50% of the coarse-grained material retained on the No. 4 sieve) were separated out to train and validate the ANN models for the SWCC fitting parameters. The 2598 plastic (PI > 0) soil data and 311 nonplastic (PI = 0) soil data were collected and used to train the neural network to predict the SWCC fitting parameters for unbound materials.

Following the architecture of Saha et al. (*23*), two three-layered ANN models consisting of one input layer, one hidden layer, and one output layer were constructed for plastic and nonplastic soils, respectively. The input variables for plastic soils included percent passing No.4 sieve, percent passing No.200 sieve, LL, plastic limit (PL), saturated volumetric water content, and local mean annual air temperature (MAAT). Input variables for nonplastic soils were particle diameters corresponding to 30%, 60%, and 90% passing of material D_{30}, D_{60}, and D_{90}, gradation scale parameter (θ), and shape parameter (Ψ), saturated volumetric water content, and local MAAT. The output of the developed ANN models were the fitting parameters (i.e., a_f, b_f, c_f and h_r) of the SWCC in the Fredlund–Xing equation, as shown in Equations 6 and 7.

The hidden layer in the ANN model was assigned 20 neurons to establish the connection between input and output layers. A total of 80% of the dataset was used for training and 20% was used for validation of the ANN models. Water contents at 0.1 bar, 0.33 bar, and 15 bar (1 bar = 100 kPa) were available in the USDA-NRCS database. In this study, the predicted SWCC parameters were input into the Fredlund–Xing equation to estimate the water content at the same suction levels. The estimated water contents were compared against the measured values to investigate the prediction accuracy.

Figure 1, *a* and *b* show the plots of measured versus predicted degree of saturation for plastic and nonplastic soils, respectively. The developed ANN models had a very good prediction accuracy generating an R^2 value of 0.91 for both plastic and nonplastic soils. Therefore, the same trained models were used to predict the SWCC fitting parameters for base materials collected from the LTPP database. Predicted a_f, b_f, c_f, and h_r parameters of the LTPP materials were input into the Fredlund–Xing equation to estimate the matric suction of the specimen at the tested water content.

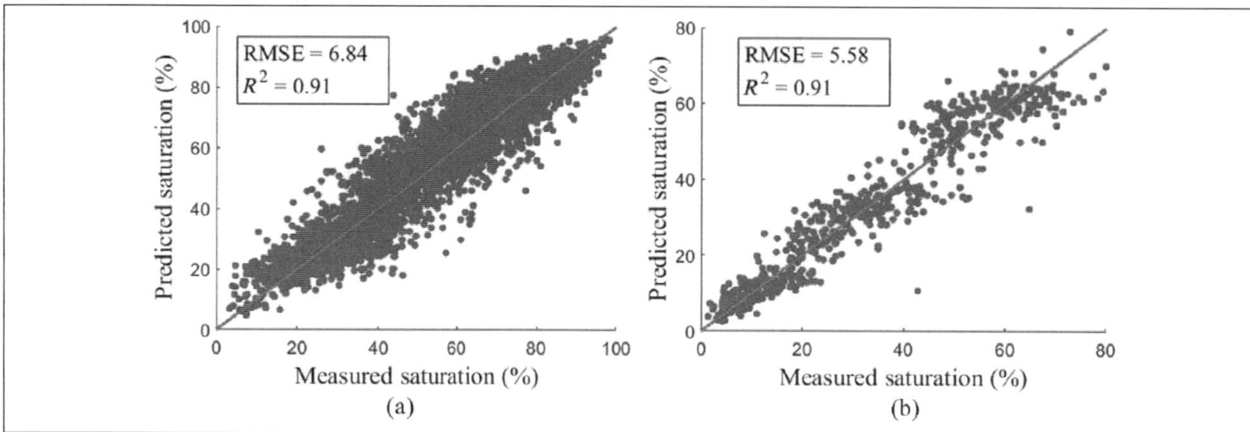

Figure 1. Comparison of measured versus predicted saturation (%) at 0.1 bar, 0.33 bar, and 15 bar suction level for unbound granular base materials using the ANN model: (*a*) plastic soils; (*b*) nonplastic soils.

Development of ANN Models for M_R Model Coefficients

The ANN approach is an adaptive information-processing technique, which allows the correlations between the input variables X_i and the output variables Y_j to be established through the inter-connected neurons (i.e., weight factor, w_{ji}). Note that the input variables X_i and the output variables Y_j are usually normalized to x_i and y_j, respectively. The correlations developed by the ANN models between the normalized input parameters x_i and the normalized output variables y_j are

$$y_j = f\left(\sum_{i=1}^{n} w_{ji}x_i\right) + b_j \quad (9)$$

where

f = a transfer function, which normally uses a sigmoidal, Gaussian, or threshold functional form;

w_{ji} = unknown weight factors; and

b_j = bias term.

A neural network model specifically adjusts the weight factors w_{ji} and bias b_j in Equation 9 based on the minimum error function. In pavement engineering, the ANN approach is usually used to develop prediction models based on large amounts of data collected from experiments and numerical analysis. ANN models have been successfully developed to predict the crack growth function (e.g., reflective cracking and top-down cracking) in asphalt concrete (*24,25*). The ANN approach was also utilized to predict the geogrid-reinforced flexible pavement performance (*26*). In general, the development of ANN models includes two critical steps: (1) data collection; and (2) construction of ANN architecture.

Data Collection

A large collection of unbound base materials was available in the LTPP database. Out of a total of 3,010 unbound base materials, 717 materials were selected that passed the gravel specifications. A total of 217 plastic (PI > 0) and 500 nonplastic (PI = 0) base materials were separated out to develop two different sets of ANN models for k_1, k_2, and k_3. Resilient modulus tests were conducted on 15 different combinations of confining pressure and nominal maximum axial stress level. In addition to the resilient modulus test data, physical properties of base materials such as gradation, Atterberg limits, moisture content, density, and specific gravity were collected from the LTPP database.

Construction of ANN Architecture

Two three-layered ANN models, one for plastic and another for nonplastic soils, were constructed as shown in Figure 2, *a* and *b*, respectively. Three layers of the ANN model consisted of a single input, a hidden layer, and an output layer. Soil physical properties, that is, percent of material passing 3/8 in. sieve, percent of material passing No. 200 sieve, PL, PI, OMC, MDD, and TMC were introduced as input parameters for plastic soils. Input parameters for nonplastic soils included the percent of material passing the 3/8 in. sieve, percent of material passing the No. 200 sieve, gradation scale parameter θ, shape parameter Ψ, OMC, MDD, and TMC. The hidden layer assigned 10 neurons to establish the connection between the output layer and the input layer. The number of neurons was selected based on the number of data points. Too many hidden neurons allow the network to memorize instead of generalizing the training set (*27*).

Both the nonplastic and plastic soil models utilized the sigmoidal transfer function (*26*):

$$f(I_i) = \frac{1}{1+\exp(-\varphi I_i)} \quad (10)$$

where

I_i = the input quantity; and

φ = a positive scaling constant.

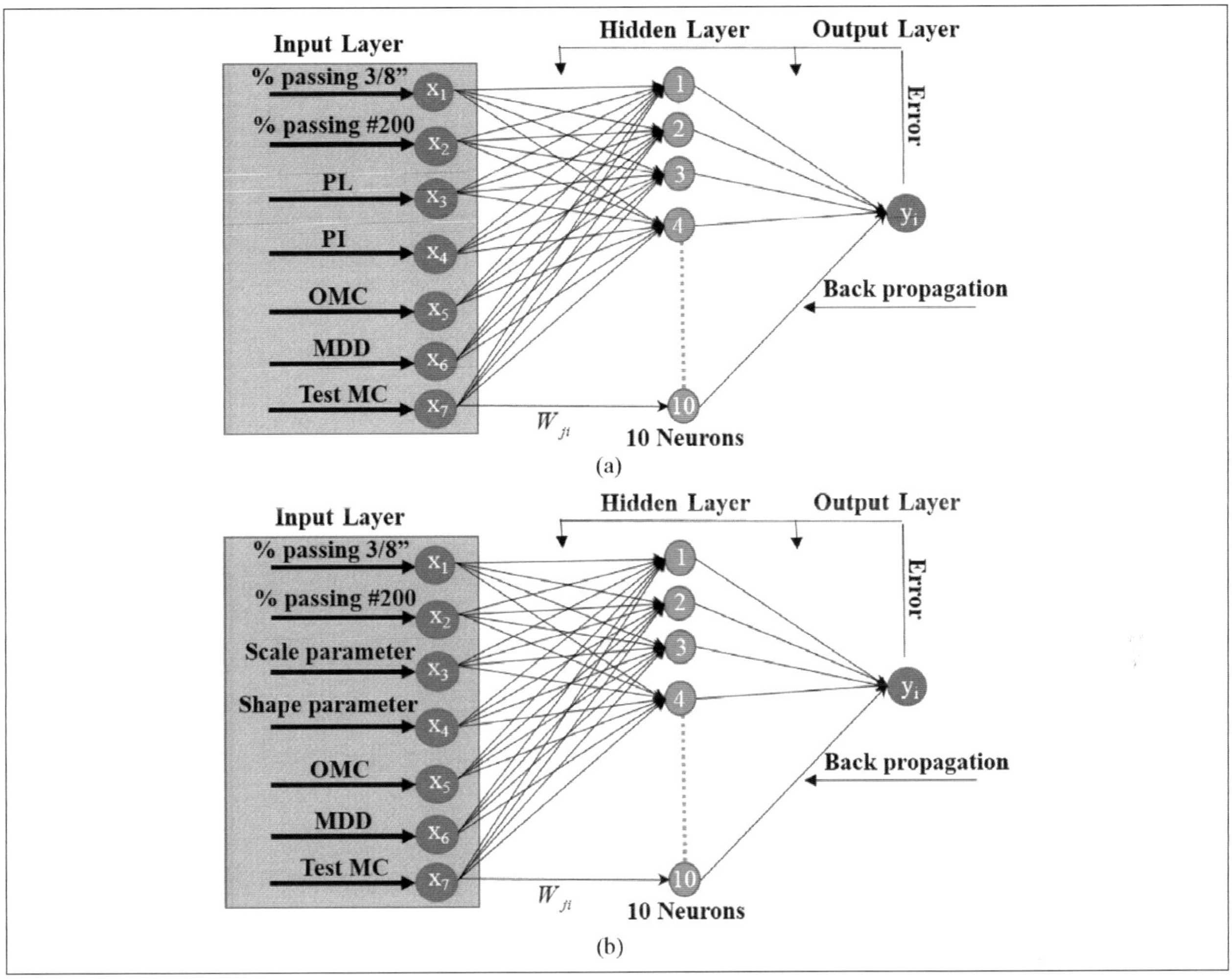

Figure 2. Illustration of the three-layered neural network architecture: (*a*) plastic soil; (*b*) nonplastic soil.

The parameter φ controls the steepness between the two asymptotic values 0 and 1. The ANN model determines these weight factors w_{ji} through the two major functions: training and validating. The training dataset is used to determine the trial weight factors, w_{ji}, and bias term, b_j; the validating dataset is employed to examine the accuracy of the model prediction. In this study, 80% of the dataset was used for training and 20% of the dataset was used for validation. The training algorithm used the Levenberg–Marquardt backpropagation method to minimize the mean squared error (MSE). The gradient descent weight function was employed as a learning algorithm to adjust the weight factors w_{ji}.

Prediction of M_R Model Coefficients Using ANN Model

In this study, the ANN model was developed to predict the M_R model coefficients (i.e., k_1, k_2, and k_3) from base physical properties. The coefficients of the M_R model were correlated with base material physical properties and used to estimate the resilient modulus. Figure 3 shows a comparison between measured and predicted M_R model coefficients for plastic soils. The predicted M_R coefficients showed high R^2 and low root mean square error (RMSE), which indicated a good prediction accuracy. The R^2 values for M_R coefficients were in the range of 0.66 to 0.73. Similarly, Figure 4 plots the comparisons between measured and predicted M_R model coefficients for nonplastic soils. The R^2 value for k_1, k_2, and k_3 coefficients were 0.78, 0.74, and 0.58 respectively. Note that a high level of variability was involved in estimated coefficients due to variations in cyclic load triaxial test results and seasonal changes in selected physical properties (*28*). Therefore, the predicted k_1, k_2, and k_3 values need to be scrutinized in the validation process. If the validation dataset provides comparable results, then the predicted coefficients can be considered as sufficiently accurate to use in the M_R model equation.

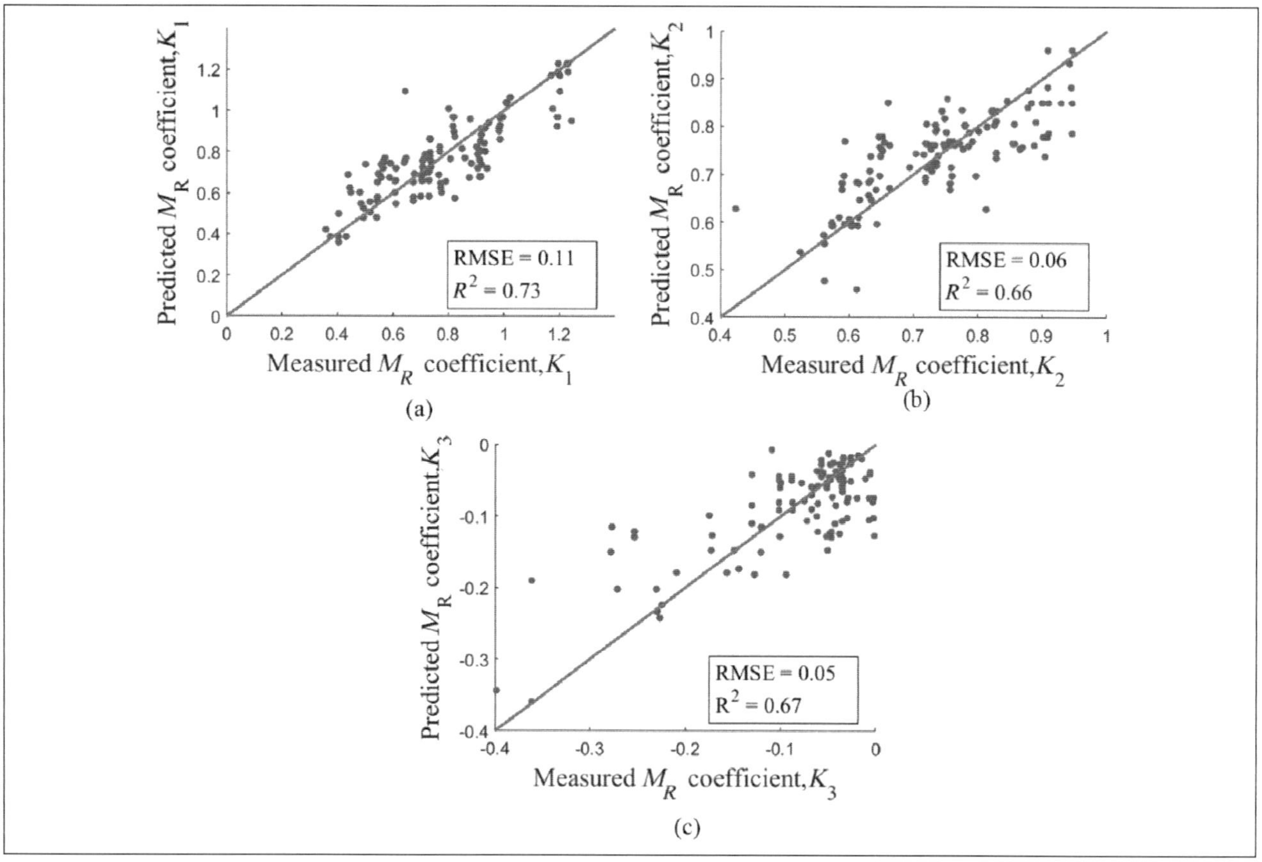

Figure 3. Predicted M_R model coefficients of plastic base materials from physical properties using an ANN approach: (*a*) k_1; (*b*) k_2; and (*c*) k_3.

Comparison of ANN Model with Other Regression Models

A number of regression models have been developed to correlate the soil physical properties with M_R model coefficients. In this study three correlation models (*16,18,29*) were selected and compared against measured values to investigate the accuracy of the M_R coefficients. They were called the Yau model (*15*), Malla model (*17*), and Soliman model (*29*), respectively. All three regression models used the generalized M_R constitutive equation shown in Equation 2.

Seven different sets of regression equations were proposed by Yau and Von Quintus (*15*) for crushed stone materials, crushed gravel, uncrushed gravel, sand, coarse-grained soil–aggregate mixture, fine-grained soil–aggregate mixture, and fine-grained soil, respectively. The physical properties used to correlate with the generalized M_R model coefficients were material percent passing 3/8 in. sieve, No. 4 sieve, No. 40 sieve, No. 200 sieve, percentage of silt (% silt), % clay, LL, PI, OMC, MDD, TMC, and TDD.

Malla and Joshi (*17*) studied the correlation between the M_R model coefficients and the gradation and compaction parameters of base materials. Regression analysis was performed on M_R model coefficients for uncrushed gravel and crushed limestone. The variables were selected based on the literature and the logical influence of gradation, compaction, and moisture content parameters on the regression constants (k_1, k_2, and k_3). These variables included: OMC, TMC, MDD, material percent passing No. 4 sieve, material percent passing No. 40 sieve, and material percent passing No. 200 sieve.

Soliman and Shalaby (*29*) presented regression equations of the generalized M_R model coefficients for six AASHTO soil types (A-1-b, A-3, A-2-4, A-4, A-6, and A-7-6). Data used in the regression analysis were collected from 19 states in New England and the nearby regions in the U.S. and two provinces in Canada. The soil properties that were considered in the regression analysis include OMC, TMC, MDD, LL, PI, TDD, percentage of coarse sand (% CSAND), percentage of fine sand (% FSAND), percentage of silt (% silt), percentage of clay (% clay), and material percent passing 3 in. sieve, 2 in. sieve, 1½ in. sieve, 1in. sieve, 3/4 in. sieve, 1/2 in. sieve, 3/8 in. sieve, No. 4 sieve, No. 10 sieve, No. 40 sieve, No. 80 sieve, and No. 200 sieve.

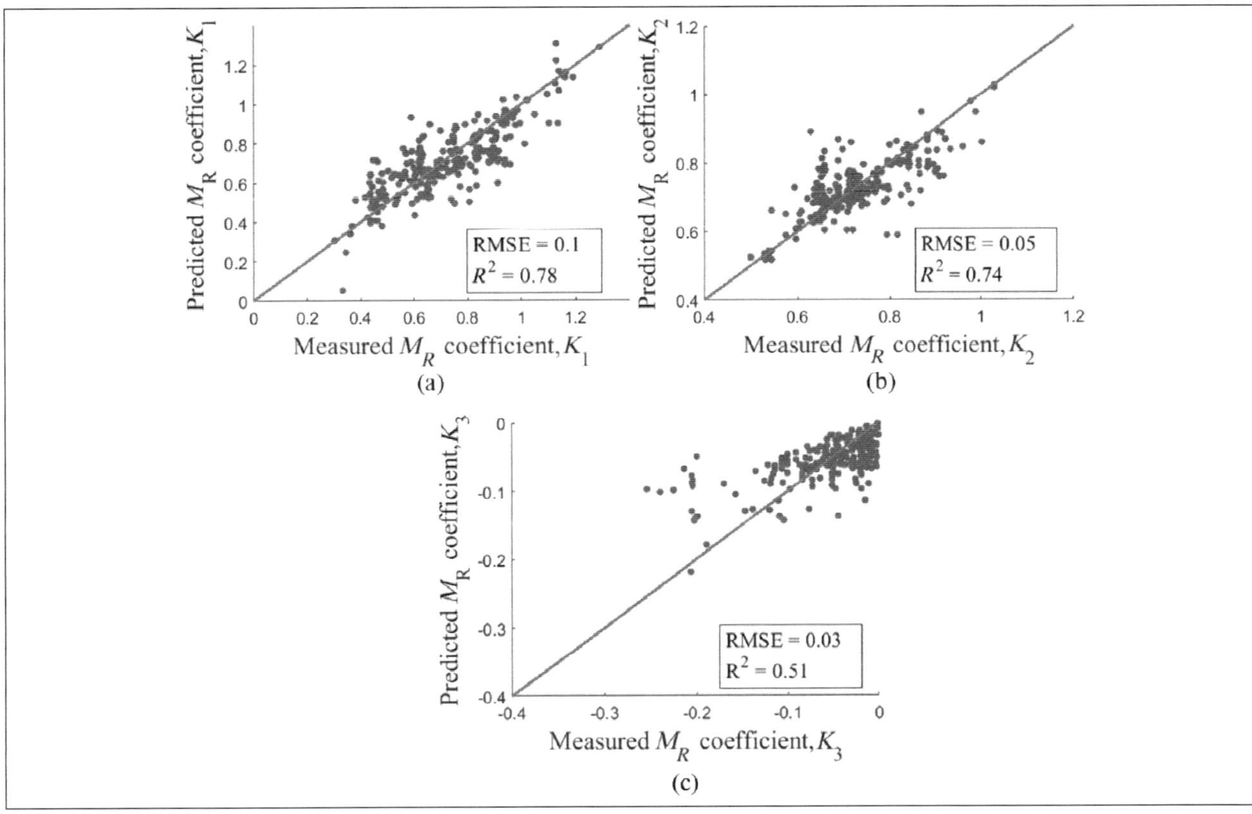

Figure 4. Predicted M_R model coefficients for nonplastic base materials from physical properties using an ANN approach: (*a*) k_1; (*b*) k_2; and (*c*) k_3.

Table 1. Prediction Accuracy of SWCC Fitting Parameter Models

	R^2			RMSE		
Model	k_1	k_2	k_3	k_1	k_2	k_3
Yau (*15*)	−4.51	−2.15	−3.06	0.31	0.25	0.54
Soliman (*17*)	−2.27	−0.16	−1.42	0.43	0.59	0.89
Malla (29)	−0.023	−0.03	−1.86	28.82	0.94	0.61
ANN	0.76	0.72	0.63	0.10	0.05	0.04

The accuracy of the predictions of the M_R model coefficients by the existing correlation models were analyzed in this study. Table 1 lists all the predicted M_R coefficients using the Yau, Soliman, Malla, and ANN models. Compared with the ANN model, the estimated RMSE values from the three regression models were high and the R^2 values were extremly low. High RMSE and low R^2 values indicated a poor prediction accuracy. The R^2 values for k_1, k_2, and k_3 from all three regression models were less than zero, whereas R^2 values from ANN model were in the range of 0.63–0.76. The negative R^2 values indicated the slope of the best fit line relating the predicted to the measured coefficients was negative. Therefore, the developed ANN models had a significant improvement in predicting the M_R model coefficients from base physical and compaction properties.

The accuracy of M_R values using the predicted coefficients are examined in the next section. Predicted coefficients from the three regression models were input into Equation 2 and the ANN model coefficients were input into Equation 5 to calculte M_R values, which were compared with the measured values. Figure 5 shows the comparisons between measured and estimated M_R values using the ANN predicted coefficients.

Clearly the M_R values estimated using the ANN coefficients matched well with the test data. The R^2 value estimated for both plastic and nonplastic base materials were 0.91, which demonstrated a high accuracy. Figure 6, *a–c* present the results from the Yau, Soliman, and Malla models, respectively. The R^2 value estimated from the Yau, Soliman and Malla models were 0.32, −1.53, and -1.62×10^6, respectively, which indicates that the regression models failed to predict

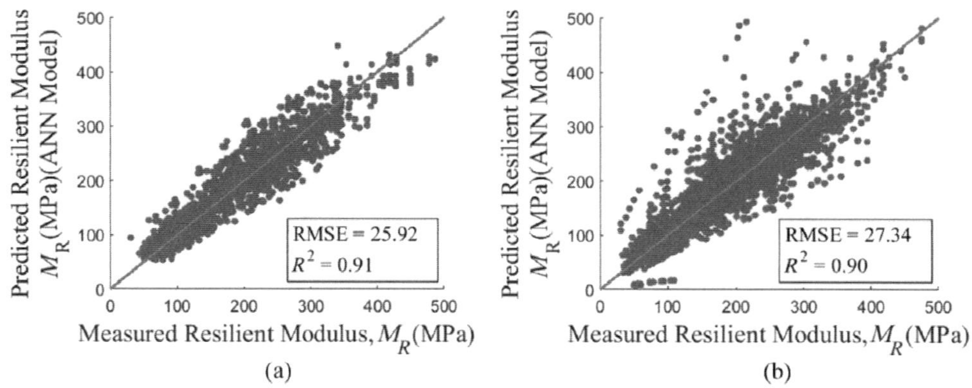

Figure 5. Comparison of ANN model predicted resilient moduli against measured values: (*a*) plastic base materials; (*b*) nonplastic base materials.

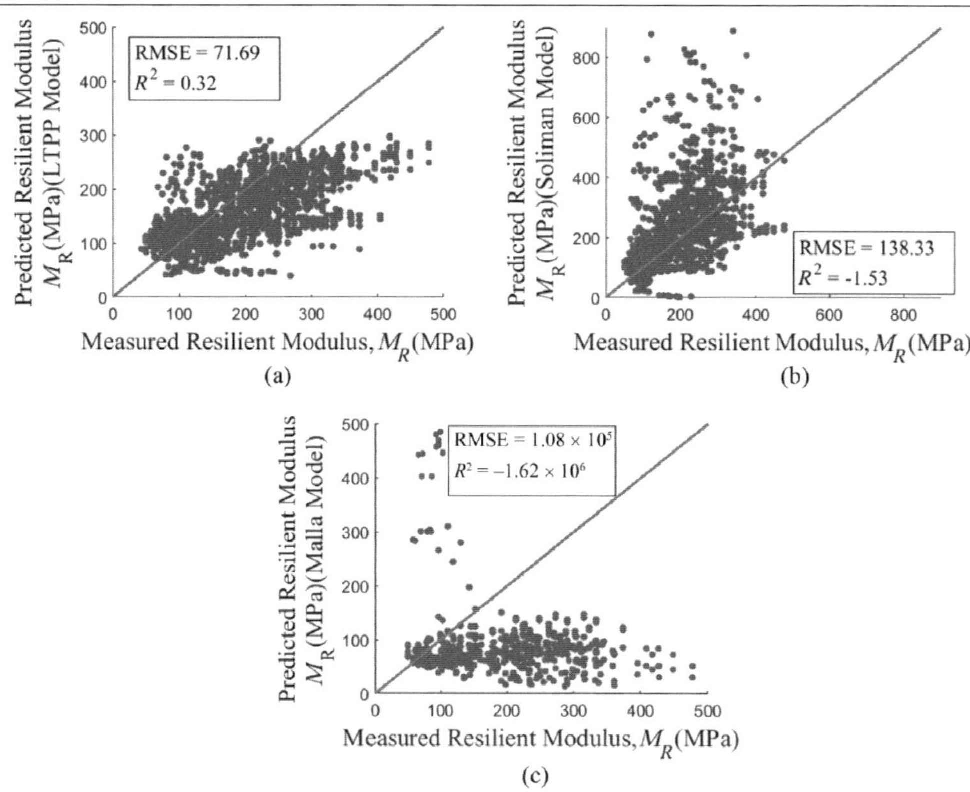

Figure 6. Comparison of measured versus predicted resilient moduli using regression models: (*a*) Yau model; (*b*) Soliman model; (*c*) Malla model.

the coefficients. The negative R^2 value from the Soliman and Malla models indicates that the best fit line relating the predicted to the measured values had a negative slope.

Validation of the Developed ANN Models

To validate the developed ANN models for the M_R model coefficients, test data were collected from the TTI (*30,31*) and UIUC (*32*). Epps et al. (*30*) conducted triaxial repeated load tests to evaluate the laboratory performance of unbound granular materials. In addition to the M_R test results at different combinations of confining pressure and deviator stresses, aggregate physical, strength and moisture properties were also collected. Titus-Glover and Fernando (*31*) performed the compressive creep and recovery tests for collected unbound base materials in the TTI laboratory. Tutumluer

Table 2. Input Parameters Collected from the Literature for Model Validation

Data Source	Soil source	3/8 in. sieve	No. 200 sieve	PL	PI	MDD	OMC	TMC
Epps et al. (*30*)	E-02-2-3-2	45.9	12.36	10	5	136.2	7.2	7.2
	E-08-2-1-6	49.2	6.3	9	5	140.4	6.5	6.5
Titus-Glover and Fernando (*31*)	Limestone (–2% OMC)	64.71	17.61	12.8	7.9	127.4	9.045	7.045
	Limestone (OMC)	64.71	17.61	12.8	7.9	127.4	9.045	9.045
Tutumluer et al. (*32*)	Dolomite FC 4% (OMC)	69.9	4	NA	NA	133	10.5	10.5
	Dolomite FC 8% (OMC)	71.1	8	NA	NA	133	8.48	8.48

NA = not available.

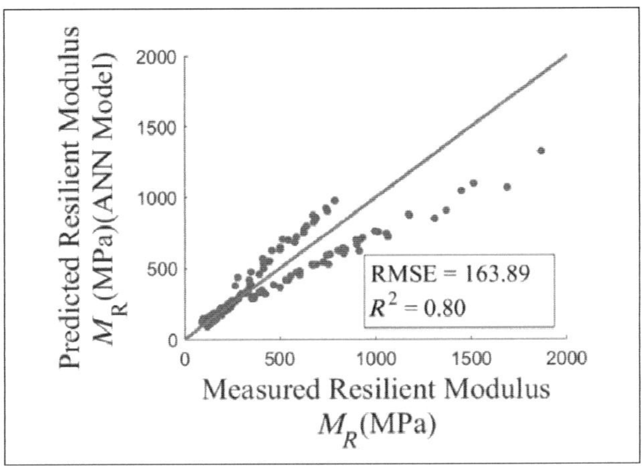

Figure 7. Validation of measured versus ANN predicted M_R at various stress levels for collected unbound materials.

et al. (*32*) conducted resilient modulus tests at the 15 various stress states. As no suction values were provided by the UIUC research team, the suction was estimated using the developed ANN models from the base physical properties. Table 2 lists all the collected physical properties used as the ANN inputs for the validation dataset.

The collected base properties were input into the trained ANN model to predict k_1, k_2, and k_3. Resilient modulus values were calculated using the predicted coefficients from the ANN models and compared with the test results. Figure 7 shows a comparison between measured and estimated M_R values using the ANN predicted coefficients. The data points plotted above the line of equality were from the UIUC dataset and those plotted below were from the TTI. The TTI data were all from plastic base materials whereas UIUC base materials were nonplastic. The overpredictions in estimated suction for nonplastic base materials contributed to the overpredicted M_R values for UIUC dataset. Overall, the M_R values estimated using the predicted coefficients from the ANN models produced a good fit with the test results. Figure 7 depicts a R^2 value of 0.8 between measured and predicted dataset. Therefore, the developed ANN model can be used as a prediction tool to accurately estimate the M_R model coefficients that are used in pavement design and analysis.

Conclusions

This paper developed ANN models to predict the coefficients of a stress- and moisture-dependent resilient modulus model. A good prediction accuracy of the developed models results in better estimation of the resilient modulus of base materials. The major findings of this study are summarized as follows.

- Two three-layered ANN models were developed for plastic and nonplastic base materials. Input variables for plastic and nonplastic soil included material percent passing 3/8 in. sieve, material percent passing No. 200 sieve, PL, PI, MDD, OMC, and TMC. The output variables were the three coefficients k_1, k_2, and k_3 of the resilient modulus model.
- The developed ANN models showed a higher prediction accuracy compared with the three regression models selected from the existing literature. The ANN models can accurately estimate the resilient modulus of base materials at any stress level. The obtained R^2 values are 0.91 and 0.90 for plastic and nonplastic base, respectively.
- The ANN models do not provide any insight into complex relationship between the M_R model coefficients and base physical properties. Thus, it is not recommended for use as a prediction tool for the values that are out of the range of training dataset. In this study, a large dataset of 779 base materials was collected from the LTPP database and, therefore, provided a wide range of input properties. A sensitivity analysis will be conducted in the future study to evaluate the influence of each input parameter on M_R model coefficients.
- The developed ANN models were validated using the resilient modulus test data from different sources. The R^2 value between the measured and predicted validation resilient modulus values was 0.8.

Acknowledgment

The authors acknowledge the financial support provided by the National Cooperative Highway Research Program (NCHRP).

References

1. AASHTO. *Mechanistic-empirical pavement design guide: A manual of practice*. AASHTO Designation: MEPDG-1, Washington, D.C., 2008.
2. Witczak, M. W. *NCHRP Report 1-28A: Harmonized test methods for laboratory determination of resilient modulus for flexible pavement design*. Transportation Research Board, Washington, D.C., 2003.
3. Lekarp, F., U. Isacsson, and A. Dawson. State of the Art. I: Resilient Response of Unbound Aggregates. *Journal of Transportation Engineering*, Vol. 66, No. 1, 2000, pp. 66–75.
4. Heath, A. C., J. M. Pestana, J. T. Harvey, and M. O. Bejerano. Normalizing Behavior of Unsaturated Granular Pavement Materials. *Journal of Geotechnical and Geoenvironmental Engineering*, Vol. 130, No. 9, 2004, pp. 896–904.
5. Liang, R. Y., S. Rabab'ah, and M. Khasawneh. Predicting Moisture-Dependent Resilient Modulus of Cohesive Soils using Soil Suction Concept. *Journal of Transportation Engineering*, Vol. 134, No. 1, 2008, pp. 34–40.
6. Cary, C., and C. Zapata. Resilient Modulus for Unsaturated Unbound Materials. *Road Materials and Pavement Design*, Vol. 12, No. 3, 2011, pp. 615-638.
7. Zapata, C. E. *NCHRP Project 9-23A: A national database of subgrade soil–water characteristic curves and selected soil properties for use with the MEPDG*. Transportation Research Board, Washington, D.C., 2010.
8. Oloo, S. Y., and D. G. Fredlund. The Application of Unsaturated Soil Mechanics Theory to the Design of Pavements. *Proceedings of the 5th International Conference on the Bearing Capacity of Roads and Airfields*, Trondheim, Norway, 1998, pp. 1419–1428.
9. Wolfe, W., and T. Butalia. *Report No. FHWA/OH-2004/007: Continued Monitoring of SHRP Pavement Instrumentation Including Soil Suction and Relationship with Resilient Modulus*. U.S. Department of Transportation, Federal Highway Administration, Washington, D.C., 2004.
10. Khoury, N. N., and M. M. Zaman. Correlation between Resilient Modulus, Moisture Variation, and Soil Suction for Subgrade Soils. *Transportation Research Record: Journal of the Transportation Research Board*, 2004. 1874: 99–107.
11. Luo, X., F. Gu, Y. Zhang, R. L. Lytton, and D. Zollinger. Mechanistic–Empirical Models for Better Consideration of Subgrade and Unbound Layers Influence on Pavement Performance. *Transportation Geotechnics*, Vol. 13, 2017, pp. 52-68.
12. Gu, F., X. Luo, Y. Zhang, R. L. Lytton, and H. Sahin. Modeling of Unsaturated Granular Materials in Flexible Pavements. E3S Web of Conferences, Vol. 9, 2016, p. 20002.
13. ARA, Inc, *Final Report, NCHRP Project 1-37A: Guide for Mechanistic-Empirical Design of New and Rehabilitated Pavement Structures*. Transportation Research Board, Washington, D.C., 2004.
14. Lytton, R. L. Foundations and pavements on unsaturated soils. Proceedings of the 1st International Conference on Unsaturated Soils, International Society of Soil Mechanics and Foundation Engineering, Paris, Vol. 3, 1995, pp. 1201–1220.
15. Yau, A., and H. Von Quintus. Predicting Elastic Response Characteristics of Unbound Materials and Soils. *Transportation Research Record: Journal of the Transportation Research Board*, 2004. 1874: 47–56.
16. Nazzal, M. D., and L. N. Mohammad. Estimation of Resilient Modulus of Subgrade Soils for Design of Pavement Structures. *Journal of Materials in Civil Engineering*, Vol. 22, No. 7, 2010, pp. 726-734.
17. Malla, R. B., and S. Joshi. Resilient Modulus Prediction Models based on Analysis of LTPP data for Subgrade Soils and Experimental Verification. *Journal of Transportation Engineering*, Vol. 133, No. 9, 2007, pp. 491-504.
18. Gu, F., H. Sahin, X. Luo, R. Luo, and R. L. Lytton. Estimation of Resilient Modulus of Unbound Aggregates using Performance-Related Base Course Properties. *Journal of Materials in Civil Engineering*, Vol. 27, No. 6, 2014, 04014188.
19. Coleri, E., M. Guler, A. Gungor, and J. Harvey. Prediction of Subgrade Resilient Modulus using Genetic Algorithm and Curve-Shifting Methodology: Alternative to Nonlinear Constitutive Models. *Transportation Research Record: Journal of the Transportation Research Board*, 2010. 2170: 64–73.
20. Kim, S. H., J. Yang, and J. H. Jeong. Prediction of Subgrade Resilient Modulus using Artificial Neural Network. *KSCE Journal of Civil Engineering*, Vol. 18, No. 5, 2014, pp. 1372-1379.
21. Nazzal, M. D., and O. Tatari. Evaluating the Use of Neural Networks and Genetic Algorithms for Prediction of Subgrade Resilient Modulus. *International Journal of Pavement Engineering*, Vol. 14, No. 4, 2013, pp. 364-373.
22. Fredlund, D. G., and A. Xing. Equations for the Soil–Water Characteristic Curve. *Canadian Geotechnical Journal*, Vol. 31, No. 3, 1994, pp. 521-532.
23. Saha, S., F. Gu, X. Luo, and R. L. Lytton. Prediction of Soil–Water Characteristic Curve for Unbound Material Using Artificial Neural Network Approach. *Journal of Materials in Civil Engineering*, 2017, In Press.
24. Ceylan, H., K. Gopalakrishnan, and R. L. Lytton. Neural Networks Modeling of Stress Growth in Asphalt Overlays due to Load and Thermal Effects During Reflection Cracking. *Journal of Materials in Civil Engineering*, Vol. 23, No. 3, 2011, pp. 221-229.
25. Ling, M., X. Luo, S. Hu, F. Gu, and R. L. Lytton. Numerical Modeling and Artificial Neural Network for Predicting J Integral of Top Down Cracking in Asphalt Pavement. *Transportation Research Record: Journal of the Transportation Research Board*, 2017. 2631: 83–95.
26. Gu, F., X. Luo, Y. Zhang, Y. Chen, R. Luo, and R. L. Lytton. Prediction of Geogrid-Reinforced Flexible Pavement Performance using Artificial Neural Network Approach. *Road Materials and Pavement Design*, 2017, pp. 1-17, DOI: 10.1080/14680629.2017.1302357
27. Lawrence, M., and A. Petterson. *Brainmaker Professional: Neural Network Simulation Software User's Guide and Reference Manual*. California Scientific Software, Nevada City, Calif, 1993.
28. Ji, R., N. Siddiki, T. Nantung and D. Kim. Evaluation of resilient modulus of subgrade and base materials in Indiana and

its implementation in MEPDG. *The Scientific World Journal*, 2014.

29Soliman, H. and A. Shalaby. Validation of Long-Term Pavement Performance Prediction Models for Resilient Modulus of Unbound Granular Materials. *Transportation Research Record: Journal of the Transportation Research Board*, 2016. 2578: 29–37.

30. Epps, J. A., S. Sebesta, B. Hewes, H. Sahin, R. Luo, J. W. Button, R. L. Lytton, C. A. Herrera, R. Hatcher, and F. Gu. *Report No. FHWA/TX-13/0-6621: Development of a specification for flexible base construction*. Texas A & M Transportation Institute, 2014.

31. Titus-Glover, L., and E. G. Fernando. *Evaluation of Pavement Base and Subgrade Material Properties and Test Procedures*. Texas A & M Transportation Institute, 1995.

32Tutumluer, E., D. Mishra, and A. Butt. Final Report, *Illinois Center for Transportation (ICT) R27-1 Project: Characterization of Illinois Aggregates for Subgrade Replacement and Subbase*, University of Illinois Urbana-Champaign, Urbana, IL, 2009.

The Standing Committee on (AFP70) Aggregates peer-reviewed this paper (18-05237).

Article

TRR
JOURNAL OF THE TRANSPORTATION RESEARCH BOARD

Transportation Research Record
2018, Vol. 2672(52) 34–45
© National Academy of Sciences:
Transportation Research Board 2018
Article reuse guidelines:
sagepub.com/journals-permissions
DOI: 10.1177/0361198118756886
journals.sagepub.com/home/trr
SAGE

Soaking Effects on Strength Characteristics of Crushed Gravel and Limestone Unbound Aggregates

Abdolreza Osouli[1], Goran Othmanawny[1], Erol Tutumluer[2], Sheila Beshears[3], and Heather Shoup[3]

Abstract

Strength characteristics of unbound aggregate materials critically affect base and sub-base thickness designs of construction working platforms and flexible pavements. Unsoaked California Bearing Ratio (CBR) is commonly used by state transportation agencies for quality control and design. However, depending on the amount of fines content (i.e., passing No. 200 sieve), the strength characteristics of unbound aggregate layers may severely be affected after soaking. This study investigates any correlations that may exist between soaked and unsoaked strengths of crushed gravel and limestone aggregates commonly used in the State of Illinois. A test matrix was established to consider the effects of varying fines content at 5% and 12%, plasticity index (PI) at 5% and 9%, and dust ratio (DR) at 0.4, 0.6, and 1.0 on aggregate strength. DR is the ratio of percent passing No. 200 sieve to percent passing No. 40 sieve. Most of the aggregate strength characterizations with 5% fines content were not sensitive to soaking in terms of CBR index. However, significant reductions in soaked and unsoaked CBR values were observed when fines content increased from 5% to 12%. The crushed limestone aggregate strengths were more affected by soaking than the crushed gravel aggregates. A prediction model was developed to predict soaked CBR values from unsoaked CBR tests considering material type, fines content, PI and DR.

Unbound aggregates are routinely constructed as base and sub-base layers in conventional flexible pavements to transfer the induced traffic loads from pavement surface to subgrade. Mineralogy, particle size distribution, fines content (FC), dust ratio (DR), particle shape, surface texture, angularity, degree of compaction, moisture content, and plasticity of fines are the most important physical properties of these aggregate materials that affect quality, strength, and performance of the pavement. DR is the ratio of percent passing No. 200 sieve to percent passing No. 40 sieve.

The empirical American Association of State Highway and Transportation Officials (AASHTO) design approach and the new Pavement ME Mechanistic-Empirical Pavement Design Guide (MEPDG) are the most commonly implemented procedures to design pavement structures (*1,2*). Based on these procedures, pavement structure can be designed using either strength or modulus layer inputs. The empirical strength-based approach focuses on the maximum load that can be carried out by the subgrade, aggregate base/sub-base, and the surface course before failure. A strength-based approach using California Bearing Ratio (CBR) test results is more commonly used in pavement design than a stiffness-based method (*3*).

Gradation and FC (i.e., passing No. 200 sieve) are important properties to consider when determining strength of unbound aggregate layer (*4–6*). Increase in FC reduces the strength of base and sub-base layers (*7*). Gray suggested a maximum allowable FC of 8% and 6% for aggregates with maximum particle sizes of 25.4 mm (1.0 in.) and 50.8 mm (2.0 in.) to achieve higher dry density and strength, respectively (*8*). A FC of 8% was proposed to achieve the highest strength in crushed limestone aggregates with a DR of 0.4 (*9*). It is worth noting that American Society for Testing and Materials (ASTM) D1241 and AASHTO M 147 standards for base and sub-base applications limit the FC to 15% and 20%, respectively. Unbound aggregate strength is also highly affected by aggregate angularity (*10–14*). Base and sub-base mixtures with uncrushed aggregate showed lower strength values than mixtures with crushed aggregate (*15*).

Liquid limit (LL), plastic limit (PL), and plasticity index (PI) also have influences on strength and rutting performance

[1]Civil Engineering Department, Southern Illinois University Edwardsville, Edwardsville, IL
[2]Department of Civil and Environmental Engineering, University of Illinois at Urbaana-Champaign, Urbana, IL
[3]Illinois Department of Transportation, Springfield, IL

Corresponding Author:
Address correspondence to Abdolreza Osouli: aosouli@siue.edu

of unbound aggregate layers. Higher plastic fines in the aggregate matrix resulted in a reduction in aggregate shear strength (*10*). Samples with lower PI values were found to have higher CBR strength indices and less permanent deformation (*16–18*). Consequently, AASHTO M147 and ASTM D1241 for base/sub-base construction have limited aggregate material PI and LL values to 6% and 25%, respectively.

Soaking has a negative effect on strength properties of soils and unbound aggregates. Baditala and Sreedhar reported up to 36% reduction in CBR values for different soils after soaking for 4 days (*19*). This was attributed to the lack of suction in soaked samples and the lubrication effect of water, which decreases the interparticle friction in the soil and aggregate matrix (*20*). Higher CBR values were found when soil specimens were compacted at Optimum Moisture Content (OMC) (*21*). Evaluation of strength behavior on the wet side of OMC is considered to be the most critical (*22*). This is in line with the common practice of treating unsoaked CBR values at the wet side of OMC similar to soaked CBR data (*22*).

AASHTO M147-65 and ASTM D1241 standards suggest agencies set their own aggregate property limits considering material availability. A survey of 46 state transportation agencies conducted by Tutumluer (*3*) concluded that 33 agencies capped maximum allowable particle size at 38 mm (1.5 in.) and the rest allowed aggregates with maximum particle sizes up to 102 mm (4 in.) in base layer. For the sub-base layer, 14 agencies capped maximum allowable particle size at 38 mm (1.5 in.) and the rest allowed aggregates with maximum particle sizes of 152 mm (6 in.) (*3*). For FC, 33 agencies limited maximum allowable FC at 12% and the rest allowed aggregate mixtures with FC up to 20% in unbound base applications. For sub-base applications, 18 agencies limited maximum allowable FC at 12% and the rest allowed aggregate mixtures with FC up to 34% (*3*). In terms of PI, 27 agencies limited maximum allowable PI at 6% and the rest allowed aggregate mixtures with PIs up to 15% in base and sub-base applications (*3*). Currently, some state standards have set 0.66 as the maximum allowable DR to be used in base and sub-base construction (*23–26*), although this criterion is not required by others (*27–30*).

There are differences in standard specifications and transportation agency practices on quality control of unbound aggregates, which directly impact aggregate behavior with respect to soaking and unsoaking. Furthermore, many studies have developed less time-consuming methods to estimate soaked CBR value for soils (*31–33*); however, there is a lack of knowledge for aggregates in this regard. Accordingly, this paper has the objective to investigate the effects of material type (i.e., crushed gravel vs. crushed limestone) and properties, such as FC, PI, and DR, on soaked and unsoaked strength and performance behavior of unbound aggregates. A prediction model is also developed and proposed to determine soaked CBR values from unsoaked CBR tests based on FC, PI, and DR of unbound aggregates.

Research Plan and Test Matrix

To study the soaked and unsoaked behavior of aggregates, a test matrix was developed to evaluate effects of material type, FC, PI and DR (see Figure 1). CA 6 gradation, which is commonly used for base and sub-base applications in Illinois, was selected for this study. The CA 6 gradation is more conforming to gradation A of AASHTO M 147 and ASTM D 1241 except for sizes greater than 12.5 mm (0.5 in.). The strength properties of unbound aggregates were determined by conducting soaked and unsoaked CBR tests following AASHTO T 193-99 on compacted samples using standard Proctor per AASHTO T 99-10. To determine maximum dry density (MDD) and OMC pairs, four to five samples were prepared with different moisture contents. For soaked CBRs, the compacted samples were soaked in a bath of water for 4 days, and for unsoaked CBRs the samples were tested after compaction. It is worth mentioning that 20% of the tests were repeated for validation purposes.

To cover a wide range, FC values of 5% and 12% were included in the test matrix. In terms of PI, many standards and state specifications have limited PI to 6%; however, some states, such as Illinois, allow higher PI values up to 9% (*34*). To cover typical ranges, PI values of 5% and 9% were included. Crushed gravel and crushed limestone fines (passing No. 200 sieve) were quite low in plasticity. Accordingly, passing No. 40 sieve fractions of crushed gravel and crushed limestone were blended with high-plasticity fines such as ball clay and kaolinite with PI values of 32% and 12%, respectively, to achieve the target PI values specified in the test matrix per AASHTO T 90. For example, PI of 9% was achieved for crushed limestone sample with 5% FC and DR of 1.0 by mixing 35% mineral filler and 65% of crushed limestone fines. Since many standards and state specifications have used a maximum allowable DR value of 0.66, DR values of 0.4 and 0.6 were selected to cover variety of DR values. However, DR of 1.0 was targeted to study the effect of skip-grading on aggregate strength.

Figure 1 shows the summery of test matrix studied in the laboratory. In the test matrix designations, aggregate samples were grouped and labeled using both letters and numbers. The letters represent DR and PI values and the numbers represent FC values of engineered samples. For example, Group B-5 in Figure 1 represents samples engineered for PI of 5%, DR of 0.6, and FC of 5%.

Aggregate Gradation

The upper and lower limits of CA 6 gradation are shown in Figure 2. To prepare targeted samples, different gradations of CA 6 were engineered for material passing No. 16 sieve

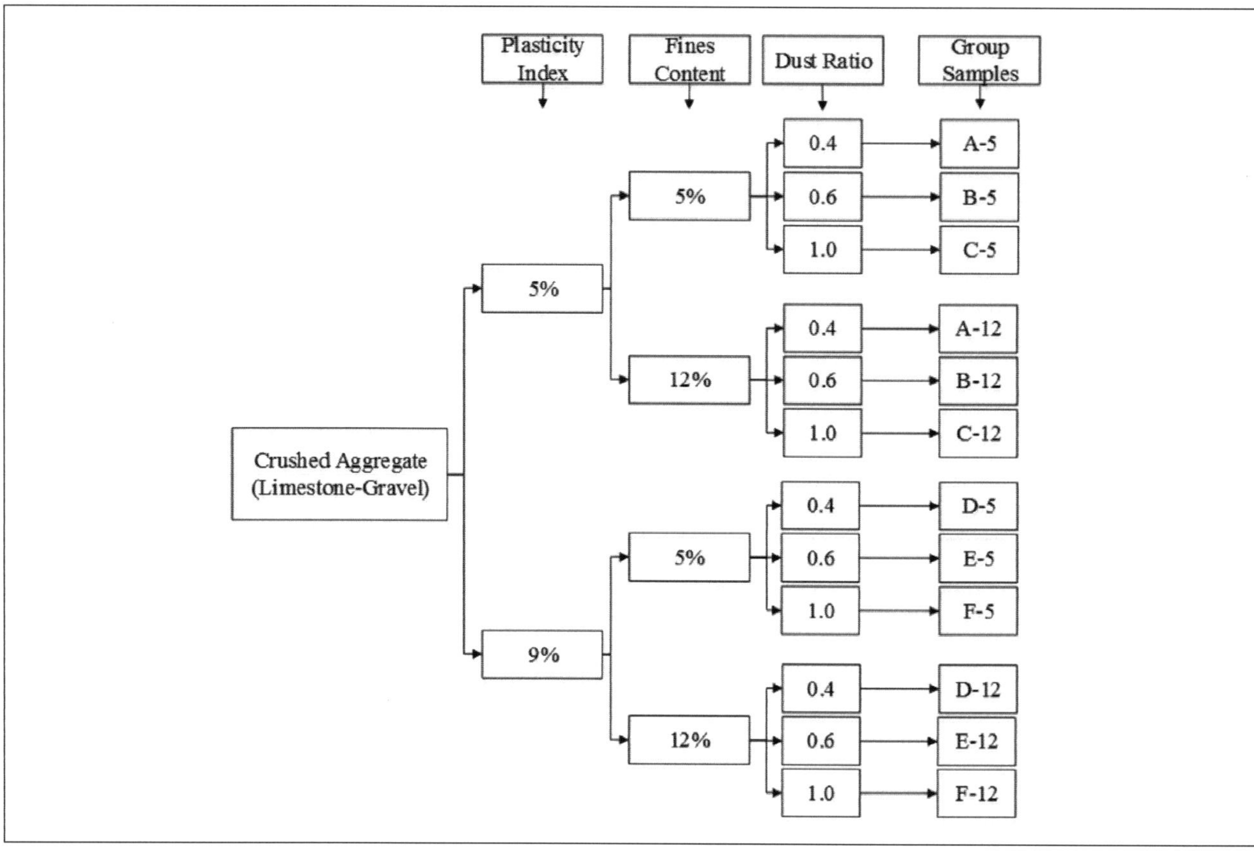

Figure 1. Test matrix.

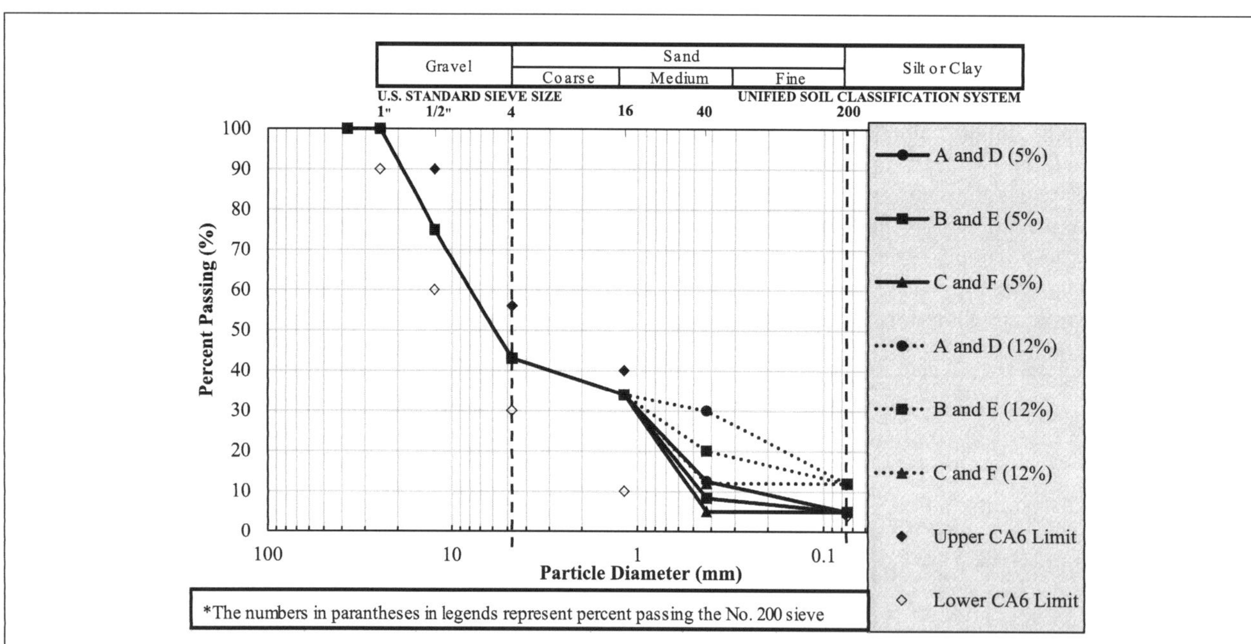

Figure 2. Crushed limestone and crushed gravel CA 6 target gradations studied.

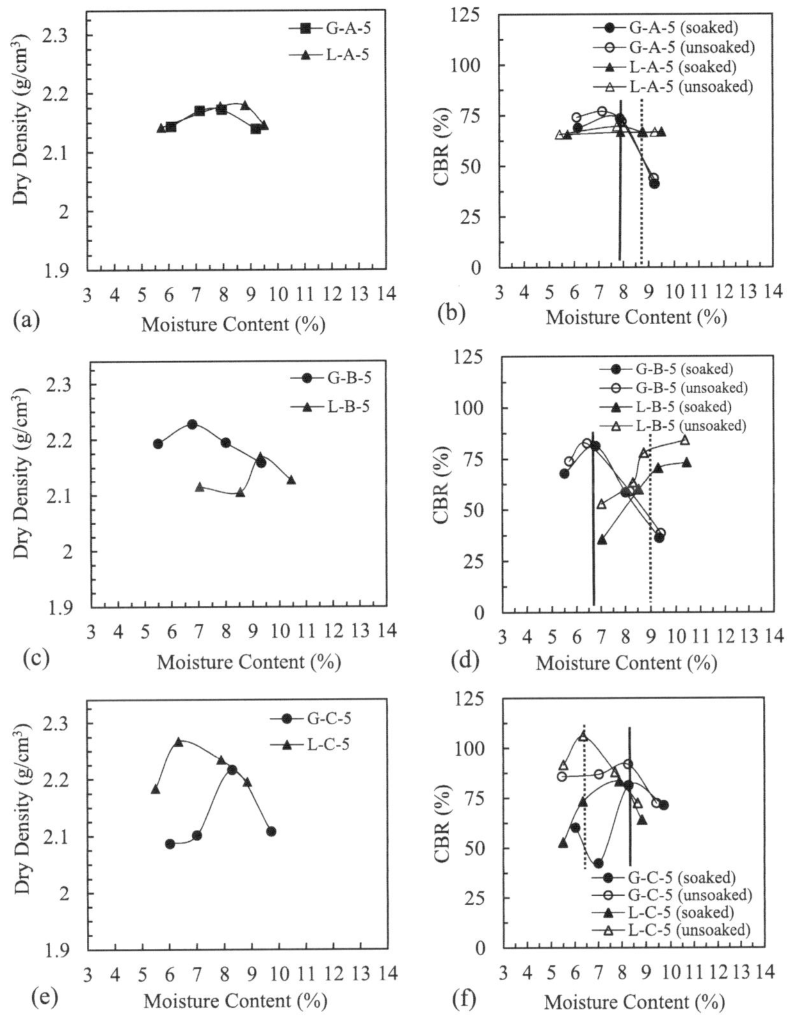

Sample Label	PI (%)	FC (%)	DR
A-5	5	5	0.4
B-5	5	5	0.6
C-5	5	5	1

Notes:
G: Crushed gravel
L: Crushed limestone
Vertical dashed line: OMC for limestone
Vertical solid line: OMC for gravel

Figure 3. Crushed limestone and crushed gravel dry density (*a, c, e*) and CBR values (*b, d, f*) vs. moisture content for A-5, B-5 and C-5 samples.

(1.19 mm size). Materials greater than No. 16 sieve size were kept constant for all the group samples. To have a dense sample with high strength, Talbot's equation (see Equation 1) was utilized for engineering gradations of materials greater than the No. 16 sieve size (*35*):

$$P=\left(\frac{d}{D}\right)^{n}\times 100 \quad (1)$$

where d is the percent passing sieve size, D is the maximum size of the aggregate, and n is a constant. For optimum packing, a value of n equal to 0.5 was used in this study to produce samples with high densities (*36*). Figure 2 shows the targeted gradations for the crushed gravel and crushed limestone samples, respectively.

Maximum Dry Density for Engineered Group Samples

The compaction curves for crushed gravel and crushed limestone samples with FC of 5% and 12% are presented in Figure 3, *a, c,* and *e*, and Figure 4, *a, c,* and *e*, respectively, for samples with the low PI of 5%. Figure 5, *a, c,* and *e*, and Figure 6, *a, c,* and *e* show compaction curves of the high PI (i.e., 9% PI) samples with 5% and 12% FC, respectively. The label letter G refers to samples of crushed

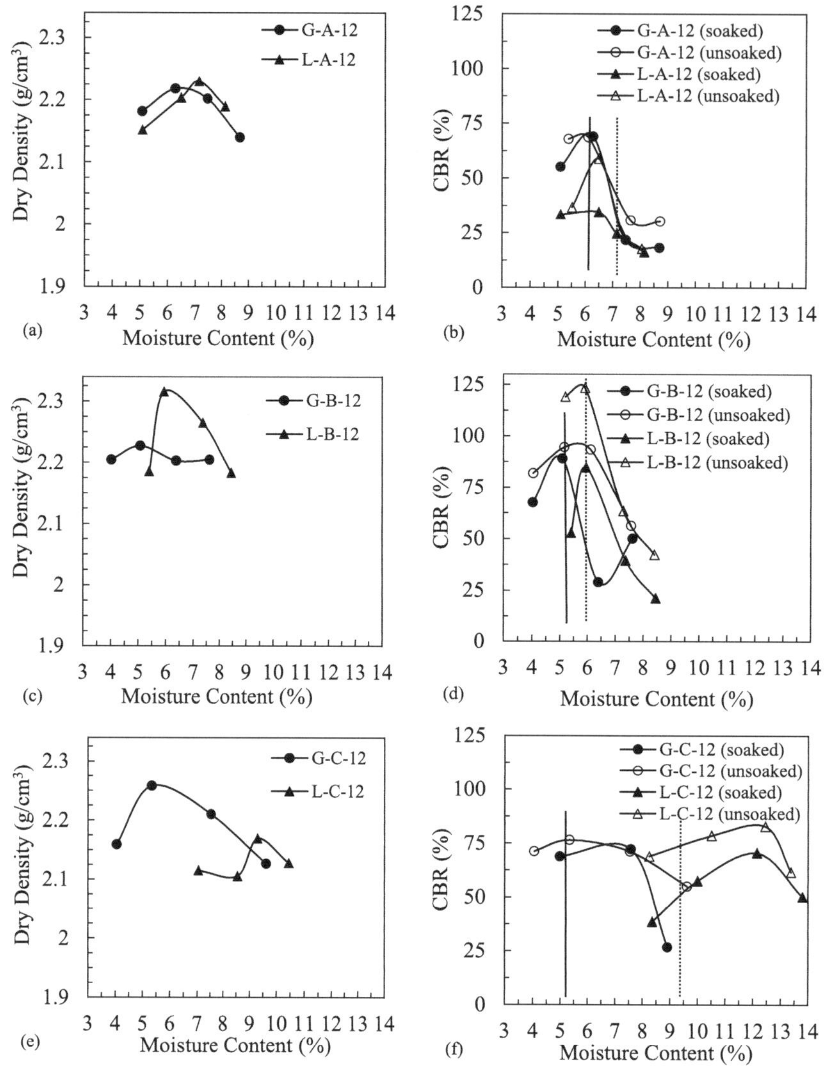

Sample Label	PI (%)	FC (%)	DR
A-12	5	12	0.4
B-12	5	12	0.6
C-12	5	12	1

Notes:
G: Crushed gravel
L: Crushed limestone
Vertical dashed line: OMC for limestone
Vertical solid line: OMC for gravel

Figure 4. Crushed limestone and crushed gravel dry density (*a, c, e*) and CBR values (*b, d, e*) vs. moisture content for A-12, B-12 and C-12 samples.

gravel, and samples labeled with letter L are crushed limestone aggregate.

Physical properties of the compacted aggregate such as moisture content, gradation, and type of the aggregate all have significant influences on the shapes of compaction curves. Lee and Suedkamp (*37*) observed four types of compaction curves: bell shape, one-and-one-half peaks, double peak, and odd shape curves for soil samples (see Figure 7). In soils, each of these compaction curve shapes represent particular LL ranges. However, for aggregates, the compaction curves for the tested samples were bell-shaped and one-and-one-half peak shaped. All samples with 12% FC were found to have bell-shaped compaction curves except for one sample of the crushed limestone and two samples of the crushed gravel, which were one-and-one-half peak shape.

The results show that the optimal DR to achieve a higher MDD value differs. The MDD for crushed gravel has the least density variability. The variability in MDD decreases as the PI increases from 5% to 9%. In general, higher MDD and lower OMC values were achieved when more fine materials were added to the aggregate mixture and fines acted like filler among the large aggregate particles.

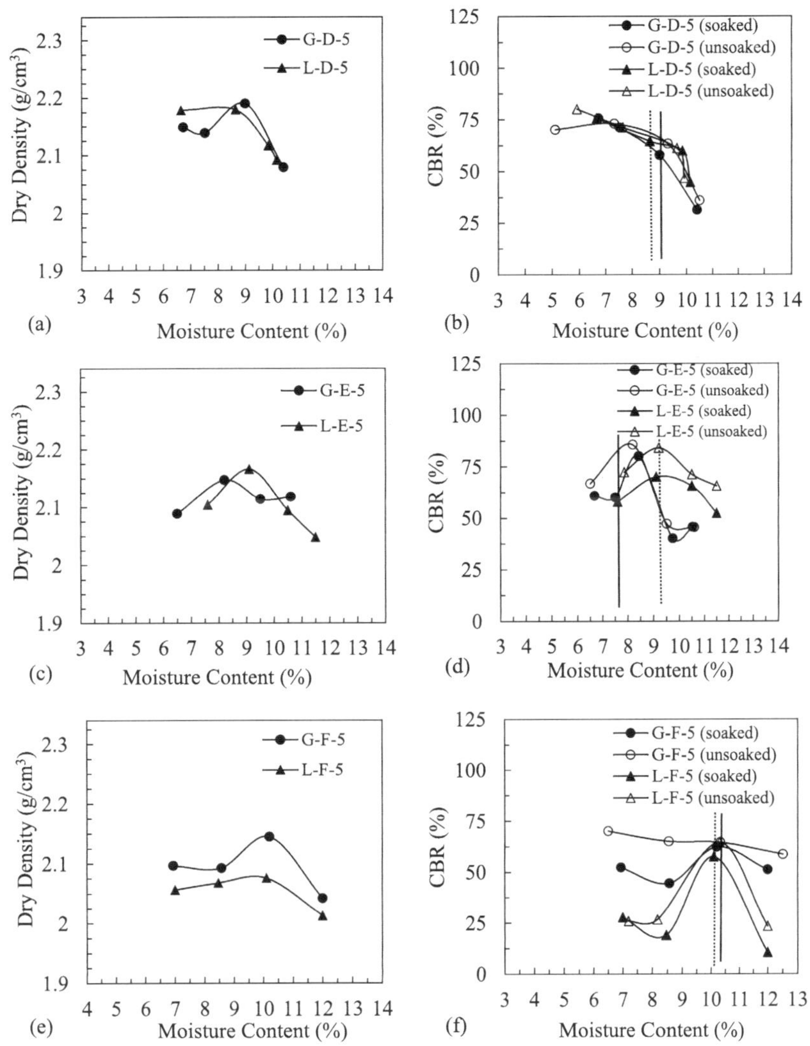

Sample Label	PI (%)	FC (%)	DR
D-12	9	5	0.4
E-12	9	5	0.6
F-12	9	5	1

Notes:
G: Crushed gravel
L: Crushed limestone
Vertical dashed line: OMC for limestone
Vertical solid line: OMC for gravel

Figure 5. Crushed limestone and crushed gravel dry density (*a, c, e*) and CBR values (*b, d, f*) vs. moisture content for D-5, E-5 and F-5 samples.

Soaked and Unsoaked Cbr Results

The soaked and unsoaked CBR curves graphed with moisture content are presented in Figure 3 through Figure 6. The soaked and unsoaked curves followed the same trend, and some overlapped one another, such as the ones for Groups A-5 and D-5. Some of the soaked and unsoaked CBR curves approached each other just on the wet side of OMC, such as Groups C-5 and A-12. Some of the crushed gravel soaked and unsoaked CBR curves were found to approach each other just around OMC, such as Groups F-5 and C-12. The remaining soaked and unsoaked CBR curves were parallel to each other. The differences between soaked and unsoaked CBR curves were found to be larger as FC increased from 5% to 12%.

The constructed base and sub-base might have moisture contents above or below OMC. Therefore, average CBR values at OMC and OMC ±1.5% for samples engineered with 5% and 12% FC were used to identify the strength characteristics, as shown in Figure 8 and Figure 9, respectively.

According to Figure 8, the difference between average soaked and unsoaked CBR values was the lowest when a DR of 0.4 was used for crushed gravel and crushed limestone

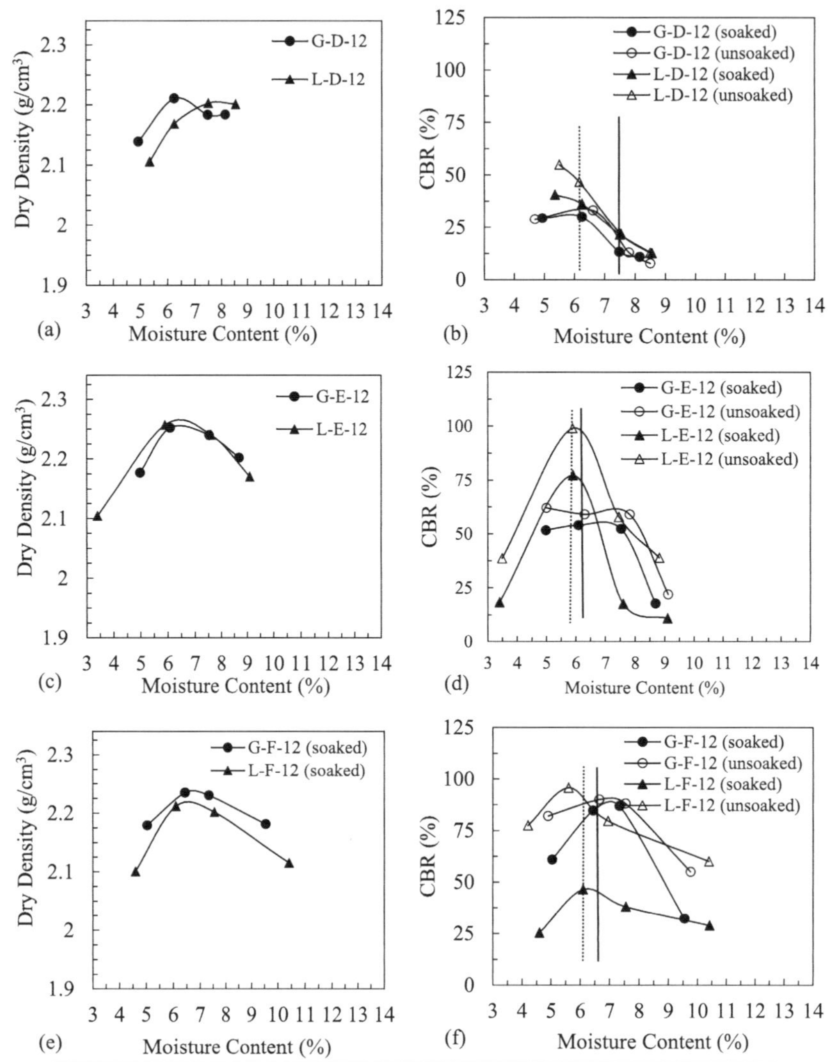

Sample Label	PI (%)	FC (%)	DR
D-12	9	12	0.4
E-12	9	12	0.6
F-12	9	12	1

Notes:
G: Crushed gravel
L: Crushed limestone
Vertical dashed line: OMC for limestone
Vertical solid line: OMC for gravel

Figure 6. Crushed limestone and crushed gravel dry density (*a, c, e*) and CBR values (*b, d, f*) vs. moisture content for D-12, E-12 and F-12 samples.

samples (i.e., Groups A-5 and D-5) because of less FC. The largest difference between average soaked and unsoaked CBR values was found for samples with DR of 1.0.

According to Figure 9, although crushed gravel and crushed limestone samples engineered with a DR of 0.4 showed the lowest average soaked and unsoaked CBR values, their strength properties were less sensitive to soaking. The effect of soaking was found to be more pronounced for crushed limestone samples engineered with DR of 0.6 and 1.0. In general, the crushed limestone samples were more influenced by soaking than crushed gravel samples when they were prepared with higher FC such as 12% and higher DR values such as 0.6 and 1.0.

A comparison of Figure 8 and Figure 9 shows:

1. Major reductions in average soaked and unsoaked CBR values are seen when FC increases from 5% to 12% for both materials engineered with DR of 0.4. It is noteworthy to mention that DR of 0.4 is well below the AASHTO M 147 standard limitation (i.e., 0.66).
2. By increasing FC from 5% to 12%, higher average soaked and unsoaked CBR values were found for

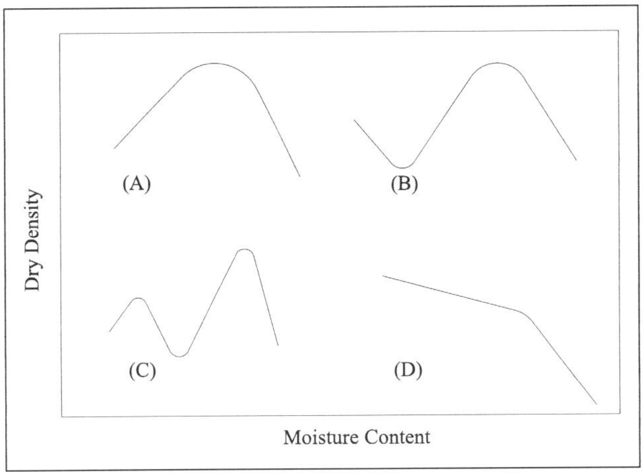

Figure 7. Compaction curve shapes: (*a*) bell shape; (*b*) one-and-one-half peak; (*c*) double peak; and (*d*) odd shape (*37*).

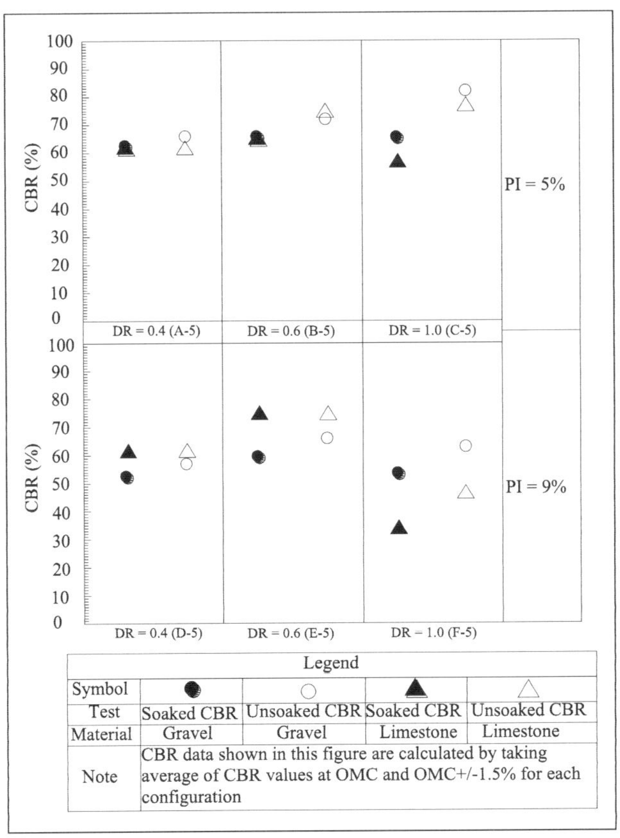

Figure 8. Soaked and unsoaked CBR results for 5% FC group samples.

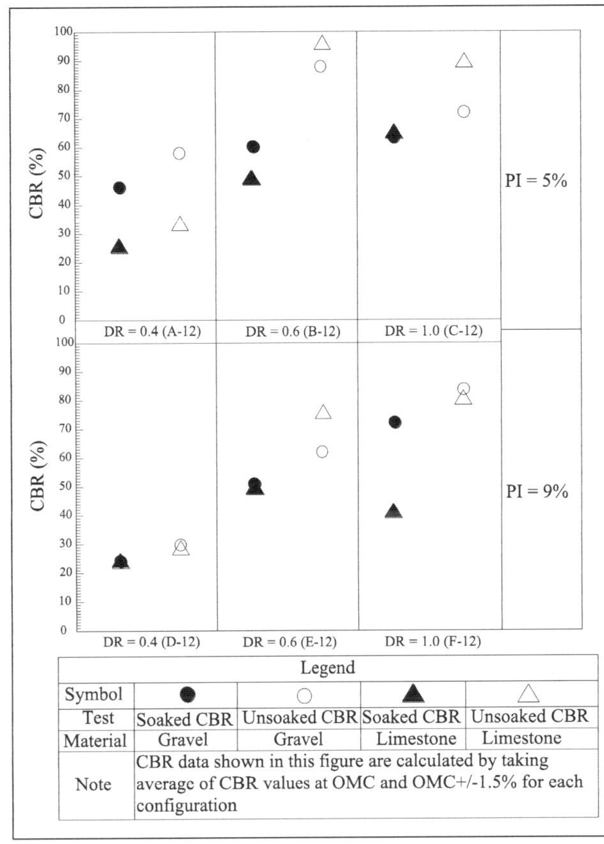

Figure 9. Soaked and unsoaked CBR results for 12% FC group samples.

crushed gravel samples with DR of 0.6 and 5% PI (i.e., B-5 and B-12); however, lower average soaked and unsoaked CBR values are seen for crushed gravel samples with DR of 0.6 and 9% PI (i.e., E-5 and E-12). Concerning crushed limestone samples, lower average soaked CBR values were generally observed when FC increased from 5% to 12% for samples with DR value of 0.6 (i.e., B and E).

3. For both materials with DR of 1.0, an increase in FC from 5% to 12% results in higher average soaked and unsoaked CBR values.
4. Very close average soaked and unsoaked CBR values were obtained for each material with 5% FC. However, for samples with 12% FC, the difference of soaked and unsoaked CBR values for each material was more pronounced.
5. The unsoaked CBRs for both materials were higher than their soaked CBR values for samples with 12% FC. However, for samples with 5% FC, the difference between soaked and unsoaked CBRs was minimal.

Prediction Model

Performing the soaked CBR test is more time consuming than the unsoaked CBR test. Many transportation agencies therefore use unsoaked CBRs for aggregate testing and pavement design needs. Therefore, developing a model to estimate soaked CBR values from available unsoaked CBR tests

Table 1. Correction Factors of the Soaked CBR Prediction Model (see Equation 2).

Plasticity Index (%)	α_p	Fines Content (%)	α_s	Dust Ratio	α_d	Material	α_m
5	0.793	5	1.079	0.4	1.096	Gravel	1.018
9	0.797	12	0.844	0.6	0.953	Limestone	1.108
–	–	–	–	1	0.794	–	–

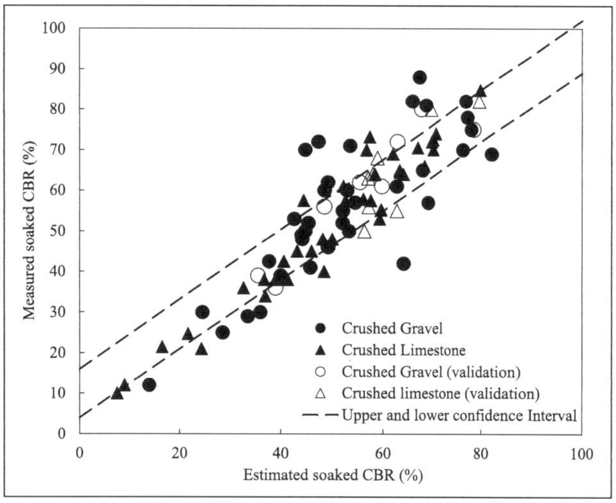

Figure 10. Predicted soaked CBR values compared with the measured ones.

with respect to aggregate index properties and material type is useful. A statistical analysis was conducted on 72 pairs of CBR test results of both materials at OMC and OMC ±1.5% to develop a prediction model. The model, which is presented in Equation 2, predicts soaked CBR values for crushed gravel and limestone from available unsoaked CBR tests and index property values of unbound aggregate:

$$\text{Soaked CBR}_{(\text{Predicted})} = \alpha_p \times \alpha_f \times \alpha_d \times \alpha_m \times \text{Unsoaked CBR} \quad (2)$$

where α_p is the PI correction factor, α_f is the percent passing the No. 200 sieve correction factor, α_d is the DR correction factor, and α_m is the material correction factor. The optimized correction factors are provided in Table 1, and Figure 10 shows the performance of the prediction model. The solid circle symbols in Figure 10 represent crushed gravel data, and triangle symbols represent crushed limestone data. Figure 10 indicates that the prediction model provides a reasonable agreement between the measured and the predicted soaked CBR values. The mean absolute percentage error (MAPE) and symmetric mean absolute percent error (SMAPE) of the prediction model were 11.8% and 12.0%, respectively. Furthermore, the *p*-value of Equation 2 was calculated and found to be less than 0.05. In addition, the results of analysis of variance (ANOVA) showed that in order of priority, FC, DR, PI, and material type were statistically significant parameters.

The prediction model was validated using a separate set of test results that were not used in development of the model. For this purpose, samples with FC of 8%, DR of 0.4, and PI of 5%, and samples with FC of 8%, DR of 0.6, and PI of 9% from crushed limestone and crushed gravel aggregates were prepared and tested. The developed model was used to predict soaked CBR values as shown in Figure 10 with hollow symbols. It can be concluded that the prediction model reasonably predicts the measured values. It should be noted that the model is only valid within the ranges of the aggregate material index used in this study.

Recommended Strength Zones

Figure 11 shows the minimum and maximum CBR values in respect to the material type, FC and DR within the OMC ±1.5% range. CBR values less than 40% are defined as the low-strength zone, while CBR values from 40% to 55% and those greater than 55% represent medium and high-strength zones, respectively.

Most of the soaked and unsoaked CBR ranges for samples with DR of 0.4 and FC of 5% are located in the high-strength zone (see Figure 11*a*). Increasing FC from 5% to 12% was found to minimally reduce maximum soaked and unsoaked CBR for crushed gravel samples with DR of 0.4. However, for crushed limestone samples, maximum soaked and unsoaked CBR values ended up shifting to the low-strength zone when FC increased from 5% to 12% (see Figure 11).

All the soaked and unsoaked CBR ranges of crushed gravel and crushed limestone samples with 5% FC and DR of 0.6 were in the high-strength zone. For both materials with a DR of 0.6, the increase in FC from 5% to 12% resulted an in an increase in the soaked and unsoaked CBRs, except for soaked crushed limestone samples (see Figure 11*d*).

Both the soaked and unsoaked CBR ranges for crushed gravel samples with DR of 1.0 were in the high-strength zone regardless of FC. For crushed limestone samples, the soaked CBRs were in the low to medium range and unsoaked CBRs were in the medium to high range (see Figure 11, *e* and *f*).

Osouli et al. (*22*) have proposed strength zones for crushed limestone with respect to FC and DR using average soaked CBR values within 1.5% of OMC, as shown in Figure

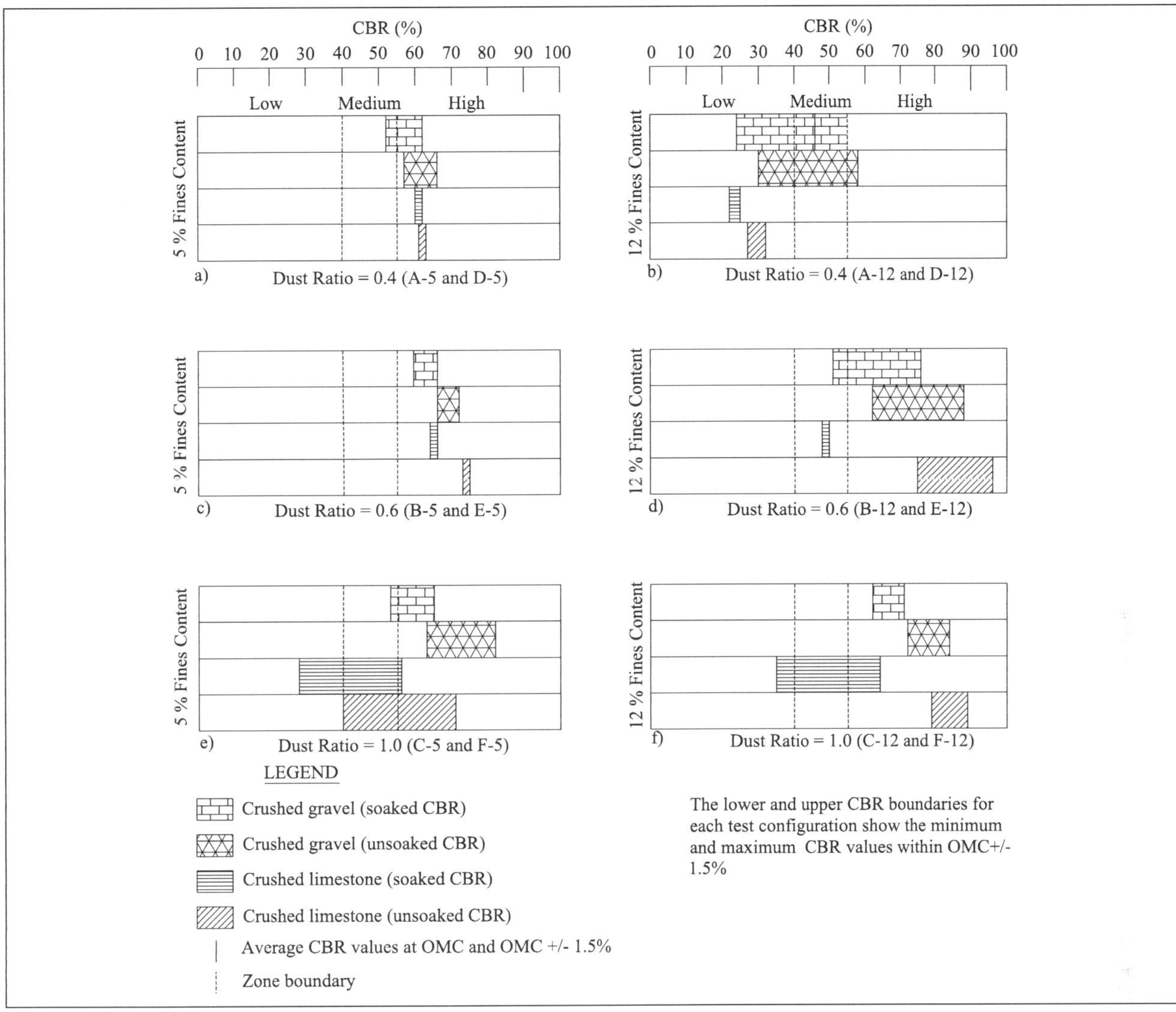

Figure 11. Soaked and unsoaked CBR zones for crushed limestone and gravel aggregates.

12 with hollow symbols. To populate this strength zone plot with more data, the unsoaked CBR values of this study were converted to soaked CBRs using the prediction model (i.e., Equation 2) and were added to this Figure and shown by solid symbols. It is observed that the additional data are also in agreement with proposed strength zones. It should be mentioned that the estimated soaked CBR values will be more accurate if additional data are added in developing the model.

Summary and Conclusions

The soaked and unsoaked CBR strength properties of crushed gravel and crushed limestone aggregates were investigated in this paper with a specific focus on the effects of FC, plasticity of fines and DR. The test matrix targeted different index property values such as FC of 5% and 12%, PI of 5% and 9%, and DR values of 0.4, 0.6 and 1.0. According to standard Proctor test results, an increase in FC results in higher maximum dry densities. In samples with 12% FC, MDD for crushed gravel has the least variability.

Results of soaked and unsoaked CBRs on crushed gravel and crushed limestone samples showed that strengths were the least and most affected by soaking when engineered with DR value of 0.4 and 1.0, respectively. The interaction of FC, DR, PI, and material type influences the CBR values within OMC ±1.5%. It is observed that for high FC values, the increase in DR would reduce the CBR values within OMC ±1.5% and DR of 0.4 with high PIs should be avoided. The average soaked and unsoaked CBR values were closer to each other for both crushed limestone and crushed gravel samples when they were engineered with 5% FC.

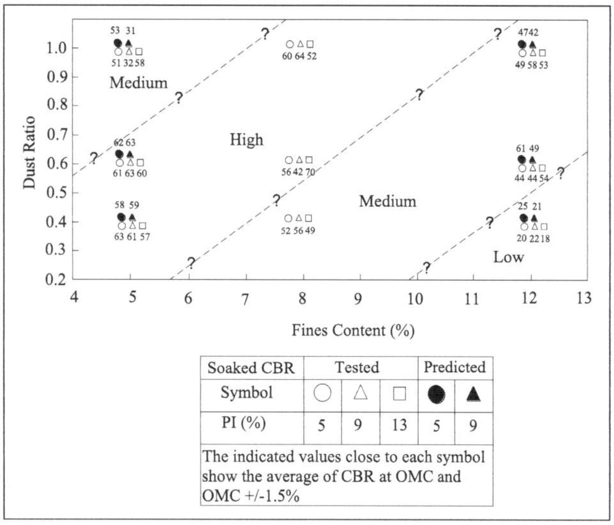

Figure 12. Strength zones corresponding to different DR and FCs (the question marks on the line show that these boundaries may change with additional data).

A prediction model was developed and validated for estimating the soaked CBR values of unbound aggregate materials from available unsoaked CBR tests for crushed gravel and crushed limestone aggregates by knowing the index properties of aggregates. Strength zones for crushed gravel and crushed limestone aggregates were also developed, showing three zones of high, medium, and low-strength properties. These strength zones may help in quality control of aggregates considering the different FC and DR properties.

Acknowledgments

This publication is based on the results of R27-157, Plasticity Requirement of Aggregate Used in Based and Sub-base. ICT R27-157 was conducted in cooperation with the Illinois Center for Transportation; the Illinois Department of Transportation, Office of Program Development, and the US Department of Transportation, Federal Highway Administration. The authors would like to acknowledge the members of IDOT Technical Review Panel (TRP) for their useful advice at different stages of this research.

References

1. American Association of State Highway, & Transportation Officials. *AASHTO Guide for Design of Pavement Structures*, 1998.
2. Xiao, Y., E. Tutumluer, and J. Siekmeier. Mechanistic-Empirical Evaluation of Aggregate Base and Granular Subbase Quality Affecting Flexible Pavement Performance in Minnesota. *Transportation Research Record: Journal of the Transportation Research Board*, 2014. 2227: 97–106.
3. Tutumluer, E. Practices for Unbound Aggregate Pavement Layers. *NCHRP Synthesis 445*. Transportation Research Board of the National Academies, Washington, D.C., 2013.
4. Dawson, A. R., N. H. Thom, and J. L. Paute. Mechanical characteristics of unbound granular materials as a function of condition, flexible pavements. In *Proc. European Symposium Euroflex* (A. G. Correia, ed.), Balkema, Rotterdam, Netherlands, 1996, pp. 35–44.
5. Barksdale, R. D. Laboratory Evaluation of Rutting in Base Course Materials. Presented at Third International Conference on the Structural Design of Asphalt Pavements, Grosvenor House, Park Lane, London, England, Sept. 11–15, 1972.
6. Kamal, M. A., A. R. Dawson, O. T. Farouki, D. A. B. Hughes, and A. A. Sha'at. Field and Laboratory Evaluation of the Mechanical Behavior of Unbound Granular Materials in Pavements. *Transportation Research Record: Journal of the Transportation Research Board*, 1993. 1406: 88–97.
7. Saeed, A., J. W. Hall, Jr., and W. Barker. *Performance-Related Tests of Aggregates for Use in Unbound Pavement Layers* (No. Project D4-23 FY'96), 2001. NCHRP Report 453, National Cooperative Highway Research Program, p. 55.
8. Gray, J. E. (1962). Characteristics of graded base course aggregates determined by triaxial tests, National Crushed Stone Association, Washington, DC.
9. Osouli, A., S. Salam, E. Tutumluer, S. Beshears, and S. Flynn. Effect of Dust Ratios on the Strength of Aggregates With Low Plasticity Fines. In *Geo-Chicago*, 2016, pp. 253–265. American Society of Civil Engineers.
10. Tutumluer, E., D. Mishra, and A. A. Butt. Characterization of Illinois Aggregates for Subgrade Replacement and Subbase. FHWA-ICT-09-060, 2009. Illinois Center for Transportation.
11. Lekarp, F., U. Isacsson, and A. Dawson. State of the Art. I: Resilient Response of Unbound Aggregates. *Journal of Transportation Engineering*, Vol. 126, No. 1, 2000, pp. 66–75.
12. Mishra, D., E. Tutumluer, and A. Butt. Quantifying Effects of Particle Shape and Type and Amount of Fines on Unbound Aggregate Performance Through Controlled Gradation. *Transportation Research Record: Journal of the Transportation Research Board*, 2010. 2167: 61–71.
13. Barskale, R. D., and S. Y. Itani. Influence of Aggregate Shape on Base Behavior. *Transportation Research Record: Journal of the Transportation Research Board*. 1989. 1227: 173–182.
14. Hicks, R. G., and C. L. Monismith. Factors Influencing the Resilient Response of Granular Materials. *Highway Research Record*, 345, 1971, 15–31.
15. Tutumluer, E., and T. Pan. Aggregate Morphology Affecting Strength and Permanent Deformation Behavior of Unbound Aggregate Materials. *Journal of Materials in Civil Engineering*, Vol. 20, No. 9, 2008, pp. 617–627.
16. Osouli, A., S. Salam, and E. Tutumluer. Effect of Plasticity Index and Dust Ratio on Moisture-Density and Strength Characteristics of Aggregates. *Transportation Geotechnics*, Vol. 9, 2016, pp. 69–79.
17. Faiz, A. The Effect of Skip-Grading on Stability of Soil-Aggregate Mixtures. 1971. Faiz, A. (1971). " The effect of skip-grading on stability of soil-aggregate mixtures. " FHWA/IN/JHRP-71/10, Joint Highway Research Project, Purdue Univ., West Lafayette, IN.
18. Chaulagai, R., A. Osouli, S. Salam, E. Tutumluer, S. Beshears, H. Shoup, and M. Bay. Maximum Particle Size, Fines Content and Dust Ratio Influencing Behavior of Base and Subbase Course Aggregates. *Transportation Research Record: Journal of the Transportation Research Board*, 2017. 2655: 20–26.
19. Baditala, S., and M. V. S. Sreedhar. Development of Correlation between Unsoaked and Soaked CBR Values. *International*

Journal of Emerging Technology and Advanced Engineering. 2016. Volume 6, Issue 11, November 16. p. 37-41

20. Onitsuka, K., S. Yoshitake, and M. Nanri. Mechanical Properties and Strength Anisotropy of Decomposed Granite Soil. *Soils and Foundations*, Vol. 25, No. 2, 1985, pp. 14–30.
21. Mahabubuzzaman, M. D. *California Bearing Ratio (CBR) and Strength Behaviour of Compacted Dhaka Clay as Road Subgrade*. MS thesis. 2001. Bangladesh University of Engineering and Technology
22. Osouli, A., S. Salam, G. Othmanawny, E. Tutumluer, S. Beshears, H. Shoup, and M. Eck. Results of Soaked and Unsoaked California Bearing Rate Tests on Unbound Aggregates With Varying Amounts of Fines and Dust Ratios. *Transportation Research Record: Journal of the Transportation Research Board*, 2017. 2655: 13–19.
23. *Standard Specifications for Roads and Bridges*, South Dakota Department of Transportation (SDDOT), 2015.
24. *Standard Specifications for Highway Construction*, Oklahoma Department of Transportation (ODOT), 2009.
25. *Standard Specifications for Road, Bridge, and Municipal Construction*, Washington State Department of Transportation (WSDOT), 2014.
26. *Standard Specification for Highway Construction*, Arkansas State Highway and Transportation Department (ARDOT), 2003.
27. *Standard Specification for Road and Bridge Construction*, Colorado Department of Transportation (CODOT), 2011.
28. *Standard Specifications*, California Department of Transportation (CADOT), 2004.
29. *Missouri Standard Specification for Highway Construction*, Missouri Department of Transportation (MODOT), 1999.
30. *Standard Specifications*, Indiana Department of Transportation (INDOT), 2014.
31. Talukdar, D. K. A study of Correlation Between California Bearing Ratio Value with Other Properties of Soil. *International Journal of Emerging Technology and Advanced Engineering*, Vol. 4, No. 1, 2014, pp. 559–562.
32. Ramasubbarao, G. V., and G. Siva Sankar. Predicting Soaked CBR Value of Fine Grained Soils Using Index and Compaction Characteristics. *Jordan Journal of Civil Engineering*, Vol. 7, No. 3, 2013, pp. 354–360.
33. Lim, S. M., D. C. Wijeyesekera, and I. Bakar. Correlations of Soil Classification and Compaction Parameters With Soaked and Unsoaked CBR of Soils. Presented at South East Asia Conference on Soft Soils Engineering and Ground Improvement. Advancement of Research and Practice for Geotechnical Solutions, Bandung, Indonesia, 20–23 October, 2014.
34. *Standard Specifications for Road and Bridge Construction*, Illinois Department of Transportation (IDOT), 2016.
35. Talbot, A. N., and F. E. Richart. The Strength of Concrete, its Relation to the Cement Aggregates and Water. *Bulletin*, No. 137, 1923. Publisher: University of Illinois at Urbana Champaign, College of Engineering. Engineering Experiment Station. Bulletin No. 137, Vol XXI, No. 7 – https://www.ideals.illinois.edu/handle/2142/4483
36. Thom, N. H., and S. F. Brown. The Effect of Grading and Density on the Mechanical Properties of a Crushed Dolomitic Limestone. Presented at 14th Australian Road Research Board Conference, Canberra, 1988.
37. Lee, P. Y., and R. J. Suedkamp. Characteristics of Irregularly Shaped Compaction Curves of Soils. 1972. Publisher: Highway Research Board, Issue Number 381, pp 1–9.

The Standing Committee on Aggregates (AFP70) peer-reviewed this paper (18-04732).

The contents of this paper reflect the views of the authors only.

Article

TRR
JOURNAL OF THE TRANSPORTATION RESEARCH BOARD

Modeling Stress-Dependent Anisotropic Elastoplastic Unbound Granular Base in Flexible Pavements

Transportation Research Record
2018, Vol. 2672(52) 46–56

DOI: 10.1177/0361198118758318
journals.sagepub.com/home/trr
SAGE

Yuqing Zhang[1], Fan Gu[2], Xue Luo[3], Bjorn Birgisson[3], and Robert L. Lytton[4]

Abstract

Unbound granular base (UGB) has a cross-anisotropic and nonlinear (stress-dependent) modulus with a plastic behavior. Existing UGB models address nonlinear cross-anisotropy and plasticity separately. It is unknown how the two characteristics are coupled into a finite element model (FEM) and how this will affect the pavement responses. This study presents a coupled nonlinear cross-anisotropic elastoplastic (NAEP) constitutive model for the UGB and implements it in a weak form equation-based FEM. No material subroutine is needed to address the circular dependence between the stress-dependent anisotropic modulus, structural stress responses, and elastoplastic deformation. The NAEP model was calibrated by triaxial resilient modulus and strength tests and validated using laboratory measurements in a large-scale soil-tank pavement structural test. It is found that the NAEP model is valid and effective in predicting the UGB responses in flexible pavements. The model predicted less horizontal tensile stresses at the base bottom and introduced compressive stresses in the middle and top of the base course. This is caused by an increasing confinement resulting from a horizontal plastic dilation in the base course, which cannot be modeled without considering plasticity. The stress-dependent modulus for the UGB material decreases with depth and the distance from loading centerline. Compared with a nonlinear anisotropic elastic model, the NAEP model predicted the same tensile strain at asphalt layer bottom, a higher base modulus, and a higher subgrade compressive strain. Thus, the nonlinear anisotropic elastic UGB model results in the same fatigue life as the NAEP model but may riskily under-predict rutting damage.

An unbound granular base (UGB) provides the foundational support to a flexible pavement and distributes the stresses induced by traffic load to the underlying subgrade. Understanding the constitutive behavior of the UGB materials is crucial to the accurate performance prediction of the pavement structures. In current pavement design, the UGB material is assumed as a linear (stress-independent) elastic material. The linear isotropic model predicted unexpected and high tensile horizontal stresses at the bottom of the base course, which cannot be sustained by the UGB material in reality. Several studies have revealed that the UGB material exhibits a nonlinear cross-anisotropic behavior, which means that the modulus of the UGB is stress-dependent (nonlinearity) and the elastic material properties (modulus and Poisson's ratio) in the vertical plane differ from those in the horizontal plane (cross-anisotropy). For example, Tutumluer and Seyhan (*1*), Adu-Osei et al. (*2*) and Gu et al. (*3*) successfully determined the nonlinear cross-anisotropic properties of the UGB using triaxial resilient modulus test at different stress states. Tutumluer and Thompson developed a finite element program to model the nonlinear cross-anisotropic behavior of UGB in flexible pavements (*4*). They found that a nonlinear cross-anisotropic model predicted much less horizontal tensile stress in base course compared with the linear elastic model. A residual compressive stress of 21 kPa resulting from compaction was suggested to compensate the tensile stresses, leading to a compressive horizontal stress in the base course. Oh et al. (*5*) and Al-Qadi et al. (*6*) reported that modeling the pavement structures using a nonlinear cross-anisotropic approach predicted greater pavement responses, including tensile strain at the bottom of asphalt layer and compressive strain on top of subgrade, than a linear model. Tutumluer et al. (*7*) and Wang and Al-Qadi (*8*) concluded that the nonlinear

[1]Aston Institute of Materials Research, and Engineering System & Management Group, School of Engineering and Applied Science, Aston University, Aston Triangle, Birmingham, UK
[2]National Center for Asphalt Technology, Auburn University, Auburn, AL
[3]Center for Infrastructure Renewal, Texas A&M University System, College Station, TX
[4]Fred J. Benson Chair, Zachry Department of Civil Engineering, Texas A&M University, College Station, TX

Corresponding Author:
Address correspondence to: Yuqing Zhang: y.zhang10@aston.ac.uk

cross-anisotropic model provided better agreement with the field measurements. Therefore, modeling UGB as a nonlinear cross-anisotropic material should be considered for pavement performance predictions.

However, some studies also demonstrated that, when subjected to a heavy load, the UGB material exhibits a significant plastic behavior in addition to the stress-dependent anisotropic behavior in flexible pavements (*9, 10*). Even at a low load level, the responses of a UGB still include the recoverable and unrecoverable strains (*11*). The recoverable strain is characterized by the nonlinear anisotropic resilient behavior of UGB, while the unrecoverable strain is reflected by the plastic behavior of UGB. Currently, few studies have been able to simultaneously model the nonlinear stress-dependence, cross-anisotropy, and plastic behavior of the UGB materials. No studies have implemented a coupled nonlinear anisotropic elastoplastic (NAEP) model for UGB in a numerical program to evaluate the material responses and pavement performance. The potential reasons include: (1) a circular dependence exists between UGB material properties (anisotropic modulus), stress distribution in pavement and the evolution of strains including elastic and plastic strains, which has not been well understood; (2) programming and utilizing a material subroutine in a finite element model (FEM) to solve the circular dependence is very complicated and time-consuming due to issues such as high nonlinearity, low-efficient iteration, and non-convergence, requiring significant training and knowledge of numerical programming for the ultimate users; and (3) difficulties exist in determining model coefficients using effective and efficient laboratory tests.

This study aims to characterize the cross-anisotropic, nonlinear, and plastic properties of UGB simultaneously using a NAEP model. The NAEP model is then implemented in a weak form equation-based FEM to demonstrate how the circular dependence is effectively modeled without needing a material subroutine. The NAEP model is validated by pavement structural loading tests in a laboratory large-scale tank. The distribution of stress-dependent modulus, horizontal stress, and plastic strain in the base course are also evaluated by comparing the NAEP model predictions with the predictions from linear (and nonlinear) anisotropic elastic models. Critical pavement responses are also determined by various UGB models to evaluate the effects of the NAEP model on the flexible pavement performance.

Anisotropic Nonlinear Elastoplastic Constitutive Model for UGB

Cross-Anisotropy of UGB Materials

A well-established cross-anisotropic constitutive model is employed for the UGB materials, as shown in Equation 1, in which the z-axis is the axis of symmetry (*12*), thus the UGB is anisotropic in z–x and z–y planes and becomes isotropic in the x-–y plane,

$$\begin{bmatrix}\sigma_{xx}\\ \sigma_{yy}\\ \sigma_{zz}\\ \sigma_{xy}\\ \sigma_{yz}\\ \sigma_{zx}\end{bmatrix}=\frac{E_z}{\alpha_0\beta_0}\begin{bmatrix} n\left(1-nv_{zx}^2\right) & n\left(v_{xyy}+nv_{z}^2\right) & nv_{zx}\alpha_0 & 0 & 0 & 0\\ n\left(v_{xy}+nv_{zx}^2\right) & n\left(1-nv_{zx}^2\right) & nv_{zx}\alpha_0 & 0 & 0 & 0\\ nv_{zx}\alpha_0 & nv_{zzx}\alpha_0 & 1-v_{xy}^2 & 0 & 0 & 0\\ 0 & 0 & 0 & \frac{1}{2}n\beta_0 & 0 & 0\\ 0 & 0 & 0 & 0 & m\alpha_0\beta_0 & 0\\ 0 & 0 & 0 & 0 & 0 & m\alpha_0\beta_0\end{bmatrix}\begin{bmatrix}\varepsilon_{xx}\\ \varepsilon_{yy}\\ \varepsilon_{zz}\\ 2\varepsilon_{xy}\\ 2\varepsilon_{yz}\\ 2\varepsilon_{zx}\end{bmatrix} \quad (1)$$

where σ_{ij} = stress tensor; ε_{ij} = strain tensor; and $i, j = x, y, z$; $n = E_x/E_z$; $m = G_{zx}/E_z$; $\alpha_0 = 1 + v_{xy}$; $\beta_0 = 1 - v_{xy} - 2nv_{zx}^2$; E_x = horizontal modulus in x-direction; E_z = vertical modulus in z-direction; G_{zx} = shear modulus in the vertical (z–x) plane; v_{xy} = Poisson's ratio in the horizontal (x–y) plane, characterizing the effect of a horizontal strain (x-direction) on the orthogonal horizontal strain (y-direction); v_{zx} = Poisson's ratio in the vertical (z–x) plane, characterizing the effect of the vertical strain (z-direction) on the horizontal strain (x-direction). Note that E_x, E_z, v_{xy}, v_{zx}, and G_{zx} are the five independent material properties characterizing the cross-anisotropic behavior of the UGB. In Equation 1, the five independent material properties are represented by E_z, n, m, α_0, and β_0.

Stress-Dependent Nonlinearity of UGB

The vertical modulus of the UGB is modeled by a nonlinear stress-dependent model as follows (*13*):

$$E_z = k_1 P_a \left(\frac{I_1}{P_a}\right)^{k_2} \left(\frac{\tau_{\text{oct}}}{P_a}\right)^{k_3} \quad (2)$$

where $I_1 = \sigma_{kk}$ is the first invariant of the stress tensor, σ_{ij}; P_a = the atmospheric pressure, 101.3 kPa; $\tau_{\text{oct}} = \sqrt{S_{ij}S_{ij}/3} = \sqrt{2J_2/3}$ is the octahedral shear stress; S_{ij} = the deviatoric stress tensor and $S_{ij} = \sigma_{ij} - 1/3\sigma_{kk}\delta_{ij}$ where δ_{ij} is the Kronecker delta, and k_1, k_2, and k_3 = regression coefficients to be determined by laboratory tests. Note that the horizontal modulus becomes stress dependent as well by the relationship of $n = E_x/E_z$ that is used in Equation 1.

Elastoplastic Models for UGB

A Generalized Drucker-Prager (GD-P) yield surface model was developed by the authors to allow a smooth and convex yield surface when frictional angle varies from 0 to 90° for granular materials such as asphalt, aggregate base, sand or soil materials (*14*). This GD-P model can remove the inherent limitations of the existing models, for example, the non-smoothness for the Mohr–Coulomb yield surface or the non-convexity for the extended Drucker-Prager yield surface when the frictional angle is greater than 22°. Furthermore, the GD-P model can be reduced to Drucker-Prager model (when parameter d = 1) or Matsuoka–Nakai model (when parameter $\kappa = 0$), which have been extensively used for modeling the yield surface of the granular materials. The GD-P model is expressed as

$$f = \sqrt{J_2}\rho(\theta') - \alpha I_1 - \kappa \tag{3}$$

where $J_2 = 1/2S_{ij}S_{ij}$ is the second invariant of the deviatoric stress tensor, S_{ij}; θ' = Lode angle that is defined as

$$\theta' = \frac{1}{3}\arccos\left[\frac{3\sqrt{3}}{2}\frac{J_3}{(J_2)^{3/2}}\right] \tag{4}$$

where $J_3 = \det(S_{ij})$ is the third invariant of the deviatoric stress tensor. θ' is zero for compression and $\pi/3$ for extension. $\rho(\theta')$ = a function defining the yield surface shape on the octahedral plane and determining the convexity of the yield surface, which is expressed as

$$\rho(\theta') = \mu\cos\left[\tfrac{1}{3}\arccos(\gamma\cos 3\theta')\right] \tag{5}$$

where μ and γ = parameters determining the size and shape of the yield surface on the octahedral plane, respectively, to ensure a smooth and convex yield surface. They are fundamentally related to material friction angle as

$$\mu = \frac{2\sqrt{1-d+d^2}}{\sqrt{3}d},\quad \gamma = -\frac{3\sqrt{3}}{2}\frac{(1-d)d}{(1-d+d^2)^{1.5}},\quad d = \frac{3-\sin\phi}{3+\sin\phi} \tag{6}$$

where d = extension ratio that is the ratio of the yield strength in extension, to that in compression. ϕ = internal friction angle, which also directly determines the slope of the yield surface on a meridian plane by

$$\alpha = \frac{2\sin\phi}{\sqrt{3}(3-\sin\phi)} \tag{7}$$

The parameter κ in the GD-P model represents the material cohesion characteristic and can be determined by material cohesion (C) and internal friction angle (ϕ):

$$\kappa = \frac{6C\cos\phi}{\sqrt{3}(3-\sin\phi)} \tag{8}$$

It should be noted that a yielding end-cap surface can be added to the GD-P model to address the yield surface softening at a high hydrostatic pressure, which is not considered in this study. Therefore a limited confining pressure will be achieved in base course, as shown in the result analysis. The increment of the plastic strain of the UGB occurs when the current state of the stress satisfies the yield surface function (i.e., $f = 0$). With hypothesizing small plastic strain, the direction of the plastic strain increment is defined by a non-associated plastic flow rule:

$$\dot{\varepsilon}_{ij}^{p} = \lambda\frac{\partial g}{\partial\sigma_{ij}} \tag{9}$$

where $\dot{\varepsilon}_{ij}^{p}$ = the rate at which the plastic strain changes with respect to $\partial g/\partial\sigma_{ij}$; λ is a plastic multiplier which depends on the current state of stress and the load history. Here λ is determined by the complementarity or Kuhn–Tucker conditions:

$$\lambda \ge 0,\ f \le 0 \text{ and } \lambda f = 0 \tag{10}$$

The variable g in the plastic flow rule is a plastic potential function. The non-associated flow rule applies when $g \neq f$, which is appropriate for shear dilating granular materials such as asphalt, UGB and soils. It is demonstrated that the plastic potential surface has the same linear form as the yield surface but with a smaller slope which affects the volumetric dilation of the material:

$$g = \sqrt{J_2}\rho(\theta') - \beta I_1 \tag{11}$$

Several studies have indicated that $\beta < \alpha$ is valid for geomaterials such as soils, sands, and asphalt mixtures (*15, 16*). Here β is derived to be a function of anisotropy (*14*):

$$\beta = 0.5889\Delta' - 0.0122 \tag{12}$$

where Δ' is a microstructural parameter quantifying anisotropy of the UGB that can be determined by (*17*)

$$\Delta' = \frac{3\sqrt{1/n} - 3}{4 - \sqrt{1/n}} \tag{13}$$

where n is defined as the anisotropic modulus ratio and $n = E_x/E_z$ that is used in Equation 1. Equations 12 and 13 were originally derived for asphalt mixtures; however, they still apply to the UGB. This is because the inherent anisotropy of both asphalt mixtures and UGB is caused by the preferentially oriented aggregate particles in the horizontal directions and can be quantified by the microstructural parameter using the same model (*17, 18*). The validation of the UGB plastic responses by laboratory tests in this study will also verify the applicability of the models for the UGB materials.

Experiments and Material Characterization

Flexible Pavement Structures and Models

Table 1 describes the structure of a flexible pavement to be modeled including the layer thickness, materials, their constitutive models and corresponding parameter determination methods. The flexible pavement consists of three layers,

Table 1. Pavement Structures, Material Models, Parameters and Characterization Tests

Pavement layers (thickness)	Materials	Constitutive models	Model parameters	Laboratory test
Layer 1 (15 cm)	Asphalt concrete	Viscoelastic	$E\infty$, v_0, E_i, τ_i	Dynamic modulus test
Layer 2 (25 cm)	Unbound granular base	(*a*) LAE (*b*) NAE (*c*) NAEP	k_1, k_2, k_3 n, m, v_{zx}, v_{xy}, C, ϕ	Resilient modulus test at different stress levels Strength test at confining pressures of 0, 20.7, and 103.4 kPa
Layer 3 (140 cm)	Subgrade	Elastic	Young's modulus	California bearing ratio test

Table 2. Model Parameters for Asphalt Concrete, UGB and Subgrade

Asphalt concrete (long term equilibrium modulus $E\infty$ = 42 MPa, Poisson's ratio v_0 = 0.35)

i	1	2	3	4	5	6
E_i (MPa)	6564	6582	3200	1342	299	103
τ_i (s)	4.09E-6	2.56E-4	7.71E-3	2.10E-1	3.88E+0	6.63E+1

UGB materials

Material models	k_1	k_2	k_3	n	m	v_{zx}	v_{xy}	C (kPa)	ϕ (°)
(*a*) LAE	E_z = 150 MPa is used by averaging NAE results			0.45	0.35	0.38	0.43	NA	NA
(*b*) NAE	1281	0.81	−0.08	0.45	0.35	0.38	0.43	NA	NA
(*c*) NAEP	1281	0.81	−0.08	0.45	0.35	0.38	0.43	20.2	51.3

Subgrade (Elastic modulus E_s = 69 MPa, Poisson ratio v_s = 0.4)

namely, 15 cm asphalt course, 25 cm base course, and 140 cm subgrade. The layer thickness was obtained based on a laboratory large-scale tank test. The asphalt concrete is modeled as a viscoelastic material. To evaluate the pavement responses predicted by different UGB constitutive models, the UGB materials are modeled as an NAEP model, a nonlinear anisotropic elastic (NAE) model, and a linear anisotropic elastic (LAE) model, respectively. The subgrade is modeled as an elastic material and its elastic modulus is estimated based on California bearing ratio test (*19*).

Asphalt Concrete Characterization

In this study, the asphalt concrete is modeled as a viscoelastic material. A solid-like generalized Maxwell model is used to model the material's relaxation modulus (*20*):

$$E(t) = E_\infty + \sum_{i=1}^{M} E_i \exp\left(-\frac{t}{\tau_i}\right) \tag{14}$$

where E_∞ = long-term equilibrium modulus; E_i = components of the relaxation modulus; τ_i = components of relaxation time; t = time; and M = the total number of the Maxwell elements (one Maxwell element is composed of one elastic spring and one viscous dashpot connected in series). Dynamic modulus tests following ASTM standard (*21*) were employed to determine the dynamic modulus and phase angle of the asphalt concrete, which are then used to determine the coefficients in the generalized Maxwell model of the relaxation modulus according to a conversion method presented previously (*22*). Note that a constant Poisson's ratio was assumed for the asphalt concrete to simplify the simulation though it is a time-dependent variable as well (*23*). The determined Maxwell model coefficients for the asphalt concrete are shown in Table 2.

UGB Material Characterization

Granite crushed aggregates are used to construct the base course. The aggregate gradation and physical properties can be found in a previous study (*11*). It has a maximum dry density of 2.16×10^3 kg/m^3 and an optimum water content of 6.7%. Repeated load triaxial tests were performed on the cylindrical aggregate specimens using a universal testing machine (UTM) with a rapid triaxial test (RaTT) cell to determine the cross-anisotropic stress dependent resilient modulus of the UGB. A test protocol with ten stress states (with different axial stresses and confining pressures) was used to produce three stress modes (compression, shear and extension) in the specimen (*2*). At each stress state, every loading cycle of the dynamic stress consisted of 1.5 seconds of loading and 1.5 seconds of unloading. The vertical and horizontal deformations of the specimen were recorded to calculate the anisotropic properties of the aggregate specimen using systematic identification method and the results are presented in Table 2.

A triaxial compressive strength test was used to determine the shearing resistance of the base materials according to Tex-117-E (*24*). The axial load was applied at a constant strain rate of 2% on the UGB specimen under three confining pressures (i.e., 0, 20.7, and 103.4 kPa) until the specimen collapsed. The initial yielding stress was identified from the curve of deviatoric stress and axial strain at a specific confining pressure. Then the material cohesion and friction angle were determined based on Mohr's yielding envelope to be 20.2 kPa and 51.3°, respectively. Details of the UGB material characterization and model coefficient determination can be found in the literature (*11, 25*). The NAE model parameters remained unchanged from the NAEP model except that the plastic model component was disabled. The LAE model parameters remained unchanged from the NAE except that a constant vertical modulus of 150 MPa was used based on the average vertical modulus of the NAE predictions.

Finite Element Modeling of UGB in Flexible Pavements

Weak Form Equation-Based Finite Element Modeling in COMSOL Program

The viscoelastic constitutions of asphalt concrete have been successfully modeled using a weak-form partial differential equation (PDE)-based FEM technique in COMSOL Multiphysics program (*22*). The viscoplastic and viscofracture properties of the asphalt concrete have also been coupled in the viscoelastic constitutive models based on the weak-form PDE based FEM method (*26*). It demonstrates that accurate viscoelastic and viscoplastic responses can be modeled and predicted for the asphalt concrete. Unlike using Abaqus, there is no need for this method to program a user-defined material subroutine. In this study, the authors continue modeling the asphalt concrete using the PDE FEM method while only considering its viscoelastic responses, details of which are referred to the above two publications (*22, 26*).

The weak-form PDE-FE is particularly advantageous in solving a coupling problem with circular dependence between the model inputs and the dependent variables to be solved. Modeling a UGB material with a stress-dependent modulus is a typical coupling problem which involves a circular dependence between modulus, stress and deformation. This circular dependence can be broken by creating two additional ***Dependent Variables*** (DVs) in the weak-form PDE FEM to define the stress-dependent modulus. The two DVs will be defined in ***Weak Expressions*** and solved simultaneously with the other three defined DVs (i.e., displacement components in spatial *x*-, *y*-, and *z*-directions). Specifically, the below modeling steps are followed. Note a bold and italic expression represents a module or a defined variable in the COMSOL program

Step 1. Define the material properties from Table 2 as ***Parameters*** to be used as model inputs or to define variables.
Step 2. Create a new physics using ***Weak Form PDE*** module and define its two ***Dependent Variables*** as ***u11*** and ***u12*** with a unit of N/m^2. In addition, global DVs have been defined for the UGB materials using a ***Linear Elastic Material*** module that includes the displacement components in three directions, that is, *u, v*, and *w*.
Step 3. Define the constitutive equation for the dependent variables in ***Weak Expressions*** as below and COMSOL will perform a numerical integration of the ***Weak Expressions*** over a volumetric region (e.g., the FEM of the base course) and find the solutions for all DVs:

$$\begin{cases} -(u11-\text{solid}.I1s)*\text{test}(u11) \\ -(u12-\text{sqrt}(\text{solid}.II2s*2/3))*\text{test}(u12) \end{cases} \quad (15)$$

where solid.$I1s$ and solid.$II2s$ are respectively the first invariant of the stress tensor (I_1) and the second invariant of the deviatoric stress tensor (J_2), which are pre-defined by COMSOL and can be directly used in the weak expressions.
Step 4. Define the stress-dependent vertical modulus of the UGB as a ***Variable*** using the DVs as below.

$$E_Z = k_1 * P_a * (\text{abs}(u11)/P_a)\wedge k_2 * (u12/P_a)\wedge k_3 \quad (16)$$

Note that ***u11*** is always negative for the UGB materials as the first stress invariant in the base course is always in compression and COMSOL defines compression as negative. Thus, an absolute value of ***u11*** is used to calculate the stress-dependent vertical modulus. Here k_1, k_2, k_3, and P_a are ***Parameters*** defined in Step 1.
Step 5. Employ the variable E_Z in the ***Linear Elastic Material*** module for the UGB material together with other cross-anisotropic model parameters defined in Step 1 as the model inputs in the elasticity matrix (the ***D matrix***) based on Equation 1. A ***Plasticity*** module was brought into the ***Linear Elastic Material*** module by utilizing a user-defined yield surface where the ***Effective stress*** is defined as $\text{sqrt}(\text{solid}.II2s)*\rho(\theta')-\alpha*(-\text{solid}.I1s)$ and the ***Initial yield stress*** can be defined as κ based on Equation 3. A user-defined ***Plastic potential*** (***Q***) is defined as $\text{sqrt}(\text{solid}.II2s)*\rho(\theta')-\beta*(-\text{solid}.I1s)$ according to Equation 11. Note that the α and β are defined in Step 1 as ***Parameters*** using Equations 7 and 8. Here $\rho(\theta')$ is defined as a ***Variable*** using Equation 5.

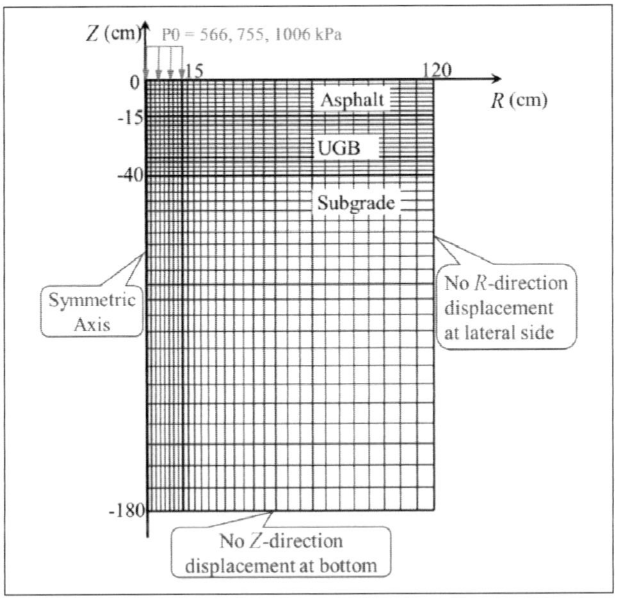

Figure 1. Axisymmetric finite element model of an asphalt pavement structure in a large-scale tank test.

FEM of Flexible Pavement Structure

A pavement structure shown in Figure 1 was modeled as an axisymmetric structure in COMSOL Multiphysics program with a width of 120 cm and three layers including an asphalt course of 15 cm, a UGB base course of 25 cm, and a subgrade of 140 cm. A ramp load was applied in 0.05 s on a circular loading area with a radius of 15 cm. Three load levels were tested, including 566, 755, and 1006 kPa. The mesh of the model used a resolution of 1.875 cm and a biased mesh distribution was used, as shown in Figure 1. The dimension and loading conditions were selected to simulate the asphalt pavement structure and the applied load used in a laboratory large-scale tank test, the results of which are employed to validate the FEM predictions. The material properties and model coefficients are shown in Table 2.

Model Validation and Result Discussions

Model Validations by Large-scale Tank Test

To validate the FEMs of the UGB in the flexible pavement, surface deflections and vertical stresses at different locations in base and subgrade were measured in the large-scale tank test. The testing procedure and measurement details of the large-tank test are described elsewhere (*27*). Figure 2*a* compares the FEM surface deflection predictions using the NAEP UGB model with the measured values at different distances (0, 30, 60, 90, and 110 cm) from the loading center under the three loading levels. It is seen that the FEM predicted surface deflections increased with loading level and matched the measured surface deflections in general, which demonstrates the robustness of the NAEP model. The measured surface deflections at the loading center ($R = 0$) under all three loading levels are greater than that predicted by FEM, which could be because a dynamic load was used in the test whereas a quasi-static ramp load was used in FEM simulations. The dynamic loading effect may need to be considered in a future FEM simulation.

Figure 2, *b, c* and *d* compare the vertical stresses at different depths ($Z = -21.25$ cm in base, -33.75 cm in base, and -55 cm in subgrade) and two distances from loading ($R = 0$ and 15 cm) under three load levels, respectively. It is observed that the FEM predicted vertical stresses are all in compression (a negative stress is in compression) and decrease with depth, which is consistent with the measured vertical stresses. The FEM predicted vertical stresses at the distance of $R = 15$ cm match well with the laboratory measurements, while those at the loading center ($R = 0$) depart from the measured stresses. For instance, the measured vertical stresses in subgrade ($Z = -55$ cm) and at loading center ($R = 0$) are always less than that predicted from FEM, which was also observed previously (*27*). This may be caused by soil arching effect over the pressure sensor during construction or not accounting for the stress-dependent behavior for soil subgrade in the FEM simulations. Nevertheless, the FEM results are in good agreement with the large-scale tank test measurements in general. The proposed NAEP model and the weak form equation-based FEM techniques are valid and effective in predicting accurate responses of the UGB materials in flexible pavements.

Pavement Response Analysis using Different UGB Models

Further analysis based on the validated FEMs was conducted to obtain the pavement responses using different UGB models including NAEP model, NAE model and LAE model. Figure 3, *a, b* and *c* show the distribution of the horizontal stresses at load centerline along base course depth with increasing load levels predicted using LAE, NAE, and NAEP models, respectively. Figure 3*a* illustrates that, when LAE model is used for UGB, the horizontal stress is mainly in tension (in positive) and increases with depth and load level. Note that if a linear isotropic elastic (LIE) model is used, the maximum tensile horizontal stress will become even higher (e.g., 50 kPa), the distribution of which is not shown here but is similar to that in Figure 3*a*.

The high tensile stress at the base course bottom predicted by LAE or LIE model is unexpected as UGB has a very limited tensile strength. The high tensile stress can be substantially reduced when a stress-dependent anisotropic modulus model is used for the base course as indicated from the literature (*4*). The same conclusion is derived from Figure 3*b* that, using an NAE model, the horizontal stresses are reduced and

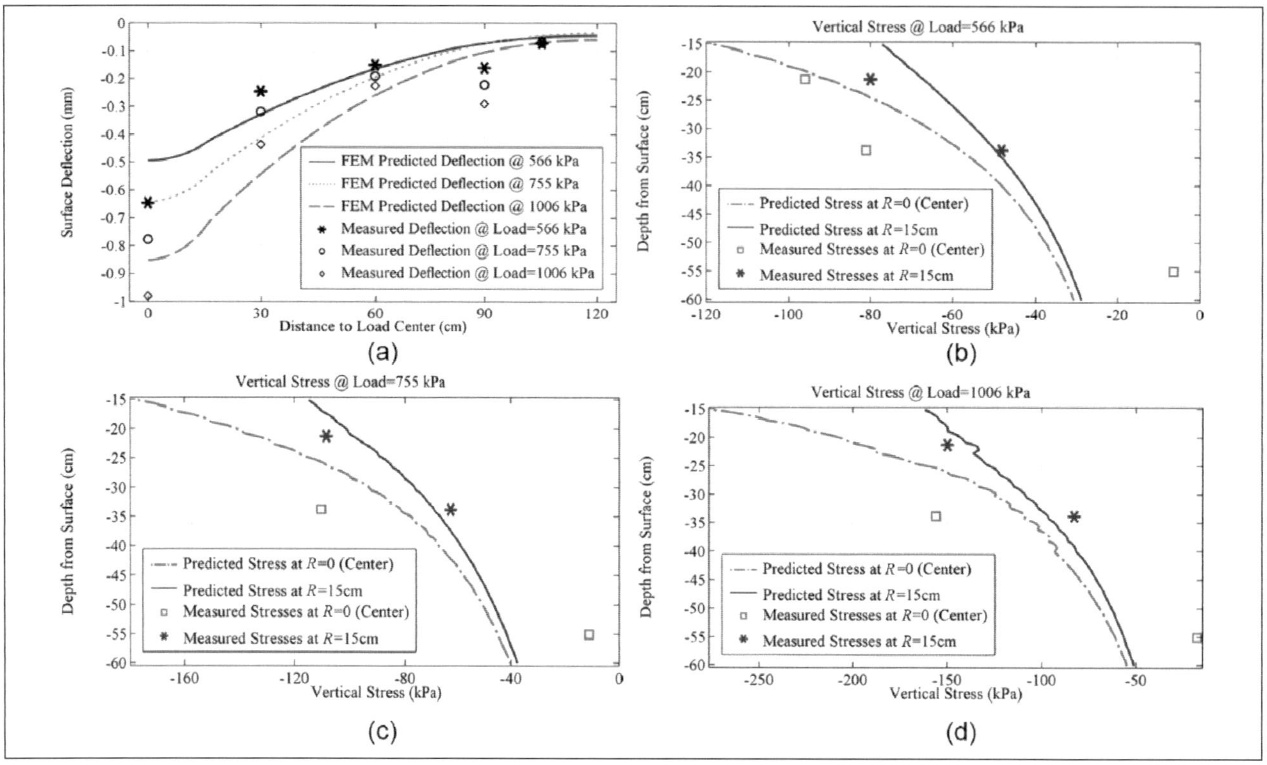

Figure 2. Validation of FEM by surface deflections and vertical stresses measured in large-scale tank tests of an asphalt pavement structure.

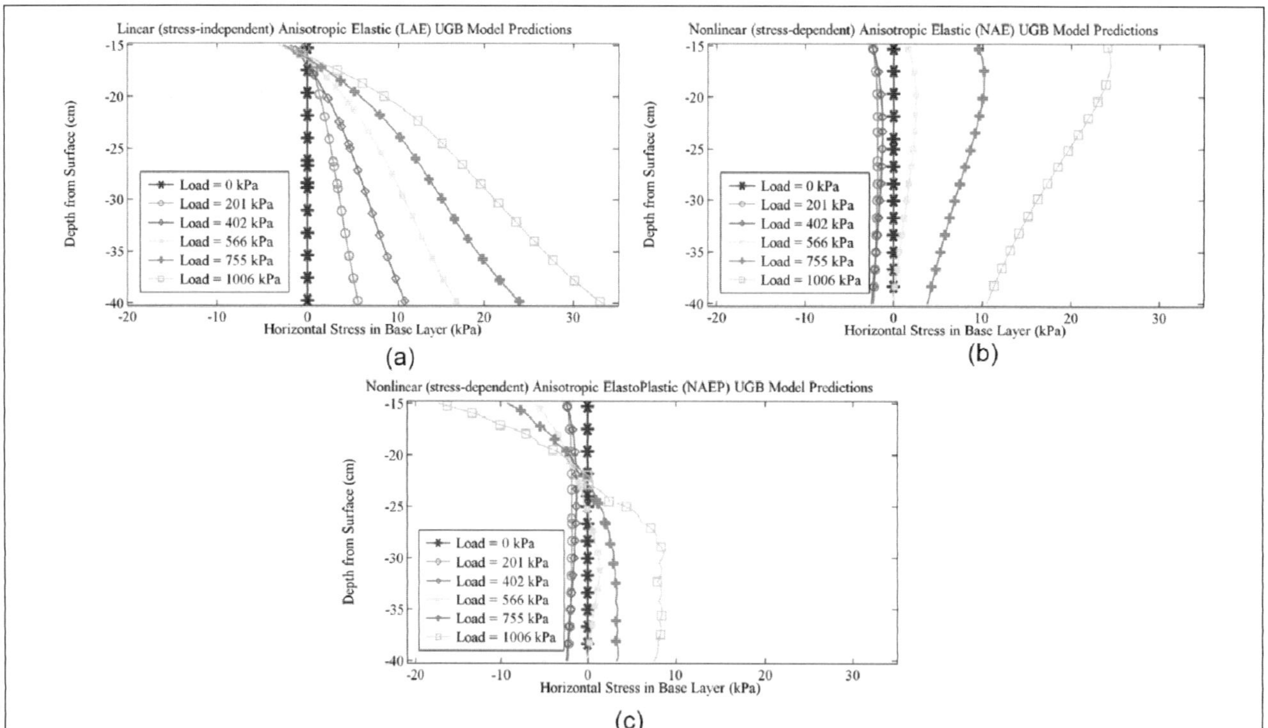

Figure 3. Horizontal stresses (kPa) at loading center along depth in base course with increasing load levels in which (*a*) LAE model, (*b*) NAE model, and (*c*) NAEP model were used for UGB (tension is positive).

vary from compression to tension as load level increases. The tensile horizontal stress values (7–10 kPa at the load level of 755 kPa) and its decreasing distribution along depth are consistent with the findings from the literature, in which it is explained that the tensile horizontal stresses are accommodated by compressive residual stresses (about 21 kPa) due to compaction (*4, 13*).

The compressive compensation from the compaction-induced residual stresses is valid, however, it may be insufficient to ensure that the base course is always in compression especially when the load level goes higher. For example, Figure 3*b* shows that the horizontal stress became as high as 25 kPa when the load level reached 1006 kPa. It could go higher when the load level kept increasing, when the NAE model is used. However, this cannot happen in the field. The UGB materials will have been yielded and plastic deformation will occur before the vertical stress reaches 250 kPa in compression (seen from Figure 2*d*, when $R = 0$, $Z = -15$cm) while the horizontal stress is 25 kPa in tension. The plastic deformation in the base course will lead to an altered stress distribution.

Figure 3*c* shows the horizontal stress distribution in the base course when the NAEP model is used for the UGB. The horizontal stresses at the base bottom were further reduced (less than 10 kPa) compared with that calculated by the NAE model (25 kPa). These low tensile stresses can be fully compensated by the compaction-induced compressive residual stresses or sustained by suction-related tensile strength. Sensitivity analysis indicates that a thicker asphalt course or a higher asphalt modulus will also further reduce this horizontal tensile stress or even produce a completely compressive horizontal stress in the base course. Sensitivity analysis also shows that this tensile stress is hardly increased under a higher load due to the UGB's saturated plastic hardening. This means a further higher load will not lead to a significant increase of the bottom tensile stress but result in an over-stress softening and final failure of the UGB material, which is out of the modeling scope of this study but will be modeled in a future study. Figure 4*c* also shows that the horizontal stress from the middle to the top along the centerline of the base became compression, which will be explained using the plastic strain and horizontal displacement presented in Figure 4.

Figure 4*a* shows the plastic strains occurring from the middle to the top of the base course when the NAEP model was used. The high plastic strains lead to a significant horizontal displacement (dilation) as shown in Figure 4*b*. This expansive deformation in the horizontal direction has been constrained by the surrounding UGB materials, resulting in an increase of the confining pressure that is in compression and produced by the surrounding UGB material. This compressive confining pressure leads to a reduction of the horizontal tensile stress at base course bottom and an increase of the horizontal compressive stress from the middle to the top of the base course. Note that the horizontal dilation is more significant and the induced confining pressure is higher in the middle and top of the base course (due to higher plastic strain) than that at the bottom of the base course. Therefore, the horizontal stress became compression in the middle and top of the base course compensated by the higher confining pressure, whereas it still remained in tension resulting from less confining pressure at the bottom of the base course.

In comparison, the horizontal displacement predicted by the NAE model (shown in Figure 4*c*) was much smaller than (only 1/1400 of) that predicted by the NAEP model (shown in Figure 4*b*). Therefore, little horizontal dilation occurred and no confining effect was introduced in the base course when the NAE model (no plastic component) was used, which leads to unreasonably high tensile stress in the horizontal direction of the base course.

The significant horizontal displacement shown in Figure 4*b* also explains an experimental finding from the authors' previous study (*27*), that is, placing the geogrid in the middle of the base course is much more effective in reducing the rutting damage of the base course than placing the geogrid at the bottom of the base course. The high plastic strain on the base top and the significant horizontal displacement in the base middle may have been effectively constrained and reduced by placing the geogrid in the middle of the base course. Placing the geogrid at the base bottom cannot effectively limit the plastic strain occurring mainly in the top area of the base course. This explanation will be further verified in a continuing NAEP modeling of the UGB material by including the geogrid in base course.

Figure 5, *a* and *b* show the distribution of the stress-dependent modulus determined by NAEP model and the NAE model, respectively. Both show the highest modulus on the top center area of the base course and a decreasing modulus with base depth and the distance to the load centerline. The results are consistent with that obtained in the literature (*4, 6*). However, compared with the modulus obtained from the NAE model (Figure 5*b*), the NAEP model predicted a higher modulus for the base course (Figure 5*a*). This is because the NAEP predicts less horizontal tensile stresses and more compressive stresses as indicated in Figure 3*c*, leading to a higher I_1 and lower τ_{oct} and eventually a higher modulus according to Equation 2.

Effect of UGB Models on Critical Pavement Responses

Table 3 compares the critical pavement responses predicted by LAE, NAE, and NAEP models, including the tensile strain at the bottom of the asphalt course, the average vertical compressive strain in the base course and the compressive strain in the top of the subgrade. It is evident that at all load levels the linear anisotropic model (LAE) under-predicted the critical pavement responses compared with the nonlinear models (NAE and NAEP). When the stress-dependent anisotropic modulus was considered for the UGB, the elastic model (NAE) predicted the

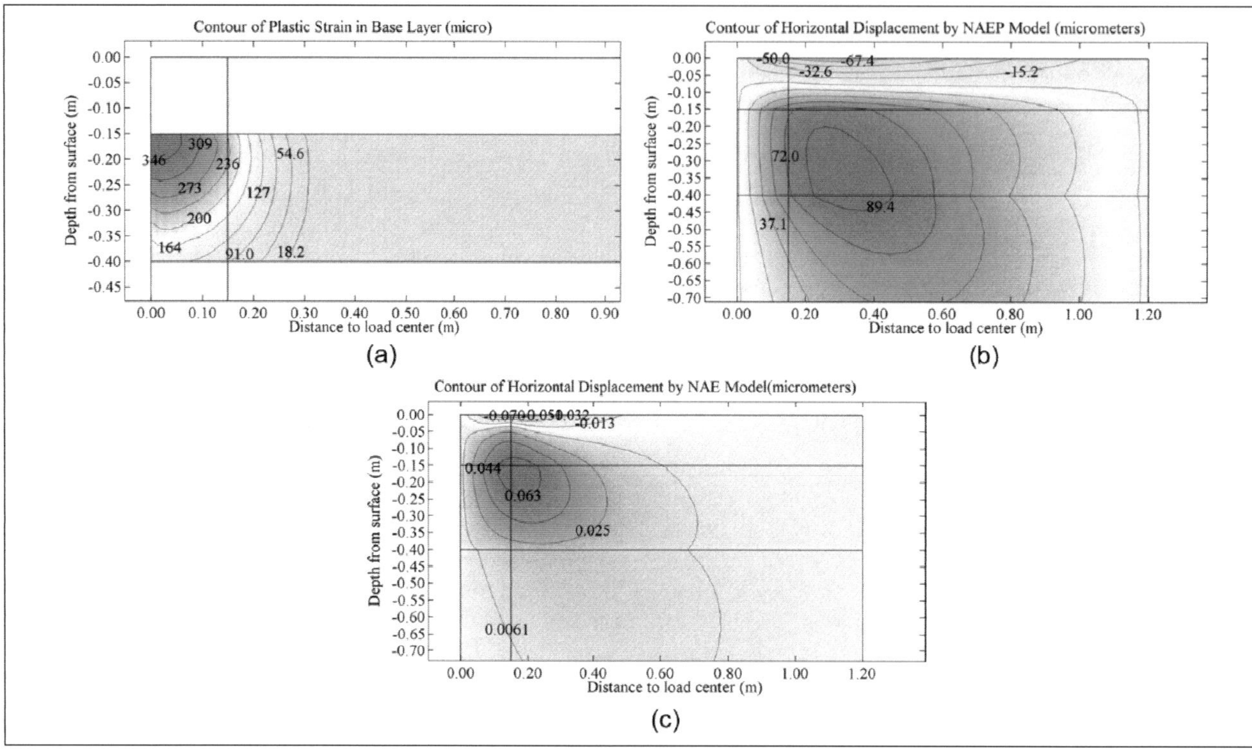

Figure 4. Contour of plastic strain (με) and horizontal displacement (μm) in base course at a load level of 1006 kPa in which (*a* and *b*) NAEP model and (*c*) NAE model were used for UGB.

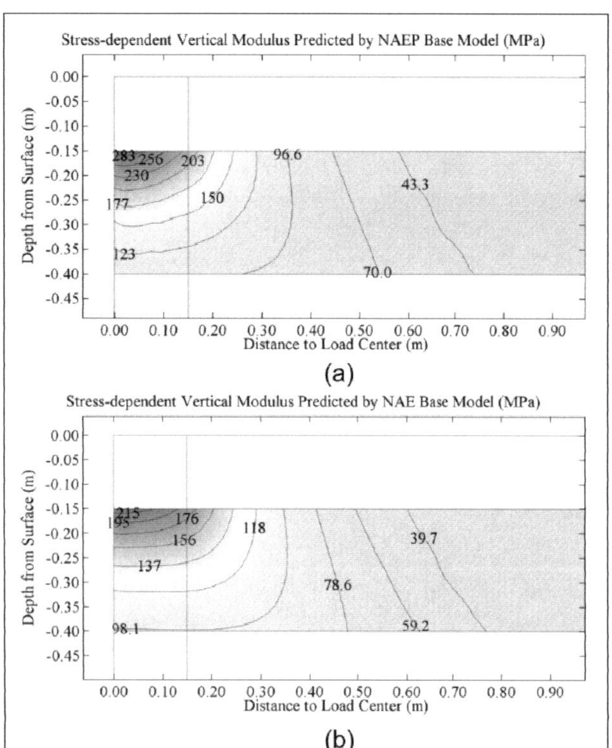

Figure 5. Contour of stress-dependent modulus (MPa) in base course at a load level of 1006 kPa in which (*a*) NAEP model and (*b*) NAE model were used for UGB.

same asphalt bottom tensile strain as that calculated by the plastic model (NAEP). However, the NAE model predicted higher than average compressive strain in the base course and lower compressive strain on the top of the subgrade compared with the NAEP model. The prediction difference between the two models became more significant when the load level increased. The lower compressive strain in the base course obtained by NAEP model was caused by the higher modulus of the UGB materials when plasticity is considered, as shown in Figure 5*a*. The higher compressive strain in the top of subgrade when NAEP model was used for UGB is because higher stress was introduced in subgrade when plastic deformation occurs in base course. One can conclude that including plasticity with a nonlinear anisotropy for UGB alters the pavement responses in the base and subgrade of a flexible pavement.

Furthermore, since a lower tensile strain at asphalt bottom leads to a longer fatigue life and a lower subgrade top compressive strain results in less rutting damage, it may be concluded that the linear anisotropic base model will riskily over-predict the fatigue life and under-predict the rutting damage in asphalt pavements. The NAE base model will result in the same fatigue life for asphalt course as the NAEP base model, but it will riskily under-predict the rutting damage in the pavements. Thus, a complete nonlinear (stress-dependent) anisotropic elastoplastic model is required for accurately modeling the UGB material in the base course.

Future research continuing from this study includes evaluating the effects of a dynamic load and a repeated load on

Table 3. Predicted Critical Pavement Responses by Different UGB Models

Load levels	UGB model	Tensile strain ($\mu\varepsilon$) at bottom of asphalt course	Average vertical compressive strain ($\mu\varepsilon$) along base course centerline	Compressive strain ($\mu\varepsilon$) on top of subgrade
566 kPa	LAE	274.1	514.4	604.4
	NAE	309.5	708.5	625.3
	NAEP	309.8	707.5	627.2
755 kPa	LAE	387.7	718.3	834.1
	NAE	421.3	888.8	876.5
	NAEP	422.6	870.7	911.8
1006 kPa	LAE	545.6	1000.1	1156.1
	NAE	564.9	1122.8	1217.1
	NAEP	568.2	1047.3	1317.8

the pavement responses and performance using the NAEP model. Further studies are also needed to predict the moisture diffusion in the pavement structure and couple the moisture effect into the nonlinear anisotropic elastoplastic model for the UGB materials.

Summary and Conclusions

1) A nonlinear stress-dependent anisotropic elastoplastic model was developed for UGB materials and successfully implemented in a weak form equation-based finite element program. A laboratory large-scale tank pavement loading test was used to validate the model using measured surface deflections and the vertical stresses in base and subgrade, proving that the model is valid and effective in predicting accurate responses of the UGB materials in flexible pavements.
2) The NAEP model predicted less horizontal tensile stresses at the base bottom and introduced compressive stresses in the middle and the top of the base course. This is caused by an increasing confinement resulting from a horizontal plastic dilation in the base course, which cannot be modeled by a NAE model. This also explains a laboratory finding that placing the geogrid in the middle of the base course is much more effective in reducing rutting than placing the geogrid at the base bottom.
3) The stress-dependent modulus for the unbound granular material in base course decreases with base depth and the distance from loading centerline. The NAEP model predicted higher modulus in the base course than the NAE model.
4) A linear anisotropic base model will riskily over-predict fatigue life and under-predict the rutting damage of flexible pavements. The NAE base model will result in the same fatigue life as the NAEP base model, but it will riskily under-predict the rutting damage. Therefore, the nonlinear (stress-dependent) anisotropic elastoplastic model is required for accurately predicting the performance of the UGB material in asphalt pavements.

Acknowledgments

Special thanks from the authors are given to Professor Raj V Siddharthan and Professor Elie Hajj from the University of Nevada, Reno, for providing valuable measuring data from the large-scale tank pavement tests for model validation.

References

1. Tutumluer, E., and U. Seyhan. Laboratory Determination of Anisotropic Aggregate Resilient Moduli Using an Innovative Test Device. *Transportation Research Record: Journal of the Transportation Research Board*, 1999. 1687: 13–21.
2. Adu-Osei, A., D. Little, and R. Lytton. Cross-Anisotropic Characterization of Unbound Granular Materials. *Transportation Research Record: Journal of the Transportation Research Board*, 2001. 1757: 82–91.
3. Gu, F., H. Sahin, X. Luo, R. Luo, and R. L. Lytton. Estimation of Resilient Modulus of Unbound Aggregates Using Performance-Related Base Course Properties. *Journal of Materials in Civil Engineering*, Vol. 27, No. 6, 2015, p. 04014188.
4. Tutumluer, E., and M. R. Thompson. Anisotropic Modeling of Granular Bases in Flexible Pavements. *Transportation Research Record: Journal of the Transportation Research Board*, 1997. 1577: 18–26.
5. Oh, J.-H., R. L. Lytton, and E. G. Fernando. Modeling of Pavement Response Using Nonlinear Cross-Anisotropy Approach. *Journal of Transportation Engineering*, Vol. 132, No. 6, 2006, pp. 458–468.
6. Al-Qadi, I. L., H. Wang, and E. Tutumluer. Dynamic Analysis of Thin Asphalt Pavements by Using Cross-Anisotropic Stress-Dependent Properties for Granular Layer. *Transportation Research Record: Journal of the Transportation Research Board*, 2010. 2154: 156–163.
7. Tutumluer, E., D. Little, and S.-H. Kim. Validated Model for Predicting Field Performance of Aggregate Base Courses. *Transportation Research Record: Journal of the Transportation Research Board*, 2003. 1837: 41–49.

8. Wang, H., and I. L. Al-Qadi. Importance of Nonlinear Anisotropic Modeling of Granular Base for Predicting Maximum Viscoelastic Pavement Responses under Moving Vehicular Loading. *Journal of Engineering Mechanics*, Vol. 139, No. 1, 2013, pp. 29–38.
9. Uzan, J. Permanent Deformation of a Granular Base Material. *Transportation Research Record: Journal of the Transportation Research Board*, 1999. 1673: 89–94.
10. Chazallon, C., P. Hornych, and S. Mouhoubi. Elastoplastic Model for the Long-Term Behavior Modeling of Unbound Granular Materials in Flexible Pavements. *International Journal of Geomechanics*, Vol. 6, No. 4, 2006, pp. 279–289.
11. Gu, F., Y. Zhang, C. V. Droddy, R. Luo, and R. L. Lytton. Development of a New Mechanistic Empirical Rutting Model for Unbound Granular Material. *Journal of Materials in Civil Engineering*, Vol. 28, No. 8, 2016, p. 04016051.
12. Lai, M., E. Krempl, and D. Ruben, *Introduction to Continuum Mechanics*. Elsevier Inc., Burlington, Mass., 2010.
13. Uzan, J. Characterization of Granular Material. In *Transportation Research Record*: Journal of the Transportation Research Board, 1985. 1022: 52–59.
14. Zhang, Y., M. Bernhardt, G. Biscontin, R. Luo, and R. Lytton. A Generalized Drucker–Prager Viscoplastic Yield Surface Model for Asphalt Concrete. *Materials and Structures*, Vol. 48, No. 11, 2014, pp. 3585–3601.
15. Tashman, L., E. Masad, D. Little, and H. Zbib. A Microstructure-Based Viscoplastic Model for Asphalt Concrete. *International Journal of Plasticity*, Vol. 21, No. 9, 2005, pp. 1659–1685.
16. Oda, M., and H. Nakayama. Yield Function for Soil with Anisotropic Fabric. *Journal of Engineering Mechanics*, Vol. 115, No. 1, 1989, pp. 89–104.
17. Zhang, Y., R. Luo, and R. L. Lytton. Microstructure-Based Inherent Anisotropy of Asphalt Mixtures. *Journal of Materials in Civil Engineering*, Vol. 23, No. 10, 2011, pp. 1473–1482.
18. Masad, E., D. Little, and R. Lytton. Modeling Nonlinear Anisotropic Elastic Properties of Unbound Granular Bases Using Microstructure Distribution Tensors. *International Journal of Geomechanics*, Vol. 4, No. 4, 2004, pp. 254–263.
19. ASTM. *Standard Test Method for California Bearing Ratio of Laboratory-Compacted Soils*. D1883-14, West Conshohocken, Pa., 2014.
20. Zhang, Y., R. Luo, and R. L. Lytton. Characterizing Permanent Deformation and Fracture of Asphalt Mixtures by Using Compressive Dynamic Modulus Tests. *Journal of Materials in Civil Engineering*, Vol. 24, No. 7, 2012, pp. 898–906.
21. ASTM. *Standard Test Method for Dynamic Modulus of Asphalt Mixtures*. D3497-79, West Conshohocken, Pa., 2003.
22. Zhang, Y., B. Birgisson, and R. L. Lytton. Weak Form Equation–Based Finite-Element Modeling of Viscoelastic Asphalt Mixtures. *Journal of Materials in Civil Engineering*, Vol. 28, No. 2, 2015, p. 04015115.
23. Zhang, Y., R. Luo, and R. L. Lytton. Anisotropic Viscoelastic Properties of Undamaged Asphalt Mixtures. *Journal of Transportation Engineering*, Vol. 138, No. 1, 2012, pp. 75–89.
24. TxDOT. *Test Procedure for Triaxial Compression for Disturbed Soils and Base Materials*. Texas Department of Transportation (TxDOT), Austin, 2010.
25. Gu, F., Y. Zhang, X. Luo, R. Luo, and R. L. Lytton. Impact of Geogrid on Cross-Anisotropy and Permanent Deformation of Unbound Granular Materials. *Transportation Research Record: Journal of Transportation Research Record*, 2016. 2580: 34–46.
26. Zhang, Y., F. Gu, B. Birgisson, and R. L. Lytton. Viscoelasticplastic–Fracture Modeling of Asphalt Mixtures under Monotonic and Repeated Loads. *Transportation Research Record: Journal of Transportation Research Record*, 2017. 2631: 20–29.
27. Gu, F., X. Luo, R. Luo, R. L. Lytton, E. Y. Hajj, and R. V. Siddharthan. Numerical Modeling of Geogrid-Reinforced Flexible Pavement and Corresponding Validation Using Large-Scale Tank Test. *Construction and Building Materials*, Vol. 122, 2016, pp. 214–230.

The Standing Committee on Aggregates (AFP70) peer-reviewed this paper (18-01217).

Article

TRR
JOURNAL OF THE TRANSPORTATION RESEARCH BOARD

Transportation Research Record
2018, Vol. 2672(52) 57–66

DOI: 10.1177/0361198118758683
journals.sagepub.com/home/trr
SAGE

Optimizing the Gradation of Fine Processed Reclaimed Asphalt Pavement and Aggregate Blends for Unbound Base Courses

Saad Ullah[1], Burak F. Tanyu[1], and Edward J. Hoppe[2]

Abstract

The purpose of this research was to investigate the effect of changes in grain size distribution to the permanent deformation of two different fine processed reclaimed asphalt pavement (RAP) blended with base course virgin aggregate (VA). Grain size distribution of the RAP-VA blends were created following two different approaches. The first approach was based on mixing RAP and VA to have one grain size distribution, regardless of how much RAP was added to VA (here referred to as the engineered mixture design). The second approach was based on mixing RAP and VA with as-is gradation from the plants to proportions determined by weight and not controlling the outcome of the specific grain size of the mixture. This approach resulted in various grain size distributions (here referred as the as-is mixture design). The engineered mixture design was useful to quantify the effect of adding RAP to the blends, but was not a realistic approach to create blends that may be achieved in the field. The as-is mixture design could not only be achieved in the field, as demonstrated in this study, but also resulted in better performance in terms of permanent deformation. This manuscript describes the comparison of these two approaches and a methodology to optimize the gradation and develop thresholds for RAP-VA blends that may result in similar or better performance than the 100% as-is VA that is used to construct base course in pavement systems.

The production of reclaimed asphalt pavement (RAP) is constantly increasing in the U.S. with time, as the U.S. possesses the world's largest infrastructure network (*1*). Early studies report that in the late 1990s, the production of RAP in the U.S. was around 50 million tons (*2*). In the late 2000s, the annual production rate had increased to be around 100 million tons (doubling the amount in nearly 10 years) (*3*). Maintenance and reconstruction produces large amount of demolished debris. The RAP generated as a result of construction and demolition of asphalt pavements is reclaimed by the asphalt industry as much as possible for the production of asphalt mixtures (AM). Although this approach is relevant and useful, it is not enough to deplete all of the excess RAP created and alternative approaches to recycle this material are needed. Previous research studies have already suggested that one of the ways that this problem may be overcome would be to incorporate RAP into the construction of new base layers for pavement structures (*3*). Being able to use RAP in base course also has potential benefits towards conservation of natural resources, reduction of exposed industrial waste stockpiles, and sustainable development.

Previous research has been conducted that has made attempts to incorporate RAP into the base course of flexible pavements by combining RAP with virgin aggregate (VA) (*2–8*). Previous literature indicates that the use of 100% RAP is not suitable for base course because of its high strain accumulation potential, temperature susceptibility, and creep. Much of the research recommended blending RAP with high quality VA to achieve improved consistency and the quality of the blend (*2, 4*). Previous research studies have reported that the increase in RAP content in the RAP-VA blends improve the resilient modulus of the blend, which is one of the main parameters required in the Mechanistic Empirical Pavement Design Guideline (MEPDG) (*4, 9*). The findings from the literature have been consistent and it appears that there is a consensus among the previous findings relating to the resilient modulus.

However, when it comes to permanent deformation (PD), the previous research findings do not always agree with each other. There are some studies that report that the addition of RAP into RAP-VA blends decreases the ability of the blend to resist permanent strain and therefore makes the blend prone to rutting and deformations (*2, 4, 5, 10*). As a solution to this problem, prior studies recommended using a thicker asphalt layer on top of the base layer created from

[1]Volgenau School of Engineering, George Mason University, Fairfax, VA
[2]Virginia Transportation Research Council (VTRC), Charlottesville, VA

Corresponding Author:
Address correspondence to Burak F. Tanyu: btanyu@gmu.edu

RAP-VA blends to prevent unsuitable deflections (*5*). Other studies suggested the use of additives to stabilize the properties of RAP-VA blends or recommended the use of geocells and geosynthetics to provide additional confinement to the base course (*7, 11, 12*). However, there are also previous studies where the researchers reported an improvement in the permanent strain of RAP-VA blends compared with the 100% VA (*13, 14*). Although these studies did not describe why such improvement was observed, compared with the opposite findings by other researchers. The discrepancy in the literature requires further evaluation and clarification. These differences could result from different aspects, but for the purposes of this study, the focus has been to evaluate the effects of different grain size distributions to create RAP-VA blends as described in detail in the subsequent sections.

Experimental Methodology

The experimental methodology was designed based on the evaluation of the effects of creating RAP-VA blends with two different mixture design approaches. One approach (here referred to as the engineered mixture design) was based on determining a target grain size distribution that is used for both 100% VA and all of the RAP-VA blends. This resulted in a unique grain size distribution regardless of any RAP percentage. The second approach (here referred to as the as-is mixture design) was based on mixing RAP and VA by weight based on their individual as-is gradation. This approach resulted in different grain size distributions for each blend. To compare and validate the findings, RAP was collected from two different sources, but VA was collected from one source. The RAP-VA blends were created to result in 20%, 30%, 40%, 50%, and 60% RAP mixed with 80%, 70%, 60%, 50%, and 40% VA, respectively, for both of the mixture design approaches. The RAP content was limited to 60% in this study based on the findings in previous research (*2, 4, 5, 14*). With five different RAP contents, two different RAP samples, and two different gradations (engineered versus as-is), the experimental program included a total of 20 different combinations and with duplicate tests on each specimen resulted in a total of 40 performance tests.

Material Characterization

The VA used in this study was collected from a quarry in northern Virginia. The aggregate is specifically produced in this quarry to meet the requirements of the Virginia Department of Transportation (VDOT) agency's base course gradation requirement for the aggregate referred to by the agency as 21 aggregate (*15*). The upper and lower limits for the 21 aggregate gradation as outlined by VDOT are presented in Figure 1.

RAP samples were also collected from northern Virginia from different asphalt plants. Two different plants owned by

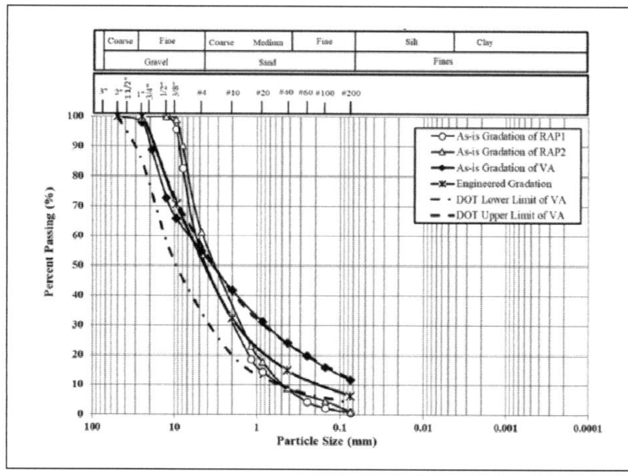

Figure 1. Grain size distribution of as-is RAP and VA, and gradation used for the engineered mixture design approach.

two different companies were used for the sampling and all RAP samples were obtained immediately after they were processed at the plant. The asphalt plants in northern Virginia produce two different types of RAP gradation, which are referred to as coarse and fine processed RAP. The threshold between the two gradations is defined by a 1/2-inch sieve size. Fine processed RAP (particles finer than the 1/2-inch sieve size) is the most abundant gradation and is primarily used for recycling asphalt for the production of new AM. This RAP type contains more consistent asphalt (binder content) than the coarse RAP. Even though fine processed RAP is used by the plants to create AM, there is still approximately close to 5 million tons of excess RAP in Virginia (*16*). Therefore, determining an alternative re-use application for this material is very relevant. For this reason both of the RAP samples used in this study are fine processed.

All samples of VA and RAP were characterized following the conventional index property tests. Dry and washed sieve analyses were performed following ASTM D 6913 and ASTM D1140 procedures, respectively. Apparent specific gravity tests were conducted according to ASTM D854 for VA and in the case of RAP, as per ASTM D2041. Atterberg limit tests were performed in accordance to ASTM D4318, and ASTM D2487 was used to determine the Unified Soil Classification. Binder contents of the RAP samples were determined from ASTM D6307. In addition to the index property tests, samples were compacted following the vibratory compaction method described in ASTM D7382 Method A to determine the optimum moisture content and maximum dry density. Proctor compaction tests following ASTM D698 were also performed, however they were not reported in this study as all samples for the PD test were prepared using a vibratory compactor at 100% relative compaction.

The results of the sieve analyses of all samples based on their as-is gradation are shown in Figure 1. The gradations of

Table 1. Properties of As-Is RAP and VA Samples and VA Gradation Created with Engineered (Eng.) Mixture Design Approach

Properties	As-Is RAP 1	As-Is RAP 2	As-Is VA	VA (Eng. GSD)
D_{10} (in)	0.020	0.020	0.003	0.008
D_{30} (in)	0.078	0.070	0.030	0.071
D_{50} (in)	0.120	0.830	0.140	0.170
D_{60} (in)	0.213	0.180	0.280	0.244
Coefficient of uniformity	10.65	9.00	93.00	31.00
Coefficient of curvature	1.43	1.36	1.10	2.60
Gravel (%)	45.0	40.0	46.0	48.0
Sand (%)	53.5	57.8	42.0	45.7
Fines (%)	1.5	2.2	12.0	6.5
Plastic limit (%)	Non-plastic	Non-plastic	17.5	15.1
Liquid limit (%)	Non-plastic	Non-plastic	19.2	16.8
USCS classification	SW	SW	SW-SM	SW-SM
AASHTO classification	A-1-a	A-1-a	A-1-a	A-1-a
Binder content (%)	5.60–5.80	4.50–4.70	N/A	N/A
Specific gravity	2.43	2.60	2.85	2.81
Max dry density (pcf)	121	124	151	151
Optimum moisture content (%)	5.5	5.5	6.2	6.2

both of the RAP samples are relatively close to each other with particles finer than 1/2-inch and a number 200 sieve. The VA samples had more gravel size particles compared with the RAP samples and also had higher fines content (~12%). The properties of all of the collected samples are tabulated in Table 1. RAP 1 and RAP 2 are non-plastic and therefore Atterberg limits could not be determined, while VA appeared to be plastic in nature. According to the Unified Soil Classification System (USCS), RAP 1 and RAP 2 are classified as well-graded sand (SW) while VA is classified as well graded sand with silt (SW-SM). Based on the AASHTO M145 classification system, all materials, RAP 1, RAP 2 and VA, are classified as gravel and sand (A-1-a).

The maximum dry density (MDD) of VA was determined as 151 pcf and the optimum moisture content (OMC) as 6.2%. The MDD and OMC values for RAP 1/VA and RAP 2/VA blends were relatively close to each other, ranging from 137 to 148 pcf and 6.15–5.85% for the 20–60% RAP content, respectively.

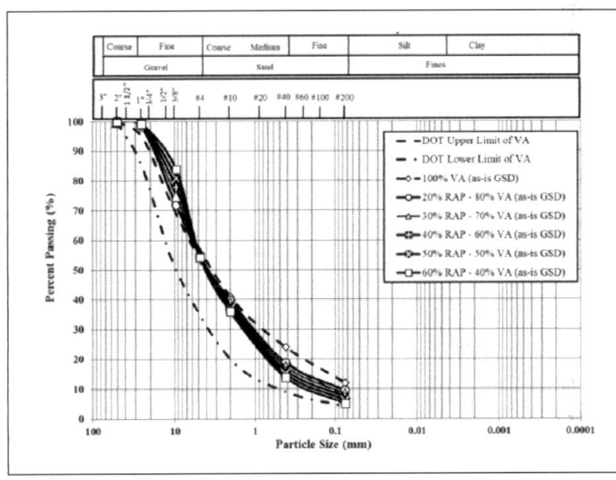

Figure 2. Grain size distribution (GSD) of RAP-VA blends created from blending RAP and VA with the as-is gradation mixture design approach (as-is GSD).

Gradations Created for Engineered and As-Is Mixture Designs

The RAP-VA blends created based on as-is gradations were created by mixing targeted percent RAP with targeted percent VA by weight. For example, to obtain 20% RAP and 80% VA, from the RAP and VA samples with as-is gradations, 20% of the whole mass of RAP-VA blend contained RAP and the rest contained VA. This approach resulted in a different grain size distribution for each RAP-VA blend, as shown in Figure 2.

The RAP-VA blends created based on the engineered mixture design approach were developed based on the previously suggested method, referred to as Fuller's curves in the literature (*17*). This method is based on creating a gradation with optimum gravel content where the gravel particles end up having particle to particle contact, as opposed to floating in a matrix of sand and finer particles. The VA gradation created based on this approach is shown in Figure 1, which fulfills the VDOT requirements for the base course and fits very well within the upper and lower bounds of the VDOT requirements for 21 aggregate in Virginia. The properties of the VA created based on this gradation are tabulated in Table 1. Once the engineered gradation was established, the RAP-VA blends were created by adding RAP and VA percentages, following each of

Table 2. Selection of Stress Levels for PD Tests

Pavement sections used in multi-layer analyses	Depth of interest (in)	Stress conditions determined at mid-depth of base layers		$\sigma_1 = \sigma_d + \sigma_3$ (psi)	References
		σ_3 (psi)	σ_d (psi)		
4-in AM and 8-in base	8	5	16	21	Calculations performed in this study based on typical road sections in Virginia
5-in AM and 8-in base	9	4	12	16	Calculations performed in this study based on typical road sections in Virginia
2-in AM and 8-in base	6	6	30	36	As presented by Barksdale and Itani (*20*)
3-in AM and 8-in base	7	5	25	30	As presented by Chow (*21*)

Note: σ_1 and σ_3 represent the vertical and horizontal (confinement) stresses and σ_d represents the deviator stress.

the specific sieve separations outlined in Table 1. For example, for the 20% RAP/80% VA blend, particles that were retained in each sieve were carefully mixed to result in 20–80% ratio by weight. Therefore, this approach resulted in all RAP-VA blends having the same grain size distribution, as shown in Figure 1 (referred to as engineered gradation) regardless of the percentage RAP content.

Permanent Deformation Testing Protocol

Currently there are no standards for performing PD tests in the U.S. This requires each researcher to develop a specific protocol to perform these tests. To develop a scenario that represents a realistic and conservative approach, the details of the testing protocol have been carefully evaluated in this study.

The sample size of all specimens was selected to be 6 inches in diameter and 12 inches in height. This sample size is consistent with samples used for resilient modulus testing (AASHTO T307) and the aspect ratio referred to in typical triaxial testing in the U.S. (ASTM D4767). All samples were compacted to satisfy 100% relative compaction and OMC as determined from vibratory compaction tests, which is consistent with VDOT requirements.

Number of loading cycles for PD tests were limited to 10,000 cycles on the basis of the prior literature and the total deformation obtained at the end of 10,000 cycles was extrapolated to obtain strains at 100,000 load cycles (*18, 19*). Bennert et al. suggested that the log N model is one of the best models to predict permanent strains at higher loading cycles, therefore samples were loaded to 10,000 loading cycles, and strains at higher loading cycles were predicted using the log N model.

The stress levels used for the PD tests during loading cycles represents traffic loading. Due to the lack of a standard, different researchers perform PD tests with different applied loading magnitudes. In this study, the magnitude of the applied stresses used for the PD tests were determined

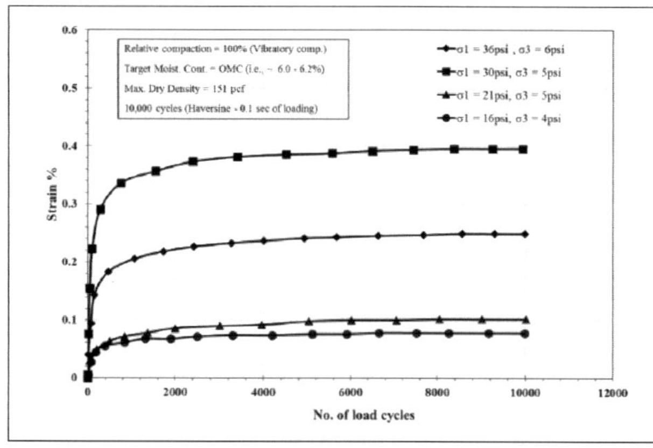

Figure 3. PD tests on 100% VA (as-is gradation) with different stress levels based on different pavement profiles.
Note: See Table 3 for the source of σ_1 and σ_3 conditions simulated in these tests.

from the multi-layer and finite element model (FEM) recommended in the literature (*4, 20*). Typical flexible pavements in Virginia were modeled to analyze the stress conditions in the middle of the base layer and the results were compared to the literature. Overall, the comparison showed that flexible pavements designed for low level traffic conditions experience the highest levels of stresses in the middle of the base layer because of the thinner asphalt layer present in these profiles. The findings of multi-layer solutions and the comparison of the findings in this study with the prior literature are presented in Table 2. Based on this comparison, PD tests were performed on 100% VA (as-is gradation) using these stress levels. The results of these tests are presented in Figure 3, which were then used to assess the worst-case scenario in generating permanent strains at the end of 10,000 loading cycles. The findings showed that tests performed with a major principal stress (σ_1) of 30 psi on the base course aggregate and confinement (σ_3) of 5 psi yields the highest amount of deformation. Therefore, all PD tests were performed in this study following these stress conditions.

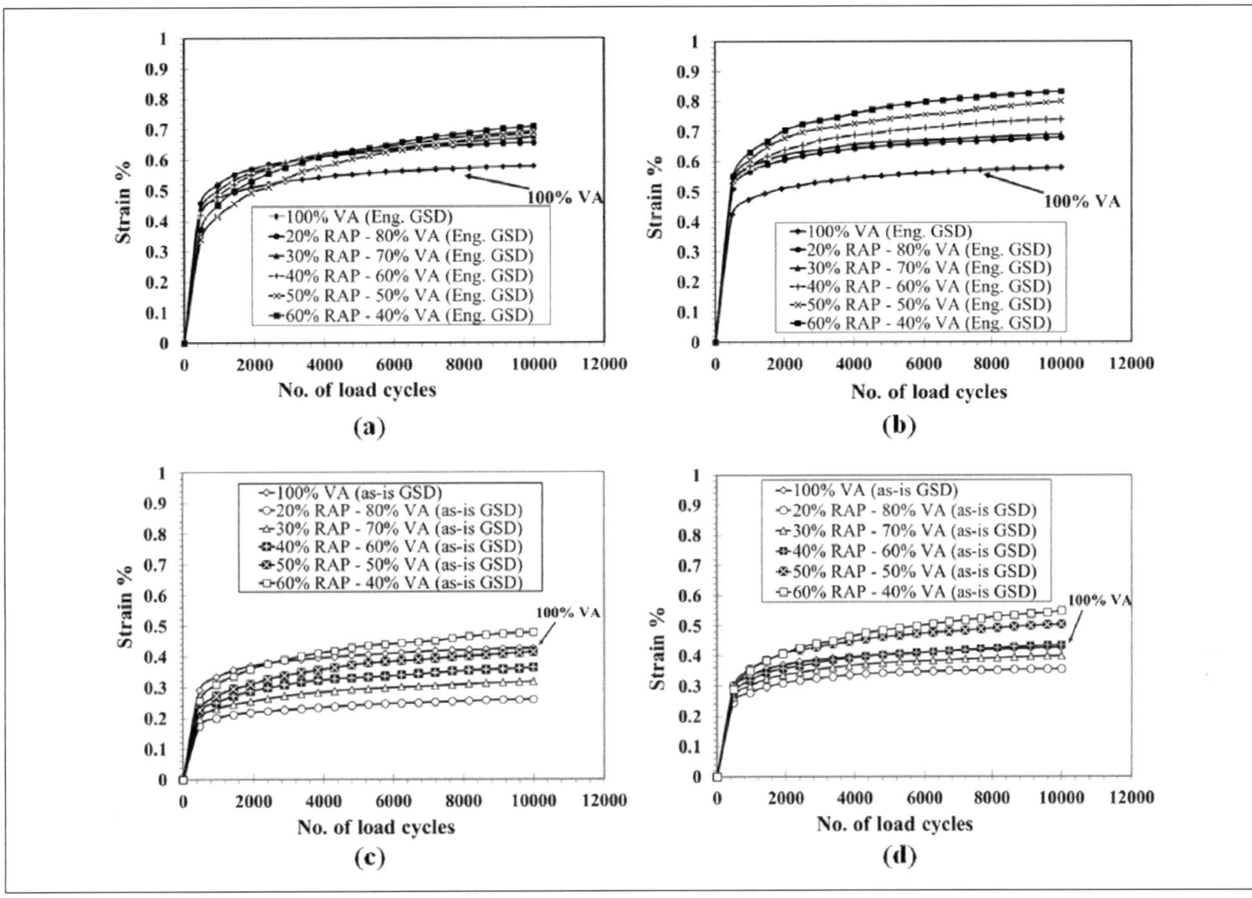

Figure 4. PD Test results of RAP1 and RAP 2 blends with VA based on blends created with (*a*) and (*b*) with engineered (eng. GSD) and (*c*) and (*d*) with as-is (as-is GSD) mixture design approaches.

Test Results and Analyses

Permanent Deformation (PD) from Engineered Mixture Design Gradation

The PD test results for all samples created based on the engineering mixture design approach are presented in Figures 4*a* and 4*b* for each RAP type used in this study. The total accumulated strains at the end of 10,000 cycles increases by increasing the RAP content (i.e., 20–60%) regardless of the characteristics of RAP. In all cases adding RAP to the blend results in permanent strains that are higher than the permanent strain of the 100% VA alone. This describes the down side of adding RAP to VA.

When the cumulative permanent strains from the RAP-VA blends were compared against the cumulative strains accumulated from the 100% VA, for blends created with RAP1, the increased strains were within the range 11.1–21.4% for 20–60% RAP content, respectively. For blends created with RAP2, the difference with the strains accumulated from 100% VA were even higher with a range of 16.2–43% for 20–60% RAP.

When the cumulative strains at 10,000 loading cycles were compared, even with the 20% RAP, the cumulative permanent strain is higher than the permanent strain from 100% VA. In fact, at 100,000 loading cycle (as determined from the log N model), the difference between the permanent strains from 100% VA and RAP-VA blends become even larger. For RAP1, this difference increases from 11.1% (at 10,000 cycles) to 19.9% (at 100,000 cycles) and for RAP2, from 16% to 20.1% for the 20% RAP-VA blend. For the 60% RAP-VA blends, these differences become 21.4% (at 10,000 cycles) to 43.7% (at 100,000 cycles) for RAP1 and 43% to 56% for RAP 2.

The differences in cumulative permanent strains from RAP1 and 2 show that the performance of RAP-VA blends is also dependent on the properties of the RAP as all samples were created with the same grain size distribution. When the difference between RAP1 and RAP2 were evaluated, it was observed that both RAP samples contained same parent rock, but different binder contents.

Permanent Deformation (PD) from As-Is Mixture Design Gradation

The PD test results for all samples created based on the as-is mixture design approach are presented in Figures 4*c* and *d*

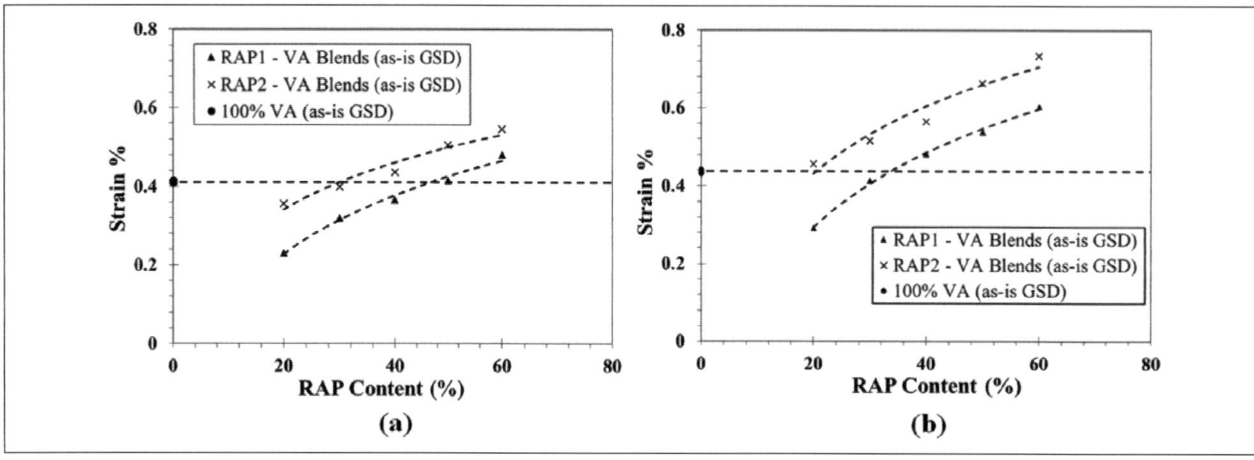

Figure 5. Cumulative permanent strains (*a*) observed from PD tests at 10,000 load cycles and (*b*) predicted with log N model at 100,000 load cycles.

for each RAP type used in this study. In both sets of tests, the results show that adding RAP to the VA tends to have a positive impact on the performance of the blends to a certain threshold. For blends created with RAP1, the cumulative permanent strain of the blends up to 50% RAP is less than the cumulative permanent strain observed from 100% VA at the end of 10,000 load cycles (Figure 5*a*). For blends created with RAP2, this threshold is observed up to 30% RAP.

The addition of RAP content tends to decrease the total accumulated permanent strains at the end of 10,000 load cycles, yet there is a contrast in the behavior of the RAP-VA blends versus 100% VA. As described by the approach defined by Werkmeister et al., RAP-VA blends exhibit creep shakedown behavior, whereas VA exhibits plastic shakedown (*22*). In both cases, the addition of RAP slows down the accumulation of permanent strains, but the strain-loading curve does not approach asymptote. Therefore, evaluating these results based on the log N model at 100,000 loading cycles is an important comparison. Figure 5 shows the predicted cumulative strains based on this approach. When values in Figures 5*a* and *b* were compared against the strains estimated for the 100% VA, at 100,000 loading cycle, the threshold where RAP-VA blend results in equal or less cumulative strain than the VA decreases to 30% for the RAP1 (from 50%) and 20% for the RAP2 blends (from 30%). The difference between the RAP 1 and RAP 2 are believed to be because ofthe difference in binder contents of the RAP samples.

Discussion of PD Results from Engineered and As-Is Mixture Designs

It is a known fact that the performance of materials is a function of gradation and therefore it is necessary to understand the preparation of RAP-VA blends (*23*). As stated previously, there is a difference in gradation between the engineered and as-is mixture design approaches. For the engineered gradation, RAP-VA blends had single gradation regardless of the percentage of RAP content and possesses 6.5% fines. Whereas in the case of as-is gradation, the percentage of fines decreases with the increase in RAP content from 12% (100% VA) to 5.5% (60% RAP–40% VA). This is because the RAP itself possesses only 1.5% fines. Increasing RAP content would result in lesser fines content in as-is RAP-VA blends, whereas in the case of engineered gradation the addition of RAP would result in the replacement of the VA. For example, 60% RAP–40% VA blend prepared with engineered gradation, will contain 4% (60% of 6.5%) RAP fines and 2.5% (40% of 6.5%) VA fines to make it 6.5% total fines. Whereas in the case of the as-is gradation approach, similar RAP-VA blend resulted in 0.9% RAP fines (60% of 1.5%) and 4.9% VA fines (40% of 12%) to make it 5.8% of total fines.

PD tests indicate that 100% VA samples prepared using engineered mixture design gradation result in less permanent strains than any of the RAP-VA blends, regardless of the RAP type used in this study. However, similar behavior is not observed in blends created with as-is mixture design gradation. In these blends, there appears to be a threshold, where adding RAP improves the overall performance up to a specific RAP content, which varies from source to source (RAP1 versus RAP2). Nevertheless, the difference in behavior between the engineered mixture design and the as-is mixture design samples is very important. In one scenario, there is an opportunity to use RAP to create blends (i.e., with the as-is mixture design approach) and in another scenario, RAP should not be added to VA if the design criterion is based on creating a blend that results in very similar or better cumulative strains as 100% VA (i.e., with the engineered mixture design approach).

The difference between the results from these two approaches presented in this study could also be used to understand the discrepancy within the literature as it relates to PD tests from RAP-VA blends. Several studies that focused on the RAP-VA blends reported that the addition of RAP decreases the performance of the blend as a base course compared to the 100% VA (higher cumulative permanent strains compared to 100% VA) (*2, 4, 24, 25*) and others reported an improved performance with blends (*13, 14*). When these studies were carefully evaluated it was evident that the improvements were only noted when the RAP-VA blends were created following a similar approach to that described in this study, with the as-is mixture design approach. All studies using the single gradation approach (similar to the engineered mixture design approach described in this study) resulted in adverse effects after adding RAP. This study demonstrates the reason between the contradictory results in the past research.

Practical Implications

In all cases (both those observed in this study and from the literature), an increase in RAP content increases the permanent strains, however with a proper mixture design, RAP in combination with VA may be used as a base course. In construction, quality control is very important, but so is the time (and associated cost) required to achieve the goal. To create an engineered mixture design gradation in the field, the asphalt plants must process RAP and VA to create many different stockpiles with a range of different particle sizes and then mix each stockpile with each other. This approach will most likely not only increase the overall production cost, but also require significant time management and is certainly not realistic to recycle a material. The alternative approach, here referred as the "as-is mixture design" gradation resulted in scenarios where depending on the RAP properties, it may be possible to create RAP-VA blends that may show similar or better permanent strains than the 100% VA. However, the main question associated with this approach is, can this approach be achieved in the field to create blends that will match the gradations. To test this, a parametric study was performed in the field, as described below.

Creation of RAP-VA Blends in the Field Based on the As-Is Mixture Design Approach

After the completion of the laboratory study, RAP-VA blends were prepared in the field to confirm the results achieved in the laboratory for the grain size distribution and controlling the percentage RAP content. The field study took place at an asphalt plant located in central Virginia that was different from the ones where the RAP samples were collected for the laboratory study. Fine processed RAP was blended with 21 VDOT aggregate by weight at the same proportions used in the laboratory (i.e., 20%, 30%, 40%, 50%, and 60% RAP). The mass needed to obtain proper blends for each RAP percentage was determined from the scale that was built-in to the front-end loader that was used for the study. For each blend, one stockpile was created by adding the proper amount of RAP and VA. For example, for the 20% RAP and 80% VA blend, 2 tons of RAP was placed on top of the 8 tons of VA to create one stockpile. The material in each stockpile with RAP and VA was tossed and turned 6 to 8 times by the bucket of the front-end loader to roughly mix the material before transporting all of the material in each stockpile to the pugmill. The pugmill was used primarily to mix RAP-VA blend and also to add water based on pre-determined target moisture contents. The mixing procedure followed for this application is consistent with the mixing requirements set forth by VDOT for preparing unbound base course aggregate (*15*). After the material was blended in the pugmill, random samples were obtained and bagged for further evaluation in the laboratory.

While at the field site, samples of 100% RAP and 100% VA were also obtained to be used later on in the laboratory to create blends following the as-is mixture design gradation approach. These samples were treated as the control and were then compared with the samples obtained from the pugmill in the field.

Comparison of the Field and Laboratory RAP-VA Blends Created Based on the As-Is Mixture Design Approach

The as-is grain size distributions of the 100% RAP and 100% VA used for the field and laboratory comparisons are shown in Figure 6*a*. These gradations were very similar to the gradations of the RAP and VA used to create blends for the PD tests in the laboratory. Figures 6*b–f* show the comparison for the RAP-VA blends. As can be seen from this figure, the laboratory and field results are very much the same for each blend. This finding shows not only that the as-is mixture design approach results in better PD results in the laboratory, but also the method used to blend the laboratory samples can easily be achieved in the field.

Conclusions

The PD test was used to determine the performance of the RAP-VA blends created at two different set of gradations (engineered and as-is gradations). The PD test was performed under representative traffic type loading and with RAP obtained from two different sources. The two main conclusions that can be drawn from the results observed in this study are:

1. The results of PD tests for the samples prepared with the as-is mixture design gradations indicate that addition of RAP to VA reduces the permanent strains up to a level when compared with the permanent strains accumulated from 100% VA. The amount of RAP that may be added to RAP-VA blends without compromising the performance of VA in the permanent strains is termed as the threshold RAP content. Although the threshold RAP content was different for RAP1 and RAP2, in both cases a threshold could be determined.

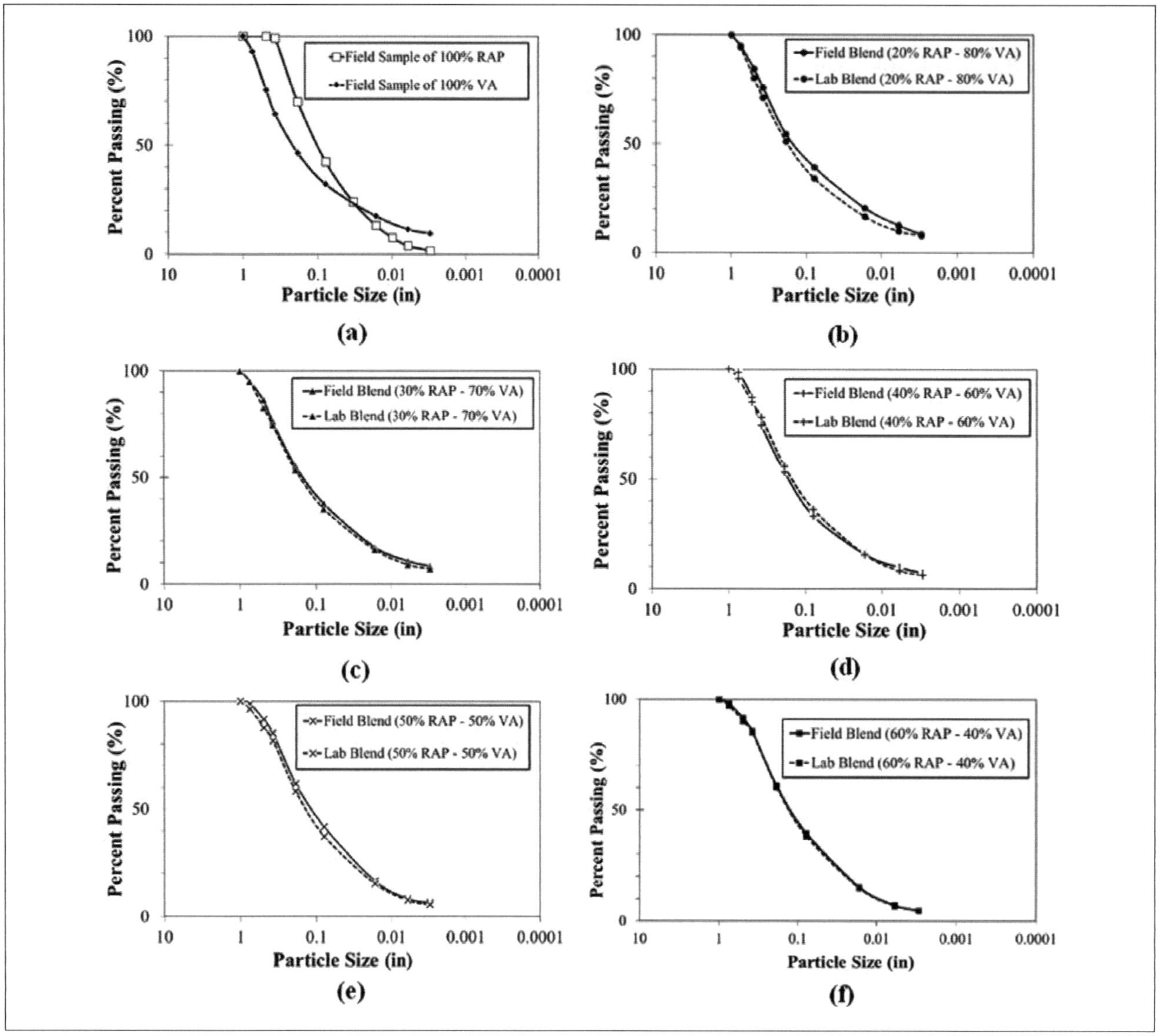

Figure 6. Comparison of grain size distributions of (*a*) 100% RAP and VA used in the field study and the blends obtained in the laboratory and field with (*b*) 20% RAP–80% VA, (*c*) 30% RAP–70% VA, (*d*) 40% RAP–60% VA, (*e*) 50% RAP–50% VA and (*f*) 60% RAP–40% VA proportions.

Although the results not shown here, the repeat tests confirmed these findings.

2. The thresholds determined based on the RAP contents were affected by the number of applied loading cycles during the PD tests because the materials exhibit creep shakedown behavior and strain-loading curves do not reach asymptote by the end of the test. The prediction of strains at a higher number of loading cycles show that the percentage of RAP allowed in RAP-VA blends must be carefully evaluated to develop appropriate thresholds.

The following are also supplementary conclusions that are important to note:

1. PD tests with RAP 1 and 2 showed similar trends but different results. This showed that each RAP that may be considered to create blends with VA must be properly characterized before any implementation in the field. The properties of RAP play a role in the outcome of the results.
2. RAP-VA blends prepared with engineered mixture design gradation accumulated higher permanent strains

than 100% VA. Although the results are not shown here, the repeat tests confirmed this observation.

3. The results of the field study showed that preparation of RAP-VA blends with an engineered mixture design gradation is not only difficult and time consuming, but is not realistic for field application. On the contrary, the as-is mixture design approach not only provided better results, but could also be implemented in the field.

The PD results presented are based on the materials used in this study. However, the approach used in this study to evaluate the RAP-VA blends both in the laboratory and field represent a methodology that may be applied to evaluate any other RAP-VA blends. The effects of temperature susceptibility and creep were not evaluated in this study. In areas where extreme temperature fluctuations occur within the base course, these properties must be carefully evaluated. The next phase of this study will include a field demonstration to confirm the validity of the thresholds determined for the RAP-VA blends.

Acknowledgments

The research presented in this paper was funded by the Virginia Transportation Research Council of the Virginia Department of Transportation. The authors greatly appreciate the in-kind support of Branscome Paving, Cedar Mountain Stone Quarry, Dalrymple Companies, and Chantilly Crushed Stone for donating materials used in this study. Several individuals were involved in this research during sample collection from the field, preparation of samples in the laboratory, and field study. The authors would like to acknowledge Mr. Ed Dalrymple, Mr. David Branscome, Mr. Joe Fitterrer, Mr. Francisco Cascante, Mr. Hamza Masud, Mr. Arslaan Khalid, and Ms. Binte Zainab, for their assistance and support.

References

1. Central Intelligence Agency. The World Factbook 2009. https://www.cia.gov/library/publications/the-world-factbook/fields/2085.html. Accessed June 10, 2017.
2. Taha, R., G. Ali, A. Basma, and O. Al-Turk. Evaluation of Reclaimed Asphalt Pavement Aggregate in Road Bases and Subbases. *Transportation Research Record: Journal of the Transportation Research Board*, 1999. 1652: 264–269.
3. Copeland, A., C. L. Jones, and J. Bukowski. Reclaiming Roads. *Public Roads*, Vol. 73, No. 5 (March/April). Publication FHWA-HRT-10-001, 2010.
4. Bennert, T., W. J. Papp, A. Maher, Jr., and N. Gucunski. Utilization of Construction and Demolition Debris Under Traffic-Type Loading in Base and Subbase Applications. *Transportation Research Record: Journal of Transportation Research Board*, 2000. 1714: 33–39.
5. Bilodeau, J.-P., G. Doré, and J. Depatie. Mitigation of Permanent Deformation in Base Layer Containing Recycled Asphalt Aggregates. *Canadian Journal of Civil Engineering*, Vol. 40, No. 2, 2013, pp. 181–187.
6. Hoppe, E. J., D. Stephen Lane, G. Michael Fitch, and S. Shetty. *Feasibility of Reclaimed Asphalt Pavement (RAP) Use as Road Base and Subbase Material*. Final report no. VCTIR 15-R6, Virginia Center for Transportation Innovation and Research, Charlottesville, 2015.
7. Hoyos, L. R., A. J. Puppala, and C. A. Ordonez. Characterization of Cement-Fiber-Treated Reclaimed Asphalt Pavement Aggregates: Preliminary Investigation. *Journal of Materials in Civil Engineering*, Vol. 23, No. 7, 2011, pp. 977–989.
8. Maher, M. H., N. Gucunski, and W. J. Papp, Jr. Recycled Asphalt Pavement as a Base and Sub-Base Material. *ASTM STP 1275*, ASTM, West Conshohocken, PA, 1997, pp. 42–53.
9. Alam, T. B., M. Abdelrahman, and S. A. Schram. Laboratory Characterisation of Recycled Asphalt Pavement as a Base Layer. *International Journal of Pavement Engineering*, Vol. 11, No. 2, 2010, pp. 123–131.
10. Edil, T. B., J. M. Tinjum, and C. H. Benson. *Recycled Unbound Materials*. Final report no. MN/RC 2012-35, Minnesota Department of Transportation, St. Paul, 2012.
11Han, J., S. K. Pokharel, X. Yang, C. Manandhar, D. Leshchinsky, I. Halahmi, and R. L. Parsons. Performance of Geocell-Reinforced RAP Bases over Weak Subgrade under Full-Scale Moving Wheel Loads. *Journal of Materials in Civil Engineering*, Vol. 23, No. 11, 2011, pp. 1525–1534.
12. Tanyu, B. F., A. H. Aydilek, A. W. Lau, T. B. Edil, and C. H. Benson. Laboratory Evaluation of Geocell-Reinforced Gravel Subbase over Poor Subgrades. *Geosynthetics International*, Vol. 20, No. 2, 2013, pp. 47–61.
13. Attia, M. I. E.-S. *Characterization of the Structural Behavior of Reclaimed Asphalt Pavement as Pavement Base Layer*. North Dakota State University, Fargo, A ProQuest dissertation publications, 2010. Document can be accessed through https://search.proquest.com/docview/837466339?pq-origsite=gscholar, Accessed March 30, 2017.
14. Jeon, E.-J., B. Steven, and J. Harvey. Comprehensive Laboratory Testing and Performance Evaluation of Recycled Pulverized Hot-Mix Asphalt Material. *Transportation Research Record: Journal of the Transportation Research Board*, 2009. 2104: 42–52.
15.Virginia Department of Transportation (VDOT) Road and Bridge Standards. 2016. Document can be accessed through http://www.virginiadot.org/business/locdes/2016_road_and_bridge_standards.asp, Accessed March 30, 2017.
16. Hoppe, E. J., D. Stephen Lane, G. Michael Fitch, and S. Shetty. *Feasibility of Reclaimed Asphalt Pavement (RAP) Use as Road Base and Subbase Material*. Final report no. VCTIR 15-R6, Virginia Center for Transportation Innovation and Research, Charlottesville, 2015.
17Skermer, N. A., and S. F. Hillis. Gradation and Shear Characteristics of Four Cohesionless Soils. *Canadian Geotechnical Journal*, Vol. 7, No. 1, 1970, pp. 62–68.
18Muhanna, A., M. Rahman, and P. Lambe. Model for Resilient Modulus and Permanent Strain of Subgrade Soils. *Transportation Research Record: Journal of the Transportation Research Board*, 1998. 1619: 85–93.
19. Puppala, A., S. Chomtid, and V. Bhadriraju. Using Repeated-Load Triaxial Tests to Evaluate Plastic Strain Potentials in Subgrade Soils. *Transportation Research Record: Journal of the Transportation Research Board*, 2005. 1913: 86–98.

20. Barksdale, R. D., and S. Y. Itani. Influence of Aggregate Shape on Base Behavior. *Transportation Research Record: Journal of Transportation Research Board*, 1989. 1227: 173–182.
21. Chow, L. C. *Permanent Deformation Behavior of Unbound Granular Materials and Rutting Model Development*. MS thesis. University of Illinois at Urbana-Champaign, Urbana, 2014. Can be accessed through https://www.ideals.illinois.edu/handle/2142/50674. Accessed March 30, 2017.
22. Werkmeister, S., A. Dawson, and F. Wellner. Permanent Deformation Behavior of Granular Materials and the Shakedown Concept. *Transportation Research Record: Journal of the Transportation Research Board*, 2001. 1757: 75–81.
23. Mishra, D., and E. Tutumluer. Aggregate Physical Properties Affecting Modulus and Deformation Characteristics of Unsurfaced Pavements. *Journal of Materials in Civil Engineering*, Vol. 24, No. 9, 2012, pp. 1144–1152.
24. Wu, M., H. Wen, B. Muhunthan, and K. Manahiloh. Influence of Recycled Asphalt Pavement Content on Air Void Distribution, Permeability, and Modulus of Base Layer. *Transportation Research Record: Journal of the Transportation Research Board*, 2012. 2267: 65–71.
25. Kim, W., J. Labuz, and S. Dai. Resilient Modulus of Base Course Containing Recycled Asphalt Pavement. *Transportation Research Record: Journal of the Transportation Research Board*, 2007. 2005: 27–35.

The Standing Committee on Aggregates (AFP70) peer-reviewed this paper (18-04673).

The conclusions and recommendations are those of the authors and do not reflect the opinions and policies of VDOT.

Article

TRR
JOURNAL OF THE TRANSPORTATION RESEARCH BOARD

Transportation Research Record
2018, Vol. 2672(52) 67–77

DOI: 10.1177/0361198118758688
journals.sagepub.com/home/trr
SAGE

Leaching Assessment of Eco-Friendly Rubberized Chip Seal Pavement

Ahmed Gheni[1], Xuesong Liu[1], Mohamed A. ElGawady[1], Honglan Shi[2], and Jianmin Wang[1]

Abstract

Companies in the United States need to mine billions of tons of raw natural aggregate each year. At the same time, billions of scrap tires are stockpiled every year. As a result, replacing the natural aggregate with recycled aggregate is beneficial to the construction industry and the environment. This paper is part of a comprehensive project that developed, and field implemented, a new eco-friendly rubberized chip seal where the mineral aggregate in chip seal is partially or totally replaced with crumb rubber made of recycled tires. This paper presents an extensive study of the environmental impact of using rubber aggregate in chip seal pavement in terms of leaching under different pH conditions, including simulated acid rain. The results are compared with those of conventional chip seal. Leaching from the constituents of chip seal, that is, rubber aggregate and emulsion, was investigated. Two types of rubber and two types of asphalt emulsions were studied. The leaching performance of rubberized chip seal was also investigated. This study revealed that the toxic heavy metals leached from the rubberized chip seal, for pH ranging from 4 to 10, were below that of the EPA drinking water standards. In addition, a significant reduction of heavy metal leaching was recorded when rubber was used with emulsion in the form of chip seal pavement under different pH conditions. Finally, the metal leaching in all types of samples (including rubber, asphalt emulsion, and chip seal) decreased with the increase in pH value.

As natural resources are depleting, the construction industry is forced to find replacements for virgin construction materials, and using recycled material is one of the main options. At the same time, the world is facing a severe problem dealing with scrap tires. More than four million tons of scrap tires were dumped in the United States during 2015 alone, taking up valuable space in landfills and wasting valuable resources in the form of the rubber material and textile and metal cord (*1*). Crumb rubber obtained from scrap tires can be used to replace mineral aggregate, leading to an eco-friendly construction industry (*2–4*). Recycled crumb rubber has been widely used in concrete to improve durability, seismic behavior, sustainability, sound absorption, noise reduction coefficient, and heat transfer properties (*5–8*).

Transportation infrastructure contributes 23% of global carbon dioxide (CO_2) emissions, making it the second largest contributor, only behind electricity generation (*9*). As a result, using recycled material in the construction of the transportation infrastructure will reduce CO_2 emissions significantly. One application is using scrap tire material in roadway construction. Previous studies have shown that using crumb rubber as an asphalt binder modifier can improve the overall performance of the binders in terms of temperature susceptibility, viscosity, and stiffness (*10*). However, this application does not consume a significant amount of scrap rubber.

Chip seal is a type of pavement treatment that consists of a single application of asphalt binder directly on existing pavement, followed by an application of aggregate chips that are rolled with a pneumatic roller. In this process, the aggregate chip layer is typically one stone thick. Chip seal with two layers of aggregate has been used (Figure 1*b*). Recycled rubber has been used with chip seal before as an asphalt binder modifier. According to the authors' best knowledge, there has been no research on using crumb rubber aggregate in chip seal.

This research aims to investigate the potential use of crumb rubber aggregate in chip seal, producing what is called rubberized chip seal, which would significantly increase the sustainability of chip seal. This project studied the overall

[1]Department of Civil, Architectural & Environmental Engineering, Missouri University of Science and Technology, Rolla, MO
[2]Department of Chemistry, Missouri University of Science and Technology, Rolla, MO

Corresponding Author:
Address correspondence to Mohamed A. ElGawady: elgawadym@mst.edu

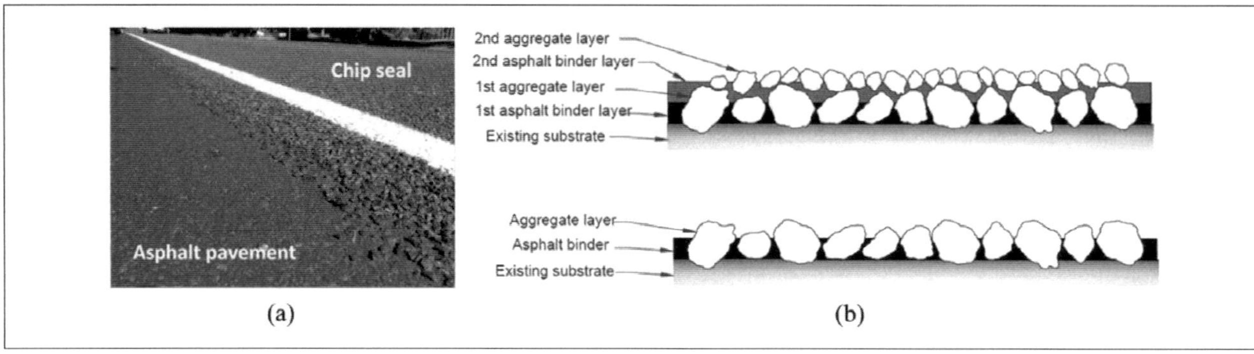

Figure 1. (*a*) Chip seal versus asphalt pavement, and (*b*) types of chip seal.

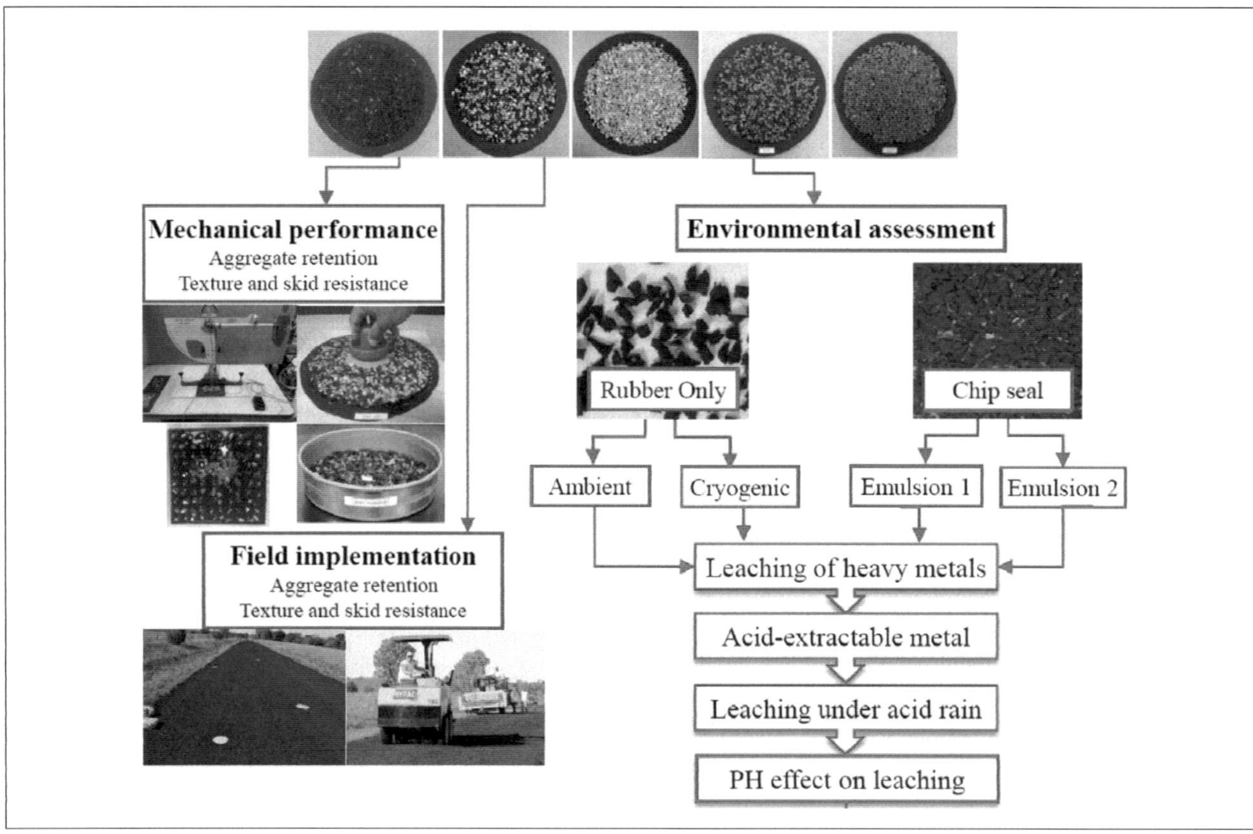

Figure 2. Schematic overview of the project.

performance of rubberized chip seal pavement (Figure 2). The aggregate retention was investigated using three standard tests. Furthermore, the micro and macro texture have been investigated, as well as the skid resistance under ambient and elevated temperatures. The results were compared with conventional chip seal using two types of asphalt emulsion and two types of mineral aggregate (*2*).

In addition to the comprehensive laboratory testing and assessment, two different locations in the State of Missouri were selected for field-testing of rubberized chip seal. The main objective of the first location was to ensure the feasibility of constructing rubberized chip seal using the same set of equipment, tools, and procedures that are currently used by an average contractor to apply a conventional chip seal; in the second location, the performance of the rubberized chip seal will be monitored for 1 year. The first location consisted of two chip seal sections with different crumb rubber volume-replacement ratios of 50 and 100%. Each section was about 350 ft long. The second location consisted of five chip seal sections with different crumb rubber volume-replacement

Figure 3. Field implementation of rubberized chip seal with 100% crumb rubber replacement ratio.

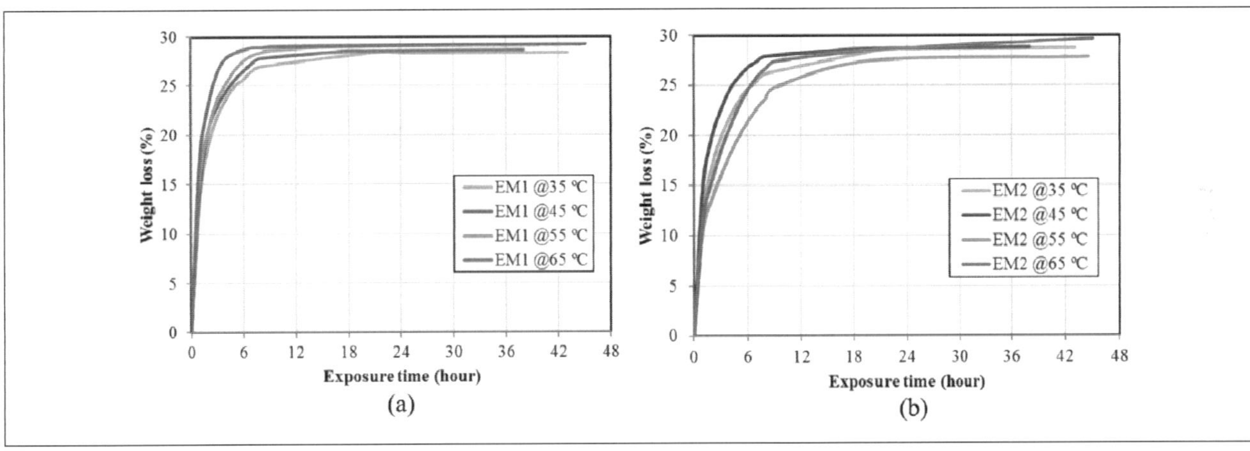

Figure 4. Emulsion weight loss due to water breakout, (*a*) Emulsion 1, and (*b*) Emulsion 2.

ratios of 0, 25, 50, 75, and 100%. Without any modification, the traditional chip seal procedure was used to apply the rubberized chip seal (Figure 3). An ambient processed crumb rubber with a size of 2.3–6.0 mm and 6.0–12.5 mm was used in location 1 and 2, respectively. This was similar in size to the creek gravel aggregate and trap rock used in the blend, respectively. Emulsion type CRS2P with a temperature of 54°C was used at an air temperature of 26°C in both locations. During the compaction, it was noticed that rubber particles were adherent to the tires of the rubber tires compactors (Figure 3), because the flexibility of the rubber particles allowed the rubber tires to penetrate and squeeze the crumb rubber layer and reach the emulsion. As a result, the rubber tire compactors were replaced by steel roller compactors which compacted the material appropriately.

Although the laboratory work and field implementation showed the high performance of rubberized chip seal, the leaching of toxic metals in rubberized chip seal could be a concern, because of the absence of any research on this issue. This paper presents a comprehensive study to determine the metal leaching behavior of rubberized chip seal under various conditions.

Experimental Program

Material Characterization and Properties

Two types of emulsion, CRS-2P and CHFRS-2P, were used during this study. Both emulsions were rapid setting, high viscous, and cationic, which is defined as the migration of asphalt particles under an electric field toward the cathode (negative electrode). Emulsion 2, CHFRS-2P, was a high-float type, and emulsion 1, CRS-2P, was a conventional type. The water breakout of the emulsions was examined through weight loss measurements after exposing the emulsion to different temperatures for different periods (Figure 4). Approximately 81% of the water breakout occurred after 6 h for both types of emulsion, whereas there was almost no evaporation after 24 h of exposure.

Two types of recycled rubber, ambient and cryogenic, were used during this study. The former type was produced by tearing and shredding the scrap tires in the cutting mills at ambient temperature. The latter type was produce by reducing the temperature of the scrap tires to −80°C using liquid nitrogen, followed by grounding the glass-like frozen tires to the required sizes using a hammer mill.

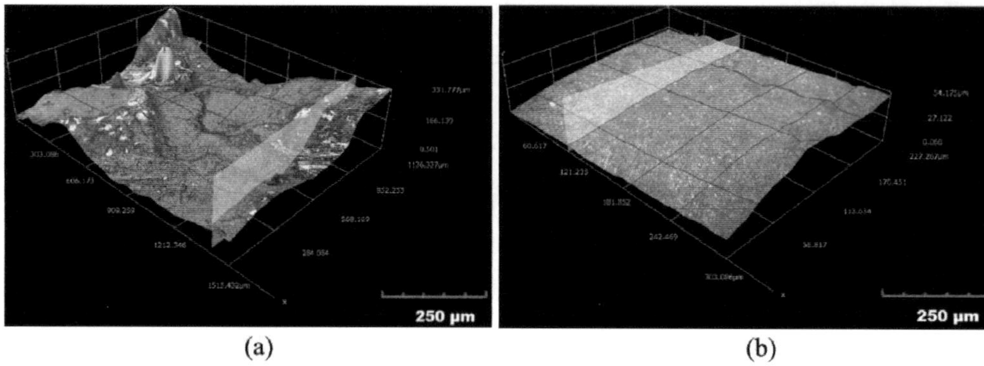

Figure 5. Microscopic results of the surface of the crumb rubber aggregates in range of 250 µm (*a*) ambient, and (*b*) cryogenic.

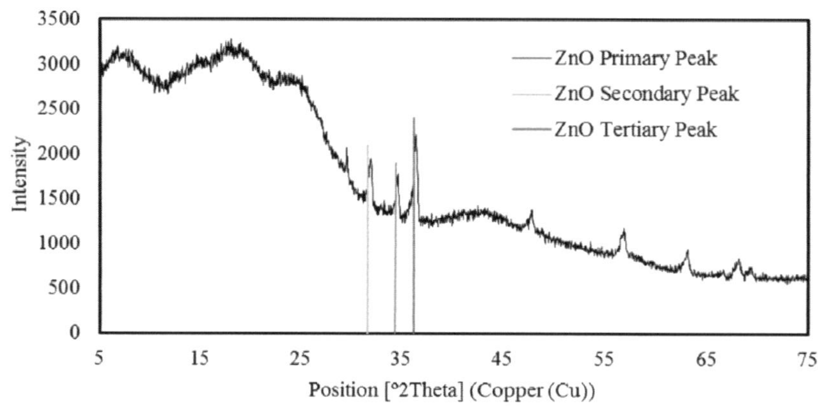

Figure 6. X-ray diffraction of crumb rubber.

The microtexture surface conditions of the mineral and crumb rubber aggregates were examined using a digital 3D microscope KH-8700. The ambient crumb rubber had a very rough surface with many crests and troughs, which improved its retention with the asphalt emulsions and binders, whereas the cryogenic crumb rubber had a smooth surface (Figure 5).

Environmental Assessment

The first step in the environmental assessment was to perform an X-ray diffraction test to determine the chemical composition of the rubber aggregate. Zinc (Zn) and zinc oxide (ZnO) had the highest concentration compared with other elements (Figure 6).

The leaching of heavy metals was studied for bare rubber particles and for chip seal (rubber and emulsion). The specimens were exposed to drinking water, acid, and simulated acid rain. The effect of pH in a pH range between 4 and 10 on metal leaching was also investigated. This part of the study consisted of the following four tasks:

1. Leaching behavior of heavy metals from bare crumb rubber under different pH conditions;
2. Acid-extractable metal contents from chip seal samples. This was to determine the maximum leachable heavy metals from these materials;
3. Leaching behavior of heavy metals from chip seal samples under a simulated acid rain condition in Missouri (west of Mississippi river);
4. Leaching behavior of heavy metals of different chip seal specimens with pH ranging between 4 and 10.

Preparation of Chip Seal Specimens

To study the difference between the leaching of crumb rubber particles and chip seal specimens, the amount of aggregate and asphalt emulsion in chip seal specimens should be calculated properly. There is no consensus in the United States on how to design a chip seal. A recent survey covering 54 cities and entire states showed that only 18% of respondents use the McLeod, Kearby, and modified Kearby methods to

Table 1. Sample Matrix for Leaching Behavior from Bare Crumb Rubber

Sample ID	Rubber (5 g)	Acid/base
F+5	Cryogenic	1.0 mL 6 M HNO_3
F+4	Cryogenic	0.8 mL 6 M HNO_3
F+3	Cryogenic	0.6 mL 6 M HNO_3
F+2	Cryogenic	0.4 mL 6 M HNO_3
F+2D[a]	Cryogenic	0.4 mL 6 M HNO_3
F+1	Cryogenic	0.2 mL 6 M HNO_3
F	Cryogenic	0
F-1	Cryogenic	0.2 mL 10 M NaOH
F-2	Cryogenic	0.4 mL 10 M NaOH
F-3	Cryogenic	0.6 mL 10 M NaOH
F-4	Cryogenic	0.8 mL 10 M NaOH
F-5	Cryogenic	1.0 mL 10 M NaOH
MQ	0	0
MQ+HNO_3	0	1.0 mL 6 M HNO_3
MQ+NaOH	0	1.0 mL 10 M NaOH
UF+5	Ambient	0.6 mL 1 M HNO_3
UF+4	Ambient	0.4 mL 1 M HNO_3
UF+3	Ambient	0.2 mL 1 M HNO_3
UF+2	Ambient	0.1 mL 1 M HNO_3
UF+1	Ambient	0.2 mL 0.1 M HNO_3
UF	Ambient	0
UF-1	Ambient	0.2 mL 0.1 M NaOH
UF-2	Ambient	0.1 mL 1 M NaOH
UF-2D[a]	Ambient	0.1 mL 1 M NaOH
UF-3	Ambient	0.2 mL 1 M NaOH
UF-4	Ambient	0.4 mL 1 M NaOH
UF-5	Ambient	0.6 mL 1 M NaOH
MQ	0	0

Note: F = cryogenic rubber; UF = ambient rubber.
[a]D represents sample duplication.

design chip seal, while 26% of the respondents do not use a formal design method. The remaining 56% of the respondents use their local, empirical, or past experience design approach (*11*). The ASTM D7000-11 provides Equation 1 for determining the aggregate application rate for sweep test:

$$\text{Aggregate weight}\left(\frac{kg}{m^2}\right)=\left(\frac{A(202.1X-15.8)}{100}+\frac{B(146.4X-47)}{100}\right)*\frac{1}{61.6} \quad (1)$$

where A is the percentage of aggregate grade 1 from 9.5 mm to 6.3 mm, B is the percentage of aggregate grade 2 from 6.3 mm to 4.75 mm, and X is the bulk specific gravity.

Equation 1 resulted in an application rate of 2.7 kg/m^2 (5.0 lb/yd^2) which was used throughout this research.

To avoid cross-contamination, special tools and cleaning procedures were used. The cutting tool used to prepare the specimens was a ceramic knife instead of a metal knife. Before cutting each specimen, the ceramic knife was cleaned with gasoline to dissolve asphalt residual, followed by deionized (DI) water rinsing. Then, the knife was cleaned again with acetone to remove gasoline residual, and then rinsed with DI water. This procedure was followed exactly for each specimen during the experiment.

Task 1: Leaching Behavior of Heavy Metals from Bare Crumb Rubber. It was found through the particle size distribution analysis that the cryogenic crumb rubber had a representative particle size between 2.0 mm and 4.75 mm, and the ambient crumb rubber had a representative particle size between 4.75 mm and 9.52 mm. As a result, these two samples were collected and used for the leaching experiment. The leaching experiments were conducted under different pH conditions by following EPA method 1313. A solid-to-liquid ratio of 1:10 was used. Pure water (MQ water) was used as the leaching liquid. The pH of different leaching bottles was adjusted by adding different volumes of stock nitric acid or sodium hydroxide solution at the beginning of the leaching experiment. Several types of control experiments were also performed. Even though EPA 1313 recommends 2 N/M HNO_3 and 1 N/M KOH/NaOH for the test, 6 M HNO_3 and 10 M NaOH were used to control the pH because the solution is National Institute of Standards and Technology traceable, so we did not dilute to make it a lower concentration (2 M HNO_3 or 1 M NaOH). Besides this, use of a high concentration acid/base can also minimize the sample volume change caused by acid/base addition. The experimental matrix is shown in Table 1.

The experimental procedure started with adding 5 g of the sample into each of the 125 mL pre-acid cleaned plastic leaching bottles (except the blanks) and then adding 50 mL of MQ water into each of the leaching bottles. After that, different volumes of nitric acid or sodium hydroxide solution were added into different bottles to adjust pH; then, all bottles were sealed tightly. All bottles were then shaken for 24 h at 180 rpm at room temperature. After shaking, all bottles were settled for 1 h. The supernatant was then filtered through 0.22 μm nylon membrane filter. A partial filtrate was then used to measure pH and total dissolved solids (TDS). The remaining filtrate was acidified with trace metal grade nitric acid to a pH <2. Finally, thirteen elements were analyzed using inductively coupled plasma mass spectrometry (ICP-MS), graphite furnace atomic absorption (GFAA), or flame atomic absorption (flame AA) after appropriate dilution with 1% HNO_3 solution if needed.

Task 2: Acid-Extractable Metal Contents from Chip Seal Samples. In this task, chip seal specimens made of ambient crumb rubber and two asphalt emulsions were digested with acid for heavy metal availability testing. The digestion experiment was conducted in a microwave digester (ETHOS E, MILESTONE) using EPA method 3051A with slight modifications. A mixture of concentrated trace metal grade

Table 2. Sample Matrix for Acid Digestion Experiment

Sample ID	Sample component (0.2 g)
T	Ambient rubber
T(D)*	Ambient rubber
Tspike	Ambient rubber with spike
A1	Emulsion 1
A2	Emulsion 2
A2(D)	Emulsion 2
A2 spike	Emulsion 2 with spike
Blank	N.A.

hydrochloric acid and nitric acid was used as the digestion solution. Several types of control experiments, spike recovery, and blank were also performed (Table 2).

The digestion procedure started with weighing 0.2 g of sample into a digestion vessel and then adding 3 mL of trace metal grade hydrochloric acid and 9 mL of nitric acid. Sample duplicate, sample spike, and blank were included using different vessels. The digestion vessels were sealed and loaded into the microwave digester. The digestion program included a 10-min ramp to 180°C with a 10-min hold, followed by a 10-min cooling time. The digestion vessels were then removed from the digester and cooled for 1 h. The solution was then transferred to pre-acid cleaned 50 mL centrifuge tubes to dilute to 50 mL with MQ water. Finally, thirteen selected elements in the digestion solution were measured using ICP-MS or flame AA after appropriate dilutions with 1% HNO_3 (if needed) and filtration through 0.22 μm nylon membrane filter.

Task 3: Leaching of Chip Seal Samples under a Simulated Acid Rain Condition. In this task, the leaching behavior of different chip seal samples was tested at a pH of 5 following EPA method 1312. The tests were performed by adding extraction fluid with a pH of 5 (containing sulfuric acid and nitric acid at a ratio of 60:40) to the chip seal at a solid-to-liquid ratio of 1:20 followed by mixing at a rate of 180 rpm on a mechanical shaker for 24 h. Several control bottles were also included. The sample information is shown in Table 3.

The test procedure started by preparing the chip seal specimens according to the aggregate and emulsion amounts mentioned in the description of the preparation of chip seal specimens. Chip seal specimens containing ambient crumb rubber were mixed with each type of asphalt emulsion. Other specimens were prepared using mineral aggregate with each type of asphalt emulsion. A control group included each of the asphalt emulsions, crumb rubber, and mineral aggregates. The sample size was reduced by cutting the specimens with a ceramic knife and putting them through standard size sieves: 3/8", 1/4", and No.4. The retained portion of the sample on No.4 sieve was used for this experiment (size range 4.75–6.30 mm). The pH 5 extraction fluid was then prepared by adding 16.5 μL acid mixture (acid mixture prepared by adding 0.15 g H_2SO_4 and 0.10 g HNO_3 in 20 mL MQ water) into 100 mL MQ water in 125 mL pre-acid cleaned plastic bottles. After that, 5 g of sample was added to each leaching bottles (except the blanks). Bottles were sealed tightly after the addition of extraction fluid and sample. The leaching bottles were then shaken for 24 h at 180 rpm at room temperature and then settled for 1 h. The supernatant was then filtered through a 0.22 μm nylon membrane filter. The partial filtrate was then used to measure pH and TDS. The remaining filtrate was acidified with trace metal grade nitric acid to a pH <2. Finally, the heavy metal concentrations in the acidified filtrate samples were determined using ICP-MS or flame AA. Appropriate dilution with 1% HNO_3 was needed before ICP-MS measurement.

Task 4: Effect of pH on Metal Leaching of Chip Seal Samples. In this task, the effect of pH on the leaching behavior of different specimens was tested with pH ranging between 4 and 10 following EPA method 1313. A solid-to-liquid ratio of 1:10 was used in this experiment. Several control experiments were also performed. The experiment procedures were similar to the acid rain leaching experiment, except the pH of the leaching bottles was adjusted individually. The detailed experimental information is shown in Table 4.

Quality Assurance/Quality Control. To ensure high-quality data, most of the recommended QA/QC by the EPA was followed. For analysis using ICP-MS, US EPA method 200.8 QC guidelines were closely followed. ICP-MS was calibrated with standard solutions diluted from a calibration standard mixture. The linear ranges of the calibration were determined and used for the quantitative analysis of the samples. Detailed control samples are listed in Tables 1, 2, 3, and 4. Laboratory reagent blank was tested to check any procedural contamination. The blank sample was prepared and measured using the same procedures as for the samples except no solid sample was used. The method detection limits (MDLs) for the leaching test were determined by instrumentation detection limits (IDLs) where MDLs were ten times the IDLs. The MDLs for the screening test and digestion test were twenty times the IDLs. To ensure good reproducibility, duplicated samples were performed for some samples. The precision of the duplication was expressed as the relative percentage difference (RPD) and was calculated using Equation 2 below.

$$RPD(\%) = 100 \times (C_h - C_l) / C_{av} \quad (2)$$

where

C_h is detected high concentration of duplicated sample,
C_l is detected low concentration of duplicated sample, and
C_{av} is the average of the C_h and C_l

Sample spike was tested by adding known concentration standards into the leached sample solution before performing the analysis. The spike recoveries (%) were calculated by the following equation:

Table 3. Sample Matrix for Simulated Acid Rain Experiment

Sample ID	Sample (5 g)	Liquid volume
T+A1	Rubberized chip seal with emulsion 1	100 mL extraction fluid
T+A2	Rubberized chip seal with emulsion 2	100 mL extraction fluid
T+A2(D)*	Rubberized chip seal with emulsion 2	100 mL extraction fluid
A+A1	Conventional chip seal with emulsion 1	100 mL extraction fluid
A+A2	Conventional chip seal with emulsion 2	100 mL extraction fluid
A1	Emulsion 1	100 mL extraction fluid
A2	Emulsion 2	100 mL extraction fluid
T	Ambient rubber	100 mL extraction fluid
A	Mineral aggregate	100 mL extraction fluid
Acid	0	100 mL extraction fluid
MQ	0	100 mL MQ water

Table 4. Sample Matrix for Effect of pH on Metal Leaching Experiment

Sample ID	Sample component (5 g in 50 mL MQ)	Acid/base
TA1-1	Chip seal with emulsion 1	0.20 mL 0.1 M HNO_3
TA1-2	Chip seal with emulsion 1	0.14 mL 0.1 M HNO_3
TA1-2(D)*	Chip seal with emulsion 1	0.14 mL 0.1 M HNO_3
TA1-3	Chip seal with emulsion 1	0.07 mL 0.1 M HNO_3
TA1-4	Chip seal with emulsion 1	0
TA1-5	Chip seal with emulsion 1	0.05 mL 0.1 M NaOH
TA1-6	Chip seal with emulsion 1	0.14 mL 0.1 M NaOH
TA1-7	Chip seal with emulsion 1	0.20 mL 0.1 M NaOH
TA2-1	Chip seal with emulsion 2	0.20 mL 0.1 M HNO_3
TA2-1(D)	Chip seal with emulsion 2	0.20 mL 0.1 M HNO_3
TA2-2	Chip seal with emulsion 2	0.14 mL 0.1 M HNO_3
TA2-3	Chip seal with emulsion 2	0.08 mL 0.1 M HNO_3
TA2-4	Chip seal with emulsion 2	0
TA2-5	Chip seal with emulsion 2	0.06 mL 0.1 M NaOH
TA2-6	Chip seal with emulsion 2	0.14 mL 0.1 M NaOH
TA2-7	Chip seal with emulsion 2	0.20 mL 0.1 M NaOH
T1	Ambient crumb rubber	0.40 mL 0.1 M HNO_3
T2	Ambient crumb rubber	0.30 mL 0.1 M HNO_3
T3	Ambient crumb rubber	0.20 mL 0.1 M HNO_3
T3(D)	Ambient crumb rubber	0.20 mL 0.1 M HNO_3
T4	Ambient crumb rubber	0.08 mL 0.1 M HNO_3
T5	Ambient crumb rubber	0.06 mL 0.1 M NaOH
T6	Ambient crumb rubber	0.14 mL 0.1 M NaOH
T7	Ambient crumb rubber	0.20 mL 0.1 M NaOH
A1-1	Emulsion 1	0.05 mL 0.1 M HNO_3
A1-1(D)	Emulsion 1	0.05 mL 0.1 M HNO_3
A1-2	Emulsion 1	0
A1-3	Emulsion 1	0.03 mL 0.1 M NaOH
A1-4	Emulsion 1	0.06 mL 0.1 M NaOH
A1-5	Emulsion 1	0.09 mL 0.1 M NaOH
A1-6	Emulsion 1	0.12 mL 0.1 M NaOH
A1-7	Emulsion 1	0.30 mL 0.1 M NaOH
A2-1	Emulsion 2	0.08 mL 0.1 M HNO_3
A2-2	Emulsion 2	0.05 mL 0.1 M HNO_3
A2-3	Emulsion 2	0.02 mL 0.1 M HNO_3
A2-4	Emulsion 2	0.03 mL 0.1 M NaOH
A2-5	Emulsion 2	0.06 mL 0.1 M NaOH
A2-5(D)	Emulsion 2	0.06 mL 0.1 M NaOH
A2-6	Emulsion 2	0.15 mL 0.1 M NaOH
A2-7	Emulsion 2	0.30 mL 0.1 M NaOH
Acid	N.A.	0.4 mL HNO_3
Base	N.A.	0.3 mL NaOH
MQ	Blank	N.A.

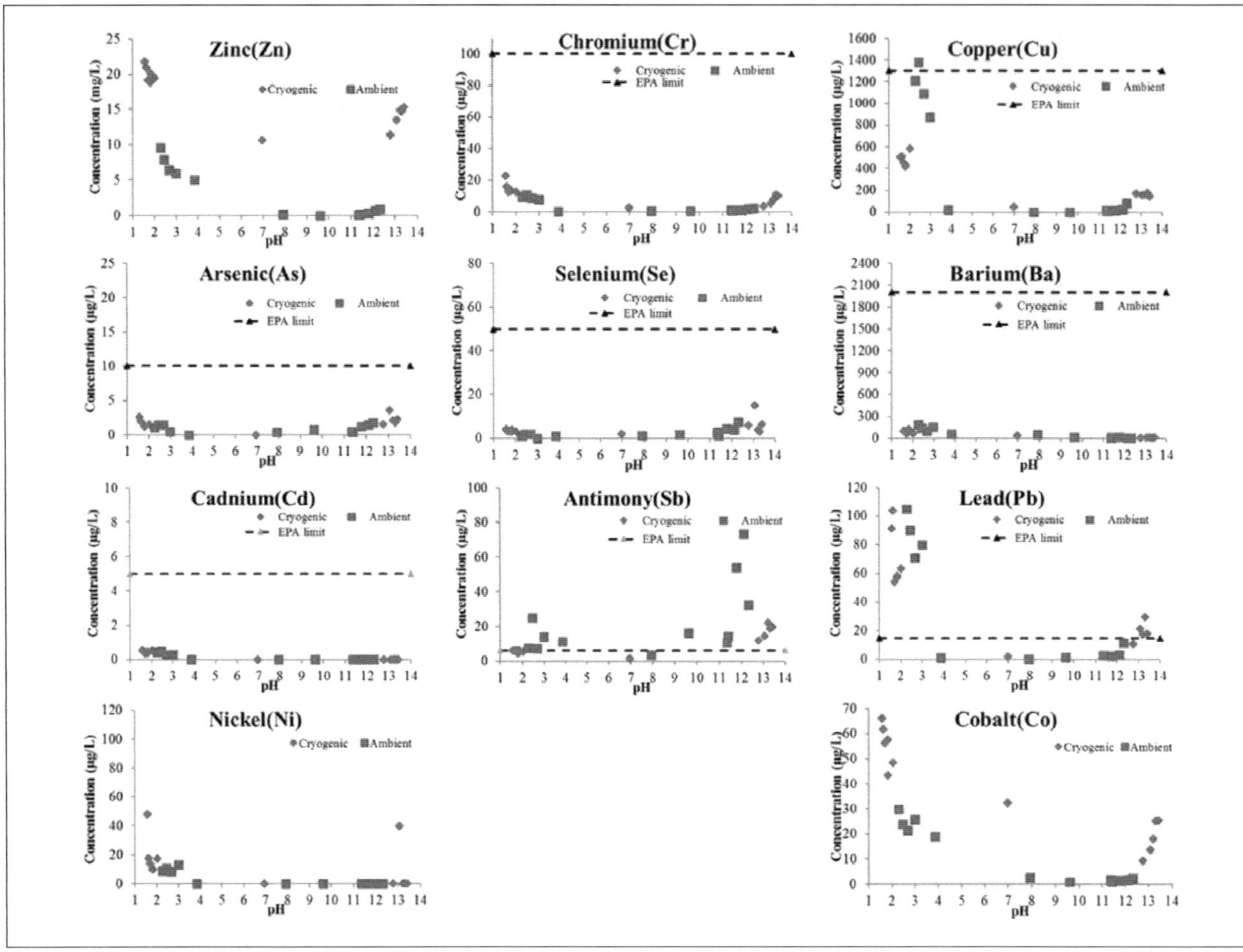

Figure 7. Metal concentration in leaching solution as function of pHs for rubber particles only.

$$\text{Spike recovery } (\%) = 100 \times (\text{detected conc. of spiked sample} - \text{control sample}) / \text{Spiked concentration} \quad (3)$$

Results and Discussion

Task 1: Leaching Behavior of Heavy Metals from Bare Crumb Rubber

The metal leaching concentration from the rubber particle specimens as a function of pH is shown in Figure 7; most of the rubber specimens were tested in very high or low pH conditions to determine which rubber, ambient or cryogenic, had more leachable metal. The major heavy metal leached out from the bare rubber is zinc (Zn), followed by copper (Cu), and barium (Ba). Note that Zn has the highest concentration in rubber (Figure 6). In the leaching solution, Zn concentration is in parts per million, which is significantly higher than any other metals. The leaching of most metals from both the cryogenic crumb rubber (2–4.75 mm) and the ambient crumb rubber (4.75–9.32 mm) was consistent, although the sizes were different. Furthermore, generally the ambient crumb rubber heavy metal leaching was significantly less than that of the cryogenic rubber and therefore was selected for further research in this project.

Task 2: Acid-Extractable Metal Contents from Chip Seal Specimens

The acid-extractable metal content in chip seal specimens was converted from metal concentration in digestion solution based on the final volume of digestion solution and mass of the sample. Table 5 shows the metal contents of chip seal specimens. The results show the major metal content in the rubber was Zn, which was about 1.6% of the total rubber weight. The major metal content in both types of asphalt was Ni, but generally, the content of all metals was very low (some of them were below the MDL). The

Table 5. Metal Concentration in Leaching Solution of Rubber and Emulsions

Sample ID	Be (μg/g)	Cr (μg/g)	Co (μg/g)	Ni (μg/g)	Cu (μg/g)	As (μg/g)	Se (μg/g)	Cd (μg/g)	Sb (μg/g)	Ba (μg/g)	Tl (μg/g)	Pb (μg/g)	Zn* (mg/g)
T	<MDL	4.97	286	3.28	244.9	24.78	2.68	0.73	2.22	4.42	<MDL	1.73	17.06
T(D)	<MDL	5.52	227.7	3.16	59.63	26.65	6.58	0.96	2.44	4.97	<MDL	2.95	15.94
T spike	2.54	6.13	30.41	31.48	78.29	24.05	7.11	2.81	3.73	5.88	1.74	20.47	21.5
Spiked conc.	2.5	2.5	10	25	50	2.5	5	2.5	2.5	2.5	2.5	10	5
T recovery (%)	101.8	35.11	−209.6	113	−147.9	−66.79	49.55	78.33	55.95	47.49	69.54	181.2	100
A1	<MDL	4.4	<MDL	65	4.73	21.72	0.87	<MDL	0.98	1.58	<MDL	0.57	0.09
A2	<MDL	5.91	<MDL	72.96	<MDL	26.57	0.96	<MDL	1.2	1.28	<MDL	<MDL	0.12
A2(D)	<MDL	5.21	<MDL	67.05	<MDL	23.62	0.7	<MDL	0.78	1.28	<MDL	<MDL	0.11
A2 spike	2.62	8.19	1.94	125	30.65	29.6	4.93	2.64	3.41	3.88	1.7	5.85	7.05
Spiked conc.	2.5	2.5	2.5	50	25	2.5	5	2.5	2.5	2.5	2.5	5	10
A2 recovery (%)	104.7	105.1	77.62	110	122.6	180.2	82.01	105.4	96.92	103.8	68.00	117	70.5
Blank	<MDL	2.61	<MDL	<MDL	<MDL	19.57	1.05	<MDL	0.8	0.44	<MDL	<MDL	<MDL
MDL	0.125	0.25	0.625	2.5	2.5	0.125	0.5	0.125	0.125	0.125	0.125	0.5	0.005
T RPD (%)	N.A.	10.49	22.73	3.88	121.7	7.28	84.22	27.09	9.33	11.82	N.A.	52.12	6.73
A2 RPD (%)	N.A.	12.67	N.A.	8.44	N.A.	11.77	31.67	N.A.	42.12	0.34	N.A.	N.A.	6.56

Table 6. Metal Concentration in Leaching Solution Under Simulate Acid Rain of Rubberized Chip Seal

Sample ID	pH	TDS (mg/L)	Be (μg/L)	Cr (μg/L)	Co (μg/L)	Ni (μg/L)	Cu (μg/L)	As (μg/L)	Se (μg/L)	Cd (μg/L)	Sb (μg/L)	Ba (μg/L)	Pb (μg/L)	Tl (μg/L)	Zn** (mg/L)
T+A1	7.10	21.20	<MDL	<MDL	3.25	<MDL	37.52	<MDL	<MDL	<MDL	<MDL	11.94	<MDL	<MDL	0.46
T+A2	6.51	16.10	<MDL	<MDL	4.22	<MDL	<MDL	<MDL	9.72	<MDL	<MDL	11.89	<MDL	<MDL	0.51
T+A2(D)	7.35	17.90	<MDL	<MDL	3.54	<MDL	<MDL	<MDL	<MDL	<MDL	0.61	12.41	<MDL	<MDL	0.29
A+A1	7.59	17.40	<MDL	<MDL	0.64	<MDL	28.16	<MDL	<MDL	<MDL	<MDL	5.83	<MDL	<MDL	<MDL
A+A2	6.95	16.90	<MDL	<MDL	0.72	<MDL	17.29	<MDL	<MDL	<MDL	<MDL	5.17	<MDL	<MDL	<MDL
A1	4.40	24.00	<MDL	<MDL	0.97	10.73	49.75	<MDL	<MDL	<MDL	<MDL	<MDL	<MDL	<MDL	0.24
A2	4.28	26.20	<MDL	<MDL	1.00	<MDL	<MDL	<MDL	8.34	<MDL	<MDL	<MDL	<MDL	<MDL	0.18
T	7.10	12.60	<MDL	<MDL	9.50	<MDL	16.08	<MDL	<MDL	<MDL	3.13	30.68	2.11	<MDL	0.40
A	7.26	5.48	<MDL	<MDL	<MDL	<MDL	<MDL	<MDL	<MDL	<MDL	<MDL	0.74	<MDL	<MDL	<MDL
Acid	4.60	N.A.	<MDL	<MDL	<MDL	<MDL	<MDL	<MDL	<MDL	<MDL	<MDL	<MDL	<MDL	<MDL	<MDL
MQ	N.A.	N.A.	<MDL	<MDL	<MDL	<MDL	<MDL	<MDL	<MDL	<MDL	<MDL	<MDL	<MDL	<MDL	<MDL
MDL			0.50	1.00	0.50	10.00	10.00	0.50	2.00	0.50	0.50	0.50	2.00	0.50	0.10
EPA limits	6.5–8.5	500	4	100	N.A.	N.A.	1300	10	50	5	6	2000	15	2	5
T+A1 spike recovery (%)			104.00	94.79	99.34	85.25	8.94	103.88	104.79	99.60	104.64	102.08	95.95	104.41	N.A.
T+A2 RPD (%)	12.12	10.59	N.A.	N.A.	17.63	N.A.	N.A.	N.A.	N.A.	N.A.	N.A.	4.27	N.A.	N.A.	54.95

reproducibility and spike recovery of the rubber sample were not good through the microwave digestion method, probably because of the incomplete digestion of rubber sample. After microwave digestion, there were some solids found in the rubber digestion vessel, but not in the asphalt digestion vessel.

Task 3: Leaching Under Simulated Acid Rain Condition

The experimental results using simulated acid rain as the leaching solution are shown in Table 6. From the test results, trace amount heavy metal was leached from the sample after 24 h and all the concentrations are below EPA drinking water standard.

Task 4: Effect of pH on Metal Leaching from Chip Seal

The soluble metal concentration as a function of pH for Task 4 (pH effect) is shown in Figure 8. It indicated that when rubber was encapsulated in emulsion for the chip seal, there was a significant reduction of heavy metal, especially Zn, leaching under different pH conditions. Approximately 50% reduction of Zn leaching was found when compared with the bare rubber. The leaching values of Co, Cu, and Ba were also reduced when rubber was used in asphalt emulsion. The leached concentrations of Be, Cr, As, Cd, Sb, Tl, and Pb were very low, that is, below or close to the MDL in the cases of crumb rubber, emulsion, and chip seal samples. Besides the effect of using asphalt

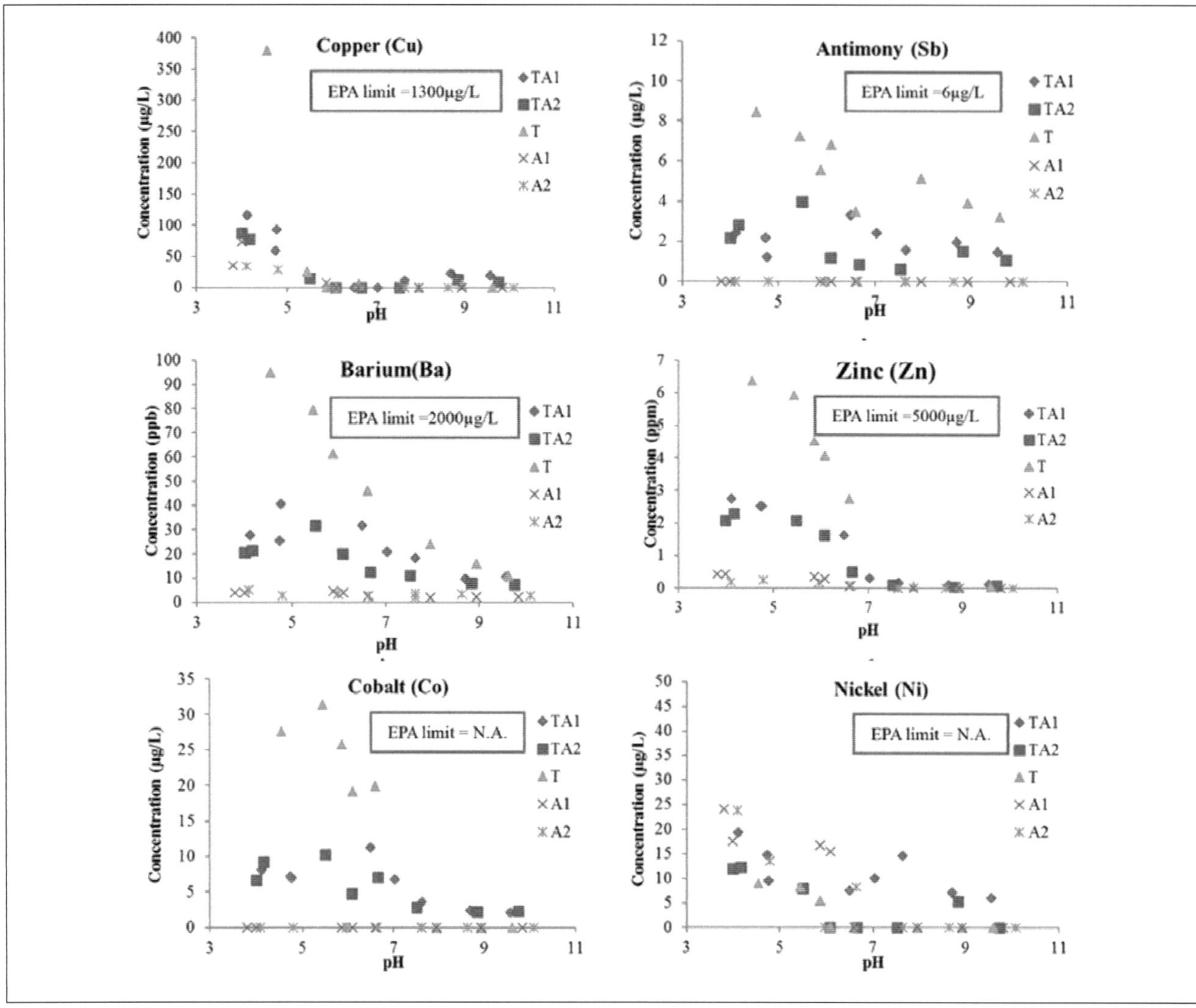

Figure 8. Metal concentration in leaching solution as a function of pHs for rubberized chip seal.
Note: Other elements are close or below MDLs.

emulsion to cover the rubber particles' surface, the elevated pH condition also reduced the metal leaching in all types of samples. From the comparison between the control group and chip seal, the rubber contributed to most of the metal leaching, except Ni, which is the major metal element in asphalt emulsion.

Conclusions

This study revealed the following findings and conclusions:

1. Using the crumb rubber that comes from scrap tires as a mineral aggregate replacement in chip seal pavement does not have a negative environmental impact in terms of heavy metal leaching. The toxic heavy metals leached from the recycle rubber or rubberized chip seal are below EPA drinking water standards.
2. The major leached heavy metal from the recycled bare rubber particles is Zn, which is consistent with the tire component. However, Zn is not regulated in primary drinking water regulations.
3. Under different pH conditions, a significant reduction of heavy metal leaching was recorded when rubber is used with emulsion in the form of chip seal pavement because asphalt is hydrophobic and prohibited the contact of tire and solution. About a 50% reduction of Zn leaching was recorded with chip seal specimens compared with the leaching from bare crumb rubber.
4. The metal leaching in all types of samples including rubber, asphalt emulsion, and chip seal decreased with the increase in pH value.

5. The cryogenic crumb rubber has a different metal leaching behavior than the ambient crumb rubber, depending on the pH value.

Acknowledgments

This research was conducted at the Missouri University of Science and Technology and supported by the Missouri Department of Natural Resources and Missouri Department of Transportation. Donations of the asphalt cement and emulsions by Vance Brothers are appreciated.

References

1. RMA. *2015 U.S. Scrap Tire Management Summary, in Wasington, DC*, 2016.
2. Gheni, A. A., O. I. Abdelkarim, M. M. Abdulazeez, and M. A. ElGawady. Texture and Design of Green Chip Seal Using Recycled Crumb Rubber Aggregate. *Journal of Cleaner Production*, Vol. 166, 2017, pp. 1084–1101.
3. Gheni, A. A., M. A. ElGawady, and J. J. Myers. Mechanical Characterization of Concrete Masonry Units Manufactured with Crumb Rubber Aggregate. *ACI Materials Journal*, Vol. 114, No. 1, 2017. P 65–76
4. Moustafa, A., and M. A. ElGawady. Strain Rate Effect on Properties of Rubberized Concrete Confined with Glass Fiber–Reinforced Polymers. *Journal of Composites for Construction*, Vol. 20, No. 5, 2016, p. 04016014.
5. Siddique, R., and T. R. Naik. Properties of Concrete Containing Scrap-Tire Rubber: An Overview. *Waste Management*, Vol. 24, No. 6, 2004, pp. 563–569.
6. Moustafa, A., A. Gheni, and M. A. ElGawady. Shaking-Table Testing of High Energy–Dissipating Rubberized Concrete Columns. *Journal of Bridge Engineering*, Vol. 22, No. 8, 2017, p. 04017042.
7. Turgut, P., and B. Yesilata. Physico-Mechanical and Thermal Performances of Newly Developed Rubber-Added Bricks. *Energy and Buildings*, Vol. 40, No. 5, 2008, pp. 679–688.
8. Gheni, A., M. A. ElGawady, and J. J. Myers. Thermal Characterization of Cleaner and Eco-Efficient Masonry Units Using Sustainable Aggregates. *Journal of Cleaner Production*, Vol. 165, 2017, pp. 980–993.
9. Ang, G., and V. Marchal. Mobilising Private Investment in Sustainable Transport: The Case of Land-Based Passenger Transport Infrastructure. *OECD Environment Working Papers*, No. 56, 2013, p. 0_1.
10. Presti, D. L. Recycled Tyre Rubber Modified Bitumens for Road Asphalt Mixtures: A Literature Review. *Construction and Building Materials*, Vol. 49, 2013, pp. 863–881.
11. Gransberg, D., and D. James. *Chip Seal Best Practices, National Cooperative Highway Research Program Synthesis 342*. Transportation Research Board, National Academies, Washington, D.C., 2005.

The Standing Committee on Geo-Environmental Processes (AFP40) peer-reviewed this paper (18-04559).

Any opinions, findings, conclusions, and recommendations presented in this paper are those of the authors and do not necessarily reflect the views of the sponsors.

Article

TRR
JOURNAL OF THE TRANSPORTATION RESEARCH BOARD

Transportation Research Record
2018, Vol. 2672(52) 78–85

DOI: 10.1177/0361198118767416
journals.sagepub.com/home/trr
SAGE

Assessment of Properties of Lateritic Gravel with a View to Enhancing Sustainable Use: A Case Study

Andrew Otto[1], Robert Tumwesige[2], and Alemayehu Ayele Endale[3]

Abstract

A number of authors have reported several properties of laterite that are not exhibited by other kinds of weathered tropical soils. Most of these properties are attributed to the presence of hydrated oxides of iron and aluminum. The combination of hydrated oxides is referred to as sesquioxides. It has been reported that the strength of soils that contain sesquioxides, such as laterite, increases due to development of self-cementation that occurs on repeated cycles of soaking and drying. The primary objective of this study was to explore the behavior of lateritic gravel under various conditions to develop methods for more beneficial and sustainable use. The scope of this study was a laboratory-based assessment of lateritic gravel obtained from a major source that supplies many construction sites in Kampala City, Uganda. The strength of the gravel was assessed on soaking after initial drying, after stabilization with lime, on drying after initial soaking, and after repeated cycles of soaking and drying. The results showed that for this particular lateritic gravel, the strength was highest when stabilized with lime. The strength development after cycles of soaking and drying was negligible; in fact, there was a reduction in strength. However, at moisture content below its optimum moisture content, the strength developed by the gravel was sufficient for use as base course or sub-base at various traffic levels provided certain measures are taken. Thus, repeated cycles of soaking and drying for this lateritic gravel may in fact be detrimental in the short term.

This study explores the behavior of lateritic gravel in the context of strength development and loss under various moisture conditions and cycles of soaking and drying. It explores the purported self-cementing characteristic of laterites and the potential for exploiting this characteristic.

The main objective of this study was to ascertain whether a major source of lateritic gravel in Kampala City, Uganda had the potential of hardening irreversibly on drying, or under repeated cycles of soaking and drying. The second objective of this study was to establish a methodology of how best to harness any potential strength gain of the lateritic gravel. Owing to the long duration of testing, the study was limited to samples taken from a single but major source of lateritic gravel in Kampala City. Based on the results, the study then explores how this laterite can be used sustainably to perform well in road construction and last as long as possible to avoid over-exploitation. It is hoped that other practitioners elsewhere may replicate this study to understand their lateritic gravels better.

The study is laboratory based and therefore such findings should be verified on site before it is applied. The site conditions are likely to experience longer dry spells than soaked spells since it is very uncommon to encounter 4 days of constant soaking. Moreover, wind action allows for faster drying conditions on site than in the laboratory.

Background to the Study

The term laterite was first derived by Buchanan in 1807 during his journey through Malabar, India. He used the term to describe a material "soft enough to be readily cut into blocks by an iron instrument, but which upon exposure to air quickly becomes as hard as brick and is reasonably resistant to the action of air and water." "Subsequently the term 'laterite' has been used or misused indiscriminately to identify various red tropical soils varying in hardness from indurated crusts to soft red earths" (*1*, p. 9). Surely Buchanan's definition of laterite raises a number of questions to the pavement engineer: How long does the material take to harden "upon exposure to air?" What does "reasonably resistant to the action of air and water" amount to strengthwise? Since, to the engineer, Buchanan's definition is vague, it calls for more elaborate definitions for such materials to be developed.

The American Geological Institute cited in (*2*, p. 5) defines laterite as "a highly weathered, red subsoil or material rich in

[1]Transport Research Laboratory, Nine Mile Ride, Wokingham, UK
[2]TecLab, Uganda, Nalukulongo, Kampala, Uganda
[3]Road Research Centre, Ethiopian Roads Authority, Ras Abebe Aregay Street, Addis Ababa, Ethiopia

Corresponding Author:
Address correspondence to Andrew Otto: aotto@trl.co.uk

secondary oxides of iron, aluminum, or both, nearly void of bases and primary silicates, and maybe containing large amounts of quartz and kaolinite. It develops in a tropical or forested warm to temperate climate, and is a residual or end product of weathering." The geological definition of lateritic soil according to the American Geological Institute cited in (*2*, p. 6) is: "a soil containing laterite; also any reddish tropical soil developed from much weathering."

Morin and Todor (*2*, p. 5) provide the following definition from the Soil Map of the World: "If a soil is named a ferralsol, no other distinguishing term is needed. The term plinthite is also recommended for soils which harden on exposure to repeated wetting and drying. Laterite gravel should be called concretionary gravel. Hardened crusts should be termed ironstone."

Krinitzsky et al. (*1*) define laterite as "varied reddish, highly weathered soils that have concentrated oxides of iron and aluminum and may contain quartz and kaolinite." Krinitzsky et al. (*1*) further go into detail in sub-dividing laterite into various groups, in which plinthite is defined as "a specific form of laterite rock, Buchanian laterite. In situ, it can be soft enough to cut with metal tools but will harden irreversibly when removed from the ground and dried." Of laterite gravel they say: "This material consists of coarse-grained, hard concretionary nodules or pisolites. These nodules are of varying strengths and may be partially cemented into larger masses or may be loose in a matrix of silty and/or clayey, lateritic soil." These definitions offer helpful distinctions between the materials; in particular, the definition of plinthite helps to explain the Buchanian laterite. That is, it hardens when removed from the ground and dried, and it is not that it simply "upon exposure to air quickly becomes as hard as brick." Krinitzsky et al. (*1*) do not clarify whether the irreversible hardening is for one or more drying–soaking cycles.

Charman (*3*, p. 18) subdivides laterites into five classes. The two most commonly used in road construction are: (1) plinthite—"soil fabric containing significant amount of laterite material. Hydrated oxides present at the expense of some soil material. Unhardened, no nodules present, but may be slight evidence of concretionary development"; (2) nodular laterite—"distinct hard concretionary nodules present as separate particles." Equivalent terms in the literature are, "laterite gravel, ironstone gravel, pisolitic gravel, and concretionary gravel." The definitions proposed by Charman (*3*) broadly agree with those proposed by Krinitzsky et al. (*1*). Charman goes a step further by stating the types of laterite most commonly used in road construction. In this paper whenever the term laterite is used it refers to the broad definition as stated by Krinitzsky et al. (*1*), and when the term lateritic gravel is used it is equivalent to laterite gravel as defined by Krinitzsky et al. (*1*), and nodular laterite as defined by Charman (*3*). Lateritic soil is used to mean plinthite as defined by Charman (*3*).

It has also been reported that when soils containing sesquioxides dry, strong bonds develop between the soil particles. The strong bond leads to high strength that is irreversible upon re-wetting. Tropically-weathered materials such as laterite contain hydrated oxides of iron and aluminum. The combination of hydrated oxides is referred to as sesquioxides. It has been reported that the strength of soils that contain sesquioxides, such as laterite, increases due to development of self-cementation that occurs on repeated cycles of wetting and drying. Charman (*3*, p. 39) indicates that "The property of self-hardening under cycles of wetting and drying could be important for the performance of certain lateritic soils."

Regarding the self-hardening concept, Charman (*3*, p. 39) further states: "At present self-hardening is not fully understood and cannot easily be predicted; indeed, there are strong differences of opinion as to whether it happens at all." Charman (*3*, p. 39) further adds that "some evidence now exists to suggest that certain nodular laterites exhibit this property to a sufficient extent that traditional requirements governing the selection of materials may be relaxed." He quotes two studies that show that materials normally considered mechanically unstable and too plastic can be satisfactory. The Brazil approach of using the silica–sesquioxide ratio to differentiate laterites and then relaxing certain requirements is also noted as support that plasticity and grading requirements may be relaxed for laterites as compared with other natural gravels. However, one must be careful when comparing specifications from different countries for differences in test methods for California bearing ratio (CBR) and Atterberg limits may also lead to false conclusions about selecting suitable materials in different parts of the world. Pinard and Netterberg (*4*) point out differences in CBR and plasticity index (PI) tested by different methods.

At this point it is worth noting two things: 1) that there is strong difference in opinion as to whether or not some laterites exhibit the property of self-hardening; and 2) that for the laterites that may exhibit self-hardening, none of the authors gives a time frame over which the self-hardening to a substantial strength takes place and whether the strength gain as a result of self-hardening is substantial. The period at which self-hardening occurs, if any, is not known and the number and frequency of cycles of repeated wetting or soaking and drying is unknown and likely to be variable for laterites that exhibit self-hardening.

Based on this, this study aims at building the body of knowledge on the investigation of the self-hardening concept.

To use naturally occurring road construction materials sustainably, it is important to understand their behavior under varying climatic conditions. This enables preventive measures to be undertaken to minimize rapid deterioration in situ. The importance of this is underlined by Greenstein (*5*) in a study of the failure of two pavements in the semi-arid zone of Peru that had performed well for more than 15 years but suddenly failed due to flash flooding. Rollings and Rollings (*6*) also cite another example of a failure of one of two runways. The first runway was constructed with base course material of plasticity index 4, and had unsoaked CBR and soaked CBR greater than 100% at both 100% and 95% compaction density. The second runway was constructed with base course material of plasticity index 12; at 100% of the maximum compaction density the unsoaked CBR was 100% but the soaked CBR decreased to 62%, and at 95% of the maximum compaction density the unsoaked CBR was 65% but the soaked CBR decreased to 11%. The latter pavement failed with severe rutting and cracking after three years in operation whereas the former runway was performing well when inspected after 7 years operation. Obviously, the strength of the failed pavement was very sensitive to moisture increase.

Luis (*7*) states: "It is possible to admit the use of lateritic materials with PI values up to 18% and expansibility up to 12%. It does not seem, as some authors (Fenzy, 1970) also recognize, that these parameters do not restrict very much the good behavior of lateritic materials. It has been recognized that a study of CBR values based on the variation of water contents is fundamental."

At this point, it is clear that it is important to study the behavior of lateritic gravel and other natural gravels when subject to varying moisture conditions. It is even more important to study this behavior under repeated cycles of wetting or soaking and drying so that if for lateritic material there is strength gain, then this can be exploited. If, however, the material is sensitive to moisture then protective measures can be undertaken to ensure the longevity of the pavement.

Given the reported failures and the fact that good laterites are not always available, how can we make lateritic gravels more sustainable when used on roads? Re-analysis of a study by Evans (*8*), of six trunk roads constructed with lateritic gravel bases in Uganda, shows that the practice was to use gravels that exhibited high soaked CBRs above 80% in any pavement layer without modification. Lateritic gravels that were marginal in quality (in terms of CBR and PI) were stabilized before they were used as base course. Lateritic gravels of low quality were used on gravel roads only. In support of this "conservative" approach, Millard (*9*, p. 16) and Clare (*10*, p. 12) note that lateritic gravels used in bases under thin bituminous surfacings may soften rapidly in wet weather due to the fact that thin bituminous surfacings are not impervious. They recommend consideration of stabilization with lime. The issue of bituminous surfacings not being completely impervious was supported by Patrick and Dias (*11*, p. 25) who found that, whilst sprayed seals are used as waterproofing treatments for pavements, they do not completely halt the ingress of water and remain somewhat permeable. In 1972 a road was constructed with a lime (2–4%) stabilized laterite base from Lichinga to Maniamba. The neat laterite had a soaked CBR of 85%, expansibility (free swell) of 14%, and plasticity index of 10% (*7*). Nevertheless, it was still stabilized for some reason (perhaps because the swell was 14% against a maximum specification of 10%). In 2012 (40 years later), this road was studied (*12*, p. 99), and the laterite base was still performing well with no structural failures. Laboratory tests on base material taken from a trial pit showed a CBR of 62%, plasticity modulus of 271, and plasticity index of 6.6% (*12*). The section was described as being on an embankment and having good drainage. We note here the added advantage of lime stabilization and of having good drainage in order to ensure good performance. Based on studies in Africa, Brazil, India, and Australia, Rodrigues et al. (*13*) have proposed a new classification system for the selection and use of lateritic gravels and soils. They recommend treatment of some sort whenever the lateritic soils or gravels do not meet CBR or shrinkage requirements. This practice is vital in roads carrying heavy traffic and for roads that are prone to soaking.

Figure 1. Nakabugo-Bira borrow-pit.

This study is divided into three sets of experiments: one, to investigate the potential for this lateritic gravel to retain the strength gained upon air-drying; two, to assess the best method of harnessing the strength gained by air-drying; and three, to investigate the potential for strength gain under repeated cycles of soaking and drying.

Materials

The materials testing was conducted in Uganda on lateritic gravel samples obtained from a major deposit that is used in construction and maintenance of roads around the capital city Kampala. The borrow-pit name is known as Nakabugo-Bira (coordinates E442879/N37955 using the UTM system). A photograph of the highly exploited borrow-pit is shown in Figure 1 and that of a sample of the lateritic gravel is shown in Figure 2.

A summary of the properties of the lateritic gravel from Nakabugo-Bira that was used in the study is shown in Table 1. The particle size distribution of the laterite gravel as delivered from site is shown in Table 2. The tests were conducted according to British Standards test method number 1377 Part 4 of 1990 (BS 1377-4:1990) for compaction related tests (including CBR) and British Standard test method number 1377 Part 2 of 1990 (BS 1377-2:1990) for soil index tests. BS

Figure 2. Photograph of a sample of the lateritic gravel from Nakabugo-Bira.

Table 1. Summary of the Properties of the Lateritic Gravel Used in the Study

Properties	Values
MDD (g/cc) at 65 blows BS 1377-4 method of compaction	2.179
OMC (%)at 65 blows BS 1377-4 method of compaction	10
CBR@65 blows BS 1377-4 method of compaction, after 4 days of soaking (%)	60
MDD (g/cc) at 56 blows BS 1377-4 method of compaction	2.175
OMC (%) at 56 blows BS 1377-4 method of compaction	11
CBR@56 blows BS 1377-4 method of compaction, after 4 days of soaking (%)	49
Liquid limit	36
Plastic limit	20
Plasticity index	16
Plasticity modulus after compaction	464
% passing 425 μm sieve	27
% passing 75 μm sieve	23

Note: MDD = maximum dry density; OMC = optimum moisture content; CBR = California bearing ratio; BS = British standard; g/cc = gram per cubic centimeter; μm = micrometer.

1377-4:1990 is very similar to American Association of State Highway and Transportation Officials (AASHTO) test method number T180-D of 1993 (AASHTO T180-D:1993) whereas BS 1377-2:1990 is similar to AASHTO T89 and AASHTO T90. In the text OMC stands for optimum moisture content, and MDD for maximum dry density.

The CBR of the lateritic gravel was measured under various density and moisture conditions. The CBR was also measured after stabilization with 4% lime, curing for 4 days and then soaking for 4 days. A surcharge weight was placed on

Table 2. Particle Size Distribution of the Lateritic Gravel Used in the Study

Sieve size (mm)	% age by mass passing
37.5	100
20	92
10	66
6.3	50
5.00	44
2.00	34
1.18	30
0.60	28
0.425	27
0.075	23

Note: mm = millimeter.

samples subjected to soaking and the swell was measured at the end of the soaking period. The maximum swell recorded on specimens prepared at 56 rammer blows per layer was 0.02%, whereas for samples prepared at 65 rammer blows the maximum recorded swell was 0.01%.

Resistance to Soaking after Air-Drying

Procedure Used to Investigate the Potential of the Lateritic Gravel to Resist Soaking after Hardening upon Air-Drying: "Buchanian Behavior"

For this stage, it was important to use clean plastic sheeting to prepare the specimens to ensure that the finer fractions of the material were not lost during handling. This is because from the literature [Charman (*3*, pp. 28 and 39)], the hardening of lateritic gravels is governed mainly by the finer fractions of the gravel.

Four specimens were prepared for this stage of the study - each compacted at OMC, 65 blows per layer for five layers, were prepared. These were Specimens 1 to 4. An additional two specimens were also prepared by treating with 4% lime, curing for 4 days and then soaking for 4 days. These were Specimens 5 and 6. Specimens 1 and 2 were soaked for 4 days and then their CBR values were determined. These specimens were used for control and comparison purposes. Specimens 3 and 4 were air-dried for 14 days first (before soaking), then after the 14 days, they were soaked for 4 days, and then their CBR values were determined. The moisture content of both specimens after drying for 14 days but before soaking was 4.7%. The moisture content of the specimens after soaking was determined after the CBR test and is shown in Table 3. Each full specimen was oven-dried to determine its moisture content.

Results on Investigation of Resistance to Soaking upon Air-Drying

Comparing the CBR values of the specimens shown in Table 3, we can draw the following conclusions.

Table 3. Results of Investigation of Resistance to Soaking upon Air-Drying

Moisture/soaking conditions	Specimen	CBR of specimen (%)	Average CBR (%)	Moisture content at time of CBR test (%)
4 Days of soaking	Specimen 1	59	61	12.4
	Specimen 2	62		12.4
14 Days drying then 4 days of soaking	Specimen 3	81	81	13.1
	Specimen 4	80		12.2
4% (by mass) lime-stabilized specimens, 4 days curing and then 4 days of soaking	Specimen 5	112	110	NM
	Specimen 6	108		NM

Note: CBR = California bearing ratio; NM = not measured.

Drying this lateritic gravel for a period of 14 days led to strength gain that is retained to an extent upon subsequent soaking for 4 days. This is demonstrated by the fact that the average CBR (81%) of Specimens 3 and 4 (that were dried for 14 days then soaked for 4 days) is higher than the average CBR (61%) of Specimens 1 and 2 (that were soaked for 4 days without prior drying). This lateritic gravel is therefore a "Buchanian laterite." This behavior, manifested in the laboratory, if proven to occur on site for a particular lateritic gravel, has a very important application in pavement construction, for it indicates that construction should take place in the dry season and the pavement should be allowed to dry for 2 weeks or more before a bituminous seal is applied.

Stabilization of this laterite with 4% lime (Specimens 5 and 6), led to an increase in the 4 days soaked CBR at 65 blows, from 60 to 110% after 4 days of curing and 4 days of soaking. The use of lime therefore results in a much higher strength gain compared with allowing the neat compacted gravel to dry for 14 days.

Investigation of a Methodology to Harness the Strength Gain from Air-Drying

Procedure Used to Investigate the Best Method for Harnessing the Strength Gained from Drying

Ten CBR specimens, each compacted at OMC, 65 blows per layer for five layers, were prepared. The prepared specimens were treated in the following ways:

1. For Specimens 7 and 8, the CBR values were determined at OMC immediately after they were prepared.
2. Specimens 9 and 10 were air-dried to 0.75 of OMC; that is after preparation, they were not soaked, but their top surfaces were left exposed for the specimen to dry until the mass of the individual compacted specimens had reduced to an extent such that the mass of moisture retained was 0.75 of OMC. This required checking the mass of the specimens on a daily basis. When this mass was attained, then the top surface of the material in the mold was tightly wrapped with cling film to allow the moisture to equilibrate throughout the specimens. In addition, the whole mold was tightly wrapped in cling film. The specimens remained wrapped for 4 days before being subjected to the CBR test. A spreadsheet was developed to calculate the mass of moisture in each mold. This was important for the subsequent tests since the moisture quantity in the specimens needed to be estimated before the specimens were tested. These specimens together with the specimens in treatment 1 were used for control and comparison purposes.
3. Specimens 11 and 12 were soaked for 2 days, then removed from the soaking basin, and allowed to air-dry. The procedure in treatment 2 above was followed until the specimens had achieved their original masses before soaking (that is the mass at OMC), wrapped in cling film, and then subjected to the CBR test.
4. Specimens 13 and 14 were soaked for 4 days after preparation, after which they were removed from the soaking basin, and allowed to air-dry. The procedure in treatment 2 above was followed until the specimens had achieved their original masses before soaking (that is mass at OMC), wrapped in cling film, and then subjected to the CBR test.

Results of Methodology to Harness Strength Gain from Air-Drying

Comparing the CBR values of the specimens shown in Table 4, we can draw the following conclusions:

1. Soaking this lateritic gravel for 2 days then drying it back to OMC led to significant strength gain. This is demonstrated by Specimens 11 and 12 having an average CBR of 97% compared with Specimen 7 having a CBR of 85% at the same moisture content. In fact the CBR (97%) value obtained after drying Specimens 11 and 12 back to OMC is higher than the CBR (90% for Specimens 9 and 10) obtained by simply drying the specimen from OMC to 0.75 of OMC. This property can be applied at the time of construction by first compacting the gravel layer at OMC to

Table 4. Results of Investigation of the Methodology to Harness Strength Gain

Moisture/soaking conditions	Specimen	CBR of specimen (%)	Average CBR (%)	Moisture content at time of CBR test (%)
OMC	Specimen 7	85	72	9.8
	Specimen 8	59		9.8
0.75*OMC	Specimen 9	91	90	7.4
	Specimen 10	89		7.4
2 Days of soaking, then drying back to OMC	Specimen 11	96	97	9.8
	Specimen 12	98		9.8
4 Days of soaking, then drying back to OMC	Specimen 13	85	85	9.8
	Specimen 14	85		9.8

Note: OMC = optimum moisture content; CBR = California bearing ratio.

the required density, then applying a light water spray after compaction before allowing the layer to dry. This may be repeated the following day but not continued on the third day.

2. However soaking the lateritic gravel excessively then allowing it to dry back did not lead to strength gain. This is demonstrated by the same CBR (85%) values shown by Specimens 13 and 14 (4 days soak then dry back to OMC) compared with Specimen 7 (at OMC).

Each full specimen was oven-dried to determine its moisture content.

Strength Assessment under Repeated Cycles of Wetting and Drying

Procedure for Investigating the Potential for Strength Gain under Repeated Cycles of Soaking and Drying

Two CBR specimens (Specimens 15 and 16) were prepared in accordance with the BS 1377-4:1990 test method previously mentioned but at a compaction level of 65 blows per layer administered in five layers. A further eight specimens (Specimens 17 to 24) were prepared at a compaction level of 56 blows per layer administered in five layers per specimen. The eight specimens were prepared at a lower density (56 blows) to assess if density influences the outcome of the results obtained. A lower density allows more water to penetrate the material and thus subjects it to a greater magnitude of soaking and drying cycles. Four specimens (Specimens 15, 16, 19 20) were treated as shown in Table 3. Specimens 17 and 18 were tested at 4 days soak to determine the standard soaked CBR to be used for comparison purposes. Testing was also conducted to investigate whether starting the procedure by drying instead of soaking would yield significantly different results. Four specimens (Specimens 21, 22, 23, 24) at the same density (56 blows) were used for this purpose. Two of the specimens (Specimens 21 and 22) were subjected to the procedure as in Table 5 (but for 19 days, to allow for 8 days of soaking). For the other two specimens (Specimens 23 and 24), a similar procedure was used but in this case the specimens were first subjected to 2 days of drying before the cycling between soaking and drying began. This also took 19 days (to allow for a total of 8 days of soaking). Each full specimen was oven-dried to determine its moisture content.

Table 5. Schedule for Soaking and Drying Cycles

Day	Activity
1	Prepare specimens and soak
2	Sun-dry or air-dry if day is cloudy
3	Sun-dry or air-dry if day is cloudy
4	Soak
5	Sun-dry or air-dry if day is cloudy
6	Sun-dry or air-dry if day is cloudy
7	Soak
8	Sun-dry or air-dry if day is cloudy
9	Sun-dry or air-dry if day is cloudy
10	Soak
11	Sun-dry or air-dry if day is cloudy
12	Sun-dry or air-dry if day is cloudy
13	Soak
14	Soak
15	Soak
16	Soak
17	CBR testing (penetrate top and bottom and determine moistures in the top and bottom halves of the specimens)

Results of Assessment under Repeated Cycles of Soaking and Drying

Comparing the CBR values of the specimens shown in Table 6, we can draw the following conclusions:

1. Repeated cycles of soaking and drying led to a significant decrease in strength as shown by the CBR (44%) of Specimens 15 and 16 compared with the CBR (61%) of Specimens 1 and 2.

Table 6. Results for Soaking and Drying Cycles

Moisture/soaking conditions	Specimen	CBR of specimen (%)	Average CBR (%)	Moisture content at time of CBR test (%)
4 days of soaking (65 blows)	Specimen 1	59	61	12.4
	Specimen 2	62		12.5
17 days of soaking and drying cycles (65 blows) as per Table 4	Specimen 15	41	44	12.7
	Specimen 16	46		11.9
4 days soaked (56 blows)	Specimen 17	48	49	11.6
	Specimen 18	51		11.3
17 days of soaking and drying cycles (56 blows) as per Table 4	Specimen 19	42	41	13.3
	Specimen 20	41		13.3
19 days of soaking and drying cycles (1 day soaking first then start cycle by 2 days drying)	Specimen 21	33	34	11.0
	Specimen 22	35		12.7
19 days of soaking and drying cycles (2 days drying first then start cycle by 1 day soaking)	Specimen 23	32	32	12.5
	Specimen 24	33		12.5

Note: CBR = California bearing ratio.

2. A reduction in the compaction density did not aggravate the problem of loss in strength due to soaking and drying cycles significantly. This is deduced from a change in comparison between the CBR (41%) of Specimens 19 and 20, compacted at 56 blows, and the CBR (44%) of Specimens 15 and 16, compacted at 65 blows. The moisture content of the lower density specimens (13.3% for Specimens 19 and 20) was not significantly higher than that of the higher density specimens (12.3% for Specimens 15 and 16), all cycled for the same length of time. It must be remembered, however, that there was still a significant reduction in strength of the cycled specimens (CBR 41% for Specimens 19 and 20) compared with the simple 4 days soaked specimens (CBR 49% for Specimens 17 and 18). The moisture content of the cycled specimens (13.3% for Specimens 19 and 20) compared with the simple 4 days soaked specimens (11.5% for Specimens 17 and 18) was also significantly different.
3. Increasing the total duration of the soaking and drying cycles led to further reduction in strength. This is evident from the comparison of the CBR (41%) of Specimens 19 and 20, cycled for 17 days, and the CBR (34%) of Specimens 21 and 22, cycled for 19 days. Moreover, the moisture content of Specimens 21 and 22 was lower than that of Specimens 19 and 20.
4. For the same total duration of the soaking and drying cycles, there was little difference between the strength (CBR 34% for Specimens 21 and 22) of the specimens that were first dried before soaking and the strength (CBR 32% for Specimens 23 and 24) of the specimens that were soaked immediately after preparation. The moisture contents were similar: 11.9 % for Specimens 21 and 22, and 12.5% for Specimens 23 and 24.

The loss in strength as a result of soaking and drying cycles could be due to the soluble sesquioxides being "leached" out of the specimens during repeated soaking. This theory is supported by the fact that leaching, transport and re-deposition/precipitation of the soluble minerals is one of the mechanisms of laterite formation. The loss in strength upon repeated cycles of soaking and drying observed for this lateritic gravel is the opposite of what literature suggests. In fact, this indicates what could happen to lateritic pavements subject to periods of soaking and drying such as in the rainy seasons in the tropics—loss of strength. Practical methods of preventing this in pavement layers and especially embankments should be devised—perhaps through providing adequate crown height and cross-fall. One could argue that the strength loss was as a result of water saturating in the specimens, but the moisture content of the specimens taken after the CBR test is very similar to that of the 4 days soak as seen in Table 5. The reason for the strength reduction is therefore not moisture saturation in the specimens.

Conclusions

Given that properties of laterites are variable from place to place, one must recall that not all of them will behave in the same way. It is therefore important that each source of lateritic gravel be tested rigorously to fully understand the behavior of the material and how best to use it.

1. The study has demonstrated a method to be used in assessing possible ways of making better use of lateritic gravel. This was done through assessing its potential to retain its strength gained by hardening by air-drying, how to enhance this strength gain further, and the effect of repeated cycles of soaking and drying on lateritic gravel that shows this behavior.

2. This particular lateritic gravel is a "Buchanian Laterite"; it hardens irreversibly upon drying. It is possible to take advantage of this behavior, manifested in the laboratory, if proven to occur on site. It indicates that construction should take place in the dry season and the pavement should be allowed to dry for 2 weeks or more before a bituminous seal is applied. For this reason, for lower category roads, this gravel can be used without stabilization.
3. It was found that moderately soaking the lateritic gravel immediately after final compaction at optimum moisture content then allowing it to dry back increased the strength of the gravel significantly. Excessive soaking and then drying back did not lead to any significant strength gain.
4. Repeated cycles of soaking and drying led to significant reduction in strength of this particular lateritic gravel contrary to what literature suggests; this could be due to the soluble sesquioxides being "leached" into the water in the soaking basin during repeated soaking. Even if this lateritic gravel showed resistance to loss of strength gained by air-drying (as described by Buchanan), the strength gain on repeated cycles of soaking and drying as suggested by some authors did not occur—not in the short term and the strength seemed to worsen with a longer test duration. This is important because thin bituminous surfacings are not impervious and therefore in the rainy season in tropical regions, lateritic gravel bases under thin bituminous surfacings will experience cyclic soaking and drying. This will be followed by significant reduction in strength and subsequent failure. It therefore follows that if this lateritic gravel is to be used in road bases in higher category roads, then it should be stabilized with lime.

It therefore follows that since road making materials such as lateritic gravels are getting increasingly scarce, we should explore ways to make them perform as well as possible for as long as possible. Therefore, to use lateritic gravel sources sustainably, methods that enhance their strength and durability should be explored as has been done in this study. This will lead to a reduction in exploitation of the gravel.

References

1. Krinitzsky, E. L., D. M. Patrick, and F. C. Townsend. *Geology and Geotechnical Properties of Laterite Gravel*. U.S. Army Engineer Waterways Experiment Station, Vicksburg, Miss., 1976.
2. Morin, W. J., and P. C. Todor. *Laterite and Lateritic Soils and Other Problem Soils of the Tropics: Volume II Instruction Manual*. Lyon Associates, and DNER Rio De Janeiro, Baltimore, Md., 1975.
3. Charman, J. H. *Laterite in Road Pavements*. CIRIA, London, 1988.
4. Pinard, M. I., and F. Netterberg. Comparison of Test Methods and Implications on Materials Selection for Road Construction. In *2nd AFCAP Practitioners Conference* (AFCAP, ed.), Maputo, Mozambique, 2012, pp. 72–91.
5. Greenstein, J. Using Nondestructive Testing in the Semi-Arid Zone of Peru. *Transport Research Record: Journal of the Transportation Research Board*, 1987. 1137: 59–70.
6. Rollings, R. S., and M. P. Rollings. Pavement Failures: Oversights, Omissions and Wishful Thinking. *Journal of Performance of Constructed Facilities*, Vol. 5, No. 4, 1991, pp. 271–286.
7. Luis, A. S. Latiritic Materials in the North of Mozambique. *9th Regional Conference for Africa on Soil Mechanics and Foundation Engineering*, Lagos, Rotterdam, Balkema, 1987, pp. 31–39.
8. Evans, E. A. *A Laboratory Investigation of Six Lateritic Gravels from Uganda: Research Note No. RN/3241/EAE*. Road Research Laboratory of the Department of Scientific and Industrial Research, London, 1958.
9. Millard, R. S., R. C. Calvert, R. Marwick, H. Grace, and L. W. Hatherly. Particular Problems of Lightly-Trafficked Roads. *PIARC XIITH Congress: Section 1. Construction and Maintenance of Roads and Runways*. PIARC, New Delhi, 1963.
10. Clare, K. E. *Road-Making Gravels in Central Africa*. Road Research Laboratory of the Department of Scientific and Industrial Research, London, 1959.
11. Patrick, S., and M. Dias. *Permeability of Sprayed Seals: Literature Review*. Austroads Ltd., Sydney, Australia, 2016.
12. Rolt, J., K. Mukura, F. Dangare, and A. Otto. *Back Analysis of Previously Constructed Low Volume Rural Roads*. DFID/AFCAP, London, 2013.
13. Rodrigues, J. K., P. Reiffsteck, J. C. Auriol, and M. Ndiaye. Universal Classification for the Use of Lateritic Soils in Low Cost Durable Pavements. *Soils and Rocks*, Vol. 36, No. 3, 2013, pp. 241–249.

The Standing Committee on Geo-Environmental Processes (AFP40) peer-reviewed this paper (18-00305).

Article

TRR
JOURNAL OF THE TRANSPORTATION RESEARCH BOARD

Transportation Research Record
2018, Vol. 2672(52) 86–95
© National Academy of Sciences:
Transportation Research Board 2018
Article reuse guidelines:
sagepub.com/journals-permissions
DOI: 10.1177/0361198118772960
journals.sagepub.com/home/trr
SAGE

Data-Based Real-Time Moisture Modeling in Unsaturated Expansive Subgrade in Texas

Asif Ahmed[1], Sahadat Hossain[1], Mohammad Sadik Khan[2], and Aya Shishani[1]

Abstract
Moisture variations significantly influence the strength and stiffness of expansive subgrade soils, shortening the service lives of pavements and increasing the associated maintenance costs. Accurate measurements of soil moisture can be obtained through soil sampling and testing, but the process can be extensive and costly. Empirical models can accurately predict the moisture variations in an expansive subgrade in a shorter period of time, with lower accompanying costs. The objective of the current study was to develop moisture models, using real-time field monitoring data from two hot mix asphalt roads in North Texas. The collected data were analyzed in a statistical environment to solve two first degree Fourier series. The solution produced a moisture variation model that captured variations associated with seasonal effects and temporary variations associated with rainfall. The outputs of this model were within 90% of the values measured on site. Application of the developed models will facilitate noninvasive estimations of the response of soil strength and stiffness properties to variations in moisture.

Expansive soil is a worldwide problem and constitutes approximately 25% of the soil in the United States (*1, 2*). Volume changes of these soils occur through adsorption or desorption of moisture, leading to cyclic swelling and shrinkage. On average, the nation's infrastructure incurs more financial losses from expansive soils than from earthquakes, floods, hurricanes, and tornadoes combined (*3*). Nelson and Miller (*1*) estimated 9 billion USD of damages annually, and Jones and Jefferson (*4*) calculated a total financial loss of 15 billion USD. The Texas Department of Transportation (TxDOT) spends 25% of its annual outlay on maintenance and repair of damaged pavements (*5*). Addressing the effects of expansive subgrade soils in both the design and construction stages could reduce future maintenance costs and extend the service lives of pavement systems.

The strength and performance of a pavement built on expansive subgrade is most influenced by moisture (*6*). Hedayati reported moisture content variation, which typically induces pavement cracking, as the prime cause of subgrade deformation in expansive soil (*7*). Fluctuations in moisture content are influenced by the pavement subgrade compaction site, environmental site conditions, location of roadside trees, and the presence of nearby drainage ditches (*3*). Moisture variations cause significant changes in the shear strength, resilient modulus, and permeability of subgrade soils (*8, 9*). They also affect hydraulic conductivity, chemical diffusivity, specific heat, and thermal conductivity (*10*). The success or failure of a pavement system is dependent on the support provided by the subgrade layers. An increase in moisture content has been shown to decrease the resilient modulus, which quantifies the support that the subgrade can offer (*11*).

Researchers have developed different subgrade moisture prediction models over the years. Their methods ranged from simple to complex in nature, that is, a single input of soil property to statistical and numerical modeling. Swanberg et al. estimated moisture variation by employing only the plastic limit value (*12*), and some studies categorized subgrade and climatic conditions by subdividing geographic areas into regions per the Thornthwaite's Moisture Index (TMI) (*13, 14*). With the

[1]Department of Civil Engineering, The University of Texas at Arlington, Arlington, TX
[2]Department of Civil and Environmental Engineering, Jackson State University, Jackson, MS

Corresponding Author:
Address correspondence to Asif Ahmed: asif.ahmed@uta.edu

advancement of technology, researchers have utilized complex mathematical relationships between soil properties and environmental conditions to predict moisture variation. Currently, the Mechanistic-Empirical Pavement Design Guide (MEPDG) recommends using the Enhanced Integrated Climatic Model (EICM) to predict moisture changes (*15*).

Other moisture variation estimation methods include soil sampling at regular time intervals (*16*), assuming seasonal variations or constant equilibrium moisture content several years after construction (*15*). Hall et al. estimated the upper and lower equilibrium moisture content of specific sites rather than predicting the real-time variations (*17*). Recent research has incorporated numerical modeling in estimating moisture variations. For example, Abed (*18*) used 2D PLAXFLOW analysis to examine the swelling and shrinkage behavior of expansive soil caused by moisture variations, and Puppala et al. (*19*) performed 3D static analysis in Abaqus to study swelling behavior.

Neither sort of developed method, from soil properties or from computational techniques, can accommodate the temporary moisture variations that are caused by rainfall. Various field studies have reported that following rainfall, increases in moisture content can be up to 20% in amplitude from the baseline moisture content (*20, 21*, and *22*). Hedayati et al. reported that this additional increase in moisture can severely deteriorate the pavement structure, a factor which is overlooked in the current models (*21*).

Therefore, it is highly recommended that moisture prediction models incorporate both seasonal and temporal variations to reflect the real-time scenario. Although moisture content can accurately be measured, soil sampling and the consequent testing required can be lengthy, destructive, and expensive (*23*). The objective of the current study was to develop moisture and temperature models, using real-time field monitoring data from two hot mix asphalt roads in North Texas. After 2 years of monitoring, the collected data were analyzed in a statistical environment to develop moisture prediction models that incorporated both seasonal and sudden increases in moisture associated with climatic factors. The predicted models were compared against the measured values of current study sites, and the models were checked against a previous study to validate their accuracy.

Site Selection

Based on the recommendations of TxDOT engineers, two sites were selected to determine the causes of roadway cracking and to provide possible remedial measures. The first site is situated in Kaufman County, 1.80 miles from the intersection of FM 2757 and I-20. The second site is located on SH 342 in Lancaster, Texas.

Several structural distresses were observed on the roadway during the initial field investigations. The sites had experienced continuous edge and surface cracks, as well as bumps, punch outs, and surface unevenness. Edge cracks up to 3 in. wide and several feet long were observed. In some cases, the depth of the crack was more than 1 ft, which resulted in complete separation of the edge from the pavement structure.

Site Description

The test site in Kaufman County is located on a farm-to-market road identified as FM 2757 in Forney, Texas. The low-volume road consists of two lanes, each measuring 11 ft wide, with no shoulder. The side slopes on both sides are covered with grass and dense trees. Small bodies of water are present to the east and west of the road. Edge cracks of 3 in. in thickness and extending up to 12 in. in depth, as well as other surface pavement distresses, indicate the presence of expansive subgrade soil and rainwater intrusion.

The SH 342 test site is situated in Lancaster, Texas, at the border between Dallas and Ellis counties. The two lanes are each 11 ft wide, and an 11 ft shoulder is present on both sides of the road. The pavement is fairly level with the ground and is flanked by grass and dense trees on both sides. To the west, a rail line runs parallel to the roadway. An edge drop up to several inches in depth was observed in the pavement shoulder. No water bodies were observed nearby.

Instrumentation Plan

Both sites were instrumented with 5TM soil moisture sensors manufactured by Decagon Devices (currently METER), a 100ECRN high-resolution tipping bucket rain gauge, and 85 mm horizontal inclinometer casings to continuously monitor moisture variations, rainfall recordings, and vertical deformations of pavement. In each 5TM sensor, dielectric permittivity is used to measure the moisture content of soil. An electromagnetic field is created to determine the dielectric permittivity of the surrounding material. Based on the manufacturer's calibration, dielectric permittivity is converted to volumetric moisture content which is further converted to gravimetric moisture content relating with soil properties. Data loggers were programmed to take hourly readings of moisture content and rainfall data. The typical sensor locations for both sites are illustrated in Figure 1*a*, and Table 1 gives the notation of sensors for both FM 2757 and SH 342 sites. For example, K 1/2 corresponds to the sensor of Kaufman (FM 2757) site located in borehole 1 and is sensor number 2 in that borehole. The table shows K 1/2 sensor was installed at a depth of 8 ft. Figure 1*b* shows some photos of field installations.

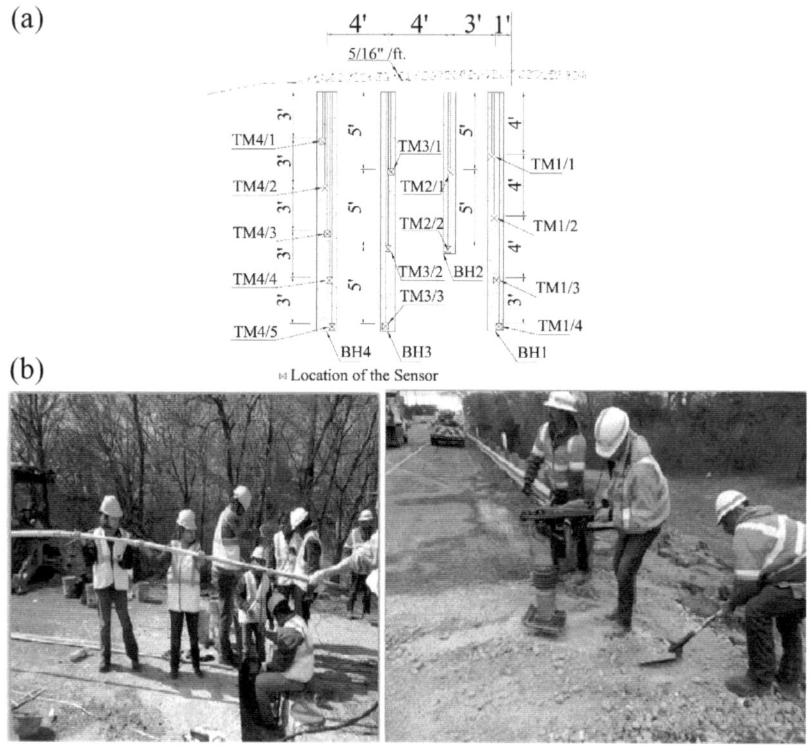

Figure 1. (*a*) Instrumentation plan for two field sites, and (*b*) instrumentation at SH 342.

Table 1. Sensor Notation for Both Sites

Tag for Kaufman Site (FM 2757)	Tag for SH 342 Site	Borehole	Sensor number	Depth (ft)
K 1/1	S 1/1	1	1	4
K 1/2	S 1/2	1	2	8
K 1/3	S 1/3	1	3	12
K 1/4	S 1/4	1	4	15
K 2/1	S 2/1	2	1	5
K 2/2	S 2/2	2	2	10
K 3/1	S 3/1	2	1	5
K 3/2	S 3/2	3	2	10
K 3/3	S 3/3	3	3	15
K 4/1	S 4/1	4	1	3
K 4/2	S 4/2	4	2	6
K 4/3	S 4/3	4	3	9
K 4/4	S 4/4	4	4	12
K 4/5	S 4/5	4	5	15

Note: In addition, there is a sensor at SH 342 at BH-2 at 15 ft depth designated as S 2/3; Number 3 sensor at Borehole 2 at 15 ft depth.

During installation, soil samples were collected from several boreholes to determine the basic soil properties. Atterberg limits, grain size distribution, specific gravity, and unit weight tests were conducted to aid in understanding the soil behavior.

Soil Properties

According to particle size distribution, samples from both sites were composed of over 85% clay, indicating the presence of very fine subgrade soil. The liquid limit of the samples varied between 50% and 64%, and the plasticity index ranged between 28% and 42%. The soil was classified as high plastic clay (CH) according to the Unified Soil Classification System (USCS) and based on the results of sieve analysis and the Atterberg limits. Specific gravity ranged between 2.68 and 2.72, with an average of 2.70. Optimum moisture content was 22%, and dry density at the optimum moisture content was 121.4 lb/ft^3 obtained from standard proctor test.

Moisture Variations in Subgrade

Because expansive clay renders pavement susceptible to edge cracking caused by moisture variations (*6*), moisture sensors were installed in four boreholes at both sites to investigate the effects of moisture variations on expansive subgrade. Only the results from the edge borehole of the Kaufman County site and center borehole of SH 342 are presented below, as the data mirror what was recorded by the remaining sensors.

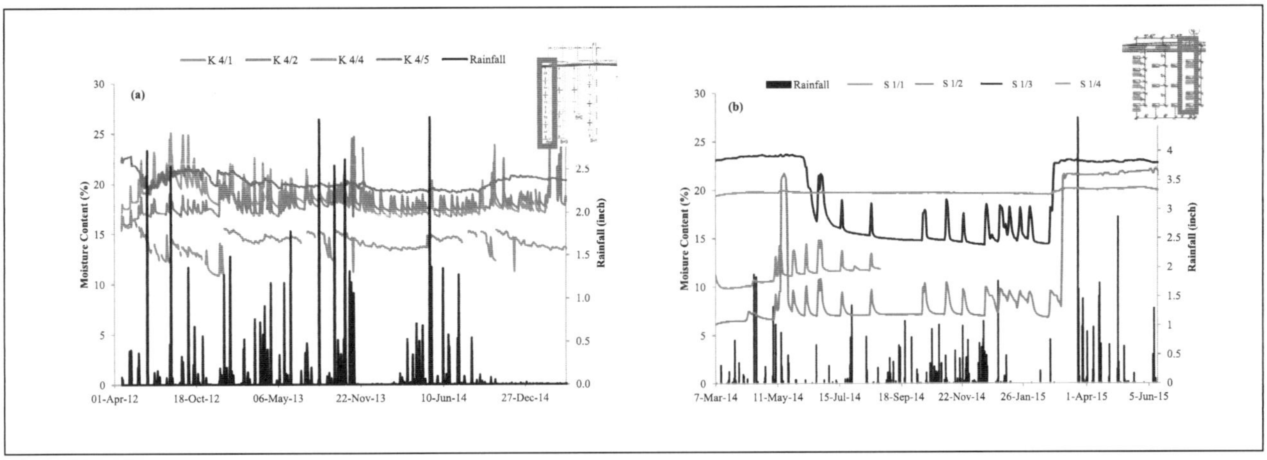

Figure 2. Moisture variation with time at (*a*) Kaufman (FM 2757) and (*b*) State Highway 342.

Figure 2*a* represents data collected from April 2012 to December 2014 at the Kaufman site. Sensors K 4/1 and K 4/2, at respective depths of 3 ft and 6 ft, exhibited both seasonal effects and temporary saturation caused by rainfall. Wetter seasons yielded a rise in the moisture variation curve, which then dipped during drier periods. Additionally, rainfall events registered instantaneous spikes in moisture content ranging from 1% to 8% in amplitude. The peaks suggested that the soil became temporarily saturated before returning to equilibrium. At greater depths, moisture content remained stable. K 4/4 and K 4/5, located at 12 ft and 15 ft depths, maintained respective moisture contents of 16% and 20%.

Data collected at SH 342 from March 2014 to June 2015 produced the curve shown in Figure 2*b*. All sensors maintained stable moisture content that responded only to individual rainfall events. Previous studies confirmed that soil properties determine whether or not a seasonal effect on moisture variation exists at a given location (*24, 25*). Furthermore, the shoulder at SH 342 might have delayed the movement of moisture, deferring the seasonal contribution. Nevertheless, SH 342 experienced temporal variations associated with rainfall. S 1/1, S 1/2, and S 1/3 (located at respective depths of 4 ft, 8 ft, and 12 ft) yielded average moisture contents of 12%, 6%, and 16%, respectively. Moisture content rose to 14% at 8 ft and 8% in amplitude at 12 ft because of temporary saturation induced by continuous rainfall. At 15 ft, S 1/4 registered no appreciable changes in moisture content and remained just below 20%.

Theory of Moisture Modeling

Annual variation of subgrade moisture can be described by the following one-dimensional nonlinear diffusion function (*21, 23*):

$$\frac{\partial \theta}{\partial t} = \frac{\partial}{\partial z}(D(\theta))\frac{\partial \theta}{\partial z}$$

where θ = volumetric moisture content at any time t at depth z, and $D(\theta)$ = Soil moisture diffusivity.

Assuming a constant D, the previous equation can be solved as a first degree Fourier series:

$$\theta(z,t) = \theta_0 + \theta_a \sin\left(\omega t - \frac{z}{d} + C_0\right)$$

where

$\theta\ (z,t)$ = volumetric moisture content at depth z at time t,

θ_0 = average moisture content over time at depth z,

θ_a = domain of moisture variation,

ω = angular frequency (equal to $2\pi/365$),

z = damping depth, and

C_0 = phase correction factor.

The first term of the solution captures the yearly average moisture content at any depth, whereas the second term accounts for the seasonal variations of moisture. Temporal variations associated with rainfall must also be addressed to capture all of the moisture variations in subgrade soil (*21*). Previous studies showed that soil and site conditions affect the seasonal and temporal variations of subgrade moisture (*24, 25*). Some soil (e.g., low plasticity clay in Ohio) experiences a sinusoidal seasonal variation (*24*), whereas other soils (expansive subgrades on high plasticity clay in Texas) tend to maintain equilibrium moisture content, not changing with the season (*22*). Therefore, soil characteristics and specific site conditions are the most important elements in determining the moisture variations. Seasonal and temporal variations of moisture content of the subgrade were statistically analyzed and incorporated in the presented model.

Moisture Data Analysis

The moisture prediction model was developed using the data from sensors with the most interpretable readings over the course of 2 consecutive years (i.e., K 1/2, K 4/2). Four sets of data were used for model development, and one sensor was randomly chosen from both the Kaufman (K 3/2) and SH 342 sites (S 1/3) for model validation.

Seasonal Trend Analysis

As discussed earlier, moisture sensors at the Kaufman County site exhibited seasonal variations, while sensors at the SH 342 site maintained an equilibrium moisture content that displayed instantaneous responses to rainfall. To determine the seasonal trends, moisture peaks associated with rainfall were removed from the Kaufman site. The seasonal output is shown in Figure 3*a*. After several trials, the seasonal trend was found to follow the first degree Fourier series. The variables were found by solving the following equation (*26*):

$$f(t) = a_0 + \sum_{n=1}^{\infty}\left(a_n \cos\frac{2n\pi x}{T} + b_n \sin\frac{2n\pi x}{T}\right)$$

The result of the series followed the form:

$$f(x) = a_0 + a_1{*}\cos(x{*}w) + b_1{*}\sin(x{*}w)$$

where, a_0 is the average value of the dataset, a_1 and b_1 are real numbers independent of the variable x, which accounts for the amplitude of the dataset, and w is the frequency (day^{-1}). As 1 year of data was used to develop the model, and seasonal trends follow annual variations, the frequency was set equal to $2\pi/365$ (0.0172 day^{-1}).

Variations of the values of a_0, a_1, b_1, and w are shown in Figure 3*a*. The moisture content of the sensors remained around 16–17%, which can be attributed to the soil's field capacity (*18*). Modeling the seasonal variation was simplified by incorporating average values of the different parameters comprising the Fourier series. For example, values of a_0 were found to be 18.85, 17.64, 16.22, and 16.42 in the four graphs in Figure 3. The

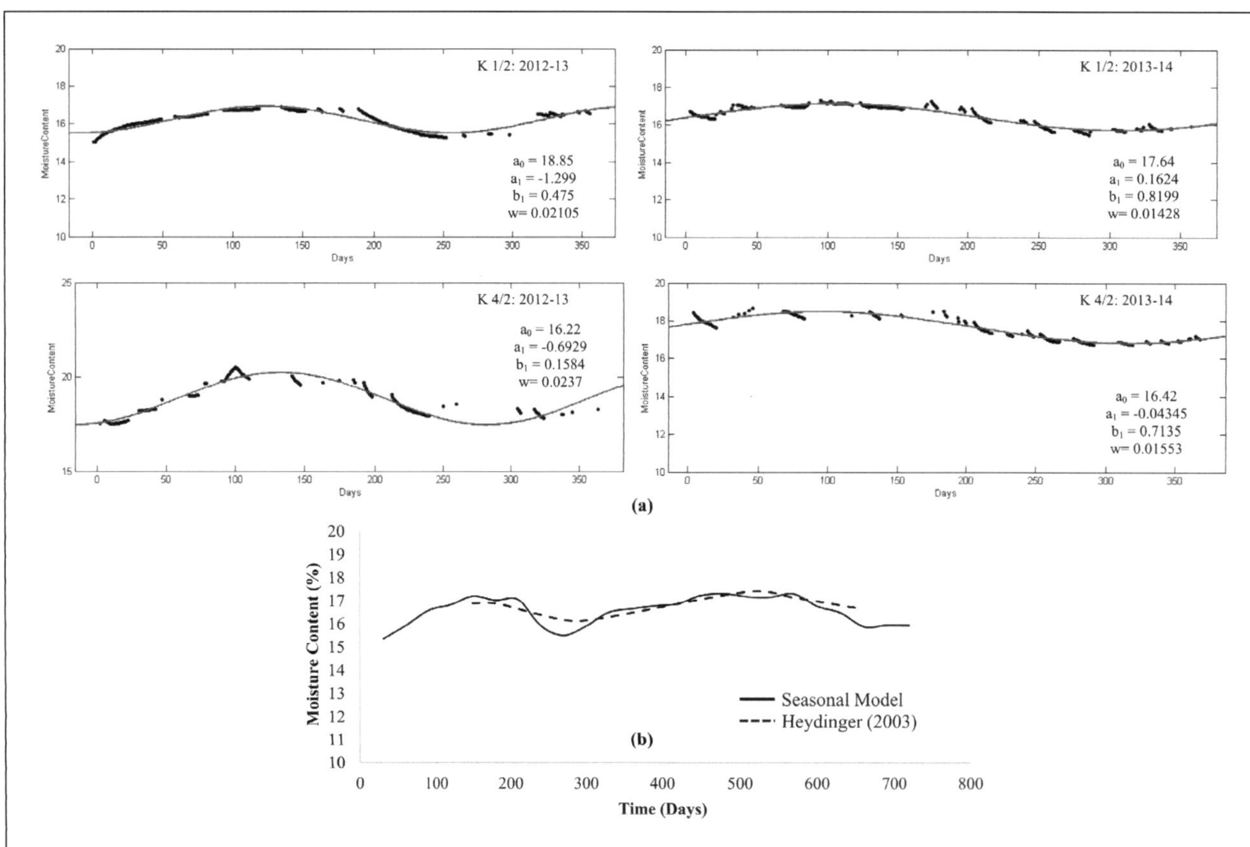

Figure 3. (*a*) Seasonal trend of moisture variation and (*b*) comparison of seasonal model with Heydinger (*24*).

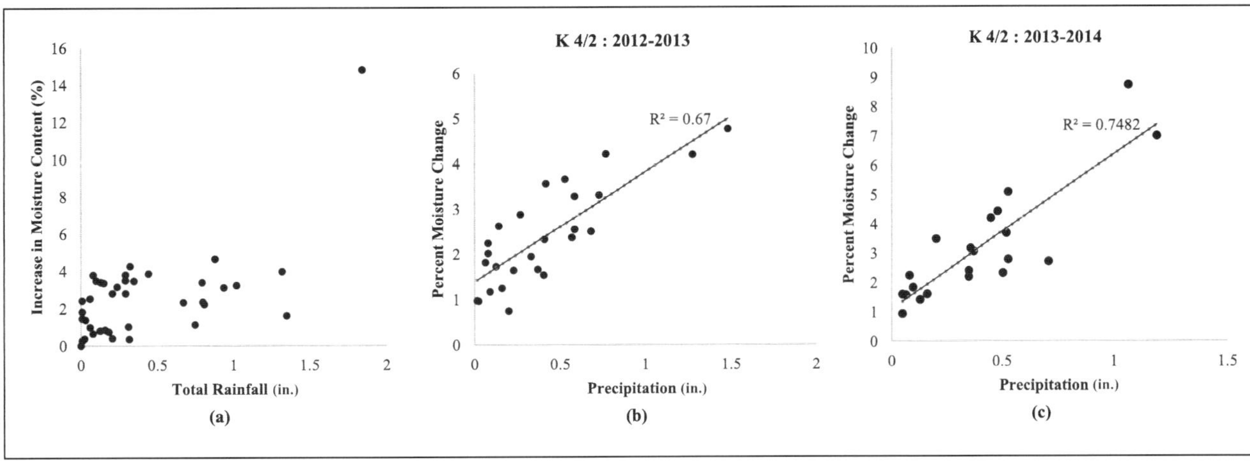

Figure 4. (*a*) Increase of moisture content after rainfall event for a random sensor; Relation between increase in moisture content and precipitation for a particular sensor in (*b*) 2012–2013, and (*c*) 2013–2014.

average of these four values, 17.2825, was selected for the seasonal variation model. Values for a_1, b_1, and ω were similarly obtained by calculating the average. The completed model followed the form:

$$\text{Seasonal M.C.} = 17.2825 - 0.46828 * \cos(x * 0.01864) + 0.5417 * \sin(x * 0.01864)$$

where M.C. = moisture content.
As the sensor began recording data in April 2012, the first day of April was set as day one (i.e., $x = 1$). It followed that March 31 was set as $x = 365$.

After developing the seasonal model, it was compared with a prominent previous study by Heydinger (*24*). It can be observed that the developed and the previous result by Heydinger overlap considerably (Figure 3*b*). The next phase of the study was to find out the effect created by rainfall.

Moisture Fluctuations Associated with Rainfall

Peaks in the moisture data (Figure 2, *a* and *b*) depicted temporary rises caused by precipitation. The increase of moisture content was analyzed separately to determine the relationship between rainfall and an increase in moisture content. Peaks were separated if the change in moisture content displayed a difference of 1% or more from the previously recorded rainfall event. A preliminary inspection of the plot showed that, in spite of an increasing magnitude in precipitation, the resulting moisture content of the soil typically only rose to a specific value that corresponded to the saturation point of the soil (Figure 4*a*). It can be observed from Figure 4*a* that in spite of different rainfall events, the moisture increase remains the same at some points. At this point, all of the medium's voids were filled with water. However, the current status of moisture content in the soil is the most determining factor for the increase of moisture the soil can experience. Thus, the more reliable plot was obtained by including only those points that exhibited an increase in moisture content caused by rainfall from the equilibrium moisture content. It was concluded that the linear plot best described the trend of moisture increase caused by rainfall, as shown in Figure 4, *b* and *c*, for a selected sensor for consecutive 2 years. Similar results were reported by previous researchers (*21, 27*). Parameters (i.e., slope and intersection) of the four sensors were averaged to find the increase in moisture content caused by rainfall. Average values of 1.39 for the intersection and 2.2085 for the slope were used in the final moisture model.

Final Moisture Model

Based on the overall analysis (i.e., seasonal trends and temporary increases associated with rainfall), moisture content at different depths could be explained as

$$\begin{aligned} \text{M.C.} &= [\text{Seasonal Variation}] + \{\text{Variation due to rainfall}\} \\ &= a_0 + a_1 * \cos(x * w) + b_1 * \sin(x * w) + f(\text{rainfall}) \\ &= [17.2825 - 0.46828 * \cos(x * 0.01864) \\ &+ 0.5417 * \sin(x * 0.01864)] \\ &+ \{(1.39 + 2.2085 * \text{Rainfall})\} \end{aligned}$$

where x = days (April 01 as day 1), and rainfall is in units of inches.

The average moisture content was generally observed to remain at the soil's field capacity of 17.5% (*21*). Monitoring of moisture variations in plastic clay in Delaware County, Ohio for a period of 5 years (1996–2001) exhibited an average moisture content of around

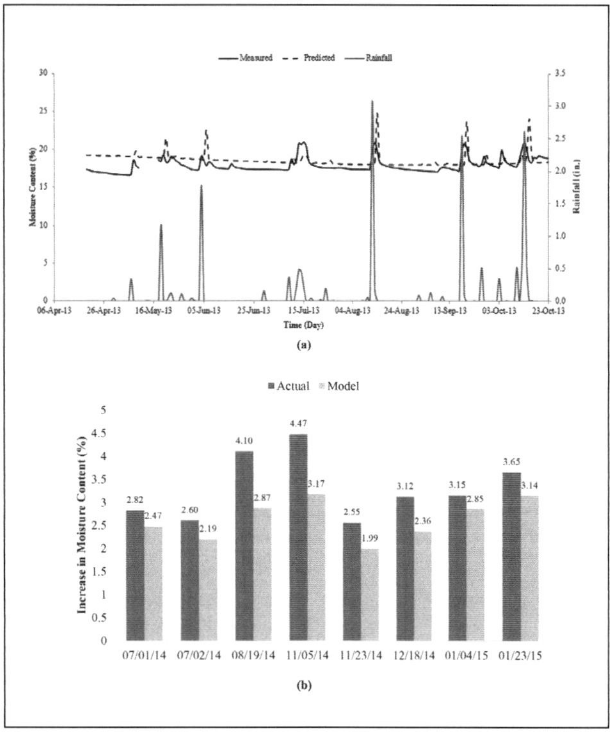

Figure 5. Validation of moisture model at (*a*) Kaufman Site and (*b*) SH 342 site.

15–17% (*24*). Similar field monitoring conducted in high plastic clay in Houston and Fort Worth, Texas for 2 years (2007–2008) showed average moisture content of 15–18% (*25*). Therefore, the applicability of the model is limited to specific soil conditions. In addition, it also considers the specific region in which the study sites are located.

Validation of the Developed Moisture Model

Sensor K 3/2 was randomly selected and tested for a period of 6 months (from April 2013 to October 2013) to determine the accuracy of the developed model. A sensor from the SH 342 site (S 1/3) was selected for model validation.

Figure 5*a* illustrates that the developed model captured both the seasonal and temporal variations associated with rainfall. A comparison of the predicted and measured values in Figure 6*a* showed that more than 90% of the data fell in the 90% confidence bands. In case of SH 342, only the increase in moisture associated with rainfall was compared against the actual increase in moisture content. A comparison of the modeled and actual increases in moisture content after randomly selected rainfall events is presented in Figure 5*b*. The model's results were compared with results obtained from previous studies (Figure 6*b*) after converting gravimetric values to volumetric moisture contents (*23*). Including the temporal variation significantly improved the model's predictions, as is depicted in Figure 6*b*.

Limitations of the Models and Further Development

The models have limitations that need to be addressed, which are that the study was based on the assumption of a homogeneous soil layer, and evaporation was not considered. To further increase the acceptance of the model, it was checked against some sensor results installed in the grassy side slope, as shown in Figure 8*a*. The rationale of selecting the grassy side was to include the effect of evaporation, as the sensors installed beneath the pavement did not include evaporation. It was observed that the sensors exhibited both seasonal and temporal variations similar to the sensors installed beneath pavement, but the average moisture content data was different. Therefore, the first version of the developed moisture model was replaced by the average moisture content of the selected sensor, and the rest of it remained the same. As such, the modified moisture model is

$$\begin{aligned} MC = {} & [X - 0.46828 * \cos(x * 0.01864) \\ & + 0.5417 * \sin(x * 0.01864)] \\ & + \{(1.39 + 2.2085 * \text{Rainfall})\} \end{aligned}$$

where X = average moisture content of the location.

Using the developed moisture model requires manually measuring the average moisture content of the selected site.

The modified moisture model was tested against the sensors installed in the grassy side of the slope. Figure 7 depicts the measured and modeled moisture content, and it is clear that the model was the most accurate at recording the changes in moisture associated with rainfall. The measured and modeled data were further plotted against a 45 degree line because that is where most of the data were clustered (Figure 8*b*), and hypothesis testing was conducted to verify the modified model.

An independent two-sample *t*-test with unequal variance was performed to determine whether there were significant differences between the actual and predicted moisture increase values. The rationale behind the independent test was the non-dependency of the two set values. In the *t*-test, the mean of the predicted values was compared with the mean of the observed values. The risk level was taken as 0.1, meaning that the test was conducted at 90% confidence level. The two-tailed test directed the significance level to 0.05 from 0.1. The basic hypothesis of the two-sample t-test, containing

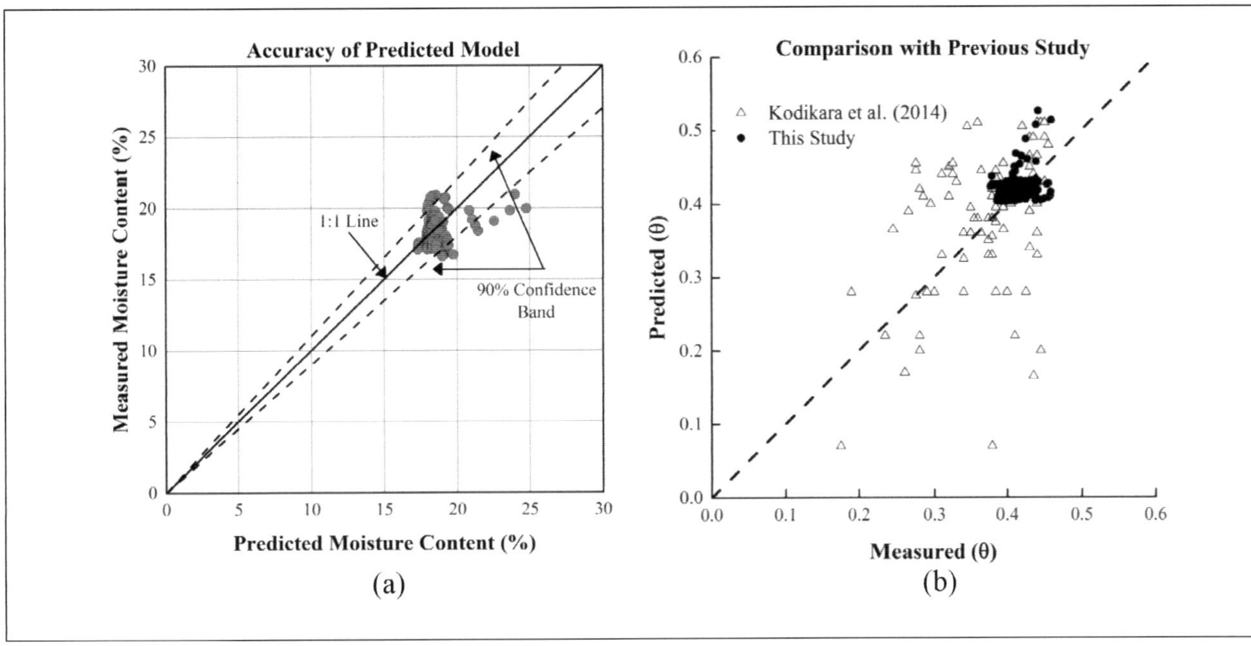

Figure 6. (*a*) Accuracy of the predicted model, and (*b*) comparison with previous study (*23*).

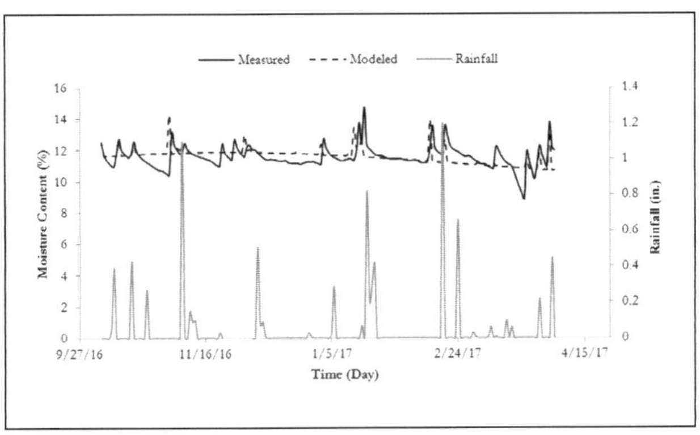

Figure 7. Comparison of modeled and measured moisture content in grassy side slope.

null (H_0) and alternative hypothesis (H_a), can be described as

$$H_0 : m_1 - m_2 = 0$$
$$H_a : m_1 - m_2 \neq 0$$

where $m_1 =$ mean of the actual moisture, and $m_2 =$ mean of the predicted moisture.

The test summary is provided in Table 2. The *t*-value found from the analysis was lower than the critical value of the two-tailed *t*-test. The *p*-value was higher than the significance level ($\alpha/2$). Based on the analysis, it was concluded that the null hypothesis could not be rejected,

Table 2. Summary of Two-Tailed *t*-Test

	Mean	SD	SE Mean	*t*-value	*p*-value
Actual	11.559	0.7	0.052	1.22	0.223
Predicted	11.639	0.523	0.039		

Note: SD = standard deviation; SE Mean = standard error of the mean.

indicating there was no significant difference between the means of the actual and predicted values. Thus, the developed prediction model is justified.

Conclusion

Results obtained from the two instrumented pavement sites were used to develop real-time moisture models. The summary of the findings is presented below:

- Moisture data showed that equilibrium moisture content varied from 6% to 20% at different depths of subgrade.
- Temporal increases of moisture caused by rainfall ranged between 1% and 15% in amplitude. Shallower sensors recorded higher variation while deeper sensors experienced less.
- Instantaneous spikes in moisture content were indicative of temporary saturation induced by rainfall. As the free water moved downwards, moisture content readings returned to equilibrium.

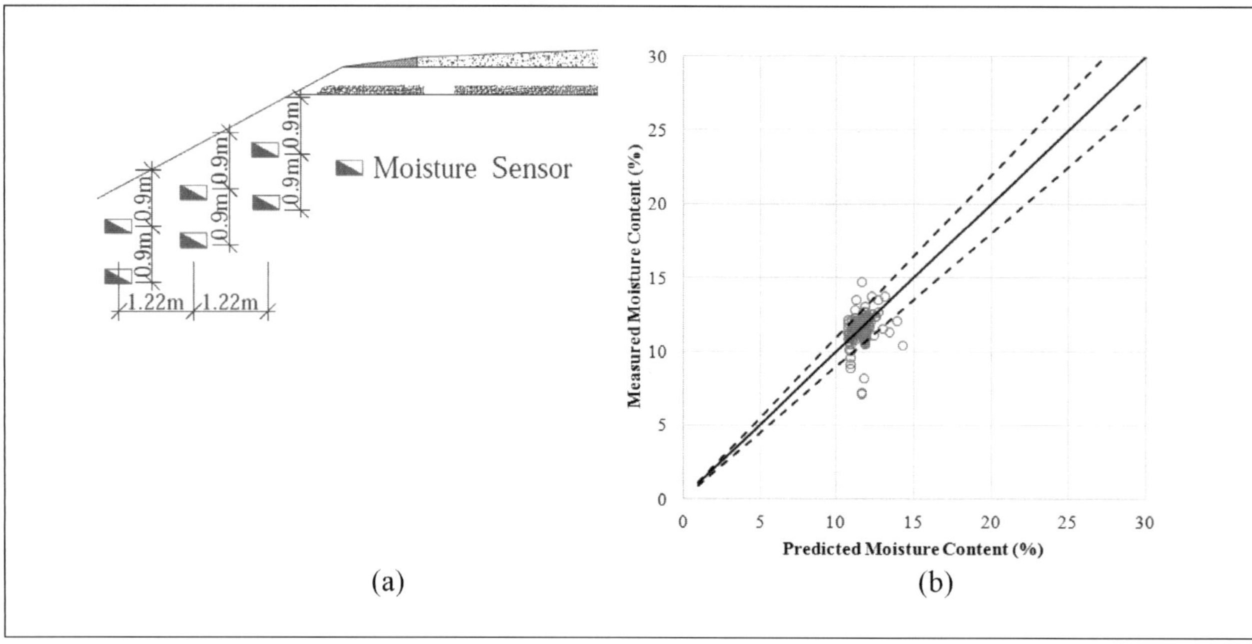

Figure 8. (*a*) Location of sensors at grassy side slope, and (*b*) predicted versus measured moisture content in grassy side slope.

- Real-time moisture model was developed incorporating both seasonal and temporal variation of moisture by solving a first degree Fourier series.
- Validation of the model indicated that the outputs of the model were within 90% of the in situ measured values. In addition, comparison with previous studies indicated that the model can capture both seasonal and temporal variation.
- The moisture variation model was studied in a homogenous single layer subgrade and specific type of soil, that is, high plastic clay. Moisture variations in different types of soils with different stratigraphy are considered outside the scope of the current case study, limiting the applicability of the model.
- It is strongly recommended that the developed models be further modified through additional analysis of different types of soil and various geometric configurations of roads.

References

1. Nelson, J., and J. D. Miller. *Expansive Soils: Problems and Practice in Foundation and Pavement Engineering*. John Wiley & Sons, Inc., New York, 1992.
2. Buhler, R. L., and A. B. Cerato. Stabilization of Oklahoma Expansive Soils Using Lime and Class C Fly Ash. *Proc., Geotechnical Special Publication No. 162 – Problematic Soils and Rocks and In Situ Characterization*, ASCE, Denver, CO, 2007, pp. 1–10.
3. Puppala, A. J., M. Thammanoon, S. Nazarian, and L. R. Hoyos. Threshold Moisture Content and Matric Suction Potentials in Expansive Clays Prior to Initiation of Cracking in Pavements. *Canadian Geotechnical Journal*, Vol. 48, No. 4, 2011, pp. 519–531.
4. Jones, L. D., and I. Jefferson. Expansive Soils. In *ICE Manual of Geotechnical Engineering, Volume 1: Geotechnical Engineering Principles, Problematic Soils and Site Investigation*, ICE Publishing, London, 2012, pp. 413–441. Edited by Burland John, Chapman Tim, Skinner Hilary and Brown Michael.
5. Sebesta, S. *Investigation of Maintenance Base Repairs over Expansive Soils: Year 1 Report*. Publication FHWA/TX-03/0-4395-1. FHWA, Texas Department of Transportation, 2002.
6. Chandra, D., R. L. Lytton, and W. Yang. *Effects of Temperature and Moisture on Low-Volume Roads*. Publication FHWA/TX-89/473-2. FHWA, Texas Department of Transportation, 1988.
7. Hedayati, M. *Rainfall Induced Moisture Distress in Low Volume Pavements*. PhD dissertation. University of Texas at Arlington, 2014.
8. Fredlund, D. G., J. Rahardjo, and M. D. Fredlund. *Unsaturated Soil Mechanics in Engineering Practice*. John Wiley & Sons, Inc., New York, 2012.
9. Khoury, N., R. Brooks, S. Y. Boeni, and D. Yada. Variation of Resilient Modulus, Strength, and Modulus of Elasticity of Stabilized Soils with Postcompaction Moisture Contents. *Journal of Materials in Civil Engineering*, Vol. 25, No. 2, 2012, pp. 160–166.
10. Lu, Y. *Temperature Effect on Unsaturated Hydraulic Properties of Two Fine-Grained Soils and Its Influence on Moisture Movement under an Airfield Test Facility*. MS thesis. Arizona State University, Tempe, 2015.

11. Mehrotra, A. *Evaluating the Influence of Moisture Variation on Resilient Modulus for Unsaturated Pavement Subgrades*. MS thesis. Louisiana State University, Baton Rouge, 2011.
12. Swanberg, J. H. Temperature Variations in a Concrete Pavement and the Underlying Subgrade. *Highway Research Board Proceedings*, Vol. 25, 1946, pp. 169–180.
13. Russam, K. The Prediction of Subgrade Moisture Conditions for Design Purposes. In *Moisture Equilibria and Moisture Changes in Soils Beneath Covered Areas*. Butterworth, Sydney, Australia, 1965, pp. 233–236.
14. Thornthwaite, C. W. An Approach toward a Rational Classification of Climate. *Geographical Review*, Vol. 38, No. 1, 1948, pp. 55–94.
15. Zapata, C. E., and W. N. Houston. Calibration and Validation of Enhanced Integration Climatic Model for Pavement Design. *Transportation Research Record: Journal of the Transportation Research Board*, 2008. 602: 70.
16. Zhang, X., and J. L. Briaud. Coupled Water Content Method for Shrink and Swell Predictions. *International Journal of Pavement Engineering*, Vol. 11, No. 1, 2010, pp. 13–23.
17. Hall, K., and R. Shreenath. Predicting Subgrade Moisture Content for Low-Volume Pavement Design Using In Situ Moisture Content Data. *Transportation Research Record: Journal of the Transportation Research Board*, 1999. 1652: 98–106.
18. Abed, A. Numerical Simulation of a Trial Wall on Expansive Soil in Sudan. *Plaxis Bulletin*, Vol. 21, 2007, pp. 14–18.
19. Puppala, A. J., T. Manosuthikij, and B. Chittoori. Swell and Shrinkage Strain Prediction Models for Expansive Clay. *Engineering Geology*, Vol. 168, 2014, pp. 1–8.
20. Nguyen, Q., D. G. Fredlund, L. Samarasekera, and B. L. Marjerison. Seasonal Pattern of Matric Suctions in Highway Subgrades. *Canadian Geotechnical Journal*, Vol. 47, No. 3, 2010, pp. 267–280.
21. Hedayati, M, and S. Hossain. Data Based Model to Estimate Subgrade Moisture Variation Case Study: Low Volume Pavement in North Texas. *Transportation Geotechnics*, Vol. 3, 2015, pp. 48–57.
22. Hossain, S., A. Ahmed, M. S. Khan, A. Aramoon, and B. Thian. Expansive Subgrade Behavior on a State Highway in North Texas. *Proc., Geotechnical and Structural Engineering Congress*, ASCE, Phoenix, AZ, 2016, pp. 1186–1197.
23. Kodikara, J., P. Rajeev, D. Chan, and C. Gallage. Soil Moisture Monitoring at the Field Scale Using Neutron Probe. *Canadian Geotechnical Journal*, Vol. 51, No. 3, 2014, pp. 332–345.
24. Heydinger, A. G. Evaluation of Seasonal Effects on Subgrade Soils. *Transportation Research Record: Journal of the Transportation Research Board*, 2003. 1821: 47–55.
25. Manosuthikij, T. *Studies on Volume Change Movements in High PI Clays for Better Design of Low Volume Pavements*. PhD dissertation. University of Texas at Arlington, 2008.
26. Sastry, S. S. *Introductory Methods of Numerical Analysis*. PHI Learning Private Limited, New Delhi, India, 2012.
27. Xu, Q, S. Liu, X. Wan, C. Jiang, X. Song, and K. Wang. Effects of Rainfall on Soil Moisture and Water Movement in a Subalpine Dark Coniferous Forest in Southwestern China. *Hydrological Processes*, Vol. 26, No. 25, 2012, pp. 3800–3809.

The Standing Committee on Engineering Behavior of Unsaturated Geomaterials (AFP60) peer-reviewed this paper (18-04285).

Article

TRR
JOURNAL OF THE TRANSPORTATION RESEARCH BOARD

Transportation Research Record
2018, Vol. 2672(52) 96–107

DOI: 10.1177/0361198118797484
journals.sagepub.com/home/trr
SAGE

The Effect of Bottom Ash on Soil Suction and Resilient Modulus of Medium-Plasticity Clay

Arian Asefzadeh[1], Leila Hashemian[2], and Alireza Bayat[2]

Abstract

The effect of adding bottom ash to medium-plasticity clay as a soil stabilizer was evaluated in this study by performing triaxial resilient modulus tests. Two log–log resilient modulus prediction equations found in the literature (MEPDG and NCHRP 1-28) were selected and calibrated for mixtures of bottom ash and clay at different moisture contents. It was found that, with 25% bottom ash in the mixture, the resilient modulus increased between 5% and 23% under different stress states. The soil total and matric suctions of the mixtures were indirectly measured using the filter paper method, and the total suction parameter was incorporated in the two log–log prediction models. The nonlinear regression analysis revealed that the average goodness-of-fit statistics for one of the modified models showed highly satisfactory performance in predicting the resilient modulus values.

The subgrade acts as the foundation of the pavement structure and plays a crucial role in supporting the upper layers of pavement and in transferring the repeated traffic-induced loadings onto the ground. The elastic stiffness of subgrade soils to cyclic loading is defined as the resilient modulus (M_r), which is numerically the ratio of the deviatoric stress to resilient or recoverable strains of the material under a constant confining pressure in repeated load triaxial tests. M_r is an important parameter for calculating resilient stresses, strains, and deformations in flexible pavements. The subgrade stress–strain behavior is reflected in pavement surface deflections and rutting failures (*1*, *2*). Therefore, having knowledge of M_r for subgrade materials is essential for a successful pavement design, which satisfies the performance expectations and provides ease of ride for commuters.

Clay is a fine-grained weak soil, which generally has unsuitable engineering properties, including low values of M_r. These properties become worse during wetting because of the ingress of rainfall through the subgrade layer or rising of the groundwater table (*3*). Studies have showed that the M_r of fine-grained soils drastically decreases as the moisture content increases, which itself leads to a decrease in soil matric suction (*4–6*). One of the other environmental parameters that adversely affects the M_r of clayey subgrade soils is the freeze–thaw cycle in cold regions. Previous studies found that the M_r of cohesive soils decreases as the freeze–thaw cycle increases (*7*, *8*). As such, using weak soils as a subgrade layer can potentially lead to immature loss of life and unsatisfactory performance of the pavement structure. In road projects where excavating and replacing the local weak soil with stronger material from other borrow pits is not feasible, amelioration of the soil through different additives is considered to stabilize the soil, improve its strength and stiffness, and provide a reliable foundation for construction projects, such as road embankments (*9*, *10*). Several researchers have studied the feasibility of using different materials, such as lime, cement, husk ash, and fly ash as soil stabilizers in clayey soils (*3*, *11–15*). Bottom ash has also been used in clayey soil stabilization in other studies, and yielded an improvement in the stiffness and elasticity of the clay (*15*, *16*). It is worth noting that bottom ash contains high amounts of silica–aluminous compounds, and can enhance pozzolanic reactions. Therefore, in other studies, clayey soil was stabilized with a mixture of bottom ash and Portland cement (*17*, *18*). However, mixing clayey soil with pure

[1]Department of Civil & Environmental Engineering, University of Alberta, Edmonton, Alberta, Canada
[2]Donadeo Innovation Center for Engineering, University of Alberta, Edmonton, Alberta, Canada

Corresponding Author:
Address correspondence to Alireza Bayat: abayat@ualberta.ca

bottom ash would result merely in mechanical stabilization of the soil, and there is no chemical reaction involved.

Bottom ash is the by-product of coal-fired power generation plants. Generally, around 4–30% of coal burned in a furnace turns into ash, and approximately 15–20% of the total generated ash collects in the form of bottom ash at the bottom of the furnace (*19*, *20*). According to the Alberta Utilities Commission, 55% of the electricity generated in the province of Alberta, Canada in 2014 stemmed from 33.8 million tons of coal consumption (*21*). Considering the ratios given above, it can be estimated that roughly one million tons of bottom ash was generated in 2014 in Alberta. Utilization of bottom ash in construction projects, such as subgrade stabilization in road embankments, can be deemed a sustainable alternative to landfilling the waste material in repositories, which can potentially create pollution risks and exothermic hazards to the environment adjacent to the landfill site (*22*, *23*).

The goal of this study is to evaluate the applicability of using bottom ash to improve the M_r of a medium-plasticity clay for use as a subgrade material through performing repeated load triaxial tests and obtaining the optimum ratio of soil and bottom ash. The total and matric suctions of soil mixtures at different moisture contents within the dry and wet sides of the optimum moisture content were also measured to integrate them into the M_r prediction equations.

Materials and Experimental Procedure

In this study, a medium-plasticity clay (with a plasticity index of 18) was used, which is found in great amounts in Edmonton, Alberta, Canada. The bottom ash was sourced from a local coal-fired power plant in Alberta. The gradation of the clay and bottom ash materials is represented in Figure 1. The Atterberg limits of the mixtures are tabulated in Table 1. Initially, three different nominal bottom ash:clay mixture ratios of 15:85, 25:75, and 35:65 were selected based on the total dry weight of the mixture.

Compaction Test

To obtain the maximum dry density and the optimum moisture content of the untreated clay and the three mixtures, modified compaction tests as per ASTM D1557 were conducted (*24*). The dry materials were first blended well by hand to achieve uniform mixing. After adding and mixing distilled water with the dry material, the mixture was pulverized through sieve No. 4 to minimize clumpy particles and ensure uniform distribution of moisture. Each mixture was then kept in a sealed plastic bag for at least 2 h before compaction. Figure 2 depicts the compaction test results for clay, bottom ash, and the three mixtures. The curves in Figure 2 are the regression lines. It can be observed that with 15% bottom ash added to the clay, the optimum moisture content of the mixture decreased to 13.5%, compared to 15.8% for the untreated clay. For each increment of bottom ash in the next two mixtures, the optimum moisture contents increased by 0.5%, reaching 14% and 14.5%, respectively. The bottom ash material did not have a conventional compaction curve, and the optimum moisture content was found to be 21%.

Specimen Preparation

To study the effect of moisture variation on M_r, three different moisture contents were considered: optimum − 2%, optimum, and optimum + 2%. The specimens for clay and each clay–bottom ash mixture were then mixed with a respective amount of water, pulverized through sieve No. 4, and placed and sealed in plastic bags for at least 2 h before compacting. A brass split mold with a diameter of 100 mm and a height of 200 mm was used. To negate the effect of varying dry densities, all the specimens were made at the maximum dry density, corresponding to each mixture's compaction curve. To this end, the exact amount of moist material was calculated based on the compaction data, and the compactive effort was modified for five layers through a trial procedure to achieve maximum dry density at three predetermined moisture contents. The top of each compacted layer was scarified with a lab claw to ensure sufficient interlocking action between layers. To stay on the conservative side, the specimens were then sealed in plastic bags and placed in a secured enclosure for one day before conducting the M_r tests.

Resilient Modulus Test

The repeated load triaxial tests for measuring the M_r values were conducted using an IPC Universal Testing Machine (UTM-100) in accordance with the AASHTO T307-99 standard, which simulates the field stress condition by applying five different deviatoric stresses to the specimen at three different confining pressures (*25*). The confining pressures used in the test were in the order of 13.8, 27.6, and 41.4 kPa, and deviatoric stresses applied on the specimens were 13.8, 27.6, 41.4, 55.2, and 69 kPa. These confining pressures and deviator stresses made overall 15 stress states. Table 2 depicts the 15 test sequences and their corresponding bulk and deviatoric stresses. A haversine-shaped load was implemented to apply cyclic deviatoric stresses on specimens with 0.1 s of load duration, and a rest period of 0.9 s. Two external

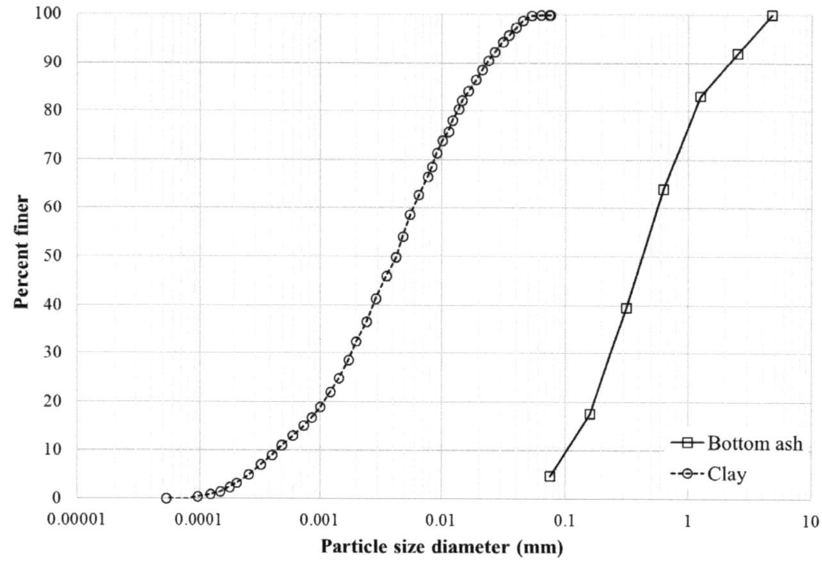

Figure 1. Particle size distribution of clay and bottom ash materials.

Table 1. Atterberg Limits of Different Mixtures (%)

Mixture	Liquid limit	Plastic limit	Plasticity index
Clay	43	25	18
15% bottom ash, 85% clay	45	27	18
25% bottom ash, 75% clay	42	26	16
35% bottom ash, 65% clay	38	30	8

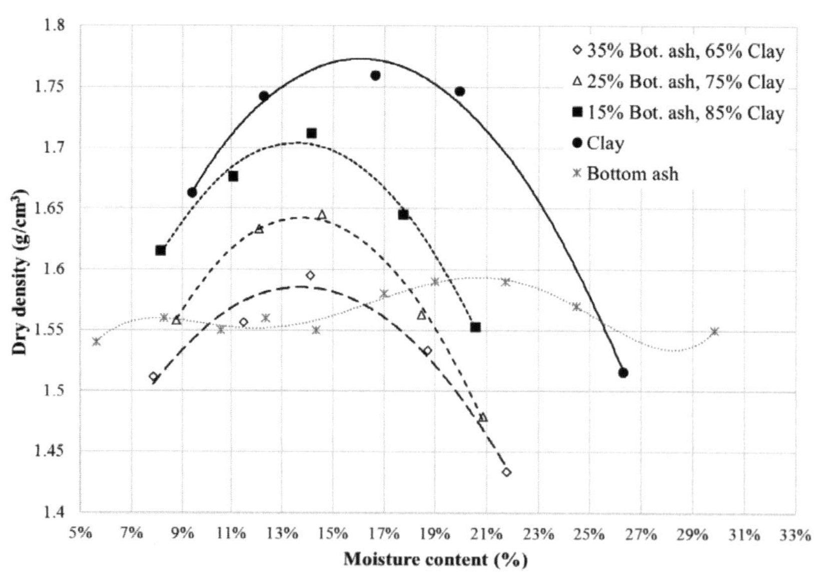

Figure 2. Modified compaction tests results.

Table 2. Resilient Modulus Test Sequences

Test sequence	Confining pressure (kPa)	Deviatoric stress (kPa)	Bulk stress (θ) (kPa)	Octahedral stress (τ) (kPa)
1	41.4	13.8	138.0	6.5
2	41.4	27.6	151.8	13.0
3	41.4	41.4	165.6	19.5
4	41.4	55.2	179.4	26.0
5	41.4	69.0	193.2	32.5
6	27.6	13.8	96.6	6.50
7	27.6	27.6	110.4	13.0
8	27.6	41.4	124.2	19.5
9	27.6	55.2	138.0	26.0
10	27.6	69.0	151.8	32.5
11	13.8	13.8	55.20	6.50
12	13.8	27.6	69.00	13.0
13	13.8	41.4	82.80	19.5
14	13.8	55.2	96.60	26.0
15	13.8	69.0	110.4	32.5

linear variable displacement transducers, with stroke lengths of 5.08 mm, were mounted on the pedestal attached to the axial loading shaft of the triaxial cell to measure the vertical resilient deformation of the specimens. Before beginning the test, each specimen was put under 500 conditioning cycles at a deviatoric stress of 13.8 kPa, and a confining pressure of 41.4 kPa, to ascertain the settling of the specimen and provide sufficient contact between the top of the specimen and the loading plate. M_r was calculated as the average M_r of the last five cycles at each stress state. Two replicates were used for each mixture under the condition that the M_r values between the two replicates had no more than a 5% difference. Otherwise, a third specimen was made and Grubbs' statistical method, with a 95% confidence level (*26*), was implemented to ensure that no specimen was an outlier, which means that a measured M_r of a specimen could potentially have a significant deviation from the average value of all three replicates. As per Grubbs' test, three outliers in the untreated clay mixture at optimum moisture content, and mixtures with 15% and 25% bottom ash contents with a moisture content of optimum − 2% were rejected. Therefore, 27 specimens were made in total.

Matric Suction Measurement

Total suction in soil is defined as the free potential energy of soil water that can be measured as the vapor pressure of the soil water, or the relative humidity of the soil (*27, 28*). Total suction is composed of two components: the matric suction and the osmotic suction. Matric suction refers to the capillary phenomenon stemming from liquid surface tension generated in the soil water media. It is defined as the difference between the pore gas pressure (surrounding gas pressure in the voids, which are partially filled with water) and pore water pressure. Osmotic suction refers to a decrease in relative humidity of soil because of the presence of dissolved salts in the soil water (*27, 28*).

The filter paper test is one of the most commonly referred to testing methods in the literature. It is capable of accurately measuring the soil matric and total suctions for engineering practices. If the filter paper's moisture reaches an equilibrium with the vapor flow, it will indirectly measure the total suction of the soil. Matric suction can be measured when the filter paper's absorbed moisture is brought to equilibrium through liquid flow and direct contact with the soil (*29*). In this study, the indirect measurement of the suctions of the soil mixtures was conducted in accordance with the ASTM D5298 standard. Soil samples were compacted in a 75 mm split mold to achieve the maximum dry density at respective moisture contents. For each test, two half-samples, with approximate thicknesses of 25 mm, were prepared. The tops of the samples were trimmed with a knife to achieve smooth surfaces. Whatman No. 42 ash-free quantitative Type II filter papers, with 55 mm and 70 mm diameters, were used. To measure the matric suction, a 55 mm filter paper was placed between two 70 mm filter papers to protect it from dirt and contaminants. The stacked filter papers were then placed between the two half-soil samples and the two halves were then secured with electrical tape to cover the gap between them and to prevent the filter paper from absorbing vapor moisture (*29*). Next, the samples were placed in 10 oz, clear, wide-mouth jars. To measure the total suction, O-rings were placed on top of the samples and two 55 mm filter papers were placed on the O-rings without making any contact with the surface of the soil. The jars' lids were closed and secured by electrical tape. The jars were then placed in an environmental chamber at 20 ± 0.5°C for 1 week. The calibration curve

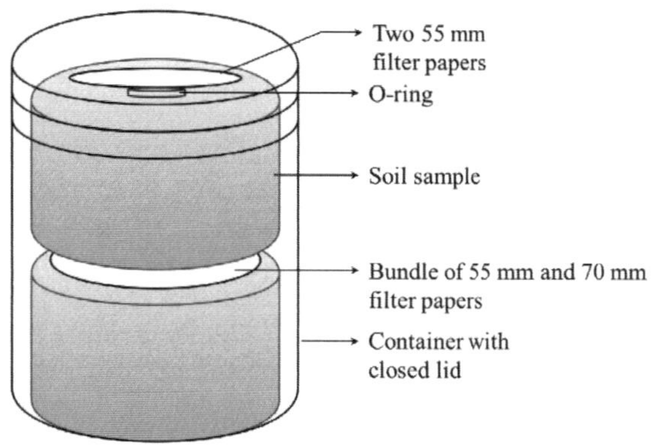

Figure 3. Schematic setup of the filter paper tests as per ASTM D5298.

of the filter paper was adopted from the ASTM D5298 standard. The weights of the filter papers were calculated using an analytical scale with a resolution of 10^{-4} g. Figure 3 schematically shows the test setup for matric and total suction measurements.

Data Representation and Discussion

Resilient Modulus

Over the course of the past five decades, several M_r constitutive prediction equations, which possess their own merits and demerits, were suggested in the literature (*30–33*). The generalized model, which is a log–log model adopted by the new AASHTO 2002 Mechanistic Empirical Pavement Design Guide (MEPDG), is as follows (*34*):

$$M_r = k_1 P_a \left(\frac{\theta}{P_a}\right)^{k_2} \left(1 + \frac{\tau_{oct}}{P_a}\right)^{k_3} \quad (1)$$

where,

M_r = resilient modulus (MPa);
k_1, k_2, k_3 = regression coefficients;
P_a = reference atmospheric pressure (101.3 kPa);
θ = bulk stress (sum of three principal stresses: σ_1, σ_2,σ_3) (kPa);
τ_{oct} = octahedral shear stress which is defined as $2^{0.5}/3(\sigma_1–\sigma_3)$ (kPa).

Another log–log M_r prediction model was suggested by NCHRP 1-28 (*35*) for stabilized soils:

$$M_r = k_1 P_a \left(\frac{\sigma_3}{P_a}\right)^{k_2} \left(\frac{\sigma_d}{P_a}\right)^{k_3} \quad (2)$$

where,

M_r = resilient modulus (MPa);
k_1, k_2, k_3 = regression coefficients;
P_a = reference atmospheric pressure (101.3 kPa);
σ_3 = confining pressure (kPa); and
σ_d = deviatoric stress (kPa).

In this study, Equations 1 and 2 were adopted to calibrate the test results. The average results of two specimens for each mixture are tabulated in Table 3. It can be observed, in all mixtures, that specimens showed higher M_r on the dry side of the optimum moisture content. The highest increase in M_r, with decrease in the optimum moisture content, was observed in 85% clay and 15% bottom ash mixtures, with an average of 65%, while the smallest increase was recorded for 65% clay and 35% bottom ash. The untreated clay specimens showed a 24% increase in the M_r. Additionally, as the moisture content of the specimens increased by 2%, a decrease in the M_r of all mixtures was noted. The mixture with 15% bottom ash showed the smallest reduction in M_r by 12%, while the clay mixture showed a 23% decrease, and both mixtures with 25% and 35% bottom ash contents yielded a 34% decrease. Additionally, all specimens showed reductions in M_r as the deviatoric stress increased. This observation can be attributed to the stress softening behavior of fine-grained soils as opposed to what is normally observed in granular materials, where M_r shows a direct relationship with bulk stress.

The effect of adding bottom ash on the M_r of mixtures, compared to that of the untreated clay, is shown in Figure 4. For brevity, the results are presented only at optimum moisture contents. Adding bottom ash at 15% and 35% dry weight ratios drastically decreased the M_r of the mixtures at an average of 27% and 22%, respectively. Mixing clay with 25% bottom ash amended the clay engineering properties, which led to the observation of an average of 15% increase in M_r, and an average of 23% was the maximum enhancement observed in the second sequence of the test. This observation relates to how the bottom ash is incorporated in the soil particles' structure of the mixture at different ratios. At 25%, maximum interlocking of particles between the clay and the bottom ash was achieved, which resulted in a significant increase in the M_r of the mixtures. Adding 15% or 35% bottom ash softened the soil mixture, which yielded lower M_r values. It is worth noting that the increase or decrease in M_r of the mixtures at different stress states were found to be variable. This can be attributed to the stress dependency of the material. The same behavior is observed in the M_r of granular materials at different stress states.

To calibrate the prediction equations and find the respective regression coefficients for each mixture, a nonlinear curve fitting, using the least squares method in Wolfram Mathematica V.10.3, was implemented. Tables 4 and 5 show the derived regression coefficient R^2

Table 3. Resilient Modulus Values of the Test Mixtures (MPa)

	Clay			15% bottom ash			25% bottom ash			35% bottom ash		
Test sequence	Optimum −2%	Optimum	Optimum + 2%	Optimum −2%	Optimum	Optimum + 2%	Optimum −2%	Optimum	Optimum + 2%	Optimum −2%	Optimum	Optimum + 2%
1	269	213	186	203	101	88	308	237	197	196	150	111
2	123	104	82	121	82	76	187	129	97	111	87	55
3	107	87	65	111	74	69	162	106	70	89	74	41
4	103	83	62	105	70	66	151	99	60	78	69	37
5	99	82	60	102	69	64	144	92	53	73	67	35
6	245	196	173	184	87	73	245	208	164	168	118	114
7	135	103	80	129	73	65	176	121	76	101	75	57
8	104	86	61	105	69	62	156	101	61	84	68	43
9	100	82	57	00	67	61	143	92	55	75	64	37
10	98	81	54	99	67	60	140	88	52	71	64	36
11	218	175	133	178	78	69	226	202	165	158	114	103
12	121	98	79	118	67	57	164	110	77	91	81	53
13	104	86	66	102	65	53	145	102	61	79	69	41
14	100	80	60	97	64	51	135	94	54	70	66	38
15	98	76	58	94	64	53	131	90	51	67	65	36

Table 4. Equation 1 Regression Coefficients and Statistics for Different Mixtures

	Clay			15% bottom ash			25% bottom ash			35% bottom ash		
Regression coefficients	Optimum − 2%	Optimum	Optimum + 2%	Optimum − 2%	Optimum	Optimum + 2%	Optimum − 2%	Optimum	Optimum + 2%	Optimum − 2%	Optimum	Optimum + 2%
k_1	3.1616	2.4988	2.3576	2.1739	0.9596	0.8466	3.1003	2.7276	2.6699	2.2887	1.5099	1.5541
k_2	0.1520	0.1512	0.2484	0.1066	0.2328	0.3376	0.2526	0.1482	0.1969	0.2167	0.1943	0.0559
k_3	−5.6661	−5.5196	−6.9731	−3.6686	−1.8829	−1.9967	−3.8309	−5.1588	−7.7809	−5.4667	−4.0163	−6.8790
R^2	0.9705	0.9720	0.9690	0.9873	0.9973	0.9978	0.9888	0.9789	0.9736	0.9845	0.9819	0.9770
S_e/S_y	0.9722	0.7611	0.6573	0.5568	0.1495	0.1191	0.7507	0.7507	0.6166	0.5211	0.4462	0.3667

Table 5. Equation 2 Regression Coefficients and Statistics for Different Mixtures

	Clay			15% bottom ash			25% bottom ash			35% bottom ash		
Regression coefficients	Optimum − 2%	Optimum	Optimum + 2%	Optimum − 2%	Optimum	Optimum + 2%	Optimum − 2%	Optimum	Optimum + 2%	Optimum − 2%	Optimum	Optimum + 2%
k_1	0.7814	1.5911	0.4330	0.8228	0.7282	0.7259	1.3600	0.7101	0.3754	0.6083	0.5728	0.2289
k_2	0.1141	0.1146	0.1848	0.0812	0.1525	0.2314	0.1759	0.1080	0.1491	0.1596	0.1316	0.0490
k_3	−0.6686	−0.6534	−0.7765	−0.4504	−0.1898	−0.1739	−0.4287	−0.6105	−0.8592	−0.6205	−0.4704	−0.8022
R^2	0.9887	0.9897	0.9871	0.9959	0.9984	0.9990	0.9951	0.9929	0.9919	0.9960	0.9909	0.9944
S_e/S_y	0.6016	0.4611	0.4168	0.3184	0.1161	0.0800	0.4875	0.4373	0.3425	0.2642	0.3168	0.1817

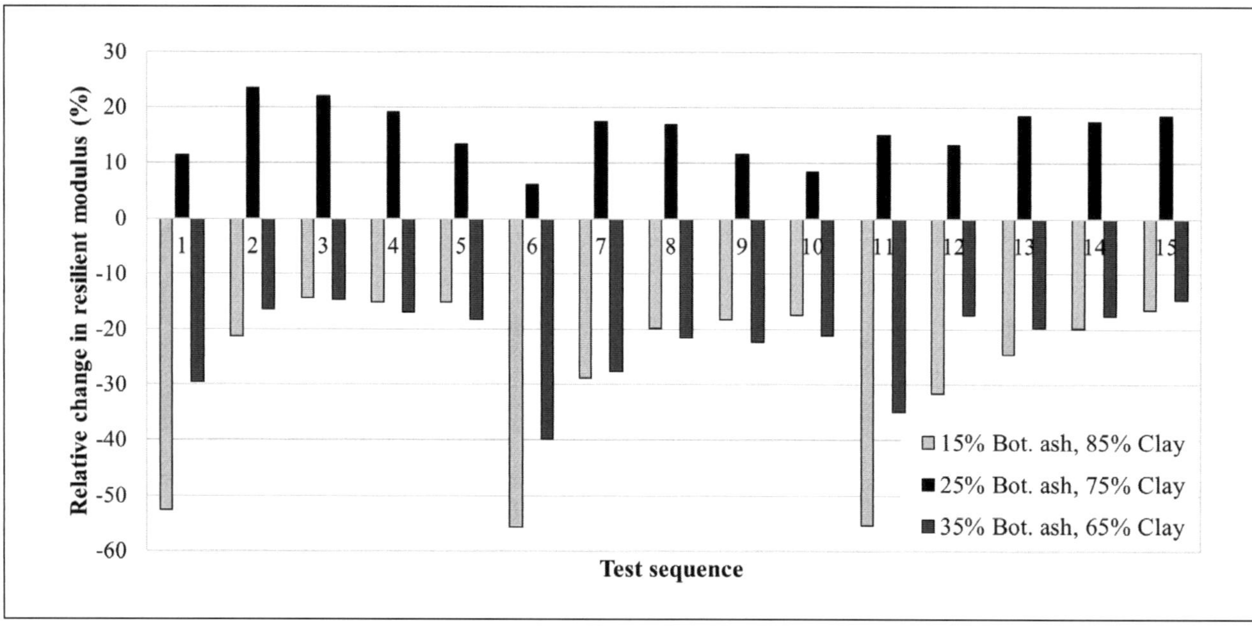

Figure 4. Relative change in resilient modulus in different mixtures compared to untreated clay.

and S_e/S_y statistic (S_e = standard error of estimate, and S_y = standard deviation within the samples) values for Equations 1 and 2, respectively. Comparing the R^2 and S_e/S_y for the two models revealed that Equation 2 outperforms Equation 1 in predicting M_r values. It was observed that in eight out of 12 mixtures, the S_e/S_y ratio for Equation 1 was above 0.50, whereas, except for the clay mixture at the moisture content of Optimum − 2%, this ratio for Equation 2 was below 0.50 for the rest of the mixtures. This indicates that the second model has significantly better goodness of fit than the first model.

Soil Suction

Figure 5 shows the variation of total, matric, and osmotic suctions in different mixtures at different moisture contents of Optimum − 2%, optimum, and optimum + 2%. The values of total and matric suctions decreased as the percentage of bottom ash in the mixture increased. This observation was expected as per the fundamentals of soil suction. The osmotic suction in the mixture with 35% bottom ash always showed higher values compared to that of the other mixtures. This observation could be attributed osmotic suction of soils being a function of ion concentration, and it increases as the concentration of electrolyte increases in the solution (*36*). To evaluate the effect of total, matric, and osmotic suctions on M_r, linear regression analysis was conducted on all data points for each test sequence. Table 6 shows the results of linear regression analysis for three soil suction components corresponding to three moisture contents. It should be mentioned that low R^2 values were not unexpected, since the regression was conducted on all the data. Since the matric suction generally decreases with the increase of the bottom ash in the mixtures, the total suction was considered and incorporated in the M_r prediction models (*37*). Table 7 shows the R^2 values for linear regression analysis between the total suction, and the M_r for each individual mixture. The overall high average R^2 values refer to a moderately high, and significant, correlation between M_r and the total suction within different mixtures.

To incorporate total suction into the M_r prediction equations, Equation 1 was modified in the form of Equation 3:

$$M_r = k_1 P_a \left(\frac{\theta}{P_a}\right)^{k_2} \left(\frac{\psi}{P_a}\right)^{k_3} \left(1 + \frac{\tau_{oct}}{P_a}\right)^{k_4} \quad (3)$$

where,

k_i = the regression coefficients; and

ψ = the total suction (kPa), and the other parameters are as defined before.

Table 8 depicts the calibrated model for the mixtures and its respective statistics. It should be mentioned that all three moisture contents were used to calibrate the models. All S_e/S_y ratios found to be lower than 0.50 indicates strong goodness of fit for the model. Additionally, the positive exponent of the total suction term in the equations refers to the direct relationship between M_r and the total suction of soil.

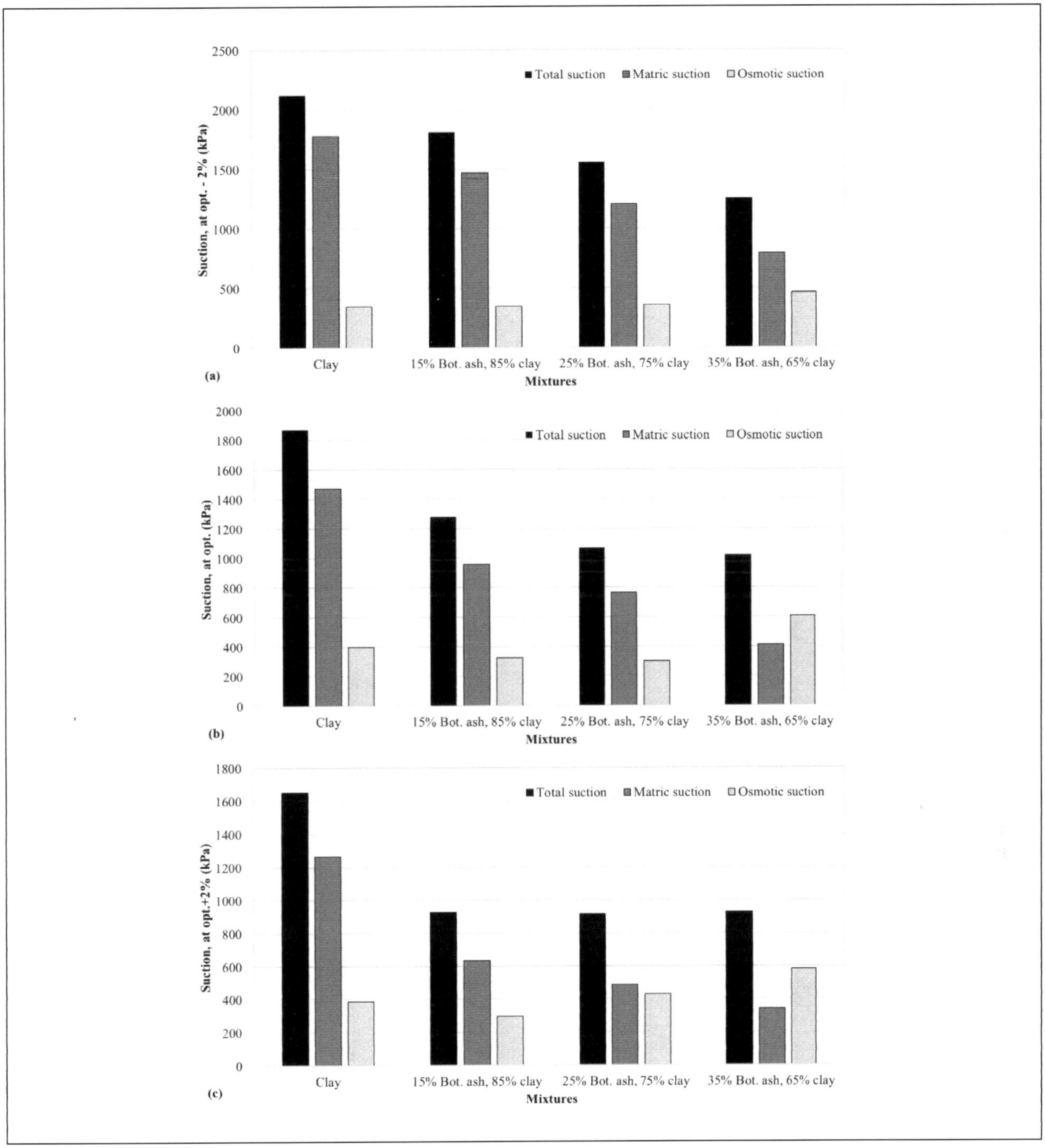

Figure 5. Mixtures suction values at different moisture contents (*a*) Optimum − 2%; (*b*) optimum; (*c*) optimum + 2%.

Equation (2) was also modified, in the form of Equation 4, incorporating the total suction term as below:

$$M_r = k_1 P_a \left(\frac{\sigma_3}{P_a}\right)^{k_2} \left(\frac{\sigma_d}{P_a}\right)^{k_3} \left(\frac{\psi}{P_a}\right)^{k_4} \tag{4}$$

The regression coefficients of the calibrated model and its statistics are reflected in Table 9.

Comparing Equations 3 and 4 reveals that the latter showed better performance in predicting M_r. Even though the R^2 values are not significantly different

Table 6. Linear Regression Coefficients of Determination for Resilient Modulus and Soil Suction for All Mixtures

Test sequence	Total suction	Matric suction	Osmotic suction
1	0.39	0.36	0.21
2	0.26	0.29	0.27
3	0.31	0.36	0.27
4	0.36	0.41	0.30
5	0.39	0.44	0.31
6	0.48	0.46	0.06
7	0.45	0.39	0.22
8	0.36	0.44	0.26
9	0.39	0.45	0.29
10	0.41	0.31	0.30
11	0.35	0.43	0.04
12	0.43	0.42	0.17
13	0.41	0.45	0.20
14	0.43	0.46	0.22
15	0.43	0.45	0.24
Average	0.39	0.41	0.22

Table 7. Linear Regression Coefficients of Determination for Resilient Modulus and Total Suctions of Each Mixture at Different Moisture Contents

Test sequence	Clay	15% bottom ash	25% bottom ash	35% bottom ash
1	0.99	0.91	0.93	0.96
2	0.98	0.92	0.94	0.90
3	0.98	0.91	0.92	0.81
4	0.98	0.91	0.89	0.71
5	0.97	0.92	0.89	0.68
6	0.99	0.91	0.80	0.95
7	1.00	0.92	0.88	0.98
8	0.96	0.93	0.90	0.88
9	0.95	0.94	0.90	0.79
10	0.94	0.95	0.90	0.69
11	0.99	0.90	0.74	0.99
12	1.00	0.94	0.92	0.78
13	0.98	0.97	0.85	0.78
14	0.99	0.98	0.84	0.64
15	1.00	0.98	0.84	0.57
Average	0.98	0.93	0.88	0.81

between the two models, the S_e/S_y ratio showed a decrease between 5% and 15% in Equation 4. Overall, the modified NCHRP model, incorporating the total suction term, outperformed the other models with the smallest average S_e/S_y of 0.30 and an average R^2 of 0.990 within different soil mixtures.

To validate the models, four extra samples from each mixture were made at their respective optimum moisture contents and their M_r values were measured. Figure 6 presents the performance of Equations 3 and 4. As expected, Equation 4 outperformed Equation 3 in predicting M_r values with higher R^2 values. It is also worth mentioning that the over- and under-predicted values by the models that were evenly distributed within the range of data still indicate a satisfactory performance of the models. This validation showed compelling outcomes for use of the developed models to predict M_r of low-plasticity clay soil treated with bottom ash.

Table 8. Equation 3 Regression Coefficients and Statistics for Different Mixtures

Regression coefficients	Clay	15% bottom ash	25% bottom ash	35% bottom ash
k_1	0.0161	0.0699	0.3848	0.0377
k_2	0.1708	0.1680	0.2160	0.1808
k_3	1.7443	1.1458	0.7777	1.6292
k_4	−5.8727	−2.9463	−4.7817	−5.2476
R^2	0.9696	0.9791	0.9692	0.9746
S_e/S_y	0.4118	0.4071	0.4233	0.3900

Table 9. Equation 4 Regression Coefficients and Statistics for Different Mixtures

Regression coefficients	Clay	15% bottom ash	25% bottom ash	35% bottom ash
k_1	0.0036	0.0347	0.1253	0.0102
k_2	0.1279	0.1203	0.1551	0.1319
k_3	−0.6851	−0.3463	−0.5497	−0.6077
k_4	1.7421	1.1542	0.7689	1.6235
R^2	0.9879	0.9843	0.9798	0.9867
S_e/S_y	0.2603	0.3529	0.3431	0.2822

Conclusions

This study was undertaken to evaluate the effect of bottom ash on the M_r of a medium-plasticity clay. The clay mixtures were made at 15%, 25%, and 35% bottom ash in resilient modulus specimens. While adding bottom ash at 15% and 35% of total dry weight of the mixture led to drastic decreases in M_r at Optimum − 2%, optimum, and optimum + 2% moisture contents, the 25% bottom ash significantly increased M_r, with a range of 5–23% among the 15 test sequences. Additionally, two well-referred resilient modulus prediction equations were calibrated for the mixtures. The NCHRP prediction equation showed better statistics compared to the log–log model adopted in MEPDG.

The total and matric suctions at different moisture contents in the mixtures were indirectly measured using the filter paper test. The total suction and matric suction of the mixtures showed a descending trend with the increase

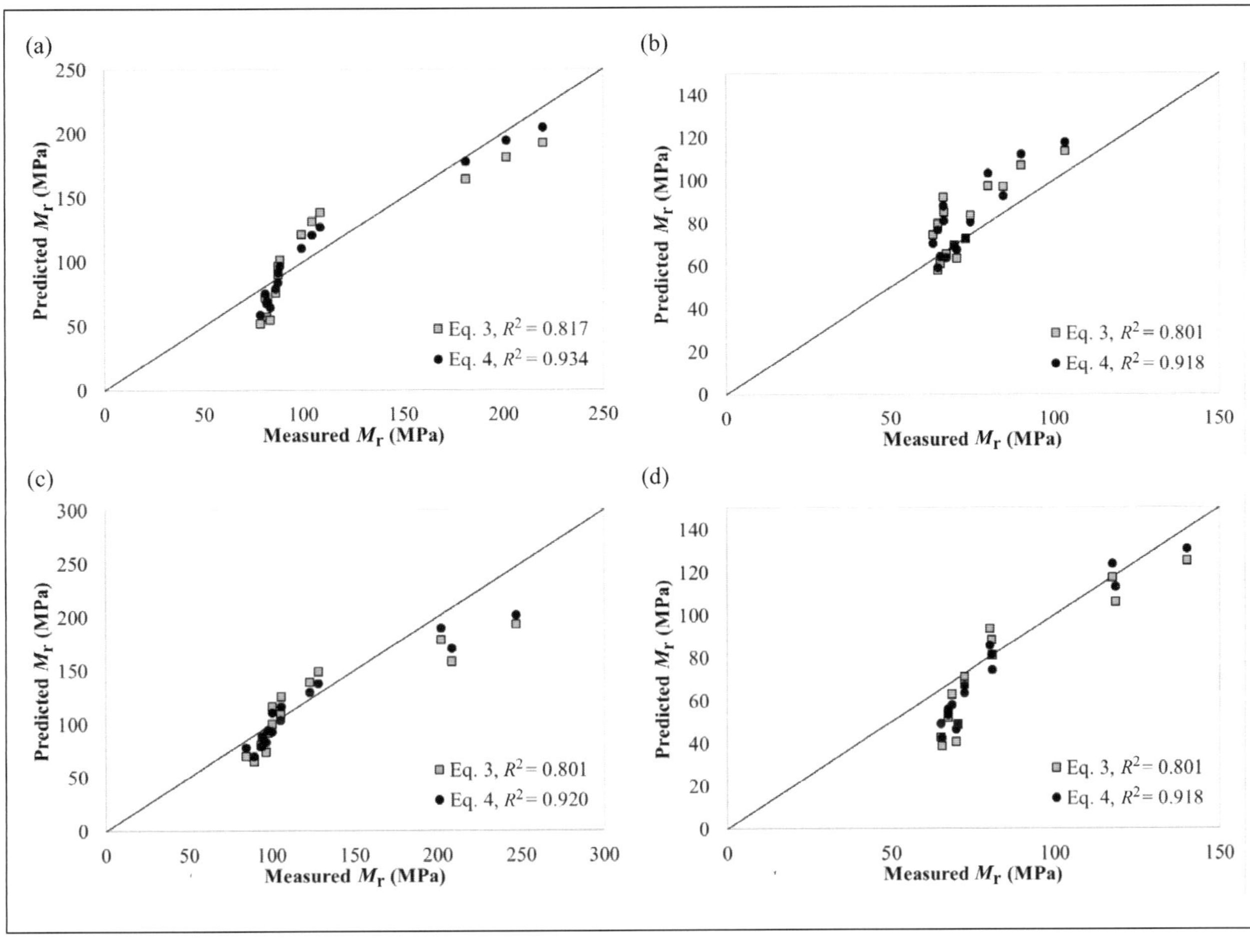

Figure 6. Validation of Equations 3 and 4 for samples made at optimum moisture contents: (*a*) untreated clay; clay with (*b*) 15%; (*c*) 25%; (*d*) 35% bottom ash.

of bottom ash in the mixtures. Incorporating the total suctions in the two M_r equations led to a higher R^2 and a lower S_e/S_y ratio. Overall, the modified NCHRP model outperformed the other models with the smallest average S_e/S_y of 0.30 and an average R^2 of 0.990 within different soil mixtures. The two models were validated for one sample from each mixture at their respective optimum moisture content. While the results showed compelling performance of both models, Equation 4 outperformed Equation 3 by showing higher R^2 values.

The authors would recommend performing the tests on different soil types within a wider range and more increments of moisture content to better refine the prediction models and enhance their corresponding goodness of fits.

Acknowledgments

The authors would like to thank Mr. Vinicius Velasco for assisting in performing the lab tests and experimental works. Ms. Delaina Lawson is also greatly appreciated for proofreading the paper.

Author Contributions

The authors confirm contribution to the paper as follows: study conception and design: AA, LH, AB; data collection: AA; analysis and interpretation of results: AA, LH, AB; draft manuscript preparation: AA, LH. All authors reviewed the results and approved the final version of the manuscript.

References

1. Li, D., and E. T. Selig. Resilient Modulus for Fine-grained Subgrade Soils. *Journal of Geotechnical Engineering*, Vol. 120, No. 6, 1994, pp. 939–957.
2. Burczyk, J. M., K. Ksaibati, R. Anderson-Sprecher, and M. J. Farrar. Factors Influencing Determination of a Subgrade Resilient Modulus Value. *Transportation Research Record: Journal of the Transportation Research Board*, 1994. 1462: 72–78.

3. Brooks, R. M. Soil Stabilization with Fly Ash and Rice Husk Ash. *International Journal of Research and Reviews in Applied Sciences*, Vol. 1, No. 3, 2009, pp. 209–217.
4. Khoury, N., and M. Zaman. Correlation Between Resilient Modulus, Moisture Variation, and Soil Suction for Subgrade Soils. *Transportation Research Record: Journal of the Transportation Research Board*, 2004. 1874: 99–107.
5. Liang, R. Y., S. Rabab'ah, and M. Khasawneh. Predicting Moisture-Dependent Resilient Modulus of Cohesive Soils Using Soil Suction Concept. *Journal of Transportation Engineering*, Vol. 134, No. 1, 2008, pp. 34–40.
6. Sawangsuriya, A., T. Edil, and C. Benson. Effect of Suction on Resilient Modulus of Compacted Fine-Grained Subgrade Soils. *Transportation Research Record: Journal of the Transportation Research Board*, 2009. 2101: 82–87.
7. Qi, J., P. A. Vermeer, and G. Cheng. A Review of the Influence of Freeze–Thaw Cycles on Soil Geotechnical Properties. *Permafrost and Periglacial Processes*, Vol. 17, No. 3, 2006, pp. 245–252.
8. Wang, D., W. Ma, Y. Niu, X. Chang, and Z. Wen. Effects of Cyclic Freezing and Thawing on Mechanical Properties of Qinghai–Tibet Clay. *Cold Regions Science and Technology*, Vol. 48, No. 1, 2007, pp. 34–43.
9. Solanki, P., M. Zaman, and J. Dean. Resilient Modulus of Clay Subgrades Stabilized with Lime, Class C Fly Ash, and Cement Kiln Dust for Pavement Design. *Transportation Research Record: Journal of the Transportation Research Board*, 2010. 2186: 101–110.
10. Güllü, H. Function Finding via Genetic Expression Programming for Strength and Elastic Properties of Clay Treated with Bottom Ash. *Engineering Applications of Artificial Intelligence*, Vol. 35, 2014, pp. 143–157.
11. Alhassan, M. Potentials of Rice Husk Ash for Soil Stabilization. *Assumption University Journal of Technology*, Vol. 11, No. 4, 2008, pp. 246–250.
12. Sezer, A., G. İnan, H. R. Yılmaz, and K. Ramyar. Utilization of a Very High Lime Fly Ash for Improvement of Izmir Clay. *Building and Environment*, Vol. 41, No. 2, 2006, pp. 150–155.
13. Misra, A. Stabilization Characteristics of Clays using Class C Fly Ash. *Transportation Research Record: Journal of the Transportation Research Board*, 1998. 1611: 46–54.
14. Prusinski, J., and S. Bhattacharja. Effectiveness of Portland Cement and Lime in Stabilizing Clay Soils. *Transportation Research Record: Journal of the Transportation Research Board*, 1999. 1652: 215–227.
15. Modarres, A., and Y. M. Nosoudy. Clay Stabilization Using Coal Waste and Lime: Technical and Environmental Impacts. *Applied Clay Science*, Vol. 116, 2015, pp. 281–288.
16. Rifai, A., N. Yasufuku, and K. Tsuji. Characterization and Effective Utilization of Coal Ash as Soil Stabilization on Road Application. In *Ground Improvement Technologies and Case Histories* (Leung, C. F, J. Chu, and R. F. Shen, eds.), Research Publishing Services, Singapore, 2009, pp. 469–474.
17. Osinubi, K. J. Stabilisation of Tropical Black Clay with Cement and Pulverised Coal Bottom Ash Admixture. In *Advances in Unsaturated Geotechnics* (Shackelford, C. D., S. L. Honston, and N.-Y. Chang, eds.), ASCE Geotechnical Special Publication, Reston, VA, 2000, pp. 289–302.
18. Ahmad, J., K. Mohamed, A. S. A. Rahman, M. I. F. Rosli, and A. R. Zulkifli. Clay Stabilization Using OPC and Bottom Ash as Additives. In *Regional Conference on Science, Technology and Social Sciences (RCSTSS 2014)*, Springer, Singapore, 2016, pp. 27–33.
19. Heidrich, C., H. Feuerborn, and A. Weir. Coal Combustion Products: A Global Perspective. *Proceedings of the World of Coal Ash Conference*, Lexington, KY, 2013, pp. 22–25.
20. Kayabal, K., and G. Buluş. The Usability of Bottom Ash as an Engineering Material When Amended with Different Matrices. *Engineering Geology*, Vol. 56, No. 3, 2000, pp. 293–303.
21. Quigley, J. By the Numbers: A Hard Look at Canada's Declining Coal Industry. *CBC News*. February 4, 2016. http://www.cbc.ca/news/business/canadian-coal-by-the-numbers-1.3408568. Accessed July 28, 2017.
22. Bethanis, S., and C. R. Cheeseman. Manufacture of Lightweight Aggregate from Incinerator Bottom Ash and Pulverised Fuel Ash. *Proceedings of the Heleco Conference*, Athens, Greece, 2005.
23. Klein, R., T. Baumann, E. Kahapka, and R. Niessner. Temperature Development in a Modern Municipal Solid Waste Incineration (MSWI) Bottom Ash Landfill with Regard to Sustainable Waste Management. *Journal of Hazardous Materials*, Vol. 83, No. 3, 2001, pp. 265–280.
24. ASTM D1557-12e1. *Standard Test Methods for Laboratory Compaction Characteristics of Soil Using Modified Effort (56,000 ft-lbf/ft^3 (2,700 kN-m/m^3))*, ASTM International, West Conshohocken, PA, 2012.
25. AASHTO T307-99. *Standard Method of Test for Determining the Resilient Modulus of Soils and Aggregates*, AASHTO, Washington, D.C., 2003.
26. Grubbs, F. E. Procedures for Detecting Outlying Observations in Samples. *Technometrics*, Vol. 11, No. 1, 1969, pp. 1–21.
27. Fredlund, D. G., and H. Rahardjo. *Soil Mechanics for Unsaturated Soils*. John Wiley & Sons, New York City, NY, 1993.
28. Leong, E-C., S. Tripathy, and H. Rahardjo. Total Suction Measurement of Unsaturated Soils with a Device Using the Chilled-Mirror Dew-Point Technique. *Geotechnique*, Vol. 53, No. 2, 2003, pp. 173–182.
29. Bulut, R., R. L. Lytton, and W. K. Wray. Soil Suction Measurements by Filter Paper. *Expansive Clay Soils and Vegetative Influence on Shallow Foundations*, Houston, TX, 2001, pp. 243–261.
30. Dunlap, W. A. *A Report on a Mathematical Model Describing the Deformation Characteristics of Granular Materials*, Texas A&M University, Texas Transportation Institute, Austin, TX, 1963.
31. Thompson, M. R., and Q. L. Robnett. *Resilient Properties of Subgrade Soils*. No. FHWA-IL-UI-160 Final Report, 1976.
32. Moossazadeh, J., and M. W. Witczak. Prediction of Subgrade Moduli for Soil that Exhibits Nonlinear Behavior.

Transportation Research Record: Journal of the Transportation Research Board, 1981. 810: 10–17.
33. Witczak, M., and J. Uzan. *The Universal Airport Design System, Report I of IV. Granular Material Characterization*. Department of Civil Engineering, University of Maryland, College Park, MD, 1988.
34. ARA, Inc., and ERES Consultants Division. *NCHRP Project 1-37A: Guide for Mechanistic-Empirical Design of New and Rehabilitated Pavement Structures*. Transportation Research Board of the National Academies, Washington, D.C., 2004.
35. *NCHRP Web Document 14 for Project 1-28: Laboratory Determination of Resilient Modulus for Flexible Pavement Design*. Transportation Research Board of the National Academies, Washington, D.C., 1997.
36. Witteveen, P., A. Ferrari, and L. Laloui. An Experimental and Constitutive Investigation on the Chemo-mechanical Behaviour of a Clay. *Geotechnique*, Vol. 63, No. 3, 2013, pp. 244.
37. Rahman, M. T., and R. A. Tarefder. Assessment of Molding Moisture and Suction on Resilient Modulus of Lime Stabilized Clayey Subgrade Soils. *Geotechnical Testing Journal*, Vol. 38, No. 6, 2015, pp. 1–11.

The Standing Committee on Engineering Behavior of Unsaturated Geomaterials (AFP60) peer-reviewed this paper (18-05098).

Article

Method for Quick Prediction of Hydraulic Conductivity and Soil-Water Retention of Unsaturated Soils

Transportation Research Record
2018, Vol. 2672(52) 108–117

DOI: 10.1177/0361198118798486
journals.sagepub.com/home/trr
SAGE

Shaoyang Dong[1], Yuan Guo[1], and Xiong (Bill) Yu[1,2]

Abstract

Hydraulic conductivity and soil-water retention are two critical soil properties describing the fluid flow in unsaturated soils. Existing experimental procedures tend to be time consuming and labor intensive. This paper describes a heuristic approach that combines a limited number of experimental measurements with a computational model with random finite element to significantly accelerate the process. A microstructure-based model is established to describe unsaturated soils with distribution of phases based on their respective volumetric contents. The model is converted into a finite element model, in which the intrinsic hydraulic properties of each phase (soil particle, water, and air) are applied based on the microscopic structures. The bulk hydraulic properties are then determined based on discharge rate using Darcy's law. The intrinsic permeability of each phase of soil is first calibrated from soil measured under dry and saturated conditions, which is then used to predict the hydraulic conductivities at different extents of saturation. The results match the experimental data closely. Mualem's equation is applied to fit the pore size parameter based on the hydraulic conductivity. From these, the soil-water characteristic curve is predicted from van Genuchten's equation. The simulation results are compared with the experimental results from documented studies, and excellent agreements were observed. Overall, this study provides a new modeling-based approach to predict the hydraulic conductivity function and soil-water characteristic curve of unsaturated soils based on measurement at complete dry or completely saturated conditions. An efficient way to measure these critical unsaturated soil properties will be of benefit in introducing unsaturated soil mechanics into engineering practice.

Hydraulic conductivity, or permeability, is a critical soil property describing the hydraulic behavior of unsaturated soils, and it is related to the water content and pore size of the soil (*1*, *2*). In the microstructural view, hydraulic conductivity depends on the structure of the soil skeleton, the connectedness of the pore space, and in essence, the intrinsic permeability of each phase of the unsaturated soil. The hydraulic conductivity of different type of soils (clay, silt, sand) varies within a wide range of several orders of magnitudes. It is challenging to determine the hydraulic conductivity because of the great variation in mineral composition and the complex texture of the soil. The wide range in hydraulic conductivity and great variation in its composition bring obstacles in analyzing hydraulic problems in geotechnical engineering.

Empirical and experimental methods are two broad method types to determine the hydraulic conductivity of unsaturated soils. The empirical approach relates the hydraulic conductivity with water content (*3*), particle size distributions (*1*), and soil texture. Table 1 lists the representative empirical equations for the unsaturated hydraulic conductivity.

The empirical equations can be applied in engineering practice when the experimental data are available for the relationship between the hydraulic conductivity and volumetric water content (or degree of saturation). Darcy's law is commonly used to determine the hydraulic conductivity from hydraulic experiments. The experimental approach is classified into two categories: one is the laboratory tests such as the transient-flow test, infiltration column tests, and deformation test (*9*, *10*); the other includes field tests such as small-scale field tests using observations of the water level in cavities in the soil, and large-scale field tests such as pump tests in wells. The experimental approach to determine the hydraulic conductivity has the advantages of being

[1]Department of Civil Engineering, Case Western Reserve University, Cleveland, OH
[2]Department of Electrical Engineering and Computer Science, Case Western Reserve University, Cleveland, OH

Corresponding Author:
Address correspondence to Xiong (Bill) Yu: xiong.yu@case.edu

Table 1. A Few Representative Empirical Equations for the Unsaturated Hydraulic Conductivity (k)

Equation†	Reference
$k = a\theta^b$	Gardner (4)
$k = k_s(\theta/\theta_s)^n$	Campbell (5)
$k = k_s exp[\alpha(\theta - \theta_s)]$	Davidson et al. (6)
$k = k_s[(\theta - \theta_r)/(\theta_s - \theta_r)]^{3.5}$	Averjanov (7)
$k = k_s S_e^{1/2}[1 - S_e(S_e^{n/(1-n)} - 1)^{(n-1)/n}]^2$ and $S_e = \frac{S - S_r}{1 - S_r}$	Mualem (8)

Note: † k_s = the saturated hydraulic conductivity; θ = the volumetric water content; θ_s = the saturated volumetric water content; n = the pore size parameter; a, b and α = the fitting parameter; θ_r = the residual volumetric water content; S_e = the effective degree of saturation; and S_r = the residual saturation.

reliable and accurate; however, it is time consuming and expensive to conduct these hydraulic experiments.

The theoretical basis for the hydraulic conductivity function (HCF) is shown by Fredlund et al. (*11*). It is proven that the HCF can be integrated from the soil-water characteristic curve (SWCC). The SWCC, or the soil-water retention curve, is another critical unsaturated soil property describing the relationship between matric suction and moisture content. The experimental approach to determine the SWCC generally includes thermocouple psychrometry and filter paper methods (*12*, *13*). Although these methods are commonly used in practice, they have the limitations of being labor intensive, time consuming and not reliable for a high suction range (over 100,000 kPa). Empirical equations for the SWCC curve are proposed by van Genuchten (*14*) and Fredlund and Xing (*15*). The fitting parameters in the equations need to be determined through laboratory or field testing.

The hydraulic properties of soil are quite sensitive to design parameters. These parameters are randomly distributed throughout the soil sample because of the complex geological process. This random process can be simulated by the random finite element method (RFEM) (*16*, *17*). The parameters include Young's modulus, Poisson's ratio, density, permeability, and a variety of other parameters (*18–20*). The RFEM is an extension of the traditional finite element method that adds randomness to material properties. Griffiths and Fenton modeled soil as a spatially random medium and represented the soil's permeability as a stochastic field to investigate the seepage beneath water retaining structures (*21*, *22*).

In the microstructural view, unsaturated soil is a three-phase material that includes soil particles, water, and air, arranged randomly as a result of complex geological processes (*23*). Researchers have built a microstructural model for soil (*24*, *25*). Tracy and Helliwell obtained the structure of soil by using X-ray computed tomography images (*26*, *27*). Zhang built 3D digital images of porous rock with porosity properties and mineral contents to study the elastic properties of the Longmaxi shale (*28*). The microstructure extracted from the high-resolution images provides a foundation for building multiscale, that is, micro-scale or pore-scale, models of soil samples.

In this research, a microstructure-based model is developed in Matlab to describe the unsaturated soil with distribution of phases based on their respective volumetric contents. The Matlab model is converted into a finite element model. The hydraulic and mechanical properties of each soil phase (soil particle, water, and air) are applied based on the microscopic structures. The flow of moisture is described with the Darcy's law (Equation 1):

$$q = -\frac{\kappa_j}{\mu}\nabla p \tag{1}$$

where the subscript j denotes different phases of the poromaterial; p is the pore-water pressure; κ_j is the intrinsic permeability; μ is the dynamic viscosity and q is the flux.

This paper aims to provide a new method to provide quick estimation of the hydraulic conductivity and soil-water characteristics curve of unsaturated soils. This new method includes a phase-coded microstructure-based RFEM model and calibration experiments conducted at completely dry and saturated conditions. These calibration experiments allow determination of the transport properties of individual phases, and account for the effects of phase interactions. By use of the RFEM and calibrated parameters for individual phase, the bulk hydraulic conductivities at different degrees of saturation are predicted. Comparison of experimental data shows that the results of hydraulic conductivity of unsaturated soils predicted by this new approach match very well with the experimental data. The approach is further extended to estimate the SWCC of unsaturated soils and demonstrates promising results.

Microstructure-Based Model Building

The mechanical and hydraulic properties of soil specimens (Denver claystone, BALT silt, Hopi silt, and

Table 2. Hydromechanical Parameters Deduced from Transient-Flow Test

Soil	n	k_s (m/s)	Porosity
Denver claystone	1.40	8.58E-9	0.55
BALT silt	1.38	1E-6	0.47
Hopi silt	1.70	3E-8	0.48
Ottawa sand	6.00	2.56E-5	0.38

Ottawa sand) are referenced from the literature (*10*), and their parameters are listed in Table 2.

To simulate the microstructure of unsaturated soils, first the volumetric content of each phase is calculated from their porosity and specific gravity. The next step is to generate a $p \times q$ matrix in Matlab. To simplify the model, for soils with finer particles such as clay and silt, the solid particles are assumed to have the same diameters. For the dimension of the matrix, p is equal to the height of the soil specimens divided by the diameter of the soil particle, and q is equal to the radius of the soil specimens divided by the diameter of the soil particle. Thus, $p \times q$ equals the total number of elements in the 2D soil model image. Each element in the matrix is assigned with a value to represent a particular phase of the soil specimens. Three different grayscale numbers (0, 0.5, and 1) are assigned to represent different phases (soil particles, water, and air) within the soil specimens. These numbers are randomly assigned, and their probability of occurrence is equal to the volumetric fraction of each phase. The final step is to generate an image from the matrix in Matlab. Each pixel of the image represents each element in the matrix, and the color in the pixel corresponds to the number in the element. Three different colors (black [0], gray [0.5], and white [1]) represent the soil particles, water, and air, respectively (Figure 1). For sand, the image in the Matlab is generated based on the particle size distribution of the sand (Figure 2).

The microstructure image is imported to COMSOL and material properties are assigned based on the color scale of the image. Figures 3 shows the 2D silt images at different volumetric water content after conversion from Matlab to COMSOL model.

Hydraulic Conductivity Simulation

With the microstructure-based RFEM model, the hydraulic parameters can be assigned to each pixel based on the phase distribution obtained from the microstructure-based image. Here the hydraulic parameters refer to those of individual phases, which is different from the bulk properties measured. Therefore, the intrinsic permeability of these phases (i.e., solid particles, water, and air) should first be calibrated. The calibration process is shown in Figure 4. Soil samples under two phased conditions are used for this purpose, using experimental data from Lu et al. (*10*). The two-phase images of totally dried soil specimens and totally saturated soil specimens are generated in Matlab. The saturated condition is used to calibrate the intrinsic hydraulic properties of water. The dry condition is used to calibrate the intrinsic hydraulic properties of air. The intrinsic properties of mineral are assumed to be magnitude smaller so that they are impermeable compared with other phases. Next, the images are converted into COMSOL, and proper physical and hydraulic parameters (Table 3) are applied to the model. No flow boundary conditions are applied on the sides, with a hydraulic pressure of 25 kPa applied to the surface to produce one-dimensional flow (Figure 2). Sensitivity studies are conducted to determine the intrinsic

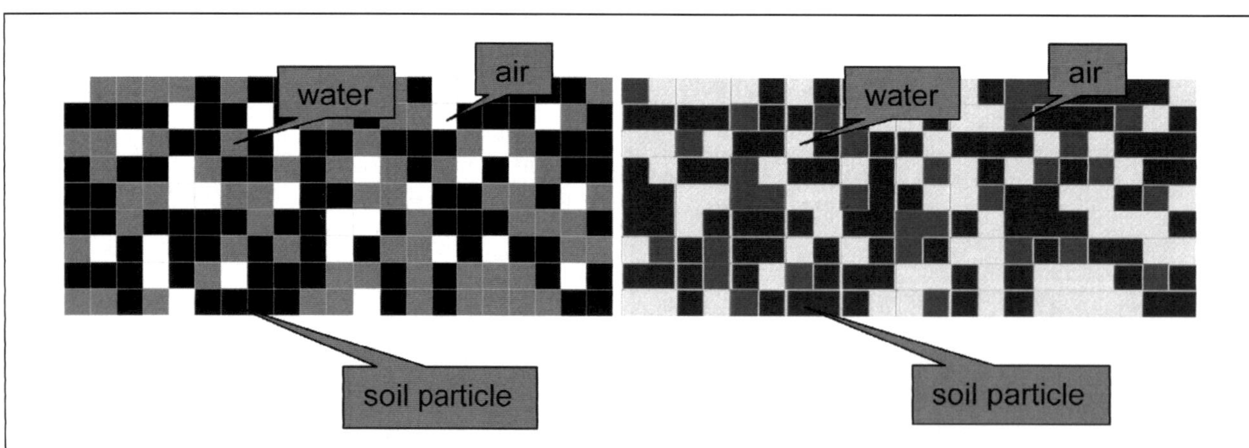

Figure 1. The enlarged image of the 2D unsaturated clay model generated in Matlab (*left*), and the corresponding image converted into COMSOL (*right*).

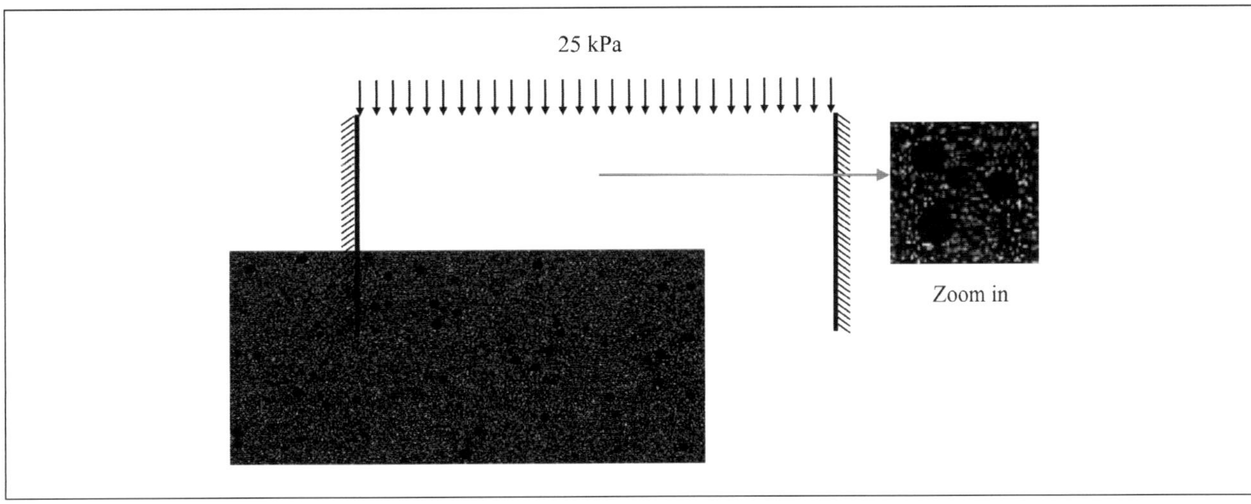

Figure 2. 2D microstructure-based image of sand generated in Matlab based on particle size distribution (the white, gray, and black pixel represents air, water, and solid particle, respectively).

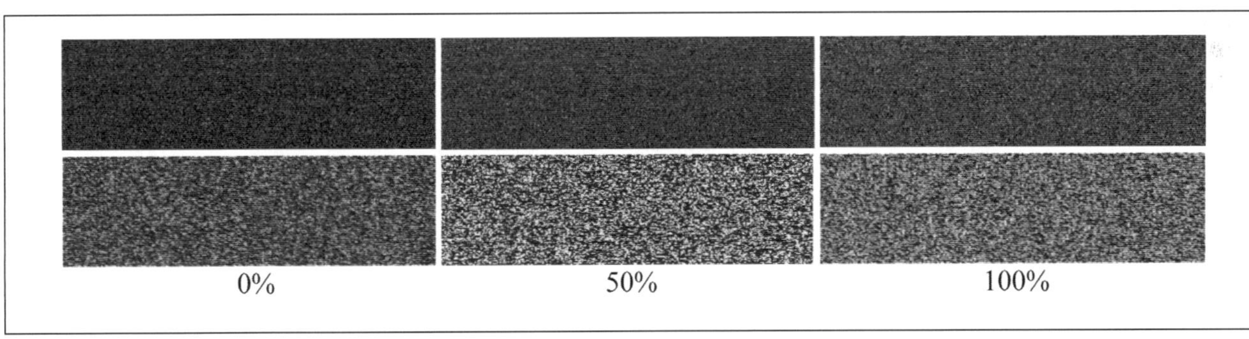

Figure 3. 2D silt microstructure-based image with different extent of saturation (the gray image is generated from Matlab, the color image is converted into COMSOL).

permeability values of solid particles, water, and air by matching the bulk hydraulic conductivities with experimental data. The calibration results for each phase of different kinds of soils are listed in Table 3.

The flowlines are not vertical in the simulation because of the microstructure of the model. As shown in Figure 5, the flowlines in the clay specimen are mostly zigzagged because the directions of the water flow are altered by the solid particles and the numbers of clay particles are magnitudes higher than those of silt and sand. In addition, clay contains more tortuous pores that end abruptly, so clay has the least permeability. For the silt specimen, the directions of the water flow are less zigzagged as there are fewer solid particles in silt. The moisture flows in the sand specimen are most vertically distributed because sand is a material with very open pores that pass completely and directly through the sand. Besides that, the numbers of sand particles are magnitudes smaller than those of clay and silt, which causes decreased impedance to the moisture flow. To summarize, the more connected and the larger the pore network is, the fewer solid particles in the soil, the greater the permeability it will be.

To further illustrate the impact of the microstructure in the simulation, the contour plots of the hydraulic head distribution with and without considering the microstructure in the clay specimen are shown in Figure 6. The hydraulic head contours are horizontally distributed in the left-hand figure because the mechanical and hydraulic properties of clay are uniformly distributed, whereas the hydraulic head contours are more complicated in the right-hand figure because randomly distributed microstructures of clay are introduced in the new model to make the prediction closer to reality.

With the calibrated hydraulic properties, the bulk hydraulic properties at different extents of saturation can be predicted via the microstructure-based model. The

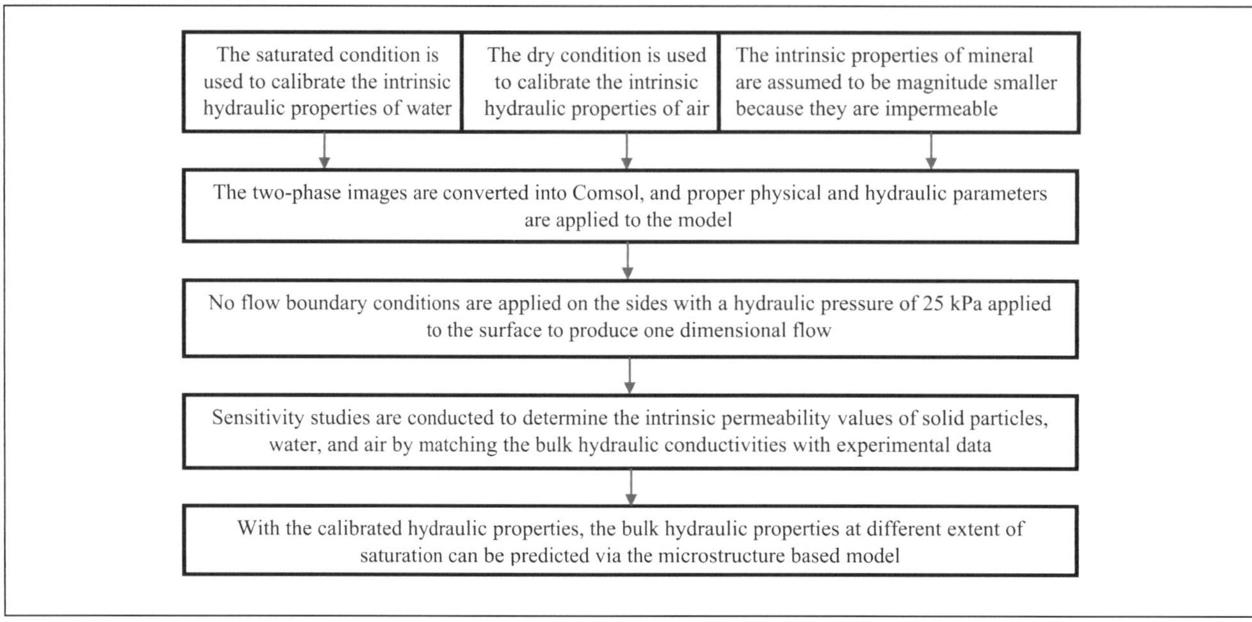

Figure 4. Flow chart of the calibration process of the intrinsic permeability and simulation process of the bulk hydraulic conductivity.

Table 3. Parameters Used for Hydraulic Conductivity Simulations

Constant	Value	Units	Description
D	100	mm	Diameter of soil specimens
H	21	mm	Height of soil specimens
E_s	12.7	GPa	Young's modulus of solid particle
E_w	3.9×10^{-5}	Pa	Young's modulus of water
E_a	0	Pa	Young's modulus of air
ρ_s	2600	kg/m^3	Density of solid particle
ρ_w	1000	kg/m^3	Density of water
ρ_a	1.29	kg/m^3	Density of air
μ_s	0.3	1	Poisson's ratio of solid particle
μ_w	0.5	1	Poisson's ratio of water
μ_a	0	1	Poisson's ratio of air
μ	8.9×10^{-4}	Pa·s	Dynamic viscosity of water
χ_f	4.6×10^{-10}	1/Pa	Compressibility of water
d_c	2	µm	Diameter of clay particle
d_s	0.01	mm	Diameter of silt particle
D_{10}	0.181	mm	The particle size where 10% of the particles in the sand are smaller
D_{60}	0.347	1	The particle size where 60% of the particles in the sand are smaller
κ_s	10^{-19}	m^2	Intrinsic permeability of solid particle for Denver claystone
κ_w	8.08×10^{-16}	m^2	Intrinsic permeability of water for Denver claystone
κ_a	8.92×10^{-19}	m^2	Intrinsic permeability of air for Denver claystone
κ_s	10^{-17}	m^2	Intrinsic permeability of solid particle for BALT silt
κ_w	3.08×10^{-13}	m^2	Intrinsic permeability of water for BALT silt
κ_a	1.31×10^{-19}	m^2	Intrinsic permeability of air for BALT silt
κ_s	10^{-18}	m^2	Intrinsic permeability of solid particle for Hopi silt
κ_w	8.08×10^{-15}	m^2	Intrinsic permeability of water for Hopi silt
κ_a	1.31×10^{-19}	m^2	Intrinsic permeability of air for Hopi silt
κ_s	10^{-16}	m^2	Intrinsic permeability of solid particle for Ottawa sand
κ_w	3.08×10^{-10}	m^2	Intrinsic permeability of water for Ottawa sand
κ_a	8.92×10^{-16}	m^2	Intrinsic permeability of air for Ottawa sand

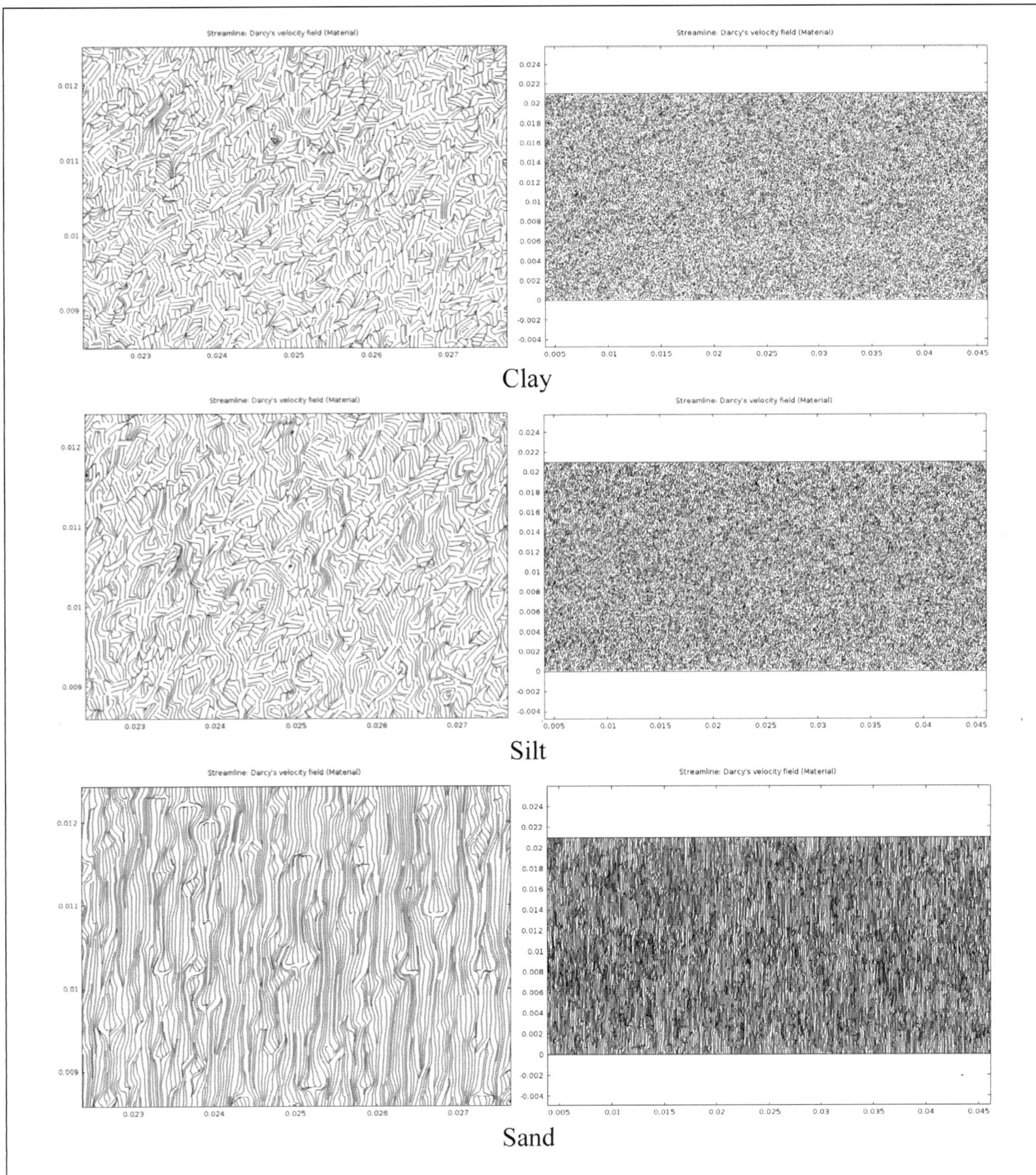

Figure 5. Flowlines simulated in the hydraulic conduction experiment in three different type of soils (the left figures are the enlarged images of the flowlines).

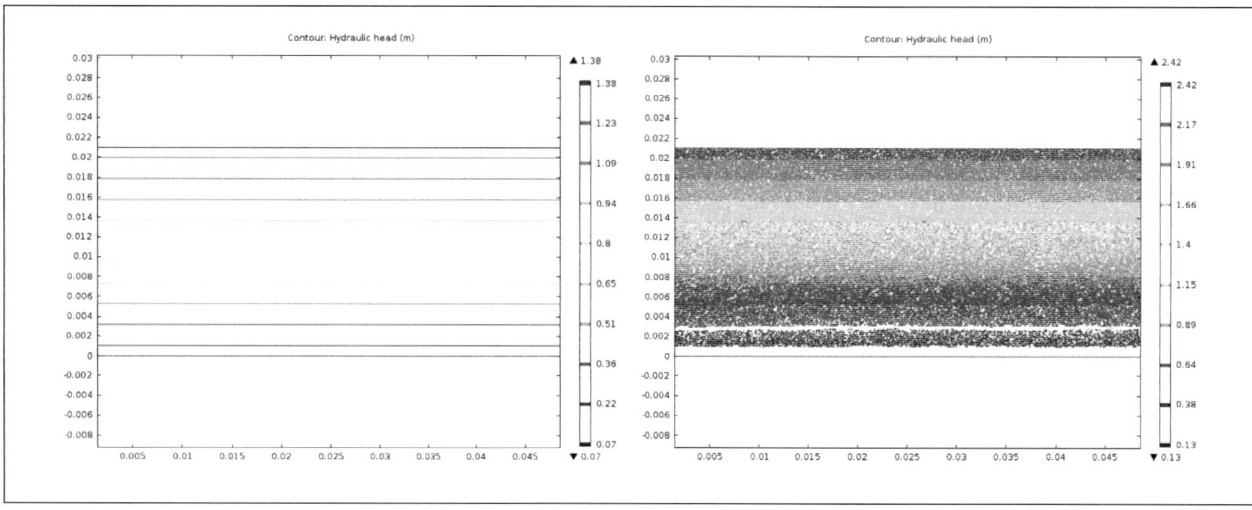

Figure 6. Contour plots of the hydraulic head distribution in the clay specimen (the microstructure is not considered in the left figure).

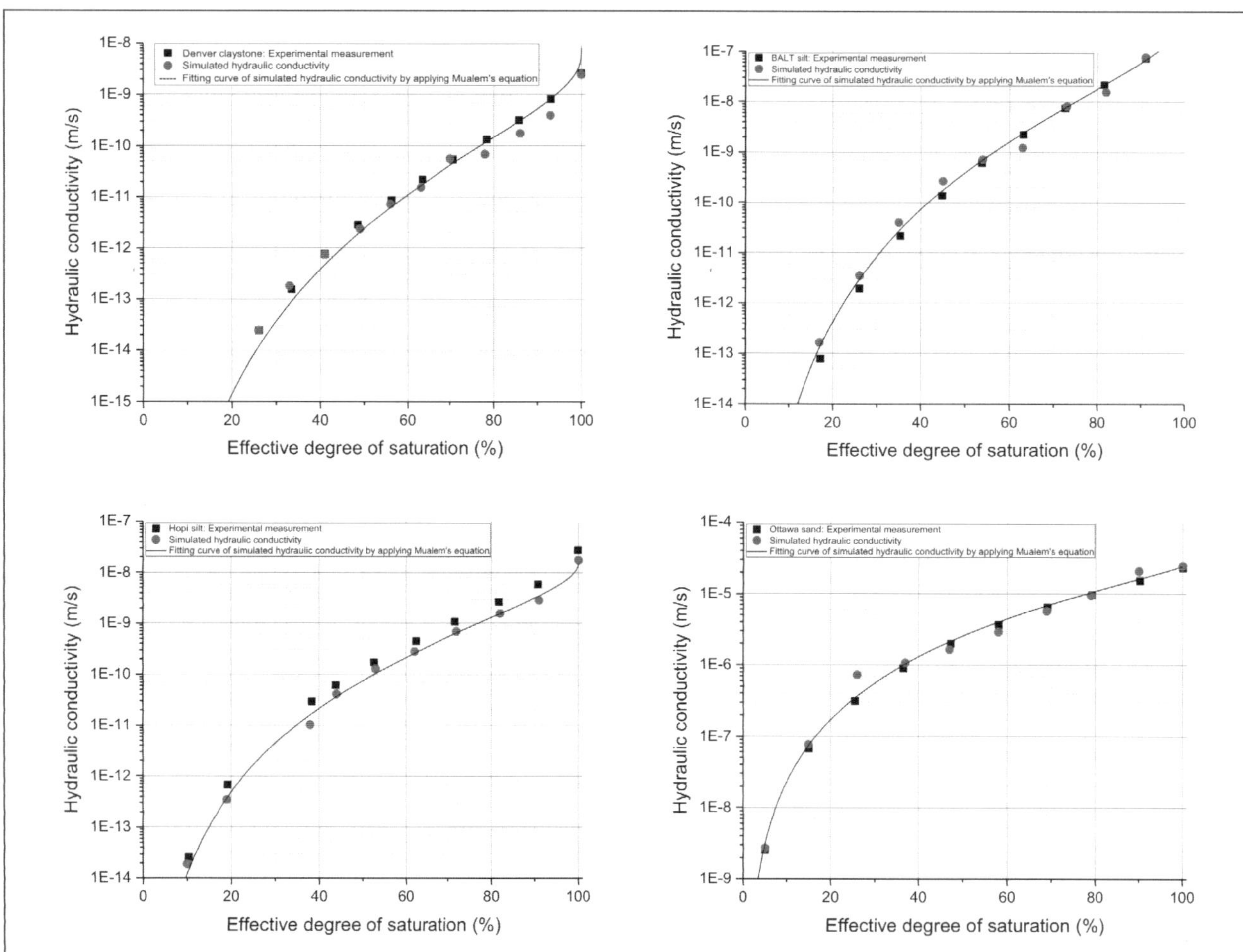

Figure 7. Comparison of the simulated hydraulic conductivity with the measured data from the experiments for the four soils and Mualem's equation.

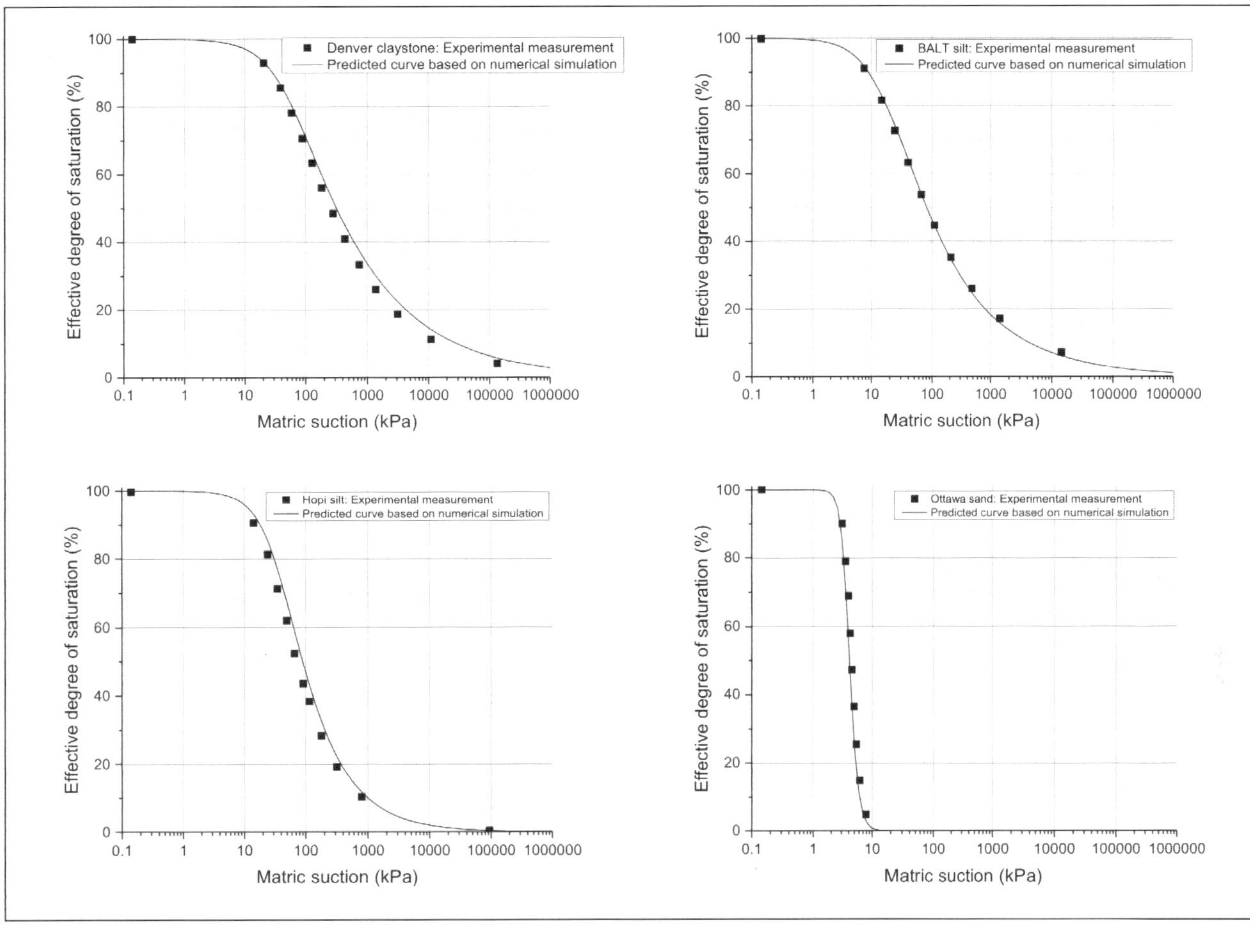

Figure 8. Comparison of the predicted SWCC curve with the measured data for the four soils.

results compared favorably with experimental results for clay, silt, and sand, as shown in Figure 7.

Mualem (*8*) proposed the following model to describe the variation of hydraulic conductivity K with the effective degree of saturation S_e:

$$K = K_s S_e^{1/2}[1 - S_e(S_e^{n/(1-n)} - 1)^{(n-1)/n}]^2 \qquad (2)$$

where n represents pore size parameter; K_s represents saturated hydraulic conductivity; and the effective degree of saturation can be defined as

$$S_e = \frac{S - S_r}{1 - S_r} \qquad (3)$$

where S_r represents residual saturation.

As shown in Figure 7, hydraulic conductivity of soils generally decreased abruptly (as much as seven orders of magnitude) from its saturated value, K_s, with decreasing saturation. The simulated hydraulic conductivities generally agree well with the measured hydraulic conductivities of Denver claystone, BALT silt, Hopi silt, and Ottawa sand. By fitting Mualem's equation to the simulated hydraulic conductivity, the pore size parameters n can be back-calculated as 1.38, 1.37, 1.69, and 6.10 for Denver claystone, BALT silt, Hopi silt, and Ottawa sand, respectively.

SWCC Prediction

From Figure 7, the fitted parameter n is calculated for different type of soils. Based on the soil-water retention equation raised by Genuchten (*14*): $S_e = (\alpha\psi^n + 1)^{\frac{1-n}{n}}$, we can plot the SWCC curve of soil specimens based on the simulation. In the equation, Ψ is the matric suction. For Denver claystone, BALT silt, Hopi silt, and Ottawa sand, the inverse values of the air-entry suction α are 0.02 kPa−1, 0.07 kPa−1, $\alpha = 0.03$ kPa−1 and $\alpha = 0.23$ kPa−1, respectively (*10*, *29*).

In Figure 8, the SWCC curve predicted by the numerical simulation agrees well with the SWCC curve obtained from flow test for different type of soils (*10*), which provides an inexpensive and efficient way to predict the SWCC curve.

Conclusion

This research developed a microstructure-based random finite element simulation model which provided a physics-based simulation of the hydraulic conductivity and soil-water retention of unsaturated soils. The advantage of this model is to simulate the bulk behaviors based on the behaviors of individual phases. This therefore could direct efforts to understand the behaviors of individual phases, which represent a more uniform material. Demonstrations were given on the calibration of hydraulic conductivity with experimental data, which was subsequently used to predict the bulk hydraulic conductivity of soil and SWCC at different extents of saturation. The results were consistent with experimental observations. The modeling approach combined the advantages of the discrete element model in simulating the behaviors of particulate system and the computational efficiency of the finite element model. It provided a new and reliable simulation tool to predict unsaturated soil behaviors, and also reduced the testing time and costs to incorporate unsaturated soil mechanics into engineering practice.

Acknowledgments

This research is partially supported by the Ohio Department of Transportation and U.S. National Science Foundation.

Author Contributions

The authors confirm contribution to the paper as follows: study conception and design: Xiong (Bill) Yu, Shaoyang Dong, Yuan Guo; data collection: Xiong (Bill) Yu, Shaoyang Dong; analysis and interpretation of results: Xiong (Bill) Yu, Shaoyang Dong; draft manuscript preparation: Xiong (Bill) Yu, Shaoyang Dong. All authors reviewed the results and approved the final version of the manuscript.

References

1. Jaafar, R., and W. J. Likos. Pore-scale Model for Estimating Saturated and Unsaturated Hydraulic Conductivity from Grain Size Distribution. *Journal of Geotechnical and Geoenvironmental Engineering*, Vol. 140, No. 2, 2013, p. 0401-3012.
2. Arya, L. M., F. J. Leij, P. J. Shouse, and M. T. Genuchten. Relationship Between the Hydraulic Conductivity Function and the Particle-Size Distribution. *Soil Science Society of American Journal*, 1999, 63, pp. 1063–1070.
3. Liu, Z., Y. Sun, and X. B. Yu. Theoretical Basis for Modeling Porous Geomaterials under Frost Actions: A Review. *Soil Science Society of America Journal*, Vol. 76, No. 2, 2012, pp. 313–330.
4. Gardner, W. R. Some Steady-State Solutions of the Unsaturated Moisture Flow Equation with Application to Evaporation from a Water Table. *Soil Science*, Vol. 85, No. 4, 1958, pp. 228–232.
5. Campbell, J. D. *Pore Pressures and Volume Changes in Unsaturated Soils*. PhD thesis. University of Illinois at Urbana Champaign, Urbana-Champaign, IL, 1973.
6. Davidson, J. M., L. R. Stone, D. R. Nielsen, and M. E. Larue. Field Measurement and Use of Soil-Water Properties. *Water Resources Research*, Vol. 5, 1969, pp. 1312–1321.
7. Averjanov, S. F. About Permeability of Subsurface Soils in Case of Incomplete Saturation. In English Collection, Vol. 7. As Quoted by P. Ya Palubarinova, 1962. The Theory of Ground Water Movement (English Translation by I. M. Roger DeWiest. Princeton University Press, Princeton, NJ, 1950, pp. 19–21.
8. Mualem, Y. Hysteretical Models for Prediction of the Hydraulic Conductivity of Unsaturated Porous Media. *Water Resources Research*, Vol. 12, No. 6, 1976, pp. 1248–1254.
9. McCartney, J. S., L. F. S. Villar, and J. G. Zornberg. Estimation of the Hydraulic Conductivity Function of Unsaturated Clays Using Infiltration Column Tests. *Proc., Sixth Brazilian Symposium on Unsaturated Soils*, Salvador, Brazil, 2007, pp. 321–328.
10. Lu, N., M. Kaya, and J. W. Godt. Interrelations Among the Soil-Water Retention, Hydraulic Conductivity, and Suction-Stress Characteristic Curves. *Journal of Geotechnical and Geoenvironmental Engineering*, Vol. 140, No. 5, 2014, pp. 04014007.
11. Fredlund, D. G., A. Xing, and S. Huang. Predicting the Permeability Function for Unsaturated Soils Using the Soil-Water Characteristic Curve. *Canadian Geotechnical Journal*, Vol. 31, No. 4, 1994, pp. 533–546.
12. Likos, W., and N. Lu. Filter Paper Technique for Measuring Total Soil Suction. *Transportation Research Record: Journal of the Transportation Research Board*, 2002. 1786: 120–128.
13. Likos, W. J., and N. Lu. Automated Humidity System for Measuring Total Suction Characteristics of Clay. *Geotechnical Testing Journal*, Vol. 26, No. 2, 2003, pp. 1–12.
14. Van Genuchten, M. T. A Closed-Form Equation for Predicting the Hydraulic Conductivity of Unsaturated Soils. *Soil Science Society of America Journal*, Vol. 44, No. 5, 1980, pp. 892–898.
15. Fredlund, D. G., and A. Xing. Equations for the Soil-Water Characteristic Curve. *Canadian Geotechnical Journal*, Vol. 31, No. 4, 1994, pp. 521–532.
16. Fenton, G. A., and D. V. Griffiths. Bearing Capacity Prediction of Spatially Random c-φ Soils. *Canadian Geotechnical Journal*, Vol. 40, No. 1, 2003, pp. 54–65.
17. Griffiths, D. V., J. Huang, and G. A. Fenton. Probabilistic Infinite Slope Analysis. *Computers and Geotechnics*, Vol. 38, No. 4, 2010, pp. 577–584.
18. Bharrucha-Reid, A. T. ed. *Probabilistic Methods in Applied Mathematics*. Academic Press, New York City, NY, 1968.
19. Sobczyk, K. *Wave Propagation in Random Media*. Elsevier, New York City, NY, 1985.
20. Chamis, C. C. Probabilistic Structural Analysis Methods for Space Propulsion System Components. *Probabilistic Engineering Mechanics*, Vol. 2, No. 2, 1987, pp. 100–110.

21. Griffiths, D.V., and G. A. Fenton. Seepage Beneath Water Retaining Structures Founded on Spatially Random Soil. *Géotechnique*, Vol. 43, No. 4, 1993, pp. 577–587.
22. Fenton, G. A., and D. V. Griffiths. Three-dimensional Probabilistic Foundation Settlement. *ASCE Journal of Geotechnical and Geoenvironmental Engineering*, Vol. 131, No. 2, 2005, pp. 232–239.
23. Andersland, O. B., and B. Ladanyi. *Frozen Ground Engineering*. Wiley, Hoboken, NJ, 2004.
24. Chang, C. S., and P. Y. Hicher. An Elasto-Plastic Model for Granular Materials with Microstructural Consideration. *International Journal of Solids and Structures*, Vol. 42, 2005, 42, pp. 4258–4277.
25. Ferber, V., J. C. Auriol, J. P. Magnan, Y. J. Cui, E. De Laure, and C. A. Gerente. Microstructural Model for the Volume Changes of Unsaturated Clayey Soils due to Wetting. *Fourth Proceedings of the International Conference on Unsaturated Soils (Unsaturated Soils)*, Carefree, AZ, 2006, pp. 861–872.
26. Tracy, S. R., K. R. Daly, C. J. Sturrock, N. M. J. Crout, S. J. Mooney, and T. Roose. Three-Dimensional Quantification of Soil Hydraulic Properties Using X-ray Computed Tomography and Image-Based Modeling. *Water Resources Research*, Vol. 51, No. 2, 2015, pp. 1006–1022.
27. Helliwell, J. R., C. J. Sturrock, K. M. Grayling, S. R. Tracy, R. J. Flavel, I. M. Young, W. R. Whalley, and S. J. Mooney. Applications of X-ray Computed Tomography for Examining Biophysical Interactions and Structural Development in Soil Systems: A Review. *European Journal of Soil Science*, 2013, 64, pp. 279–297.
28. Zhang, W. H., L. Y. Fu, Y. Zhang, and W. J. Jin. Computation of Elastic Properties of 3D Digital Cores from the Longmaxi Shale. *Applied Geophysics*, Vol. 13, No. 2, 2016, pp. 364–374.
29. Likos, W., and N. Lu. Filter Paper Technique for Measuring Total Soil Suction. *Transportation Research Record: Journal of the Transportation Research Board*, 2002. 1786: 120–128.

The Standing Committee on Engineering Behavior of Unsaturated Geomaterials (AFP60) peer-reviewed this paper (18-05995).

Article

TRR
JOURNAL OF THE TRANSPORTATION RESEARCH BOARD

Transportation Research Record
2018, Vol. 2672(52) 118–128

DOI: 10.1177/0361198118758625
journals.sagepub.com/home/trr
SAGE

Forensic Investigations into Recurrent Pavement Heave from Underlying Expansive Soil Deposits

Bhaskar C. S. Chittoori[1], **Debakanta Mishra**[1], **and Kazi Moinul Islam**[1]

Abstract

Pavement sections constructed over expansive soil deposits often exhibit excessive distresses due to volume changes in the underlying soil strata. Differential movements within these deposits resulting from fluctuations in the moisture content manifest themselves in the form of localized heaves on the pavement surface. A pavement section near the western border of Idaho has experienced recurrent damage due to volume changes in the underlying expansive soil layer; traditional treatment methods such as lime stabilization and moisture barrier installation have provided partial relief over the years. A recently concluded forensic research study at Boise State University investigated the causes for failure of earlier treatment methods. This study involved extensive laboratory characterization of expansive soil samples collected from underneath this pavement section to identify location of the problematic soil strata, and to propose suitable rehabilitation measures. Laboratory characterization included tests such as moisture content, Atterberg limits and One-Dimensional swell test to determine the potential vertical rise (PVR) and establish approximate active zone. Laboratory test results indicated that the most expansive soil deposits were at a depth of at least 1.83 m from the pavement surface. PVR values calculated closely matched with the surface profile trends observed in the field. In addition, the soluble sulfate tests performed on various soil samples indicated that sulfate heaving could be a problem for calcium-based stabilizer. Based on the findings, the research team proposed that the pavement section be reinforced using a flexible mechanical system that dissipates the swell pressures originating from the underlying clay layers.

Expansive soil is a term used to describe soils that undergo volume changes due to fluctuations in the moisture content (*1*). These soils experience significant swelling on increase in the moisture content, whereas removal of moisture from the soil leads to shrinkage. Expansive soils typically exhibit moderate to high plasticity, low to moderate strength, and high swell and shrinkage characteristics (*2–5*). Their shrink/swell characteristics have deteriorated many overlying structures, such as roads, residential buildings and other lightly loaded structures (*1, 6, 7*). The damage is particularly severe for lightly loaded structures such as pavements as the weight of these structures is often not sufficient to counter the upward heave forces exerted by the expansive soil deposits. A study sponsored by the National Science Foundation (NSF) reported that expansive soils cause more damage to light buildings and pavements than any other natural disaster including earthquakes and floods (*8*). Through a detailed review of the effects of expansive soils on civil engineering infrastructure, Gromko (*9*) estimated that the annual cost of damage from these soils in the United States alone is $2.3 billion. This number is likely much higher now. Puppala and Cerato (*10*) have estimated the cost to be about $13 billion in 2009. Vanapalli and Adem (*11*) present similar annual costs for several regions of the world including the United States, United Kingdom, France, Saudi Arabia, China, and Australia.

Volume changes observed in expansive soils on fluctuations in moisture content can be attributed to the presence of Montmorillonite, which has an expanding lattice, and undergoes significant expansion when exposed to water (*12*). Soils rich in these minerals are abundant in arid and semi-arid regions such as Australia, Canada, China, India, Israel, South Africa, and the United States (*1, 13*). In these regions, the evaporation rate is higher than the annual rainfall, thus leading to moisture deficiency in the soil. Subsequent rain increases the moisture content of the soil layers resulting in swelling behaviors. Similarly, in semi-arid regions a pattern of short periods of rainfall followed by long dry periods (drought) can develop, resulting in seasonal cycles of swelling and shrinkage.

[1]Department of Civil Engineering, Boise State University, Boise, ID

Corresponding Author:
Address correspondence to Bhaskar C. S. Chittoori: bhaskarchittoori@boisestate.edu

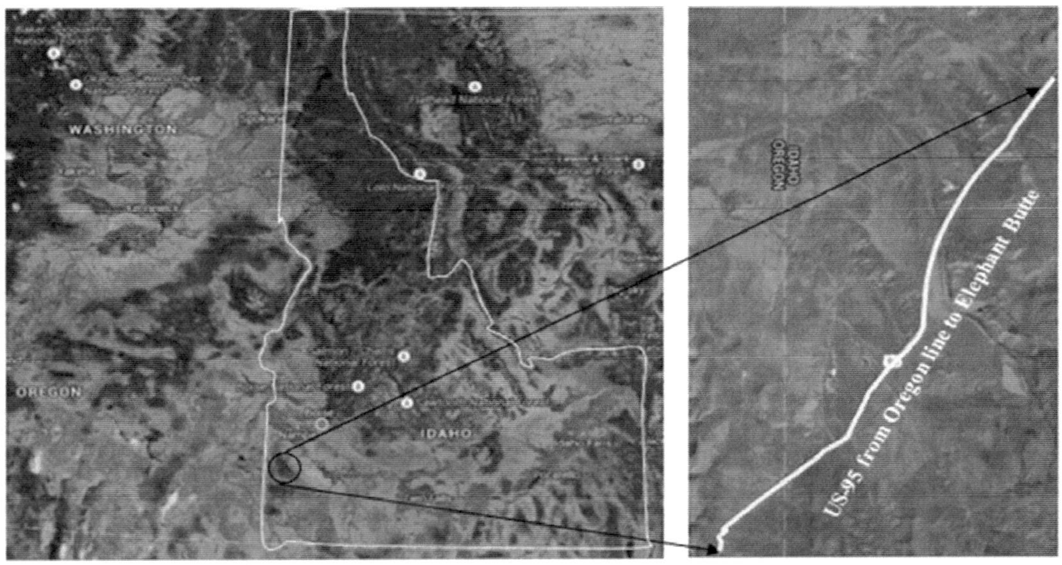

Figure 1. Aerial image showing project location (Source: Google Maps).

In the United States, expansive soil deposits can be found in several regions of states like Texas, Colorado, and California. This has led to extensive soil characterization and stabilization efforts by the respective state transportation agencies (*14–16*). Several guidelines have also been developed by these agencies specifying recommended construction practices in regions containing expansive soil deposits. As expansive soil deposits are not abundant in the state of Idaho, the Idaho Transportation Department (ITD) currently does not have specific guidelines concerning rehabilitation measures for pavement structures constructed over expansive soil deposits. However, a section of US Highway 95 (US-95) just east of the Idaho-Oregon border has experienced recurrent pavement damage due to volume changes in the underlying expansive soil deposit. Several reconstruction and rehabilitation measures have been implemented in the past with only partial success (*17*). A recently concluded research study at Boise State University focused on identifying the reasons behind the failed rehabilitation measures, and recommended suitable rehabilitation alternatives. Findings from this research study are presented here.

Objective and Scope

The primary objective of this study was to identify the primary mechanism governing the recurrent pavement damage for a section of US-95 near the Idaho-Oregon border. An extensive laboratory test matrix was designed and performed to map the exact location of expansive soil deposits underlying this stretch of the pavement. A total of 10 boreholes were drilled to collect soil samples corresponding to heave and non-heave zones in the pavement. A conceptual rehabilitation measure involving placement of a hybrid geosynthetic system was proposed. Details of the tasks undertaken to achieve these objectives are presented below.

Project Background

The problematic stretch of US-95 under consideration lies just east of the Idaho-Oregon border, and is 29.8 Km (18.5 mi) long. Figure 1 shows an aerial map of the location highlighting the segment of interest. Hardcastle (*17*) presented a brief construction history for this highway segment, that was originally constructed in the 1940s and was subsequently subjected to several reconstruction/rehabilitation efforts that resulted in the current alignment.

Chronology of Reconstruction

Sections of the highway exhibited signs of expansive soil-related problems a few years into the service life, and high-plastic soils with Liquid Limit (LL) values exceeding 125% were detected. Remedial measures attempted involved: (1) compaction at moisture contents slightly above the Optimum Moisture Content (OMC), (2) use of an impermeable asphalt membrane, and (3) increased surcharge loads through the construction pavement layers with greater than usual thicknesses. The first extensive geotechnical study for this problematic pavement section was performed in 1979 (*18*). Subsequently, Smith in 1986 (*19*) and Nottingham in 1988 (*20*) also reported the presence of moderate to high amounts of expansive soils in the subgrade, and recommended lime stabilization and/or membrane encapsulation as the most suitable alternatives to arrest the swell-shrink behavior. Nottingham (*20*) advocated,

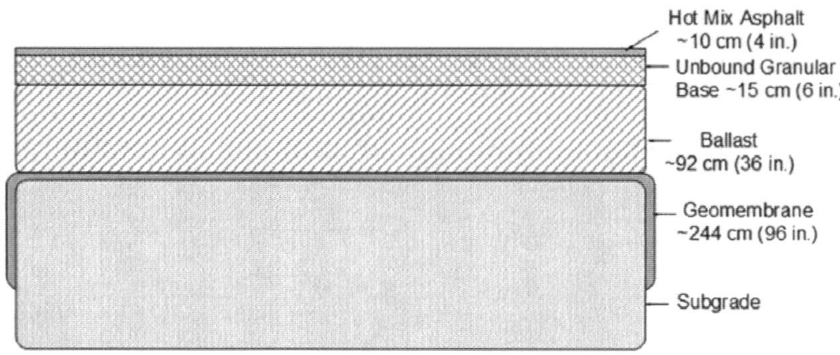

Figure 2. Schematic representation of the pavement section configuration proposed by Hardcastle (*17*).

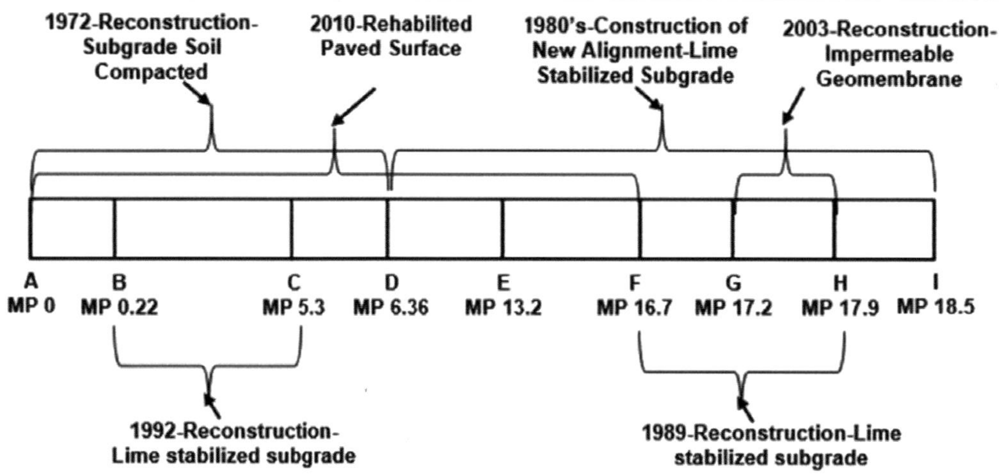

Figure 3. Schematic representation of different sections of the problematic highway segment highlighting different reconstruction/rehabilitation efforts (1 mi = 1.6 km).

and subsequently implemented, lime stabilization of the subgrade along with placement of an impermeable asphalt on top of the treated subgrade to prevent future ingress of moisture. However, by 1999, some of the reconstructed sections started exhibiting heaving-related distresses, which led to an extensive laboratory testing effort by Hardcastle (*17*).

Hardcastle's study comprised conventional geotechnical tests to establish the Atterberg limit and One-Dimensional swelling properties, along with tests such as X-Ray Diffraction (XRD) and Cation Exchange Capacity (CEC) aimed at identifying the soil mineralogical composition. Hardcastle (*17*) also evaluated candidate stabilization methods to remediate the problem of recurrent pavement heave, and recommended pavement reconstruction using well-graded aggregates with non-plastic fines and low hydraulic conductivities ($< 10^{-6}$ Cm/s). In circumstances where such materials were not available, the use of conventional free-draining base materials was recommended accompanied by the placement of an impervious asphalt or geosynthetic membrane on top of the subgrade. Figure 2 illustrates a schematic of this section configuration proposed by Hardcastle (*17*).

Figure 3 graphically identifies different sections of this 29.8 km long segment along with corresponding reconstruction/rehabilitation efforts. Different points along the segment have been identified using Mile Post (MP) numbers, as is the convention for transportation agencies in the United States. A chronological list of the reconstruction/rehabilitation efforts is presented in Table 1.

Subsurface Investigation to Identify Problematic Soil Layers

As already mentioned, past reconstruction and rehabilitation efforts implemented along this problematic section of US-95 have not been successful in eliminating the problem of recurrent differential heave at the pavement surface. Figure 4 shows photographs of the problematic highway segment

Table 1. Construction/Reconstruction Activities over the Life of the Project

Year	Segment ID	Details of reconstruction/rehabilitation efforts
1972	AD (MP 0.0–6.36)	Reconstruction - subgrade soils compacted wet of OMC
Early 1980s	DI (MP 6.36–18.5)	Construction of new alignment incorporating a 38 cm (15 in.) thick lime-stabilized subgrade layer
1989	FH (MP 16.7–17.9)	Reconstruction incorporating a 38 cm (15 in.) thick lime-stabilized subgrade layer
1992	BC (MP 0.22–5.3)	Reconstruction incorporating a 38 cm (15 in.) thick lime-stabilized subgrade layer
2003	GH (MP 17.2–17.9)	Reconstruction incorporating an impermeable geomembrane
2010	AF (MP 0.0–16.7)	Rehabilitated paved surface

(a)
(b)

Figure 4. Photographs taken in May 2015 showing pavement surface irregularities along the problematic section of US-95 between MP 16.0 and 17.0: (*a*) looking south on US-95; (*b*) looking north on US 95.

taken in May 2015. As clearly seen from the photographs, the pavement surface resembles a rolling pattern leading to significant discomfort for drivers, and raising safety concerns at relatively high operating speeds (88 km/h).

Drilling and Sample Collection

A total of 10 boreholes were drilled between MP 16.3 and 18.0, a segment of the highway that exhibited the most severe pavement distresses. The locations for these boreholes were strategically determined with boreholes drilled on and adjacent to surface heaves. The primary objective was to collect soil samples from underneath heave and non-heave portions of the highway to differentiate between the soil stratigraphy, and identify the cause of differential heave patterns. The boreholes were designated as AH-1, AH-2, through AH-10, where AH stands for 'Auger Hole'. Boreholes AH-1 through AH-8 corresponded to segments of the highway that exhibited noticeable surface heave. AH-9 and AH-10 on the other hand, were drilled outside this zone to identify any differences in the soil stratigraphy that could potentially contribute to the differential heaves.

Drilling and sample collection efforts were carried out in May–June 2015. Hollow stem augers were used for drilling, and sample collection was carried out using a continuous sample tube system manufactured by Central Mine Equipment Company (CME). This sampling system retains the soil inside a clear Polyvinyl Chloride (PVC) sampler tube with ~8 cm inside diameter and ~76 cm length. Standard Penetration Tests (SPT) were also performed at different depths during the drilling, and split-spoon samples were collected. Figure 5 shows photographs of the drilling and sample collection steps. The boreholes were drilled to different depths ranging between 366 cm and 793 cm, and were terminated when strong layers in the substructure made sampling difficult. The plastic tubes containing the soil samples were sealed, and transported back to the laboratory for detailed characterization.

All soil samples were stored in a moisture-controlled chamber until further testing. Table 2 lists the boreholes with their location specifics and termination depths measured from the pavement surface.

Laboratory Characterization of Soil Samples

Soil samples collected from the 10 boreholes were tested in the laboratory to establish the following physical, mechanical, and chemical characteristics: (a) Atterberg's limits; (b) One-Dimensional Swell Characteristics; (c) Soluble Sulfate Contents. Significant findings from the laboratory testing efforts are presented in the following sections along with relevant inferences.

In-Place Moisture Content

In-place moisture content values were established for the samples collected from the 10 boreholes. This was used to assess the degree of moisture access to each layer, and to make

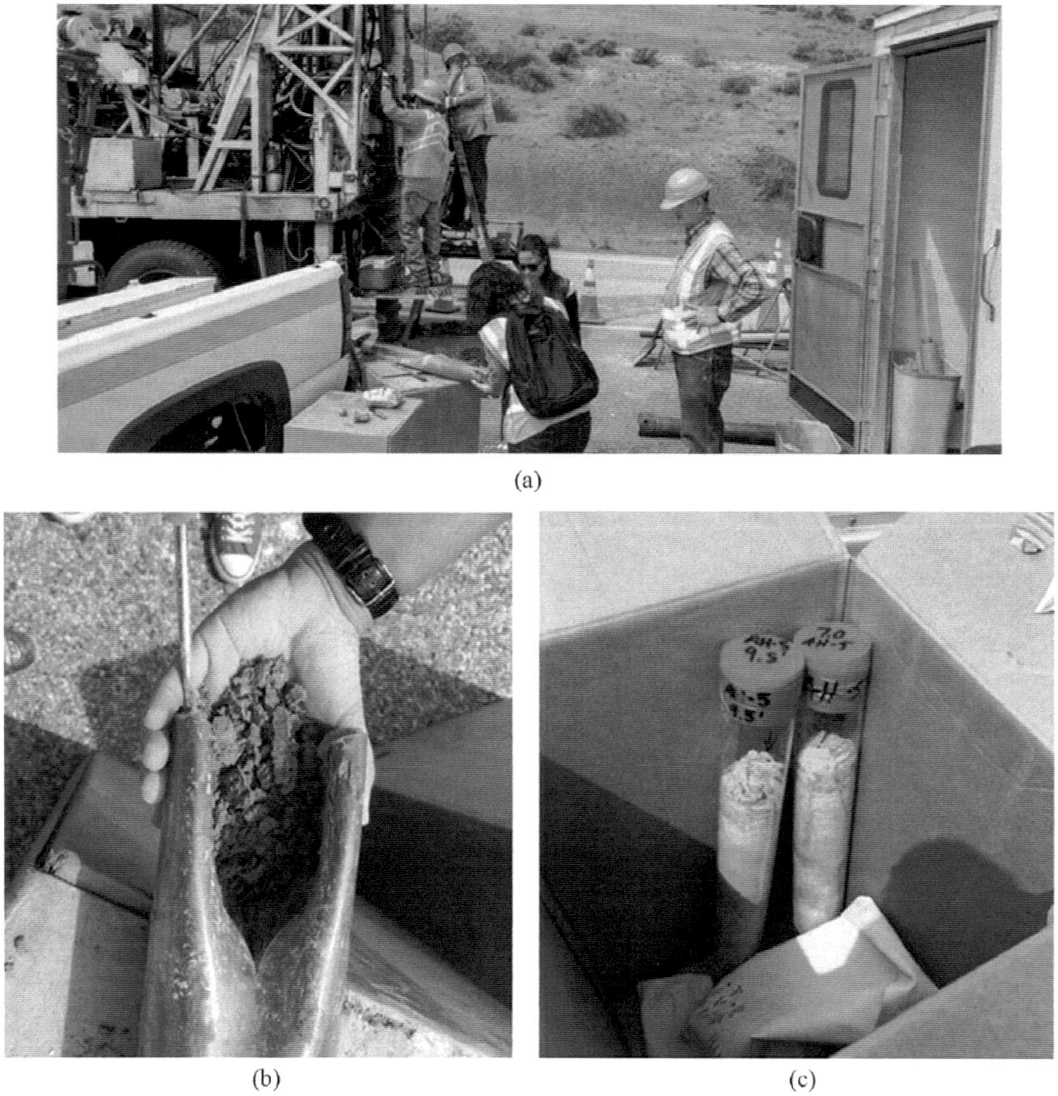

Figure 5. Photographs showing different steps in the drilling and sample collection operations: (*a*) view of the overall field operations; (*b*) end of the plastic tube sampler deformed on encounter with gravelly layers; (*c*) plastic tube samplers containing soil samples sealed and ready for transportation back to the laboratory.

Table 2. List of Boreholes with Location Specifics, and Termination Depths (Measured from the Pavement Surface)

Borehole ID	Location specifics	Termination depth (cm)	Borehole ID	Location specifics	Termination depth (cm)
AH-1	On the heave	793	AH-6	Next to heave	454
AH-2	Next to heave	637	AH-7	On the heave	564
AH-3	On the heave	686	AH-8	Next to heave	594
AH-4	Next to heave	357	AH-9	Non-heave zone	497
AH-5	On the heave	546	AH-10	Non-heave zone	396

inferences regarding the possible sources for moisture ingress. A total of 169 samples were tested corresponding to different depths within the 10 boreholes. Ground Water Table (GWT) was not encountered during the drilling operation at any of the boreholes. This indicated that the GWT was more than 7.9 m (26 ft) deep from the pavement surface. This indicated that

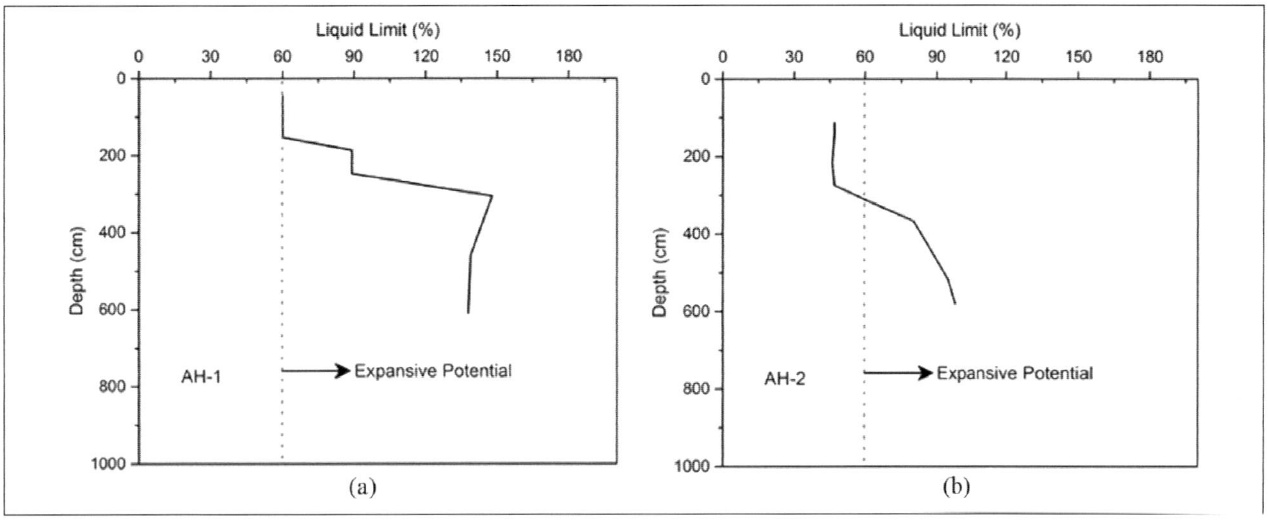

Figure 6. Variation in LLwith depth for (*a*) AH-1 and (*b*) AH-2 boreholes.

suction was most likely not the primary mechanism for moisture movement within the soil, and therefore, moisture ingress from other sources (such as cracks in the pavement, percolation from drainage ditches, etc.) was probably responsible for volume changes in the expansive soil deposits.

Atterberg Limits

Atterberg limit tests were conducted as per ASTM D4318 (*21*) test procedure. A total of 68 soil samples corresponding to different depths within the 10 boreholes were tested to establish the LL and Plasticity Index (PI) values. Note that LL and PI values are often used as measures of the swelling potential in clays. For example, LL values exceeding 60 are believed to correspond to highly expansive soils, whereas the corresponding value for PI is 35 (*22*). Establishing the LL and PI values of soil samples from different depths within the pavement substructure would help determine location of the highly expansive clay layers. LL values for the tested samples ranged from 25 to 165, whereas the PI values varied between 20 and 75. The LL values were plotted with depth, and a threshold value of 60 was used to differentiate between expansive and non-expansive soil layers. As an example, Figure 6 shows the variation in LL values with depth for AH-1 and AH-2. As seen from the figure, all samples tested from AH-1 exhibited severe swelling potential (LL > 60), with the highest LL value exceeding 150. Samples tested from AH-2 showed relatively lower expansion potential. Moreover, the soil samples corresponding to the first 274 cm (9 ft) within AH-2 had LL values lower than 60; the LL value gradually increased with depth, with a maximum value (LL ~100) observed at a depth of approximately 610 cm. This clearly demonstrates that although shallow soil layers at the AH-2 location were relatively non-expansive, the expansion potential increased significantly with depth. Moreover, samples from AH-1 were significantly more expansive in nature than those from AH-2. Similar trends of increasing expansion potential with depth were observed for the other boreholes. However, plots for all the boreholes have not been included in this article for brevity. It should be noted that all samples tested for AH-4, AH-9, and AH-10 exhibited LL values lower than 60, thus indicating no expansive layer at those locations. This observation will be used in the next section during the discussion of potential vertical rise (PVR) values.

Establishing the Active Zone using Liquidity Index Values. Establishing an approximate active zone for a given soil stratum will greatly benefit the comprehension of the expansive potential of a particular soil stratum. The "active zone" is defined as the zone of soil that is either contributing to, or has the potential to contribute to, heaving in expansive soils (*23*). In other words, the active zone corresponds to the depth over which significant fluctuations in the moisture content are likely to be observed; this, in conjunction with the presence of expansive soils, can lead to volume changes. The depth of the active zone at a particular location is influenced by the soil type, soil structure, topography, and climate (*4*). Nelson and Miller (*1*) reported that in soil clays and clay shales in the western United States, the depth of the active zone can be close to 1,500 cm (49 ft). To establish the active zone for the given pavement section, a method based on the soil Liquidity Index (LI) value, proposed by O'Neill and Poormoayed (*24*), was used. However, this method requires the LI data to be obtained over several seasons. Seasonal variation in LI indicates the moisture fluctuation which leads to shrink-swell behavior in expansive soils. Based on the data obtained for the 10 boreholes, a set of moisture data were selected over different years (instead of seasons). The moisture data for

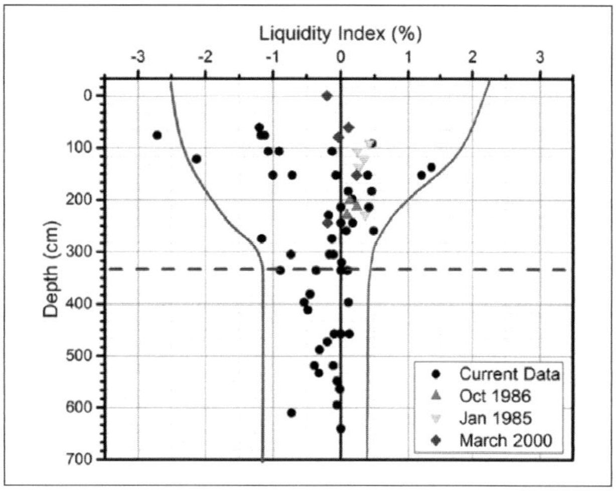

Figure 7. Active zone depth established using LI profile.

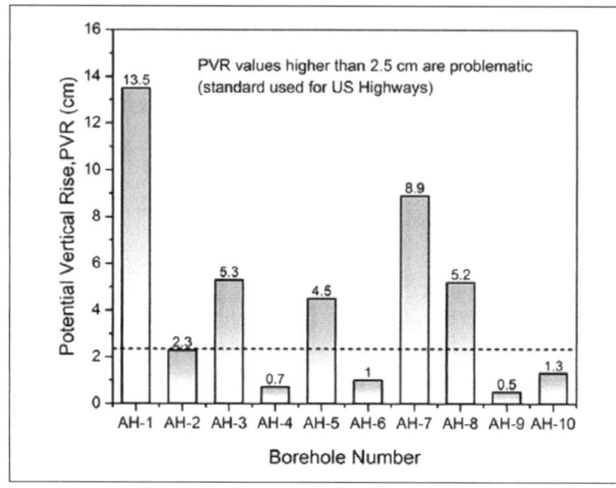

Figure 8. PVR values calculated for the ten boreholes drilled during the current study.

different years was collected by Nottingham (*20*) and Hardcastle (*17*) studies. Figure 7 shows the variation of LI values with depth for the pavement section under investigation. Note that data from all boreholes have been merged to prepare this plot. Moreover, data from previous studies have been used to represent temporal variation in the LI values. As seen from Figure 7, large variation in the LI values was observed up to a depth of 335 cm (11 ft). For depths greater than 335 cm, no significant temporal change in LI (and hence moisture content) was observed. This indicates that the active zone underneath the pavement section extends to a depth of least 335 cm (11 ft). It is important to highlight that this inference was made based on data collected from the current study, and limited data extracted from previous study. Therefore, it is possible that the active zone may extend to greater depths than indicated by the available data. Nevertheless, the LI data indicates that soils to a depth of at least 335 cm from the pavement surface may be contributing to the pavement heave problem. Accordingly, traditional soil stabilization/treatment options such as chemical stabilization, soil removal and replacement, etc. will not be suitable to remediate the recurrent pavement heave. Once the depth of the problematic zone was established, the next task involved quantifying the swell potential of the soil stratum. This was carried out using the concept of Potential Vertical Rise (PVR).

PVR Calculations. Originally developed by McDowell (*25*) in 1956, PVR is an index property linked to the expansion potential of individual layers in a soil stratum when subjected to moisture fluctuations under given density and loading conditions. Texas Department of Transportation (TxDOT) procedure Tex-124-E (*26*) was developed based on McDowell's original work, and has undergone several modifications since then. Although several other advanced methods to predict the swelling of expansive soil deposits have been developed over the years (*27–29*), Tex-124-E represents the simplest and most applicable method to predict the volume change in expansive soil deposits. The PVR value for a particular layer is calculated based on the following information: (1) layer depth, (2) LL, (3) percent moisture, (4) percent soil binder (material finer than 0.42 mm, or passing the No. 40 sieve), and (5) PI. Once the PVR value for each layer has been calculated, the overall PVR value for the stratum is calculated by summing the corresponding values for each layer. As far as pavements are concerned, the PVR value can be thought of as the maximum surface heave that a particular soil stratum will cause if unlimited access to moisture was ensured. According to Jayatilaka, et al. (*30*) PVR values greater than 2.54 cm are considered problematic for US Highways, and hence remedial measures need to be adopted in such scenarios.

PVR values for all 10 boreholes drilled within the problematic section of US-95 were calculated, and are shown in Figure 8. The objective was to estimate the theoretical swelling potentials for the soil layers corresponding to each borehole location. This would also help compare the trends in the PVR values to the observed pavement surface heaves. As seen from Figure 8, the maximum PVR value (13.5 cm) was observed for AH-1, followed by AH-7 (8.9 cm) and AH-3 (5.3 cm). Only four boreholes (AH-2, AH-4, AH-9, and AH-10) exhibited PVR values less than the threshold value of 2.5 cm specified for US Highways. Interestingly, it can be seen that all boreholes drilled through the heave zone (AH-1, AH-3, AH-5, and AH-7; refer to Table 2) exhibited higher PVR values compared to those drilled next to the heave zones (AH-2, AH-4, AH-6, and AH-8; refer to Table 2). AH-9 and AH-10 were drilled outside the heave zone, and accordingly exhibit the lowest PVR values (0.5 cm and 1.3 cm, respectively). It can be seen that the trend in PVR values (Figure 8) closely corresponds to the visual assessment

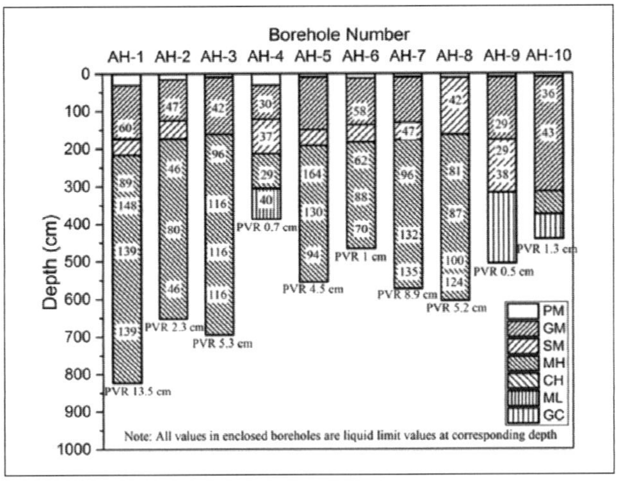

Figure 9. PVR for different boreholes along with LL values established at different depths.

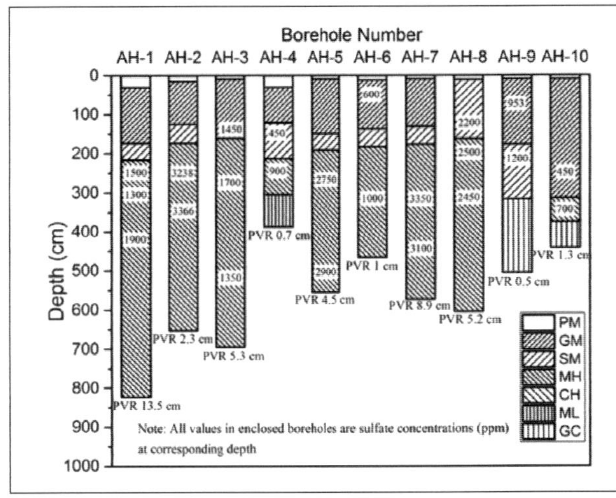

Figure 10. PVR for different boreholes along with sulfate content values established at different depths.

of the surface heaves at the borehole locations (Table 2). Note that the low PVR values for AH-4, AH-9, and AH-10 directly correspond to the low LL values reported in the previous section. Therefore, it appears that PVR values can be used as representative indicators for the likelihood of pavement surface heaves along this section of US-95. Note that the PVR is primarily an index value, and assumes a case where the soil stratum has unlimited access to moisture. However, this may not be a reasonable assumption for all scenarios as the moisture access to a particular soil stratum is largely dependent on site conditions. Therefore, PVR values established for a particular borehole may not necessarily correspond to the actual amount of surface heave observed at that location. Nevertheless, establishing the PVR values for all boreholes in the section can help identify locations likely to exhibit the most swelling, and hence can be useful during the planning of rehabilitation methods. Figure 9 shows the PVR values for the different boreholes along with LL values established at different depths. Different fill patterns in the profiles correspond to different soil types encountered during the drilling process. As shown in the figure, LL values as high as 164 were observed in the subgrade soil, accounting for the severe differential heave problems.

In an effort to compare the PVR trends with actual heave amounts observed along the pavement surface, the research team collected International Roughness Index (IRI) values along the problematic pavement section from ITD's inertial pavement profiler database. IRI values corresponding to the exact locations (in terms of latitude and longitude) of the boreholes were observed to follow a trend similar to those exhibited by the PVR values. Therefore, PVR values for the boreholes can be reasonably assumed to be indicative of actual expansion potential for the soil layers at those locations. Detailed discussions on this comparison effort have been eliminated from this article for the sake of brevity and will be presented in future publications.

Soluble Sulfate Content

Soluble sulfate measurement in subgrade soils is an integral part of geotechnical investigations due primarily to sulfate-induced heave problems experienced by certain chemically treated sulfate soils. The presence of sulfates in expansive soils can lead to the formation of high-swelling minerals like ettringite and thaumasite when treated with calcium-based stabilizers such as lime and cement. It is therefore important to establish the sulfate content of soils requiring chemical stabilization. Sulfate content tests were conducted on soil samples obtained from the 10 boreholes to assess the viability of treatment using calcium-based stabilizers such as lime and cement. It should be noted that LL and LI profiles established the presence of clay layers with high expansive potential up to 6 m underneath the pavement surface. Although chemical stabilization of soil layers that far below the pavement surface is not possible (due to construction feasibility related issues), quantifying the sulfate contents would indicate the suitability of lime/cement stabilization for shallow soil layers. Hayes (*31*) prescribed threshold values for sulfate contents to classify soils based on their likelihood for sulfate attack. Based on Hayes' threshold values, soil samples with sulfate contents higher than 2,000 parts per million (ppm) can be marked as candidates for "severe" degree of sulfate attack, whereas soils with sulfate content values between 1,000–2,000 ppm can be marked as candidates for "considerable" sulfate attack.

42 soil samples were selected from the 10 boreholes to determine the sulfate content using a test method proposed by Puppala et al. (*32*). The sulfate content values ranged from 450 to 3,366 ppm. Figure 10 shows a plot similar to that in Figure 9, except this time the LL values in the soil profiles have been replaced with sulfate content values. As shown in Figure 10, soil samples from different depths in AH-1 exhibited sulfate content values up to 1,900 ppm, therefore

classified as candidates for "considerable" sulfate attack. Samples from AH-2, AH-5, and AH-8 indicated "severe" potential for sulfate attack. Lastly, boreholes AH-9 and AH-10 do not pose any considerable risk for sulfate attack. From Figure 10, it is apparent that sulfate attack may be a considerable problem for the problematic pavement section, if calcium-based stabilization alternatives are pursued.

One-Dimensional Swell Test Results

The swell pressures for the soil samples collected from the 10 boreholes were established following ASTM D 4546-90, "One Dimensional Swell or Settlement of Cohesive Soils"; both Methods A and B were used in the current study. In Method A the test specimen is allowed to expand after being inundated under a 1.0 kPa vertical stress. The change in the specimen height expressed as a percentage of the original height is defined as the "free swell." After free swell is complete, the specimen is loaded incrementally as in the conventional One-Dimensional compression test until the specimen height is equal to or smaller than its original height (usually 25 mm). The swell pressure is defined as the vertical stress required to bring the specimen back to its original thickness. Method B is similar to Method A except that the initial vertical stress under which the specimen undergoes its free swell is selected by the operator. A total of 46 samples were tested from the 10 boreholes to determine the swell pressure of the soil. This was important to quantify the heave pressures exerted by different soil layers within the pavement substructure. Samples from AH-1 exhibited the highest swell pressure values (168 kPa). This again corroborates the high swelling observed in the field at AH-1 borehole and the PVR values reported in the previous sections. Samples from AH-3, AH-6, AH-9, or AH-10 did not demonstrate significant swell pressure values.

Inferences Based on Laboratory Test Findings

The following inferences can be drawn based on findings from the laboratory experimental matrix:

1. The Atterberg limit values established that the soil layers corresponding to most boreholes (except for AH-4, AH-9, and AH-10) exhibit expansive characteristics at depths greater than 183 cm (6 ft) from the pavement surface.
2. The LI profile exhibited that that active zone could extend up to a depth of at least 335 cm (11 ft) below pavement surface. Accordingly, shallow stabilization efforts may not be suitable to mitigate the differential heave problems.
3. Investigation of pavement roughness profiles using IRI values corroborated trends predicted from PVR calculations.
4. Soluble sulfate content tests indicated that soils from several of the boreholes are susceptible to high levels of sulfate attack if chemical stabilization using calcium-based stabilizers is pursued.
5. Results from One-Dimensional swell tests were not conclusive for all borehole locations. However, samples from AH-1 exhibited the highest swell pressure values (168 kPa), which corresponds to the fact that the pavement section overlying AH-1 exhibited the highest level of swelling.

Potential Rehabilitation Approach Comprising a Hybrid Geosynthetic System

Based on findings from the laboratory test matrix, it can be concluded that chemical stabilization or shallow treatment alternatives are not likely to be successful in mitigating the recurrent differential heave problems. Considering the significant depth (from the pavement surface) of expansive soil layer deposits, and also the fact that complete prevention of moisture fluctuation within the subsurface layers is not practical, alternatives to improve the structural response of the pavement section against heaving need to be explored. In other words, the pavement section needs to be reinforced using a flexible mechanical system that dissipates the swell pressures originating from the underlying clay layers, thus eliminating the differential heaving. One such alternative involves the use of a hybrid geosynthetic system that is strong enough to sustain the traffic-induced stresses, and at the same time is flexible enough to dissipate subgrade-generated swell pressures.

A hybrid geosynthetic system combines two or more geosynthetic types to address particular engineering problems. One particular instance of using a hybrid geosynthetic system (comprising geocells and geogrids) to mitigate problems related to expansive soils was reported by Kief (2015) (*33*). In this particular application, a section of a railroad track overlying an expansive soil deposit was reinforced using a combination of geogrids and geocells. A stiff biaxial geogrid was placed at the subgrade/pavement interface, and two stiff geocell layers were embedded in the granular sub-ballast layer. The semi-rigid section thus constructed, served as a "bridge" between the expansive soil and the upper track structure, thereby absorbing or smoothing the swell pressures and significantly reducing track geometry defects, as established through track monitoring measurements. A similar rehabilitation approach can be adopted for the problematic section of US-95 under investigation. Further studies are underway to explore this aspect and the results from these studies will be discussed in future publications.

Summary and Conclusions

This paper presented findings from a recently completed research study undertaken at Boise State University to investigate the problem of differential heaving in a section of

US-95 at the Idaho-Oregon border. A total of 10 different boreholes were strategically drilled at different locations along the problematic pavement section, and soil samples were collected for extensive laboratory testing. Different physical, mechanical, and chemical tests conducted on the soil samples established that the problematic soil layers were deep within the subgrade, and therefore traditional rehabilitation measures such as chemical stabilization, or soil removal and replacement are not likely to address the root cause of the problem. It was also determined that the active zone in this area could extend below 3.3 m from pavement surface. The concept of a hybrid geosynthetic system was proposed as a possible means to dissipate the excessive swell pressures originating from the underlying soil layers.

Acknowledgments

The authors are thankful to Keith Nottingham, John Arrambari, and Daris Bruce of ITD for their consistent help and support. The help by Sikha Neupane, Thomas Robbins, S M Naziur Mahmud, Barbara Santana, and Mir Md. Tamim during the laboratory testing of soils is greatly acknowledged.

Author Contributions

The authors confirm contribution to the paper as follows: study conception and design: B. C. S. Chittoori, D. Mishra; data collection: K. M. Islam; analysis and interpretation of results: B. C. S. Chittoori, D. Mishra and K. M. Islam; draft manuscript preparation: B. C. S. Chittoori, D. Mishra and K. M. Islam. All authors reviewed the results and approved the final version of the manuscript.

References

1. Nelson, J. D., and D. Miller. *Expansive Soils Problems and Practice in Foundation and Pavement Engineering*. John Wiley & Sons, Hoboken, NJ, 1997.
2. Holtz, W. G., and H. J. Gibbs. Engineering Properties of Expansive Clays. *Transaction of ASCE Vol*, Vol. 121 SRC-, 1956, pp. 641–667.
3. Aitchison, G. D. The Quantitative Description of the Stress-Deformation Behavior of Expansive Soils-Preface to Set of Papers. *Proc., 3rd International Conference on Expansive Soils Vol No 658 pp 1719 ALRawas The Factors Controlling the Expansive Nature of the Soils and Rocks of the Northern Oman Engineering Geology 53 327350*, Vol. 81 SRC-, 1999, pp. 79–82.
4. Chen, F. H. *Foundations on Expansive Soils*. Elsevier Scientific Publishing Co., Amsterdam, Netherlands, 1988.
5. Sherwood, P. T. *Soil Stabilization with Cement and Lime: State of the Art Review*. HMSO, London, UK, 1993.
6. Chittoori, B., and A. J. Puppala. Quantitative Estimation of Clay Mineralogy in Fine-Grained Soils. *Journal of Geotechnical and Geoenvironmental Engineering*, Vol. 137, No. 11, 2011, pp. 997–1008. https://doi.org/10.1061/(ASCE)GT.1943-5606.0000521.
7. Pedarla, A., S. Chittoori, and A. Puppala. Influence of Mineralogy and Plasticity Index on the Stabilization Effectiveness of Expansive Clays. *Transportation Research Record: Journal of the Transportation Research Board*, 2011. 2212: 91–99.
8. Jones, D. E., and W. G. Holtz. Expansive Soils- The Hidden Disaster. *Civil Engineering-ASCE*, Vol. 43, No. 8, 1973, pp. 49–51.
9. Gromko, G. J. Review of Expansive Soils. *Journal of the Geotechnical Engineering Division*, Vol. 100 SRC-, 1974, pp. 667–687.
10. Puppala, A., and A. Cerato. Heave Distress Problems in Chemically-Treated Sulfate-Laden Materials. *Geo-Strata*, Vol. 10, No. 2, 2009, p. 28.
11. Adem, H. H., and S. K. Vanapalli. Constitutive Modeling Approach for Estimating *1-D* Heave with Respect to Time for Expansive Soils. *International Journal of Geotechnical Engineering*, Vol. 7, No. 2, 2013, pp. 199–204. https://doi.org/10.1179/1938636213Z.00000000024.
12. Mitchell, J. K., and K. Soga. *Fundamentals of Soil Behavior*. John Wiley & Sons Inc., New York, NY, 2013.
13. Al-Rawas, A. A. The Factors Controlling the Expansive Nature of the Soils and Rocks of Northern Oman. *Engineering Geology*, Vol. 53, No. 3, 1999, pp. 327–350. https://doi.org/http://dx.doi.org/10.1016/S0013-7952(98)00085-4.
14. Ardani, A. *Expansive Soil Treatment Methods in Colorado*. Colorado Department of Transportation, Denver, CO, 1992.
15. Transportation, T. *Guidelines for Modification and Stabilization of Soils and Base for Use in Pavement Structures*. Texas Deparment of Transportation, Austin, TX, 2005.
16. Jones, L. D., and I. Jefferson. Expansive Soils. In *ICE manual of geotechnical engineering. Volume 1, geotechnical engineering principles, problematic soils and site investigation* (J. Burland, ed.), ICE Publishing, London, UK, 2012, pp. 413–441.
17. Hardcastle, J. H., and I. T. D. Report. *Evaluation and Treatment of Expansive Volcanic Soils US95 Owyhee County, Idaho*. National Institute for Advanced Transportation Technology, University of Idaho, Moscow, ID, 2003, pp. 2000–2185.
18. Ulrich, M. 1979. *Soils and Geotechnical Investigations for a Portion of Elephant Butte South*. Project Idaho FLH 11-1(3). September, 1979.
19. Smith, R. M. Expansive Subgrade Investigation. In *IntraDepartmental Correspondence to District 3 Materials Engineer*. Idaho Transportation Department, Boise, ID, 1986, p. 2.
20. Nottingham, K. N. *Swelling Soils-Oregon Line to Elephant Butte*. Intra-Departmental 1 Correspondence to District 3 Materials Engineer, Idaho Transportation Department, Boise, ID, 1988, p. 23.
21. American Society for Testing and Materials. *ASTM D4318 Test Methods for Liquid Limit Plastic Limit and Plasticity Index of Soils*. Annual Book of Sec Vol 0408 Soil and Rock, American Society for Testing and Materials, West Conshohocken, PA, 2001.
22. Department of the Army. *Foundation in Expanisve Soils*. Technical Manual 5-818-7. Department of the Army, 1984.
23. Nottingham, K. N. Swelling Soils-Oregon Line to Elephant Butte. In *Intra-Departmental Correspondence to District 3 Materials Engineer*. Idaho Transportation Department, Boise, ID, 1988, p. 23.
24. Neill, M. W., and N. Poormoayed. O'and Methodology for Foundations on Expansive Clays. *Journal of the Geotechnical Engineering of Civil Engineers Vol No GT12*, Vol. 106 SRC-, 1980, pp. 1345–1367.

25. McDowell, C. Interrelationship of Load, Volume Change, and Layer Thickness of Soils to the Behavior of Engineering Structures. *Proceedings of the Highway Research Board*, No. 35 SRC-, 1956, pp. 754–772.
26. Texas Department of Transportation (TxDOT). *Test Procedure for Determining Potential Vertical Rise*. Designation Tex124E. TxDOT, Austin, TX, 1999, p. 11.
27. Lytton, R. L. *The Characterization of Expansive Soils in Engineering. Presentation at the Symposium on Water Movement and Equilibria in Swelling Soils*, American Geophysical Union, San Francisco, CA, 1977.
28. Mitchell, P. W., and D. L. Avalle. A Technique to Predict Expansive Soil Movements. *Proc., 5th International Conference on Expansive Soils*, Institution of Engineers, Adelaide, Australia, 1984.
29. Jayatilaka, R., and A. Texas. *A Model to Predict Expansive Clay Roughness in Pavements with Vertical Moisture Barriers*. Doctoral thesis. Texas A&M University, College Station, TX, 1999.
30. Jayatilaka, R., D. A. Gay, R. L. Lytton, and W. K. Wray. Effectiveness of Controlling Pavement Roughness due to Expansive Clays with Vertical Moisture Barriers. *Texas Transportation Institute Research Report May*, 1993, pp. 1165-2-NaN-0.
31. Hayes, C. *Test Methods for Water-Soluble Sulfate in Soils*. PCA Research No 3016. Portland Cement Association, Skokie, IL, 2007.
32. Puppala, A. J., C. Viyanant, A. P. Kruzic, L. Perrin, and G. Journal. Evaluation of a Modified Soluble Sulfate Determination Method for Fine-Grained Cohesive Soils. *Geotechnical Testing Journal*, Vol. 25 SRC-, 2002, pp. 85–94.
33. Keif, O. The Hybrid Geosynthetic Solution for Rail Track on Expansive Clay. *Proc., Geosynthetics Conference*, Portland, OR, February 2015.

The Standing Committee on Geotechnical Site Characterization (AFP20) peer-reviewed this paper (18-06304).

Article

TRR
JOURNAL OF THE TRANSPORTATION RESEARCH BOARD

Transportation Research Record
2018, Vol. 2672(52) 129–141

DOI: 10.1177/0361198118758631
journals.sagepub.com/home/trr
SAGE

Soil–Geosynthetic Interface Shear in Different Testing Scales

Gholam H. Roodi[1], Amr M. Morsy[1], and Jorge. G. Zornberg[1]

Abstract

Geosynthetics have been used to improve mechanical performance of roadway layers (e.g., geosynthetic-reinforced asphalt, geosynthetic-stabilized bases) and a wide range of transportation infrastructures (e.g., geosynthetic-reinforced soil walls). A key aspect in understanding soil–geosynthetic interaction mechanisms involved in each application includes characterization of the interface between geosynthetics and adjacent materials. This study evaluates soil–geosynthetic interface shear in various pullout test scales including standard, smaller than standard, and larger than standard scales. Experimental results obtained from tests conducted in each scale were analyzed to determine the soil–geosynthetic interface shear model. An iteration procedure, similar to that used in *t–z* analysis of pile loading, was developed to simulate incremental geosynthetic movements. Shape and parameters of the interface shear model were changed to minimize the residual error between experimental and simulated data. It was found that mobilization of the interface shear in the small-scale test differs from that in the standard- and large-scale tests. In the standard- and large-scale tests, the ultimate soil–geosynthetic interface shear mobilized at comparatively small displacements, which could be represented by a linear plastic interface shear model. In the small-scale test, however, the interface shear developed in two phases. A portion of the ultimate interface shear mobilized at comparatively small displacements while additional resistance continued to mobilize at extended displacements. Consequently, the development of interface shear resistance in the standard- and large-scale tests was found to depend on progressive increase of the geosynthetic mobilized length, whereas in the small-scale test the interface shear resistance developed by displacement of the entire geosynthetic.

Geosynthetics have been widely used to improve mechanical performance of pavement layers as well as a wide range of transportation infrastructures. Specifically, geosynthetic layers have been used to reinforce hot mix asphalt concrete (*1*), to stabilize unbound granular base course (*2,3*), to stabilize subgrade layer, and to reinforce backfill materials in retaining structures (*4,5*). Quantification of the benefits derived from the use of geosynthetics in each application requires understanding characteristics of the geosynthetic material, the surrounding materials, and the soil–geosynthetic interaction.

Soil–geosynthetic interaction has typically been evaluated using interface direct shear and pullout tests (*6,7*), which mobilize the soil–geosynthetic interface shear in different modes. Pullout test would be relevant in cases where elongation of the geosynthetic specimen is considered. Results obtained from pullout tests have been used to gain better understanding of (1) ultimate conditions in structures that are designed using limit states conditions (e.g., geosynthetic-reinforced soil walls) (*4,8*) and (2) service conditions in structures that are designed using serviceability criteria (e.g., geosynthetic-stabilized roadways) (*9,10*).

This study evaluates soil–geosynthetic interface shear in three pullout test scales including: a standard scale, a significantly smaller and a larger than standard scales. Experimental results obtained from tests conducted in the three scales were analyzed to estimate soil–geosynthetic interface shear model that governed the response in each scale. An iteration procedure, similar to that used in *t–z* analysis of pile loading, was developed to simulate incremental geosynthetic movements. The soil–geosynthetic interface shear model was optimized to minimize the residual error between experimental and simulated data.

Background

Modeling soil–geosynthetic interaction requires suitable constitutive relationships to be defined for the soil, the geosynthetic, and the soil–geosynthetic interface. Although the unit tension–strain response of geosynthetics is inherently

[1]Department of Civil, Architectural, and Environmental Engineering, The University of Texas at Austin, Austin, TX

Corresponding Author:
Address correspondence to Gholam H. Roodi: hroodi@utexas.edu

nonlinear, a linear response has often been adopted (*11–13*). Nonlinear functions have also been used, including polynomial functions (*14,15*) and hyperbolic functions (*16*).

Interface constitutive relationships relate the shear stress mobilized at the soil–geosynthetic interface to the relative displacements that are mobilized between the two materials. Models adopted in previous studies to describe the interface interaction include: linear elastic (*13*); linear elastic–perfectly plastic (*11,14*); rigid–perfectly plastic (*12*); bilinear and hyperbolic models (*15*); and more complex nonlinear multiphase models such as elasto-plastic strain hardening and softening models (*11,16*).

This study focuses on determining the soil–geosynthetic interface shear model in various scales of pullout testing. Unlike conventional approaches in evaluation of pullout test results, in which constitutive models for soil–geosynthetic interface shear and geosynthetic material were typically determined using additional tests (e.g., direct shear test, wide width tensile test), this study adopted a procedure in which the pullout test results can directly be used to identify constitutive models for the geosynthetic and the soil–geosynthetic interface shear.

Factors that may affect the pullout test results include thickness of the soil, dimensions of the geosynthetic specimen, loading rate, and friction between soil and the side and front walls of pullout box (*17–19*). The large volume of soil and extensive efforts required to conduct experiments in large scales makes the use of this scale particularly onerous in studies where production of large volume of data is essential. In addition, although avoiding boundary effects may be essential in studies on large-displacement parameters, when the focus is on ultimate condition, studies on stiffness parameters, when the focus is on the response under small displacements, may not be sensitive to boundary effects. The results obtained in this study underscore that the difference between the pullout test data in various testing scales may partially be attributed to the different soil–geosynthetic interface shear models that govern the response in each scale.

Methodology

Pullout tests were conducted and results were back-analyzed to predict the constitutive models for geosynthetic and soil–geosynthetic interface. The constitutive model for the geosynthetic was obtained using data collected from the unconfined length of the geosynthetic. The constitutive model for soil–geosynthetic interface was estimated by simulation of the pullout data in the confined length using a procedure similar to *t–z* analysis procedure originally developed for simulation of deep foundations subjected to axial loading (*20*). This procedure and specific algorithm to simulate pullout test data are presented next.

t–z *Analysis*

The *t–z* analysis approach is a method that was developed to predict the load transfer between piles and surrounding soils. This method is deemed as one of the most widely used techniques to estimate the settlement of individual axially loaded piles. This approach involves modeling the pile as a series of discrete elements connected axially by nonlinear springs, which represent the axial stiffness of the pile. In addition, these elements are connected through their skin to surrounding soil by other nonlinear springs (*t–z* relationship), which represent the resistance of the soil in skin friction. An additional nonlinear spring is assumed at the pile tip, which represents the end-bearing (*q–y* relationship). Several research studies were conducted and obtained *t–z* and *q–y* relationships empirically based on model and full-scale pile load tests (*21*). Subsequent studies proposed general recommendations for estimating *t–z* and *q–y* curves (*22*). Theoretical approaches have also been developed to relate the *t–z* and *q–y* relationships to the soil properties surrounding the pile (*23,24*). A few studies have also adopted *t–z* analysis to simulate geosynthetic pullout test results (*11, 25, 26*).

Solution Algorithm for Pullout Test Data

The general approach adopted in this study to simulate experimental pullout test data involves discretizing the geosynthetic specimen into several segments, adopting an incremental displacement to the free end of the geosynthetic, and satisfying the force equilibrium equation of each segment using an iterative procedure to eventually determine the displacement, strain, interface shear, and load profile along the geosynthetic length. The constitutive model for soil–geosynthetic interface was used to estimate the interface shear resistance mobilized along each segment. This model relates geosynthetic displacement (u) to the interface shear between soil and the geosynthetic (τ):

$$\tau = f(u) \tag{1}$$

The model adopted for geosynthetic material was used to estimate the unit tension developed in the geosynthetic. This model relates geosynthetic unit tension (T) to its tensile strain (ε):

$$T = g(\varepsilon) \tag{2}$$

Specific steps adopted in this study to implement the *t–z* analysis procedure to simulate pullout test data are as follows:

Step 1: The geosynthetic specimen is discretized into several segments from 1 (at the loading front) to n (at the free end). Definitions used in discretization of the geosynthetic specimen are presented in Figure 1. Geosynthetic nodal displacements, unit tension in the geosynthetic at the nodal locations, interface shear between soil and geosynthetic segments, and tensile strain in the geosynthetic are defined by u, T, τ, and ε, respectively.

Step 2: An incremental displacement, δu, is adopted at the free end of the geosynthetic (Segment n):

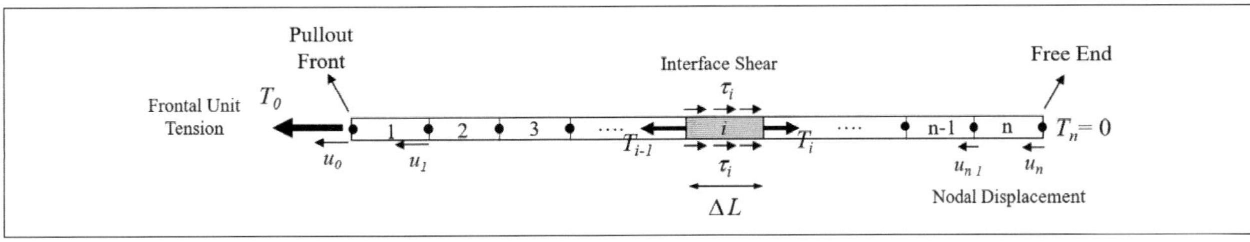

Figure 1. Discretization of geosynthetic along with definition of variables.

$$u_n = \delta u \quad (3)$$

The unit tension at the free end of geosynthetic is zero:

$$T_n = 0 \quad (4)$$

Step 3: Displacement at the front of Segment n is assumed (u_{n-1}) and the average displacement in Segment n ($u_{n(\text{avg})}$) is estimated as follows:

$$u_{n(\text{avg})} = \frac{u_n + u_{n-1}}{2} \quad (5)$$

Step 4: The interface shear along Segment n is estimated using the adopted interface shear model:

$$\tau_n = f\left(u_{n(\text{avg})}\right) \quad (6)$$

Step 5: Considering the force equilibrium in Segment n, unit tension at the front of this segment yields as follows:

$$T_{n-1} = T_n + 2.\Delta L.\tau_n \quad (7)$$

where ΔL = segment length. The average unit tension in Segment n ($T_{n(\text{avg})}$) is then obtained:

$$T_{n(\text{avg})} = \frac{T_n + T_{n-1}}{2} \quad (8)$$

Step 6: Tensile strain in Segment n is estimated using the adopted constitutive model for the geosynthetic and elongation of this segment (δ_n) is obtained:

$$\varepsilon_n = g^{-1}\left(T_{(\text{avg})}\right) \quad (9)$$

$$\delta_n = \Delta L.\, \varepsilon_n \quad (10)$$

Step 7: Displacement at the front of Segment n is updated using u_n and δ_n:

$$u_{n-1} = u_n + \delta_n \quad (11)$$

The updated displacement at the front of Segment n, obtained in this step, is compared with that assumed in Step 3. If the difference between the two was negligible, proceed to the next step. Otherwise, the updated displacement is used in Step 3 and Steps 3 to 7 are iterated to converge.

Step 8: Steps 3–7 will be repeated for Segment n-1 using the displacement and unit tension obtained at the end of this segment (u_{n-1} and T_{n-1}) to eventually determine the displacement and unit tension at the front of this segment. This procedure will then be repeated for Segments n-2 to 1 to obtain the displacement, strain, interface shear, and load profile along the geosynthetic length. Eventually, the geosynthetic frontal displacement (u_0) and frontal unit tension (T_0) will be determined.

Step 9: The incremental displacement in Step 1 is increased and the iteration is repeated to determine the updated load and displacement profiles along the geosynthetic.

Constitutive Models

The constitutive models adopted in this study to evaluate soil–geosynthetic interaction response in pullout tests are presented in this section.

Constitutive Model for Geosynthetic. The constitutive model adopted for the geosynthetic was consistent with the information typically provided by geosynthetic manufacturers. Specifically, the tensile loads correspond to 1, 2, 5, and 10% strains were identified and linear unit tension–strain relationships were considered between consecutive points.

Constitutive Model for Soil–Geosynthetic Interaction. Several interface shear models were evaluated in this study to simulate pullout test data. Ultimately, the two models finally adopted were those that could effectively explain the differences in the development of interface shear in various testing scales. Specifically, linear plastic and bilinear plastic soil–geosynthetic interface shear models were adopted, which are discussed in this section.

Linear Plastic Soil–Geosynthetic Interface Shear Model. A linear relationship between interface shear (τ) and geosynthetic displacements (u) was defined using a slope referred to as the soil–geosynthetic shear stiffness (K_τ). The interface shear was limited by a maximum value referred to as the ultimate soil–geosynthetic interface shear (τ_{ult}) at a displacement value referred to as u_{ult}:

$$K_\tau = \frac{\tau_{ult}}{u_{ult}} \tag{12}$$

The ultimate soil–geosynthetic interface was estimated using the ultimate unit tension obtained in the pullout test (T_{ult}) and the confined contact area between soil and geosynthetic (2A):

$$\tau_{ult} = \frac{T_{ult}}{2A} \tag{13}$$

It should be noted that at the ultimate condition all geosynthetic segments displace at the same rate without developing significant additional strain. At this condition, a uniform interface shear was assumed along the geosynthetic.

The soil–geosynthetic shear stiffness (K_τ), or alternatively u_{ult}, was obtained after iterative procedure to minimize the sum of the squares of the residuals (S) between the experimental data and estimations made using the t–z analysis procedure adopted in this study.

Bilinear Plastic Soil–Geosynthetic Interface Shear Model. The linear plastic model was modified by splitting its linear portion into two linear portions using different soil–geosynthetic shear stiffness values. The stiffness adopted in the first linear portion was referred to as $K_{\tau-1}$ and that adopted in the second portion was $K_{\tau-2}$.

Two ratios were utilized to facilitate investigation of the optimum interface shear model. The displacement ratio (α) defines the ratio between the limit displacement value corresponds to the first linear portion of the model (u_1) and the displacement value corresponds to the ultimate interface shear (u_{ult}):

$$\alpha = \frac{u_1}{u_{ult}} \tag{14}$$

The interface shear ratio (β) defines the ratio between the limit interface shear corresponds to the first linear portion of the model (τ_1) and the ultimate interface shear (τ_{ult}):

$$\beta = \frac{\tau_1}{\tau_{ult}} \tag{15}$$

Scope of Experimental Program

A pullout testing program was conducted using three devices of different scales. Characteristics of the materials, specifications of the equipment, and the test procedure are described in this section.

Material Characteristics

Backfill Soil. The backfill material was Monterey No. 30 sand, which is a uniformly graded clean sand classified as SP (poorly graded). This sand has rounded to sub-rounded particles and consists predominantly of quartz with a trace of feldspars and other minerals. The grain size of Monterey No. 30 sand ranges from 0.2 to 2 mm with a mean grain size of 0.7 mm. The coefficients of uniformity and curvature are 1.9 and 1.3, respectively. The specific gravity of this soil is 2.65 and its minimum and maximum dry unit weight is14.76 and 16.70 kN/m^3, respectively. In this study, the backfill materials were placed at a moisture content of approximately 2% and were compacted using a hand tamper to reach a target dry unit weight of 16.05 kN/m^3.

Geosynthetic. The geosynthetic material was a polypropylene woven geotextile. This geotextile has mono-filament yarns oriented in the machine direction and multi-filament yarns oriented in the cross-machine direction. The unconfined tensile strength of this geosynthetic in the cross-machine direction, tested in this study, is 70 kN/m at failure and 19.3 and 39.4 kN/m at 2 and 5% strains, respectively.

Equipment and Test Setup

Pullout Test Device. Figure 2 shows a schematic layout of the testing devices. The devices were instrumented to map geosynthetic displacements in the confined and unconfined zones. The box dimensions (length × width × depth) were approximately 1,500 × 750 × 450 mm, 1,500 × 600 × 300 mm, and 250 × 300 × 150 mm for the large-, standard-, and small-scale devices, respectively. The confined length of geosynthetic specimens was approximately 1000, 600, and 250 mm and their width was approximately 750, 300, and 280 mm for the large-, standard-, and small-scale tests, respectively. A normal pressure of 21 kPa was generated on the geosynthetic specimens using two different systems. Roodi (*12*) extensively evaluated (and confirmed) consistency of the normal pressures generated by the two systems. In the large- and standard-scale setups, the normal pressure was applied using a set of six pneumatic actuators installed on top of a stack of wooden plates. The plates were backed with thick neoprene rubber mats to reduce the effect of rigidity. The pneumatic actuators were positioned to allow both piston extension and retraction to accommodate soil compression and dilation,

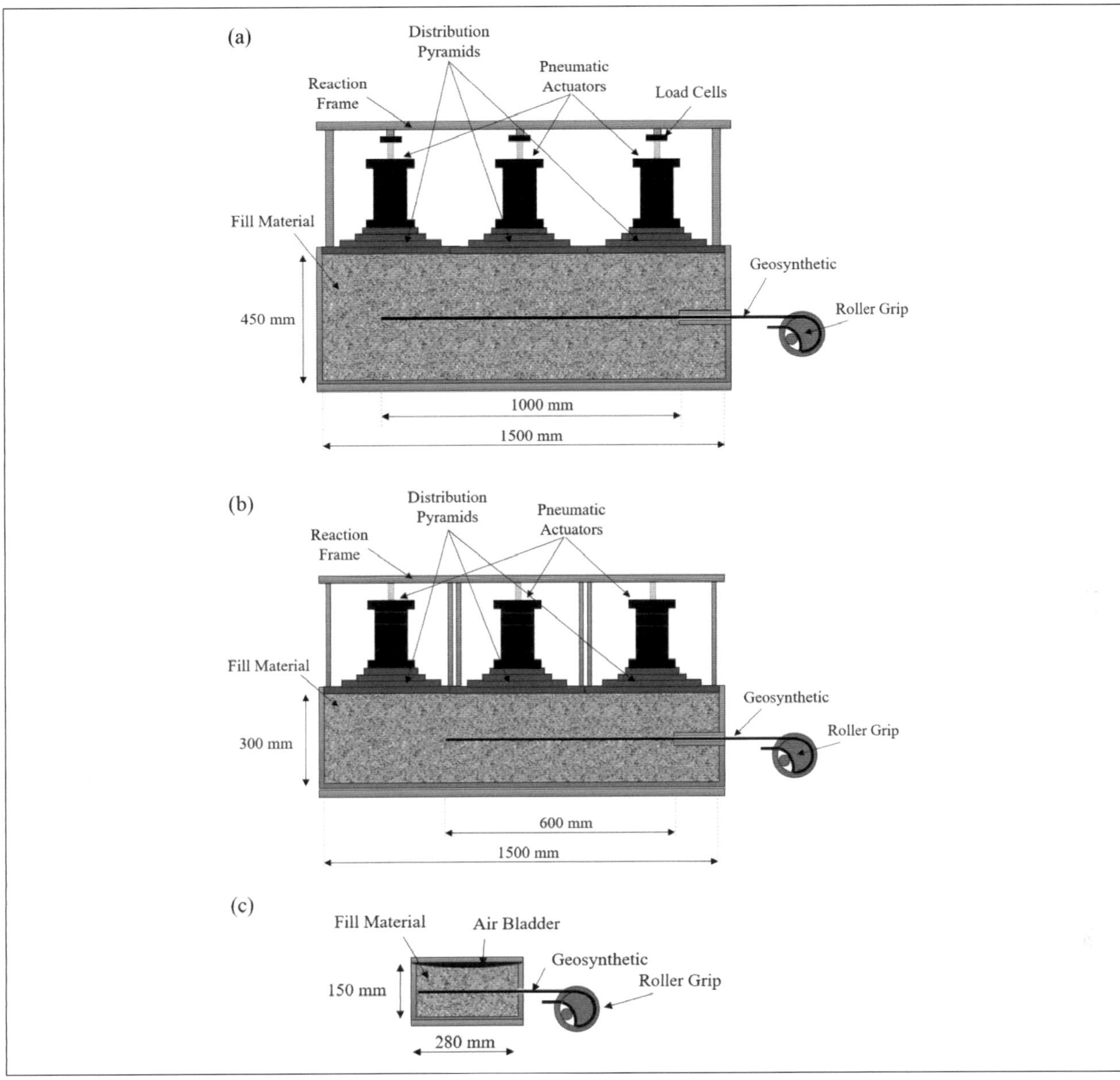

Figure 2. Schematics of pullout test equipment: (*a*) large-scale setup, (*b*) standard-scale setup, and (*c*) small-scale setup.

respectively. The load applied by the actuators was independently monitored during entire test using load cells. In the small-scale setup, the normal pressure was applied through an air bladder that was mounted in the box lid.

The geosynthetic was placed in the middle of the boxes and clamped to roller grips of different scales for the three different devices. The roller grips consisted of steel cylinders with crescent cuts. The clamp was lined with sandpaper to avoid geosynthetic slippage. The edge of the crescent cut in contact with the geosynthetic was lined with a neoprene foam strip to mitigate potential stress concentration at this location. This clamping system provides an unconfined geosynthetic section (i.e., outside the box), which allowed evaluation of the tensile behavior of the geosynthetic specimen during pullout tests. Pullout tests were conducted by pulling the grip at a rate of approximately 1 mm/min.

Results and Discussion

Experimental Test Results

The experimental data obtained from pullout tests conducted using the three scales of pullout devices are presented in Figure 3, *a–c*. The plots on the left presents the

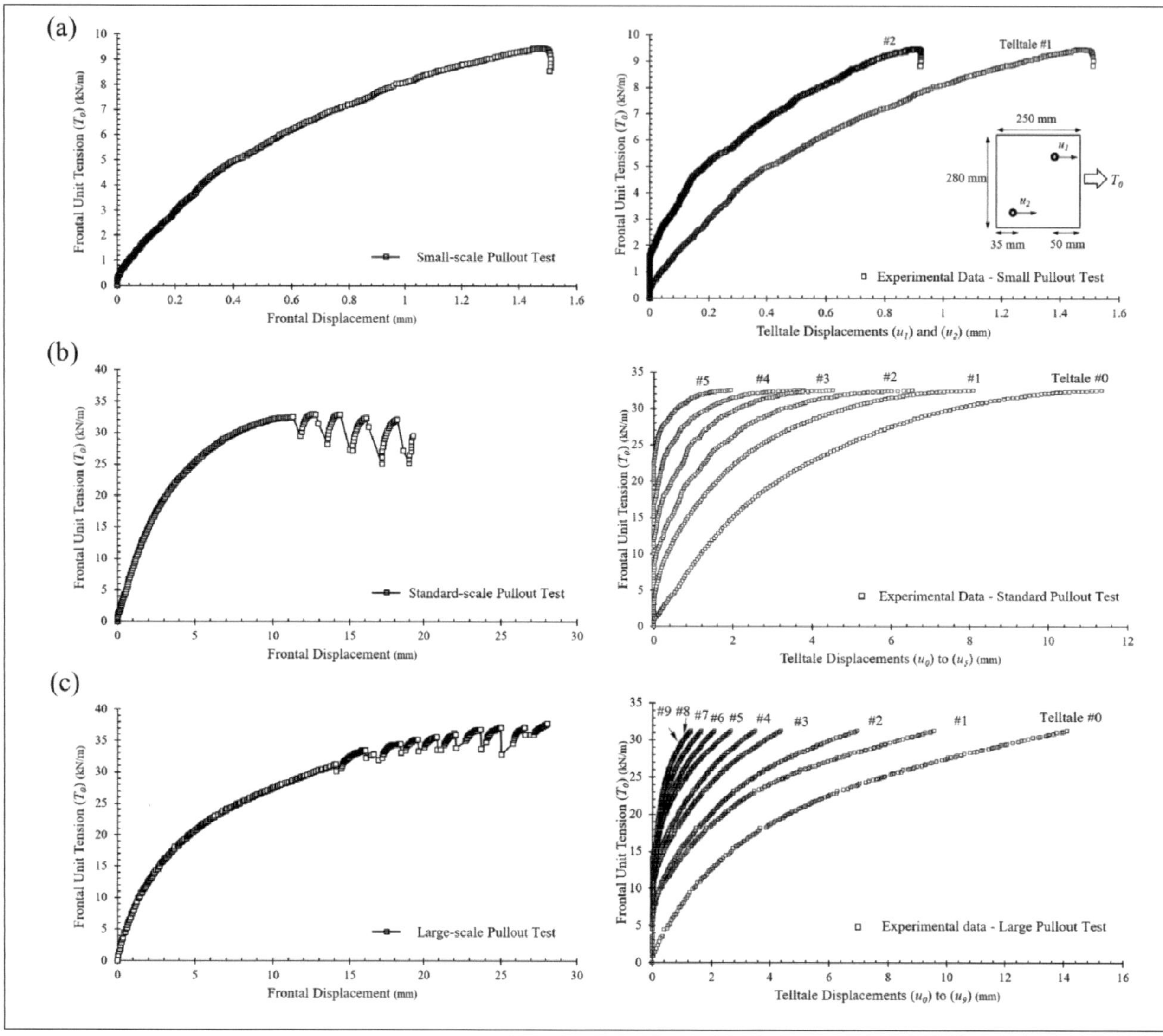

Figure 3. Frontal unit tension versus frontal displacements and versus geosynthetic displacements in the confined length: (*a*) small-scale test, (*b*) standard-scale test, and (*c*) large-scale test.

frontal unit tension versus frontal displacements and the plots on the right presents the frontal unit tension versus displacements along geosynthetic confined length recorded at telltale locations. Pullout tests stopped when (1) pullout load no longer increased and (2) displacement rate was the same along the geosynthetic. As shown in Figure 3, *b* and *c*, fluctuation in the unit tension was observed toward the end of the standard- and large-scale tests. This fluctuation, which was observed in the form of consecutive drops and recovery in the mobilized pullout resistance, is referred to as slip-stick mechanism in which temporary loss of resistance results from shear failure of soil adjacent to geosynthetic specimen. It should be noted that the plots on the right presents only the telltale data recorded before the slip-stick behavior starts.

The ultimate unit tension was found as approximately 10, 32, and 35 kN/m in the small-, standard-, and large-scale tests, respectively. Corresponding ultimate interface shear was calculated using contact areas between soil and geosynthetic in each test as 19.1, 26.0, and 18.6 kN/m^2, respectively. The slightly lower ultimate interface shear obtained in the small- and large-scale tests may be attributed to the comparatively higher coverage ratios (i.e., the ratio between geosynthetic confined area and the total soil area) and comparatively smaller clearance on the sides of the specimens to the sidewalls of the box in these scales as compared with the standard-scale test.

Understanding soil–geosynthetic interface shear response requires evaluation of the displacements recorded by telltales connected to the confined geosynthetic length. Specifically,

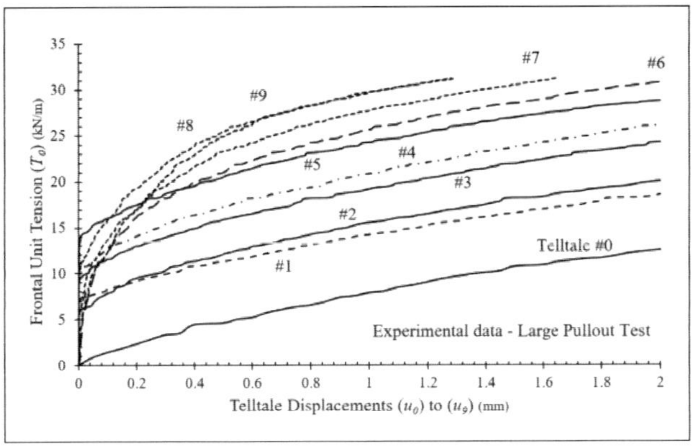

Figure 4. Close view of initial geosynthetic displacements in the confined length in the large-scale test.

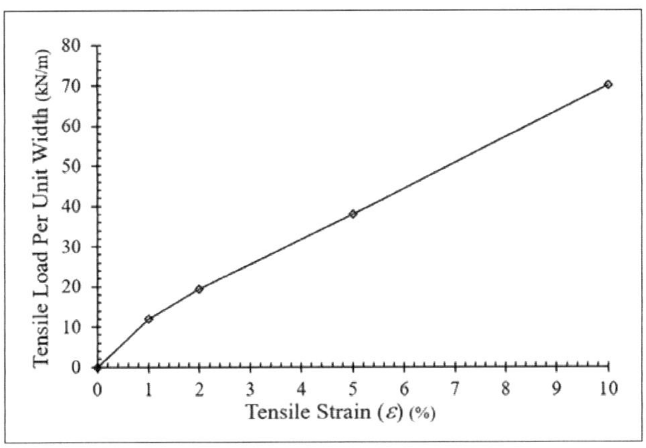

Figure 5. Geosynthetic constitutive model using data collected from geosynthetic unconfined length.

for evaluation of the data at small displacements, it is important to inspect the experimental data at the onset of the geosynthetic movement. Progressive mobilization of the interface shear resistance from pullout front toward the free end of the geosynthetic requires telltale displacements to be triggered in the order of their locations from the pulling front.

The displacement data recorded at the telltale locations versus frontal unit tension in the three pullout tests are also presented in Figure 3. As shown in Figure 3*a*, two telltales (Telltales #1 and #2) were installed in the small pullout test located 50 mm and 215 mm from the pulling front, respectively. The total number of telltales in the standard and large pullout tests was 6 and 10, respectively, out of which Telltale #0 was installed at the front of the confined geosynthetic length and the rest were equally spaced from the front to the free end of the geosynthetic. Inspection of the telltale data presented in Figure 3, *a* and *b*, indicates that the order of displacement triggering among various telltales was reasonably good in the small- and standard-scale tests. However, the data obtained from the large-scale test (Figure 3*c*) had to be studied carefully. Although telltales were attached to evenly spaced locations along the geosynthetic, it was found that Telltales #1 and #2 curves and Telltales #3 and #4 curves started at the same unit tension values.

A close view of the initial telltale displacements in the large-scale test is presented in Figure 4. Inspection of the data presented in this figure, reveals inconsistency in the expected order of triggered telltales and their movements thereafter. For example, although Telltales #6 to #9 were installed toward the free end of the geosynthetic, their displacements were triggered before those in Telltales #2 to #5. This inconsistency can be attributed to malfunction of the telltale attachments, waves in the geosynthetic, or effects from sidewalls and boundary conditions. Although this inconsistency might not be important in evaluation of pullout data at large displacements, it affects evaluations at small displacements. To minimize potential effects from misleading data, only the data represented by solid lines in Figure 4 were used in calibration of the interface shear model in this study. Displacements recorded at Telltales #1, #4, and #6 to #9 were discarded.

Geosynthetic Constitutive Model

Tensile characteristics of geosynthetic specimens were determined in each test using instrumentations installed in the unconfined portions of the geosynthetics. This information was compared with tensile characteristics reported by the geosynthetic manufacturer. Specifically, tensile loads corresponded to 1, 2, 5, and 10% strains were determined. As presented in Figure 5, the geosynthetic constitutive model was generated considering linear relationships between these data points. The difference between the tensile loads versus strain data obtained from the geosynthetics in the three testing scales was small. Therefore, the geosynthetic model presented in Figure 5 was used in simulation conducted for all testing scales.

Soil–Geosynthetic Constitutive Model

In this section, the *t–z* analysis procedure developed in this study was implemented to simulate the experimental data. Linear plastic and bilinear plastic models were used to find the optimum model for each scale. The ultimate interface shear values, calculated using the ultimate unit tensions, were used and model parameters were changed to obtain the best match to the experimental data. Specifically, u_{ult} was changed in the linear plastic interface model and u_{ult}, u_1, and τ_1 (or alternatively, u_{ult}, α, and β) were changed in the bilinear plastic interface model. The model parameters corresponded to the lowest total sum of the squared residuals between measured and simulated displacements (for the same frontal unit tension) were determined.

Displacements recorded in the confined geosynthetic length at the telltale locations were used to find the soil–geosynthetic interface shear model in each testing scale. The displacements

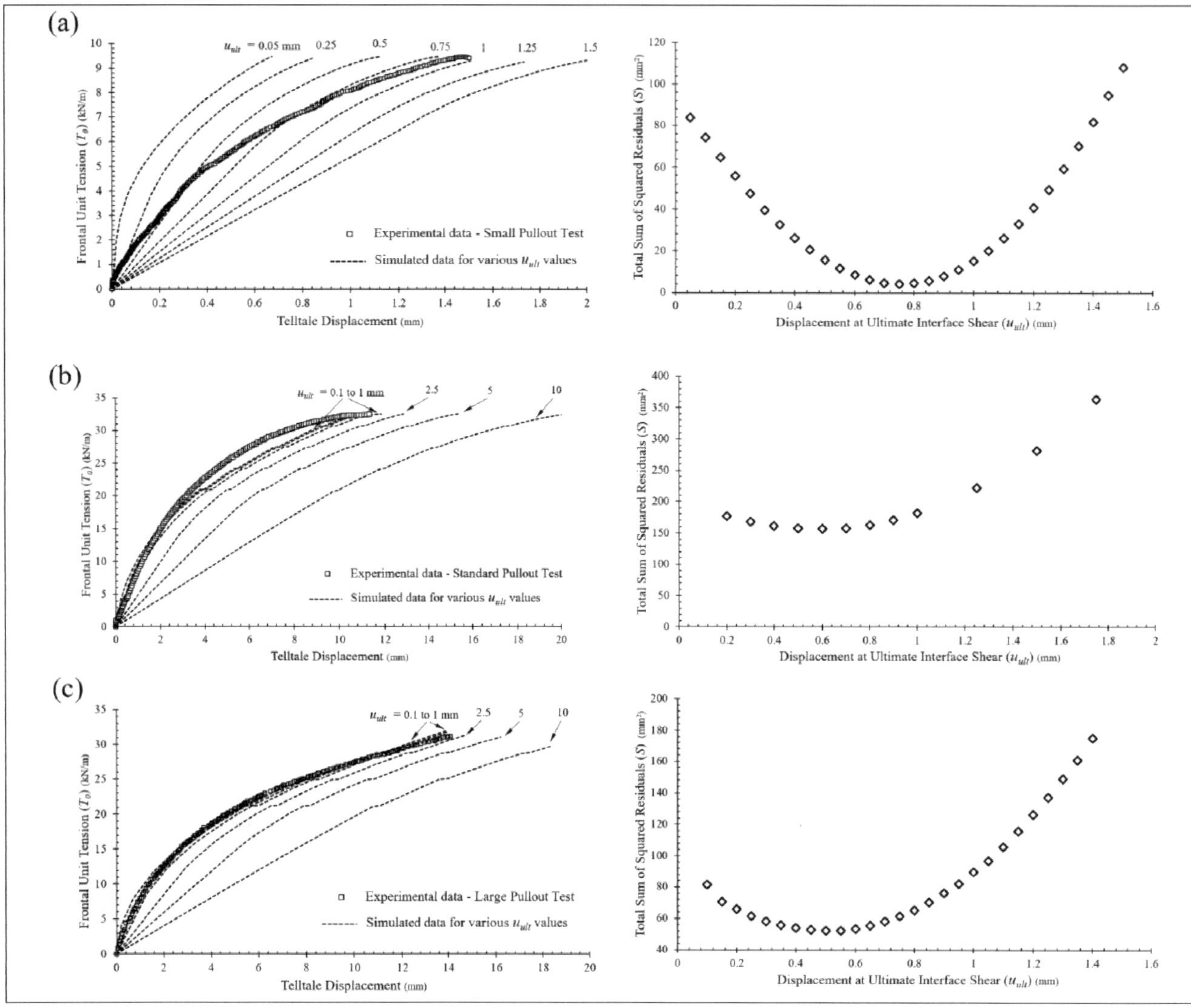

Figure 6. Simulation of the pullout test results using a linear plastic interface model and variation of the total sum of the squared residuals between the experimental and simulated data points: (*a*) small-scale test, (*b*) standard-scale test, and (*c*) large-scale test.

recorded at the end of the tests were used to estimate the displacement required to mobilize the ultimate soil–geosynthetic interface shear (u_{ult}). The ultimate pullout strength corresponds to the condition where the ultimate interface shear is mobilized over the entire interface area. At this condition, relative displacement between soil and geosynthetic is expected to exceed u_{ult} along the entire length of the geosynthetic. Therefore, the displacement recorded at the free end of the geosynthetic is expected to be a reasonably good estimate for u_{ult}. On the other hand, the displacement recorded at the pullout front corresponds to the displacement of the free end along with accumulated elongation of the entire geosynthetic.

Linear Plastic Interface Model. The limits of the displacement required to mobilize the ultimate interface shear, u_{ult}, was estimated using the experimental data presented in Figure 3. Specifically, u_{ult} was changed from 0.05 to 1.5 mm, from 0.1 to 10 mm, and from 0.1 to 10 mm to simulate the small-, standard-, and large-scale pullout test data, respectively. The *t–z* analysis was conducted using the algorithm previously detailed and the total sum of the squared residuals between the measured and simulated displacements (*S*) was determined. Figure 6 presents the simulated data as compared with the experimental data. Visual inspection of the data presented in this figure indicates that the linear plastic interface model could not reproduce the experimental data in the small-scale test (Figure 6*a*). However, the simulated data in the standard- and large-scale tests were found to be reasonably close to the experimental data for comparatively small values of u_{ult} (Figure 6, *b* and *c*).

Figure 6 also presents the total sum of the squared residual (*S*) between the experimental and the simulated data for various u_{ult} values. It should be noted that the *S* values in Figure 6 were computed considering the residual displacement values for all telltales. In the small-scale test, the lowest *S* value was

Table 1. Model Parameters and Total Sum of Squared Residuals Obtained in Simulations

Pullout Test Scale	Linear Plastic Interface Shear Model		Bilinear Plastic Interface Shear Model			
	u_{ult}	S	u_{ult}	α	β	S
Small pullout test	0.05	83.7	0.75	0.1	0.2	2.30
	0.25	47.4				
	0.5	15.6	1	0.2	0.5	0.79
	0.75	4.1				
	1	15.1	1.25	0.6	1	4.38
	1.25	49.6				
Standard pullout test	0.2	176.8	0.5	0.2	0.6	153.3
	0.5	157.4	0.6	0.1	0.5	147.3
	0.6	156.4	1.5	0.1	0.9	122.1
	1	181.9	2	0.1	0.9	122.2
	2.5	740.1	2.5	0.1	0.9	140.6
Large pullout test	0.1	81.5	0.5	1	1	52.3
	0.5	52.3	0.6	0.8	1	52.4
	0.6	53.5	0.75	0.7	1	52.4
	0.75	61.5	1.5	0.3	1	52.9
	1.5	204.0	2	0.3	1	53.8
	2	401.2	2.5	0.2	1	52.3

found to be 4.1 mm^2, which corresponded to u_{ult} = 0.75 mm. In the standard- and large-scale tests, the lowest S value was found to be 156.4 (corresponding to u_{ult} = 0.6 mm) and 52.3 mm^2 (corresponding to u_{ult} = 0.5 mm), respectively.

Bilinear Plastic Interface Model. A bilinear soil–geosynthetic interface shear model was also adopted to simulate the experimental pullout results. The simulation was aimed at refining the linear plastic model obtained in the previous section. Specifically, u_{ult} was changed from 0.75 to 1.25 mm, from 0.5 to 2.5 mm, and from 0.5 to 2.5 mm to simulate the small-, standard-, and large-scale test results, respectively. For each u_{ult} value, the displacement ratio (α) was changed from 0.1 to 1 by an increment of 0.1. The interface shear ratio (β) was then changed from α to 1 by an increment of 0.1. The optimum values for α and β were determined considering the lowest total sum of the squared residuals (S). It should be noted that interface shear ratios (β) lower than the displacement ratio (α) were not considered because the slope of the interface shear versus displacement relationship in the first linear portion of the bilinear model ($K_{\tau-1}$) was assumed to be always larger than this slope in the second linear portion of the model ($K_{\tau-2}$).

Table 1 summarizes the model parameters and resulting S values for the three testing scales and for both interface shear models. The S value in simulating the small-scale test results significantly decreased when a bilinear plastic model was used instead of a linear plastic model. The lowest S value in this testing scale was 0.79 mm^2, which corresponded to u_{ult} = 1 mm, α =0.2, and β =0.5.

Using the bilinear plastic model did not result in decreased values for S in simulating the large-scale pullout test. Specifically, the lowest S value obtained using the bilinear plastic model corresponded to cases where β =1 (or $\tau_1 = \tau_{ult}$), meaning that the stiffness of the second linear portion of the bilinear plastic model was zero ($K_{\tau-2} = 0$). In other words, the bilinear plastic models were equivalent of linear plastic models. Calculating $\alpha.u_{ult}$ for all bilinear plastic models for the large-scale test in Table 1 results approximately 0.5 mm. This is the same u_{ult} value as that corresponded to the lowest S value in this testing scale using a linear plastic model.

Evaluation of the results obtained for the standard-scale test in Table 1 indicates that the lowest value for S was obtained for a bilinear plastic model with u_{ult} =1.5 mm, α =0.1, and β = 0.9. The comparatively high value of β obtained in this case indicates that the stiffness of the second linear portion of the bilinear plastic model for the standard-scale test is comparatively small.

Relevant Interface Shear Model for Various Testing Scales. The interface shear models that resulted in the minimum S value for each testing scale and corresponding simulated telltale displacement data are presented in Figure 7. Comparison of the simulated and experimental data in Figure 7*a* indicates that using a bilinear plastic interface shear model was necessary to simulate telltale displacements in the small-scale pullout test. In contrast, the telltale displacements in the geosynthetic confined length in the large-scale pullout test could well be simulated using a linear plastic interface shear model (Figure 7*c*). Although the optimum model obtained to simulate the standard-scale pullout test data (Figure 7*b*) was bilinear plastic, this model was similar to an equivalent linear plastic model because the stiffness of the second linear portion was found to be very small.

Comparison of the slopes of the interface shear versus displacement relationships (and corresponding displacements) between the optimum models obtained for various testing scales underscores the different mechanisms involved in the development of interface shear. In the standard- and large-scale tests, the ultimate interface shear developed following a single slope (K_τ = 156 and 31 (kN/m^2)/mm, for the standard- and large-scale tests, respectively) at a geosynthetic displacement of 0.15 and 0.5 mm, respectively. On the other hand, in the small-scale test, half of the ultimate interface shear developed at comparatively smaller displacement of u_1 = 0.1 mm compared with that in the standard- and large-scale tests. However, additional shear resistance continued to develop up to a comparatively larger displacement of u_{ult} = 1 mm.

The comparatively high slope of the linear relationship between geosynthetic displacements and interface shear in the standard- and large-scale tests indicates that the ultimate interface shear at each point along the geosynthetic was fully mobilized at comparatively small displacements.

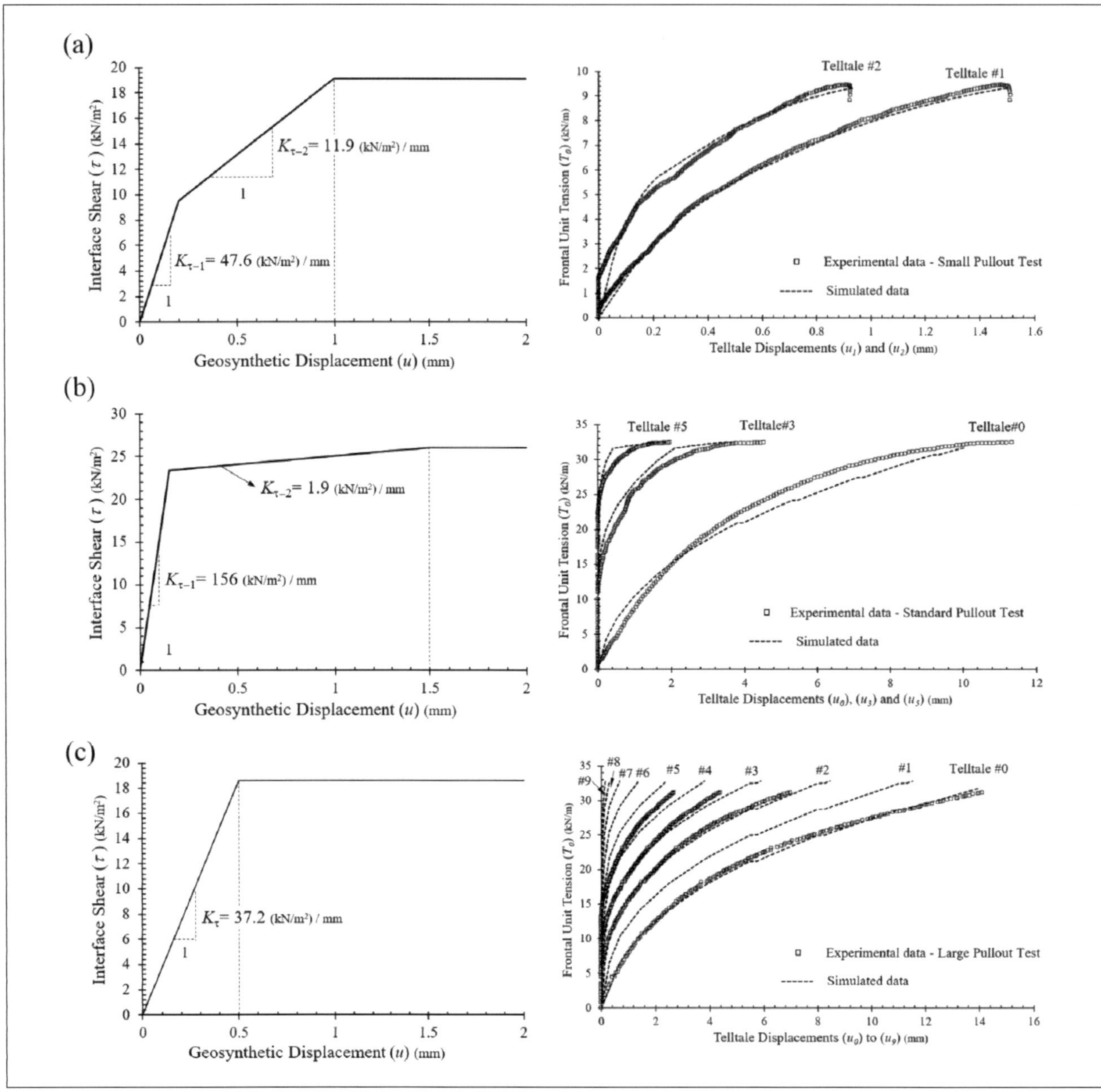

Figure 7. Optimum interface shear model obtained for various testing scales and experimental data versus simulation results using the obtained optimum interface shear model: (*a*) small-scale test, (*b*) standard-scale test, and (*c*) large-scale test.

Therefore, additional displacements would not necessarily increase the pullout resistance. Instead, additional pullout resistance was developed as the mobilized length of the geosynthetic increased. Consequently, the ultimate pullout resistance was reached shortly after the entire length of the geosynthetic was mobilized. In contrast, in the small-scale test, shear resistance was mainly developed by increased displacements. Consequently, unlike the larger scale tests, in the small-scale test significant additional resistance was developed even after the entire length of the geosynthetic was mobilized.

Displacements, Strains, and Interface Shear along Geosynthetics

The geosynthetic displacements measured in the confined portion of the geosynthetics were compared with the simulated data obtained for each testing scale. The simulated displacements are presented for the optimum models obtained for each scale at various levels of frontal unit tension. In addition, variation of the tensile strain in the geosynthetic confined portion and the interface shear values estimated along geosynthetics are also discussed.

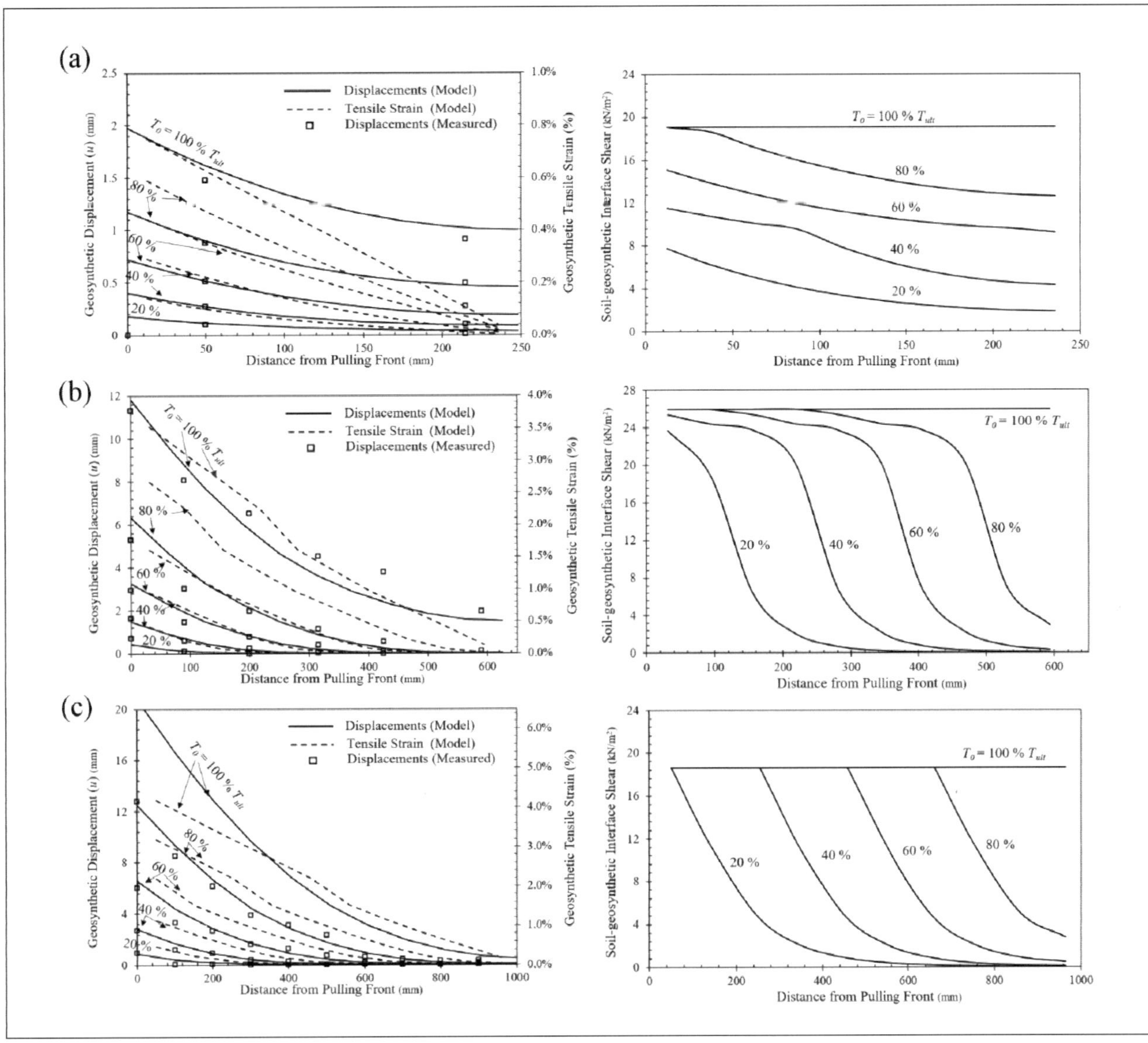

Figure 8. Displacements, tensile strains, and soil–geosynthetic interface shear along geosynthetics at various levels of ultimate frontal unit tension: (*a*) small-scale test, (*b*) standard-scale test, and (*c*) large-scale test.

Figure 8 presents displacements, tensile strains, and interface shear estimated along the geosynthetics at 20, 40, 60, 80, and 100% of the ultimate frontal unit tension. Comparison of the measured displacements with simulated displacements underlines suitability of the models obtained in various testing scales. In addition, comparison between the mobilized lengths of the geosynthetics at various levels of frontal unit tension further underscores the different interface shear mobilizations among the testing scales. In the small-scale test, the total length of the geosynthetic mobilized at a comparatively low frontal unit tension (20%) and the entire geosynthetic displaces thereafter. This condition was not interpreted as the ultimate pullout condition because the specimen continued to mobilize additional pullout resistance and the displacement rate was also not the same along the geosynthetic. However, as a result of the movement of the entire specimen, comparatively small tensile strains (elongations) developed in the geosynthetic in this testing scale. In the standard- and large-scale pullout tests, however, the active length of the geosynthetic continued to increase as the interface shear resistance progressively mobilized. The ultimate frontal unit tension was reached shortly after shear was mobilized along the entire length of the geosynthetic. Therefore, comparatively large deformations (elongations) developed in the geosynthetic in these testing scales.

In the standard- and large-scale tests, the interface shear was found to be distributed differently over three sections along the geosynthetic. The interface shear was found to be comparatively uniform in the frontal portion (i.e., toward the pullout front) of the geosynthetic. Then, along a transition section, the interface shear reduces to approximately zero in the stationary length of the geosynthetic (i.e., toward the free end; Figure 8, *b* and *c*). In the small-scale test, however, the interface shear was found to have a continuous profile along the entire geosynthetic length (Figure 8*a*). The shape of this profile was found to change as the frontal unit tension increased. At comparatively large displacements, the interface shear was found to be uniformly distributed over the portion of the length within which displacements exceeded u_{ult}. The interface shear gradually reduced in the remaining length of the geosynthetic within which displacements were below u_{ult}.

Summary and Conclusions

The development of soil–geosynthetic interface shear was studied using experimental data obtained from pullout tests conducted on a woven geotextile in a uniformly graded sand under a normal pressure of 21 kPa at three different testing scales. Unlike conventional analytical approaches in evaluation of pullout test results in which constitutive models for soil–geosynthetic interface shear and geosynthetic material were typically determined using additional tests (e.g., direct shear test, wide width tensile test), this study adopted a procedure in which the pullout test results can directly be used to identify constitutive models for the soil–geosynthetic interface shear and the geosynthetic. Specifically, the constitutive model for geosynthetic specimen was estimated from data collected in the unconfined portion of the geosynthetic. The soil–geosynthetic interface shear model was obtained by simulating the displacements measured in the geosynthetic confined length. The procedure adopted for this simulation was similar to the *t–z* analysis procedure originally developed to simulate piles subjected to axial loading. As part of this study, several interface shear models were evaluated. The two models that were eventually adopted were those that could underscore the difference in the development of interface shear in various testing scales. Specifically, linear plastic and bilinear plastic soil–geosynthetic interface shear models were adopted.

Comparison of the experimental and simulated data for the standard- and large-scale tests revealed that a linear plastic model can adequately define the soil–geosynthetic interface shear responses in these scales. The slope of the linear relationship between interface shear and geosynthetic displacements was found to be comparatively high. Consequently, in the standard- and large-scale tests, the ultimate interface shear at each point along the geosynthetic developed at comparatively small displacements. Therefore, the development of additional interface shear resistance required increased geosynthetic active length. Progressive mobilization of the interface shear continued until shear was mobilized along the entire length of the geosynthetic.

The development of soil–geosynthetic interface shear in the small-scale test was found to differ from that obtained for the standard- and large-scale tests. Specifically, a bilinear plastic model was found to be necessary to define the soil–geosynthetic interface shear response in the small-scale test. The slope of the interface shear versus geosynthetic displacement at the onset of the movement was found to be particularly high. This means that significant interface shear primarily mobilizes at comparatively small geosynthetic displacements. However, the slope of the second linear portion of the bilinear interface shear model was found to be smaller than the first portion. This means that after initial mobilization of shear, the development of additional shear resistance required comparatively larger displacements. Consequently, in the small-scale test, significant additional shear resistance was found to develop even after shear was mobilized along the entire length of the geosynthetic.

As comparatively small elongation was found in the geosynthetic in the small-scale test, the interface shear resistance in this scale developed mainly by displacement of the entire geosynthetic. In the standard- and large-scale tests, however, the entire geosynthetic did not significantly displace. Instead, the interface shear resistance in these scales developed mainly by deformation (elongation) of the geosynthetic.

Findings from this study suggest that the differences among pullout test results obtained using different test scales may partially be attributed to differences in mobilization of the interface shear resistance. Gaining better understanding of the impact of testing scale on soil–geosynthetic interaction data may be particularly important in adopting each testing scale for appropriate applications. Although a large-scale device may be essential for characterization of ultimate conditions, a small-scale test may be appropriate to be used for working conditions and small-displacement responses. Specifically, using a small-scale test device maybe relevant for specification of a soil–geosynthetic composite property under small displacements, which is particularly relevant for practical applications, such as the base stabilization of roadways.

References

1. Brown, S. F., J. M. Brunton, D. A. B. Hughes, and B. V. Brodrick. Polymer Grid Reinforcement of Asphalt. *Asphalt Paving Technology*, Vol. 54, 1985, pp. 18–41.
2. Al-Qadi, I. L., S. H. Dessouky, J. Kwon, and E. Tutumluer. Geogrid in Flexible Pavements: Validated Mechanism. *Transportation Research Record: Journal of the Transportation Research Record*, 2008. 2045: 102–109.
3. Zornberg, J. G., G. H. Roodi, J. Ferreira, and R. Gupta. Monitoring Performance of Geosynthetic-Reinforced and Lime-Treated Low-Volume Roads Under Traffic Loading and

Environmental Conditions. *Proc., GeoCongress 2012: State of the Art and Practice in Geotechnical Engineering*, Oakland, Calif, ASCE, Reston, VA, 2012, pp. 1310–1319.

4. Bathurst, R. J., T. M. Allen, and D. L. Walters. Reinforcement Loads in Geosynthetic Walls and the Case for a New Working Stress Design Method. *Geotextiles and Geomembranes*, Vol. 23, No. 4, 2005, pp. 287–322.
5. Morsy, A. M., J. G. Zornberg, B. R. Christopher, D. Leshchinsky, B. F. Tanyu, and J. Han. Experimental Approach to Characterize Soil-Reinforcement Composite Interaction. *Proc., 19th ICSMGE*, Seoul, Korea, ISSMGE Open Access Online Library, London, UK, 2017, pp. 451-454.
6. Chen, Q., and M. Abu-Farsakh. Structural Contribution of Geogrid Reinforcement in Pavement. *Proc., GeoCongress 2012: State of the Art and Practice in Geotechnical Engineering*, Oakland, California, ASCE, Reston, Va., 2012, pp. 1468–1475.
7. Weldu, M. T., J. Han, S. M. Rahmaninezhad, R. L. Parsons, J. I. Kakrasul, and Y. Jiang. Effect of Aggregate Uniformity on Pullout Resistance of Steel Strip Reinforcement. *Transportation Research Record: Journal of the Transportation Research Record*, 2016. 2579: 1–7.
8. Morsy, A. M., D. Leshchinsky, and J. G. Zornberg. Effect of Reinforcement Spacing on the Behavior of Geosynthetic-Reinforced Soil. *Proc., Geotechnical Frontiers*, Orlando, Forida, ASCE, Reston, VA, USA 2017, pp. 112–125.
9. Roodi, G. H., and J. G. Zornberg. Stiffness of Soil-Geosynthetic Composite under Small Displacements: II. Experimental Evaluation. *Journal of Geotechnical and Geoenvironmental Engineering*, Vol. 143, No. 10, 2017, p. 04017076.
10. Zornberg, J. G., G. H. Roodi, and R. Gupta. Stiffness of Soil-Geosynthetic Composite under Small Displacements: I. Model Development. *Journal of Geotechnical and Geoenvironmental Engineering*, Vol. 143, No. 10, 2017, p. 04017075.
11. Juran, I., and C. L. Chen. Soil-Geotextile Pull-Out Interaction Properties: Testing and Interpretation. *Transportation Research Record: Journal of the Transportation Research Record*, 1988. 1188: 37–47.
12. Roodi, G. H. *Analytical, Experimental, and Field Evaluations of Soil-Geosynthetic Interaction under Small Displacements*. PhD dissertation. University of Texas, Austin, 2016.

13Yuan, Z. Pullout Response of Geosynthetic in Soil-Theoretical Analysis. *Proc., Geo-Frontiers 2011: Advances in Geotechnical Engineering*, Dallas, Tex., ASCE, Reston, VA, USA 2011, pp. 4388–4397.

14. Bergado, D. T., and J. C. Chai. Pullout Force–Displacement Relationship of Extensible Grid Reinforcements. *Geotextiles and Geomembranes*, Vol. 13, No. 5, 1994, pp. 295–316.
15. Wilson-Fahmy, R. F., and R. M. Koerner. Finite Element Modeling of Soil-Geogrid Interaction with Application to the Behavior of Geogrids in a Pullout Loading Condition. *Geotextiles and Geomembranes*, Vol. 12, No. 5, 1993, pp. 479–501.

16Perkins, S. W., and E. V. Cuelho. Soil-Geosynthetic Interface Strength and Stiffness Relationships from Pullout Tests. *Geosynthetics International*, Vol. 6, No. 5, 1999, pp. 321–346.

17. Farrag, K., A. B. Yalcin, and I. Juran. Pull-Out Resistance of Geogrid Reinforcements. *Geotextiles and Geomembranes*, Vol. 12, 1993, pp. 133–159.
18. Moraci, N., and P. Recalcati. Factors Affecting the Pullout Behaviour of Extruded Geogrids Embedded in a Compacted Granular Soil. *Geotextiles and Geomembranes*, Vol. 24, No. 4, 2006, pp. 220–242.
19. Palmeira, E. M. Soil–Geosynthetic Interaction: Modelling and Analysis. *Geotextiles and Geomembranes*, Vol. 27, No. 5, 2009, pp. 368–390.
20. Chin, J. T., and H. G. Poulos. A "T-Z" Approach for Cyclic Loading Analysis of Single Piles. *Computers and Geotechnics*, Vol. 12, 1991, pp. 289–320.
21. Coyle, H. M., and I. H. Sulaiman. Skin Friction for Steel Piles in Sand. *Soil Mechanics and Foundation*, Vol. 93, No. SM6, 1967, pp. 261–278.
22. Vijayvergiya, V. N. Load-Movement Characteristics of Piles. *Proc. Ports '77, 4th Annual Symposium of the American Society of Civil Engineers*, Waterway, Port, Coastal and Ocean Division, Long Beach, CA, 1977, pp. 269–284.
23. McVay, M. C., F. C. Townsend, D. G. Bloomquist, M. O. O'Brien, and J. A. Caliendo. Numerical Analysis of Vertically Loaded Pile Groups. *Proc., Foundation Engineering: Current Principles and Practices*, Evanston, Illinois, ASCE, Reston, VA, 1989, pp.675–690.
24. Randolph, M. F. Design Methods for Pile Groups and Piled Rafts. *Proc., 13th ICSMFE*, New Delhi, Oxford and IBH Publishing Co. Pvt. Ltd., New Delhi, India, Vol. 5, 1994, pp. 61–82.

25Beech, J.F. Importance of Stress-Strain Relationships in Reinforced Soil System Design. Proc., Geosynthetic '87, New Orleans, Louisiana, Industrial Fabrics Association Int., St Paul, MN, 1987, pp. 133-144.

26. Teixeira, S.H.C., B.S. Bueno, and J.G. Zornberg. Pullout Resistance of Individual Longitudinal and Transverse Geogrid Ribs. *Journal of Geotechnical and Geoenvironmental Engineering*, Vol. 133, No. 1, 2007, pp. 37-50.

The Standing Committee on Geosynthetics (AFS70) peer-reviewed this paper (18-06179).

Article

TRR
JOURNAL OF THE TRANSPORTATION RESEARCH BOARD

Transportation Research Record
2018, Vol. 2672(52) 142–151
© National Academy of Sciences:
Transportation Research Board 2018
Article reuse guidelines:
sagepub.com/journals-permissions
DOI: 10.1177/0361198118790848
journals.sagepub.com/home/trr
SAGE

Validation of the Deep Patch Embankment Repair Design Method using a Large-Scale Centrifuge

Eli V. Cuelho[1], Steven W. Perkins[2], and Brian M. Collins[3]

Abstract

The deep patch repair technique has been widely used in the western U.S.A. to address subsidence and shallow slope failures mainly within poorly performing side-cast embankments and natural slopes consisting of weak soils overlying stronger soils. This project was a multi-phase effort to examine, refine, update, and verify the design procedure associated with this repair method. An experimental design was developed to examine the performance of various deep patch repair configurations and to validate the design technique developed using FLAC (Fast Lagrangian Analysis of Continua) modeling results during Phase I. A series of unreinforced and reinforced slope configurations were tested using a large centrifuge to study their behavior and document their performance. Fourteen centrifuge models (seven unreinforced and seven deep patch reinforced) were run to verify the design chart for a 39-degree slope angle and having one-foot thick deep patch layer spacing (one of four design charts developed during Phase I). Performance of the deep patch models was evaluated based on a serviceability criterion limiting movement in the road bench atop the reinforced zone. Models reinforced with deep patches helped stabilize slope failures by acting as a cantilever and redirecting the failure path to the face of the slope beneath the slope crest, thereby reducing distresses near the top surface of the deep patch. Overall, results from this analysis indicated that the design tool created during Phase I sufficiently predicts the depth of the deep patch necessary to achieve good performance.

The deep patch repair technique was originally developed by the U.S. Forest Service for low volume roads in the mountainous terrain of the Pacific Northwest (*1*). A deep patch is used to mitigate temporarily roadway damage associated with poorly performing side-cast fill embankments and shallow landslides in native slopes. Despite the fact that many deep patches have been in service for more than a decade, the deep patch is considered a temporary repair technique used to slow damage to or redirect the shallow slope failure from the roadway until a more permanent repair can be constructed—if needed in the future. The deep patch consists of multiple layers of geosynthetic reinforcement (geotextile or geogrid) spaced 6 to 24 inches vertically between layers of compacted granular backfill. Depths typically range from 3 to 6 feet below the pavement surface.

This project was part of a multi-phase effort to examine the design requirements for repairing roadways on steep hillsides by the deep patch method. The first phase of the project involved two main study tasks (*2*). The first task involved field reconnaissance of 48 individual deep patch installations on 11 roads within four forests visited during two separate site visits in fall 2010 and spring 2011. The site visits were made to inspect visually the performance of existing deep patch installations and to document site characteristics of road distresses commonly addressed by the deep patch repair technique. The second task consisted of an analytical study using limit equilibrium and finite difference models of steep slopes incorporating the deep patch repair method. This study led to an understanding of the limitations associated with the limit equilibrium method, methods for interpreting results from finite difference models, and an understanding of the importance of depth of the deep patch and reinforcement spacing. Finite difference models were also used to examine the importance of biaxial versus uniaxial reinforcement materials (*3*) and the performance of deep patch with negative batter.

[1]TRI-Environmental, Inc., Austin, TX
[2]Montana State University, Bozeman, MT
[3]Western Federal Lands Highway Division, Vancouver, WA

Corresponding Author:
Address correspondence to Eli V. Cuelho: ecuelho@tri-env.com

The finite difference model was used in a parametric study to generate design guidelines for deep patch slope repairs. Results from this study were summarized in design charts giving depth of the reinforced mass for reinforcement layer spacing of one and two feet. Design charts were developed for slopes with an angle of 34 and 39 degrees, respectively. Design equations were also developed for the required tensile strength of the reinforcement.

The design recommendations developed in Phase I were based on analytical evaluations and require experimental validation. The high cost and difficulties of maintaining good quality control associated with full-scale experimental models led to the decision to conduct model experiments in a geotechnical centrifuge. Use of the centrifuge allowed construction of scaled models that were subjected to an increased gravity field in the centrifuge. The increased gravity on the model increased the stress field and effectively scaled up the size of the model. This allowed for more model configurations to be examined and for greater control over the model construction process.

The scope of the second phase (the subject of this paper) was to validate the deep patch repair design technique developed in Phase I by performing several scaled model laboratory tests using a centrifuge. To accomplish this goal, a brief literature review was conducted to obtain knowledge of the use of centrifuge testing to study slope behavior, geosynthetics, and other soil improvements. From this, an experimental design was developed to examine the performance of various deep patch repair configurations (based on depth of deep patch and geosynthetic strength), and to validate the design model developed in Phase I. A series of unreinforced and reinforced slope configurations were tested using a large centrifuge to study their behavior and document their performance.

Experimental Design

Two failure mechanisms are commonly observed on sites where deep patches have historically been constructed. The first is a rotational failure within the body of the weak zone of soils (comprised of cast material or weaker natural ground); the second is sliding along the interface between this zone of soil and the adjoining stronger natural ground (i.e., along the dashed red line extending up to the roadway bench in Figure 1). In Figure 1, the slope's original and natural grade is given by the slope angle α. The roadway surface that was created by the side-cast operation has an embankment slope defined by the angle β, which is steeper than α. The slope height is defined as H. The parameter X defines the distance from the slope crest to the interface between the fill and native soil. The parameter X_c is the distance from the slope crest to the observed failure (shown as a tension crack in Figure 1) initiating on the roadway surface. The weak soil zone is assumed to have strength properties different from and inferior to the surrounding natural ground to confine failure to be within this region. For cases where the slope consists entirely of natural ground, the geometry shown in Figure 1 is still valid, with the exception that the soil interface defined by the angle α does not intersect the roadway bench.

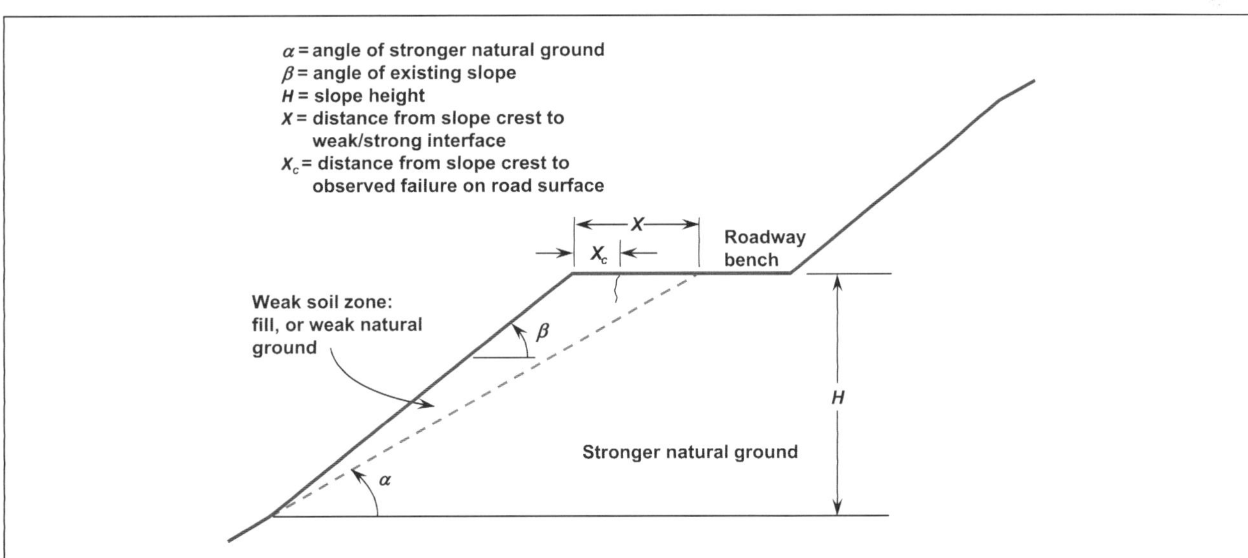

Figure 1. Deep patch slope geometry.

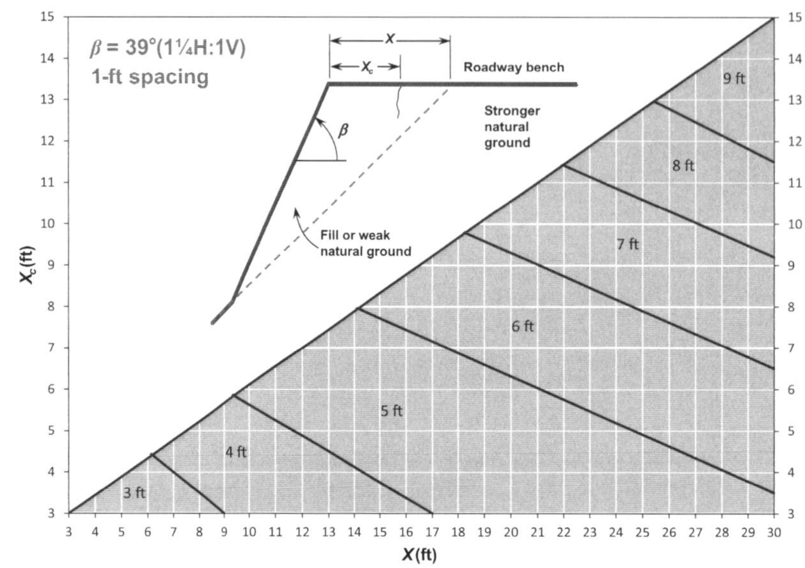

Figure 2. Design chart to determine depth of deep patch for β = 39° for one-foot layer spacing.

Based on the parametric study conducted in Phase I of this project, multiple cases were run in FLAC for several geometries where $\beta = 34°$ or 39° and 3 ft $< X <$ 30 ft. X_c corresponded to the point at which the failure surface intersected the roadway bench, and ranged from approximately 3 to 15 feet. The results of this modeling work were used to develop a design procedure in which a combination of X and X_c could be used to determine the depth and number of geosynthetic layers for the $\beta = 34°$ or 39° cases and for one- or two-foot reinforced layer spacing by determining their intersection point on a design chart (example design chart shown in Figure 2). This project focused on the $\beta = 39°$ for one-foot reinforced layer spacing design chart for verification purposes (shown in Figure 2).

The design charts were developed based on the numerical results of both rotational and wedge failures. The upper region of the design chart is primarily (although not solely) based on wedge failures because they tend to exhibit higher X and X_c values. Conversely, lower X and X_c values are more prominent in rotational failures. Rotational failures are based on the cohesion and friction angle of the deposited soil unique for a given geometry. Manufacturing a soil mass with precise values of both cohesion and friction angle to be used in the centrifuge model is very difficult. Due to these difficulties, wedge failures were solely used to induce slope failure in the centrifuge experiments. Wedge failures are more predictable because the wedge of material can be made to slip on a purely cohesive substrate, rather than a blend of cohesive and non-cohesive materials. However, because wedge failures tend to exhibit larger X and X_c values, the question became how to verify areas lower on the design chart, with lower X and X_c values. This was achieved using a clay seam that ran at a steeper α angle, which was initially investigated by numerical analyses. Failure in this case was dependent on the shear strength of a clay slip layer, and induced when a predetermined g level was reached in the centrifuge.

The centrifuge model was designed to accommodate a wide range of slope configurations, but was limited to an angle of the existing slope, $\beta = 39°$. The angle of the stronger natural ground from the horizontal, α, had a practical range from 22 degrees to about 32 degrees. Other geometric properties (X_c, and X or H) were controlled based on the g level produced by the centrifuge device (dimensions are directly proportional to the g level). By targeting particular X and X_c values, various areas of the design chart were studied and verified.

Unreinforced (no deep patch) and reinforced (deep patch) centrifuge models were constructed so that failure occurred uniformly across the entire slope face side-to-side (i.e., simulating two-dimensional failure). To accomplish this, the length of the model (corresponding to the direction of traffic on the road bench) was made relatively short and the interface between the soil embankment and the end wall support was lubricated. The stiff centrifuge frame that represents the stronger ground was made to accommodate two 2D models simultaneously, each model being 17 inches long (as described below).

Deep Patch Centrifuge Model Design

Centrifuge testing of the deep patch experiments was used to understand the basic behavior of these systems in a variety of conditions to avoid having to build full-scale experiments in the field. Rotation of a scale model via a centrifuge induces centripetal forces normal to the experiment. These additional centripetal forces increase as a function of the rotational velocity. Earth's gravitational acceleration causes a centripetal force that acts upon masses on its surface. The result of this phenomenon is that greater forces are induced on objects with greater densities. The unit weight of an object is defined as the density of the object multiplied by gravity ($\gamma = \rho g$). Materials of a certain density have defined unit weights because the gravitational acceleration, g, is relatively constant; however, when additional centripetal forces are induced (as with the centrifuge) the unit weight of the object increases. This increase is often put in terms of g, where $2g$ is equal to twice the gravitational acceleration and therefore twice the unit weight. This principle is helpful when understanding the effect of the centripetal forces acting upon the centrifuge model. Expanding the equation for unit weight illustrates how it is related to volume (Equation 1). From this, it can be recognized that the volumetric dimensions of an object (its length, l, width, w, and height, h) having a certain mass (m) are directly proportional to the gravitational acceleration. Subsequently, even though the centrifuge is actually increasing the unit weight of an object, the previously described principles can be used to understand how this can also be viewed as an increase in the volume of that same object at a constant unit weight. Therefore, the physical dimensions of the scale models used in the centrifuge tests can be linearly scaled by the increase of additional centripetal forces induced by the centrifuge (in terms of gravity, g). Predetermined stress levels are produced by controlling the rotational velocity of the centrifuge. Centrifuge models are commonly tested by gradually increasing centrifugal acceleration until they reach failure.

$$\gamma = \frac{m}{V} g \rightarrow \gamma = \frac{m}{l \cdot w \cdot h} g \rightarrow \gamma(l \cdot w \cdot h) = mg \qquad (1)$$

A scaled experiment consisting of a stiff steel frame and aluminum and acrylic boundary walls was designed to withstand a centripetal force of 100G. The dimensions, thicknesses, and support of the various pieces of the experiment were designed to minimize distortion under these loads during testing. The overall dimensions of the centrifuge experiment were approximately 48 in. deep (front to back) by 37 in. wide by 26 in. tall and weighed around 900 lb. Centrifuge testing was conducted using a 400 g-ton centrifuge located at the University of Colorado, Boulder. The swinging-basket platform on the centrifuge, located approximately 18 feet from the center of rotation, was designed to accelerate an experiment weighing up to 4000 lb to a maximum of 200G. The test platform was connected to the main body of the centrifuge by a pair of swing arms that allowed the payload to maintain its perpendicular position with respect to the direction of applied acceleration.

The rigid frame that supported the slope was made of flat plates of steel welded together at precise angles and reinforced to provide a stable platform consisting of the proper geometry for the remaining components and instrumentation to be attached. The support frame was designed to provide a stable platform by which the weak soil wedge and deep patch could be constructed and tested in the centrifuge without distorting or moving excessively under the increased gravitational forces induced by the spinning motion of the centrifuge. In this way, the scale model assembly was able to maintain a plane–strain condition throughout testing.

A separate mold assembly was designed to form the "cast" wedge of material. It was constructed of aluminum panels, other than the two side panels that were made of one-inch thick acrylic to allow viewing during testing. A center wall, made of aluminum, was used to split the model into two sections so that two centrifuge models can be run simultaneously side-by-side, each model being 17 in. wide. The wedge assembly was designed to be filled by placing it face down in a jig to hold it stationary while the soil is placed into the mold. Once full, the wedge assembly could then be lifted and placed on the support frame. The top and front confinement plates were designed to confine the soil and plaster wedge during this transition. Once placed on the support frame, these plates could be removed to expose the slope and allow the reinforced deep patch to be built on top. Extensions were made to form the sides, back, and center divider of the deep patch region, as shown in Figure 3. During the centrifuge runs, a catch basin was attached to the front of the apparatus to contain the soil from the slope during failure.

Each of the soils used to form the weak fill material, the thin slip layer, and the deep patch components of the scaled model was designed to meet predetermined strength properties associated with specific FLAC model runs consistently. It was critical that these soils were homogeneous and consistent from test to test. A naturally occurring silt (Bonney silt from Colorado) and powdered modeling clay were used to represent the weak fill zone and the weak clay seam, respectively. The properties of these soils were manipulated mainly through the addition of water. An all-purpose sand from Quikrete® was used as the fill material for the deep patch, and three different materials were used to represent the geosynthetic reinforcement. Three polymeric textile materials were

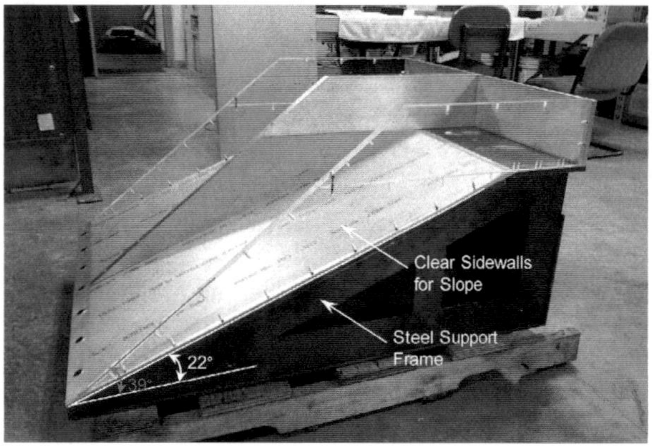

Figure 3. Deep patch extensions on top of wedge mold and support frame.

selected to simulate geosynthetic reinforcement in the deep patches—one resembled a woven grid and the other two non-woven textiles. Although these products were not true geosynthetics, their key attributes (strength, stiffness, flexural rigidity, and thickness that under gravitational accelerations of around 30*G* in the centrifuge) were similar to typical geosynthetic reinforcement products used in this application. Scaled strengths of these materials are directly proportional to the gravitational acceleration. Based on a *g* level of 30, the ultimate strengths of the three materials were 18,000 lb/ft, 6,450 lb/ft, and 2,490 lb/ft.

Construction and Testing of Centrifuge Models

Construction of the deep patch models consisted of three to four main steps, depending on the configuration. The silt wedge representing the cast material was composed of the Bonney silt. The slip plane was made of clay to facilitate a wedge failure. The slip plane was oriented at the angle α, the inclination of the stronger natural ground. A quick-dry plaster filler was used to create the angle α with respect to the angle of the steel support frame. These three components were constructed in a triangular-shaped mold and carefully placed on the support frame. The major steps in the construction process are shown in Figure 4. A photograph and illustration of the side view of an unreinforced model (i.e., no deep patch) is shown in Figure 5. The Bonney silt was prepared to a moisture content of 20% and compacted to within 95% of maximum dry density based on the Proctor mainly to ensure consistency from test to test. The clay slip plane was also prepared to the moisture content associated with a particular shear strength based on direct shear and triaxial test results to ensure failure along the slip plane at a specific *g* level. Shear strengths of the clay layer ranged from approximately 200 to 600 psf, depending on the model configuration. For the deep patch reinforced models, an extra step was added to create the geosynthetic-reinforced layers. Once constructed, the deep patch model was loaded onto the centrifuge platform, instrumented, and run.

The deep patch was constructed directly on top of the silt wedge, slip plane, and plaster surfaces. Sidewalls and a back-plate extension were attached to the top of the model to confine the deep patch. The front of the deep patch was temporarily shored during the construction process using 0.25-inch thick metal strips that spanned the length of the deep patch (from side-to-side), as shown in Figure 6. These shoring strips were held in place with two metal brackets that were bolted to the front of the deep patch sidewall extensions. Various heights of shoring were used depending on the thickness of the deep patch reinforced layer. The first layer of "geosynthetic" was installed directly on top of the silt/clay/plaster surface, and the Quikrete sand placed on top of that to form a single layer 0.35 to 0.5 inches thick (depending on the model configuration and anticipated *g* level at failure). The sand was prepared to a moisture content of approximately 7% and compacted in place with a hand-held, pneumatic, vibrating plate compactor. A small amount of sand was excavated from the front of the deep patch to accommodate the wrapping of the geosynthetic. The geosynthetic was wrapped back approximately 1.5 inches, tucked under a layer of sand, and re-compacted (Figure 6).

Once construction of the deep patch model was finished, it was loaded into the centrifuge basket. All exposed soil surfaces were lightly misted with water to keep them from drying out prior to testing. Lighting and GoPro® cameras were also set, and all wires zip-tied and/or taped in place. The front catch basin was also installed to catch the slope materials as the slope failed. The final setup of one of the tests is shown in Figure 7, showing also the positions of the cameras and linear variable displacement transformers (LVDTs). Two of the GoPro cameras and all of the LVDTs were fed live to monitors in the centrifuge control room during testing. Control of the centrifuge was managed by setting the maximum *g* level and allowing the centrifuge to attain that level as quickly as it could. It took approximately 10 minutes to accelerate to about 30*G*. Deceleration of the centrifuge was accomplished by an internal brake after the model had failed.

Centrifuge Test Results

Seven separate centrifuge runs, each consisting of two individual tests side-by-side, were made during this

Figure 4. Construction of the centrifuge model: (*a*) wedge mold in cradle at designated angle; (*b*) installation of the silt wedge; (*c*) installation of the clay layer; (*d*) filling remaining volume with plaster.

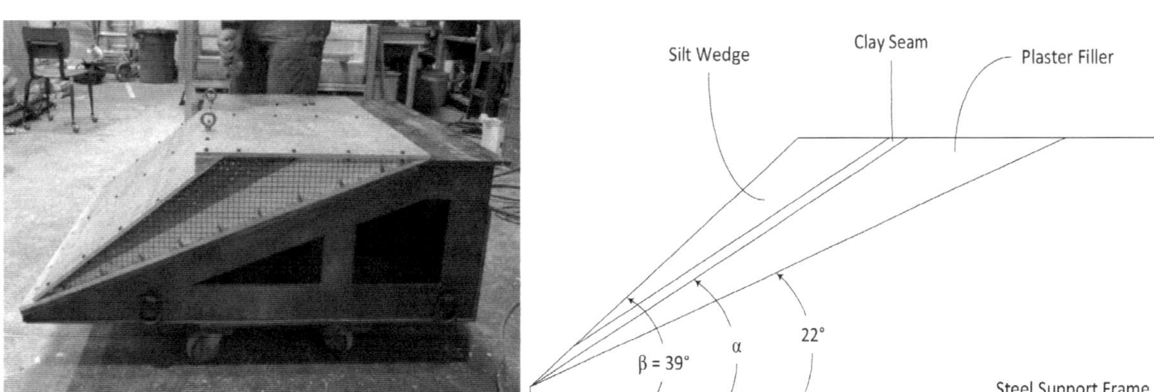

Figure 5. Side view photograph and illustration of unreinforced slope geometry and construction components.

Figure 6. Folding over geosynthetic to create wrapped face.

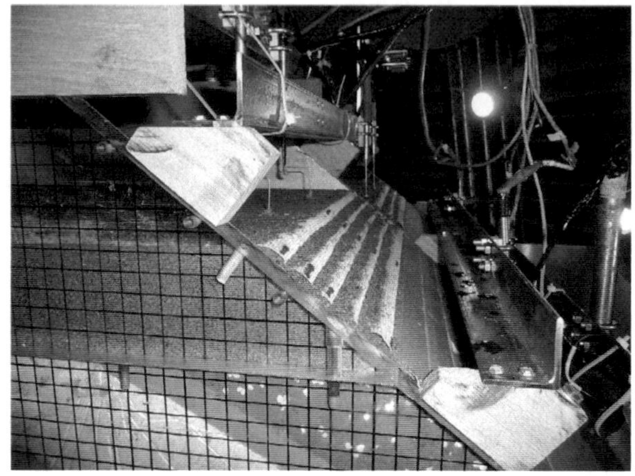

Figure 7. Final setup of deep patch models with instrumentation (five-layer model in foreground, three-layer model in background) prior to centrifuge run.

project. Design features of the centrifuge models were predetermined using FLAC so that various areas on the design chart could be verified. The two physical properties of the centrifuge models that determined where it would plot on the design chart were the angle α and the cohesive strength of the clay slip plane. Unreinforced models (i.e., no deep patch) were primarily used to verify the predictions made by FLAC. Reinforced models (i.e., deep patch) were primarily used to evaluate whether the slope could maintain adequate serviceability of the horizontal bench at the top of the deep patch, which corresponds to the location of the road in a full-scale arrangement. Reinforced models were intended to verify the design chart associated with the one-foot thick deep patch layer spacing (Figure 2). Seven unreinforced configurations and seven reinforced configurations were run. The following sections describes the results of the unreinforced and reinforced centrifuge testing.

Unreinforced Centrifuge Models

The purpose of testing the unreinforced centrifuge models was to understand the behavior of the scaled models and to verify the values of X and X_c associated with specific soil characteristics and geometric properties. The results from FLAC (namely g level at failure, X, and X_c) were used to determine the geometry of the centrifuge models and the clay shear strength that would plot in specific regions of the design chart. Failure of the unreinforced models was defined as impending downward sliding of the silt wedge along the clay seam, induced by rotation of the centrifuge. X_c was determined by measuring the distance from the slope face to the crack that formed on the top of the silt wedge. In general, the plane strain condition in the models was maintained, as evidenced by the slippage of the entire wedge from side-to-side during failure.

The g level at failure was correlated to the displacement measurements made during centrifuge testing. Failure of the slope occurred when the cast wedge of silt material slid down the slope along the clay slip plane, as evident in the displacement measurement traces. When the displacement measurements were plotted with respect to the g level, failure was apparent as an asymptotic increase in displacement at a particular g level, and that g level was defined as the failure point. Displacement measurements were scaled by multiplying the measured LVDT displacement by the g level.

FLAC was also used to verify the responses from the unreinforced centrifuge tests using material properties and failure results from the constructed centrifuge models. Two back-calculations were used in this analysis: (a) a prediction of the g level at failure based on the cohesion of the clay layer (determined from clay moisture content measurements before and after centrifuge testing), and (b) a prediction of the cohesive strength of the clay slip plane based on the g level at failure. Using the X and X_c values from the centrifuge tests and the FLAC models, the plot in Figure 8 was created to allow a direct visual comparison of the results. Referring to Figure 8, the lines identify the possible range of X and X_c as determined by FLAC, while the circles represent the results from the individual centrifuge tests. In general, the comparisons between the centrifuge results and the FLAC results for the unreinforced tests show modest agreement—the best correlation being with Test 3B and the worst correlation being with Test 5B. In general, the centrifuge test results were nearer the lower part of the range of back-

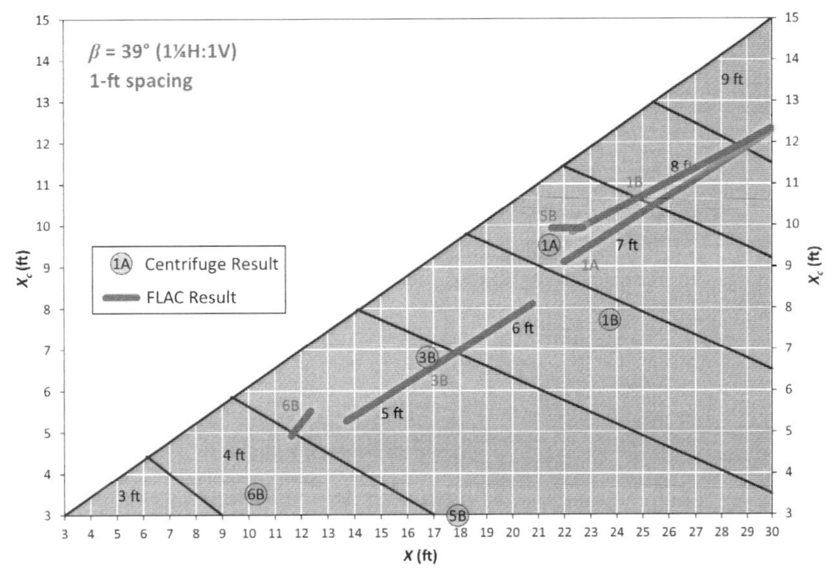

Figure 8. Comparison of centrifuge and FLAC results for unreinforced models plotted on the design chart.

calculated X and X_c from FLAC, indicating that the lower slip plane cohesion is likely more indicative of its true strength.

Reinforced Centrifuge Models

The purpose of the reinforced centrifuge models was to assess how well the reinforced earth layers forming the deep patch maintained the stability and serviceability of the top of the reinforced zone (representative of the road bench) under failure conditions. Similar to the unreinforced models, failure of the slope was defined as the impending downward slippage of the silt wedge along the clay seam. Despite the fact that this type of failure would normally warrant immediate repair, the presence of the deep patch reinforced zone created a cantilever effect and redirected the failure surface toward the face of the slope beneath the deep patch rather than propagating into the road surface. Therefore, for the purposes of the analysis, the behavior of the top of the surface (location of the roadway) was analyzed with respect to its ability to maintain a relatively smooth driving surface despite the underlying slope failure beneath the deep patch. Small depressions with no abrupt crack or differential settlement may be acceptable in many situations; however, excessive deformations or large cracks on the road surface can create potentially hazardous driving conditions.

Achieving a failure condition in the reinforced centrifuge models was accomplished similarly to the unreinforced models in that the weak clay seam provided a slip surface for the silt wedge to slide upon when g levels reached critical levels. The g level at failure was directly related to the cohesive strength of the clay slip layer. Failure of the silt wedge was determined using displacement measurements made on the unreinforced slope face beneath the deep patch. Two vertical displacement measurements were also made on the surface of the deep patch. A subjective threshold of three inches was set to facilitate comparisons of the vertical displacements at the fog line. Based on this serviceability criterion, most of the deep patches worked well, with the exception of Tests 4B and 5A whose average displacements exceeded this threshold, and Test 7B whose maximum displacement exceeded the threshold.

The target thickness of the individual reinforced layers in the centrifuge models was one foot. (based on scaled values determined by multiplying actual layer thickness by g at failure) corresponding to the one-foot design chart being validated. The anticipated g level at failure was determined using FLAC. This g level was used to back-calculate the actual thickness of the individual reinforced layers for construction of the scaled models.

To judge how well the reinforced centrifuge models performed, the material properties of the silt wedge and clay slip plane from the deep patch reinforced models were used in FLAC to determine theoretical values for X and X_c corresponding to the unreinforced condition. Results from the centrifuge models themselves were then used to determine how well they performed in terms of providing adequate serviceability and stability of the road bench.

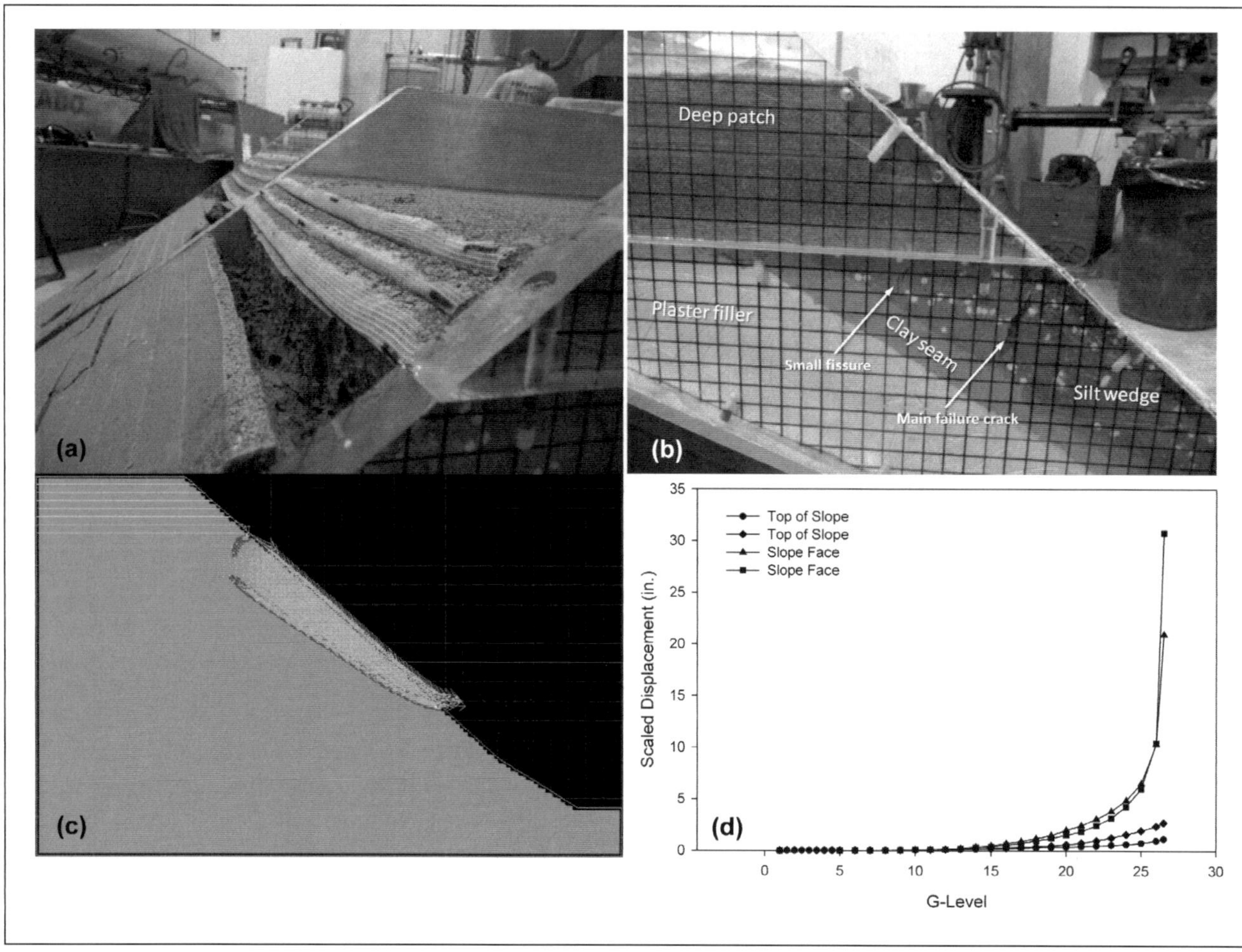

Figure 9. Representative results from reinforced centrifuge models: (*a*) failure crack beneath the deep patch; (*b*) side view of failure crack in the silt wedge; (*c*) prediction of slope failure behavior from FLAC; (*d*) typical displacement traces on the top and front face of a model.

The centrifuge tests run on deep patch reinforced models generally showed that failure of the silt wedge was intercepted by the reinforced zone and turned outward toward the face of the slope, as depicted in Figure 9*a* and *b*. This result also corresponds well to the results of the FLAC analyses (example in Figure 9*c*). Displacements on the face and top of the slope were also used to determine when failure occurred (typical displacement results are shown in Figure 9*d*).

A summary of the reinforced centrifuge tests is provided in Table 1. The bulk of the reinforced models verified the five-to-seven-foot region of the design chart. The results from most of the reinforced models generally indicated that the one-foot design chart developed during Phase I of this study was able to provide sufficient stability and serviceability of the road bench, indicative of the relatively small average displacements. The only exception to this was Test 5A, whose results were believed to be negatively affected by the flexural rigidity (stiffness perpendicular to the plane of the material) of the simulated geosynthetic material used in the experiment.

Summary and Conclusion

This project is the second phase of a study that developed a new design methodology for repairing roadways on steep hillsides by the deep patch method. This phase consisted of a comprehensive experimental program to further examine and verify the deep patch design tool created during Phase I. The high cost and difficulty of maintaining good quality control associated with multiple full-scale experimental models led to the decision to conduct experiments using a large centrifuge. Use of the centrifuge allowed construction of scaled models that were subjected to an increased gravity field. The increased *g* on the models increased the stress field and

Table 1. Summary of Reinforced Centrifuge Tests

Test no.	*G* level @ failure	Number of layers	Position on design chart[a]	Avg. scaled displacement (in.)	Scaled deep patch depth (ft.)	Scaled layer thickness (in.)
3A	27	7	5–6 ft	0.1	7.9	13.5
4A	26	5	6 ft	1.9	5.4	13.0
4B	27	3	6 ft	3.7	3.4	13.5
5A	25	6	6–7 ft	4.0	4.4	8.8
6A	25	3	5 ft	2.3	2.3	8.9
7A	27.5	4	5 ft	2.6	3.3	10.3
7B	28	2	5 ft	2.2	1.8	10.5

[a]As determined from FLAC using unreinforced conditions of the deep patch models.

effectively scaled up the size of the models. This allowed a greater number of configurations to be examined and unlimited control over the model construction process.

In general, models reinforced with deep patches helped stabilize slope failures by redirecting the failure path to the face of the slope beneath the crest of the slope. Redirection of the failure path also helped reduce the direct impact of displacements on the top surface of the deep patch. Displacements near the top were further reduced by the deep patch because it behaved as a cantilever to restrict movement in the top of the slope.

Deep patch reinforced models were analyzed with respect to: (a) the number of reinforced layers compared to where the model plotted on the design chart, (b) how much displacement took place at the surface of the deep patch, and (c) how well the model performed when compared with an unreinforced model with the same material properties. Utilizing the material properties of the deep patch reinforced centrifuge models, FLAC was used to determine where on the design chart the unreinforced conditions would have plotted. This information, coupled with the serviceability criterion limiting the displacement at the fog line to three inches, was used to determine how well the one-foot design chart worked. Overall, these assessments showed that the $\beta = 39°$ design chart for one-foot thick layers, created under Phase I of this effort, works sufficiently well to determine the depth of the deep patch necessary to suppress and redirect failures from reaching the road bench atop the reinforced zone.

Acknowledgments

The authors express their appreciation for the financial and technical support provided by the Western Federal Lands Division of the Federal Highway Administration.

Author Contributions

The authors confirm contribution to the paper as follows: study conception and design: Eli Cuelho, Steve Perkins, Brian Collins; data collection: Eli Cuelho; analysis and interpretation of results: Eli Cuelho, Steve Perkins; draft manuscript preparation: Eli Cuelho. All authors reviewed the results and approved the final version of the manuscript.

References

1. Musser, S. W., and C. Denning. *Deep Patch Road Embankment Repair Application Guide*. Report No. 0577 1204–SDTDC. USDA Forest Service and Federal Highway Administration, 2005.
2. Cuelho, E. C., S. W. Perkins, and M. S. Akin. *Deep Patch Repair, Phase I: Analysis and Design*. FHWA-WFL-TD-12-003. Final report to the Federal Highway Administration, Washington, D.C., 2012.
3. Perkins, S. W., and E. V. Cuelho. Biaxial Versus Uniaxial Geogrids for Deep Patch Embankment Repair. *Proc., Geosynthetics Conference*, Portland, OR, 2015, pp. 654–662.

The Standing Committee on Geosynthetics (AFS70) peer-reviewed this paper (18-04768).

Article

TRR
JOURNAL OF THE TRANSPORTATION RESEARCH BOARD

Transportation Research Record
2018, Vol. 2672(52) 152–163

DOI: 10.1177/0361198118769900
journals.sagepub.com/home/trr
SAGE

A Practical Approach for the Estimation of Strength and Resilient Properties of Cementitious Materials

Mohammad Rashidi[1], Reza S. Ashtiani[1], Jimmy Si[2], Richard P. Izzo[2], and Mark McDaniel[2]

Abstract

Cementitious stabilization of granular soils has been proven to be a cost-saving option for the use of materials with marginal quality in the construction and rehabilitation of pavement structures. The orthogonal load distribution capacity of the Cement-Stabilized Materials (CSM) is typically characterized by Unconfined Compressive Strength (UCS), Indirect Tensile Strength (IDT), and Resilient Modulus (M_r) tests in the laboratory. The aforementioned parameters and properties are integral components of the analysis and design of pavements with stabilized layers. Time and budget constraints make it impractical for many state agencies to complete the full laboratory characterization protocols to determine all the design input parameters. Therefore, in many cases, the design engineers opt out of laboratory testing and primarily rely on past experience and engineering judgments to assign design input parameters. Such an approach compromises the reliability of the pavement life predictions, and can potentially incur unforeseen costs to the traveling public. This study was designed to bridge this gap by developing a series of statistically robust relationships between the laboratory achived data to provide an estimate of the design input parameters. To accomplish this objective, 570 stabilized cylindrical specimens were prepared and subjected to UCS, IDT, and submaximal modulus tests at three strength ratio levels. Subsequently, the relationships between the IDT, UCS, and resilient modulus at small-strain and intermediate strain levels were developed in this study. Such relationships can serve as a valuable means for the estimation of the tensile and compressive strength of the CSM for pavement design.

In recent decades, cement stabilization technology has been successfully used as an economical means for the construction of new roads and the rehabilitation of the existing pavements. It is well established in the literature that the stabilization of the granular soils greatly improves the load bearing capacity and durability of pavement foundations. Traditionally, the compressive strength of the cement stabilized materials (CSM) is obtained by an unconfined compressive strength test, while the tensile strength is estimated based on the third point bending beam flexural or split tension tests. The resilient properties of the CSM can be obtained based on stress path tests or the submaximal modulus tests. Detailed discussions of the advantages and shortcomings of the laboratory procedures and the parameters of the stress path protocols for the estimation of the resilient properties of the CSM are provided by the authors in reference (*1*). Execution of all the laboratory tests is a costly and time consuming undertaking for many state agencies. Customarily, a lack of equipment availability, time constraints, operator concerns, budgetary issues, and so forth mean that many design engineers rely on experience for the selection of the type and the stabilizer contents in pavement projects (*1*). This study was designed to provide a series of practical and robust relationships between the routine laboratory tests for the estimation of the pavement design input parameters. The developed relationships are valuable in the hierarchical pavement design approach for incorporation in the level II design and analysis. Traditionally a third point bending beam test is used to estimate the modulus of rupture of stabilized materials in the laboratory. This test was originally developed for large concrete beams subjected to a strain-controlled loading. Several researchers reported inconsistencies with the bending beam test, and explored alternative testing methods to estimate the tensile behavior of stabilized materials. Ashtiani et al. performed several Finite Elements (FE) analysis considering different material properties and stress paths to investigate the systematic error associated with the bending beam test for stabilized materials in the laboratory (*1*). The authors concluded that at the onset of fracture, approximately 60% of the beam is still

[1]Civil Engineering Department, The University of Texas at El Paso, El Paso, TX
[2]Transportation Engineer, Texas Department of Transportation, Austin, TX

Corresponding Author:
Address correspondence to Reza S. Ashtiani: Reza@utep.edu

in compression as a result of the nature of loading of the prismatic beams. Additionally, the distributions of the stresses followed a highly nonlinear pattern in the mid-span cross section of the prismatic beam. The researchers also showed the capability of the Indirect Diametric Tensile (IDT) test to induce a relatively uniform tension along the axis of loading in the specimen. The authors finally introduced the modified IDT test with a retrofitted bracket for outside strain measurements in lieu of a traditional bending beam test.

Explorations of the relationships between the mechanical properties of the calcium-based stabilizers date back to the 1960s. Thompson, in a pioneering study, developed relationships between the resilient modulus and the unconfined compressive strength of the lime-stabilized subgrade soils (*2*). He also used the typical stress-strain curves to provide the relationships between the tangent modulus, as a measure of undamaged modulus, and the resilient modulus of lime-stabilized granular soils, as presented in Equations 1 and 2.

$$M_r = 3.49 + 1.9 \times E_t \quad (1)$$

$$M_r = 9.98 + 0.124 \times UCS \quad (2)$$

where:
M_r: Resilient modulus (ksi);
E_t: Initial tangent modulus (ksi); and,
UCS: Unconfined compressive strength (psi)

Little et al. also developed a relationship between unconfined compressive strength and the resilient modulus for lime-stabilized subgrade soils (*3*). The resilient modulus was back-calculated based on non-destructive field tests using the Falling Weight Deflectometer (FWD). The researchers determined the the compressive properties of the mixes in accordance with TxDOT (Texas Department of Transportation) test method Tex-121-E for a curing period of 28 days. The authors also found 15 correlations between the UCS and resilient modulus in the literature, and they reported the range in correlations to be extremly wide (*4*). Toohey et al. assessed Little's and Thompson's equations by performing laboratory resilient modulus and unconfined compressive strength testing on multiple specimens of three lime-stabilized soils (*5*). They revealed that Thompson's equation provides a lower-bound estimate of M_r from UCS and that Little's recommended relationship provides a better agreement to their laboratory data. They also studied the influence of sample size by establishing comparative plots of resilient modulus for 115 mm and 200 mm tall specimens. They concluded that the UCS test specimen size does noticeably influence the relationships. In a follow-up study, Hossain and Kim proposed a model to estimate the resilient modulus of fine grained cement stabilized soils from the results of unconfined compression tests (*6*). The authors subjected 26 specimens to UCS tests and resilient modulus tests according to AASHTO T 208 and AASHTO T 307 specifications, respectively. The researchers examined the role of the stress path on the resilient and strength properties of the cement treated mixes. They also explored the influence of the compaction method on the mechanical properties of the cement stabilized materials in the laboratory. The authors reported that the goodness of the fit for relationships developed for the specimen compacted using the impact hammer was superior to the relationships developed for the static compactor.

Piratheepan et al. characterized lightly stabilized materials using UCS and IDT tests with internal displacement measurements (*7*). Based on the laboratory archived data, the researchers developed correlations between the unconfined compressive strength and the tensile strength from both the monotonic and cyclic loading protocols in the IDT test. The authors concluded that the IDT strength is approximately 11% of the compressive strength for lightly stabilized materials. The authors also suggested that the developed relationship can be used as a reasonable estimate of the tensile strength of the materials without the need to incorporate the IDT test in the experiment matrix. However, it should be noted that the estimation of the tensile behavior from the compressive strength should be treated with caution. Further studies by the authors revealed that this anomaly is more pronounced in cement treated reclaimed aggregates and lightly stabilized soils.

Research Objective

The primary objective of this research was to develop relationships between the resilient modulus, compressive strength, and tensile strength of cement stabilized materials for the analysis and design of pavement structures. This objective was achieved by establishing a multi-dimensional aggregate database, consisting of approximately 570 stabilized specimens subjected to multiple mechanical and durability tests. The research team incorporated four aggregate sources, four cement contents, and moisture susceptibility tests to improve the generalization of the developed models.

Laboratory Testing Program

Four aggregate sources with distinct lithologies were incorporated in this research to account for the impact of mineralogy and surface properties on the compressive, tensile, and resilient behavior of cement treated aggregates. Four increments of cement content ranging from 2 to 5% were added to each permutation of the experiment design to cover the light to traditional stabilization spectrum. Subsequently, the variants of the experiment matrix were subjected to several destructive and non-destructive laboratory tests to characterize the resilient, tensile, and compressive behavior of

Table 1. Experiment Design

		Aggregate type															
		El Paso				San Antonio				Pharr				Paris			
		Limestone				Limestone				Gravel				Sandstone			
		Cement content (%)															
	Test	**2**	**3**	**4**	**5**	**2**	**3**	**4**	**5**	**2**	**3**	**4**	**5**	**2**	**3**	**4**	**5**
7-Day Moist Cured	UCS Test	√	√	√	√	√	√	√	√	√	√	√	√	√	√	√	√
	Indirect Diametrical Tensile (IDT) Test	√	√	√	√	√	√	√	√	√	√	√	√	√	√	√	√
	Submaximal Test @ 20% UCS Strength	√	√	√	√	√	√	√	√	√	√	√	√	√	√	√	√
	Submaximal Test @ 40% UCS Strength	√	√	√	√	√	√	√	√	√	√	√	√	√	√	√	√
	Submaximal Test @ 60% UCS Strength	√	√	√	√	√	√	√	√	√	√	√	√	√	√	√	√
	Seismic Modulus Test	√	√	√	√	√	√	√	√	√	√	√	√	√	√	√	√
	Dielectric Value Test	√	√	√	√	√	√	√	√	√	√	√	√	√	√	√	√
10-Day Capillary Soak	UCS Test	√	√	√	√	√	√	√	√	√	√	√	√	√	√	√	√
	Indirect Diametrical Tensile (IDT) Test	√	√	√	√	√	√	√	√	√	√	√	√	√	√	√	√
	Submaximal Test @ 20% UCS Strength	√	√	√	√	√	√	√	√	√	√	√	√	√	√	√	√
	Submaximal Test @ 40% UCS Strength	√	√	√	√	√	√	√	√	√	√	√	√	√	√	√	√
	Submaximal Test @ 60% UCS Strength	√	√	√	√	√	√	√	√	√	√	√	√	√	√	√	√
	Seismic Modulus Test	√	√	√	√	√	√	√	√	√	√	√	√	√	√	√	√
	Dielectric Value Test	√	√	√	√	√	√	√	√	√	√	√	√	√	√	√	√

stabilized materials. Table 1 presents the experiment design developed in this study.

Two different methods of conditioning/curing of the stabilized materials were incorporated in this research to study the influence of moisture ingress on the mechanical properties of the stabilized materials.

a. 7-Day Moist Cured: stabilized specimens were deposited in a chamber set at 23°C (73°F) with at least 95% relative humidity for seven days.
b. 10-Day Capillary Soak: stabilized specimens were placed on porous stones in a tub of water and were subjected to capillary soak for ten days before laboratory testing.

The main motivation for the inclusion of the capillary soak procedure in the experiment was to identify the sensitivity of the aggregates to hold and transport unbound moisture. The rationale is that unreacted moisture trapped in the pore structure can potentially degrade the stiffness properties of the stabilized layers and consequently jeopardize the longevity of the pavement structure.

The particle size distributions for four aggregate sources in this study is presented in Figure 1. The selection of the particle size distributions was based on Item 247 of TxDOT standard construction specifications for Grade 4 materials. The moisture-density tests were performed on untreated materials to identify the Optimum Moisture Content (OMC) and Maximum Dry Density (MDD) following Tex-113-E test method (*8*). Figure 1 also provides the OMC and $(\gamma_d)_{max}$ results. Subsequently, the molding moisture content was adjusted for each increment of cement added to the mix based on Tex-120-E procedure (*9*).

Unconfined Compressive Strength (UCS)

The unconfined compressive strength test was performed on 6 × 12 in. stabilized specimens to provide an indication of the compressive strength of stabilized systems subjected to the vertical compressive load. This test was the precursor to the submaximal test which was utilized to determine the dynamic stress level amplitudes.

Indirect Diametrical Test (IDT)

The Indirect Diametrical Tensile (IDT) tests were performed at an imposed strain rate of 1 mm/min on 6 × 4.5 in. stabilized specimens as observed in Figure 2. This test provides an indication of the tensile strength of the specimen subjected to strain-controlled stress path tests in the laboratory. Similar to the UCS tests, two curing/conditioning procedures were considered to monitor the influence of the moisture ingress on the degradation (or improvement) of tensile strength of the stabilized materials.

Submaximal Modulus Test

Figure 3 shows the stress-controlled submaximal tests (strength ratio tests) performed on the 6 × 12 in. cylindrical specimens stabilized with varying cement contents. The

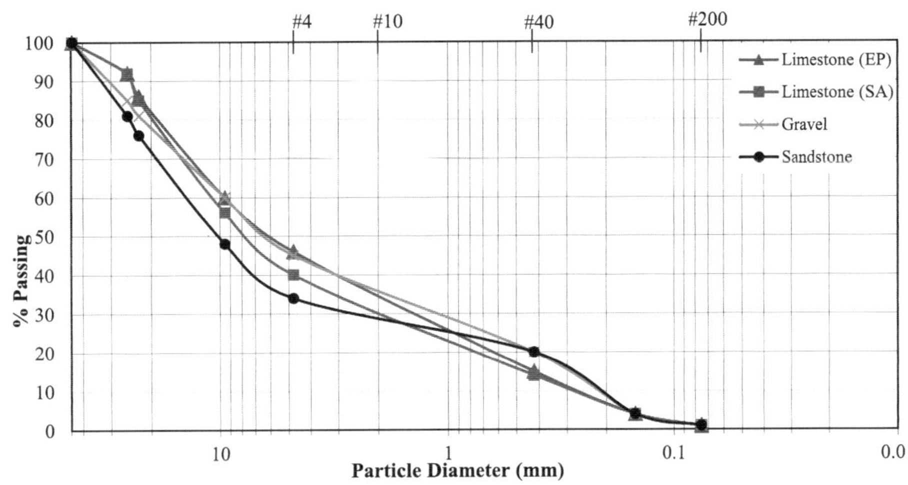

Material	OMC (%)	MDD (lb/ft^3)
El Paso Limestone	6.8	139.1
San Antonio Limestone	6.5	140.1
Pharr Gravel	8.8	134.5
Paris Sandtone	6.8	137.4

Figure 1. Particle size distributions of aggregate materials and moisture-density test results.

strength values obtained from the UCS test were the basis for the selection of the stress amplitudes applied to the permutations of the experiment design. Pre-determined fractions of the UCS-value, namely 20, 40 and 60% were cycled for 5,000 repetitions for each variant to characterize the resilient behavior of stabilized systems. The cyclic loading consists of repeated cycles of a haversine shaped load pulse at a frequency of 3 Hz. These load pulses had a 0.1-second load duration and a 0.9-second rest period. This provided valuable information on the influence of the incremental increase in cement contents on the resilient modulus of the stabilized mixes. Similar to the procedure outlined for the UCS tests, two sets of samples for each [Material Type-Stabilizer Content] combination were subjected to 7-day moist cured and 10-day capillary soak conditions. The moisture susceptibility tests provided valuable information on the deleterious (or beneficial) influence of provided moisture through capillary action on the resilient modulus of stabilized materials in the laboratory.

Labratory Test Results and Analysis

The following section provides the analysis and discussion of the models developed based on the laboratory tests outlined in the experiment design. The primary focus was to investigate the inter-relations between the resilient properties and the strength characteristics of cement stabilized materials in tension and compression.

Relationship between the UCS and IDT

Figure 4 shows the relationship between the unconfined compressive strength and the indirect diametrical tensile tests results. The ascending nature of the trendline suggests a direct correlation between the compressive and tensile behavior of the cement stabilized materials. In other words, permutations with higher compressive strength exhibited higher tensile strength compared with other counterparts in the experiment matrix.

Equation 3 presents the relationship between the compressive and tensile strength of the cement stabilized systems. The model is based on the UCS and IDT testing of 68 permutations in the laboratory. The relationship developed between the IDT strength (S_{IDT}) and the UCS is presented in Equation 3.

$$S_{IDT} = 0.12 \times UCS \qquad (R^2 = 0.68) \tag{3}$$

This relationship suggests that the tensile strength from the IDT test is approximately 12% of the unconfined compressive strength of the cement stabilized materials. The results are in good agreement with models developed by other researchers as shown in Table 2.

The proposed model has several advantages compared with the previous studies outlined in Table 2. The first advantage is associated with the range of the cement contents added to the mixes. As previously stated in this paper, all

Figure 2. Static IDT test set-up and fractured specimens.

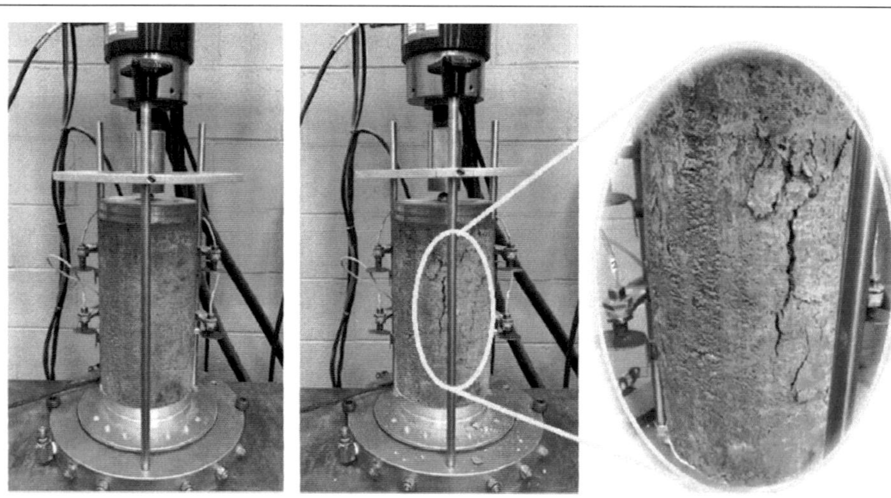

Figure 3. Submaximal modulus test set up and fractured specimens.

four aggregate sources were stabilized from 2 to 5% in this study. Therefore, the models incorporate both lightly stabilized systems as well as rigid aggregate matrices. The overly rigid aggregate systems are not desireable in pavements as they may be prone to shrinkage cracking. For this reason, several state agencies defined thresholds for the UCS to

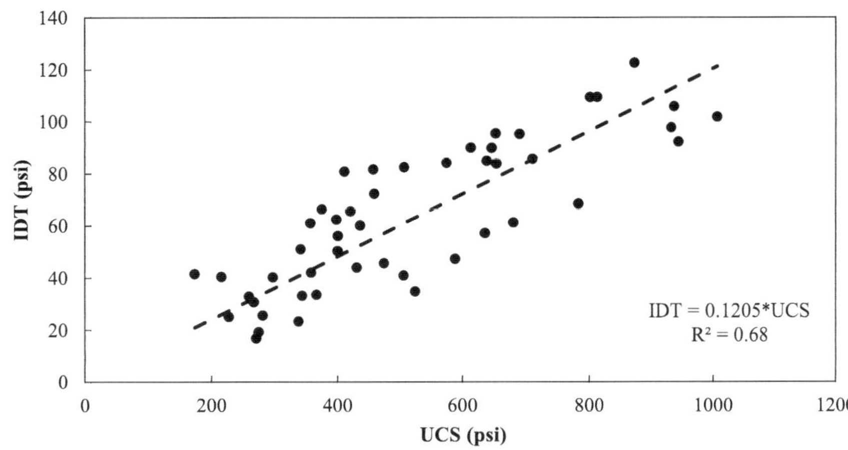

Figure 4. Relationship between UCS and IDT strength for cement stabilized base materials in the laboratory.

Table 2. Available Relationships for IDT Strength and UCS

Relationship (psi)	Researcher	Year
$S_{IDT} = 0.166 \times UCS - 11.38$	Kennedy *(10)*	1973
$S_{IDT} = 0.114 \times UCS$	Piratheepan *et al. (7)*	2010
$S_{IDT} = 0.177 \times UCS - 9.31$	Scullion *(11)*	2012
$S_{IDT} = 0.12 \times UCS$	Wen & Edil *(12)*	2014

protect the pavements against distresses associated with the reflective cracking. Another major advantage of the newly developed model is attributed to the incorporation of durability tests in this study. As previously outlined in the experiment design segment, two replicates of the stabilized specimen were subjected to 7-day moist-cured and 10-day capillary soak or the Tube Suction Test (TST) protocols. The change in the dielectric value at the top of the specimen, as well as the seismic modulus values, were determined for 10 consecutive days for the TST specimen. The simultaneous monitoring of the unbound moisture and the rate of strength gain (or loss) in the TST test provided valuable information on the moisture susceptibility of the stabilized mixes. A detailed discussion of the pair-wise plots of dielectric value and seismic modulus, and the procedure for the characterization of the moisture susceptibility of stabilized materials are provided by authors in reference (*1*). In summary, incorporation of four aggregate sources, four increments of cement content, and integration of the durability tests potentially enhanced the generalization of the relationships developed in this study.

Figure 5 underscores the influence of the stabilizer content on the compressive and tensile behavior of the cement stabilized materials. As expected, the incremental increase in the cement content in the mixes resulted in the improvements in both compressive and tensile properties of the stabilized systems. Additionally, this plot clearly shows the synergistic influence of the aggregate mineralogy and the cement content on the mechanical behavior of cement treated materials. The spread of the datapoints corresponding to the tensile and compressive behavior of stabilized specimens can be explained by the schematic data cloud analogy. As evidenced in Figure 5, the cloud of the data is relatively rounded for lightly stabilized materials. However, the increase in the cement contents resulted in a larger spread of the cloud of the data. This variation is more pronounced for 5% cement stabilized systems as depicted with an elongated ellipse. In other words, the results indicate that the distance from the mean increases with the increase in the cement content. This can potentially be attributed to the role of the mineralogy and surface properties of the aggregates, and their influence on the tensile and compressive behavior of the cement treated systems. Another plausible explanation can be attributed to the initiation of the micro-cracks in overly rigid stabilized mixes. Advanced imagining techniques such as X-Ray Computed Tomography can provide deterministic means to test the validity of these theories.

Another noteworthy observation is the distinct transition of lightly stabilized materials that resemble soil cement, to the heavily stabilized materials that behave more like concrete. This was the motivation for establishing a new relationship exclusively for the 5% cement stabilized materials and for comparing it with conventional concrete models. The general power form function used by American Concrete Institute ACI (*13*) and Oluokon (*14*) was adopted for direct comparison of the models. Equation 4 presents the model based on the 5% cement stabilized materials for four different aggregate sources.

$$S_{IDT} = 7.77 \times UCS^{0.38} \qquad (R^2 = 0.79) \qquad (4)$$

Figure 6 illustrates the comparison between the widely used concrete relationships and the model developed for the

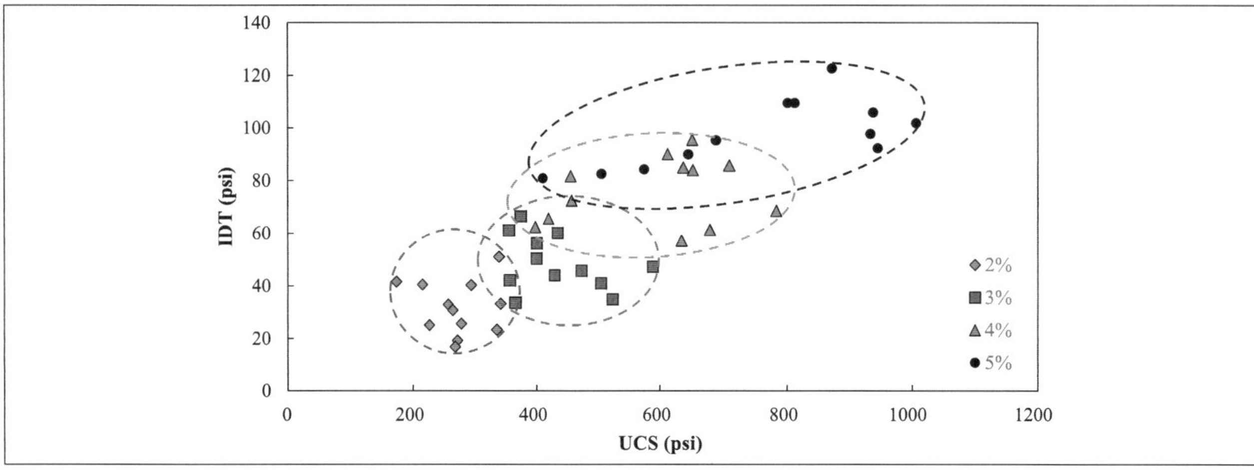

Figure 5. Relationship between UCS and idt strength for cement stabilized base materials categorized based on the cement content.

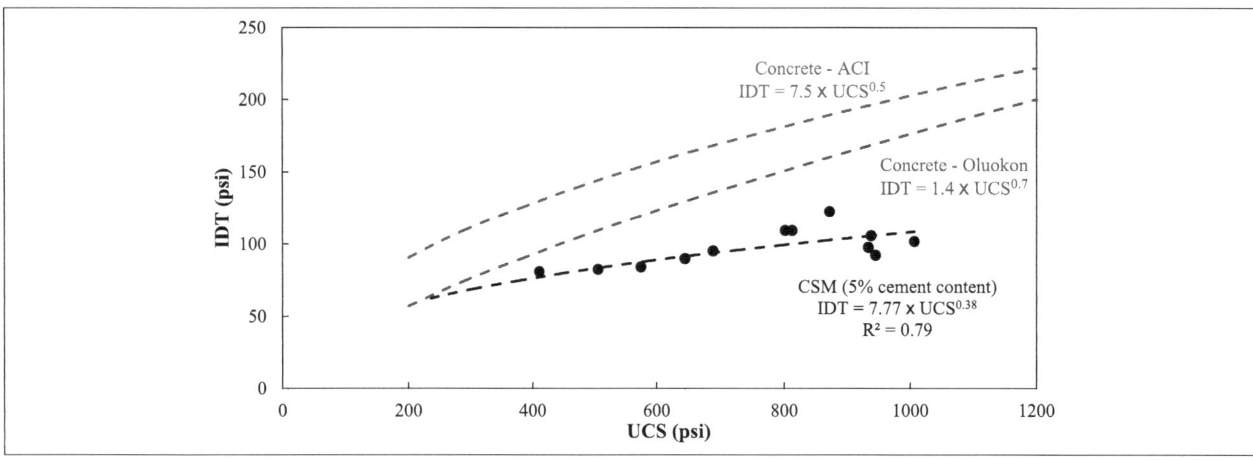

Figure 6. Comparisons of tensile and compressive strengths for heavily cement stabilized materials and conventional concrete models.

highly cement stabilized materials in this study. As expected, the envelope for the cement treated materials was contained below the traditional concrete models. Additionally, the rate of the tensile strength gain with the increase in the UCS is shown to be significantly higher for concrete. In other words, the slope of the concrete models is much higher compared with the cement stabilized materials.

Figure 7 provides the IDT and UCS results categorized based on the aggregate type. As previously stated, four aggregate sources, namely two limestone materials sourced from San Antonio and El Paso, Sandstone from Paris district in Texas, and gravel from Pharr stabilized with 2 to 5% cement in this study. The results clearly underscore the role of the aggregate mineralogy on the mechanical performance of the cement stabilized materials. As evidenced from Figure 7, the IDT strength for limestone materials from El Paso and San Antonio, and silicious sandstone from Paris is approximately 10~15% of the unconfined compressive strength. However, the gravel sourced from Pharr district benefitted less from the incremental increase in the cement contents, and therefore underperformed other variants in the experiment design. The lackluster results for Pharr materials are in agreement with other laboratory performance tests such as repeated load permanent deformation tests, and dynamic IDT tests executed by the authors (*1*).

Resilient Modulus of the Cement Stabilized Materials

The new Mechanistic-Empirical Pavement Design Guide (MEPDG) requires a measure of the resilient modulus for the analysis and design of pavements. This information is required for the calculation of the responses of pavements subjected to traffic loads. Therefore, the submaximal modulus tests at three Strength Ratios (SR) were performed on 7-day moist-cured and 10-day capillary soak specimen in the

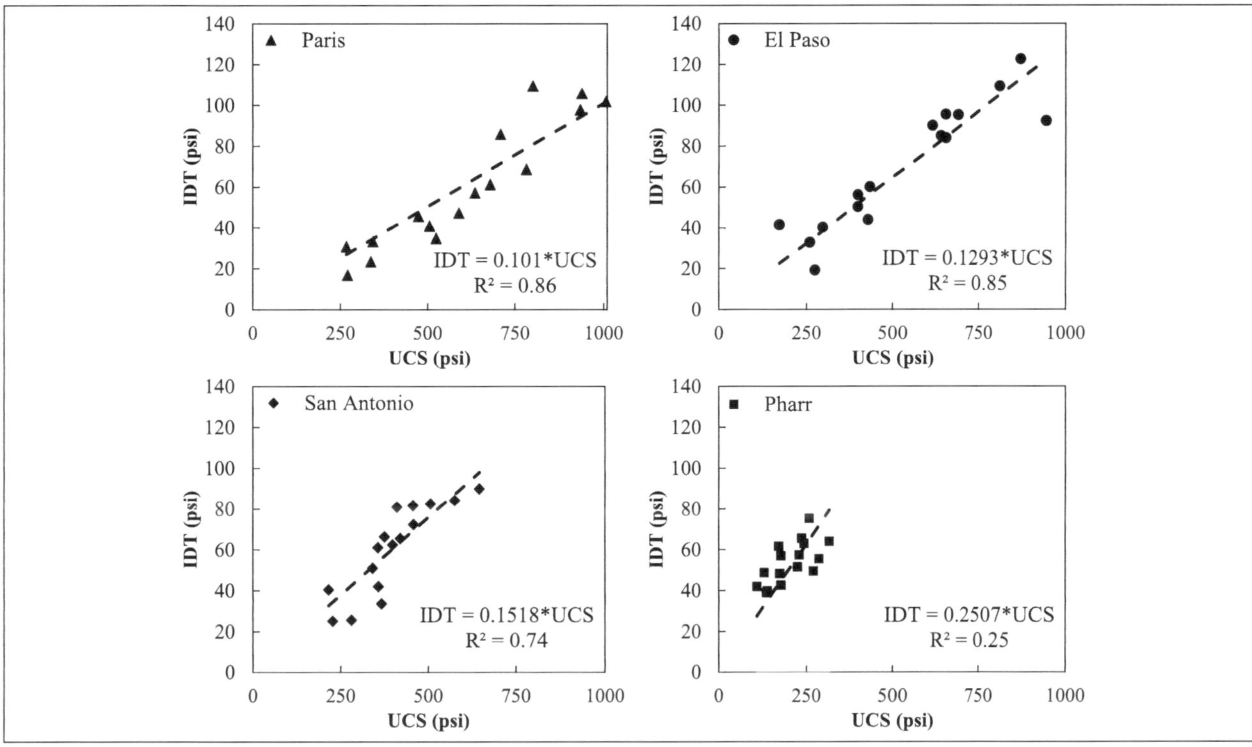

Figure 7. Relationships between the UCS and IDT categorized by the aggregate type.

laboratory. Pre-determined fractions of the UCS-value, namely 20, 40 and 60% were cycled for 5,000 repetitions to calculate the resilient modulus of the stabilized mixes.

Execution of the full spectrum of the submaximal modulus test is a costly and time-consuming undertaking for small projects. A series of relationships between the UCS and the resilient modulus tests at different strength ratios were, therefore, developed in this study. Such models can serve as a starting point to provide an estimate of the resilient modulus of cement treated layers for the practitioners and the pavement design engineers.

Figure 8 provides the resilient modulus tests results for 7-day moist-cured and 10- day capillary soak (TST) specimen. This plot is primarily based on the averages of two replicates of 3% cement stabilized aggregates subjected to 5,000 load cycles for all aggregate types in the experiment design. Based on the survey conducted by the authors, 3% cement was the predominant stabilizer content used by the design agencies (*15*). This was the motivation for the selection of the 3% stabilized systems to demonstrate the deleterious effect of unreacted moisture on the resilient properties of the laboratory specimen. As illustrated in Figure 8, a stabilized specimen subjected to the TST moisture susceptibility protocol had lower resilient modulus values compared with traditional 7-day moist-cured specimen. Such degradation of the stiffness properties is greatly influenced by the aggregate type. For instance, the 3% cement stabilized limestone materials sourced from San Antonio had approximately 50% lower resilient modulus when subjected to the moisture susceptibility protocol. Conversely, sandstone materials sourced from Paris district in Texas showed a mere 8% reduction in the resilient modulus when compared with the 7-day moist cure as the benchmark. This underscores the role of the mineralogy and the aggregates source with regard to the moisture damage potential during the service life stabilized pavement foundations. The incorporation of the moisture susceptibility protocols provides a mechanistic approach for the quantification of the softening behavior of the stabilized layers upon moisture

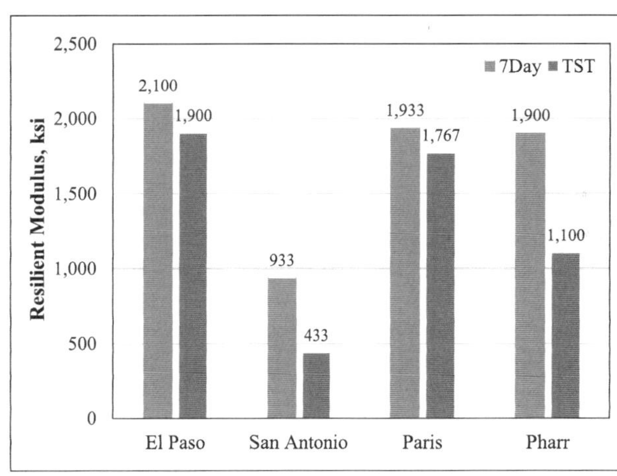

Figure 8. Comparisons of the average resilient modulus for 3% cement stabilized aggregate systems.

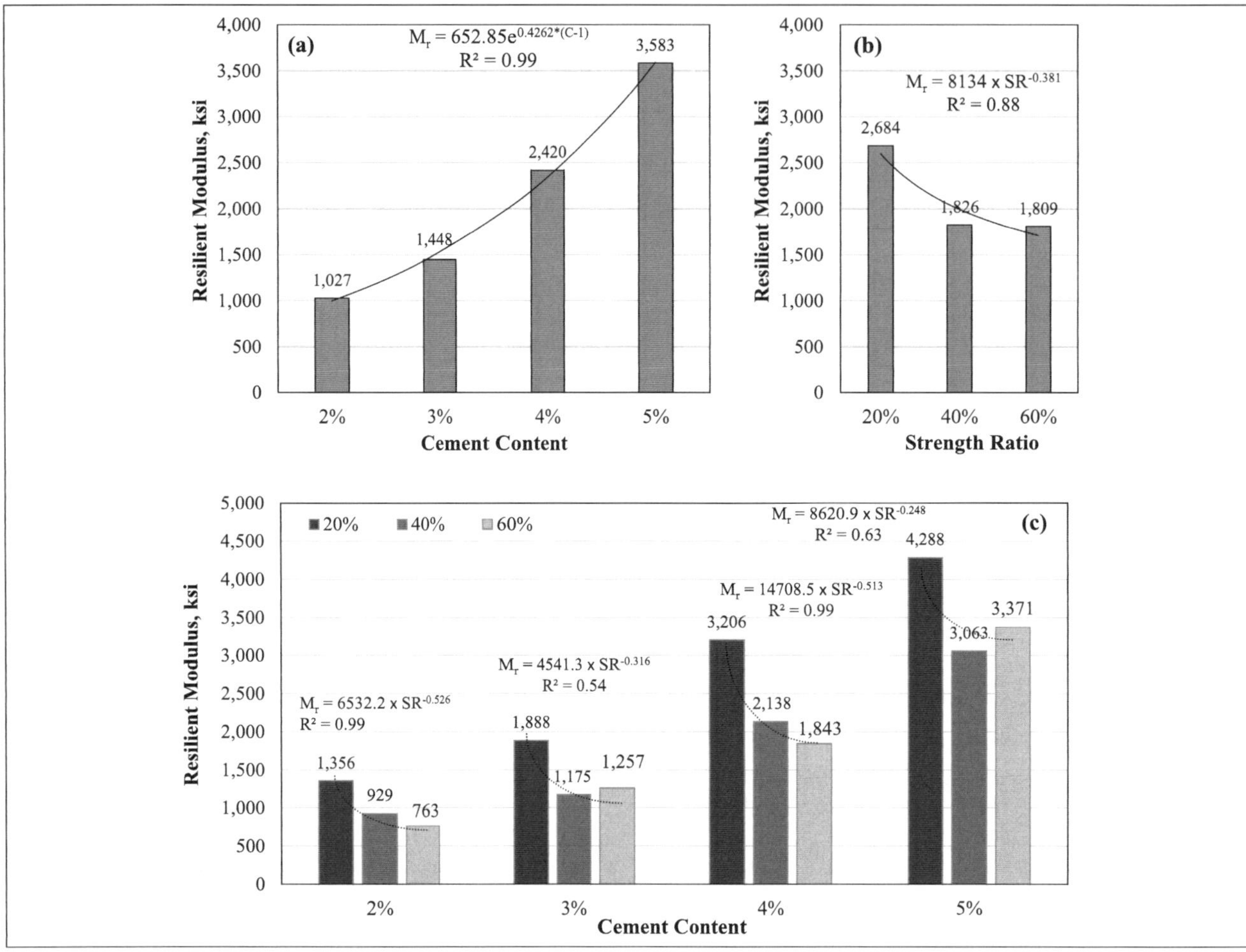

Figure 9. Effects of (*a*) cement content (*b*), and (*c*) strength ratio on resilient modulus.

intrusion. This provides a valuable means for the selection of the type and source of the aggregates for the construction of cement stabilized base layers.

Figure 9 demonstrates the submaximal modulus results for all permutations of the experiment design in this study. Figure 9*a* presents the beneficial role of the incremental increase in the stabilizer content on the resilient modulus of the treated systems. The ascending nature of the trendline is an indication of the favorable influence of the cement on the resilient properties of the stabilized specimen. Equation 5 provides the relationship between the resilient modulus and the cement content developed in this study.

$$M_r = 652.85 \times e^{04262\times(C-1)} \quad (R^2 = 0.99) \tag{5}$$

where:

M_r: Resilient modulus (ksi); and
C: Percent cement content in the mix (%).

Figure 9*b* underscores the influence of the stress path protocol on the resilient modulus of the stabilized matrials. As previously indicated, 20, 40, and 60% of the unconfined compressive strength was cycled for 5,000 repetitions to characterize the resilient properties of the cement treated materials. Figure 9*b* and *c* indicate that the variants subjected to less taxing stress paths had higher resilient modulus values. This results from the low-stress excitation levels and therefore low induced strains in the SR = 20% loading protocol. Conversely, SR = 40 and 60% stress path protocols induce higher strains and the net effect will be lower modulus values. Therefore, it is deemed necesary to consider the strain levels imposed by traffic loads on pavement layers for the realistic selection of the resilient modulus for the design and analysis of pavements.

Relationship between the UCS and the Resiliant Modulus

Figure 10 presents the relationship between resilient modulus at 20, 40 and 60% of the UCS values. As previously indicated in this paper, the resilient modulus of the stabilized bases is a stress-path dependent property. Therefore, the

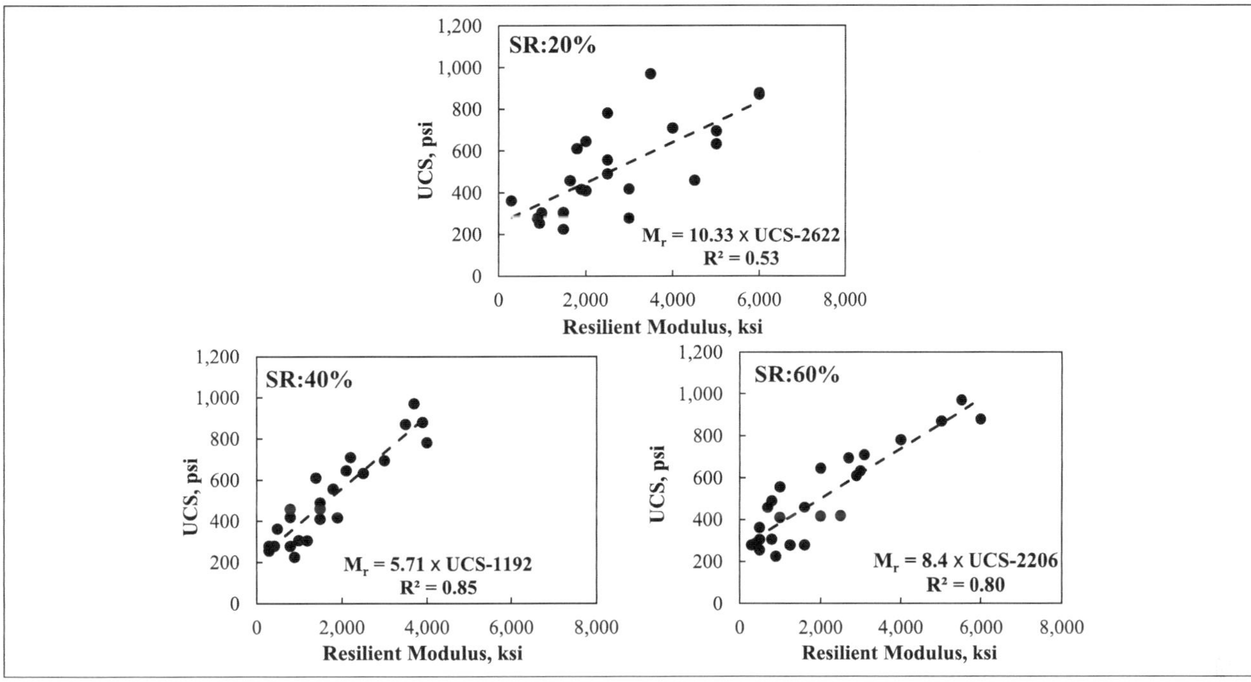

Figure 10. Relationship between resilient modulus and unconfined compressive strength at 20%, 40%, and 60% strength ratios.

stress dependency of the resilient properties of the cement stabilized materials should be considered for the selection of the design modulus values. For this reason, two relationships based on the strength ratio levels were developed in this study to cover small-strain and intermediate-strain level behavior of stabilized layers. Small-strain behavior could be associated with pavements with thick/stiff surface layers. In such pavements, most of the traffic-induced stresses dissipate within the stiff/thick surface layer, and therefore the magnitude of the stresses transmitted to the stabilized layer will be relatively small. Subsequently, the small stress excitations will result in low strain levels induced by traffic loads. A similar argument is valid for inverted pavements for which the stabilized layer is located below the asphalt and the unbound granular base layers. Such cases are plausible scenarios for the small-strain behavior of stabilized base layers in the field.

Equations 6 and 7 provide relationships between the UCS and resilient modulus for small-strain and intermediate strain levels. The small-strain behavior model was primarily developed based on SR = 20%, and the intermediate-strain model is based on the combination of the SR = 40 and 60%.

$$M_r = 10.33 \times UCS - 2622 \quad \left(R^2 = 0.53\right) \tag{6}$$

$$M_r = 7.40 \times UCS - 1940 \quad \left(R^2 = 0.81\right) \tag{7}$$

Influence of the Type of Calcium-Based Stabilizer

The resilient modulus relationships developed in this research were juxtaposed on prominent studies such as National Lime Association by Little (*4*), Barenberg (*16*), American Coal Ash Manual (*17*), and Australian Road Research Laboratory (*18*). Figure 11 provides a visual comparison of the models developed under various sample preparation procedures, base/subgrade materials, and different calcium-based stabilizers.

As previously stated in this paper, the proper selection of the type and the amount of the calcium-based stabilizers in the mix design is of paramount importance to ensure the longevity of the pavement foundations. Excessive amounts of stabilizers will result in overly rigid systems that are prone to shrinkage cracking. The propagation of the cracks to the surface of the asphalt layers can significantly compromise the structural integrity, serviceability, and life of pavements. Therefore, it is imperative to establish an upper bound limit for the strength properties of the calcium-based stabilized materials to protect the pavements against reflective cracking.

Conclusions

The focus of this effort was to investigate the inter-relations between the compressive, tensile, and resilient properties of cement stabilized base materials in the laboratory. This objective was achived by the execution of a comprehensive

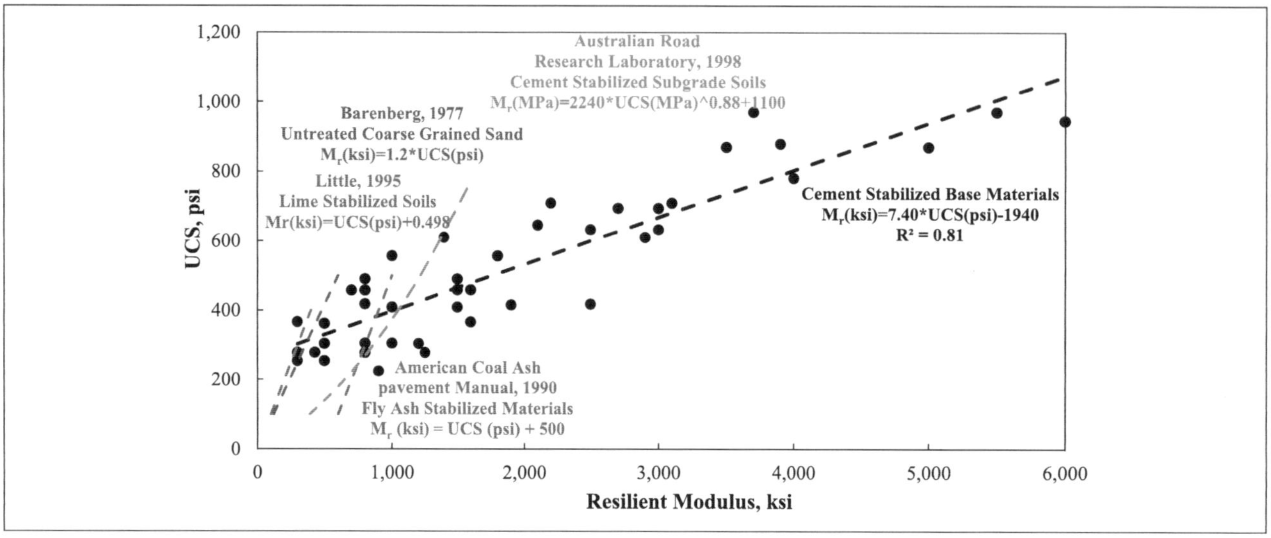

Figure 11. Comparison of the resilient modulus and UCS relationships for the cementitiously stabilized base and subgrade soils.

experiment design consisting of four aggregate sources with distinct lithologies to account for the influemce of aggregate mineralogy on the strength of cement stabilized materials. Four levels of cement content, ranging from 2 to 5% were added to the mixes to cover a wide spectrum of cement treatment from light stabilization to heavily stabilized systems. The stabilized aggregate specimen was subjected to two curing/conditioning procedures before the mechanical tests, to study the affinity of the aggregates to adsorb and transport moisture. Several mechanical tests such as UCS, IDT, and submaximal modulus tests at 20, 40, and 60% strength ratios were performed on the permutations of the experiment design. Based on the post-processed data, the tensile strength of the cement stabilized materials can be approximated as the 12% of the unconfined compressive strength. Subsequently, the spread of the data points corresponding to the tensile and compressive behavior was evaluated via the data cloud analogy. The results show that distance away from the mean for lightly stabilized materials was significantly smaller compared with heavily stabilized systems. The stretch of the datapoints in hevealy stabilized systems underscores the role of the aggregate mineralogy on the compressive and tensile behavior of cement stabilized base materials.

The relation between the resilient modulus and compressive strength properties of the stabilized mixes was also investigated in this study. The submaximal modulus tests at three strength ratios were performed to study the behavior of the systems at small-strain and intermediate-strain levels. Consequently, the relationships between the resilient modulus, UCS, strength ratio, and the cement content were developed in this study.

The incorporation of moisture susceptibility tests, the diversity of the aggregate sources, and the wide range of the cement treatments were the main prerogative of this study. This can potentially enhance the generalization of the models for improved prediction of the resilient and strength properties of the stabilized systems. The proposed models can serve the pavement design industry to provide an estimate of the tensile strength and the resilient modulus of cement stabilized materials for the analysis and design of pavement structures.

References

1. Ashtiani, R. S., and J. Tarin. *Testing Procedure for Long Life Heavy Duty Stabilized Bases*. TxDOT Project Number 0-6812, FHWA/TX-06/0-5223-2. Texas Transportation Institute, 2016.
2. Thompson, M. R. *Shear Strength and Elastic Properties of Lime Soil Mixtures*. Highway Research Board, University of Illinois, Champaign, Ill., 1966.
3. Little, D. N., T. Scullion, P. Kota, and J. Bhuiyan. Identification of the Structural Benefits of Base and Subgrade Stabilization. Report 1287-2. Texas Transportation Institute, Texas A&M University, College Station, 1994.
4. Little, D. N., A. M. Godiwalla, P. Y. Oshiro, and P. S. Tang. *Characterization of Design Properties (Compressive Strength and Resilient Modulus) of Lime, Cement, Fly Ash Stabilized Structural Recycled Concrete Base as a Function of Curing Time*. Presented for the Federal Aviation Adiminstration Airport Thechnology Transfer Conference, 2002.
5. Toohey, N. M., M. A. Mooney, and R. G. Bearce. Relationship between Resilient Modulus and Unconfined Compressive Strength for Lime-Stabilized Soils. *Journal of Geotechnical and Geoenvironmental Engineering*, Vol. 139, No. 11, 2013, pp. 1982–1985.
6. Hossein, M. S., and W. S. Kim. Estimation of Subgrade Resilient Modulus Using Unconfined Compression Test. *Transportation Research Record: Journal of the Transportation Research Board*, 2015. 2473: 126–135.

7. Gnanendran, C. T., and J. Piratheepan. Determination of Fatigue Life of a granular Base Material Lightly Stabilized with Slag Lime from Indirect Diametral Tensile Testing. *Journal of Transportation Engineering*, Vol. 136, No. 8, 2010, pp. 736–745.
8. Texas Department of Transportation. *Test Procedure for Laboratory Compaction Charactrestics and Moisture Density Relationship of base Materials* (Tex-113-E). Texas Department of Transportation, 2016.
9. Texas Department of Transportation. *Test Procedure for Soil-Cement Testing* (Tex-120-E). Texas Department of Transportation, 2013.
10. Kennedy, T. W., and W. R. Hudson. *Tensile Properties of Subbases for Use in Rigid Pavement Design*. Report No. CFHR 3-8-66-98-14F. Texas Highway Department, 1973.
11. Scullion, T., S. Sebesta, C. Estakhri, P. Harris, C. Shon, O. Harvey, and K. Rose-Harvey. *Full-Depth Reclamation: New Test Procedures and Recommended Updates to Specifications*. 2012 FHWA/TX-11/0-6271-2, Texas Transportation Institute, 2012.
12. Wen, H., and T. Edil. Characterization of Cementitiously Stabilized Layers for Use in Pavement Design and Analysis. NCHRP Project 04-36. 2014.
13. Mindess, S., and J. F. Young. *Concrete*. Prentice-Hall, Englewood Cliffs, N.J., 1981.
14. Neville, A. M. *Properties of Concrete*. John Wiley & Sons, Inc., New York, 1966.
15. Ashtiani, R., I. Abdallah, and S. Nazarian. *Updated Testing Procedures for Long Life Heavy Duty Stabilized Bases: District Survey Results*. Texas Deparment of Transportation Report 0-6812, 2014.
16. Barenberg, E. J. *Evaluating Stabilized Materials*. National Cooperative Highway Research Program. Contract No. HR 63-4-1, Unpublished Report, 1977.
17. American Coal Ash Association. *Flexible Pavement Manual*. Alexandria, Va., 1990.
18. Australian Road Research Laboratory. Unpublished Correlations between Compressive Strength and Modulus. Sydney, New South Wales, Australia, 1998.

The Standing Committee on Stabilization of Geomaterials and Recycled Materials (AFS80) peer-reviewed this paper (18-02950).

Article

TRR
JOURNAL OF THE TRANSPORTATION RESEARCH BOARD

Transportation Research Record
2018, Vol. 2672(52) 164–173

DOI: 10.1177/0361198118757439
journals.sagepub.com/home/trr
SAGE

Performance Evaluation of Cement and Slag Stabilized Expansive Soils

Masrur Mahedi[1], Bora Cetin[1], and David J. White[2]

Abstract

Swelling, shrinking, and subsequent low strength of expansive soil poses significant damage to structures if it is considered as foundation or fill material. Recently, the use of cement has become very prevalent in stabilizing these problematic soils owing to its effectiveness. However, the swelling potential of expansive soil is not always adequately resolved by cement. The presence of sulfate salts aggravates the situation impairing the effectiveness of cement, leading to the need to reassess its performance. In this study, the effectiveness of different stabilizers was investigated in stabilizing high-plasticity soil. Two types of soil with variable sulfate content were treated with slag, Type I/II, and Type V Portland cement, and their performances were evaluated based on Atterberg limits, pH, unconfined compression, and volumetric swell tests. A total of 312 samples were prepared for 18 different soil–stabilizer blends tested after 7, 28, and 90 days of curing period. Laboratory test results indicated that strength gain performance was attenuated and swelling potential increased due to the presence of sulfate salts. Adding stabilizers improved the strength of soils by a factor of 4–10 and decreased the swelling potential to < 1%. Atterberg limits decreased initially and then slightly increased with the increase of additive dosages. Additives increased the pH up to a maximum value of 11–12, which could be used as an indicator of target stabilizer addition rate. Finally, slag improved the performance of cement significantly and has proven to be a better option for treating high sulfate expansive soils.

Expansive soil is a worldwide problem causing significant structural issues to the infrastructure. In the U.S., the problem is more acute as about 20% of the land surface is composed of high-plasticity expansive soil (*1*). Mitigation of the swelling and shrinking potential of expansive soils is a very crucial design aspect of pavement construction when expansive soils are used as subgrade soil. Expansive soils undergo large amounts of heaving and shrinking due to seasonal moisture variation, which can lead to failure of the upper pavement structure. Expansive soils typically consist of fine smectite clay minerals such as montmorillonite with illite, claystones, residual, and sedimentary soils that absorb large amounts of water and swell (*2*). Desiccation of clay exacerbates the problem by causing shrinkage cracks during the dry season allowing water to penetrate even deeper. Through this cyclic process of shrinking and swelling, soil undergoes large volume changes and subsequently loses its strength (*3*). This shrink–swell behavior also induces tensile stresses on structures leading to deterioration by heaving and differential settlement (*4–6*). Under such conditions, soil improvement by using appropriate additives is a viable approach to ensure the stability of the infrastructure, which also results in substantial economical savings by reducing construction, repair, and rehabilitation costs.

Cement is one of the most common additives used as a stabilizing agent for expansive soils. Extensive evaluations have been carried out on the cement stabilization of expansive soils. However, cement stabilization usually results in high stiffness, and makes the soil brittle, which is undesirable in dynamic loading conditions such as pavement systems (*7*). Moreover, swelling tendency of cement stabilized soil is not adequately compromised which need to be addressed (*8*). Substantial challenged arise when the expansive soil contains considerable amounts of sulfate salts such as sodium sulfate, calcium sulfate, and magnesium sulfate (*9*). Sulfate minerals attack the tricalcium aluminate (C_3A) of hardened cement in the presence of calcium hydroxide and form secondary gypsum and ettringite, which lead to the deterioration of the cement–soil matrix (*10*). Mehta demonstrated the inadequacy of calcium-based cements in stabilizing high sulfate soils (*11*). Therefore, the expected performance of cement treatment is not often met. It is,

[1]Department of Civil, Construction and Environmental Engineering, Iowa State University, Ames, IA, USA
[2]Ingios Geotechnics, Inc., Ames, IA, USA

Corresponding Author:
Address correspondence to Bora Cetin: bcetin@iastate.edu

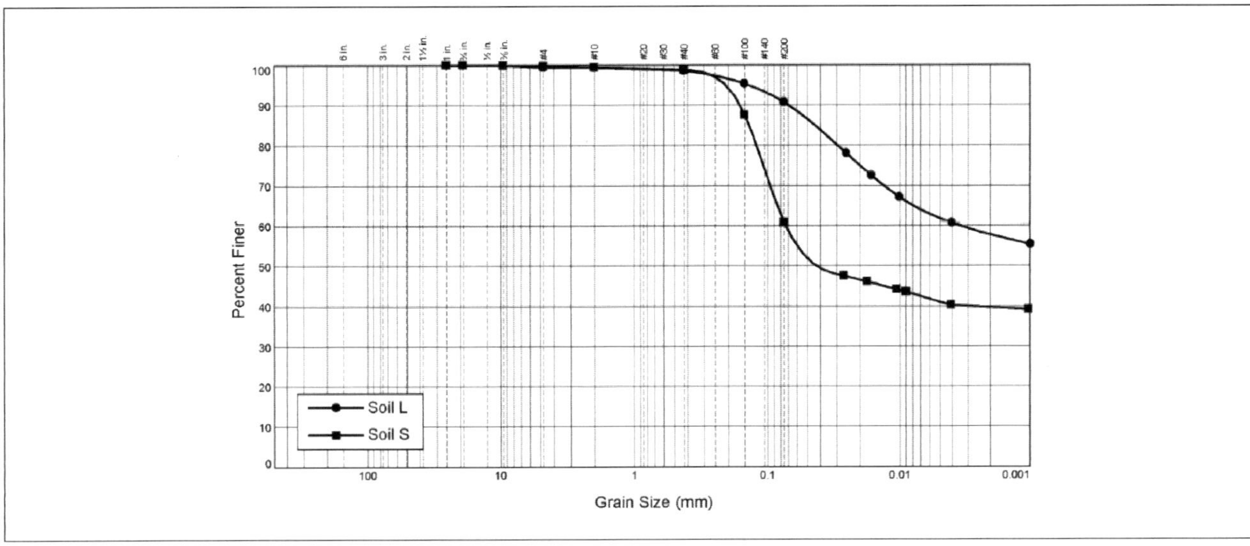

Figure 1. Particle size analysis of Soil-L and Soil-S.

thus, necessary to reassess and predetermine the appropriate cement type and application dosage for stabilizing expansive soils.

Owing to these concerns, it was decided to formulate an experimental program to evaluate the effectiveness of cement in stabilizing expansive soil, along with the investigation on the prospective use of steel slag. To achieve this goal, different quantities of Type I/II and Type V Portland cement, and Slag-Type I/II cement blends were mixed with two different types of high-plasticity soils of variable sulfate content. Atterberg limits, pH, unconfined compressive strength (UCS), and volumetric swell tests were conducted on samples after a curing period of 7, 28, and 90 days. In addition, X-ray fluorescence (XRF) analysis was performed on soil and stabilizing agents.

Materials

Soils

For the experimental program of this study, two different soils were sampled from Austin and San Antonio, Texas. Based on the nearest city from the soil sample locations, materials are designated herein as soil groups "S" (Sequin) and "L" (Lockhart). Particle size distribution of the soils were investigated following the standard test method Tex-110-E specified by Texas Department of Transportation (TxDOT) and results are presented in Figure 1. According to the Unified Soil Classification System (USCS), both soils were classified as high-plasticity clay (CH). However, the clay content of Soil-L (58%) was higher than Soil-S (40%) resulting in a higher plasticity index (PI) of Soil-L (49). Compared with Soil-S, Soil-L had a higher optimum moisture (28%) and lower maximum dry unit weight (14.5 kN/m^3) due to its higher clay content at standard Proctor compaction energy.

Moreover, soils were tested for soluble sulfate content according to Tex-145-E and Tex-146-E. Tex-145-E testing was performed using a turbidity colorimeter. Tex-146-E testing was conducted using a conductivity probe with a nominal cell constant of 1.0/cm. Previous lab testing has resulted in quantitative correlations between conductivity and sulfate concentration (*12*), which were used in this analysis. A container of Soil-L samples was found to be of high sulfate content that has been designated herein as Soil-L$_s$. pH, cation concentration, organic matter, and total carbon content measurements were conducted using the "Recommended Chemical Soil Test Procedures for the North Central Region" (*13*). The pH of the soils was analyzed using a pH meter following a 1:1 ratio technique. A total of 5 g of air-dried soil sieved through a No. 10 sieve was mixed with 5 ml of distilled water and mixed thoroughly. pH of the mixture was measured after 10 minutes using the pre-calibrated pH meter. pH of Soil-L, Soil-S, and Soil-L$_s$ were 8.7, 5.7, and 7.8, respectively. pH was relatively low for high-sulfate Soil-L (7.8), though its Ca^{2+} concentration was the highest. It was anticipated that most of the Ca^{2+} ions in Soil-L$_s$ were in the form of calcite, dolomite, gypsum, or basanite, which are neutral in terms of pH. For quantitative comparisons, all the soil test results are listed in Table 1.

Stabilizers

Two different cements, Type I/II and Type V, and steel slag were used as stabilizing agents in this study. XRF

Table 1. Properties of Soil Materials Tested

Soil designation	Soil-L	Soil-S	Soil-L_S
PI	49	31	45
LL	74	51	73
Dry unit weight (kN/m^3)	14.5	17.3	15.7
Optimum moisture content (%)	28.0	17.5	25.8
Clay content (0.002 mm) (%)	58	40	54
Volumetric swell (%)	12	18	14
Initial moisture content (%)	28.3	16.8	27.0
Final moisture content (%)	39.2	27.5	43.8
ppm K	434	185	328
ppm Ca	10,088	3,010	26,285
ppm Mg	147	670	653
ppm Na	38	396	715
pH	8.7	5.7	7.8
% carbon	2.5	0.4	0.8
% organic matter	4.5	0.7	1.4
Sulfate TX 145-E (ppm)	< 100 (low)	< 100 (low)	30,320 (high)[a]
Sulfate TX-146-E (ppm)	700 (low)[b]	< 100 (low)[b]	31,300 (high)[b]

[a]Sample diluted to 1:200, a dilution outside of standard used to obtain quantifiable amount in equipment measurement range.
[b]Using the correlation of Harris (*12*).
Note: LL = liquid limit; PI = plasticity index; ppm = parts per million.

Table 2. Summary of Type I/II and Type V Cement, and Slag

Oxide or constituent (mass %)	Type I/II cement	Type V cement	Steel slag
Na_2O	0.19	0.33	0.34
MgO	1.09	0.72	7.48
Al_2O_3	4.49	4.16	10.4
SiO_2	20.1	21.3	35.1
P_2O_5	0.07	0.06	0.02
SO_3	2.9	3.91	—
K_2O	0.59	0.22	0.27
CaO	64.4	63.3	41.2
TiO_2	0.21	0.17	0.43
Fe_2O_3	3.37	4.39	0.74
SrO	0.05	0.07	0.08
Mn_2O_3	0.04	0.05	0.31
LOI	2.52	1.03	N/A
Calculated via ASTM C 15			
C_3S	66	51	
C_2S	8	23	
C_3A	6	4	
C_4AF	10	13	

Note: — = not detected; N/A = not available; LOI = loss on ignition of cement.

measurements of the stabilizers were performed using an X-ray spectrometer with Rh X-ray tubes operated at 3,600 W. Table 2 summarizes the chemical properties of the stabilizers. However, Table 2 shows that the SO_3 content (3.91%) of Type V cement exceeded the ASTM C 150-07 designated maximum value (2.3%).

Table 3. Soil–Stabilizer Combinations

Soil Type	Stabilizers type and percentages
Soil-L and Soil-S	8% Type I/II cement
	12% Type I/II cement
	16% Type I/II cement
	4% Slag + 4% Type I/II cement
	8% Slag + 8% Type I/II cement
	8% Type V cement
	12% Type V cement
	16% Type V cement
Soil-L_S	12% Type I/II cement
	8% Slag + 8% Type I/II cement

Methods

To evaluate the effects of different stabilizers, 18 different soil–stabilizer batches were prepared. A summary of the batches is provided in Table 3. Mixtures were prepared by mixing air-dried soil with a specific amount of stabilizer by weight. The amounts and types of stabilizer were varied between batches and some batches were treated with a combination of stabilizers. For both Type I/II and Type V cement, addition rates were selected as 8%, 12%, and 16% as the cement requirement is higher for treating expansive soils. Moreover, the presence of sulfate salts impairs the performance of cement, leading to higher stabilizer requirement. Slag was used along with Type I/II Portland cement, as the CaO (41%) content of slag was low for stabilizing expansive soil. Two different combinations of slag–Type I/II cement were considered where slag and cement were used in equal proportions. No slag–Type V cement blend was utilized because of the inherent superiority of Type V cement in stabilizing sulfate rich soils. ASTM C 150-16 (Standard specification for Portland cement) classifies Type V Portland cement as having high resistance to sulfate. High-sulfate soil (Soil-L_S) batches were treated with 8% Type I/II cement and 8% slag–8% Type I/II cement mixtures due to lower amount of available soil. Three samples were prepared for each batch and all specimens were cured for 7 days, 28 days, and 90 days at ambient temperature (21°C) and 100% relative humidity. After curing, samples were tested for strength and volumetric swell tests. Owing to the low amount of high-sulfate soil, samples of Soil-L_S were cured for only 7 days before they were subjected to the strength test. In addition, the liquid limit (LL) and plastic limit (PL) tests were conducted on soil–stabilizer blends after 28 days of curing. pH tests were also performed for all samples after 90 days of curing.

Unconfined Compressive Strength

ASTM D 2166-16 (Standard test method for unconfined compressive strength of cohesive soil) was followed to conduct UCS tests on each specimen. The loading rate was 1.20 mm/min. Material proportions were measured gravimetrically and mixed thoroughly, mellowed for 15 minutes (for compaction delay), and then compacted using the Iowa State University 2 × 2 apparatus (*14*) at optimum moisture content.

Volumetric Swell Test

Volumetric swell measurements were recorded using the method based on ASTM D 559-15 (Standard test methods for wetting and drying compacted soil–cement mixtures—Method A). Samples were placed on porous stones and protected from direct water contact in a 100% relative humidity room to allow expansion/swelling. Height and diameter measurements were taken with digital calipers and average values were recorded. The final values for all three samples were combined for the final reported results.

Test Results

Atterberg Limits

Table 4 summarizes the LL and PI values of stabilized soils after 28 days of curing. Adding stabilizing agents resulted in a decrease of LL and PI in all mixtures. Li et al. also reported the decrease of LL/PI of for high-plasticity clay materials due to the addition of stabilizers (*15*). The addition of Type V cement resulted in a greater decrease in LL values, regardless of soil types. At lower cement content, a greater decrease in PI was observed for Type I/II cement stabilization, whereas at higher cement content, Type V stabilized soils experienced a greater decrease. These trends indicate that the strength of stabilized soils should be higher for Type I/II cement stabilization at lower cement content, whereas at higher cement content Type V cement-stabilized samples should have higher strength. This expected behavior has been proved in the UC strength results section. For Slag–Type I/II cement mixtures, LL and PI values were relatively higher because of lower CaO content of the mixtures (Table 2). In addition, for the same amount of cement, Soil-L showed lower plasticity, which be due to its higher initial moisture content (28.3%). Higher moisture content may have increased the pozzolanic reactions by facilitating the hydration of cement constituents and eventually decreasing the plasticity of Soil-L.

Table 4. LL and PI of Soil-L and Soil-S after 28 Days of Curing

Soil–stabilizer mixture	Soil-L LL (PI)	Soil-S LL (PI)
Natural soil	74 (49)	51 (31)
8% Type I/II cement	43 (1)	37 (5)
12% Type I/II cement	43 (2)	40 (5)
16% Type I/II cement	42 (4)	41 (12)
4% Slag + 4% Type I/II cement	47 (13)	46 (20)
8% Slag + 8% Type I/II cement	52 (9)	36 (5)
8% Type V cement	39 (2)	33 (8)
12% Type V cement	42 (1)	35 (7)
16% Type V cement	43 (1)	34 (5)

Moreover, for both soils LL/PI decreased initially with the addition of Type I/II cement and then started to increase with increasing cement content. For Type V cement, LL/PI values kept decreasing with the increase of cement percentage, though the decrease was insignificant beyond the addition rate of 8%. According to Al-Zoubi, the initial decrease in LL/PI values is attributed to the cation exchange process of soil and cement associated with minimal pozzolanic activities at lower cement content (*16*). With the increase of cement content, pozzolanic activities increase. Pozzolanic reactions increase the water-holding capacity and, thus, increase the LL/PI of the soils. For Type V cement, pozzolanic activities was lower at 28 days, which might be the reason for decreasing LL/PI values of Type V cement-treated soils with time. From UCS test results, the strength of Type V cement was lower at 28 days compared with Type I/II cement, though after 90 days of curing Type V cement had higher strength. This is also an indication of the lower pozzolanic activity of Type V cement compared with Type I/II cement after 28 days of curing period.

pH Results

Soil stabilization resulted an increase in pH of all soils within the range of 10.6–12.4 as listed in Table 5. Smaller variation of pH was observed depending on the stabilizers percentage. Therefore, pH value can be used as an indicator of the target stabilizer addition rate. For Soil-L and Soil-S, pH mostly approached 12, whereas for the high-sulfate content Soil-L_S the maximum pH was around 11. Sulfate attacks the calcium silicate hydrates, leading to a lower calcium–silicon ratio that results in buffering of pH at lower levels (*17*). Stabilized Soil-S mixtures had slightly higher (> 0.25) pH compared with Soil-L mixtures owing to its initial low pH value (5.7). Ghobadi et al. (*18*) concluded that the initial low pH of the soil causes the leaching of more Ca^{2+} ions from the cement that eventually increases the pH. Among the three different stabilizers, Type I/II cement showed higher increase in pH, whereas mixtures with Type V cement had relatively smaller increase in pH. XRF analyses showed that metal oxides' proportions (CaO, Al_2O_3, Na_2O, MgO) are higher in Type I/II cement which could be the reason for the higher pH values of Type I/II cement treated soils (Table 2). pH of slag–Type I/II cement mixtures was mostly controlled by cement, buffering at pH levels obtained by Type I/II cement exclusively. For instance, the pH of 8% slag–8% Type I/II cement treated Soil-L was 11.83 whereas, the pH of 8% Type I/II cement was 11.85 (Table 5). Higher amount of CaO in Type I/II cement compared with slag might be the reason for the greater influence of cement on the pH of slag–cement mixtures (Table 2).

Table 5. pH Test Results of Treated Soils at 90 Days

Soil type	Stabilizer	Addition rate (%)	Soil-L $pH_{(90\text{-Days})}$	Soil-S $pH_{(90\text{-Days})}$
Soil-L and Soil-S	Initial	0	8.70	5.70
	Cement I/II	8	11.85	11.79
		12	12.10	12.25
		16	12.26	12.35
	Slag + Cement I/II	4+4	11.83	11.73
		8+8	11.83	11.89
	Cement V	8	11.83	12.09
		12	10.82	12.18
		16	12.30	11.95
Soil-L_S	Initial	0	7.80	
	Cement I/II	12	11.06	
	Slag + Cement I/II	8+8	10.60	

UCS Results

The UCSs of Soil-L treated with three different stabilizers (Type I/II cement, Type V cement, slag–cement) are shown in Figure 2. As can be seen from Figure 2*a*, after 7 days of curing Soil-L treated with Type I/II cement has higher strength at every addition rate. The high amounts of C_3S (66%) are responsible for the higher strength of Type I/II cement at early stage. Termkhajornkit concluded that the compressive strength of soils increased more quickly when they were stabilized with additives rich in C_3S than additives with higher C_2S (*19*). However, the addition of Type I/II cement beyond 8% by weight does not increase the strength significantly. This trend became more evident at longer curing times. For instance, as shown in Figure 2*c*, UCS for 12% and 16% Type I/II cement treatment after 90 days of curing period are 1413 kPa and 1434 kPa, respectively, indicating only a trivial increase in strength beyond the addition rate of 8%. This indicates that the use of 8% Type I/II cement to be the optimum to stabilize Soil-L. Moreover, the effect of curing time on strength gain is not well pronounced for Type I/II cement. Strength increased with curing time, though the rate of increase in strength was not significant beyond the curing of 28 days. In contrast, UCS of Soil-L stabilized with Type V cement kept increasing with the increase of cement content and curing time. Chusilp et al. found that lower loss on ignition of cement (LOI) is associated with higher strength development (*20*). Type V cement has a LOI of 1.03%, whereas Type I/II cement has a LOI of 2.52%, which could be the reason for the incessant strength development of Type V cement treatment with the addition rate. In addition, the relatively higher amount of C_2S in Type V (23%) compared with Type I/II (8%) cement increased the strength of Type V cement-stabilized samples with time. After 90 days of curing, the strength achieved by Type V cement exceeded the strength of Type I/II cement-treated samples.

For the slag–cement combinations, the total percentages of slag and cement were taken as the stabilizer content (4% slag + 4% cement = 8% stabilizer). At higher percentage (16%), slag–cement blends had higher UCS compared with Type I/II cement when cured for 28 and 90 days. Moreover, UCS of slag–cement mixtures kept increasing with curing period unlike Type I/II cement treatment. El-Alfi et al. found that increasing SiO_2 content as silica fume increased the strength of slag–cement pastes at all curing times (*21*). According to El-Alfi et al., additional hydration of cement phases and pozzolanic reaction of silica fume with liberated lime improves the strength of slag–cement paste. From XRF analysis (Table 2) it was found that slag had higher percentage of SiO_2 (35.1%) compared with Type I/II cement (20.1%). Higher amount of SiO_2 may have contributed significantly to the strength gain performance of slag–cement mixtures with curing time. As shown in Figure 3*a*, similar trends of UCS were observed for Soil-S. UCS of Soil-S increased with the increase of Type I/II cement content, though the strength development decreased beyond the addition rate of 8%. Curing time does not also have any considerable influence on strength gain of Soil-S–Type I/II cement mixtures. Furthermore, similarly to Soil-L, 7-day cured Soil-S stabilized with Type V cement had lower UCS than that stabilized with Type I/II cement. After 90 days of curing, Type V cement exhibited higher strengths than those mixed with Type I/II cement. The strengths of 7-day cured slag–cement-stabilized Soil-S were lower than the strengths of cement-stabilized Soil-S. However, the ultimate strengths of slag–cement mixtures were higher compared with Type I/II cement–Soil-S mixtures after 90 days of curing.

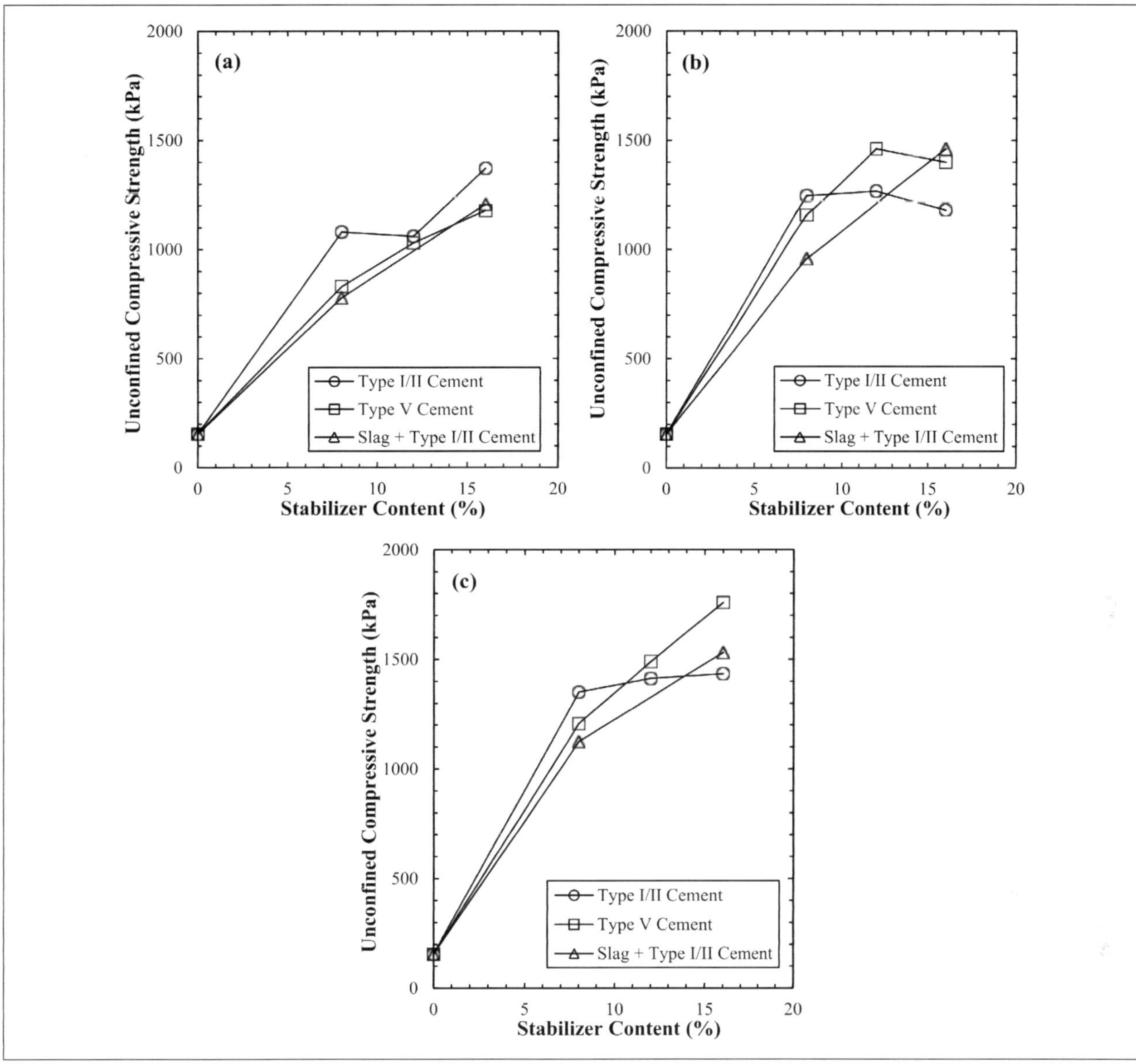

Figure 2. UCS of stabilized Soil-L at (*a*) 7 days, (*b*) 28 days, and (*c*) 90 days.

For both types of cement, the 28-day strength of Soil-L is higher compared with Soil-S. The higher moisture (Table 1) of Soil-L may have facilitated the hydration of cementitious elements such as calcium hydrates, gypsum, ettringite, and calcium silicate hydrates, which has a significant influence on strength. For the same amount of slag–cement stabilization, higher strength was observed for Soil-L compared with Soil-S. In a nutshell, Type I/II cement worked better in terms of 7-day strength, but the ultimate strength for Type V was higher than Type I/II cement. The addition of slag can overcome the problem of Type I/II cement by developing strength with time and, therefore, is proven to be a better option for stabilizing high-plasticity soils.

Volumetric Change

Untreated soil materials had high volumetric swell potential ranging from 12% to 18%. Adding stabilizers resulted in a decrease in swelling to below 1% as shown in Figure 4. For Soil-L, very small amounts of swelling (<0.7%) was observed after 7 days of curing time, except for the mixture stabilized with 12% Type V cement where shrinkage (–0.2%) was detected (Figure 4*a*). It was observed that after 28 days of curing all mixtures showed shrinkage ranging from –0.3% to –0.5% which could be attributed to the chemical shrinkage occurring over time due to hardening of cement. According to Tazawa et al., the volume of the cement hydration products

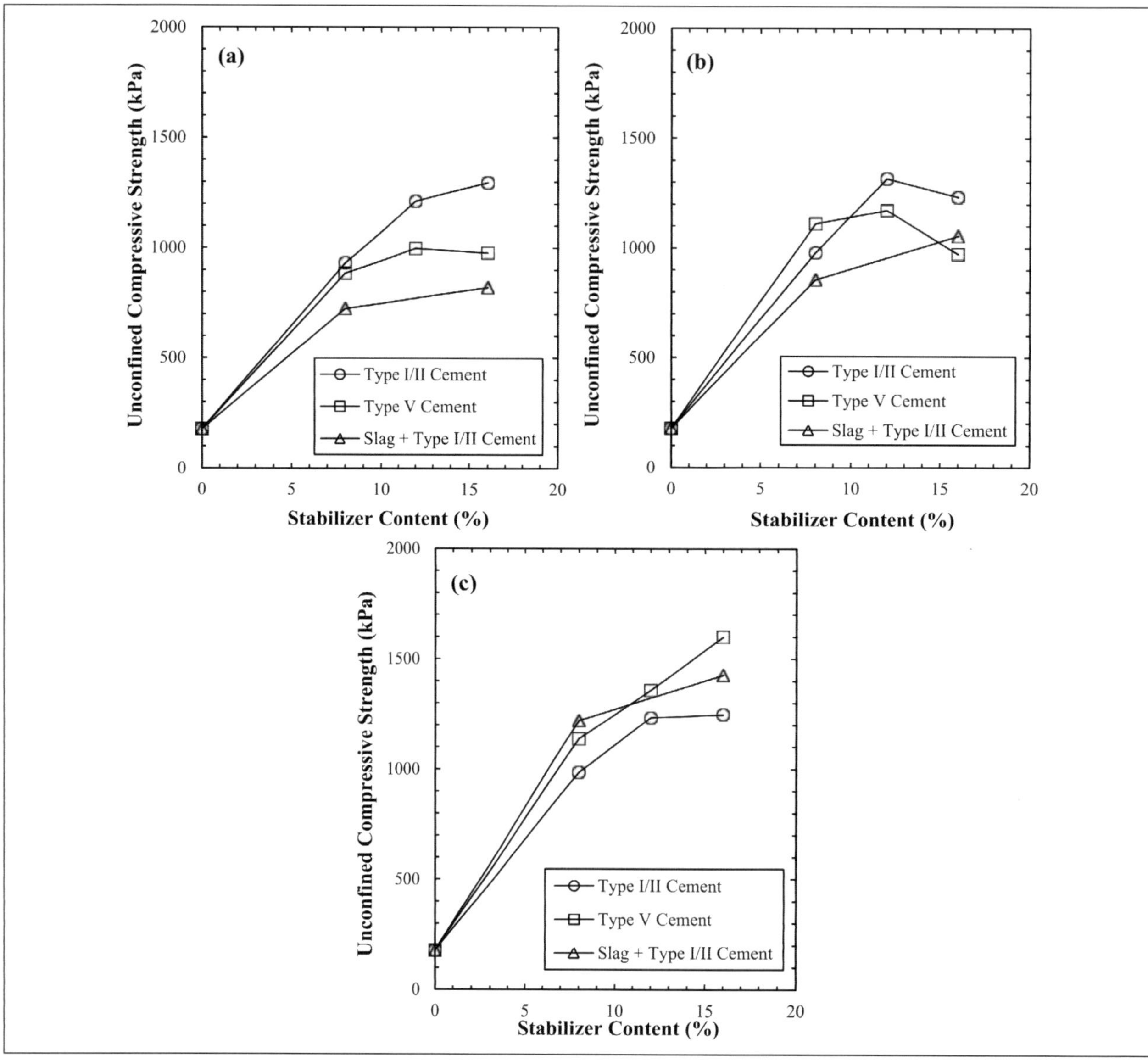

Figure 3. UC-strength of stabilized Soil-S at (*a*) 7 days, (*b*) 28 days, and (*c*) 90 days.

(C-S-H gel, CH) is smaller than the reactants (C_3S, C_2S, C_3A, and C_4AF), which results in the overall volume decrease of the cement paste (*22*). In this study, it was anticipated that the volume changes of treated soils were accompanied by two phenomena; (1) swell of clay particles; and (2) chemical shrinkage due to cement hardening. With shorter curing periods, such as 7 days, the maximum degree of hydration varied within the range of 55% to 60% (*23*). Hence, the expansive clay was not fully stabilized and the swelling of clay took over the chemical shrinkage, resulting an overall swelling of the samples. After 28 days of curing, swelling decreased due to the increase of cement hydration which altered chemical shrinkage to be dominant. Therefore, an overall volume decrease was observed after 28 days of curing time. However, after 90 days of curing 16% Type V cement and 4% slag–4% Type I/II cement-treated samples showed swelling while the others demonstrated persistent shrinkage. According to Powers (*24*), prolonged curing increases the strength and modulus of elasticity, thus making the cement paste more susceptible to cracking around the aggregates when restrained. This internal cracking is very fine and may not be visible. Internal cracking allows shrinkage to occur without reducing the volume. Ingress of water into the cracks while curing reduces the surface tension of cement gel, and further expansion takes place. Moreover, water molecules act against the cohesive forces and tend to force the gel particles further apart, which results in overall swelling (*25*). It is speculated that internal cracks may have occurred in 16% Type V cement

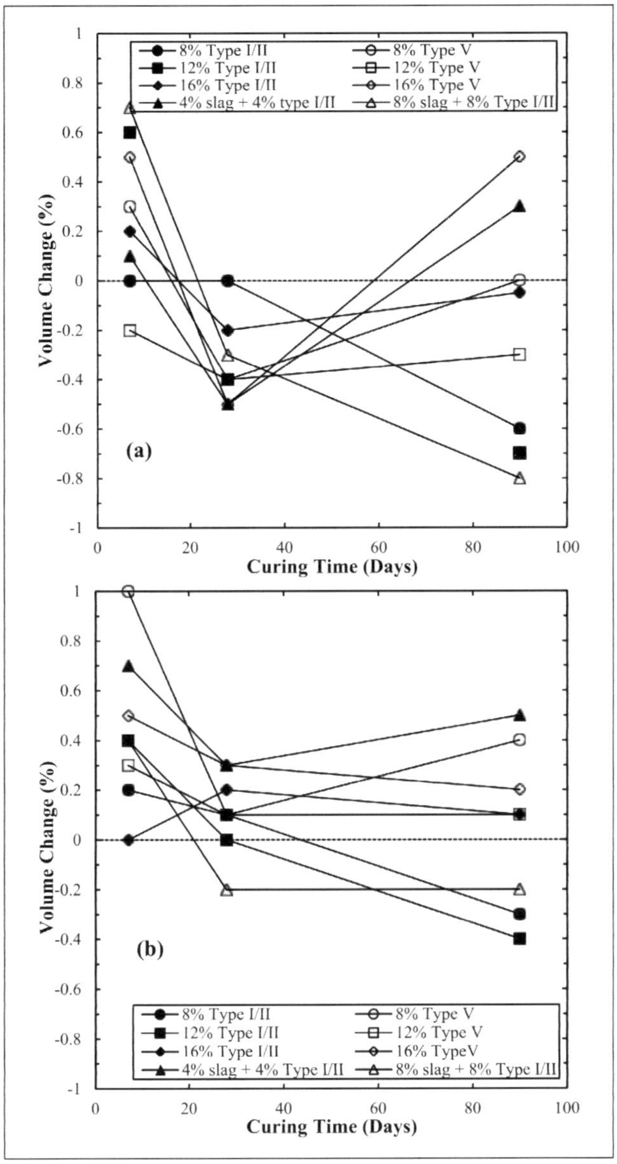

Figure 4. Volumetric swell for (*a*) Soil-L and (*b*) Soil-S.

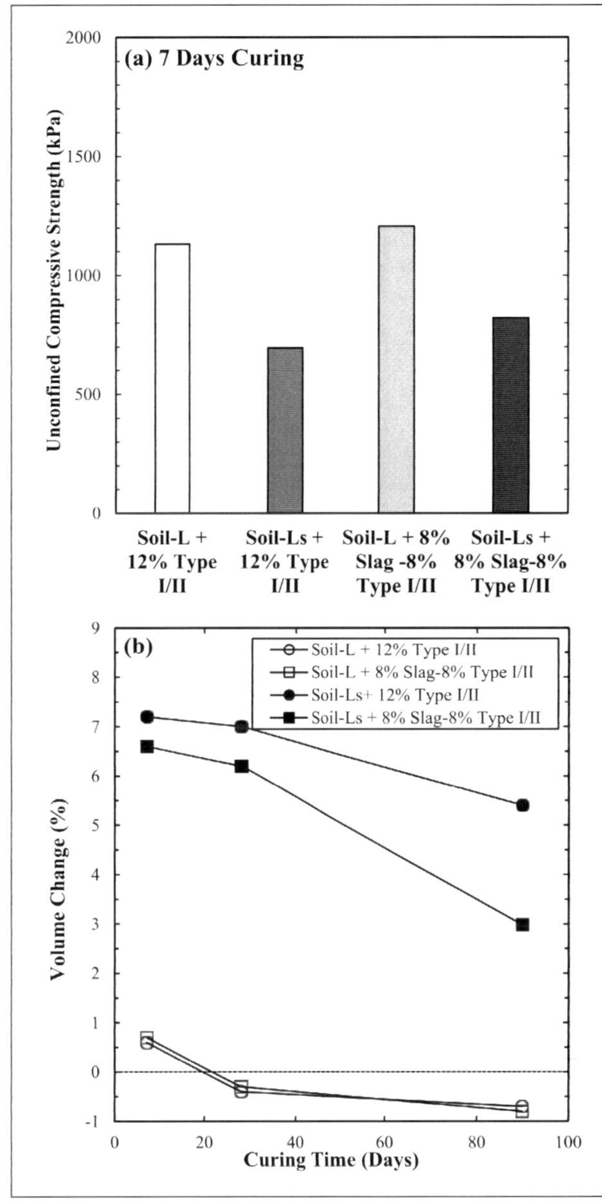

Figure 5. Effect of sulfate content on (*a*) UCS and (*b*) swelling behavior.

and 4% slag–4% Type I/II cement treated samples, and water intrusion may have caused the swelling.

Figure 4*b* shows that stabilized Soil-S after 7 days of curing experiences more swelling compared with Soil-L. In addition, swelling persisted even after 28 days of curing, though swelling percentages were decreased (Figure 4*b*). Only the 8% slag–8% Type I/II cement stabilized samples contracted after 28 days of curing. Even after 90 days, a decrease volume was only observed for three blends (8% Slag+8% Type I/II cement, 8% and 12% Type I/II cement). The low initial and optimum moisture content of Soil-S (Table 1) resulted in low water–cement ratio (w/c), which could be the reason for lower shrinkage and, hence, higher swelling behavior of Soil-S. According to Oh and Cha (*26*), the degree of hydration is largely influenced by water storage and water distribution. From mathematical modeling and lab testing, they found that the hydration reaction of cement decreased with the decrease in w/c ratio. Lin and Meyer (*27*) concluded that, in early stage of hydration, the w/c ratio has a small effect on the hydration rate, but after the middle period of hydration, the lower w/c ratio could lead to a lower hydration rate.

Effect of High Sulfate Content

As shown in Figure 5*a*, for high sulfate Soil-L_S, 7 days UCS of 12% Type I/II cement treatment is 696 kPa, which is very low compared with low-sulfate Soil-L. Sulfate reacts with the chemical constitute of cement, especially with C_3A. The product (delayed ettringite and gypsum) produced by the hydration reactions had larger volume than the reactant, which caused additional stress and weakened the soil–cement bonds (*28*). Strength of high-sulfate soil treated with 8% Slag+8% Type I/II mixture was also low (821 kPa). However, strength impairment was relatively less for slag–cement stabilization due to the low calcium content of slag. Tikalsky (*29*) suggested the use of low-calcium-based stabilizers in treating soils with high-sulfate salts.

Figure 5*b* shows the comparisons of swelling behavior of Soil-L_S with Soil-L. High-sulfate Soil-L_S, although showing a decrease from the untreated 20% swell potential, had relatively higher swelling compared with low-sulfate Soil-L. For the use of 12% Type I/II cement, swelling after 7 days of curing was 0.6% for Soil-L, whereas 7.2% swelling was observed for Soil-L_S. At prolonged curing such as 28 and 90 days, Soil-L showed shrinkage although Soil-L_S demonstrated persistent high swelling. Figure 5*b* also demonstrates that the smaller swelling was observed when Soil-L_S stabilized with 8% Slag+8% Type I/II cement. Slag does not contain C_3A (Table 2), which dilutes the total amount of C_3A in the system. In addition, the larger amount of SiO_2 in slag reacts with excess $Ca(OH)_2$ and produces additional calcium–silicate hydrate gel, thus decreasing available $Ca(OH)_2$ in the system to produce delayed ettringite. Moreover, a decrease in swelling with curing was also observed for high-sulfate soil. Higher decrease rates were detected for slag–cement stabilizing confirming its applicability to stabilize high-sulfate expansive soils.

Conclusions

In this experimental study, the performance of two types of cement (Type I/II and V), and steel slag was evaluated in stabilizing expansive clay based on strength improvement and swelling potential reduction. Atterberg limits, pH, UCS and volume change assessments were performed on two different soils, treated with variable dosages of stabilizers. Tests were conducted after 7, 28, and 90 days curing periods on 312 samples. The following conclusion were obtained based on this laboratory experiment.

- The LL and PI of soils decreased initially up to a certain amount of cement content (8% to 10%) and then started to increase slightly with an increase in cement content.
- Soil stabilization increased soil pH up to a maximum value depending on the stabilizer type. Type I/II cement showed highest pH, whereas the lowest pH was observed for Type V cement treated blends.
- The UCS of soils increased with an increase in stabilizer content and curing time. Optimum stabilizer requirement in treating expansive soils is likely to be 8–10%. Higher amounts of cement do not necessarily offer higher strength and may lead to the impairment of expected performance. However, a comprehensive laboratory setup with a wide range of soils is needed for definitive conclusions.
- Slag–cement treatment is preferable to Type I/II cement based on ultimate strength requirement. The strength of slag–cement treatment increased significantly with curing time offering significant economic advantages in stabilizing expansive soils.
- The presence of sulfate salts decreased the strength significantly by hindering the formation of calcium hydrates. The addition of slag improved the performance of cement by decreasing the calcium–silicon ratio and, therefore, has proven to be a better alternative for stabilizing high-sulfate expansive soils.
- Adding stabilizers resulted in a decrease in the swell potential of the soils depending on moisture, clay content, curing time, and stabilizer percentage. For stabilizing high–sulfate–high-plasticity soil, slag is preferred because of its low CaO content, whereas lime-based stabilizers such as cement exacerbate the situation by inducing undue swelling.

References

1. Taylor, R. K., and T. J. Smith. The Engineering Geology of Clay Minerals; Swelling, Shrinking and Mud–Rock Breakdown. *Clay Minerals*, Vol. 21, No. 3, 1986, pp. 235–260.
2. Mokhtari, M., and M. Dehghani. Swell–Shrink Behavior of Expansive Soils, Damage and Control. *Electronic Journal of Geotechnical Engineering*, Vol. 17, 2012, pp. 2673–2682.
3. Guney, Y., D. Sari, M. Cetin, and M. Tuncan. Impact of Cyclic Wetting–Drying on Swelling Behavior of Lime-Stabilized Soil. *Building and Environment*, Vol. 42, No. 2, 2007, pp. 681–688.
4. Bell, F. G., and R. R. Maud. Expansive Clays and Construction, Especially of Low-Rise Structures: A Viewpoint from Natal, South Africa. *Environmental & Engineering Geoscience*, Vol. 1, No. 1, 1995, pp. 41–59.
5. Khan, M. S., S. Hossain, S., Ahmed, A., and M. Faysal. Investigation of a Shallow Slope Failure on Expansive Clay in Texas. *Engineering Geology*, Vol. 219, 2017, pp. 118–129.
6. Hossain, S., A. Ahmed, M. S. Khan, A. Aramoon, and B. Thian. Expansive Subgrade Behavior on a State Highway in North Texas. In *Geotechnical and Structural Engineering Congress 2016*.
7. Basha, E. A., R. Hashim, H. B. Mahmud, and A. S. Muntohar. Stabilization of Residual Soil with Rice Husk Ash and Cement. *Construction and Building Materials*, Vol. 19, No. 6, 2005, pp. 448–453.
8. Stavridakis, E. I. Evaluation of Engineering and Cement-Stabilization Parameters of Clayey–Sand Mixtures under Soaked Conditions. *Geotechnical & Geological Engineering*, Vol. 23, No. 6, 2005, pp. 635–655.
9. Rajasekaran, G. Sulphate Attack and Ettringite Formation in the Lime and Cement Stabilized Marine Clays. *Ocean Engineering*, Vol. 32, No. 8, 2005, pp. 1133–1159.

10. Santhanam, M., M. D. Cohen, and J. Olek. Mechanism of Sulfate Attack: A Fresh Look: Part 2. Proposed Mechanisms. *Cement and Concrete Research*, Vol. 33, No. 3, 2003, pp. 341–346.
11. Mehta, P. K. Effect of Fly Ash Composition on Sulfate Resistance of Cement. *American Concrete Institute Journal Proceedings*, Vol. 83, No. 6, 1986, pp. 994–1000.
12. Harris, P. *Laboratory and Field Procedures for Measuring the Sulfate Content of Texas Soil* Develop Guidelines and Procedures for Stabilization of Sulfate Soils. Texas Transportation Institute, 2002.
13. Warncke, D., and J. R. Brown. Potassium and Other Basic Cations. *Recommended Chemical Soil Test Procedures for the North Central Region*, Vol. 1001, 1998, p. 31.
14. O'Flaherty, C. A., C. E. Edgar, and D. T. Davidson. The Iowa State Compaction Apparatus: A small sample apparatus for use in obtaining density and strength measurements of soils and soil-additives. *Special Report Prepared for Presentation at the 42nd Annual Meeting of the Highway Research Board, Contribution* (No. 63-5), 1963.
15. Li, S., D. J. White, and P. Vennapusa. Cement Stabilization of Embankment Materials, 2015.
16. Al-Zoubi, M. S. Undrained Shear Strength and Swelling Characteristics of Cement Treated Soil. *Jordan Journal of Civil Engineering*, Vol. 2, No. 1, 2008, pp. 53–62.
17. Cook, E. R., and B. Batchelor. Stabilization/Solidification of Hazardous Wastes in Soil Matrices. *Advances in Porous Media*, Vol. 3, 1996, pp. 307–359.
18. Ghobadi, M. H., Y. Abdilor, and R. Babazadeh. Stabilization of Clay Soils Using Lime and Effect of pH Variations on Shear Strength Parameters. *Bulletin of Engineering Geology and the Environment*, Vol. 73, No. 2, 2014, pp. 611–619.
19. Termkhajornkit, P., Q. H. Vu, R. Barbarulo, S. Daronnat, and G. Chanvillard. Dependence of Compressive Strength on Phase Assemblage in Cement Pastes: Beyond Gel–Space Ratio—Experimental Evidence and Micromechanical Modeling. *Cement and Concrete Research*, Vol. 56, 2014, pp. 1–11.
20. Chusilp, N., C. Jaturapitakkul, and K. Kiattikomol. Effects of LOI of ground bagasse ash on the compressive strength and sulfate resistance of mortars. *Construction and Building Materials*, Vol. 23, No. 12, 2009, pp. 3523–3531.
21. El-Alfi, E. A., A. M. Radwan, and H. Abu-El-Naga. Influence of Substitution of Ordinary Portland Cement by Silica Fume on The Hydration of Slag–Portland Cement Pastes. *Ceramics-Silikaty*, Vol. 55, No. 2, 2011, pp. 147–152.
22. Tazawa, E. I., S. Miyazawa, and T. Kasai. Chemical Shrinkage and Autogenous Shrinkage of Hydrating Cement Paste. *Cement and Concrete Research*, Vol. 25, No. 2, 1995, pp. 288–292.
23. Feng, X., E. J. Garboczi, D. P. Bentz, P. E. Stutzman, and T. O. Mason. Estimation of the Degree of Hydration of Blended Cement Pastes by a Scanning Electron Microscope Point-Counting Procedure. *Cement and Concrete Research*, Vol. 34, No. 10, 2004, pp. 1787–1793.
24. Powers, T. C. Causes and Control of Volume Change. *Journal of Portland Cement Association Research and Development Laboratories*, Vol. 1, No. 1, 1959, pp. 29–39.
25. Neville, A. M. *Properties of Concrete*, 4th ed. Prentice Hall, Englewood Cliffs, NJ, 1995.
26. Oh, B. H., and S. W. Cha. Realistic Models for Degree of Hydration and Moisture Distribution in Concrete at Early Age. *Fourth International Conference on Fracture Mechanics of Concrete and Concrete Structures*, 2001, pp. 279–286.
27. Lin, F., and C. Meyer. Hydration Kinetics Modeling of Portland Cement Considering the Effects of Curing Temperature and Applied Pressure. *Cement and Concrete Research*, Vol. 39, No. 4, 2009, pp. 255-265.
28. Collepardi, M. A State-of-the-Art Review on Delayed Ettringite Attack on Concrete. *Cement and Concrete Composites*, Vol. 25, No.4, 2003, pp. 401–407.
29. Tikalsky, P. J. Influence of Fly Ash on the Sulfate Resistance of Concrete. *Materials Journal*, Vol. 89, No. 1, 1993, pp. 69–75.

The Standing Committee on Stabilization of Geomaterials and Recycled Materials (AFS80) peer-reviewed this paper (18-03489).

TRR
JOURNAL OF THE TRANSPORTATION RESEARCH BOARD

Article

Transportation Research Record
2018, Vol. 2672(52) 174–184
© National Academy of Sciences:
Transportation Research Board 2018
Article reuse guidelines:
sagepub.com/journals-permissions
DOI: 10.1177/0361198118797214
journals.sagepub.com/home/trr
SAGE

Strength and Shrink/Swell Behavior of Highly Plastic Clay Treated with Geopolymer

Suraj D. Khadka[1], Priyantha W. Jayawickrama[1], and Sanjaya Senadheera[1]

Abstract

This paper presents findings from an experimental study in which a novel, non-energy intensive, environmentally-friendly stabilizer known as geopolymer was used for stabilization of a highly plastic clay. Two forms of the stabilizer were synthesized, one using metakaolin (MK) and the other using fly ash (FA) as the alumino-silicate precursor. The paper describes the process of geopolymer synthesis as well as quality control tests conducted during geopolymer synthesis. Synthesized geopolymers were analyzed using scanning electron microscopy (SEM) and energy-dispersive x-ray spectroscopy (EDX). The paper also presents results obtained from a comprehensive laboratory test program that examined the effectiveness of the stabilizer in improving strength and controlling moisture induced swelling of a highly plastic clay soil. The geopolymer content of the soil ranged from 3% to 15% by weight ratio. SEM-EDX and XRD were performed to investigate the treated specimen for uniformity and stabilization mechanism. The data obtained from unconfined compressive strength tests conducted after seven days of curing showed three- to four-fold improvement in strength for geopolymer stabilized soils with MK geopolymer consistently providing better performance than FA geopolymer. The results from one-dimensional swell tests indicated significant reduction of swell behavior in FA geopolymer treated specimens while no improvement in swell behavior was observed in MK treated soil. This study demonstrates that, while geopolymers can overcome many limitations that exist in traditional stabilizers, the strength and swell performance of geopolymer stabilized clay soil can vary significantly depending on the source of alumino-silicate used in the production of the geopolymer.

Soil subgrade provides overall support for roadway pavements and hence plays an important role in determining stability and successful performance of the pavement system. Clay soils, especially those with high plasticity, are often associated with structural problems in pavements (*1*). First, clay soils lack sufficient strength to support the pavement during construction and throughout service life of the pavement. Second, many clay soils are highly sensitive to moisture variations and tend to shrink and swell with fluctuations in the water table or changes in moisture level (*2–4*). Traditionally, lime and Portland cement have been used for stabilization of clay soil subgrades. Clay stabilization by lime and cement treatment occurs primarily due to cation exchange followed by flocculation and agglomeration (*5*). A major concern with the use of cement and lime for soil stabilization is the large amount of energy required during their production, including heating raw materials within a rotating kiln at a temperature greater than 1400°C. Production of Portland cement, as an example, requires 90 to 150 kWh per 1 ton of cement which represents 40% of total production costs. The production of cement also requires large amounts of water and results in emission of gases such as CO, CO_2, NO_x, and SO_2 which are harmful to the environment (*6, 7*). It is estimated that approximately one ton of CO_2 is emitted during the production of 1 ton of cement. Another problem that is often associated with calcium-based stabilizers such as lime and cement is sulfate-induced heave. Sulfate-induced heave occurs when an expansive mineral called ettringite forms due to reactions between the stabilizer and sulfate minerals (usually gypsum) present in the soil (*8*). Another concern with stabilization of soil with lime involves loss of calcium ions from the soil stabilizer system due to leaching. Leaching tests conducted on lime stabilized soil have shown that removal of calcium ions can occur as leaching

[1]Department of Civil, Environmental and Construction Engineering, Texas Tech University, Lubbock, TX

Corresponding Author:
Address correspondence to Suraj D. Khadka: suraj.khadka@ttu.edu

progresses resulting in gradual deterioration of engineering properties of the stabilized soil with time (*9*). In recent years, geopolymers have drawn much attention as a soil stabilizer that can potentially overcome many of the above shortcomings found in traditional stabilizers.

Production of geopolymers does not require extremely high temperature kilns with large demand for fuel, nor do they require a large capital investment in plant and equipment. Naturally occurring alumino-silicates provide necessary geopolymeric raw materials. Geopolymers have been synthesized from different alumino-silicate precursor materials such as metakaolin (MK), fly ash (FA), rice husk, furnace slag, Ground granulated blast-furnace slag (GGBS), silica fume and red mud (*10–12*). When synthesizing geopolymers, the selected alumino-silicate precursor material is allowed to react with an alkali hydroxide solution and alkaline silicate solution to form reactive silica and alumina. Then the dissolved species polycondenses into amorphous or semi-crystalline materials which further polymerize and harden to form geopolymer. The reasonable synthesis temperature of geopolymer is in the range of 25 to 80°C. Geopolymers are considered a greener alternative to Portland cement because of lower energy consumption and CO_2 emission associated with geopolymer production. It is estimated that geopolymer production results in 70% to 90% reduction in CO_2 emissions when compared with Portland cement manufacturing (*13*).

Mechanism of Geopolymerization

Geopolymers are alumino-silicate networks which consist of tetrahedral silica (SiO_4) and alumina (AlO_4) that are linked alternatively by sharing all oxygen atoms where positive ions such as Na^+, K^+, Li^+, Ca^{++} balance the negative charges (*14*). The sequencing of SiO_4 and AlO_4 tetrahedra is based on Lowenstine's avoidance principle. According to this principle, AlO_4 tetrahedra do not directly link with other AlO_4 tetrahedra, instead they require a bridging SiO_4 tetrahedron between them (*15*, *16*). Geopolymers can be represented with the following empirical formula (*14*):

$$M_n\{-(SiO_2)_z-AlO_2-\}_n \qquad (1)$$

where M is an alkali cation, n is the degree of polycondensation and z is the Si/Al molar ratio. Alumino-silicates are polymeric chains with Si^{4+} and Al^{3+} in four-fold coordination with oxygen. Three types of alumino-silicate networks have been identified: polysilialate (–Si–O–Al–O–), polysilialate–silioxo (–Si–O–Al–O–Si–O–) and polysilialate–disilioxo (–Si–O–Al–O–Si–O–Si–O–) which range from amorphous to semi-crystalline form (*14*, *17*). Properties of geopolymer vary depending on the relative amounts of silica, alumina, and alkali. Among the various constituents, Si/Al molar ratio is known to have the most significant effect (*18*, *19*). Data obtained from previous experimental studies suggest that geopolymers with significantly improved properties can be obtained when Si/Al molar ratio is in the range 1–3 and Na/Al molar ratio is maintained at unity (*17*). Solubility of these precursors is highly dependent upon pH conditions during reaction. Alkali raises the pH which in turn increases the solubility of geopolymer precursor (*20*). Alkali also provides cations to balance the negative charge in the aluminate groups (*21*). Geopolymers cured at higher temperature (60–80°C) acquire higher early-age compressive and flexural strength compared with geopolymers cured at lower or ambient temperature (10–40°C). However, the long-term strength is considerably lower in geopolymers cured at higher temperature due to the rapid formation of polymeric chains which results in weaker bonds (*22*). Regardless of initial water content, geopolymer maintains constant residual water at the end of the curing period at ambient condition and the amount of initial water does not have any effect on its typical features (*23*).

High strength geopolymer can be used as a substitute for ordinary Portland cement in infrastructure development (*24*). Different techniques such as addition of fiber to produce strain hardening geopolymer composites, distinct alumino-silicate sources activated with varying Si/Al, and Na/Al molar ratios have been investigated to determine the feasibility of geopolymer as an alternative construction material to cement (*25–29*). Previous studies have explored the use of geopolymer as a soil stabilizer to improve the mechanical properties and volume stability of artificially prepared soil (*30*, *31*). Artificially prepared soils treated with MK and FA geopolymer exhibit increased compressive and flexural strength and reduced shrinkage strains (*30*, *32*). In this research study, MK and FA were selected as the alumino-silicate precursors to produce geopolymers which were used to stabilize a natural high plastic clay with low strength and high shrink–swell potential. Since the properties of geopolymer are significantly affected by their Na/Al and Si/Al molar ratios, optimum Na/Al and Si/Al molar ratios were identified separately for both MK and FA geopolymers based on the strength of synthesized geopolymer. To determine the feasibility of using geopolymer as a viable option for stabilization of high plastic clay, unconfined compressive strength, Young's modulus, and volumetric swell were determined for geopolymer treated soil specimens. Finally, microstructural analysis was performed by using scanning electron microscopy (SEM) and mineralogical characterization was performed using x-ray diffraction (XRD) and x-ray fluorescence (XRF) on geopolymer, untreated clay, and treated clay specimens to determine the mechanism of polymerization and

Table 1. Chemical Composition of MK and FA

Chemical components	MK	FA
SiO_2	52%	41.96%
Al_2O_3	43%	20.04%
Fe_2O_3	< 2.20%	5.76%
CaO	< 0.20%	21.09%
MgO	< 0.10%	4.30%
Na_2O	< 0.05%	1.44%
K_2O	< 0.40%	0.73%
SO_4	< 0.05%	0.99%
Water content	0.33%	0.33%

effect of geopolymer treatment on the thickness, basal spacing or d-spacing of clay interlayers (*33*).

Geopolymer Synthesis and Quality Control

Geopolymer Synthesis

The chemical composition of the alumino-silicate precursors used in this study, MK and Class-C FA, is summarized in Table 1. The alkaline activator solution (AAS) was a mixture of sodium hydroxide solution (NaOH ~ 50%, H_2O ~ 50%, Na_2CO_3 = 0.3% max, Fe = 5 ppm max, PO_4 = 5 ppm max, density = 1.53 gm/cm^3 at 20°C) and technical grade sodium silicate solution (Na_2O = 10.6%, SiO_2 = 26.5%, H_2O = 62.9%, pH = 11.3, specific gravity = 1.39 gm/cm^3 at 20°C). Initially, NaOH and Na_2SiO_3 were mixed at a predetermined concentration to prepare the AAS. The AAS was then left in a sealed container for 30 min to allow the heat of the reaction to dissipate. Finally, the geopolymer precursor was added to the AAS and mixed for approximately 5 min to obtain the geopolymer slurry.

Compressive Strength Testing

High strength geopolymer is typically associated with reduced porosity, increased density, and fine-graded microstructure. In one previous research study, MK geopolymers with uniform microstructure and composition and high strength were obtained when Na_2O/Al_2O_3 molar ratio was maintained at 1.2 and SiO_2/Al_2O_3 molar ratio was maintained at between 3.5 and 3.8 (*18*). In a different study, high strength MK geopolymer was obtained when Na/Al molar ratio was maintained at between 1.0 and 1.5 and Si/Al molar ratio was maintained at between 2.0 and 2.5 (*16*). For FA geopolymer, high strength and fine-grained microstructure were obtained when Na_2O/Al_2O_3 molar ratio was maintained at 1.0 and SiO_2/Al_2O_3 molar ratio was maintained at 3.9 (*19*).

To find the optimum Na_2O/Al_2O_3 and SiO_2/Al_2O_3 molar ratios to be used in the current study, both types of geopolymers were synthesized by combining different proportions of sodium hydroxide and sodium silicate with MK and FA. Geopolymer pastes prepared in this manner were first poured into 2-inch cubical molds. These test specimens were then cured in two separate phases. During the first phase, the geopolymer mixtures were cured under ambient conditions (RH ~ 40%, temperature ~ 23°C) for two days. Test specimens were covered with a plastic film to avoid any moisture loss during the first phase. In the second phase, the specimens were extracted from the molds, sealed in plastic bags, and allowed to cure for an additional five days. Finally, UCS tests were performed on the specimens after seven days of curing. Results from the UCS tests are shown in Table 2.

The final geopolymer mix design was determined based on the strengths of tested geopolymer specimens. Based on these test results, MK geopolymer was synthesized with Si/Al molar ratio of 1.66 and Na/Al molar ratio of 1.06. Similarly, FA geopolymer was synthesized with Si/Al molar ratio of 3.14 and Na/Al molar ratio of 2.27.

Microstructural Characterization

SEM imaging was performed on geopolymer synthesized from MK and FA to observe morphological changes in material obtained using different alumino-silicate precursors. Hitachi S-4300 E/N (FESEM), with 1.5 nm resolution at 15 kV accelerating voltage was used for this purpose. Energy-dispersive x-ray spectroscopy (EDX) analyses were performed with the accessory EDAX EDX system. Sample preparation was carried out according to procedures used by previous researchers who performed SEM analyses on geopolymers (*30*, *31*). Accordingly, the cured cubical specimens were cut along the centerline with a V-shaped grooving tool and broken into two parts by bending. A hydraulic press was used to break the specimen at the desired location. Small fragments on the surface of the broken pieces were carefully removed using adhesive tapes. The specimens were then coated with palladium before performing SEM imaging.

SEM imaging was performed on MK geopolymer with Si/Al molar ratio of 1.66 and Na/Al molar ratio of 1.06 and FA geopolymer with Si/Al molar ratio of 3.41 and Na/Al molar ratio of 2.27, as shown in Figure 1. Characterization of the geopolymer matrix using EDX suggested that both the geopolymer samples had identical matrices with negligible variation in weight percentages of each of the elements. Distinct morphological differences were observed between them, however. For the MK geopolymer, surface of the specimen exhibited a polishing relief, which suggested that MK was completely dissolved in the alkaline solution, resulting in a harder and denser matrix compared with FA geopolymer. This validates the

Table 2. Compressive Strength of MK and FA Based Geopolymer with Different Na/Al and Si/Al Molar Ratio

MK based geopolymer				FA based geopolymer			
Si/Al molar ratio	Na/Al molar ratio	Compressive strength (ksi)	Standard deviation	Si/Al molar ratio	Na/Al molar ratio	Compressive strength (ksi)	Standard deviation
1.5	1.06	3.25	0.75	3.14	1.57	0.84	0.8
1.66	1.06	6.40	0.07	3.14	2.27	0.98	0.09
1.77	1.06	4.01	2.56	3.14	3.02	0.47	0.73
1.87	1.06	3.17	7.25	2.68	2.27	0.85	1.12
2.01	1.06	NA[a]	NA[a]	3.6	2.27	0.60	0.56

[a]MK geopolymer sample with Si/Al molar ratio of 2.01 and Na/Al molar ratio of 1.06 developed extreme fractures during the curing period. Hence, compressive strength could not be determined.

(a)

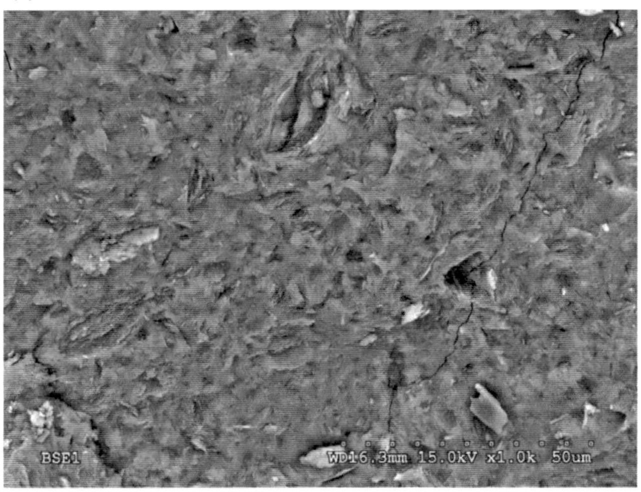

Quantification		Weight %	Atomic %
Elements	NaK	10.53	12.45
	AlK	24.28	24.46
	SiK	65.20	63.10
	Total	100	100
Oxides	Na_2O	8.95	9.67
	Al_2O_3	24.40	16.03
	SiO_2	66.65	74.30
	Total	100	100

(b)

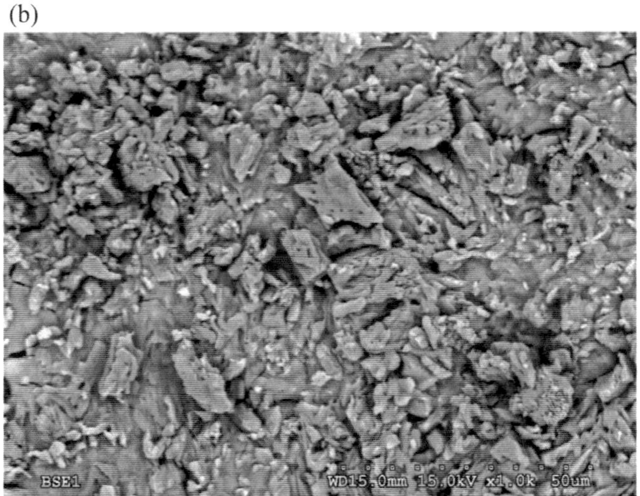

Quantification		Weight %	Atomic %
Elements	NaK	11.83	13.95
	AlK	24.33	24.44
	SiK	63.84	61.61
	Total	100	100
Oxides	Na_2O	11.06	11.99
	Al_2O_3	25.07	16.53
	SiO_2	63.88	71.48
	Total	100	100

Figure 1. SEM images of (*a*) MK and (*b*) FA geopolymers after seven days of curing.

result of strength tests performed on MK (6.4 ksi) and FA (0.98 ksi) geopolymers.

Strength of geopolymer is affected by the rate at which alumina is released from the precursor aluminosilicate (*34*). Geopolymerization reaction occurs with the formation of Al-rich gel followed by Si-rich gel. When alumina is readily available from the precursor aluminosilicate, as in FA, rapid formation of Al-rich gel occurs resulting in increased early strength. This retards the formation of Si-rich gel, however, causing a reduction in final strength. The FA geopolymer matrix was observed to gain strength faster but had lower final strength than the MK geopolymer. This observation was consistent with the hypothesis described above. In this study, both MK and FA geopolymers were considered as potential candidates for enhancing the mechanical properties of highly plastic clay. Therefore, in the next phase of the research, a highly plastic clay soil was treated with MK and FA geopolymers with a range of geopolymer contents and the strength and swelling behavior were determined.

Geopolymer Stabilized Soil: Materials and Test Methods

Soil

The highly plastic clayey soil used in this study was obtained from Atlanta, Texas, and therefore, is referred to as "Atlanta clay" throughout this paper. Preliminary processing of the soil involved the following steps. The soil was initially oven dried at 105°C for 24 hours to remove any moisture present. The oven dried soil was then passed through a grinder and then sieved to remove foreign particles and coarse aggregates that could affect the test results. Residue measuring less than 425 μm fraction (or soil passing the No. 40 sieve) was used to prepare specimens for the entire laboratory test program.

Atterberg limit tests conducted according to ASTM D4318 on the less-than 425 μm fraction of the Atlanta clay yielded liquid limit of 82.2% and plastic limit of 28.8%. Thus, the plasticity index (PI) of the soil was 53.4%. According to the Unified Soil Classification System (USCS), this soil is classified as CH (high plasticity clay). Optimum moisture content (OMC) and dry density (γ_{dry}) of the Atlanta clay were determined to be 30.8% and 112.3 lb/ft^3 respectively.

XRF was used to identify the elemental composition of the Atlanta clay. The data were also analyzed to determine the elements and oxides in the clay. Clay specimens were further characterized to determine their mineralogical composition using XRD. The predominant mineralogical constituent of the Atlanta clay was determined to be montmorillonite and quartz. Montmorillonite is commonly associated with high shrink–swell behavior (*3*, *35*, *36*). The chemical and mineralogical composition of Atlanta clay are presented in Table 3.

Table 3. Chemical and Mineralogical Composition of Atlanta Clay

Constituent	Percentage
Elements	
Si	30.11
Al	8.65
Fe	6.28
Ca	2.79
Mg	1.07
K	1.42
Ti	0.82
P	0.35
Mn	0.12
Zr	0.08
Ce	0.05
Oxides	
SiO_2	64.28
Al_2O_3	16.39
Fe_2O_3	9.03
CaO	3.92
MgO	1.78
K_2O	1.72
TiO_2	1.38
P_2O_5	0.79
ZrO_2	0.11
MnO	0.15
BaO	-
Minerals	
Montmorillonite	39.5
Quartz	50.3
Feldspar	10.2

Unconfined Compressive Strength Test

Dry soil was initially mixed at its OMC of 30.8%. The moist soil was treated with geopolymer for 30 minutes at concentrations of 3%, 6%, 9%, 12%, and 15% (by weight of dry soil). In addition, soil samples treated with lime were prepared at 4% lime content based on ASTM C977-10 to serve as reference. Two replicates of each soil stabilizer mixture were prepared to ensure consistency of treatment. All the samples were compacted with a 5.5 lbf rammer dropped from a height of 12 inches to produce a compactive effort of 12,400 ft-lbf/ft^3. The samples were compacted to a dry density (γ_{dry}) of 110.5 lb/ft^3. The samples were then placed in a sealed plastic bag and cured for seven days in an ambient condition (RH ~ 40%, temperature ~ 23°C). At the end of the curing period, diameter and height of the sample were measured to determine if any swelling had occurred during the curing period. Treated soil samples that were 2.7 in. in diameter and 4 in. in height at a height/diameter ratio of 1.5 were tested to determine unconfined compressive strength. Testing of the specimen was performed at the end of seven days at a loading rate of 0.05 in./min. Both unconfined compressive strength and Young's modulus of each sample were determined.

One-Dimensional Swell Test

One-dimensional swell test was performed in accordance with ASTM D4546. However, sample preparation in this test procedure was modified to represent samples prepared for UCS tests. Initially, water was added to soil at OMC and allowed to equilibrate in a sealed container for at least 16 hours. Soil was then treated with geopolymer for 30 minutes and then compacted into a mold 2.7 in. in diameter and 4 in. in height. A single layer of 1.5 in. height was molded with 25 blows. A circular consolidation ring 2.5 in. in diameter and 0.575 in. in height was then used to extract a specimen of 2.5 in. diameter and 0.575 in. height. The extraction was regulated by pressing the ring into the specimen with a hydraulic press. Two replicates were prepared for each test mixture to ascertain the consistency of treatment. The specimens were then stored in a sealed container for seven days under ambient conditions (RH ~ 40%, temperature ~ 23°C).

At the end of the curing period, each specimen was measured to account for any increase in volume during the curing period. This volumetric increase was added to the final volumetric increase observed in the specimen during moisture absorption. The specimen was then placed inside the ring with a porous disk on the bottom to allow moisture absorption. A metallic cap weighing 4 ± 0.2 oz was placed on top of each specimen to simulate swell under a pressure of 0.043 psi. The metallic cap was assumed to have no significant friction with the wall of the ring. Water was added between the ring and wall of the consolidation cell such that water percolated into the specimen only through the porous disc placed at the bottom. A linear variable differential transformer was used to measure the vertical swell of the specimen at intervals of 20 min. The complete test setup is shown in Figure 2. The swell test was continued until there was no significant difference in vertical swell between 10 consecutive readings. At the end of the test, the final mass and water content of each specimen were recorded.

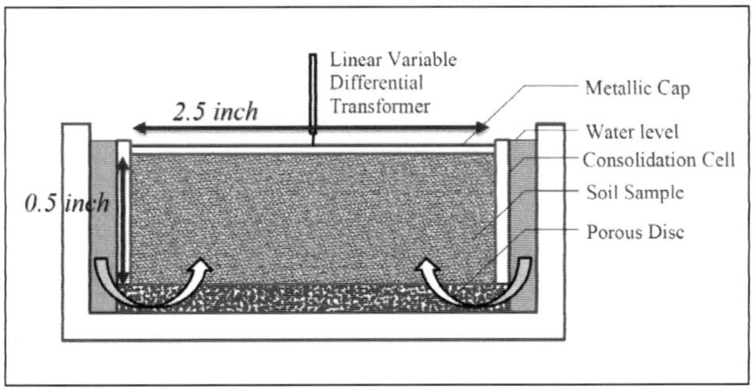

Figure 2. Schematic of the test setup used for the one-dimensional swell test.

Mineralogical Characterization

Soil samples were investigated using XRD to identify d-spacing of untreated and geopolymer treated soil samples at different moisture conditions. Increase in d-spacing of soil minerals can be observed by XRD which can then be used to determine the number of layers of structured water present within the clay sheet (*37–39*). Soil samples used for XRD analysis were primarily composed of untreated, lime treated, and geopolymer treated samples at dry, optimum moisture, and fully saturated conditions. Two different techniques were adopted while preparing the samples for XRD tests. Dry samples were initially crushed with a mortar and pestle and then loaded into a sample holder for testing. Wet samples were applied as a paste. XRD patterns were collected on the samples using Rigaku Miniflex II equipped with graphite monochromatic and scintillation detector which produced x-ray radiation (Cu Kα) with a voltage of 30 kV and a current of 15 mA. Samples were scanned in a range of 2θ from 3° to 80° at the rate of 0.02°/step. From the XRD pattern, phase abundances were analyzed with Reitveld refinement and whole-pattern fitting. Finally, d-spacing results established the swell potential of treated and untreated clay to verify the results of one-dimensional swell tests.

Microstructural Characterization

SEM imaging was performed on untreated clay and clay treated with MK and FA geopolymer using Hitachi S-4300 E/N (FESEM), with 1.5 nm resolution at 15 kV accelerating voltage used for this study (see Figure 3). EDX analyses were performed by the accessory EDAX EDX system.

Mechanical Properties of Geopolymer Stabilized Soil

Unconfined Compressive Strength

Test results for untreated, lime treated, and MK and FA geopolymer treated clay specimens after seven days of curing are presented in Figure 4. Significant improvement in strength was observed for specimens treated with both MK and FA geopolymer. Specimens prepared with MK geopolymer treatment had higher strength than FA geopolymer treated specimens at all concentrations of geopolymer. The higher strength of MK geopolymer treated specimens can be attributed to stronger interparticle bonding provided by the stronger MK geopolymer matrix. Even though FA geopolymer matrix was

(a)

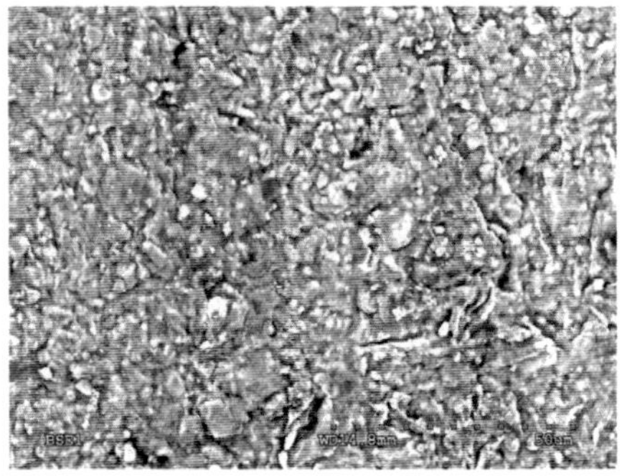

Quantification		Weight %	Atomic %
Elements	SiK	30.11	63.57
	AlK	8.65	19.01
	FeK	6.28	6.67
	CaK	2.79	4.13
	MgK	1.07	2.61
	KK	1.42	2.15
	TiK	0.82	1.02
	PxK	0.35	0.67
	MnK	0.12	0.13
	ZrK	0.08	0.05
	Total	100	100
Oxides	SiO_2	64.28	73.94
	Al_2O_3	16.39	11.11
	Fe_2O_3	9.03	3.91
	CaO	3.92	4.83
	MgO	1.78	3.05
	K_2O	1.72	1.26
	TiO_2	1.38	1.49
	P_2O_5	0.79	0.19
	ZrO_2	0.11	0.06
	MnO	0.15	0.15
	Total	100	100

(b)

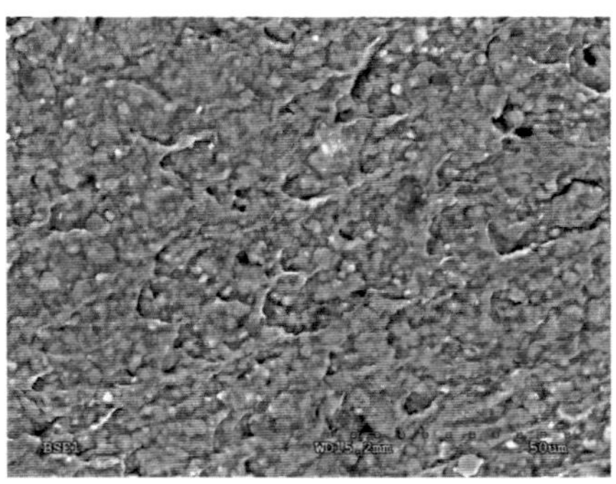

Quantification		Weight %	Atomic %
Elements	NaK	1.76	2.31
	MgK	1.64	2.04
	AlK	19.61	21.96
	SiK	55.88	60.11
	CaK	10.17	7.66
	FeK	10.94	5.92
	Total	100	100
Oxides	Na_2O	1.42	1.50
	MgO	1.56	2.54
	Al_2O_3	20.54	13.22
	SiO_2	61.84	67.51
	CaO	7.37	6.72
	FeO	7.27	8.51
	Total	100	100

(c)

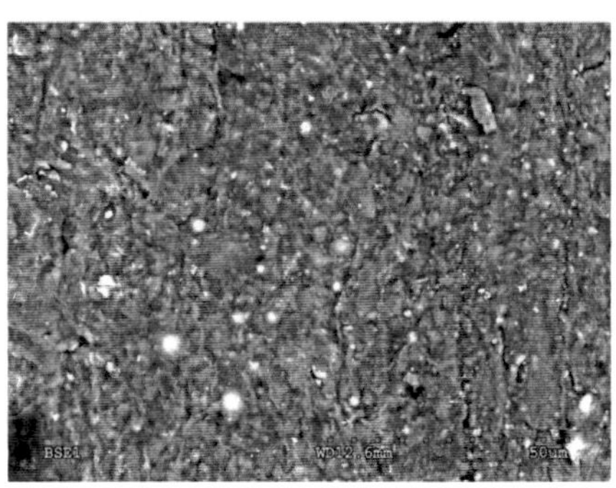

Quantification		Weight %	Atomic %
Elements	NaK	1.36	1.69
	MgK	2.60	3.05
	AlK	22.30	23.61
	SiK	62.56	63.64
	KK	2.23	1.63
	CaK	8.95	6.38
	Total	100	100
Oxides	Na_2O	1.13	1.19
	MgO	2.47	4.01
	Al_2O_3	22.90	14.71
	SiO_2	66.2	72.17
	K_2O	1.28	0.89
	CaO	6.01	7.02
	Total	100	100

Figure 3. SEM image of (*a*) untreated clay; (*b*) MK geopolymer treated; and (*c*) FA geopolymer treated clay after seven days of curing.

weaker than the MK geopolymer matrix, considerable improvement in strength was observed for FA geopolymer treated clay as well. It can be further observed that increase in unconfined compressive strength with increase in geopolymer content is minimal. This is particularly evident in the case of FA geopolymer treated clay. These

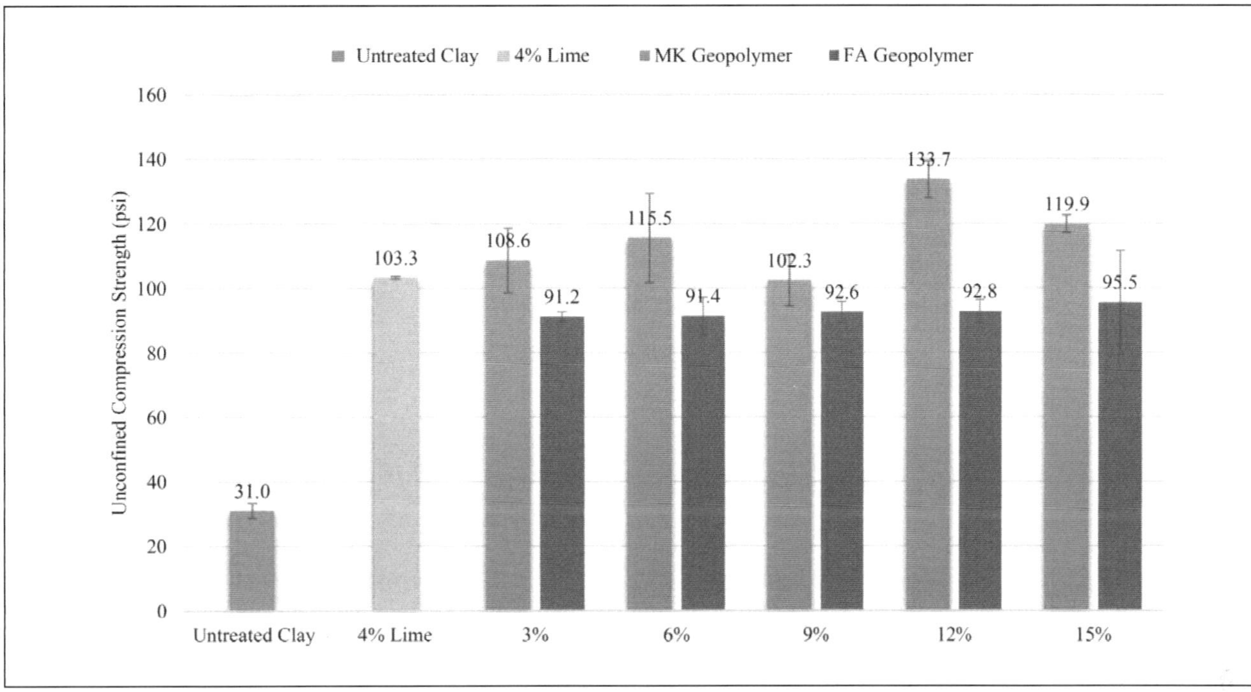

Figure 4. Unconfined compressive strength of untreated, 4% lime stabilized, and MK and FA geopolymer stabilized clay after seven days of curing.

results suggest that the lowest geopolymer concentration used (i.e., 3%) was adequate to provide necessary inter-particle bonding and that further increase in geopolymer does not contribute to further enhancement in strength.

Percentage Swell

Results obtained from swell tests performed on MK and FA geopolymer treated clay specimens are shown in Figure 5 along with percentage swell measurements made for untreated and lime treated clay soil. Clay treated with 4% lime showed dramatic reduction in swell behavior. Significant reduction in swell was also observed in 6%, 9%, and 12% FA geopolymer. Among the geopolymer treatments evaluated, 12% FA geopolymer provided the best results with over 80% reduction in overall swelling. Interestingly, MK geopolymer, which proved to be quite effective in improving the strength of the clay soil, failed to produce any improvement in swell behavior. In fact, MK geopolymer seemed to have adverse effect on the swell behavior of the clay soil.

Mineralogy of Geopolymer Treated Clay

Comparison between results from XRD tests performed on untreated and treated clay specimens did not reveal any major changes in the mineralogical make-up of the material. This suggests there were no large-scale mineral neoformations in the soil upon the treatment with geopolymers (similar to ettringite which forms during lime treatment of sulfate rich soils). There was good agreement between d-spacings obtained for soil at OMC and at full saturation and measured swell percentages. Comparisons were also made between d-spacings obtained for uncompacted soil and compacted soil with treatment. The results showed that the treatment was most effective when combined with mechanical compaction.

Conclusions

Calcium-rich stabilizers such as lime and Portland cement are commonly used to stabilize weak subgrade soils with high plasticity. However, high energy demand associated with production of these stabilizers and adverse environmental impact from emission of harmful gases has given much impetus to the search for alternative soil stabilizers. In a number of previous studies, geopolymers have been identified as a more sustainable alternative to traditional soil stabilizers. However, these previous studies primarily focused on improvement of the strength of the stabilized soil. Furthermore, the results were based on testing performed on artificially prepared siliceous materials with low PI. This research investigated the effectiveness of geopolymer on

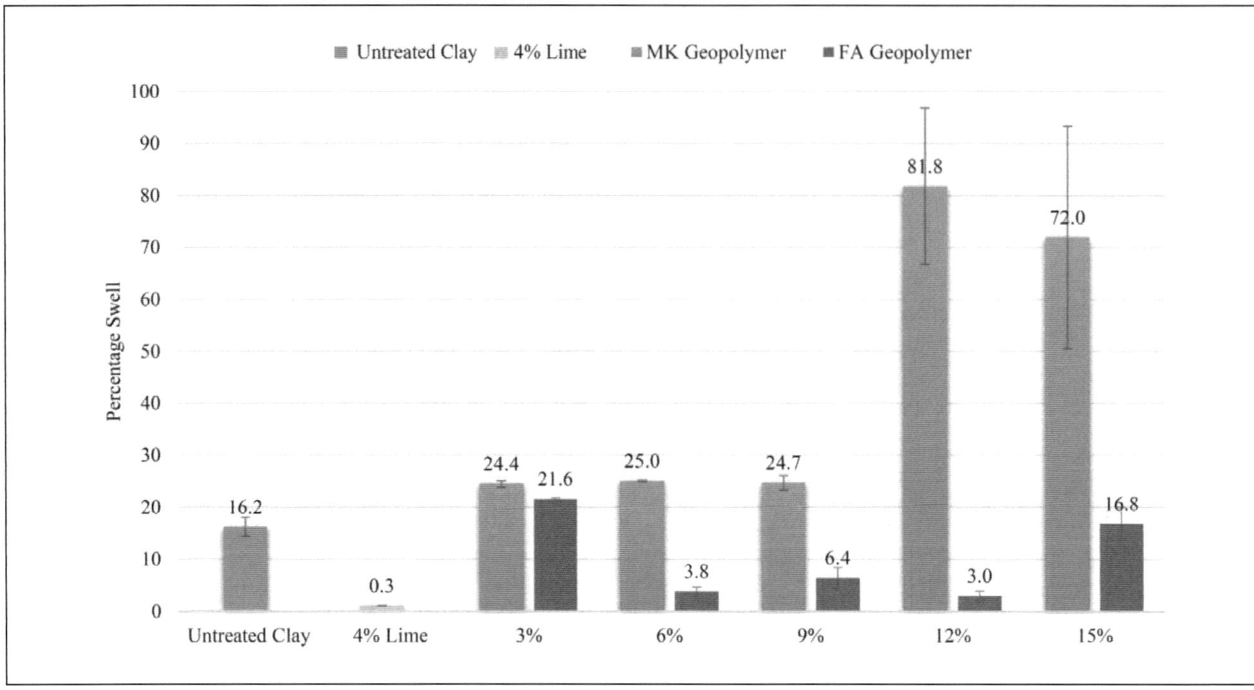

Figure 5. Swell percentage of untreated, 4% lime stabilized, and MK and FA geopolymer stabilized Atlanta clay after seven days of curing and seven days of swell test.

improving both strength and shrink/swell behavior of a high PI, natural clay soil. Two forms of geopolymer, MK and FA based geopolymer, were included in the study.

The strength measurements made on the geopolymer matrix with different alumino-silicate precursors indicated that optimum conditions are achieved at a Si/Al molar ratio of 1.66 and Na/Al molar ratio of 1.06 for the MK geopolymer. For the FA geopolymer, the optimum conditions are achieved at a Si/Al molar ratio of 3.41 and Na/Al molar ratio of 2.27. Significant improvement in strength, comparable to that achieved with the traditional stabilizer, lime, was observed for both MK and FA geopolymer treated soil. In excess of four-fold increase in strength was observed for MK geopolymer after seven days of curing while more than three-fold increase in strength was obtained for FA geopolymer treated soil. MK geopolymer treated soil demonstrated higher strength than FA geopolymer at all geopolymer concentrations. The difference in strengths achieved for MK versus FA geopolymer treated soil can be attributed to difference in the strength of the geopolymer matrix. A direct correlation between percentage of geopolymer used and strength could not be established. Based on the results obtained, 3% MK and 3% FA geopolymer can be recommended as optimum when the primary objective of stabilization is strength improvement.

The percentage swelling measured on geopolymer treated clay soil revealed that FA geopolymer was significantly more effective in controlling moisture induced swelling than the MK geopolymer. More than 80% reduction in swell percentage was achieved with FA geopolymer treatment. Among the different dosages of FA geopolymer used, 6% appears to be optimum in terms of achieving satisfactory levels of swell control. Similar results for swell reduction were observed at FA geopolymer contents of 9% and 12%. Dosages lower than 6% and higher than 12% did not produce satisfactory swell reduction. On the other hand, MK geopolymer treated specimens did not produce swell reduction at any of the geopolymer dosages tested. Neither form of geopolymer produced reduction in swell comparable to that achieved with lime. However, since geopolymers can be synthesized using alumino-silicates low in calcium, they may have an advantage over traditional stabilizers in the treatment of sulfate-bearing high PI clay soils. This aspect deserves further investigation.

Acknowledgments

The authors thank Texas Department of Transportation for funding this research project. Technical support by Dr Branimir Segvic, Dr Callum Hetherington, Dr Zhan Bo and Mahmoud Mahrous for assisting with XRD and SEM-EDX is greatly appreciated. The authors also wish to thank Sean

Starcher for procuring the clay, and Gerardo Tueme and Lan Ventura for assistance in lab work.

Author Contributions

The authors confirm contribution to the paper as follows: study conception and design: Suraj Khadka, Priyantha Jayawickrama, Sanjaya Senadheera; data collection: Suraj Khadka; analysis and interpretation of results: Suraj Khadka, Priyantha Jayawickrama; draft manuscript preparation: Suraj Khadka, Priyantha Jayawickrama. All authors reviewed the results and approved the final version of the manuscript.

References

1. Wise, J. R. *An Examination of Expansive Clay Problems in Texas*. Research Report Number 118-5. Center for Highway Research, University of Texas at Austin, Austin, TX, 1971.
2. Basma, A. A., A. S. Al-Homoud, A. I. H. Malkawi, and M. A. Al-Bashabsheh. Swelling-Shrinkage Behavior of Natural Expansive Clays. *Applied Clay Science*, Vol. 11, No. 2–4, 1996, pp. 211–227.
3. Mašin, D., and N. Khalili. Swelling Phenomena and Effective Stress in Compacted Expansive Clays. *Canadian Geotech Journal*, Vol. 53, No. 1, 2016, pp. 134–147.
4. Prime, N., S. Levasseur, L. Miny, R. Charlier, A. Léonard, and F. Collin. Drying-Induced Shrinkage of Boom Clay: An Experimental Investigation. *Canadian Geotech Journal*. Vol. 53, No. 3, 2016, pp. 396–409.
5. Prusinski, J. R., and S. Bhattacharja. Effectiveness of Portland Cement and Lime in Stabilizing Clay Soil. *Transportation Research Record: Journal of the Transportation Research Board*, 1999. 1652: 215–226.
6. Stajanča, M., and A. Eštoková. Environmental Impacts of Cement Production. *Theory and Practice of Building*, Vol. 56, 2012, pp. 296–302.
7. Chen, C., G. Habert, Y. Bouzidi, and A. Jullien. Environmental Impact of Cement Production: Detail of the Different Processes and Cement Plant Variability Evaluation. *Journal of Cleaner Production*, Vol. 18, 2010, pp. 478–485.
8. Harris, P., J. V. Holdt, S. Sebesta, and T. Scullion. *Recommendations for Stabilization of High-Sulfate Soils in Texas*. Report No. FHWA/TX-06/0-4240-3. Texas Department of Transportation, Austin, TX, 2006.
9. McCallister, L., and T. Petry. Leach Tests on Lime-Treated Clays. *Geotechnical Testing Journal*, Vol. 15, No. 2, 1992, pp. 106–114.
10. Detphan, S., and P. Chindaprasirt. Preparation of Fly Ash and Rice Husk Ash Geopolymer. *International Journal of Minerals Metallurgy and Materials*, Vol. 16, 2009, pp. 720–726.
11. He, J., and G. Zhang. Geopolymerization of Red Mud and Fly Ash for Civil Infrastructure Applications. *Geo-Frontiers*, 2011, pp. 1287–1296.
12. Fofomatasja, S. N., A. M. Mustafa, M. A. Bakri, H. Kamarudin, Y. Zarina, A. R. Rafiza, and J. Liyana. Reviews on the Different Sources Materials to the Geopolymer Performance. *Advances in Environmental Biology*, Vol. 7, 2013, pp. 3853–3842.
13. Davidovits, J. *False Values on CO_2 Emission for Geopolymer Cement/Concrete Published in Scientific Papers*. Technical Paper #24, Geopolymer Institute Library, Saint-Quentin, France, 2015, pp. 1–9. https://www.geopolymer.org/fichiers_pdf/False-CO2-values.pdf
14. Davidovits, J. Geopolymers and Geopolymeric Materials. *Journal of Thermal Analysis*, Vol. 35, 1994, pp. 429–441.
15. Loewenstein, W. The Distribution of Aluminum in the Tetrahedra of Silicates and Aluminates. *American Mineralogist*, Vol. 39, 1954, pp. 92–96.
16. Rowles, M. R., and B. H. O'Connor. Chemical and Structural Microanalysis of Alumino-silicate Geopolymers Synthesized by Sodium Silicate Activation of Metakaolinite. *Journal of American Ceramic Society*, Vol. 92, 2009, pp. 2354–2361.
17. Davidovits, J. Geopolymer: Inorganic Polymeric New Materials. *Journal of Thermal Analysis*, Vol. 37, 1991, pp. 1633–1656.
18. Steveson, M., and K. Sagoe-Crentsil. Relationships between Composition, Structure and Strength of Inorganic Polymers. Part I Metakaolin-derived Inorganic Polymers. *Journal of Materials Science*, Vol. 40, 2005, pp. 2023–2036.
19. Steveson, M., and K. Sagoe-Crentsil. Relationships Between Composition, Structure and Strength of Inorganic Polymers. Part II Fly Ash-Derived Inorganic Polymers. *Journal of Materials Science*, Vol. 40, 2005, pp. 4247–4259.
20. Van Jaarsveld, J. G. S., and J. G. S. Van Deventer. The Potential Use of Geopolymeric Materials To Immobilize Toxic Materials. Part I Theory and Applications. *Minerals Engineering*, Vol. 19, 1996, pp. 659–669.
21. Davidovits, J. *Mineral Polymers and Methods of Making Them*. US Patent 4349386 A, 1982.
22. Rovnanik, P. Effect of Curing Temperature on the Development of Hard Structure of Metakaolin-based Geopolymer. *Construction and Building Materials*, Vol. 24, 2010, pp. 1176–1183.
23. Lizcano, M., A. Gonzalez, S. Basu, K. Lozano, and M. Radovic. Effect of Water Content and Chemical Composition on Structural Properties of Alkaline Activated Metakaolin-Based Geopolymers. *Journal of American Ceramic Society*, Vol. 95, 2012, pp. 2169–2177.
24. Bakri, A. M. M., H. Kamarudin, M. Bnhussain, I. K. Nizar, A. R. Rafiza, and Y. Zarina. The Procressing, Characterization and Properties of Fly Ash Based Geopolymer Concrete. *Reviews on Advanced Materials Science*, Vol. 30, 2012, pp. 90–97.
25. Motorwala, A., V. Shah, R. Kammula, P. Nannapaneni, and D. B. Raijiwala. Alkali Activated Fly-Ash Based Geopolymer Concrete. *International Journal of Emerging Technology and Advanced Engineering*, Vol. 3, 2013, pp. 159–166.
26. Mathew, B. J., M. Sudhakar, and C. Natarajan. Strength, Economic and Sustainability Characteristics of Coal Ash-GGBS Based Geopolymer Concrete. *International Journal of Computational Engineering Research*, Vol. 3, 2013, pp. 207–212.

27. Palomo, A., M. W. Grutzeck, and M. T. Blanco. Alkali-activated Fly Ashes: A Cement for the Future. *Cement and Concrete Research*, Vol. 29, 1999, pp. 1323–1329.
28. Ohno, M., and V. C. Li. A Feasibility Study of Strain Hardening Fiber Reinforced Fly Ash-Based Geopolymer Composites. *Construction and Building Materials*, Vol. 57, 2014, pp. 163–168.
29. Temuujin, J., A. Van Riessen, and K. J. D. MacKenzie. Preparation and Characterization of Fly Ash Based Geopolymer Mortars. *Construction and Building Materials*, Vol. 24, 2010, pp. 1906–1910.
30. Zhang, M., H. Guo, T. El-Korchi, G. Zhang, and M. Tao. Experimental Feasibility Study of Geopolymer as the Next-Generation Soil Stabilizer. *Construction and Building Materials*, Vol. 47, 2013, pp. 1468–1478.
31. Zhang, M., M. Zhao, G. Zhang, P. Nowak, A. Coen, and M. Tao. Calcium-Free Geopolymer as a Stabilizer for Sulfate-Rich Soils. *Applied Clay Science*, Vol. 108, 2015, pp. 199–207.
32. Chimoye, W. Strength of Soft Bangkok Clay Improved by Geopolymer from Palm Fuel Ash. *International Journal of Engineering and Technology Research*, Vol. 2, 2015, pp. 1–10.
33. Nayak, P. S., and B. K. Singh. Instrumental Characterization of Clay by XRF, XRD and FTIR. *Bulletin of Material Science*, Vol. 30, 2007, pp. 235–238.
34. Hajimohammadi, A., J. L. Provis, and J. S. J. Van Deventer. Effect of Alumina Release Rate on the Mechanism of Geopolymer Gel Formation. *Chemistry of Materials*, Vol. 22, 2010, 5199–5208.
35. Zhen, Y., A. Zaoui, and I. Shahrour. A Theoretical Study of Swelling and Shrinking of Hydrated Wyoming Montmorillonite. *Applied Clay Science*, Vol. 51, 2011, pp. 177–181.
36. Basma, A. A., A. S. Al-Homoud, A. I. H. Malkawi, and M. A. Al-Bashabsheh. Swelling-Shrinkage Behavior of Natural Expansive Clays. *Applied Clay Science*, Vol. 11, 1996, pp. 211–227.
37. Amorim, C. L. G., R. T. Lopes, R. C. Barroso, J. C. Queiroz, and D. B. Alves. Effect of Clay-Water Interactions on Clay Swelling by X-Ray Diffraction. *Nuclear Instruments and Methods in Physics Research*, Vol. 580, 2007, pp. 768–770.
38. Mooney, R. W., A. G. Keenan, and L. A. Wood. Adsorption of Water Vapor by Montmorillonite. I. Heat of Desorption and Application of BET Theory. *Journal of American Chemical Society*, Vol. 74, 1952, pp. 1367–1371.
39. Mooney, R. W., A. G. Keenan, and L. A. Wood. Adsorption of Water Vapor by Montmorillonite. II. Effect of Exchangeable Ions and Lattice Swelling as Measured by X-Ray Diffraction. *Journal of American Chemical Society*, Vol. 74, 1952, pp. 1371–1374.

The Standing Committee on Stabilization of Geomaterials and Recycled Materials (AFS80) peer-reviewed this paper (18-04882).

TRR
JOURNAL OF THE TRANSPORTATION RESEARCH BOARD

Article

Transportation Research Record
2018, Vol. 2672(52) 185–194

DOI: 10.1177/0361198118792996
journals.sagepub.com/home/trr
SAGE

Expansive Soil Treatment with Liquid Ionic Soil Stabilizer

Shi He[1], Xinbao Yu[1], Aritra Banerjee[1], and Anand J. Puppala[1]

Abstract

Calcium-based stabilizers such as lime and cement control swell and shrinkage behavior and enhance strength properties for expansive soils through the formation of pozzolanic components. However, sulfate-bearing subgrade soils stabilized with calcium-based stabilizers might cause excessive swelling and shrinkage due to the formation of highly expansive minerals like ettringite and thaumasite. In this paper, one liquid ionic soil stabilizer (LISS) was evaluated as an alternative stabilizer used to control swelling and shrinkage behavior of expansive soils. A comprehensive laboratory experiment program including a linear shrinkage test, a one-dimensional swell test, and an unconfined compressive strength test, was designed and carried out on soils from Dallas, Texas before and after treatment. Three dosage levels of stabilizer and four different curing periods were investigated. Test results indicate that LISS is an effective stabilizing agent, which not only reduces swelling and soil plasticity but also increases soil strength. Furthermore, a similar type of LISS is utilized to treat the soil in Dallas via deep injection using a hydraulic pump. Field emission scanning electron microscopy results on the test soil showed that the stabilizing program is likely to work through clay flocculation and morphological variations in the clay particles.

Expansive soil tends to change volume when the moisture content changes (*1*, *2*). Some soils in Texas can absorb a significant amount of water after rainfall and swell substantially; however, during the drought season they lose moisture and shrink, often resulting in desiccation cracks. Moreover, swelling and shrinkage behaviors of expansive soil are not completely reversible. The recurring cracks on the ground are inevitably more severe as the soil fails to seal completely after rewetting (*1*, *3*). Therefore, the State of Texas has to spend at least US$1 billion per year to repair the damaged houses, pavements, buried utilities, and embankment constructed on expansive soils (*2*).

To prevent or mitigate the loss from expansive soil, various treatment methods have been developed over the past decades. Essentially conventional soil stabilizers, including lime, Portland cement, and fly ash, are utilized to improve workability after compaction, to inhibit swell and shrinkage behavior, and to enhance the strength and stiffness of the foundation of residential homes (*4–8*). However, calcium-based stabilizers have a negative impact on treating sulfate-rich expansive soil, which leads to excessive volume change and heaving (*4*, *5*). This phenomenon occurs when the calcium in the stabilizer reacts with sulfate and alumina in the soil to form ettringite (*4*, *6*, *9*). Ettringite tends to swell up to 250% when it comes in contact with water (*10*, *11*). Furthermore, conventional soil stabilizers are harmful to the environment. For instance, lime treatment not only dramatically raises the pH of the soil but also emits a large amount of carbon dioxide, which exacerbates the global greenhouse effect (*7*).

Hence non-traditional stabilizers without calcium content are emerging on the market to treat high sulfate soil. Most non-traditional stabilizers are classified into three categories: ionic, polymer, and enzyme (*4*, *9*). The majority of the products contain secondary additives including surfactants, catalysts, and ultraviolet inhibitors (*12*). Despite the many benefits of liquid ionic soil stabilizers (LISSs), many engineers and manufacturers are reluctant to accept this. The primary reason is the lack of literature on the mechanisms involved in the soil treatment using LISSs. Scholen indicated that LISS causes cation exchange and clay mineral flocculation (*13*). The chemical reaction makes the clay structure much more stable and dispels the excess double film water. Studies in China have attributed the stabilization effect to the reduction of the cation-exchange capacity, which compresses the thickness of double layer water and changes

[1]Department of Civil Engineering, University of Texas at Arlington, Arlington, TX

Corresponding Author:
Address correspondence to Xinbao Yu: xinbao@uta.edu

the density of the electrical charge of clay particle surface (*14*, *15*). Ali claims that the primary electrical charges of clays are negative, which attract positive molecules (*16*). During the treatment process, weaker cations such as H + are inevitably replaced. Finally, chemical compounds modify the clay particle surface and inhibit water absorption.

In this study, one type of LISS was obtained to treat one soil from Dallas by hand mixing the stabilizer and the soil in the lab. Comprehensive tests, including the Atterberg limit test, linear shrinkage test, standard compaction test, unconfined compressive strength (UCS), 1D swell, and swell pressure tests, were carried out on soil specimens before and after treatment with different dosages of LISS to evaluate the effectiveness of the treatment. Additionally, field emission scanning electron microscopy (FESEM) was utilized to study the morphological variations between untreated and treated samples (*9*).

Materials and Experiment Study

Materials

The soil used in this study was collected from a site in Caddo Mills near Dallas, Texas. Table 1 summarizes the basic properties of the soil used in this study. Because a clay fraction of soil is around 40%, LISS is suitable to treat the expansive soil with more than 25% clay content (*14*).

In this study, LISS was obtained to treat one expansive soil collected in Texas. This liquid stabilizer is composed of sulfuric acid, phosphoric acid, citric acid, water, and surfactant. Ionic stabilizers could reduce the surface charge of the soil particles, which cause cation exchange and flocculation of the clay minerals (*12*). According to the manufacturer, this LISS is environmentally friendly, non-toxic, and an effective solution to the treatment of expansive clay soil. Around a 50% reduction in the potential vertical rise can be achieved within 1 year after the treatment. In addition, the transportation cost of LISS is far less than traditional soil stabilizers (*5*).

In the field, LISS concentrate is diluted with water and injected into the ground under pressure to treat soil layers, as shown in Figure 1. The dilution ratio recommended by the supplier is to 8 gal of the chemical concentrate to 340 g of surfactant with 6,000 gal of water. On the site, the five rods are statically pressed 6 in. into the ground, and the solution is supplied at 100 gal per min to complete the chemical pass for around 12 s. Then the rods are advanced another 6 in. to repeat the above chemical pass. This process continues until the desired treatment depth is reached, typically 10 ft for residential homes. After the soils are completely saturated, the tractor is moved 3 ft away and the injection progress repeats until the site is completely covered. In the lab treatment, the field application ratio was replicated by mixing 5-ml liquid chemical concentrate and 0.057-g surfactant with 1 gal of water. In addition, two other ratios (50% dilution and double concentration) were also prepared to evaluate the LISS dosage treatment effects, which are shown in Table 2. In the lab tests of treated soils, soil samples were obtained by hand mixing dry soil powder with LISS solutions to prepare wet treated soils at optimum moisture content (OMC). The hand mixing procedure is different than the field injection and results in much more uniform chemical distribution in the treated soils. This sample preparation method represents a more favorable treatment condition. The focus of this study is to evaluate the stabilization effect of LISS under lab testing conditions.

Table 1. Basic Properties of Soil Obtained from Dallas, TX

Soil properties	Dallas soil
Liquid limit, LL (%)	76
Plasticity index, PI (%)	58
Specific gravity, G_s	2.84
Maximum dry density, (MDD) (pcf)	85.75
Optimum moisture content, (OMC) (%)	31
USCS classification	CH
% clay fraction	40
Montmorillonite content (%)	55.44
Illite content (%)	17.47
Kaolinite content (%)	27.1

Figure 1. A deep injection rake with perforated rods for pressurized injection of LISS.

Experiment Study

Atterberg Limit Test. Atterberg limit tests were conducted according to ASTM D4318-17. Treated samples were

Table 2. Three Liquid Stabilizer Dosage Designs for Soil Treatment

LISS proportion (chemical/water)	Company ratio (5 ml/gal)	Double water ratio (2.5 ml/gal)	Double chemical ratio (10 ml/gal)
Chemical concentrate (ml)	5	5	10
Surfactant (g)	0.057	0.057	0.114
Water (gal)	1	2	1

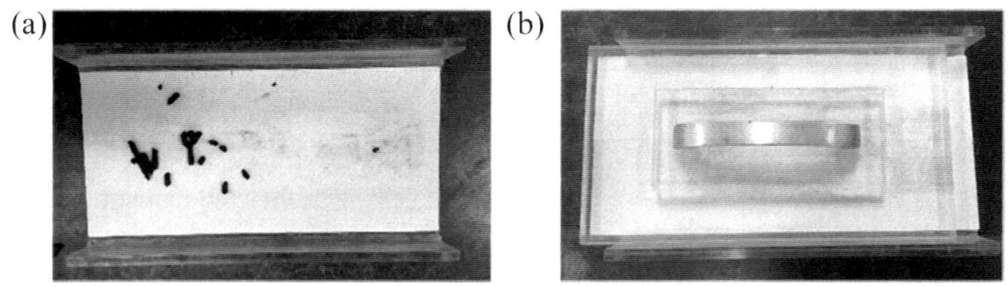

Figure 2. E-180 plastic limit rolling device.

prepared by mixing about 100 g of pulverized dry soil, passing through a No. 40 sieve, with various dosages of LISS solutions to prepare moist samples close to OMC. Samples were cured in a moisture room to allow reaction between liquid chemical and soil. During the liquid limit (LL) test, distilled water was added to enable the mixed soil to reach the liquid state; dry soil should not be added to ensure a consistent soil–chemical reaction time. After the LL test, the wet samples were air dried until the plastic limit (PL) state was obtained. An E-180 PL rolling device (Figure 2), initially developed by Texas Department of Transportation, was utilized to measure the PL. The major benefit of this device is that it can provide consistent results. The gap between the top roller plate and the bottom roller plate is set at 0.125 in. to conform to ASTM D4318–17. The linear shrinkage tests were conducted via Tex-107-E. Sufficient pulverized dry soil was mixed with enough water (control samples) or LISS (treated samples) to make the wet soil flow. Then, the wet soils were evenly transferred into the bar-shaped mold and tapped gently to let the soil flow and eliminate the entrapped bubbles. After the molds were filled completely, the samples were air dried for about 5 h and then transferred to the oven with a temperature set at 110°C.

Standard Compaction. A standard Proctor test was carried out to determine the relationship between OMC and maximum dry density (MDD) for control and treated soil samples. For treated soils, around 1.8 kg of pulverized dry soils were mixed with various proportions of LISS in separate containers. The moist soil specimens were stored in the moisture room for at least for 24 h before the Proctor test was conducted to allow proper moisture equilibration (*17*, *18*). This also allows a proper reaction between the liquid chemicals and the soil before compaction testing.

1D Free Swell Test. ASTM D 4546-14 introduces the procedures for free swell of compacted soil. The test is performed using a conventional oedometer (*19*). The sample size is 2.5 in. in diameter and 1 in. in thickness. The soil specimens were obtained by statically compressing wet mixed soil samples to achieve their optimum compaction state. All treated specimens were kept in the moisture room for 1 day, 7 days, or 28 days. During the swell tests, all the specimens were fully inundated in the oedometer with an initial seating load of 1 psi for at least 48 h. Filter papers and porous stones were placed at the top and bottom of the specimen that was confined in a metal ring. A dial gauge was set up on the loading frame to monitor the vertical deformation during water absorption. As the 1D swell test progresses, once the specimen reaches the saturation point and the dial gauge reading stops changing, loads are gradually added to return the dial gauge reading back to its original point. The total load applied on the soil specimen was calculated from the swell pressure. Figure 3 shows the test set up for 1D swell under the swell potential test.

Unconfined Compressive Strength (UCS). UCS tests were performed in accordance with ASTM D 2166-16. Both untreated and treated soil samples were mixed with

Figure 3. 1D swell and swell pressure equipment.

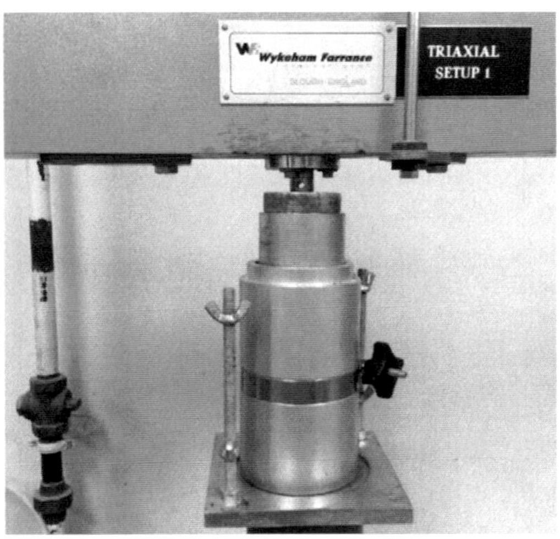

Figure 4. Static compaction machine.

additives and then compacted in three layers via a static compaction machine shown in Figure 4. The sample size was 2.8 in. in diameter and 5.8 in. in height. To simulate the field condition, all UCS and 1D swell specimens were prepared with a target of OMC and 95% MDD. For treated specimens, soil was mixed with liquid chemical by hand uniformly and then stored in a 100% humidity-controlled room overnight to ensure enough time to allow the chemical to react with the soil after initial mixing. After compaction, soil specimens were cured for 1 day, 7 days, and 28 days, respectively in the moisture room. At least three duplicate samples were prepared for UCS and 1D swell tests to avoid erroneous conclusions.

SEM Imaging Test. FESEM was used to study the microstructural changes in clay particles for untreated and treated specimens (*7, 9*). In this research, an Hitachi S-4800 II FESEM was used to obtain images at around 6,000 magnification for specimens before and after treatment. To simulate the field condition, around 50 g of the soil were soaked with LISS or water in ceramic bowls. They were then air dried for almost 28 days. A typical sample was removed from the dry soil and coated with silver via the sputter system because a sample that can conduct electricity is preferred for FESEM analysis.

Results and Discussion

Atterberg Limit Test

The LL and PL tests reflect the effect of moisture content on the mechanical properties of soil to estimate the soil swell potential (*20*). Table 3 indicates the Atterberg limits of the Dallas soil after adding various dosages of LISS. The plasticity index (PI) of Dallas soil dropped to the minimum level when the maximum dosage of LISS was applied. When compared to the untreated specimen, the PI had reduced by 50%. In summary, to some extent, the increase of LISS content in the Dallas soil helped decrease the PI value.

Bar Linear Shrinkage Test

The bar linear shrinkage bar tests were performed on untreated and treated samples to quantify the shrinkage potential of the soil. A series of the linear shrinkage bar tests from Dallas soil with LISS treatment are displayed in Figure 5. Table 4 illustrates the test results of Dallas soil, which shows a slight difference between shrinkage ratios among various treatment dosage levels and control samples. It was observed that LISS might not be effective in controlling the shrinkage potential.

Standard Proctor Compaction Test

Standard Proctor compaction tests were conducted to measure the relationship between MDD and OMC. Figure 6 reveals the standard compaction curve for untreated and treated Dallas soil. For untreated samples, the mixture should be hydrated completely prior to compaction because inadequate hydration time may cause erroneous results (*20*). Regarding treated soil, the chemical reaction between soil particles and LISS is likely to occur before compaction, causing soil particle flocculation. Hence, the void ratio decreases during the reaction, accompanied by a reduction in MDD. Generally, for chemical treatment of soil after compaction, the MDD of soils decreases and the OMC increases (*21*). The soil

Table 3. Atterberg Limit Results of Dallas Soil

LISS concentration to water (ml/gal)	LL (%)	PL (%)	PI (%)
0	76	18	58
2.5	72	15	57
5	79	29	50
10	69	43	26

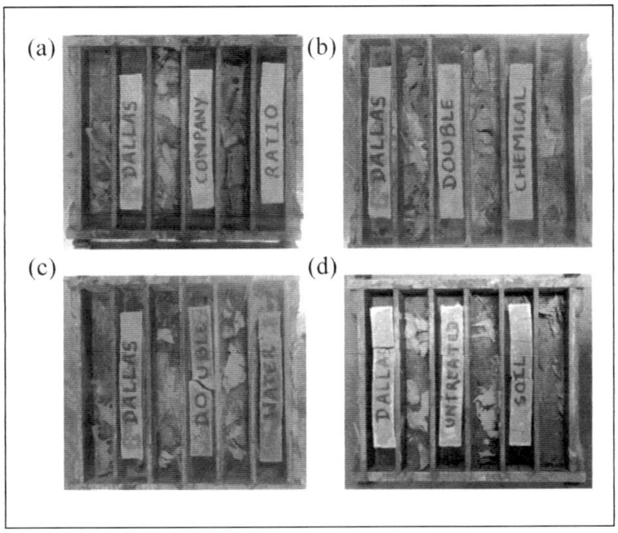

Figure 5. Bar linear shrinkage bar tests for Dallas soil with different dosages of LISS and control soil.

Table 4. Linear Shrinkage Measurement of Control and Treated Dallas Soil

LISS concentration to water (ml/gal)	Average linear shrinkage (%)
0	22.1
2.5	24.42
5	23.1
10	23.35

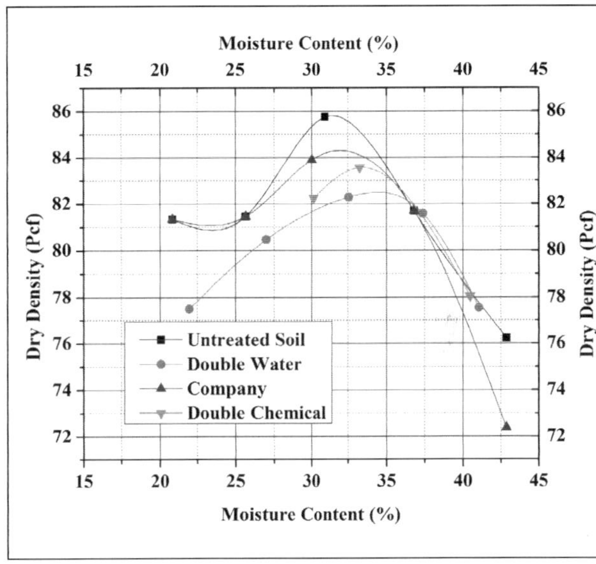

Figure 6. Standard compaction curve for Dallas soil.

chemical reaction may not finish completely without enough water, which results in OMC rising.

Unconfined Compressive Strength (UCS) Test

UCS tests were performed on the control and LISS-treated soil specimens. Three different stabilizer ratios (as shown in Table 2) were used to treat the soil. Two specimens were prepared for each stabilizer ratio. Figure 7 shows the mean UCS results for Dallas soil treated with different dosages of LISS. Figure 8 summarizes the difference in ultimate compressive strength among the three treatment ratios, the ultimate compressive strengths of control, company ratio, double water ratio, and double chemical ratio, all of which were influenced by the curing time. In most cases, more curing time resulted in higher compressive strength after treatment. In the double chemical ratio, the LISS content increased. The compressive strengths of the cured control and sample after 28 days were 16.83 psi and 28.42 psi, respectively. The vertical compressive strength rose more than 45% after treatment. This can be attributed to soil particle flocculation. The effects of the company-treated soil ratio were not clearly comparable until the results of soil before and after treatment were tallied. These results showed both the control soil and the company soil to be very similar. For double water ratio, its ultimate strength slightly dropped to 13.21 psi even though the sample was stored in the moisture room for 28 days. There is no doubt that the clay structure forms a gel after LISS treatment that provides more resistance during static compaction. However, larger compaction efforts are needed to match the corresponding desired density. Over-compacted samples become much weaker as shown by their decline in compressive strength after treatment (*22*).

One-Dimensional Swell Test

At least two samples were prepared for each dilution ratio. Figures 9 and 10 shows the average results of the one-dimensional (1D) swell test. Comparing samples before and after treatment, the original swell strain of control sample was 5.3% whereas the swell strain dropped to 0.92% after curing 28 days, treated using the

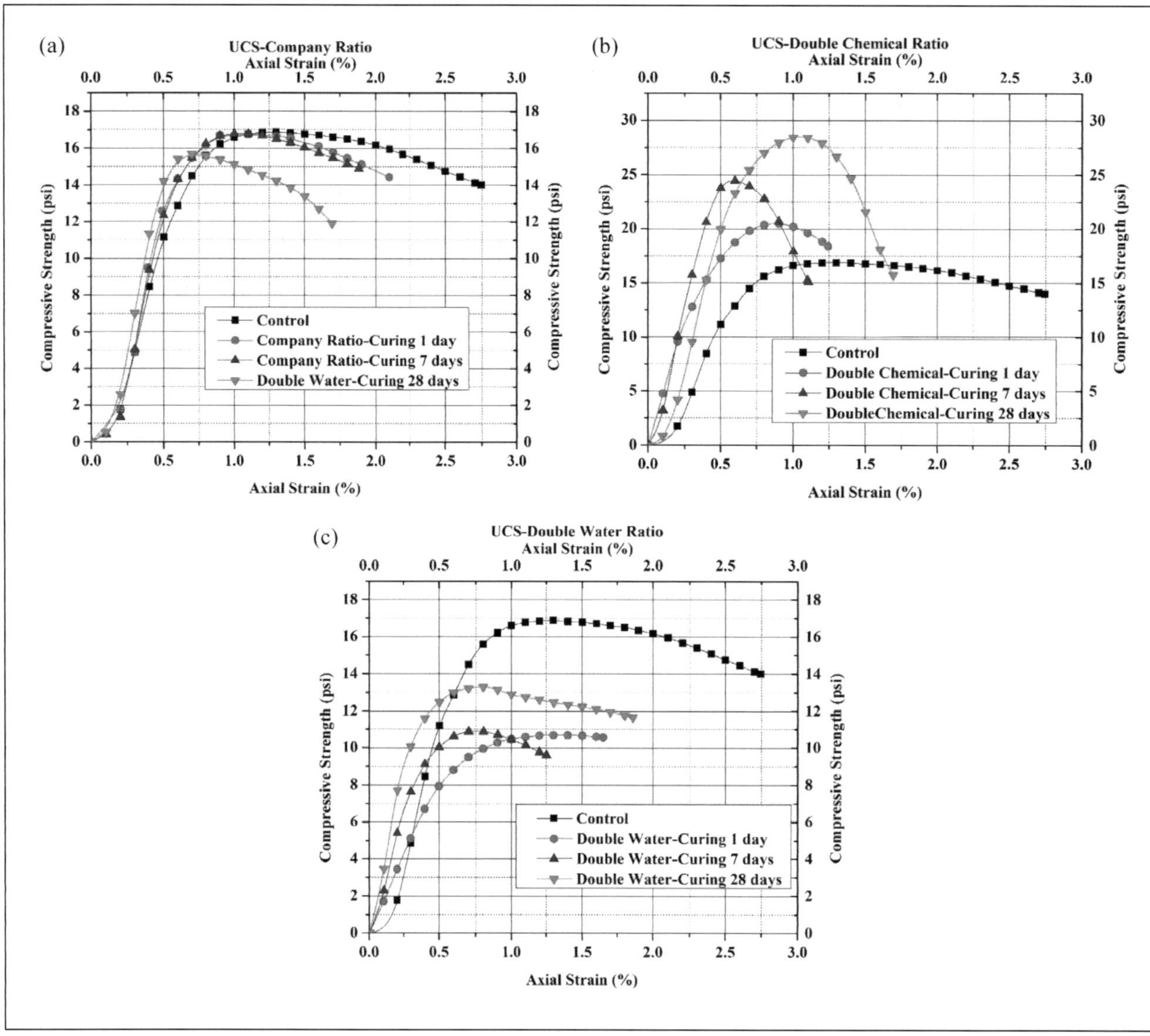

Figure 7. Unconfined compressive strength (UCS) test results.

double chemical ratio. After treatment, the swell strain reduced around 87%. Although the soil samples were treated with different dosages of LISS curing after only 1 day, the swell strains of all treated samples decreased 32%, 34%, and 51%, respectively. The swell strain reductions due to LISS treatment may be attributed to flocculation. Overall, this kind of LISS has a positive impact on inhibiting soil swell behavior.

Swell Potential Test

The pressure swell test mean results for Dallas soil are presented in the form of bar charts in Figure 11. Soil without LISS displayed large swell pressure whereas stabilized samples sustained low swell pressures. Among all the treated specimens, the longer the curing time, the lower the swell strain yields. The swell strain of soil treated by the double chemical ratio reduced about 34% compared with the control sample. This phenomenon can be explained as related to the time needed to complete soil chemical reaction. Furthermore, LISS treatment is helpful to dissipate pore pressures in the loading progress.

Field Emission Scanning Electron Microscopy (FESEM)

Figure 12 indicates the micromorphology among the soil sample with different dosages of LISS and untreated specimens. From the figure, it is obvious that microstructure

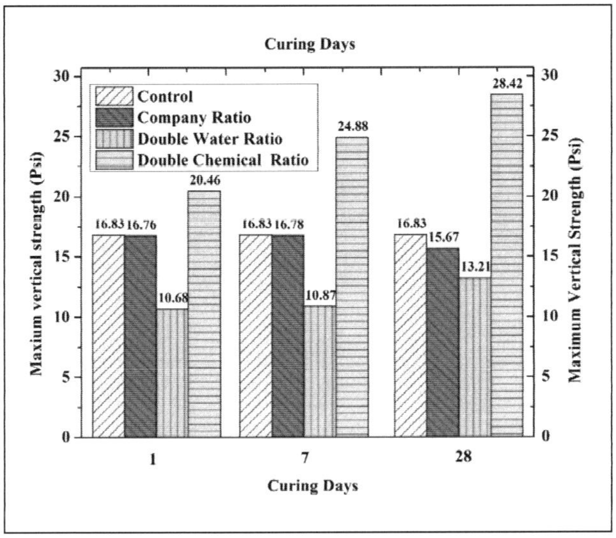

Figure 8. Effect of LISS dosages and curing time on UCS.

morphology is different. The soil particles become intact after adding LISS. The higher binding and gel formation in the microstructure may result in higher strength and stiffness of the treated soil (7).

Summary and Conclusions

LISSs are regarded as a new kind of calcium-free stabilizers, which are being applied in Texas. This kind of LISS consists of sulfuric acid, phosphoric acid, citric acid, water, and surfactant. The conclusions obtained from the laboratory tests before and after treatment are summarized as follows:

1. The PL of treated soils decreases with the increase of LISS dosage whereas LL remains relatively unchanged. Thus, PI decreases with the increase of LISS dosage. In contrast, the shrinkage ratio is nearly constant before and after treatment.

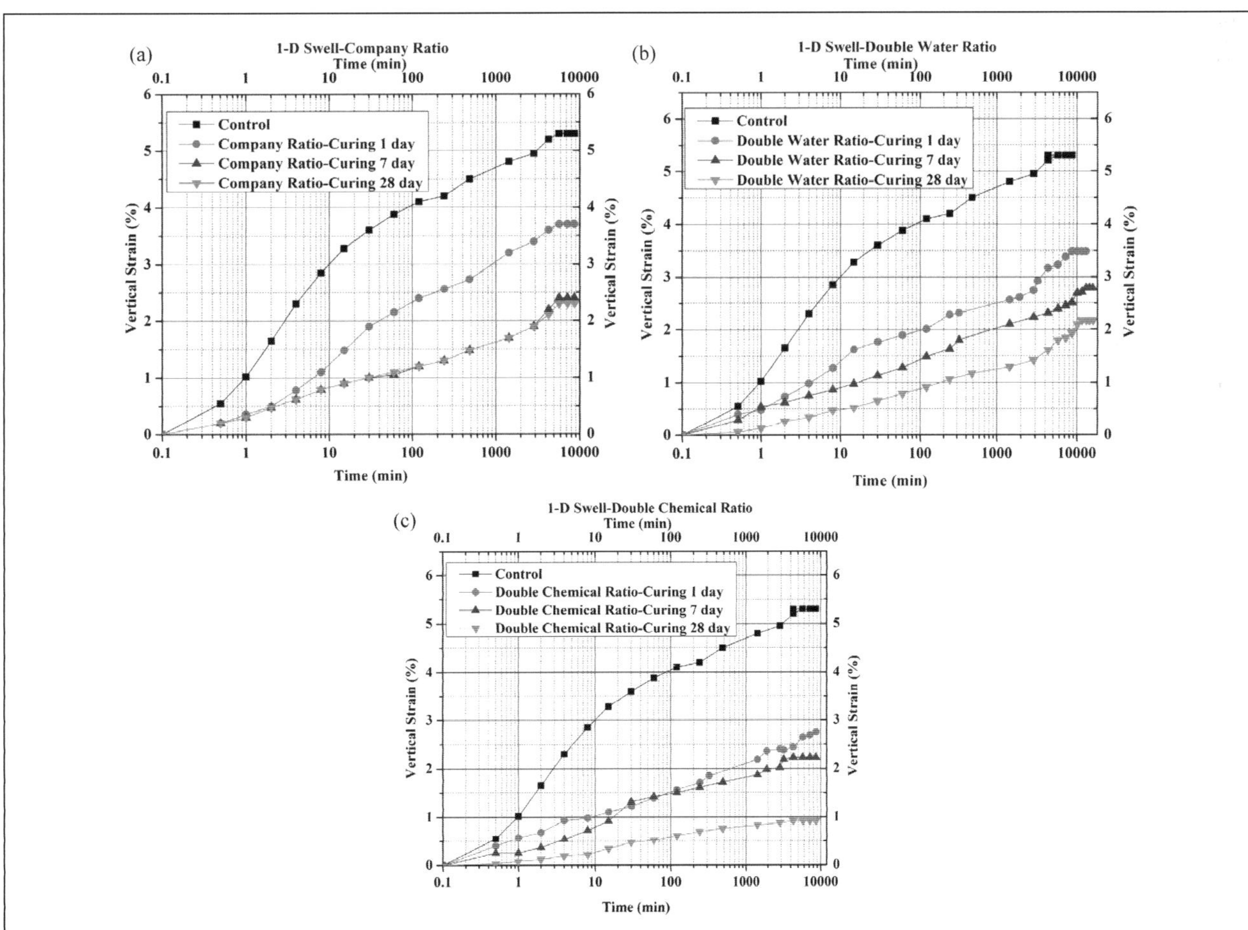

Figure 9. 1D swell test results for Dallas soil with different dosages of LISS treatment.

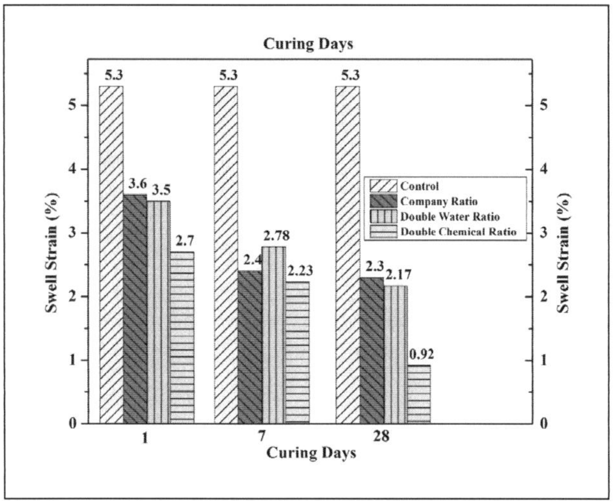

Figure 10. Swell strain percentages of Dallas soil using stabilizer ratios.

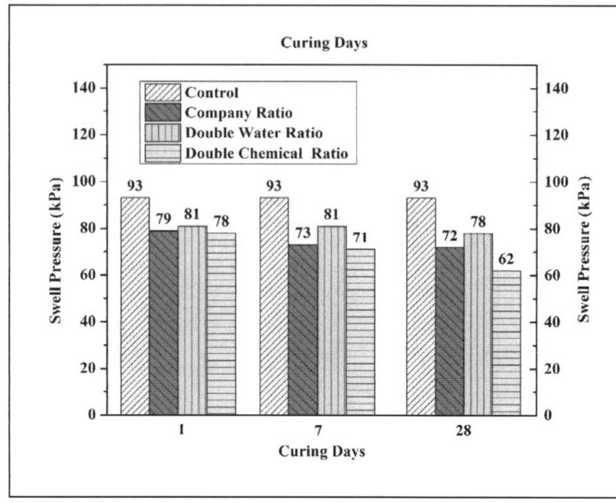

Figure 11. The FESEM images of Dallas soil: (*a*) control sample (0 ml/gallon); (*b*) double water ratio treatment sample (2.5 ml/gallon); (*c*) company ratio treatment sample (5 ml/gallon); (*d*) double chemical ratio treatment sample after 28 days curing (10 ml/gallon).

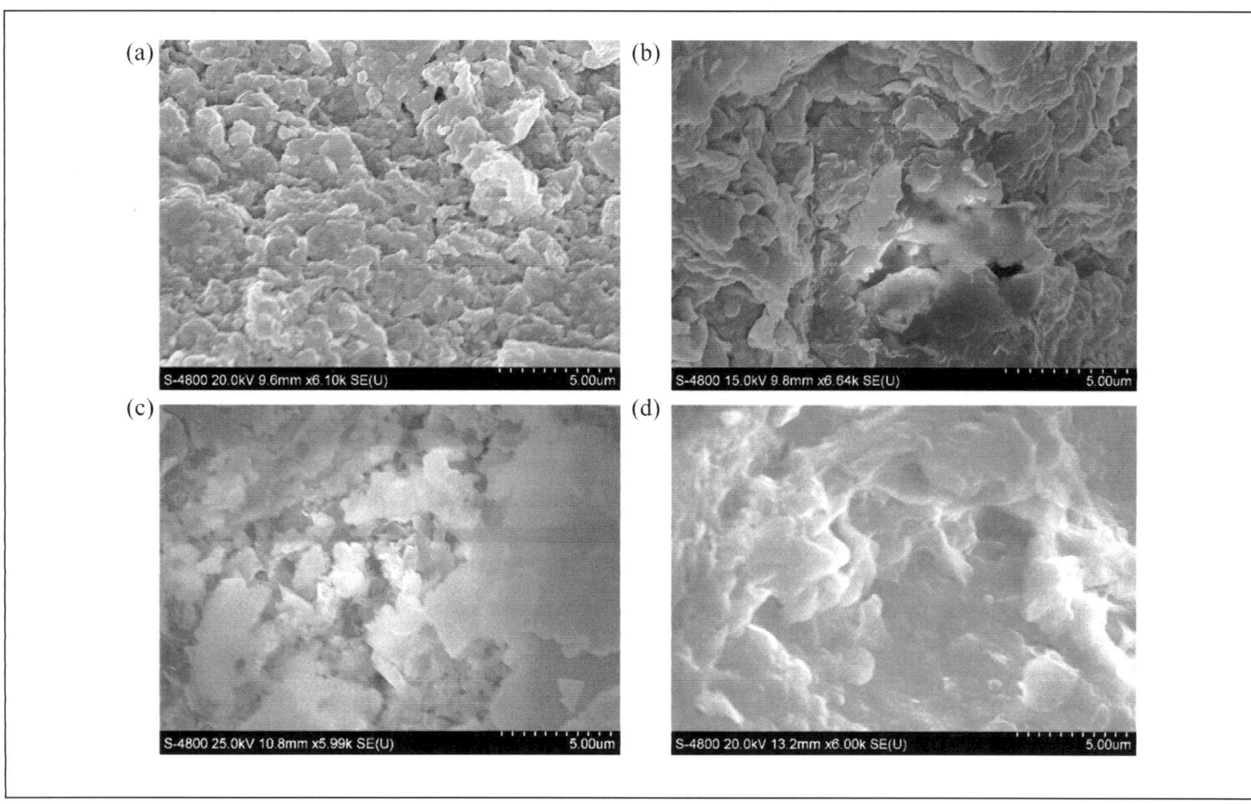

Figure 12. The FESEM images of (*a*) control sample (0 ml/gal), (*b*) double water ratio treatment sample (2.5 ml/gal), (*c*) company ratio treatment sample (5 ml/gal), and (*d*) double chemical ratio treatment sample after 28 days' curing (10 ml/gal).

2. In general, after treatment with LISS, soil strength significantly increased, whereas free swell strain decreased. This revealed that the thickness of the double water-infused layer of soil particles reduced and the bond among soil particles increased.

3. From the FESEM figures, it is evident that the soil particles became much more flocculated and formed a gel after LISS treatment. To some extent, the level of clay flocculation depends on the concentration of LISS. Among various LISS dosage treatments considered in this study, the double chemical ratio treatment not only increased the UCS but also reduced the swell pressure and swell strain. In addition, the PI value dropped to around 50% when compared with those PI values of the control specimens. Hence, for field applications, double chemical ratio treatment is recommended.

Although treatment with LISS reduces the swell strain and increases the strength of tested soils in the lab, additional research is necessary to compare the stabilized soil in the field with lab results. Furthermore, a comparison study between calcium-based stabilizers and LISS should be carried out to better understand the benefits of LISS.

Acknowledgments

The authors acknowledge their appreciation for Ben Baker of TX Prochemical Soil Stabilization, Inc., who provided the LISS and sampled the soil for this study. Technical training and guidance in FESEM from Jiechao Jiang of University of Texas at Arlington are gratefully acknowledged.

Author Contributions

The authors confirm contribution to the paper as follows: study conception and design: XY, SH, AP; data collection: SH; analysis and interpretation of results: SH, XY, AB; draft manuscript preparation: SH, XY, AB, AP. All authors reviewed the results and approved the final version of the manuscript.

References

1. Jones, L. D., and I. Jefferson. Expansive Soils. In J. Burland (ed.), *ICE Manual of Geotechnical Engineering, Vol. 1: Geotechnical Engineering Principles, Problematic Soils, and Site Investigation*. ICE Publishing, London, UK, 2012, pp. 413–441.
2. Punthutaecha, K., A. J. Puppala, S. K. Vanapalli, and H. Inyang. Volume Change Behaviors of Expansive Soils Stabilized with Recycled Ashes and Fibers. *Journal of Materials in Civil Engineering*, Vol. 18, No. 2, 2006, pp. 295–306.
3. Jones, D. E., and K. A. Jones. Treating Expansive Soils. *Civil Engineering—ASCE*, Vol. 57, No. 8, 1987, pp. 62–65.
4. Rauch, A., J. Harmon, L. Katz, and H. Liljestrand. Measured Effects of Liquid Soil Stabilizers on Engineering Properties of Clay. *Transportation Research Record: Journal of the Transportation Research Board*, 2002. 1787: 33–41.
5. Katz, L., A. Rauch, H. Liljestrand, J. Harmon, K. Shaw, and H. Albers. Mechanisms of Soil Stabilization with Liquid Ionic Stabilizer. *Transportation Research Record: Journal of the Transportation Research Board*, 2001. 1757: 50–57.
6. Puppala, A. J., N. Talluri, and B. C. S. Chittoori. Calcium-Based Stabiliser Treatment of Sulfate-Bearing Soils. *Proc., Institution of Civil Engineers-Ground Improvement*, Vol. 167, No. 3, ICE Publishing, London, UK, 2014, pp. 162–172.
7. Zhang, M., H. Guo, T. El-Korchi, G. Zhang, and M. Tao. Experimental Feasibility Study of Geopolymer as the next-Generation Soil Stabilizer. *Construction and Building Materials*, Vol. 47, 2013, pp. 1468–1478.
8. Das, J. T., A. J. Puppala, T. V. Bheemasetti, L. A. Walshire, and M. K. Corcoran. Sustainability and Resilience Analyses in Slope Stabilisation. *Proc., Institution of Civil Engineers - Engineering Sustainability*, Vol. 171, ICE Publishing, London, UK, 2016, pp. 1–12. https://doi.org/10.1680/jensu.16.00054.
9. Rauch, A. F., L. E. Katz, and H. M. Liljestrand. AN Analysis of the Mechanisms and Efficacy of Three Liquid Chemical Soil Stabilizers. Research Report 1993-1. Center for Transportation Research, University of Texas at Austin, Austin, TX, 2003, pp. 1–204.
10. Adams, A. G., O. M. Dukes, W. Tabet, A. B. Cerato, and A. Miller. Sulfate Induced Heave in Oklahoma Soils Due to Lime Stabilization. GeoCongress 2008, ASCE, Reston, VA, 2008, pp. 444–451.
11. Zhang, M., M. Zhao, G. Zhang, P. Nowak, A. Coen, and M. Tao. Calcium-Free Geopolymer as a Stabilizer for Sulfate-Rich Soils. *Applied Clay Science*, Vol. 108, 2015, pp. 199–207.
12. Tingle, J., J. Newman, S. Larson, C. Weiss, and J. Rushing. Stabilization Mechanisms of Nontraditional Additives. *Transportation Research Record: Journal of the Transportation Research Board*, 2007. 1989: 59–67.
13. Scholen, D. E. *Non-Standard Stabilizers*. FHWA-FLP-92-011, Federal Highway Administration, Washington, D.C., 1992.
14. Xiang, W., D. Cui, Q. Liu, X. Lu, and L. Cao. Theory and Practice of Ionic Soil Stabilizer Reinforcing Special Clay. *Journal of Earth Science*, Vol. 21, No. 6, 2010, pp. 882–887.
15. Liu, Q., W. Xiang, W. Zhang, and D. Cui. Experimental Study of Ionic Soil Stabilizer-Improves Expansive Soil. *Rock and Soil Mechanics*, Vol. 8, 2009, p. 16.
16. Ali, F. Stabilization of Residual Soils Using Liquid Chemical. *Electronic Journal of Geotechnical Engineering*, Vol. 17, 2012, pp. 115–126.
17. Banerjee, A. Response of Unsaturated Soils under Monotonic and Dynamic Loading over Moderate Suction States. Doctoral dissertation, University of Texas at Arlington, Arlington, TX, 2017.
18. Banerjee, A., A. J. Puppala, U. D. Patil, L. R. Hoyos, and P. Bhaskar. A Simplified Approach to Determine the Response of Unsaturated Soils Using Multistage Triaxial Test. *IFCEE* 2018: Advances in Geomaterial Modeling and Site Characterization, GSP 295, pp. 332–342.
19. Chen, F. H. *Foundations on Expansive Soils*. Elsevier, New York, NY, 2012.
20. Kalinski, M. E. *Soil Mechanics: Lab Manual*. John Wiley & Sons, Hoboken, NJ, 2011.
21. Little, D. N. Evaluation of Structural Properties of Lime Stabilized Soils and Aggregates, Vol. 3: Mixture Design

and Testing Protocol for Lime Stabilized Soils. *National Lime Association*, Arlington, VA, 2000. https://www.lime.org/documents/other/SOIL3A.pdf. Accessed May 10, 2017.

22. Sweeney, D. A., D. K. H. Wong, and D. G. Fredlund. Effect of Lime on a Highly Plastic Clay with Special Emphasis on Aging. *Transportation Research Record: Journal of the Transportation Research Board*, 1988. 1190: 13–23.

The Standing Committee on Stabilization of Geomaterials and Recycled Materials (AFS80) peer-reviewed this paper (18-05146).

Article

TRR
JOURNAL OF THE TRANSPORTATION RESEARCH BOARD

Transportation Research Record
2018, Vol. 2672(52) 195–204

DOI: 10.1177/0361198118799167
journals.sagepub.com/home/trr
SAGE

Use of Real-Time Dust Monitoring and Surface Condition to Evaluate Success of Unpaved Road Treatments

Bethany K. Kunz[1], Nicholas S. Green[1,2], Janice L. Albers[1,3], Mark L. Wildhaber[1], and Edward E. Little[1]

Abstract

Fugitive dust from unpaved roads creates human health hazards, degrades road surfaces, and increases the cost of road maintenance. As a result, many different chemical treatments are applied to unpaved roads in an attempt to control dust and stabilize the wearing course. However, investigations of the effectiveness of these treatments have often been poorly planned or executed. The objective of this study was to use a combination of real-time dust monitoring and objective road condition evaluations to assess the success of two chemical treatments for a period of 19 months post-application, to provide quantitative information in support of road management decisions. Dust production from road sections treated with calcium chloride-based durablend-C™ or the synthetic fluid EnviroKleen® was monitored on five dates using a vehicle-mounted particulate matter meter. Both products reduced dust by up to 99% relative to an untreated control section during the monitoring period, and quantitative data from the meter were consistent with qualitative observations of dust conditions. Linear models of dust production indicated that road treatment and humidity explained 69% of the variation in dust over time. Road sections treated with either product developed less rutting and fewer potholes than the untreated control. Overall, the combination of real-time dust monitoring and surface condition evaluation was an effective approach for generating quantitative data on endpoints of interest to road managers.

Fugitive dust from unpaved roads decreases visibility for drivers and can represent a health hazard for those traveling on or living near roads. Not surprisingly, unacceptable levels of dust are one of the most common complaints of the public to road managers. The loss of fine material from the road surface also contributes to the development of surface distresses and increases costs for road maintenance. For these reasons, various chemical stabilizers (e.g., chlorides, lignosulfonates, synthetic polymer emulsions, and synthetic fluids) have been applied to unpaved roads since at least the 1960s (*1*).

Although there have been numerous product trials and research studies on unpaved road chemical treatments in the past 50 years, many of these efforts have suffered from insufficient monitoring, a lack of untreated controls for comparison, and a lack of reporting (*1, 2*). Even well-planned and executed studies show great variability in methods of evaluating performance. Some studies have focused solely on engineering endpoints (dynamic cone penetrometer and falling-weight deflectometer measurements among others [e.g., *3*]), whereas other studies have reported only subjective observations. Neither of these categories is ideal for informing road management decisions and justifying those decisions to the public. In order to maximize the value of field trials and research studies, performance endpoints should reflect the major concerns of road managers—dust production and surface distresses that lead to aggregate loss and increased maintenance (*4*).

Many previous studies of road dust production have employed visual ratings of dust. Although such ratings can generate useful comparisons, they cannot estimate mass concentrations of particles to which road users or roadside organisms might be exposed. From a research standpoint, quantitative measurement of dust is a much more powerful approach. Measuring road dust in the field, however, can be problematic. Passive methods such

[1]U.S. Geological Survey, Columbia Environmental Research Center, Columbia, MO
[2]Waterborne Environmental, Inc., Leesburg, VA
[3]Michigan State University, East Lansing, MI

Corresponding Author:
Address correspondence to Bethany K. Kunz: bkunz@usgs.gov

as dust collection pans tend to suffer from extreme variability, as well as interference from humans and animals (*5*). Active methods such as vehicle-mounted techniques (*2, 6, 7*) have been used successfully, but these methods also face challenges, including lack of widespread access to custom-built equipment.

A quantitative, yet relatively simple and economical approach is needed to evaluate the success of chemical treatments for unpaved roads. Such an approach would allow road managers to assess in a reliable way whether or not a particular treatment is appropriate for their road setting, and to effectively justify their road management decisions. The objective of this project was to provide a rigorous, systematic assessment of the success of two unpaved road chemical treatments for 19 months post-application. Success was evaluated through a combination of real-time dust monitoring using a commercially available meter and an objective road condition rating procedure. Together, these methods allow road managers to quantitatively assess two of the most important endpoints for unpaved roads—dust production and surface distresses.

Methods

Test Site

The Loess Bluffs National Wildlife Refuge (NWR), formerly Squaw Creek NWR, is a 3,011-ha (7,440-acre) complex of wetlands, grasslands, and forest in northwest Missouri. Designated a Globally Important Bird Area by Birdlife International, Loess Bluffs hosts more than a million snow geese, waterfowl, and shorebirds each year. The primary road on the refuge is a 16-km (10-mi) auto tour loop that runs adjacent to managed wetland units. Historically, the refuge has had moderate to severe issues with dust on the auto tour loop, particularly associated with special events that can bring thousands of visitors to the refuge on a single day. Dust on the auto tour loop creates problems for wildlife viewing and photography, which are priority refuge uses. Refuge management has also received complaints from recreational bikers and visitors with respiratory issues.

In an effort to reduce dust and stabilize the road surface on the auto tour loop, two dust control products were applied: durablend-C™, a polymer-enhanced calcium chloride (EnviroTech Services, Greeley, CO), and EnviroKleen®, a synthetic fluid with binder (Midwest Industrial Supply, OH). Each of these products had demonstrated low aquatic toxicity in previous laboratory tests, and was expected to pose minimal risk to the aquatic habitats along the road. Prior to application, product vendors were provided a summary of road geometry, traffic patterns, weather conditions, and aggregate composition and gradation to ensure that the selected products were appropriate for installation at the refuge.

Test Design and Product Applications

The two products were each applied to two different 0.8-km (0.5-mi) sections of the Loess Bluffs NWR auto tour loop in July 2014, and an additional 0.8-km (0.5-mi) section was left untreated as a control (Figure 1). These road sections were selected for consistency of conditions that could influence dust production, namely, canopy cover, orientation relative to prevailing winds, and drainage. Based on discussions with refuge staff, several sections of the auto tour loop were excluded from consideration because of planned construction projects or vulnerability to flooding. These limitations precluded the inclusion of a second untreated control section. Because traffic on the loop is one-way and there are no entrances/exits between the selected sections, all sections should have experienced the same traffic (predominantly light passenger vehicles, with occasional heavy equipment use by refuge staff). During the test period, refuge traffic counters recorded average daily traffic of 71 vehicles, with seasonal peaks of up to 505 vehicles per day.

Prior to product applications, all sections including the untreated control received new crushed limestone surface aggregate (Missouri Department of Transportation [MoDOT] Type I Base) to ensure uniform road conditions. Both products were applied according to vendor specifications, with on-site supervision from vendor technical advisors. For durablend-C™, the road was pre-wet with a water truck, cut with the grader, and then the product was incorporated into the top ~7 cm (~3 in.) of the road surface (i.e., a mixed-in application) at a rate of 2.26 L/m^2 (0.5 gal/yd^2). The road was then shaped and compacted, and a topical application of product at 1.81 L/m^2 (0.4 gal/yd^2) was applied. For EnviroKleen®, the road was pre-wet, cut, shaped, and compacted in the same way, and EnviroKleen® was applied topically at 1.36 L/m^2 (0.3 gal/yd^2). Twenty-four hours later, the EnviroKleen section received final compaction. Applications of both products were performed by specialized trucks with computerized spray systems provided by the vendors. The EnviroKleen® section also received a lighter maintenance application (0.68 L/m^2; 0.15 gal/yd^2) 10 months after initial application. All road preparation steps, application procedures, and maintenance application procedures (or lack thereof) were specified by the product vendors to represent a "typical" application scenario. Therefore, applications of the two products were not standardized, but should be representative of conditions as used in the field.

Dust Measurement

Dust production on each treated and untreated road section was measured at 2, 3, 6, 11, and 19 months after the initial product applications with a DustTrak DRX Aerosol Monitor (Model 8533, TSI Incorporated,

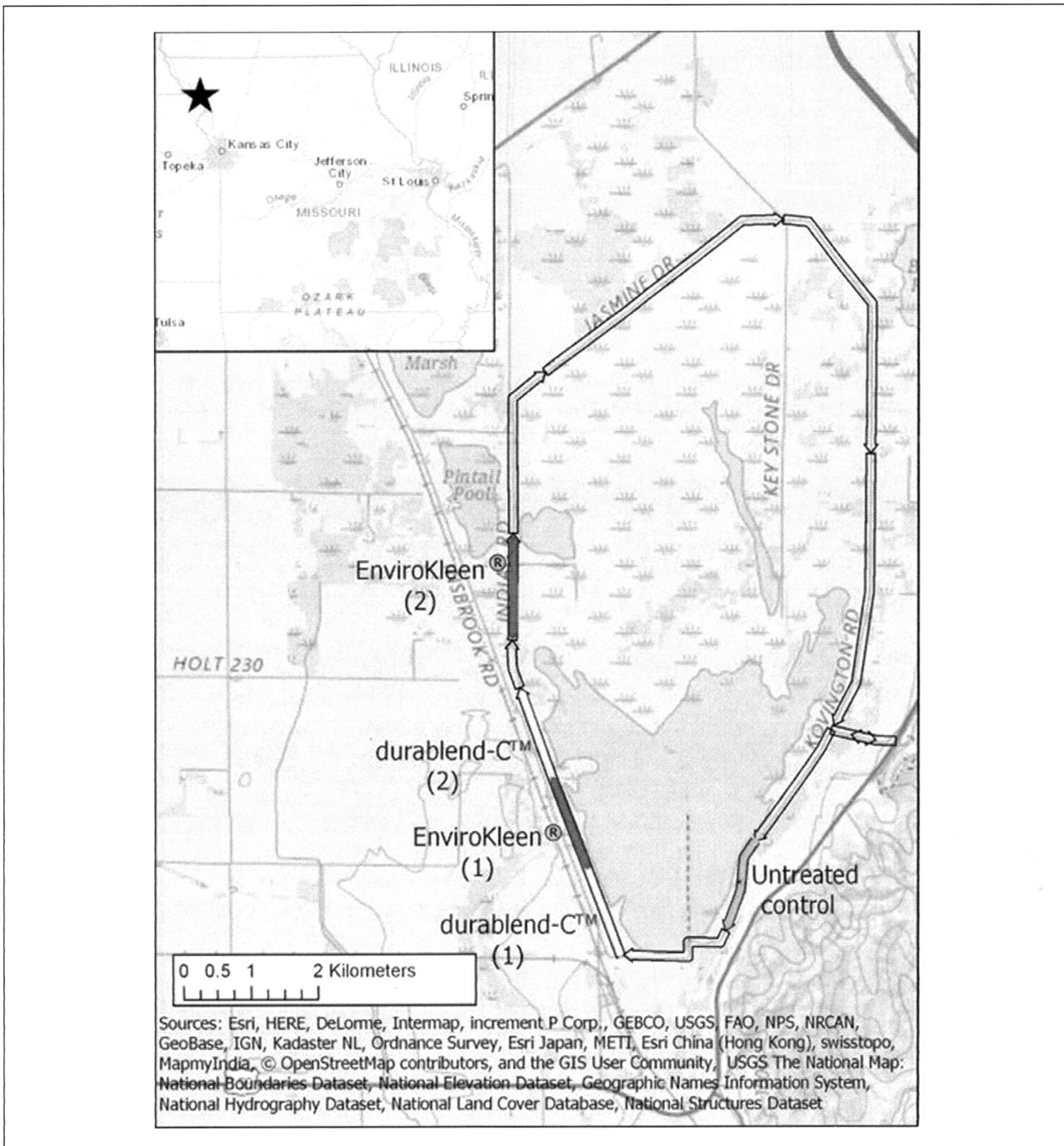

Figure 1. Layout of experimental treatment sections on the Loess Bluffs NWR one-way auto tour loop. Traffic enters from the east side and arrows show direction of travel. Each treatment section = 0.8 km (0.5 mi).

Shoreview, MN). The DustTrak DRX was chosen because of its portability, the ease with which it could be vehicle-mounted, and its ability to simultaneously infer the mass concentration (mg/m^3) of particulate matter in five size ranges from PM_1 to total particulate matter (Table 1). The same or similar models of laser photometers have been used successfully in both vehicle-mounted applications (*7–10*) and stationary applications (*7*). The DustTrak DRX was mounted on the tailgate of a refuge truck inside a weatherproof Storm Case™

Table 1. Size Ranges of Particles Reported by the DustTrak DRX Meter

Size	Particle diameter (μm)
PM_1	0.10–1.0
$PM_{2.5}$	0.10–2.5
PM_{resp}	0.10–4.0
PM_{10}	0.10–10.0
PM_{total}	0.10–15.0

(iM2450, Pelican Products, Inc., CA) with holes drilled for the intake and exhaust lines. The intake line was secured horizontally 1 m (3.28 ft) above the road surface, which is the height at which peak PM10 exposure is expected (*11*). On each sampling date, each road section was driven three times with a DustTrak sampling rate of 1 sample/second, yielding three dust profiles per section. All measurements were taken with the DustTrak mounted on the same refuge truck, with the same driver for all measurements on a sampling date. All sections were driven in the direction of prevailing traffic on the one-way tour loop. All measurements were taken at a speed of 40 kph (25 mph) according to previous recommendations (*8, 12*), with smooth accelerations and decelerations at boundaries between treatment sections. Any passing cars or other potential influences on dust measurements were recorded in field notes for each sampling run. On each sampling date, the project team also made qualitative observations and took digital photographs of dust production on each section.

Data were plotted and compared with field notes to determine quality assurance and control. In two cases, data points were removed from the beginning or end of a run to reflect delays mentioned in the field notes. On the 29 October, 2014, sampling date, the measurements from one run of two sections (one durablend-C™ and one EnviroKleen®) were excluded from analysis because of a meter error. Measurements from each run within a section were averaged to yield a value representing the run. Then, the run-level means from within each sampling date were averaged to yield a value representing that section and sampling date. Because dust data from all five particulate size ranges were strongly right-skewed, all data were $\log_{10}$-transformed prior to analysis to improve normality.

Weather Data

Weather data for the monitoring period (1 June, 2014 to 29 February, 2016) were downloaded from the closest National Oceanic and Atmospheric Administration weather station, which was located at Brenner Field Airport, Falls City, NE (WBAN:94957), 27.5 km (17.1 mi) west of the refuge. To explore the possible influence of weather conditions on dust production, three weather variables representing precipitation, wind speed, and humidity were selected. For each sampling date, precipitation was characterized by days since precipitation >0.03 cm (0.01 in.). Wind speed was characterized as the daily mean wind speed (km/h) for each sampling date, and humidity was characterized as the daily mean relative humidity (%) for each sampling date.

Objective Road Condition Evaluation

In addition to dust production over time, road condition was used as a measure of success for the chemical treatments. Road condition was assessed on 21 December, 2015, approximately 18 months after the initial product applications. At the time of evaluations, no maintenance had been required on any treated section since the product applications. On the untreated control section, approximately one-third of the section length had recently required grading. Objective road condition evaluations followed those used by the Central Federal Lands Highway Division at a road stabilization project at Seedskadee NWR (*13*). These evaluations focused on the same categories of surface distresses as other established unpaved road assessment systems, such as the U.S. Army Corps of Engineers' unsurfaced road condition index (URCI [*14*]) and Wisconsin's gravel pavement surface evaluation and rating (PASER) system (*15*), but used a rating scale optimized to distinguish among roads in relatively good condition, rather than across the full spectrum of "failed" to "excellent."

Each treated and control road section was sampled using three stops, at the 0.1-, 0.2-, and 0.3-mi mark (0.16-, 0.32-, and 0.48-km mark), measured from the beginning of each section. At each stop, a team of two observers measured the depth and extent of four categories of surface distresses—washboarding, raveling, rutting, and potholes—present in a 7.6-m (25-ft) length of road. Washboarding was measured as the depth (in mm) of six consecutive troughs. Raveling was measured as the depth (mm) of loose material at four locations across the width of the road—on the outside and inside edges of each wheel path. Rutting was measured as the depth of depression (mm) in the right and left wheel paths, and potholes were measured as the number and average depth (mm) of potholes in the 7.6-m (25-ft) length. In addition, the total number of potholes observed in each 0.8-km (0.5-mi) road section was recorded. All measurements were subsequently averaged and converted into ratings (0 to 10, with 10 = best condition) to provide an intuitive basis for comparison of sections, according to scales in Appendix B of Woll et al. (*13*). For example, a section with an average washboarding trough depth of 23 mm (0.9 in.) would be assigned a rank of 5.

Statistical Analyses

Dust monitoring data from the five sampling dates were used to generate linear models of dust production. The purpose of these models was to evaluate whether variation in dust measurements over time was associated with application of chemical treatments, with weather conditions on each day of sampling, or with some combination thereof. The analysis also allowed a rigorous assessment of whether chemically treated sections differed from each other and from the untreated control section in dust production.

Correlations. Linear correlation (Pearson's product moment correlation coefficient, *r*) was used to test for correlations between the five measures of dust production (PM_1, $PM_{2.5}$, PM_{resp}, PM_{10} and PM_{total}; Table 1), and between continuous weather covariates. Correlations between dust particle sizes were tested because if two dust production variables were highly correlated with each other (correlation coefficient near 1 or −1), separate analyses of both measures would be redundant. Correlations between potential predictor variables were also tested to ensure that models did not include redundant predictor variables.

Dust Production. Linear models were used to test for a relationship between dust production (response variable) and road treatment (control vs. treated), and four additional predictor variables: time, precipitation, wind speed, and relative humidity. Random intercept models were considered to account for variability between individual sections (*16, 17*); however, preliminary tests showed that the random effect due to section was zero and therefore random effects were excluded. All models with more than one predictor included treatment as a factor ($X_{treatment}$) and either days since 1 June, 2014 (X_{time}), days since precipitation (X_{since}), mean daily wind on sampling date in km/h (X_{wind}), or mean relative humidity on sampling date ($X_{humidity}$) as a continuous covariate (Table 2). The model with the lowest Akaike information criterion value corrected for small sample sizes (AICc [*18*]) was selected as the most parsimonious (i.e., best-fitting) model. Type-III analysis of variance (ANOVA) was used to assess model significance to ensure that comparisons were made in light of other model terms, rather than sequentially (*19*). The use of the linear model with Type-III sums of squares also accounted for the unbalanced sampling design with respect to road treatment (one untreated control section versus two treated sections for each product). Tukey's honest significant difference (HSD) post hoc test was used to test which road treatments differed from each other.

Table 2. Models of Dust Production Tested

Model	Form
0	$y = Constant$
1	$y = \beta_0 + \beta_1 X_{treatment}$
2	$y = \beta_0 + \beta_1 X_{time}$
3	$y = \beta_0 + \beta_1 X_{since}$
4	$y = \beta_0 + \beta_1 X_{wind}$
5	$y = \beta_0 + \beta_1 X_{humidity}$
6	$y = \beta_0 + \beta_1 X_{treatment} + \beta_2 X_{time}$
7	$y = \beta_0 + \beta_1 X_{treatment} + \beta_2 X_{since}$
8	$y = \beta_0 + \beta_1 X_{treatment} + \beta_2 X_{wind}$
9	$y = \beta_0 + \beta_1 X_{treatment} + \beta_2 X_{humidity}$
10	$y = \beta_0 + \beta_1 X_{treatment} + \beta_2 X_{time} + \beta_3 X_{treatment} X_{time}$
11	$y = \beta_0 + \beta_1 X_{treatment} + \beta_2 X_{since} + \beta_3 X_{treatment} X_{since}$
12	$y = \beta_0 + \beta_1 X_{treatment} + \beta_2 X_{wind} + \beta_3 X_{treatment} X_{wind}$
13	$y = \beta_0 + \beta_1 X_{treatment} + \beta_2 X_{humidity} + \beta_3 X_{treatment} X_{humidity}$

Note: In these models, y = total dust production (mg/m^3); $X_{treatment}$ = treatment (control, durablend-C™, or EnviroKleen®); X_{time} = days since June 1, 2014; X_{since}= is days since precipitation ⩾0.03 cm (0.01 in.); X_{wind} = mean wind speed on sampling date (km/h); $X_{humidity}$ = mean relative humidity on sampling date (%). β_0 = model intercept; β_1, β_2, and β_3 = regression coefficients.

Software. All analyses were performed using R version 3.3.3 (R Core Team 2017). Within R, package "MuMIn" (*20*) was used to calculate AICc values, package "car" version 2.1-4 (*19*) was used for Type-III ANOVAs and package "multcomp" version 1.4-6 was used for Tukey tests (*21*).

Results and Discussion

Dust Measurement

Overall, quantitative dust monitoring data indicated that treatment with either product reduced dust by up to 99% for the first 11 months of the study relative to the untreated control (Figure 2). Although absolute dust levels varied by sampling date, the reductions associated with treatment were relatively consistent over time, with greater dust suppression by durablend-C™ than EnviroKleen® on most dates (Figure 2). By the final monitoring visit (19 months after initial applications), dust levels on EnviroKleen® sections were 71% lower than the control, and those on durablend-C™-treated sections were 93% lower than the control. The fact that control efficacy did not decline in a consistent manner over the course of the study precludes plotting of deterioration curves (*22*) or drawing definitive conclusions about the effective lifespan of the two products. However, based on dust reductions at the end of the monitoring period and the difference in application history (one initial application for durablend-C™ versus an initial application and a maintenance dose for EnviroKleen®), durablend-C™

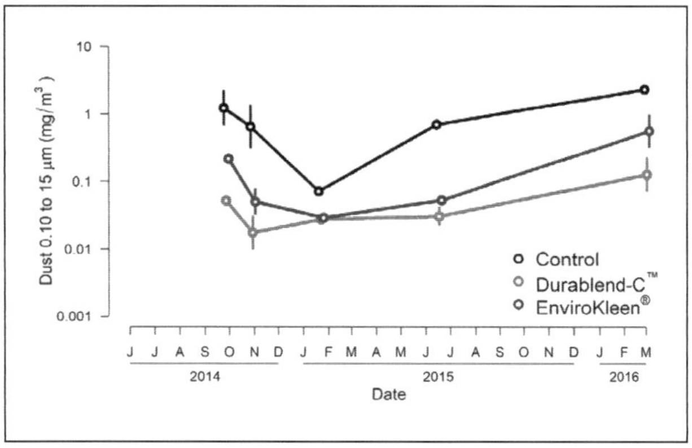

Figure 2. Dust production (PM_{total}; 0.10 to 15 µm) over time on untreated and treated road segments. Points represent means; error bars represent 95% confidence intervals. Note that the y-axis scale is logarithmic. For clarity, points for control treatment are offset 3 days to the left and points for EnviroKleen® are offset 3 days to the right.

provided longer-lasting dust control under the conditions of the current study.

Importantly, the results for dust production based on real-time dust monitoring with the DustTrak DRX are consistent with qualitative observations of dustiness by project researchers. Reductions in dust associated with road treatment were clear when driving the auto tour loop (Figure 3). Improvements in driving conditions were even more evident when compared with an adjacent section of the auto tour loop that did not receive new aggregate or chemical treatment as part of the study (i.e., business-as-usual conditions).

Objective Road Condition Evaluation

The mean depths of each type of surface distress (washboarding, raveling, rutting, and potholes) and the associated ratings are presented in Table 3. Each road section is listed individually to highlight any variation between replicate sections of the same treatment. Overall, all sections, including the untreated control, performed relatively well, with minimal washboarding and limited raveling. Relative to the treated sections, the untreated control exhibited moderately more rutting and substantially more potholes. The untreated section was characterized by both greater numbers and greater depths of potholes, as reflected in the pothole rating. In general, condition ratings for replicate sections of the same treatment were consistent. One exception was the EnviroKleen® (2) section, which developed less rutting and more potholes than the EnviroKleen® (1) section.

Figure 3. Representative dust levels 15 months after original product applications (5 months after maintenance application of EnviroKleen®): (*a*) durablend-C™- treated section; (*b*) EnviroKleen®-treated section; (*c*) untreated control section; (*d*) an adjacent road section on the auto tour loop that did not receive new surface aggregate or chemical treatment as part of the study, shown for reference. Photographs were taken from the back of a passenger truck traveling 40kph (25mph). Photo credit: B. Kunz, USGS.

Table 3. Objective Road Condition Evaluation

Section	Washboarding mean depth (mm)	Washboarding rating[a]	Raveling mean depth (mm)	Raveling rating	Rutting mean depth (mm)	Rutting rating	Pothole mean depth (mm)	Number of potholes in overall section	Pothole rating	Overall rating
Untreated control	2.4	9	8.9	8	17.0	6	36.7	19	5	7.0
durablend-C™ (1)	0.6	9	6.3	8	7.3	8	0	3	9	8.5
durablend-C™ (2)	0	10	8.8	8	7.5	8	0	4	9	8.8
EnviroKleen® (1)	0	10	9.2	8	11.0	7	0	4	9	8.5
EnviroKleen® (2)	0.7	9	4.2	9	4.3	9	70[b]	13	7	8.5

Note: 25.4 mm = 1 in.
[a]All ratings on a scale of 0 to 10, with 10 = best condition.
[b]A single deep pothole in one of the 7.6-m (25-ft) lengths. Because this depth was not characteristic, rating was based on other potholes within the overall section.

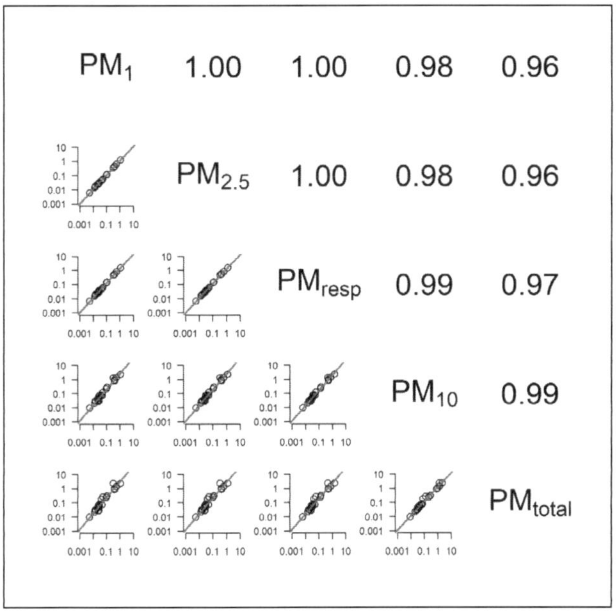

Figure 4. Correlations between five measures of dust production. Variables and associated dust size ranges are as follows: PM_1 (0.1 to 1.0 µm); $PM_{2.5}$ (0.1 to 2.5 µm); PM_{resp} (0.1 to 4.0 µm); PM_{10} (0.1 to 10 µm); and PM_{total} (0.1 to 15 µm). Numbers in panels above the diagonal show Pearson product moment correlation coefficient (r). Red lines show linear relationship between the variables. Note that all axes are on a $\log_{10}$ scale.

Statistical Analyses

Correlations. All five measures of dust production were significantly positively correlated with each other (all $p < 0.0001$; Figure 4). In particular, the correlations between total particulate matter (PM_{total}) and particulate matter <2.5 µm ($PM_{2.5}$) and <10 µm in diameter (PM_{10}) were 0.96 and 0.99, respectively. Therefore, at least for the surfacing aggregate and conditions tested in this study, PM_{total} is a useful surrogate for particulate matter in the two size ranges of greatest concern for human health. Because analyzing all five highly correlated size fractions separately would be redundant, only analyses of total dust production are presented.

All combinations of continuous covariates (time and weather variables) were significantly correlated (all $p < 0.0001$; data not shown). Therefore, no model was tested with more than one continuous covariate.

Dust Production. Over the 19 months of the study, the best model of dust production was Model 9, in which dust varied with road treatment and relative humidity on the sampling date (Table 4). In this model, treatment and humidity explained 68.5% of variation in total dust production (Tables 4 and 5). Dust production decreased with increasing humidity (Figure 5). This result is not surprising, given the effect of road bed moisture on dust production (*23*). Models that included days since precipitation were not supported in this study. This result is in contrast with an earlier study that documented a trend of increasing dust for at least 7 days after a rainfall event (*10*). The stronger relationship with humidity than days since precipitation in the current study may have been a function of seasonal differences in sampling or geographic differences in climate.

Dust production was significantly greater on control sections than on durablend-C™ or EnviroKleen® sections. Differences in dust production between durablend-C™ and EnviroKleen® sections over the entire monitoring period were not statistically significant (Table 6).

Benefits and Limitations of the Approach

The greatest benefit of this approach is the quatitative nature of the data generated. For practitioners, these data could be used to set definitive performance measures for chemical treatments (e.g., "80% reduction in average dust production for 6 months"), or make more

Table 4. Dust Production Model Selection Results

Model	F	DF_{model}	DF_{res}	P	R^2	$AICc$	$\Delta AICc$	$AICc_{wt}$
9	20.0988	3	21	<0.0001	0.6852	31.7961	0	0.9223
13	2.2931	5	19	0.0864	0.6640	38.3524	6.5564	0.0348
6	13.0844	3	21	<0.0001	0.5823	38.8688	7.0727	0.0269
1	12.6412	2	22	0.0002	0.4914	41.7939	9.9979	0.0062
8	9.9545	3	21	0.0003	0.5236	42.1561	10.3601	0.0052
7	9.4374	3	21	0.0004	0.5064	43.0439	11.2479	0.0033
10	2.2727	5	19	0.0886	0.5474	45.8032	14.0072	0.0008
12	0.9689	5	19	0.4614	0.4946	48.5586	16.7626	0.0002
11	3.9258	5	19	0.0130	0.4920	48.6876	16.8915	0.0002
5	7.4276	1	23	0.0121	0.1556	52.7248	20.9288	0
2	30.3341	1	23	<0.0001	0.0616	55.3629	23.5669	0
4	8.6603	1	23	0.0073	0.0080	56.7515	24.9554	0
3	31.3855	1	23	<0.0001	−0.0077	57.1447	25.3486	0

Note: F = f statistic in omnibus (Type-III ANOVA) test; DF_{model} = degrees of freedom in the model; DF_{res} = degrees of freedom in the residuals; P = p-value in omnibus test; R^2 = coefficient of determination; $AICc$ = Akaike's information criterion corrected for small sample sizes; $\Delta AICc$ = difference from lowest AICc; $AICc_{wt}$ = AICc weight. Level of significance for P = 0.05.

Table 5. Type-III ANOVA Table for Final Model of Dust Production

Source	SS	DF	MS	F	P
Intercept	0.7424	1	0.7424	5.0537	0.0354
Treatment	5.9788	2	2.9894	20.3497	<0.0001
Humidity	2.1364	1	2.1364	14.5434	0.0010
Residuals	5.0618	25	0.2025	na	na

Note: na = not applicable; SS = sum of squares associated with each term; DF = degrees of freedom associated with each term; MS = mean squared error; F = f statistic of ANOVA test; P = p-value. Level of significance for P = 0.05.

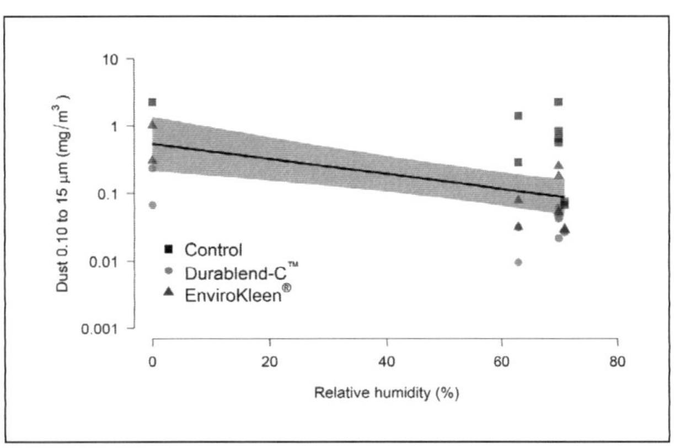

Figure 5. Total dust production (PM_{total}; 0.10 to 15 µm) decreased with increasing relative humidity (%) on untreated and treated road segments. Shaded area indicates 95% confidence interval. Note that the y-axis scale is logarithmic. Points represent samples taken on a particular road segment on a particular date.

compelling decision justifications to the public or funding sources. For researchers, these data could be used to standardize evaluations of relative chemical treatment effectiveness across a range of climatic conditions and geographic locations.

Monitoring with the commercially available DustTrak DRX as described here requires no installation of infrastructure or modification of existing vehicles. The unit can easily be tailgate-mounted and removed at the end of sampling. Although the purchase price is substantial (approx. US$11,000), units can also be rented and shipped to a field site for < US$500/week. It is important to recognize, however, that measurements using this meter provide meaningful comparisons of *relative* dust production on different sections. Because dust production varies with weather, vehicle weight and speed (*23*) among other factors, all sampling for a given set of sites must take place under consistent conditions.

The objective road condition evaluations did require more time than visual (i.e., "windshield") methods. A two-person crew required approximately 10 min to complete each 7.6-m (25-ft) section. However, this measurement-based approach generated more comprehensive data on the severity of surface distresses, while eliminating some of the subjective determinations required by visual methods (e.g., PASER [*15*]). In the

Table 6. Tukey's HSD Post Hoc Test for Differences in Means between Levels of Treatment

Level 1	Level 2	Estimate	SE	*T*	*P*
durablend-C™	Control	−1.3376	0.2099	−6.3720	<0.0001
EnviroKleen®	Control	−0.9415	0.2099	−4.4850	0.0006
EnviroKleen®	durablend-C™	0.3961	0.1714	2.3110	0.0756

Note: Estimate = estimated difference in means ($\mu_{level1} - \mu_{level2}$); SE = standard error of estimate; T = t statistic; P = p-value adjusted for multiple comparisons. Level of significance for P = 0.05.

current study, this method was sensitive enough to detect differences in surface distresses among road sections that were all in relatively good condition. For a study design in which a greater range of performance is expected, a system such as the URCI (*14*) may be more appropriate.

Conclusion

This paper describes an approach for quantitatively evaluating the success of unpaved road treatments. Based on real-time dust monitoring data, road treatment with one application of durablend-C™ suppressed dust by up to 99%, relative to the untreated control. At 19 months post-application, average dust levels remained 93% lower than those on the control section. Treatment with an initial application and a maintenance dose of EnviroKleen® suppressed dust by a similar magnitude for 11 months, with a smaller reduction (71%) relative to the control by the end of the project. Objective road surface condition evaluations were used to quantify the type and severity of road surface distresses, and determine that sections treated with either product according to the vendors' directions developed less rutting and substantially fewer potholes in the first 15 months after application.

Data generated by this quantitative approach also facilitated modeling to determine whether road treatment, weather, or some combination thereof was the most important influence on dust production. Linear models revealed that chemical treatment and relative humidity on the sampling date were the dominant influences on dust production on the different road sections in this study. Understanding the factors controlling dust is critical for designing effective dust monitoring programs.

Both of the techniques described here are relatively simple to implement, and required minimal equipment and training of staff. Although this approach would not be detailed enough for some research studies and would be too time-consuming for general unpaved road management, it provided a useful balance between ease of use and quality of data in the current study.

Overall, a combination of real-time dust monitoring and road surface condition evaluations was an effective approach for determining success of unpaved road treatments. The approach generated quantitative data for comparisons of treatments over time, while retaining more flexibility than some other quantitative methods. These data could be used to develop more definitive performance measures for chemical treatments and should also facilitate better communication regarding justification for road management decisions.

Acknowledgments

Funding was provided by the U.S. Fish and Wildlife Service through Interagency Agreement #4500054314.

Author Contributions

The authors confirm contribution to the paper as follows: study conception and design: BK, EL; data collection: BK, EL; analysis and interpretation of results: NG, JA, MW, BK; draft manuscript preparation: BK, NG. All authors reviewed the results and approved the final version of the manuscript.

References

1. Campbell, A. E., and D. Jones. Soil Stabilization in Low-Volume Roads: Obstacles to Product Implementation from Additive Supplier's Standpoint. *Transportation Research Record: Journal of the Transportation Research Board*, 2011. 2204: 172–178.
2. Jones, D. Holistic Approach to Research into Dust and Dust Control on Unsealed Roads. *Transportation Research Record: Journal of the Transportation Research Board*, 1999. 1652: 3–9.
3. Guthrie, W. S., D. O. Simmons, and D. L. Eggett. Enzyme Stabilization of Low-Volume Gravel Roads. *Transportation Research Record: Journal of the Transportation Research Board*, 2015. 2511: 112–120.
4. Jones, D. Toward Fit-for-Purpose Certification of Road Additives. *Transportation Research Record: Journal of the Transportation Research Board*, 2003. 1819: 208–217.
5. Rosbury, K. D., and R. A. Zimmer. *Cost Effectiveness of Dust Control Use on Unpaved Haul Roads, Vol. 1- Results, Analysis, and Conclusions*. Report PN-3559. U.S. Department of the Interior, Bureau of Mines, Twin Cities Research Center, Minneapolis, MN, 1983.
6. Sanders, T. G., and J. Q. Addo. Experimental Road Dust Measurement Device. *Journal of Transportation Engineering*, Vol. 126, 2000, pp. 530–535.

7. Etyemezian, V., H. Kuhns, J. Gillies, M. Green, M. Pitchford, and J. Watson. Vehicle-Based Road Dust Emission Measurement: I—Methods and Calibration. *Atmospheric Environment*, Vol. 37, 2003, pp. 4559–4571.
8. Edvardsson, K., and R. Magnusson. Monitoring of Dust Emission on Gravel Roads: Development of a Mobile Methodology and Examination of Horizontal Diffusion. *Atmospheric Environment*, Vol. 43, No. 4, 2009, pp. 889–896. https://doi.org/10.1016/j.atmosenv.2008.10.052.
9. Kunz, B. K., and E. E. Little. Dust Control Products at Hagerman National Wildlife Refuge, Texas: Environmental Safety and Performance. *Transportation Research Record: Journal of the Transportation Research Board*, 2015. 2472: 64–71.
10. Kuhns, H., V. Etyemezian, M. Green, K. Hendrickson, M. McGown, K. Barton, and M. Pitchford. Vehicle-Based Road Dust Emission Measurement—Part II: Effect of Precipitation, Wintertime Road Sanding, and Street Sweepers on Inferred PM_{10} Emission Potentials from Paved and Unpaved Roads. *Atmospheric Environment*, Vol. 37, 2003, pp. 4573–4582.
11. Muleski, G. E., A. R. Trenholm, D. L. Gebhart, D. L. Franke, and C. Cowherd. New Method to Measure Control Performance of Dust Palliatives on Unpaved Roads at Federal Facilities. *Federal Facilities Environmental Journal*, Vol. 14, 2003, pp. 23–33.
12. Thenoux, G., J. P. Bellolio, and F. Halles. Development of a Methodology for Measurement of Vehicle Dust Generation on Unpaved Roads. *Transportation Research Record: Journal of the Transportation Research Board*, 2007. 1989: 299–304.
13. Woll, J. H., R. W. Surdahl, R. Everett, and R. Andresen. *Road Stabilizer Product Performance, Seedskadee National Wildlife Refuge*. FHWA-CFL/TD-08-005. FHWA, U.S. Department of Transportation, Lakewood, CO, 2008.
14. Eaton, R. A., and R. E. Beaucham. *Unsurfaced Road Maintenance Management*. Special Report 92-26. United States Army Corps of Engineers—Cold Regions Research and Engineering Laboratory, Hanover, NH, 1992. http://www.dtic.mil/get-tr-doc/pdf?AD=ADA262015 Accessed November 10, 2017.
15. Walker, D. *Gravel PASER Manual: Pavement Surface Evaluation and Rating*. Wisconsin Transportation Information Center, Madison, WI, 1989.
16. Laird, N. M., and J. H. Ware. Random-Effects Models for Longitudinal Data. *Biometrics*, Vol. 38, 1982, pp. 963–974.
17. Zuur, A. F., E. N. Ieno, N. J. Walker, A. A. Saveliev, and G. M. Smith. *Mixed Effects Models and Extensions in Ecology with R*. Springer Science + Business Media, Berlin, Germany, 2009.
18. Burnham, K. P., and D. R. Anderson. *Model Selection and Multimodel Inference: A Practical Information-Theoretic Approach*, 2nd ed. Springer-Verlag, Berlin, Germany, 2002.
19. Fox, J., and S. Weisberg. 2011. *An {R} Companion to Applied Regression*, 2nd ed. SAGE Publications, Thousand Oaks, CA, 2011. http://socserv.socsci.mcmaster.ca/jfox/Books/Companion.
20. Bartoń, K. *MuMIn: Multi-Model Inference*. R Package Version 1.40.0. 2018. https://cran.r-project.org/web/packages/MuMIn/index.html. Accessed January 3, 2018.
21. Hothorn, T., F. Bretz, and P. Westfall. Simultaneous Inference in General Parametric Models. *Biometrical Journal*, Vol. 50, 2008, pp. 346–363.
22. Monlux, S., and M. Mitchell. Chloride Stabilization of Unpaved Road Aggregate Surfacing. *Transportation Research Record: Journal of the Transportation Research Board*, 2007. 1989: 50–58.
23. U.S. Environmental Protection Agency. *Compilation of Air Pollutant Emission Factors*. Publication AP-42. EPA, Office of Air Quality Planning and Standards, Research Triangle Park, NC, 1995. https://www.epa.gov/air-emissions-factors-and-quantification/ap-42-compilation-air-emission-factors.

The Standing Committee on Low-Volume Roads (AFB30) peer-reviewed this paper (18-06624).

Any use of trade, firm, or product names is for descriptive purposes only and does not imply endorsement by the U.S. Government.

Article

TRR
JOURNAL OF THE TRANSPORTATION RESEARCH BOARD

Transportation Research Record
2018, Vol. 2672(52) 205–215

DOI: 10.1177/0361198118781147
journals.sagepub.com/home/trr
SAGE

Applying Large-Scale Optimization to Evaluate Pavement Maintenance Alternatives for Low-Volume Roads using Genetic Algorithms

Marwan Hafez[1], Khaled Ksaibati[1], and Rebecca A. Atadero[2]

Abstract

Over the last decade, significant progress has been made to customize the maintenance policies of low-volume roads (LVRs) to local needs and available resources. Low-cost treatments and surface repairs are extensively employed to reduce annual maintenance costs. Colorado Department of Transportation (CDOT) uses chip seals and thin overlays as the available treatment options applied to LVRs. However, the effectiveness of these treatments differs depending on the existing condition of pavements. Some surface treatments and light rehabilitations provide only short-term effectiveness. Multi-year optimization techniques can support decision makers with a set of optimal maintenance activities to achieve specific pavement performance targets. This study applies large-scale optimization to compare the current CDOT maintenance policy with an alternative strategy recommended for low-volume paved roads in Colorado. Genetic algorithms were applied in the optimization models because they are capable of resolving the computational complexity of optimization problems in a timely fashion. The optimized maintenance alternatives were comprehensively investigated for a LVR network in Colorado over a specific planning horizon. The specific optimization constraints and limitations prevailing in LVRs are addressed and introduced in the problem formulation of the optimization process. The results of both performance and cost analysis emphasize the effectiveness of the proposed maintenance strategy compared with the existing one. The alternative policy provides much more benefit-cost saving while preserving the overall pavement performance of the network. This approach is expected to be efficient to quantify the mid- and long-term financial impact of different treatment policies applied to LVRs within modest resources.

Departments of Transportation (DOTs) face the challenge of maintaining the pavement condition of road networks. Although pavement deterioration rates increase over time, the financial resources for their maintenance are not raised accordingly (*1*). This challenge is more critical for low-volume roads (LVRs) which are managed under the supervision of state DOTs and local agencies. Most federal aid supports state highway agencies to improve the condition of national highways. For LVRs, state and local agencies are interested in developing lower-cost pavement maintenance policies. Low-cost treatments are applied to LVRs even if the pavement performance is poor. In the context of pavement preservation strategies, low-cost and surface treatments should be applied to roads while the pavement is still performing well. When applying preventive maintenance on severely distressed LVRs, some of the treatments applied may provide only short-term effectiveness. In today's global economy, it is realized by pavement engineers and researchers that considerable savings can be obtained by implementing an efficient pavement management system (PMS) on LVRs (*2*). Decision makers are trying to get the best value for the allocated resources by considering effective pavement treatments and taking informed maintenance decisions. Such decisions should be optimized within the limited budgets to preserve the LVR network at an acceptable level of service.

In the State of Colorado, low-volume paved roads are currently maintained with only low-cost treatments (*3*). In 2016, Colorado State University and the Wyoming Technology Transfer Center/Local Technical Assistance

[1]Department of Civil and Architectural Engineering, University of Wyoming, Laramie, WY
[2]Department of Civil and Environmental Engineering, Colorado State University, Fort Collins, CO

Corresponding Author:
Address correspondence to Khaled Ksaibati: khaled@uwyo.edu

Program (WYT^2/LTAP) conducted a comprehensive evaluation to document the effectiveness of some in-practice treatments applied to LVRs in Colorado. By analyzing historical pavement performance data and treatment records, the effectiveness of each treatment type was determined according to the existing pavement condition. An alternative maintenance strategy was proposed for Colorado Department of Transportation (CDOT) (*4*). Cost analysis was also conducted and it was found that the proposed maintenance strategy could provide cost-effective maintenance solutions. However, a comprehensive optimization analysis is required to study the long-term effectiveness of the proposed treatments compared with the current CDOT policy.

The primary objective of this study is to assist CDOT and other transportation agencies in determining the optimal allocation of available network budget, and the corresponding maintenance policy. This paper presents an approach of artificial-intelligence optimization analysis to identify the benefit-cost impact of considering maintenance activities other than those currently applied. The optimized maintenance alternatives were comprehensively investigated for a LVR network in Colorado over a specific planning horizon. This approach is expected to be efficient to quantify the mid- and long-term financial impact of different treatment policies applied to LVRs within a context of modest resources.

Background

According to the definition of the Manual on Uniform Traffic Control Devices, roads are considered low volume when the average daily traffic (ADT) is less than 400 vehicles per day (*5*). However, most DOTs consider reasonable traffic ranges of high-, medium-, and low-volume roads to keep the road network of each type in manageable sizes (*6*). According to CDOT, LVRs are defined as roads carrying ADT of less than 2,000 vehicles per day and average daily truck traffic (ADTT) of less than 100 vehicles per day, due to the high traffic distributions in Colorado (*3*). Several states have been making efforts to develop specific policies for maintaining LVRs in order to reach specific pavement performance targets. Many state DOTs recommend constructing and preserving the pavement of LVRs using low-cost thin overlays (*7*). In the State of Nevada, life-cycle cost analysis was conducted for LVRs (*8*). It was found that cold in-place recycling (CIR) with double chip seals can effectively rehabilitate a LVR at almost half the cost of placing a 2-inch overlay. The application of full-depth replacement (FDR) was also found to be an effective treatment to increase the structural capacity of LVRs. Another management effort was initiated by Minnesota DOT. A survey was conducted for county engineers in Minnesota and neighboring states to investigate any applied practices of recycling techniques on LVRs (*9*). The results of the survey revealed that CIR and FDR are extensively applied in Minnesota. A study evaluating four surface treatments was then performed using test sections. These treatments comprised combinations of CIR, FDR, chip seals, and thin overlays. It can be noticed from the literature that several DOTs are trying to find alternative surface treatments applied to LVRs to reduce the cost of maintenance. However, optimizing proposed maintenance strategies is not well investigated.

Many mathematical optimization methods have been used in the pavement maintenance and management area such as optimal control theory, linear and nonlinear programming, dynamic programing, and integer programming (*10–12*). However, these conventional optimization techniques require complex forms of mathematical formulations when optimizing pavement maintenance activities. The development of artificial intelligence allows road engineers to include applications of fuzzy systems, neural networks, and genetic algorithms to identify the optimal solution of pavement maintenance problems (*13*). Since the early 1990s, genetic algorithms (GAs) have been used widely by many researchers in civil engineering. The reason is that GAs are general-purpose stochastic optimization techniques that are able to generate and store multiple solutions of a global optimization problem with a comparable level of accuracy, especially for large size networks (*14*). Also, local pavement management officials can process GAs in optimization problems using user-friendly software. The application of GAs in road maintenance planning was introduced by Fwa et al. (*15*). Pavement performance modeling is a key element in formulating the optimization problem. Performance models are used to project future pavement conditions so that multi-year maintenance programs can be assigned. Ferriera et al. (*16*) optimized road maintenance activities using deterministic pavement performance models. Other optimization efforts were conducted by Odoki and Kerali (*17*) in HDM-4 to find the optimal set of maintenance treatments applied at the network level. Probabilistic performance models were included in bridge pavements to optimize the maintenance strategies using Markov-chain models (*14*). Most of studies focused on primary roads, and limited applications were implemented for local PMSs. Mathew and Isaac (*18*) developed a PMS optimization model based on GAs for rural roads. The strategy of the previously mentioned studies was to demonstrate the use of GAs in determining multi-year maintenance plans. Few road segments were considered in the analysis for simplicity and efficiency of optimization analysis. However, this study establishes large-scale optimization models for a whole LVR network in Colorado.

Methodology

This section provides brief definitions of the pavement performance variables used in this study. The formulation of optimization models for LVRs is also introduced in this section followed by the main steps in the GA process.

Drivability Life

Drivability life (DL) is an indication of how long a road has acceptable driving conditions in terms of years (*3*). DL is determined based on a trend analysis of five distresses that are normalized into the following condition indices:

- Fatigue Index
- Longitudinal Index
- Transverse Index
- Ride Index
- Rut Index

These indices are on a scale from 0 (pavement's worst state) to 100 (excellent or brand new pavement). According to CDOT, the pavement has a DL of 0 when any of the distress indices is less than 50. Each condition index has a specific performance curve as a function of pavement age and they were modeled in the pavement evaluation study (*4*). The overall DL value of a road is the minimum time remaining until any of the condition index reaches a value of 50.

Optimization Problem Formulation

At the network level, most of the optimization models in PMSs are segment-linked (*14*, *16*, *18*, *19*). That is, a set of road segments of the network is identified every year where maintenance actions should be taken. Therefore, a decision variable (X_{st}) should be optimized for each segment (s) and at a time (t). The following subsections describe some of the important objective functions used when optimizing LVRs in Colorado.

Pavement Performance Maximization. The first strategy is to maximize the overall pavement performance within an annual maintenance budget. This strategy is applied when the network-level funding is inadequate to repair the road network. As shown in Equation 1, the objective function is to maximize the overall weighted DL (OWDL_t) of the road network every year t. Pavement performance can be weighted by different variables such as road length, ADT, ADTT, or the associated risk of maintenance (*20*). However, the LVR network tends to show low variability of traffic volumes statewide. Also, the risk of applying low-cost treatments is not significant on LVRs which have low economic and service return, compared with the rehabilitation activities of primary roads. In the proposed optimization model, the OWDL_t considers the length of road segments (l_s) as the weighting factor so that longer segments would have maintenance priorities. The main constraint of formulating this objective function is that the total costs must be less than the available budget, see Equation 2.

$$\text{Maximize}: \mathrm{OWDL}_t = \frac{\sum_{s=1}^{S} \mathrm{DL}_{st} * l_s}{\sum_{s=1}^{S} l_s} \tag{1}$$

subject to

$$\sum_{m=1}^{M} \sum_{s=1}^{S} X_{st}\,\mathrm{Cos}\,t_{mst} \leq \mathrm{Budget_t}\,,\ \forall t = 1 \text{ to } T \tag{2}$$

$$X_{st} \in \{0, 1\}\,,\ \forall s = 1 \text{ to } S,\ \forall t = 1 \text{ to } T \tag{3}$$

The expected performance of pavement depends mainly on an optimal set of X_{st} which is an integer variable of 0 for "do nothing" and 1 when a maintenance is applied to the road, see Equation 3. Equation 4 formulates the DL_{st} value at the end of a time interval t for each segment s. For the "do nothing" option, the DL value of the previous year (e.g., $\mathrm{DL}_{s(t-1)}$) is expected to be decreased by an annual decline of DL ($d\,\mathrm{DL}$) due to pavement deterioration. When applying a maintenance activity (m) (e.g., $x_{st} = 1$), the pavement improvement is introduced by an increase in the DL value $(\Delta\mathrm{DL})_{sm}$ caused by the treatment. The amount of $d\,\mathrm{DL}$ and $(\Delta\mathrm{DL})_{sm}$ can be estimated based on expertise or it can be determined using deterministic performance and improvement models for each condition index.

$$\mathrm{DL}_{st} = \mathrm{DL}_{s(t-1)} + X_{st}(\Delta\mathrm{DL})_{sm} - (d\,\mathrm{DL} - X_{st} d\,\mathrm{DL}) \tag{4}$$

where

$(\Delta\,\mathrm{DL})_{sm}$ = Expected increase in DL due to maintenance treatment; and

$d\,\mathrm{DL}$ = Annual decline in DL due to pavement deterioration.

Maintenance Costs Minimization. In this strategy, the present worth of the total maintenance cost is minimized, as shown in Equation 5. Any future costs are discounted to the base year of the analysis period using a discount rate (r). This strategy is applied when a minimum performance target ($\mathrm{OWDL_{min}}$) is required to achieve by the end of the analysis period (T) as shown in Equation 6.

$$\text{Minimize : TMC} = \sum_{m=1}^{M}\sum_{s=1}^{S}\sum_{t=1}^{T}\frac{1}{(1+r)^t}X_{st}\,Cost_{mst} \quad (5)$$

subject to

$$\text{OWDL}_T \geq \text{OWDL}_{\min} \quad (6)$$

It can be noted that the decision variable in this strategy has a length of ($S \times T$) which is a long length at network level. That affects the computational efficiency of GA analysis in the optimization model. Thus, the objective function should be modified to have a lower number of decisions in each model. Morcous and Lounis (*14*) proposed grouping the road segments into distinctive characteristics such as type, material properties, operating loads, and environmental conditions. However, LVR networks exhibit low variability of the main characteristics so classifying roads into groups is not a practical method. Therefore, the large-scale optimization model of this strategy is formulated as shown in Equation 7 where the total annual maintenance costs in each year (TMC_t) are minimized. The modified objective function reduces the length of the decision variable to ($S \times 1$). In this case, the pavement performance target needs to be distributed linearly or nonlinearly every year according to the management needs.

$$\text{Minimize : TMC}_t = \sum_{m=1}^{M}\sum_{s=1}^{S}X_{st}\,\text{Cos}\,t_{mst}\ ,\ \forall t = 1\ to\ T \quad (7)$$

Multi-Objective Optimization. The objective functions of both strategies, mentioned above, can be combined into a single model shown in Equation 8. Since they are different in the fitness nature, the maximization problem of OWDL_t is converted into a minimization one using the transformation ($1/\text{OWDL}_t$). All of the previously mentioned constraints can be included in this model. This strategy includes multi-objective optimization problems with nonlinear constraints which requires complex GA analysis to find a solution. In some cases, the solution remains infeasible to satisfy the constraints after a number of trials defined by stopping criteria. The unsatisfied solution is then subjected to a penalty process defined by the analyst (*21*).

$$\text{Minimize : }\frac{\text{TMC}_t}{\text{OWDL}_t} \quad (8)$$

Genetic Algorithms Process

GAs are evolutionary selection algorithms that initiate solutions of the constrained maintenance program in PMSs. They then evolve the maintenance decisions toward a set of more optimal solutions. The decision variable is divided into discrete parts that can vary independently. These parts are called "genes" and they are applied with operations of reproduction, crossover, and mutation. Figure 1 depicts the main steps executed in the GA analysis. After importing the data, the encoded optimization parameters create initial population or "generation" of the potential solutions. They are selected as feasible solutions according to their fitness where all constraints are satisfied. All feasible solutions are used to produce offspring generations with more optimal features by a mating process. Each two individual solutions are assigned as parents then they are combined together to form a new set of potential solutions. The genes of parents are manipulated with a crossover process where sections of the genes, called "chromosomes," are swapped. Then a few of the produced genes are treated with a random mutation to prevent more solutions falling into a local optimum of the feasible solutions space. Developing subsequent generations stops when specific stopping criteria are met, and the best fit offspring population is considered the optimum solution.

The crossover probability (P_c) represents the amount of population strings used in the crossover operation. High P_c probabilities allow fast solutions and better performance of the analysis. However, it is recommended the the mutation rate (P_m) should be low to converge the solution. The typical P_c rates are from 0.5 to 1.0 while P_m should be in the range of $0.001 \sim 0.05$ (*22*).

Case Study: Low-Volume Paved Roads in Colorado

There are 2,022 miles of low-volume paved roads in Colorado distributed in five engineering regions. In this study, the LVR network of Region 4 was comprehensively analyzed where 85 road segments are distributed along a total of 422 miles. Region 4 has varying traffic volumes with different DL categories, as shown in Figure 2. About half of the LVRs (48% of road miles) have moderate DL values. There are 41% of road miles in the low DL category while only 11% of roads have high drivability lives of more than 10 years. Most of the LVRs have traffic volumes of not more than 800 vehicles per day. The diverse pavement conditions and traffic volumes would support the validity of the optimization models since most treatment options are included in the analysis with different deterioration amounts.

Pavement Treatments

Table 1 shows the current treatment options recommended by CDOT for LVRs. The current maintenance policy of CDOT recommends investigating rehabilitation alternatives

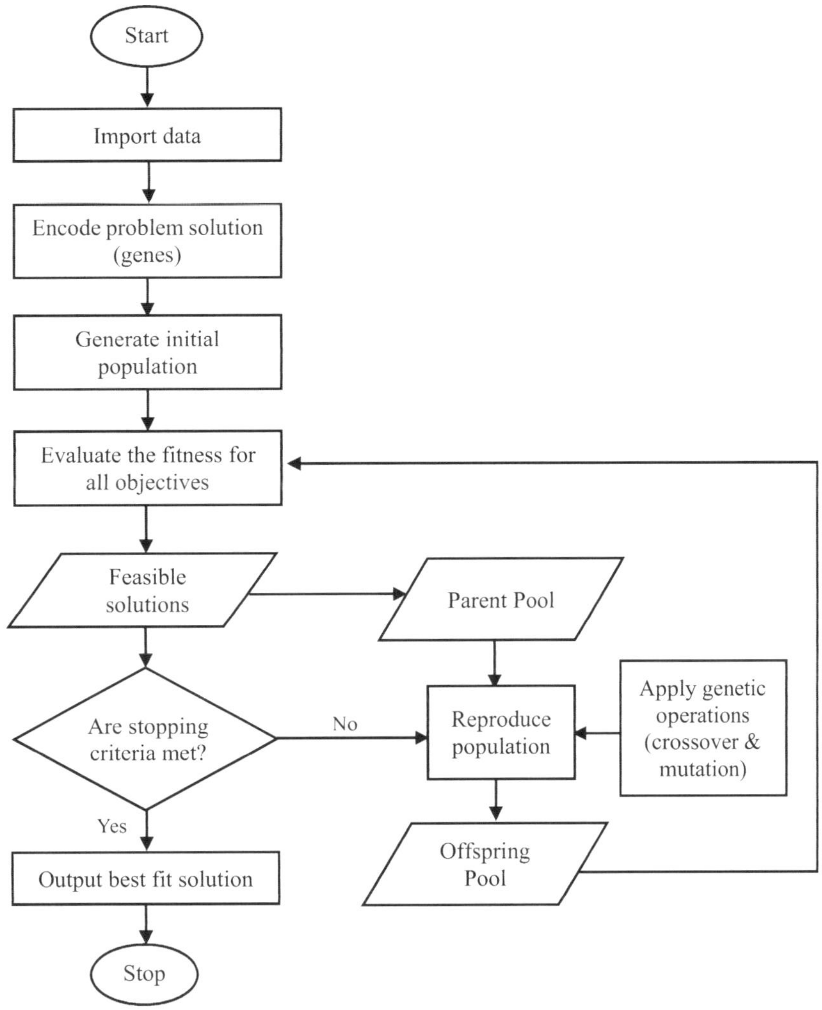

Figure 1. Main steps of problem solving using genetic algorithms.

Category	DL (Years)
Low	0–4
Moderate	4–10
High	>10

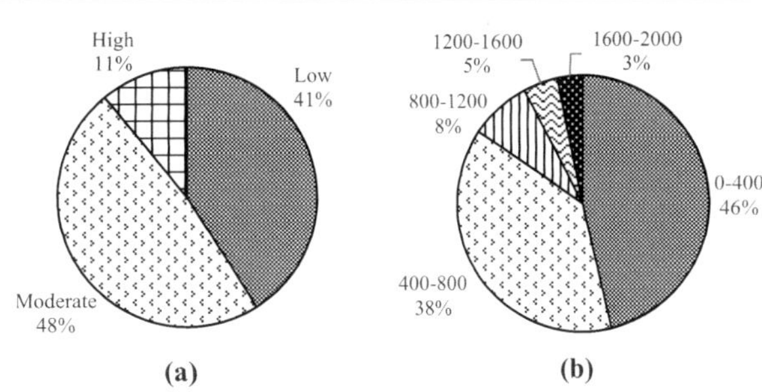

Figure 2. Mileage distribution of Region 4 low-volume paved roads: (*a*) drivability life; (*b*) average daily traffic (vehicles per day).

to balance between pavement preservation and capacity improvement on LVRs. Four pavement rehabilitation alternatives involving recycling technologies are integrated with the current policy. These treatments are:

Table 1. Current CDOT Pavement Treatment Policy on Low-Volume Roads

DL category	Treatment	Description
High	GM	General maintenance
Moderate	ACHP	Asphalt chip seal
Low	ATHO	Asphalt thin overlay < 2 in.

- CIR of 4 in, of asphalt surfaced with a fog coat (FC) surface
- CIR layer surfaced with an asphalt chip seal
- CIR layer with an asphalt thin overlay
- Full-depth replacement (FDR)

Unlike the current policy, all treatment options in the proposed strategy were included in a developed decision tree shown in Figure 3. Trigger values of fatigue and transverse cracking were considered for each treatment option. The methodology of selecting triggers was derived from the effectiveness of treatments. This term depends on the initial condition of pavements and it is not fixed. The treatment evaluation study performed a comprehensive analysis on the past performance of the applied treatments and it was able to define the lower limit of each condition index where treatments were found to be effective (*4*). Accordingly, the trigger values were assigned in the decision tree in which they do not go below the minimum effective values to ensure effectiveness of treatments.

Multi-Year Optimization Analysis

As explained in the methodology, the second strategy of minimizing costs is applied in this study to demonstrate the economic impact of maintenance alternatives so that specific performance targets are achieved at network level. Consequently, cost analysis was conducted to compare each strategy over a study time period of 10 years. The costs were obtained from CDOT and they are listed in Table 2. These costs are based on the dollar value of 2016 which is the base year of the analysis period.

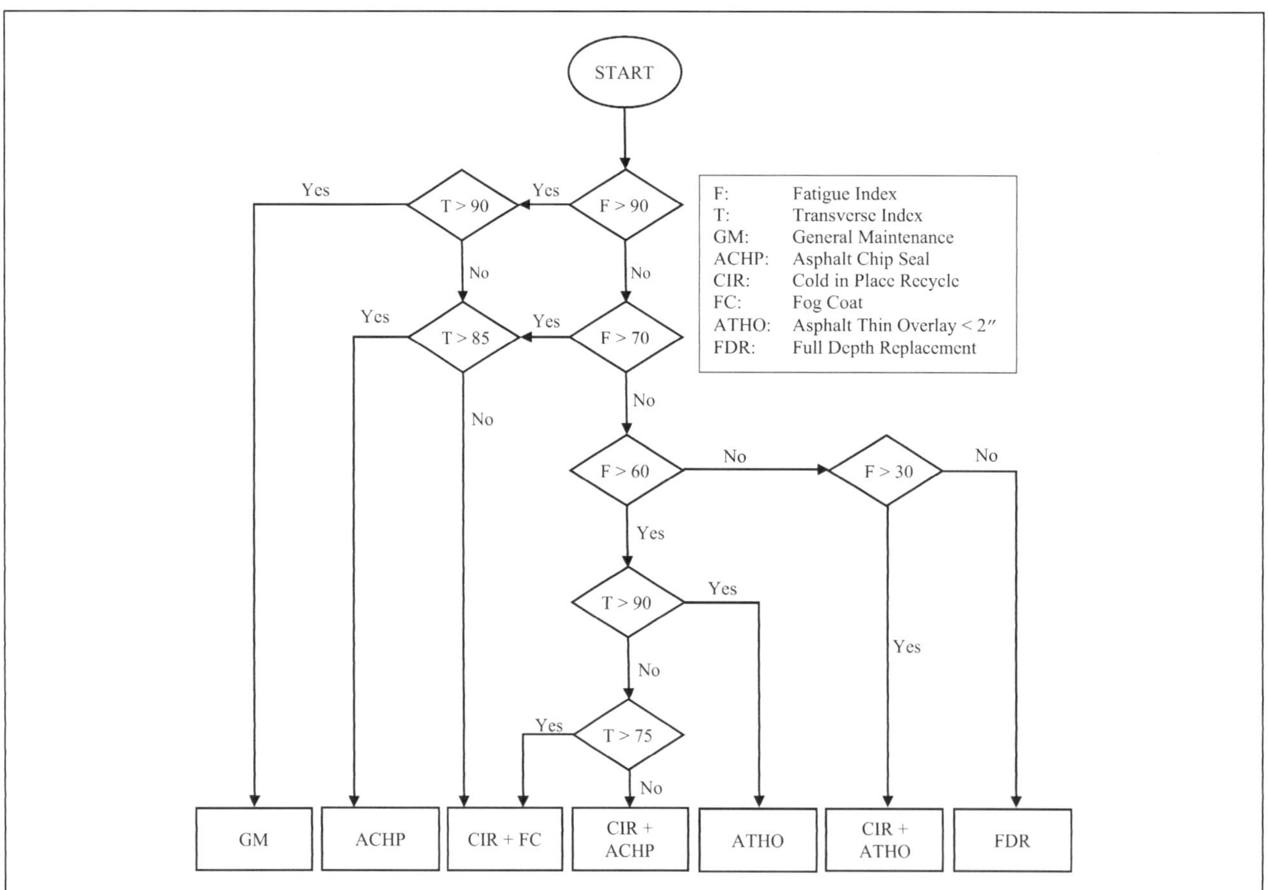

Figure 3. Decision tree of the alternative maintenance strategy.

Table 2. Estimated Maintenance Unit Costs for Colorado Department of Transportation

Maintenance	Cost (Lane – mile)
GM	\$3,000
CHP	\$42,240
CIR + FC	\$40,480
CIR + CHP	\$77,440
ATHO	\$151,360
CIR + ATHO	\$186,560
FDR	\$195,350

The multi-year optimization process for selecting treatments is first conducted at the base year where all segments are assigned proper treatments according to each maintenance policy. The decision to maintain segments is then optimized using the developed optimization models. For the subsequent years over the analysis period, the future pavement conditions are projected for both treated and untreated roads. For those roads receiving treatments, the expected improvement of each condition index was identified in the treatment evaluation study as a function of treatment type and the initial value of the condition index (*4*). Accordingly, the term $[DL_{s(t-1)} + (\Delta DL)_{sm}]$ in the optimization model was determined depending on the terminal value of each condition index after applying treatments.

For Region 4, the overall weighted DL is 4.87 years. This value was intended to change over the analysis period with different scenarios. Table 3 lists four maintenance scenarios followed in the optimization analysis. The first scenario was to keep the overall DL steady over time by eliminating the overall annual deterioration of the network. The second scenario was to achieve an overall DL target with an improvement of 10% at the end of the analysis period. Scenario III and IV optimized the maintenance activities to achieve improvement amounts of 20% and 30%, respectively. Including different scenarios in the optimization analysis allows decision makers to investigate the impact of allocating different maintenance budgets. At the end of the analysis, a summary was developed showing the costs required for each maintenance policy.

Optimization Results

Since the optimization model does not have budget constraints, all analysis periods yielded a theoretical optimized solution. The minimum annual performance targets were attained in each maintenance scenario in a reasonable time. The optimized cash flow diagrams of Region 4 are shown in Figure 4 (for Scenario I and II) and Figure 5 (for Scenario III and IV). At the base year of 2016, it can be noticed that all the allocated budgets of the alternative strategy are less than half of the budgets of the current policy. The reason is that about 50% of the LVR network has low DL values. Most of these roads were optimized with lower-cost alternatives such as CIR with fog coats or CIR with chip seals. The costs of the alternative treatments are almost half of the costs of thin overlays applied in the model of current policy.

Over the study period, it was found that the alternative maintenance strategy allocates almost uniform annual budget except the costs at 2023. When the targeted improvement increased in each scenario, the annual costs slightly increased, except Scenario IV where a significant improvement target is required to be achieved. On the other hand, the current policy requires much higher budgets at the beginning of the analysis period. The costs then decreased over time which appeared to show a good performance of the maintenance activities. However, the annual costs increased after 2020 for all scenarios, revealing the short-term effectiveness of the applied treatments. Chip seals provide short-term effectiveness if the condition indices are below specific values derived from the evaluation study (*4*). However, all roads in the moderate category were optimized with chip seals regardless the value of the condition indices. Also, thin overlays should not be applied directly on severely damaged roads without considering the appropriate pre-overlay treatments. Although the post-treatment performance of thin overlays is expected to be high, some untreated defects may reflect cracks directly to the overlaid pavements. As a result, treated roads show higher deterioration rates.

In order to compare the total costs required for each maintenance strategy, all the allocated funds were

Table 3. Maintenance Scenarios of Optimization Analysis

		Performance targets ($OWDL_t$, years)	
Scenario	Maintenance objective	From 2016	To 2025
Scenario I	Steady State	4.87	4.87
Scenario II	+10% Improvement	4.87	5.36
Scenario III	+20% Improvement	4.87	5.85
Scenario IV	+30% Improvement	4.87	6.33

Note: OWDL = Overall Weighted Drivability Life (years).

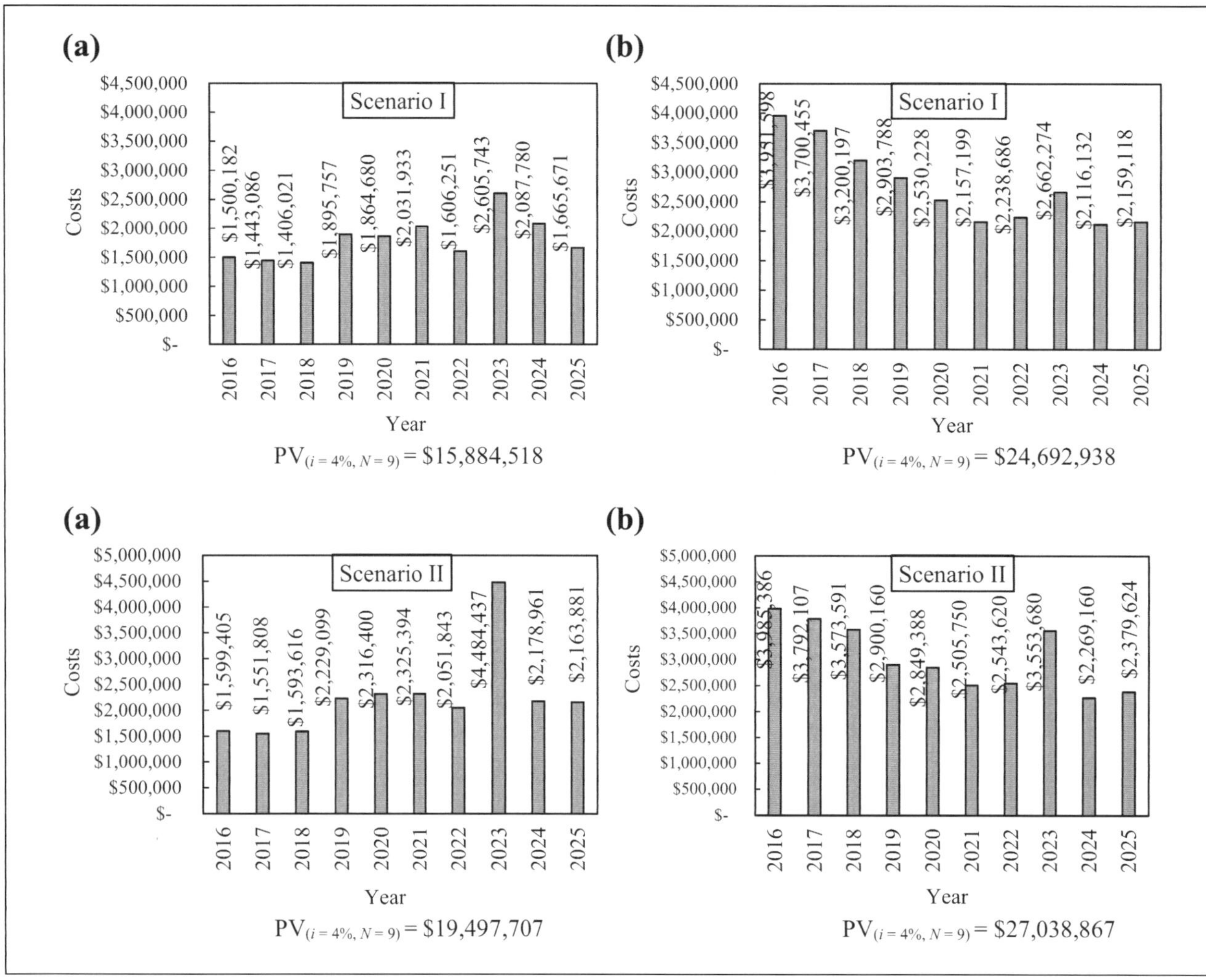

Figure 4. Region 4 cash flow diagram for scenario I and II: (*a*) alternative maintenance strategy; (*b*) current maintenance strategy.

discounted to the base year of 2016 using a discount rate of 4%. It can be seen that all present values of the alternative maintenance strategy are less than those of the currently applied policy. Also, the total discounted value increased with increasing the targeted pavement improvement in each maintenance strategy. It was found that a sensitivity analysis is recommended to incorporate a cost-benefit comparison of the two policies on each maintenance scenario.

Sensitivity Analysis

It is highly recommended to identify the maximum benefits of different allowable treatments on LVRs. Benefit-cost analysis is applied in PMSs to evaluate the economic strength of different maintenance alternatives over the lifespan of pavements (*23*). The benefit-cost ratio represents the compound benefits of treatment selection. However, the maintenance benefits should be determined in the same currency of costs. This is not practical on LVRs where a low return is expected due to the low social impact of LVRs. For simplicity, the benefit from applying the alternative maintenance strategy is introduced by the percentage of cost saving given the current budget. This amount is weighted by the corresponding condition level target, as shown in Equation 9.

$$\%\text{Benefit} = \frac{PV_{current} - PV_{altr}}{PV_{current}} \times (1 + \%\text{Improvement}) \times 100\% \quad (9)$$

where,

$PV_{current}$: present value of current maintenance costs;

PV_{altr}: present value of alternative maintenance costs; and

%Improvement: percentage of improvement target of the overall drivability.

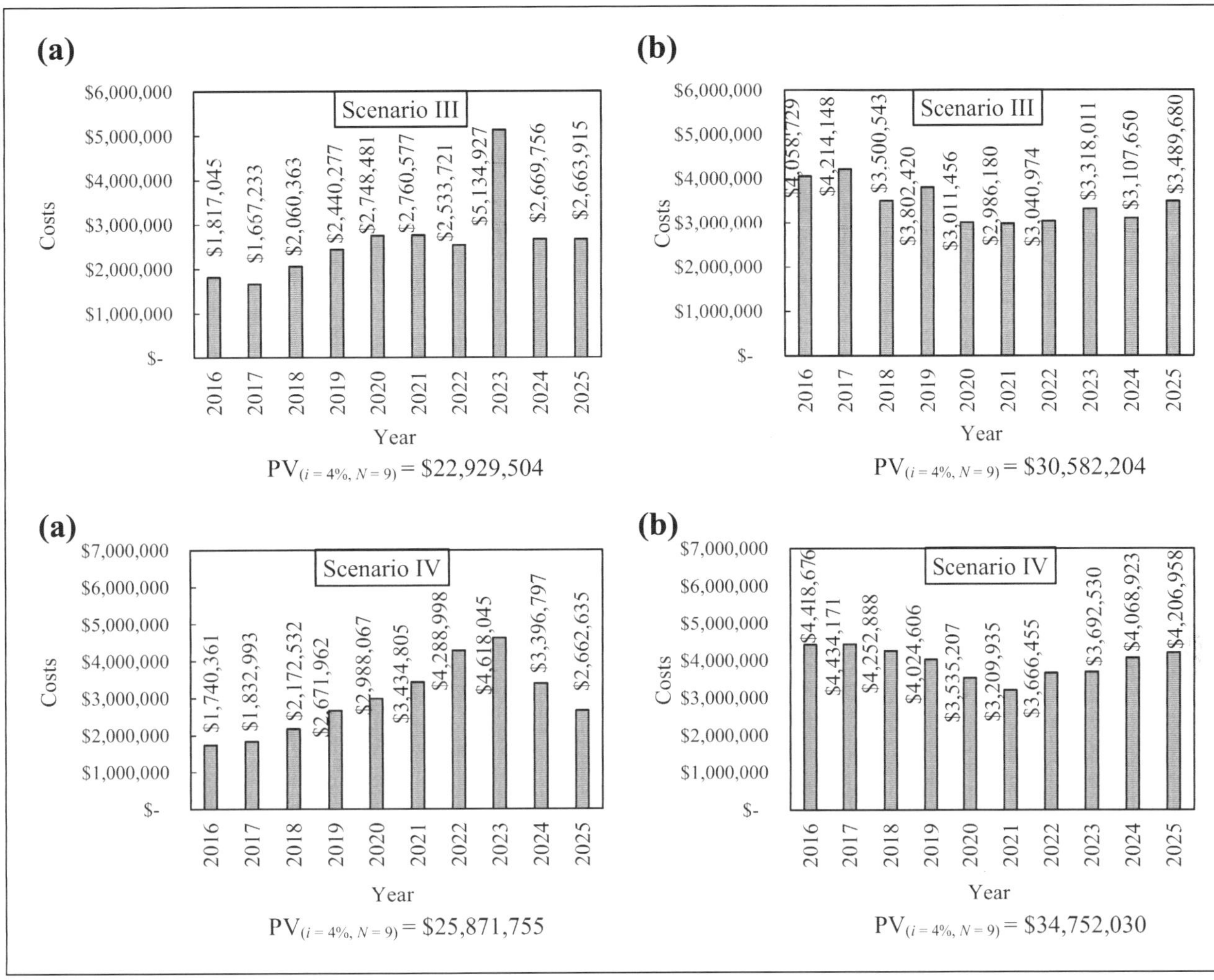

Figure 5. Region 4 cash flow diagram for scenario III and IV: (*a*) alternative maintenance strategy; (*b*) current maintenance strategy.

Figure 6 displays the percentage of benefits estimated for each maintenance scenario. The results show that the alternative maintenance strategy achieves the maximum benefit when the overall pavement performance is intended to be steady overtime. The second recommended strategy was found to be the fourth scenario, where an improvement of 30% is achieved at the end of the tenth year. It is obvious that there is no significant difference among the scenarios. Also, the optimized budget levels required to achieve the fourth scenario are much more than what can be available to LVRs. Therefore, it is recommended to apply the first maintenance scenario to keep the overall pavement performance of the network steady over the planning period.

Conclusions

A comprehensive optimization analysis was conducted to support pavement management decisions on the maintenance and rehabilitation planning of LVRs. In the State of Colorado, a LVR network of 422 miles was optimized using GAs to minimize the annual maintenance costs restricted by drivability performance targets. Based on the findings of this research study, the following conclusions are drawn:

- The optimization tool is functional to seek the long-term effectiveness of different maintenance policies among various treatment options. In this study, the key benefit of optimizing the LVR network is making more effective use of maintenance and rehabilitation strategies.
- GAs are able to solve a large-scale optimization problem for LVR networks. The maintenance decision matrix of multi-year planning can be optimized within a reasonable time. However, the decision variable of a large road network can be optimized annually to reduce the number of genes

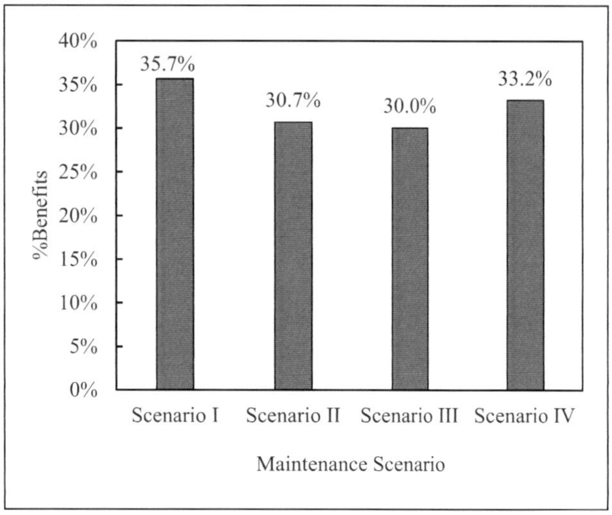

Figure 6. Percentage of benefits of applying alternative maintenance strategy.

set in the analysis. That enhances the performance and efficiency of the optimization process.

- The large-scale optimization of LVRs provides more realistic solutions to quantify the maintenance funding needs. Engineers can identify prudent treatment policies to achieve a desired target state of the network. Moreover, the impact of various condition targets can be evaluated to determine the optimal allocation of maintenance budgets.
- The current CDOT policy requires additional funding to sustain the network-level pavement condition given the application of only chip sealing and thin overlaying. Including alternative light rehabilitation strategies for poor roads can save significant amounts of money.
- CIR and FDR techniques enhance the overall drivability of LVRs with lower costs compared with the regular rehabilitation of overlays. Also, applying thin overlays over severely deteriorated pavements provides short-term effectiveness. The application of thin overlays should be addressed considering minimum threshold values of pavement distress.
- Although raising the performance target levels may add drivability value to LVRs, the benefit-cost impact is not significant. Applying the alternative maintenance policy provides maximum benefit-cost saving when the maintenance decisions are optimized to keep the overall drivability in constant performance.

Recommendations

The large-scale optimization approach is recommended to be applied when investigating alternative maintenance strategies. State DOTs should define all possible constraints that significantly provide benefit-cost effectiveness of proposed treatments. CDOT is recommended to develop statewide implementation plans of pavement maintenance activities on its LVRs to define the appropriate funding levels and capital improvement plans. Also, a comparison between deterministic and probabilistic performance models is recommended to investigate the accuracy of optimization results. The reason is that some transportation agencies do not have enough condition data on LVRs to fit deterministic performance models. Therefore probabilistic performance models could be more practical while optimizing LVRs.

Acknowledgments

The authors gratefully acknowledge CDOT for the financial support provided to this research.

References

1. The Pew Charitable Trusts. *Intergovernmental Challenges in Surface Transportation Funding*, 2014. http://www.pewtrusts.org/en/research-and-analysis/reports/2014/09/intergovernmental-challenges-in-surface-transportation-funding. Accessed February 5, 2018.
2. Landers, S., D. Mason, and J. MacNaughton. Development of a Sustainable Pavement Management Strategy for Resurfacing Low Volume Roads in New Brunswick. *Proc., 9th International Conference on Managing Pavement Assets*, Alexandria, VA, 2015.
3. Colorado DOT. *2016 M-E Pavement Design Manual*. Colorado Department of Transportation, 2016. www.codot.gov/business/designsupport/matgeo/manuals/pdm/2016-pdm/2016-pdm-final. Accessed February 5, 2018.
4. Hafez, M., K. Ksaibati, and R. Atadero. Developing a Methodology to Evaluate the Effectiveness of Pavement Treatments Applied to Low-Volume Paved Roads. *International Journal of Pavement Engineering*, 2017.
5. Manual on Uniform Traffic Control Devices. *Part 5 - Traffic Control Devices for Low-Volume Roads*. Federal Highway Administration, U.S. Department of Transportation, 2009.
6. Hafez, M., K. Ksaibati, and R. Atadero. Best Practices to Support and Improve Pavement Management Systems for Low-Volume Paved Roads. *International Journal of Pavement Engineering*, 2017.
7. Brown, E. R., and M. Heitzman. *Thin HMA Overlays for Pavement Preservation and Low Volume Asphalt Roads*. NCAT Report No. 13-05. National Center for Asphalt Technology. Auburn, AL, 2013.
8. Maurer, G., S. Bemanian, and P. Polish. Alternative Strategies for Rehabilitation of Low-Volume Roads in Nevada. *Transportation Research Record: Journal of the Transportation Research Board*, 2007. 1989: 309–320.
9. Jahren, C., J. Yu, and R. Williams. *Alternate Design Methods to Renew Lightly Traveled Paved Roads*. Final Report,

No. MN/RC 2016-14. Minnesota Department of Transportation, St. Paul, MN, 2016.

10. Golabi, K., R. Kulkarni, and G. Way. A Statewide Pavement Management System. *Interfaces*, Vol. 12, No. 6, 1982, pp. 5–21.
11. Jiang, Y., and K. C. Sinha. A Dynamic Optimization Model for Bridge Management Systems. *Transportation Research Record: Journal of the Transportation Research Board*, 1989. 1211: 92–100.
12. Markow, M. J., and W. S. Balta. Optimal Rehabilitation Frequencies for Highway Pavements. *Transportation Research Record: Journal of the Transportation Research Board*, 1985. 1035: 31–43.
13. Sundin, S., and C. Braban-Ledoux. Artificial Intelligence–Based Decision Support Technologies in Pavement Management. *Computer-Aided Civil and Infrastructure Engineering*, Vol. 16, No. 2, 2001, pp. 143–157.
14. Morcous, G., and Z. Lounis. Maintenance Optimization of Infrastructure Networks Using Genetic Algorithms. *Automation in Construction*, Vol. 14, No. 1, 2005, pp. 129–142.
15. Fwa, T. F., C. Y. Tan, and W. T. Chan. Road-Maintenance Planning Using Genetic Algorithms. II: Analysis. *Journal of Transportation Engineering*, Vol. 120, No. 5, 1994, pp. 710–722.
16. Ferreira, A., L. Picado-Santos, and A. Antunes. A Segment-Linked Optimization Model for Deterministic Pavement Management Systems. *International Journal of Pavement Engineering*, Vol. 3, No. 2, 2002, pp. 95–105.
17. Odoki, J. B., and H. R. Kerali. *Analytical Framework and Model Descriptions.* Highway Development and Management Series, Vol. 4 (HDM-4). International Study of Highway Development and Management, World Road Association, Paris, France, 2006.
18. Mathew, B. S., and K. P. Isaac. Optimisation of Maintenance Strategy for Rural Road Network Using Genetic Algorithm. *International Journal of Pavement Engineering*, Vol. 15, No. 4, 2014, pp. 352–360.
19. Ferreira, A., A. Antunes, and L. Picado-Santos. Probabilistic Segment-Linked Pavement Management Optimization Model. *Journal of Transportation Engineering*, Vol. 128, No. 6, 2002, pp. 568–577.
20. Saha, P., and K. Ksaibati. A Risk-Based Optimisation Methodology for Pavement Management System of County Roads. *International Journal of Pavement Engineering*, Vol. 17 No. 10, 2016, pp. 913–923.
21. Goldberg, D. E. *Genetic Algorithms in Search, Optimization, and Machine Learning.* Addison-Wesley Publishing Company, Inc., New York, 1989.
22. Lin, W. Y., W. Y. Lee, and T. P. Hong. Adapting Crossover and Mutation Rates in Genetic Algorithms. *Journal of Information Science and Engineering*, Vol. 19, No. 5, 2003, pp. 889–903.
23. Sallman, D., E. Flanigan, K. Jeannotte, C. Hedden, and D. Morallos. *Operations Benefit/Cost Analysis Desk Reference.* FHWA-HOP-12-028. FHWA, U. S. Department of Transportation, Washington, D.C., 2012.

The Standing Committee on Low-Volume Roads (AFB30) peer-reviewed this paper (18-01478).

Article

TRR
JOURNAL OF THE TRANSPORTATION RESEARCH BOARD

Transportation Research Record
2018, Vol. 2672(52) 216–225

DOI: 10.1177/0361198118787372
journals.sagepub.com/home/trr
SAGE

Performance-Based Design Method for Gradation and Plasticity of Granular Road Surface Materials

Cheng Li[1,2], Jeramy C. Ashlock[2], Bora Cetin[2], Charles T. Jahren[2], and Vanessa Goetz[3]

Abstract

Granular-surfaced roads frequently experience severe surface damage and degradation, which adversely affects traffic safety and significantly increases maintenance costs. The importance of the gradation and plasticity of granular-surfaced road materials have long been recognized. However, very few studies have focused on quantifying the effects of gradation and plasticity on the resulting mechanical performance of granular roadways. In this study, laboratory tests and statistical analyses were conducted to quantify the effects of variations in gradation and plasticity on the mechanical performance of granular road surface materials. A performance-based design method was developed using Fuller's model to replace the commonly used arbitrary gradation band specifications. To validate the performance of the proposed method, field test sections were constructed then tested through a seasonal freeze–thaw period. Compared with existing gradation band specifications, the laboratory and field test results demonstrated that the proposed method is more closely tied to performance and can be used to develop specifications with more precise targets. To help secondary roads agencies implement the method while also recycling existing surface materials, a gradation optimization program was developed to determine the mixing ratios of existing roadway aggregate and two to three new quarry materials to come closest to the theoretical gradation for optimum performance. A complete set of testing, design, and construction procedures is also recommended to provide more cost-effective solutions to building or reconstructing granular-surfaced roads.

Unpaved roads comprise 35% of the total 6.7 million kilometers of public roadways in the United States (*1*). Compared with paved roads, granular-surfaced roads more frequently experience severe damage including rutting, potholes, and washboarding, which adversely affect traffic safety and significantly increase maintenance costs (*2–4*). Current practice for repairing such damage typically involves covering the damaged roadway surface with fresh quarried aggregate. However, most state department of transportation (DOT) specifications for gradation and plasticity of granular surface materials are neither performance based nor strictly executed. Consequently, considerable variation exists in the performance and durability of granular surface roads, and substantial amounts of the freshly placed material for maintenance rapidly degrades to smaller particles and dust (*5, 6*).

The importance of the gradation and plasticity of granular surface materials has long been recognized (*5, 7–11*). However, very few studies have focused on quantifying the effects of gradation and plasticity on the performance of granular road surface materials (*12*). Granular surface materials are quite different from pavement-base materials, which usually have a larger top size and a very small percentage of fines, which provides better drainage but would result in a surface that is unstable and difficult to maintain when used on granular-surfaced roads (*7*). Granular-surfaced roads therefore typically require a smaller top size for better stability and ride quality, and a small amount of plastic fines to bind the aggregate together and reduce aggregate loss. Specifications for gradation and plasticity index of unpaved road surface materials vary between different state DOTs. However, most of the existing specifications were established based on arbitrary gradation bands with less than six control points, and many do not specify a plasticity index.

Paige-Green developed a performance-related material selection chart for unpaved road surfaces based on testing and monitoring of 110 unpaved road sections for more than three years in Southern Africa (*12*). In this chart, a grading

[1]Key Laboratory for Special Area Highway Engineering, MOE, School of Highway, Chang'an University, Xi'an, China
[2]Department of Civil Construction and Environmental Engineering, Iowa State University, Ames, IA
[3]Iowa Department of Transportation Office of Research and Analytics, Ames, IA

Corresponding Author:
Address correspondence to Cheng Li: cli@chd.edu.cn

Figure 1. Photograph of (*a*) UCS test and (*b* and *c*) slaking test after 1 minute and 35 minutes, on the control (C) and 4% bentonite (B) treated 2-by-2 specimens.

coefficient and shrinkage product, which are calculated using results of sieve analyses and bar linear shrinkage tests, are related to the observed performance of granular roads. To use the chart, however, requires calibrations for local subgrade conditions and local surface material properties.

Previous studies also found that sodium bentonite could be used to increase stability and reduce dust for crushed limestone-surfaced roads, because the negatively charged surfaces of the clay particles interact with positively charged limestone surfaces to effectively bond the finer particles to the larger limestone particles (*13, 14*). Field observations also showed that an appropriate amount of plastic fines can help to obtain a tighter and smoother roadway surface (*15, 16*). However, the presence of large quantities of plastic fines could lead to a slippery surface and swelling-related issues (*4, 5, 8*).

In this study, extensive laboratory soaked California bearing ratio (CBR), unconfined compressive strength (UCS), and slaking tests were conducted to quantify the effects of gradation and plasticity on the shear strength and slaking characteristics of granular surface materials. Based on the laboratory tests and statistical analysis results, a performance-based method is proposed for designing and specifying the gradation and plasticity of materials for new granular roadway construction, as well as recycling of existing surface materials by mixing with virgin aggregate and plastic fines.

To validate the performance of the proposed method, field test sections were constructed, tested, and monitored in Pottawattamie County, Iowa. Dynamic cone penetrometer (DCP) and light weight deflectometer (LWD) tests were performed after construction and after one seasonal freeze–thaw period to compare the performance of the sections. In this study, a gradation optimization program was also developed and implemented in a Microsoft Excel spreadsheet to enable secondary roads agencies to implement the proposed design method. This paper presents the laboratory and field test results, details the proposed design method, and recommends complete testing, design, and construction procedures to more cost-effectively build, reconstruct, and maintain unpaved granular-surfaced roadway systems.

Laboratory and Field Testing Methods

Unpaved road surfaces are most prone to damage during spring thaws and rainy seasons, when the thawing or infiltrated water cannot drain efficiently, and the nearly saturated surface materials can easily lose strength under heavy agricultural traffic loads (*15*). Therefore, the laboratory evaluations in this study were focused on quantifying the gradation and plasticity effects on the post-saturation performance of surface materials. DCP and LWD tests were conducted to compare the performance of the various test sections.

California Bearing Ratio Test

Soaked CBR tests were conducted in accordance with ASTM D1883 (*17*). The specimens were soaked for 48 h before testing and were submerged in a water tub during the test. The gradations of test specimens were adjusted manually, and sieve analysis tests were performed after each test in accordance with ASTM C136 (*18*).

Unconfined Compressive Strength and Slaking Tests

To evaluate the effects of plasticity on the undrained shear strength and slaking behavior of granular surface materials, the UCS and slaking tests were conducted on specimens consisting of the minus No. 40 fraction of the samples, as shown in Figure 1. The plasticity of the specimens was adjusted by incorporating different percentages of bentonite powder (sodium montmorillonite). The bentonite ($Na_{0.3}(Al, Mg)_2Si_4O_{10}(OH)_2\ 4H_2O$) used in this study was purchased

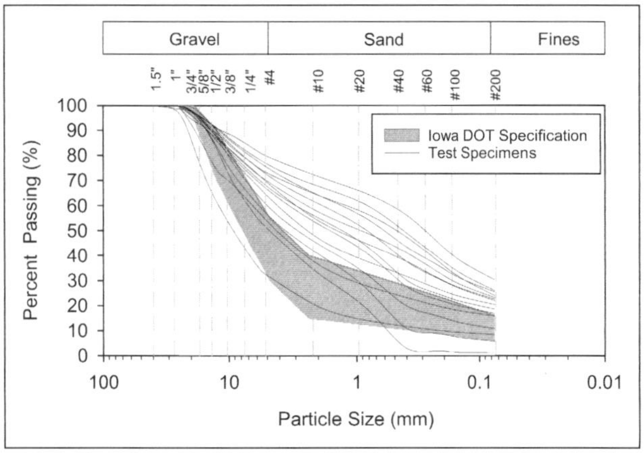

Figure 2. Particle size distribution curves of the 14 laboratory CBR test specimens.

from a company in Wyoming. The liquid limit and plastic limit of the material were found to be 297% and 35%, respectively.

For the UCS and slaking tests, cylindrical specimens with a diameter and height both equal to 50 mm (2 in.) were prepared using the 2-by-2 compaction apparatus (O'Flaherty et al. 1963). Previous studies have demonstrated that the 2-by-2 compaction device can achieve moisture-density results similar to those obtained by standard Proctor compaction tests (*19*). The slaking test was used to evaluate the long-term water susceptibility of the specimens by placing them on a No. 4 sieve and submerging them in a water tub, as shown in Figure 1*b*. The time for the specimens to disintegrate and pass through the No. 4 sieve was then recorded (Figure 1*c*). Atterberg limits of the specimens were determined in accordance with ASTM D4318 (*20*).

Dynamic Cone Penetrometer Test

DCP tests were performed in accordance with ASTM D6951 to estimate the in situ CBR of the surface aggregate and subgrade materials of the test sections (*21*). The in situ CBR values for USCS CL soils were estimated using empirical correlations recommended by the ASTM standard Equations 1 and 2, wherein the DCP Index (DCPI) is defined as the penetration distance (mm) per blow;

$$CBR = \frac{292}{DCPI^{1.12}} \left(\begin{array}{c} \text{for all soils, excluding CH} \\ \text{soils and CL soils with CBR>10} \end{array} \right) \quad (1)$$

$$CBR = \frac{1}{(0.017019 \times DCPI)^2} \left(\begin{array}{c} \text{for CL soils} \\ \text{with CBR<10} \end{array} \right) \quad (2)$$

The thickness of each material layer was estimated using DCP test data, with boundaries between the layers identified by sudden changes in the slope of the cumulative blows versus depth profile, as well as sudden jumps in the CBR versus depth profile. The depth-weighted average CBR of each material layer was calculated using Equation 3;

$$\begin{array}{c}\text{Weighted-average} \\ DCP-CBR\end{array} = \frac{(CBR_i \times H_i) + (CBR_{i+1} \times H_{i+1}) \cdots (CBR_n \times H_n))}{\sum_i^n H_i} \quad (3)$$

The notation $DCP\text{-}CBR_{AGG}$ and $DCP\text{-}CBR_{SG}$ will be used to denote the weighted average *DCP-CBR* of the surface aggregate layer and the subgrade to a depth of 450 mm below the surface of the subgrade, respectively.

Light Weight Deflectometer Test

The LWD test was used to evaluate the composite stiffness of the test sections. The test involves dropping a 10-kg weight onto a rigid 300-mm diameter loading plate and measuring ground surface deflections using an accelerometer embedded in the loading plate. The composite elastic modulus (E_{LWD}) can be calculated using Equation 4;

$$E_{LWD} = \frac{(1 - v^2)\sigma_0 A}{d_0} \times f \quad (4)$$

where d_0 is the measured deflection at the center of the loading plate (mm); v is the Poisson's ratio (assumed to be 0.4); σ_0 is the normalized applied peak stress (MPa); A is the radius of the plate (mm); and f is the shape factor (assumed to be 2).

Laboratory Evaluations of Gradation and Plasticity Effects

Gradation Effects on the Soaked Bearing Capacity

To evaluate the influence of gradation on the soaked bearing capacity, a total of 14 crushed limestone specimens with various gradations were manually prepared and tested using the soaked CBR test. The particle size distribution (PSD) curves of the testing specimens are compared with the current Iowa DOT gradation specification for unpaved road surface materials in Figure 2.

The gradation and CBR test results are summarized in Table 1 in descending order of fines content (percent passing No. 200) of the specimens. Previous studies have investigated the influence of fines content on the mechanical properties of aggregate materials. However, in this study, statistical analysis conducted on the test data showed that both the fines content and gravel content (percent retained on

Table 1. Gradations, Fuller's Model Parameters, and Soaked CBR Test Results of the 14 Specimens

Specimen no.	Gradation			Parameters of Fuller's model			Water content after soaking (%)	As-compacted dry unit weight (lb/ft^3)	Soaked CBR (%)
	% Gravel	% Sand	% Fines	D_{max} (in.)	n	R^2			
1	20.8	48.9	30.3	0.82	0.17	0.97	11.1	19.6	8.9
2	25.5	49.0	25.5	1.00	0.20	0.98	9.0	20.4	30.5
3	27.5	48.5	24.0	0.93	0.20	0.98	10.8	20.5	26.0
4	30.9	45.9	23.2	0.92	0.23	1.00	9.4	20.6	31.7
5	31.7	45.7	22.6	0.91	0.25	1.00	8.8	20.9	56.0
6	28.5	49.5	22.0	0.81	0.24	0.99	8.5	21.1	41.0
7	37.1	42.5	20.4	0.92	0.29	1.00	9.0	20.7	63.3
8	33.2	48.2	18.6	0.85	0.28	0.99	10.1	21.3	41.2
9	34.4	48.5	17.1	0.84	0.30	0.99	9.4	21.6	51.3
10	47.3	36.8	15.9	1.04	0.36	1.00	8.4	20.4	66.0
11	41.0	48.2	10.8	0.86	0.36	0.99	9.0	21.4	50.7
12	68.7	22.8	8.5	1.02	0.64	0.98	9.2	19.2	42.2
13	44.5	49.7	5.8	0.84	0.43	0.98	8.2	20.6	37.8
14	50.1	48.7	1.2	0.83	0.51	0.98	8.8	20.5	20.7

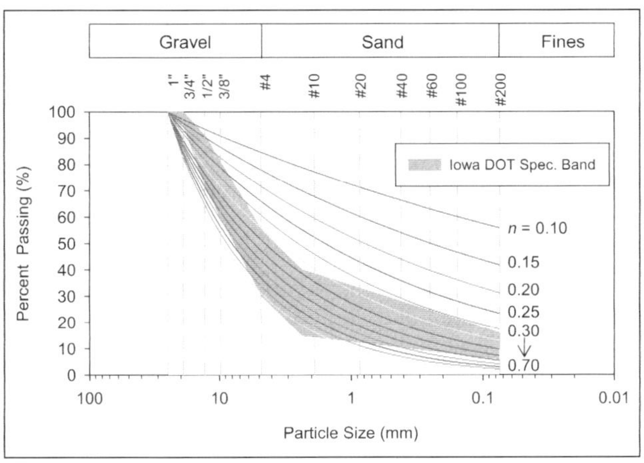

Figure 3. PSD curves generated using Fuller's model with D_{max} equal to 25.4 mm and n value increasing from 0.1 to 0.7.

No. 4 sieve) had a statistically significant influence on the soaked CBR values. The sand content was determined to have a statistically insignificant influence in this study, but this could be because the sand contents of the specimens varied within a small range. These results suggest that the soaked strength of a granular material depends more strongly on its gradation and associated particle packing than whether the particles fall within a specific range, such as a gradation specification band. To determine the optimal gradation that will give the best strength and performance of a granular road surface material, parameters or models that can quantify the material's PSD and packing are needed.

Based on comparisons among various parameters and models in this study, it was found that Fuller's model Equation 5 is the simplest mathematical model for quantifying PSDs of well-graded granular materials (*22*);

$$p_i = \left(\frac{D_i}{D_{max}}\right)^n \times 100 \qquad (5)$$

where subscript "i" represents a particular sieve; p_i is percentage passing sieve number i; and D_i is the corresponding opening size.

In Fuller's model, the gradation of a material can thus be represented by only two parameters: the maximum aggregate size (D_{max}) and the shape factor of the PSD curve (n). Previous studies demonstrated that the coefficient of determination (R^2) values of Fuller's model are usually higher than 0.97 for well-graded materials, but the model may not be appropriate for gap-graded materials (*22, 23*).

The PSD curves of the laboratory CBR test specimens in this study were fit by Fuller's model, giving the D_{max}, n, and R^2 values listed in Table 1. The maximum aggregate size varies from 20.7 to 26.4 mm, and the n value varies from 0.17 to 0.64.

For granular road surface materials, 19 mm (3/4 in.) and 25.4 mm (1 in.) are the most commonly used top sizes, and n values between 0.1 and 0.7 can cover a wide range of gradations as demonstrated in Figure 3. The n value increases as the coarseness of the material increases.

A multiple regression analysis was performed on the data of Table 1 to quantify how changes in gradation influence the soaked CBR strength of the tested material. The resulting regression model equation, shown in Figure 4, has an R^2 of 0.80 and all variables of the model are statistically significant with a 95% confidence interval (p-values < 0.05 and t-values < -2 or > 2). A 3D regression surface plot is compared with the lab testing results in Figure 4*a*. In general, the soaked CBR value increases as D_{max} increases, and a bell-shape trend can be observed as n varies. This result indicates

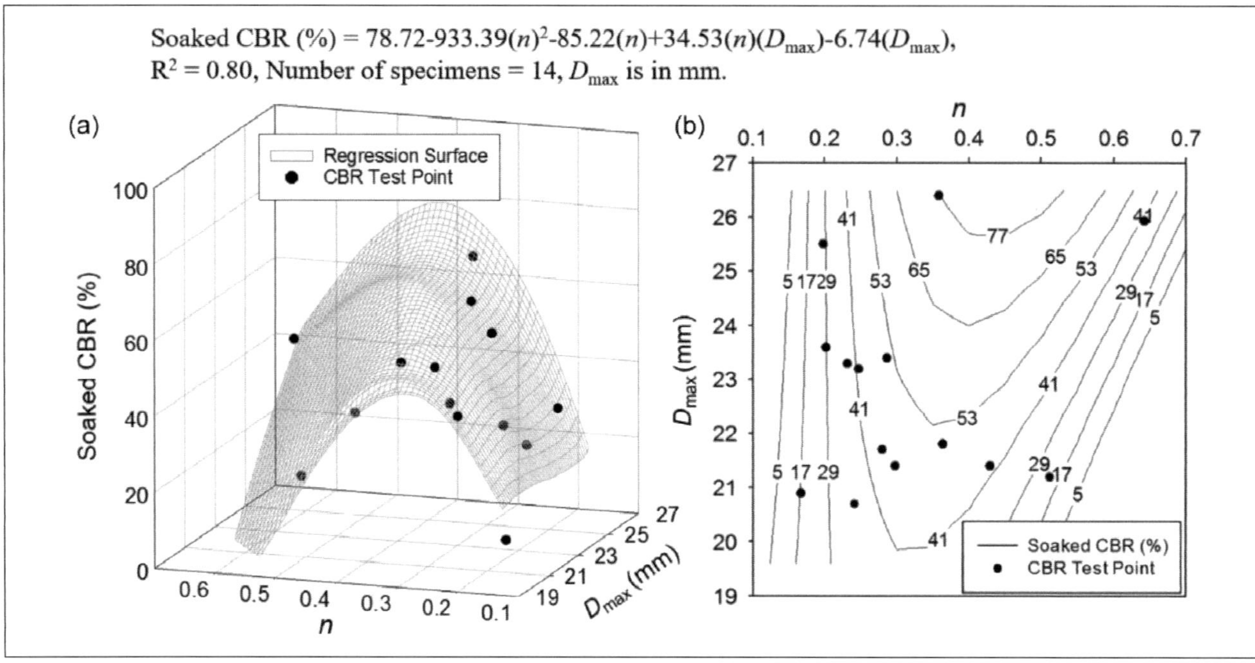

Figure 4. Multiple regression analysis results on soaked CBR of specimens: (*a*) 3D regression surface and (*b*) 2D contour map of regression surface.

that the undrained bearing capacity of a well-graded granular material increases as top size increases, and an optimal gradation in terms of the soaked CBR can be quantified by the shape parameter (n) for a given top size.

The regression surface is also plotted as a 2D contour map in Figure 4*b*. Since CBR is commonly used for thickness design of granular surface layers, and each CBR data point in Figure 4*b* represents a specific PSD curve, this contour plot can be used to develop performance-based design criteria and specify acceptable or target gradation ranges for granular surface materials.

From Figure 3, it can be seen that materials with a top size of 25.4 mm (1 in.) and n values between 0.3 and 0.7 can meet the current DOT specification band. However, Figure 4*b* reveals that the corresponding soaked CBR values reduce by an order of magnitude (from 65% to 5%) as n increases from 0.3 to 0.7. This result clearly demonstrates that an arbitrary gradation band is not sufficient to ensure strength and performance of granular road surface materials, and it may also explain the considerable variations in observed performance and durability of granular-surfaced roads.

For the materials tested in this study, Figure 4*b* also shows that the optimal n value increases from 0.28 to 0.4 as D_{max} increases from 19.5 mm to 25.4 mm. For a material with a given top size, as the n value decreases beyond the optimal value, the soaked CBR decreases rapidly and becomes practically independent of D_{max}. This can be explained by particle packing characteristics; when the coarseness of a material decreases (n decreases) beyond a certain level, the coarse aggregates eventually become floating in a matrix of finer particles, and the particle interlocking between coarse aggregates is thereby greatly reduced, significantly lowering the shear strength of the material. The saturated mechanical properties of the material are then governed by those of the fine matrix. Conversely, when the coarseness of the material increases (n increases), the material becomes less well graded as indicated by a steeper PSD (Figure 3), and the CBR for a given top size also decreases as n increases beyond the optimum value (Figure 4*b*), but CBR remains somewhat proportional to top size.

It is well known that particle packing is also influenced by the material morphology. The statistical model developed in this study (the Equation in Figure (4)) is based on the test results of crushed limestone materials with an average sphericity of 0.6 as reported in Li et al. (*6*). Therefore, the statistical model needs to be validated for a wider range of materials having sufficiently different morphologies and geologic origins. Based on the authors' visual observations, most granular surface materials except for those consisting of rounded river gravel have very similar morphologies to those tested in this study. The statistical model is therefore likely to be useful for typical crushed limestone materials with top sizes of 19 mm to 25.4 mm.

Plasticity Effects on Undrained Shear Strength and Slaking Characteristics

Previous studies have demonstrated that the plasticity of granular road surface materials is related to their surface deterioration, stability, potholes, aggregate loss, and dust

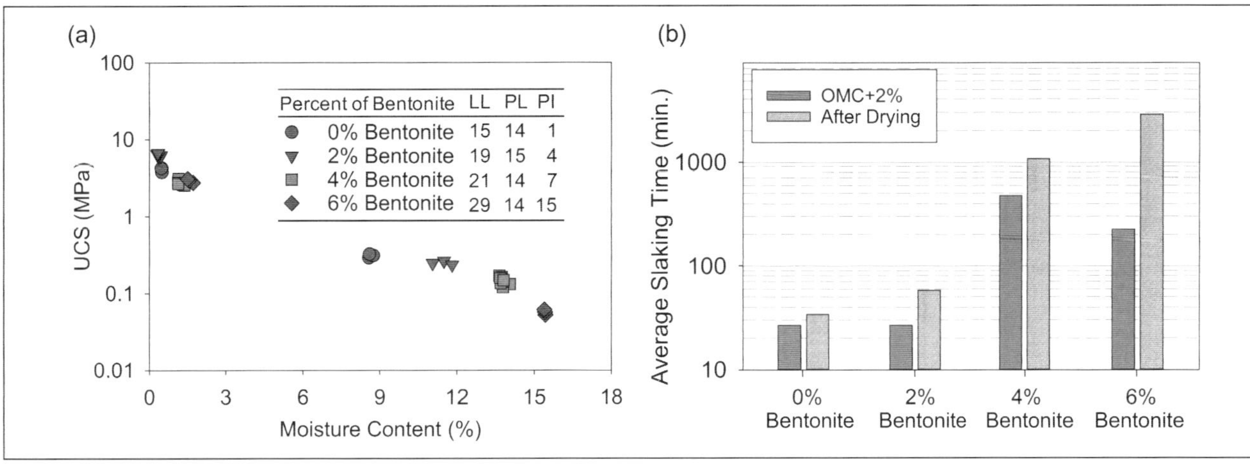

Figure 5. Results of (*a*) UCS and (*b*) slaking tests conducted on 2-by-2 specimens of minus No. 40 materials treated with different percentages of bentonite.

emissions (*5, 12, 15*). In the present study, laboratory tests were conducted on specimens of minus No. 40 material with various amounts of added bentonite clay, to determine the optimal plasticity index for granular road surface materials with respect to shear strength and slaking characteristics.

The UCS and slaking test results are shown in Figure 5*a* and *b*, respectively, which both have logarithmic scales on the *y*-axes. The tests were conducted on specimens that were compacted approximately 2% wet of optimum moisture content, and also on specimens that were similarly compacted then oven-dried for 24 h at 40°C. These two moisture contents are approximately the upper and lower bounds of the moisture content range observed in the field. Figure 5*a* shows that the UCS of the dried specimens with a range of bentonite percentages are approximately the same, but the UCS can reduce by more than one order of magnitude as moisture content increases. In addition, the specimens with 6% bentonite (by dry mass of minus No. 40 material) generally resulted in the highest reductions in UCS. These test results suggest that adding too much plastic fines such as bentonite may lead to rutting and skidding issues under wet conditions.

For slaking behavior, Figure 5*b* shows that the oven-dried specimens yielded longer disintegration times than the specimens with as-compacted (OMC+2%) water contents, and the 4% bentonite-treated specimens yielded the best slaking behavior among the specimens with as-compacted water contents. Based on these results, a bentonite content between 4% and 6% by dry mass of particles passing the No. 40 sieve may be the optimal range for treating crushed limestone surface materials. The corresponding plasticity index range of the bentonite-treated materials is between 7 and 15.

Construction and Performance of Field Test Sections

Based on the laboratory and statistical analysis results, three test sections were designed and constructed in Pottawattamie County, Iowa. One existing control section next to the test sections was also included for comparison. Pre-construction DCP tests were conducted to determine the thickness of the existing surface layer and the in situ CBR of the subgrade material. Sieve analysis and plasticity tests were also performed to determine the soil index properties of the existing surface material. The selected road had an annual average daily traffic count of 70 vehicles the year 2015 (*24*). According to the AASHTO aggregate-surfaced road design catalog (*25*), a surface thickness of 178 mm (7 in.) is recommended for roads in climatic Region 3 with fair or good quality subgrades and low traffic volumes. The as-built cross-section profiles of the four test sections are shown in Figure 6, which were determined based on the DCP test results performed after construction.

For test Section 1, the existing surface material was recycled by mixing in virgin materials and bentonite to reach to the optimal gradation ($n = 0.4$, $D_{max} = 25.4$ mm) using Figure 4*b*. As discussed in the previous section, the optimal range of the bentonite content is 4% to 6% by dry mass of particles passing the No. 40 sieve. For Section 1, 6% bentonite for the portion passing the No. 40 sieve equals 2% of the dry mass of the entire surface material gradation, and an additional 1% was added to compensate for material loss during construction. Therefore, 3% bentonite by dry mass of the entire surface material gradation was used to adjust the plasticity. The material labeled Virgin Aggregate 1 was uniformly blended with the existing road surface aggregate by windrowing using a motor grader. A tractor-powered soil reclaimer was used to incorporate the 3% bentonite. The surface layer was then compacted using four passes of a 7,000 kg vibratory smooth-drum roller, and the compaction water content was controlled to be close to the laboratory-determined optimum water content. For Section 2, the county's conventional reconstruction method was followed, except compaction was performed using the vibratory roller instead of a rubber tire roller. For Section 3, the Virgin Aggregate 2 and existing

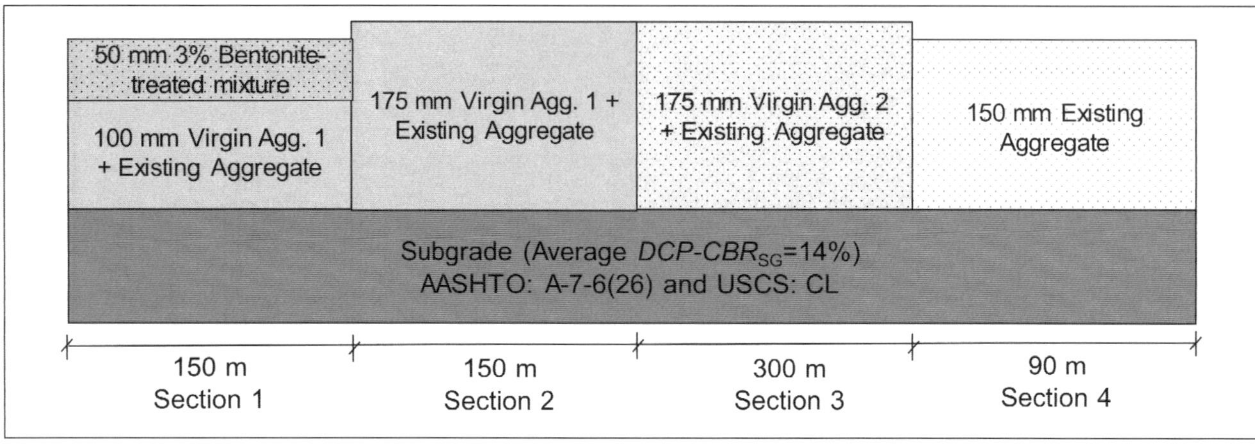

Figure 6. As-built cross-section profiles of the field test sections (not to scale).

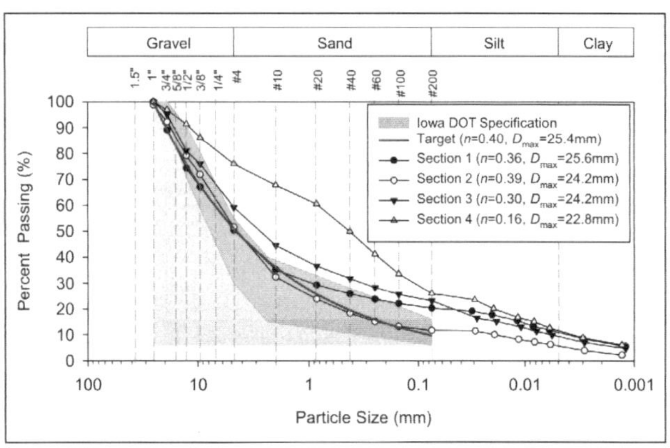

Figure 7. PSD curves of the representative surface materials of the test sections.

aggregate were mixed using the soil reclaimer and compacted using the vibratory roller. Section 4 was a control section featuring existing aggregate and current maintenance practices. The test sections were constructed on July 5th and 6th, 2016.

Sieve analysis tests were performed on representative samples collected from the test sections during construction, and their resulting PSD and Fuller's model gradation parameters are compared in Figure 7. The n value of the test section surface materials ranges between 0.16 and 0.39, and the gradation of the Section 2 material is closest to the theoretical optimal gradation ($n = 0.40$) determined using Figure 4.

Field DCP and LWD tests were conducted on the four test sections after construction and again after the 2016–2017 seasonal freeze–thaw period. The soaked CBR values predicted using the statistical model shown in Figure 4 are compared in Figure 8*a* to the in situ CBR values estimated using the DCP correlations. The trends of the in situ DCP test results generally agree with the predictions of the statistical model. However, the in situ CBR values of the surface aggregate layers of Sections 3 and 4 are similar, but do not agree with the predicted results. The statistical model predicts a much lower CBR for Section 4 than measured in the field (note the different y-axis scales for the predicted and in situ values). The higher in situ CBR values can be explained by the fact that the statistical model was developed based on laboratory soaked CBR test results, whereas both of the two groups of field DCP tests were conducted under dry conditions. Field observations also revealed more severe surface damage for the control Section 4 after heavy rain events, which indicates that the bearing capacity of the existing surface material with a low shape factor of $n = 0.16$ can greatly decrease under wet conditions.

The LWD test results show that the as-built composite modulus (E_{LWD}) values of the three test sections were approximately the same, and were all greater than that of the control Section 4 (Figure 8*b*). After one seasonal freeze–thaw period, the E_{LWD} values of all sections decreased noticeably except for Section 2, for which the gradation was the closest to the theoretical optimal target gradation as shown in Figure 7. This observation supports the hypothesis that the granular roadway will perform best the closer the material is to the optimum gradation determined by the statistical model developed in this study.

Recommended Testing, Design, and Construction Procedures

Based on the laboratory and field evaluations of the proposed method, a complete set of testing, design, and construction procedures is recommended below. In addition, to help secondary roads agencies implement the proposed method and also recycle existing degraded surface materials, a Microsoft Excel-based program was developed to optimize proportions of existing surface materials and two or three available quarry materials to reach the target "optimal" gradation (Figure 9). The program can be obtained from the second author. The recommended steps are as follows:

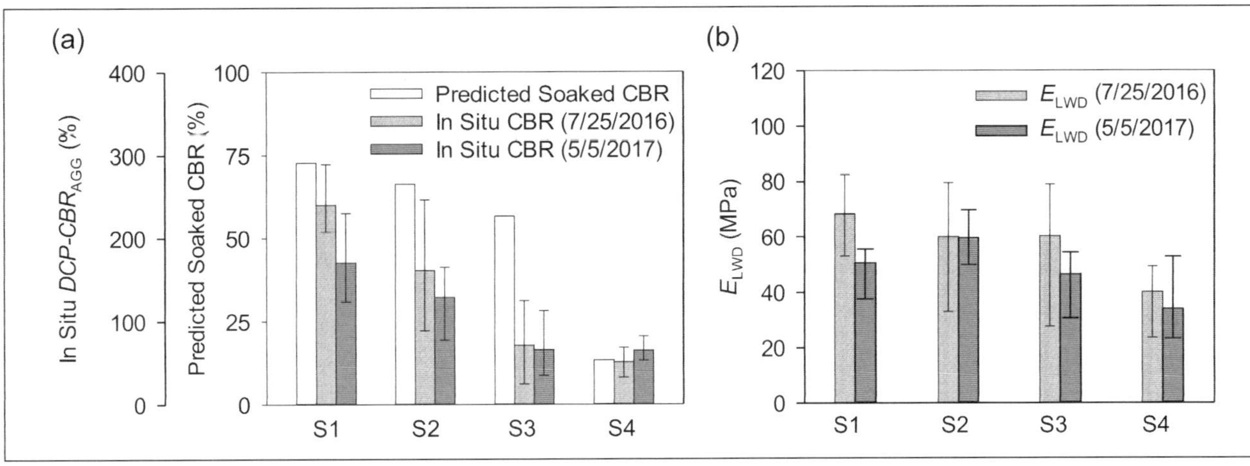

Figure 8. (*a*) Predicted soaked CBR versus in situ DCP-CBR values and (*b*) in situ composite modulus values of the four test sections.

1. Determine the road geometries and the thickness, gradation, and plasticity of the existing surface aggregate layer. To collect the design inputs, DCP tests can be used to rapidly determine the average thickness of the existing surface aggregate layer as well as the in situ DCP-CBR values of the surface and subgrade materials.
2. Design the thickness of the aggregate surface layer based on the in situ DCP-CBR values and traffic information. The thickness of the surface layer can be designed following the AASHTO (*25*), Giroud and Han (*26*), or other methods.
3. Determine the optimal shape factor (n) based on the top size of the material using Figure 4. Note, the statistical model was developed based on crushed limestone materials, and needs further validation for other materials having significantly different particle shapes and origins.
4. Calculate the design proportions for mixing the existing and two or three available virgin quarry materials using the Microsoft Excel-based program. Most quarries either do not produce custom gradations or else they charge extra for them, so blending of two or three available quarry materials with the existing surface material is a more feasible solution to achieve the target gradation. The program will optimize the proportions of the quarry materials and calculate the needed quantities in tons per mile for each type material. Note that the quarry gradations used for the optimization should be representative and may therefore need to be verified.
5. For construction, the aggregate materials can be mixed in a quarry or onsite using motor graders. Conventional equipment is sufficient to implement the proposed method, although compaction at optimum moisture content by a smooth-drum vibratory roller is highly recommended. Quality control tests may be required to check the gradations of the quarry materials and the blended mixtures.
6. If bentonite or a local clay will be incorporated to increase the plasticity of the surface material, Atterberg limits tests should be performed to determine the plasticity index of the treated mixture. In addition, it is recommended that the plastic fines be mixed only into the top 50–75 mm of the roadway, because plastic fines can greatly reduce shear strength of granular materials under prolonged wet conditions. By mixing the fines into the top portion of the surface course, they perform the desired function of binding the larger aggregates to reduce material loss, while preserving the shear strength of the deeper aggregates in the lower part of the surface course.

Summary and Conclusion

This study focused on evaluating the effects of gradation and plasticity on the mechanical properties of granular road surface materials. Based on results of laboratory tests, field tests, and statistical analyses, a performance-based method for designing or specifying the gradation and plasticity of granular road surface materials was presented.

The theoretical optimal gradation in terms of the predicted soaked CBR for a given material with known top size can be determined using the statistical model presented here. The model is applicable to crushed limestone materials, and may require further validation for other aggregate types or morphologies. Comparing the proposed method with current DOT specifications in the form of an arbitrary gradation band, the proposed method is more performance related than the gradation band, and can be used to develop specifications with more precise targets. For the plasticity effect, laboratory test results showed that adding plastic fines can reduce the shear strength under wet conditions but greatly improve the

Gradation Optimization for Unpaved Road Surface Materials

District		**Project**	TR-685 Test Section 2	**Date**	5/19/2016
County	Pottawattamie	**Note**	This is a trial version of the program	**Designer**	Cheng Li

Road Length	500	ft
Average Road Width	30	ft
Thickness of Existing Aggregate	2.30	in.
Target Final Thickness	6.00	in.
Compacted Dry Unit Weight	130	pcf
Maximum Aggregate Size (D_{max})	1.00	in.
PSD Shape Factor (*n*)	0.40	

INSTALL SOLVER

RUN

Sieve No.	Sieve size (mm)	Target Gradation (%)	Existing Material Gradation (%)	Target Virgin Material Gradation (%)	Quarry Material A	Quarry Material B	Quarry Material C	Optimized Virgin Material Gradation (%)	Actual Final Gradation (%)
2	50.8	100.0	100.0	100.0	100.0	100.0	100.0	100.0	100.0
1.5	38.1	100.0	100.0	100.0	100.0	100.0	100.0	100.0	100.0
1	25.4	100.0	100.0	100.0	60.0	95.0	100.0	95.9	97.5
3/4	19.00	89.0	97.5	85.9	23.0	84.0	100.0	92.2	94.2
1/2	12.70	75.8	89.8	71.4	6.3	70.0	83.0	75.2	80.8
3/8	9.51	67.5	81.8	64.1	6.0	60.0	71.0	64.4	71.1
#4	4.76	51.2	66.6	48.8	5.6	37.0	45.0	41.0	50.8
#8	2.38	38.8	55.9	36.1	5.1	21.0	27.0	24.8	36.7
#16	1.19	29.4	48.4	25.5	4.6	14.0	20.0	18.4	29.9
#30	0.595	22.3	41.9	17.5	4.1	11.0	16.0	14.8	25.2
#50	0.297	16.9	35.7	12.0	3.4	8.4	14.0	12.9	21.7
#100	0.149	12.8	29.3	8.6	2.9	6.8	12.0	11.1	18.1
#200	0.075	9.7	25.2	5.4	2.3	5.6	10.2	9.4	15.5
			Proportion (%)	100	10	0	90		
			Quantity (tons)	300.6	30.5	0.0	270.1		

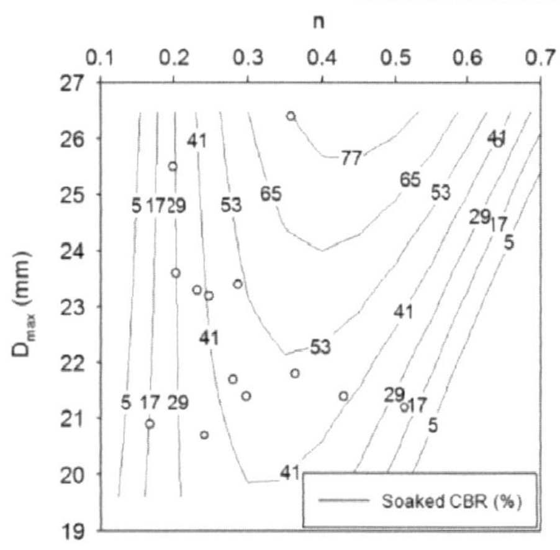

The optimal *n* value can be selected using the chart developed based on lab CBR test results in the IHRB Project TR-685.

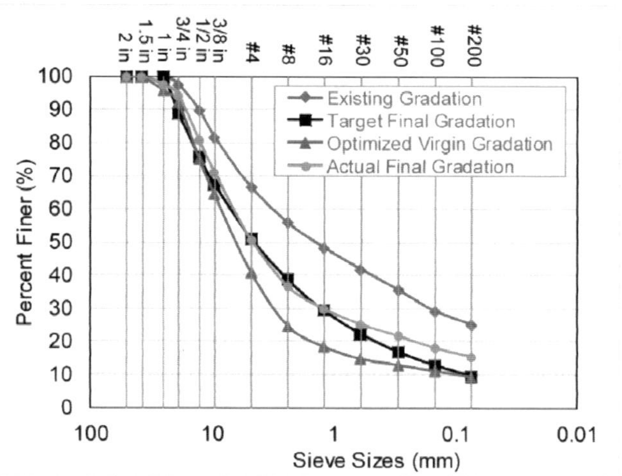

Figure 9. Screenshot of the gradation optimization program.

slaking performance. Based on the results of this study, the plastic index of granular surface materials should be between 7 and 15.

Field test sections were designed and constructed using the proposed method, and field test results conducted following construction and after one seasonal freeze–thaw period further validated the performance of the proposed method. The set of testing, design, and construction procedures described above are recommended to provide state secondary roads departments with more cost-effective solutions to build or reconstruct granular road systems with improved performance and durability.

Acknowledgments

The research project was sponsored by the Iowa DOT through the Iowa Highway Research Board. The authors would like to acknowledge the support and assistance provided by the Pottawattamie County Secondary Roads Department in constructing the test sections. The assistance of research assistants Yijun Wu, HaSung Kim, and Sajjad Satvati is also appreciated.

Author Contributions

The authors confirm contribution to the paper as follows: study conception and design: CL, JA, BC, CJ, and VG, data collection: CL, JA, BC; analysis and interpretation of results: CL, JA, BC; draft manuscript preparation: CL, JA. All authors reviewed the results and approved the final version of the manuscript.

References

1. FHWA. *Highway Statistics 2016*. FHWA, Washington, D.C., https://www.fhwa.dot.gov/policyinformation/statistics/2016/. Accessed July, 2017.
2. Jahren, C. T., D. Smith, J. Thorius, M. Rukashaza-Mukome, D. White, and G. Johnson. *Economics of Upgrading an Aggregate Road*. MN/RC-2005-09, Minnesota DOT, Research Services Section, Iowa State University, Saint Paul, 2005.
3. De Vries, S. W. *Financial Needs of Iowa's County Roads*. Project TR-608. Iowa Department of Transportation, Ames, 2012.
4. White, D., and P. Vennapusa. *Low-Cost Rural Surface Alternatives: Literature Review and Recommendations*. Project TR-632. Iowa Department of Transportation, Ames, 2013.
5. Jones, D., and P. Paige-Green. Limitations of Using Conventional Unpaved Road Specifications for Understanding Unpaved Road Performance. *Transportation Research Record: Journal of the Transportation Research Board*, 2015. 2474: 30–38.
6. Li, C., J. C. Ashlock, D. J. White, C. T. Jahren, and B. Cetin. Gyratory Abrasion with 2D Image Analysis Test Method for Evaluation of Mechanical Degradation and Changes in Morphology and Shear Strength of Compacted Granular Materials. *Construction and Building Materials*, Vol. 152, 2017, pp. 547–557.
7. Skorseth, K., and A. A. Selim. *Gravel Roads: Maintenance and Design Manual*. U.S. Department of Transportation Federal Highway Administration, Washington, D.C., 2000.
8. Paige-Green, P. *Material Selection and Quality Assurance for Labour-Based Unsealed Road Projects*. Technical Brief No. 9. International Labour Organisation, Advisory Support, Information Services, and Training, Geneva, 1998.
9. Hudson, S. W., B. F. McCullough, and R. F. Carmichael. *Aggregate and Paved Surface Design and Rehabilitation Manual for Low-Volume Roads*. FHWA Report FH40/3. Federal Highway Administration U.S. Department of Transportation, Washington, D.C. 1986.
10. Li, Y., Yang, Y., Yu, H., and Roberts, G. Principal Stress Rotation under Bidirectional Simple Shear Loadings. *KSCE Journal of Civil Engineering*, Vol. 22, No. 5 , 2017, pp. 1651–1660.
11. Van Zyl, G., M. Henderson, and R. Uys. Applicability of Existing Gravel-Road Deterioration Models Questioned. *Transportation Research Record: Journal of the Transportation Research Board*, 2007. 1989: 217–225.
12. Paige-Green, P. *The Influence of Geotechnical Properties on the Performance of Gravel Wearing Course Materials*. PhD dissertation. University of Pretoria, South Africa, 1989.
13. Bergeson, K. L., J. W. Waddingham, S. G. Brocka, and R. K. Lapke. *Bentonite Treatment for Economical Dust Reduction on Limestone Surface Secondary Roads*. Project HR-351. Iowa DOT, Ames, 1995.
14. Bergeson, K. L., and A. M. Wahbeh. *Development of an Economic Dust Palliative for Limestone Surfaced Secondary Roads*. Project HR-297. Iowa DOT, Ames, 1990, p. 64.
15. Li, C., P. K. R. Vennapusa, J. Ashlock, and D. J. White. Mechanistic-Based Comparisons for Freeze-Thaw Performance of Stabilized Unpaved Roads. *Cold Regions Science and Technology*, Vol. 141, 2017, pp. 97–108.
16. Li, C., J. C. Ashlock, D. J. White, and P. K. R. Vennapusa. Mechanistic-Based Comparisons of Stabilised Base and Granular Surface Layers of Low-Volume Roads. *International Journal of Pavement Engineering*, 2017, pp. 1–13. doi:10.1080/10298436.2017.1321417
17. ASTM International. *Standard Test Method for California Bearing Ratio (CBR) of Laboratory-Compacted Soils*. ASTM D1883-06. ASTM International, West Conshohocken, Pa., 2014.
18. ASTM International. *Standard Test Method for Sieve Analysis of Fine and Coarse Aggregates*. ASTM C136/C136M-14. ASTM International, West Conshohocken, Pa., 2014.
19. O'Flaherty, C. A., C. E. Edgar, and D. T. Davidson. Iowa State Compaction Apparatus for Measurement of Small Soil Samples. *Highway Research Record*, No. 22, 1963.
20. ASTM International. *Standard Test Methods for Liquid Limit, Plastic Limit, and Plasticity Index of Soils*. ASTM D4318-10. ASTM International, West Conshohocken, Pa., 2013.
21. ASTM International. *Standard Test Method for Use of the Dynamic Cone Penetrometer in Shallow Pavement Applications*. ASTM 6951-09. ASTM International, West Conshohocken, Pa., 2013.
22. Fuller, W. B., and S. E. Thompson. The Law of Proportioning Concrete. *Transactions of ASCE*, Vol. 59, No. 2, 1907, pp. 67–143.
23. Xiao, Y., E. Tutumluer, Y. Qian, and J. A. Siekmeier. Gradation Effects Influencing Mechanical Properties of Aggregate Base-Granular Subbase Materials in Minnesota. *Transportation Research Record: Journal of the Transportation Research Board*, 2012. 2267: 14–26.
24. Iowa DOT. *Iowa County Traffic Map*. Iowa DOT. https://iowadot.gov/maps/digital-maps/traffic-maps/county. Accessed April, 2015.
25. AASHTO. *AASHTO Guide for Design of Pavement Structures*. Washington, D.C.: AASHTO, 1993, 48–63.
26. Giroud, J. P., and J. Han. Design Method for Geogrid-Reinforced Unpaved Roads. I. Development of Design Method. *Journal of Geotechnical and Geoenvironmental Engineering*, Vol. 130, No. 8, 2004, pp. 775–786.

The Standing Committee on Low-Volume Roads (AFB30) peer-reviewed this paper (18-05314).

Article

TRR
JOURNAL OF THE TRANSPORTATION RESEARCH BOARD

Transportation Research Record
2018, Vol. 2672(52) 226–237

DOI: 10.1177/0361198118799035
journals.sagepub.com/home/trr
SAGE

Safety Performance Functions for Rural Two-Lane County Road Segments

Steven Y. Stapleton[1], Anthony J. Ingle[1], Meghna Chakraborty[1], Timothy J. Gates[1], and Peter T. Savolainen[1]

Abstract

Safety performance functions (SPFs) were developed for rural two-lane county roadway segments in Michigan. Five years of crash data (2011 to 2015) were analyzed for greater than 6,500 mi of rural county roadways, covering 29 of Michigan's 83 counties and representing all regions of the state. Three separate models were developed to estimate annual deer-excluded total and injury crashes on rural county roadways: 1) paved federal-aid segments, 2) paved non-federal-aid segments, and 3) paved and gravel non-federal-aid segments with fewer than 400 vpd. To account for the unobserved heterogeneity associated with differing county design standards, mixed effects negative binomial models with a county-specific random effect were utilized. Not surprisingly, the county segment SPFs generally differed from traditional models generated using data from state-maintained roadways. County federal-aid roadways general showed greater crash occurrence than county non-federal-aid roadways, the *Highway Safety Manual* (HSM) two-lane rural roadways model, and rural state highways in Michigan. County non-federal-aid paved roadways showed crash occurrence rates that were remarkably similar to the HSM base rural two-lane roadway model, whereas gravel roadways showed greater crash occurrence rates. The presence of horizontal curves with design speeds below 55 mph had a strong association with the occurrence of total and injury crashes across all county road classes. Increasing driveway density was also found to be associated with increased crash occurrence. However, lane width, roadway surface width, and paved shoulder width had little to no impact on total or injury crashes.

Safety on rural highways continues to be a serious concern throughout the United States. Nationwide, approximately one-half of motor vehicle fatalities occur in rural areas, although only approximately 20% of the U.S. population lives in rural areas. In 2015, the rural highway fatal crash rate (per vehicle miles traveled) in the United States was 2.6 times higher than for urban areas, providing further evidence of an overrepresentation of crashes in rural areas (*1*). Several factors contribute to the elevated rural crash risk, including speeds, geometry, lack of lighting, and other factors, each of which contribute to an elevated risk for lane departure crashes, including head-on, sideswipe, or run-off-road, which are among the most severe types.

In many states, the majority of rural highways are maintained by the state department of transportation (DOT). However, several states, particularly those in the Midwest and Great Lakes regions, possess a substantial rural county highway network. Michigan has approximately 120,000 mi of public roadways, nearly 75% of which are owned by one of the 83 county road agencies across the state, with the remainder owned by the state (8%) or city/village (17%). Given the size of Michigan's county roadway network, it is not surprising that 60% of the 71,402 traffic crashes in rural areas in 2015 occurred on county-owned facilities (*2*). Thus, determination of factors affecting crashes on rural county highways is important to support highway safety programs in Michigan.

The first edition of the *Highway Safety Manual* (HSM) includes separate families of safety performance functions (SPFs) to estimate annual crash occurrence for three specific roadway facility types: rural two-lane/two-way roads, rural multilane highways, and urban and suburban arterials (*3*). Separate SPFs exist for intersections and road segments for the base conditions within each facility type, whereas crash modification factors (CMFs) are provided to account for deviations from the base condition of the particular facility type. Because the SPFs contained in the HSM were developed based on a

[1]Department of Civil and Environmental Engineering, Michigan State University, East Lansing, MI

Corresponding Author:
Address correspondence to Timothy J. Gates: gatestim@egr.msu.edu

limited sample of data collected from select states, direct application of the SPFs from the HSM does not tend to provide accurate results unless the models are calibrated using local data (*4–12*). Although the HSM provides details related to local calibration of the models, the fact remains that the HSM's original SPFs were generated based on data obtained from select state highways, and therefore assumptions made on the general effect of characteristics, such as traffic volume or lane width, may not apply to low-volume, county-maintained highways.

In Michigan and elsewhere, county-maintained highways typically possess characteristics that differ considerably from those maintained by the state DOT (MDOT), which limits the usefulness of SPFs and CMFs generated based on rural state highways. These differences include traffic characteristics (e.g., lower traffic volumes, shorter trip lengths, driver familiarity, etc.) and design characteristics (e.g., lower design speeds, gravel surfaces, extreme curvature, narrow lanes and shoulders, reduced sight distances, reduced clear zones, etc.). Although Michigan-specific SPFs currently exist, they are limited to urban, state highways (*13*, *14*). Furthermore, the SPFs contained in the HSM are only applicable to paved roads (*15*), and additional research related to the safety performance of unpaved highways is also limited. Thus, there is a clear need for development of SPFs that are unique to rural county highways. To that end, the primary objective of this research was to develop SPFs unique to county highways, including both paved and unpaved roadways. To accomplish this objective, county highway data, including traffic volumes, roadway characteristics, and traffic crashes were collected from 29 counties across Michigan and subsequently analyzed. The methods and results of this analysis are described in the sections that follow.

Literature Summary

Prior research has explored the safety effects of various cross-sectional and geometric factors for rural highways. The following subsections summarize the extant research literature regarding the relationship between the primary characteristics of rural highways and safety performance.

Lane Width

Narrow travel lanes on rural two-lane highways have been associated with increases in single-vehicle run-off-the-road, head-on, and sideswipe-type crashes (*3*, *16*), and the effect is most pronounced at lane widths of 9 ft or less. However, a recent study in rural Pennsylvania found a lower occurrence of total crashes and fatal and injury (FI) crashes on segments with narrower lanes (*17*). Yet another study found that narrower lane widths were associated with reductions in same-direction crashes, and fatal and incapacitating injury crashes, but an increase in single-vehicle crashes as well as total crashes, non-incapacitating injury, and property damage only (PDO) crashes (*18*).

Shoulder Width

Though the effect of lane width on crashes on rural two-lane highways is mixed, research has consistently found that wider shoulders on rural highways are associated with fewer crashes (*3*, *16–18*), due to the increased recovery and vehicle storage space and increased separation from roadside hazards. Although the magnitude of this effect depends on traffic volumes, the frequency of traffic crashes tends to increase as paved shoulder widths are reduced below 6 ft. Further, safety performance tends to degrade substantially as the paved shoulder width decreases below 2 ft on roadways with greater than 2,000 vpd (*3*).

Access Points

Several prior studies have demonstrated that increasing access point density leads to an increase in crash occurrence, particularly for multi-vehicle crashes (*3*, *19*, *20*). This is at least partially due to driving errors caused by intersections, driveways, or both intersections and driveways, which may result in rear-end, sideswipe, or both rear-end and sideswipe-type crashes (*3*). Specifically, NCHRP Report 420 concluded that increasing the access point density from 10 to 20 access points per mile led to a 40% increase in crashes, whereas increasing access points to 40 access points per mile was associated with a doubling of crash occurrence (*20*).

Alignment

Horizontal curves are among the most critical geometric design elements related to the influence of driver behavior and crash risk (*21*). Specifically, horizontal curves with radii less than 2,600 ft tend to cause a drop in highway running speeds below that of adjacent tangent sections, with substantial speed declines observed for curves with radii less than 800 ft (*22*). It is generally understood that crash occurrence tends to increase as the degree of curvature, length of curvature, or both degree and length increases along a rural highway segment (*5*, *12–15*). On two-lane rural highways, horizontal curves increase crash risk, particularly if operating speeds through the curve are reduced by more than 3 mph from the adjacent tangent section (*9*). A recent analysis of rural two-lane state highways in Pennsylvania found locations with curve radii less than 1,008 ft, which is the approximate radius of a curve designed for 55 mph with a superelevation of

7%, to have 43% greater total crashes and 48% greater injury crashes compared with tangent sections (*18*).

Prior research has also demonstrated that steeper vertical grades are associated with higher crash rates (*5, 11*). Total crash rates generally increase with the degree of vertical curvature (*5*), particularly where hidden horizontal curves, intersections, or driveways are present (*21*).

Pavement Surface

Rural unpaved roads include a wide variety of design standards, design speeds, and surface characteristics, which can be greatly affected by the effects of weather and heavy traffic loads. The safety of these roads may also face issues associated with a lack of pavement markings and insufficient signage, narrow road widths, and the absence of shoulders. A limited amount of research has investigated the safety effects of paved versus unpaved surfaces for low-volume roadways. Differing design standards between primary and local roadways make it difficult to compare safety performance between paved and unpaved roadways without constraining such analyses to roadways with lower traffic volumes and lower functional classes. Nevertheless, research has found that at the lowest of volumes, little to no difference in crash occurrence is observed between paved and unpaved roads. However, at higher volumes, paved roads were found to have fewer crashes than unpaved roads (*23*).

Methods

To provide a better understanding of the relationship between various roadway characteristics and safety performance on rural two-lane county roadways in Michigan, roadway and crash data were obtained for a statewide sample of county highway segments. These data were obtained from a variety of sources for the 5-year period of 2011 through 2015. To provide adequate representation across all classes of rural county roadways, it was necessary to collect data for both federal-aid and non-federal-aid county roadways. Details pertaining to the identification of county highway segments and collection of the relevant data are provided in the sections that follow.

Roadway Segmentation

The Michigan Geographic Framework All Roads (MGF-AR) shapefile, which represents a digital base map for the state, was used to provide the spatial basis for collection of the necessary roadway and traffic-related attributes for each county highway segment (*24*). The file represents a digital base map for the state, consisting of all public road segments, in addition to census boundaries and other relevant political boundaries, and other spatial characteristics across the state. Among the attributes contained within the shapefile are the physical road (PR) reference number, the beginning and ending milepoint of each road segment, jurisdictional ownership, and the urban/rural classification of the segment. Segment begin/end milepoints within the MGF-AR are based on a change in one or more primary characteristics, including pavement surface, annual average daily traffic (AADT), major junction, jurisdictional boundary, and numerous other features. Thus, the PR and mileposts from the MGF-AR file effectively partition each roadway into unique homogeneous segments, which provided the roadway segmentation basis for data collection performed during this study.

U.S. census boundaries were then used to isolate rural segments for use in the analysis. Rural areas are typically defined as locations that fall outside of urban boundaries with populations of greater than 5,000 (*2, 3*). However, this research sought to isolate high-speed sections of county highways where the statutory rural speed limit of 55 mph would apply. Thus, road segments falling inside any incorporated census area boundary were excluded from this sample, including small cities and villages with populations of less than 5,000 and unincorporated census designated places. This step was important, as speed limit signs are not required on roadways utilizing the statutory speed limit, making speed limit verification difficult.

It was also necessary to define a minimum segment length for the analysis to lessen the effects of segmentation bias due to features like freeway interchanges bridges, offset intersections, and other features that would result in short segmentation and atypical crash occurrence or injury severity. Although the HSM recommends a minimum segment length of 0.1 mi, a random sampling of segments between 0.1 and 0.2 mi revealed that most of these segments were located on bridges, freeway interchange overpasses, or had other atypical features that would potentially bias the results. Therefore, a minimum segment length of 0.2 mi was selected for this analysis.

Data Collection

Traffic Volumes. Development of SPFs for any roadway facility is critically dependent on the availability of traffic volume data. Thus, the initial step was to identify counties for which traffic volume data were available across all classes of county roadways. Though traffic volume data were available for all federal-aid county roadways statewide, a review of transportation agency websites along with direct agency contact found that only a limited number of counties possessed traffic volume data for non-federal-aid roadways. Ultimately, it was determined that an adequate sample of traffic volume data for non-federal-aid county roadways were available for 29 of 83 counties in Michigan. Thus, these 29 counties represented the sampling frame for selection of county highway segments included in this study and are displayed in Figure 1 along with the number of total roadway centerline miles for which data were collected within each county.

Traffic volume data were assembled from a variety of sources. Data for federal-aid county roadways were obtained from an MDOT-maintained database, whereas data for non-federal-aid county roadways were obtained directly from county road commissions or regional planning commissions. In all cases, these traffic volume data represented actual counts taken along the particular highway segment, and were not estimated based on data from other roadways. Annual traffic growth factors based on regional and road-type factors were obtained from MDOT and applied to the traffic volumes to adjust to the appropriate analysis year, where necessary. The AADT values were then spatially matched to the appropriate segment via the PR and milepoint values.

Horizontal Curvature. The radius and length of all horizontal curves with radii up to 2,640 ft (0.5 mi) were extracted from the MGF-AR shapefile. Curve data were converted to binary indicator values based on several radius thresholds, ranging from a maximum radius of 465 ft up to curves with maximum radii of 2,640 ft. The curve data were then merged with the roadway inventory data for the respective segment. To account for segment breaks across curves, the curve data were aggregated for each segment in the following manner: length of the curved portion of the segment, proportion of the segment on a curve, and the average radius of curves on the segment. Ultimately, horizontal curves with radii less than 1,008.5 ft, which corresponds to a design speed of 55 mph at 7% superelevation, were selected for use in the analyses. This curve design speed was chosen because 55 mph was the statutory speed limit for rural county highways in Michigan during the analysis period, and

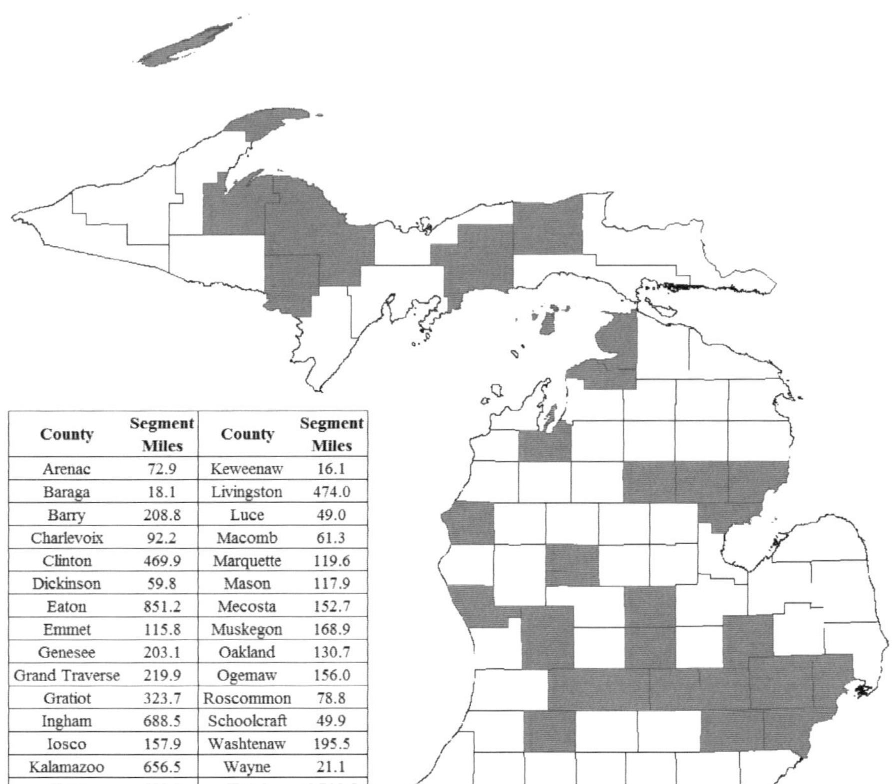

County	Segment Miles	County	Segment Miles
Arenac	72.9	Keweenaw	16.1
Baraga	18.1	Livingston	474.0
Barry	208.8	Luce	49.0
Charlevoix	92.2	Macomb	61.3
Clinton	469.9	Marquette	119.6
Dickinson	59.8	Mason	117.9
Eaton	851.2	Mecosta	152.7
Emmet	115.8	Muskegon	168.9
Genesee	203.1	Oakland	130.7
Grand Traverse	219.9	Ogemaw	156.0
Gratiot	323.7	Roscommon	78.8
Ingham	688.5	Schoolcraft	49.9
Iosco	157.9	Washtenaw	195.5
Kalamazoo	656.5	Wayne	21.1
Kent	580.2	**Total**	**6,510.3**

Figure 1. Represented counties and corresponding segment mileage.

curves designed below this speed would typically possess curve warning signage.

Crash Data. Crash data were collected from the annual statewide crash databases maintained by MDOT for the sample of county highways included in the analysis. Data were obtained from 2011 to 2015, which was the most recent 5-year period available at the time of the study. Only crashes with a roadway area type coded as "midblock segment," which excluded crashes occurring at a public road intersection, were included. The crash data were aggregated annually and merged with the roadway inventory data for each segment. All relevant crash related fields (i.e., severity, crash type, etc.) were retained within the crash database. Because deer crashes accounted for greater than 63% of all crashes occurring on the study segments, for analytical purposes, it was decided that only non-deer crashes would be included in the models. Due to the unpredictability of deer crashes, leaving such crashes in the analysis would likely have provided misleading SPF results with respect to the various roadway characteristics.

Other Data. Each segment was also manually reviewed in Google Earth based on the PR and begin/end milepoints to extract additional roadway information. Data collected manually included driveway counts, surface type (i.e., paved or gravel), surface width, lane width, number of lanes, shoulder width, and the presence of centerlines, edgelines, or both centerlines and edges. Cross-sectional characteristics, including surface width, lane width, and shoulder width, were measured using the ruler tool in Google Earth. The precision of this measurement method was validated by comparing the Google Earth lane and shoulder width measurements to known measurements for a sample of segments. Lane and paved shoulder width were recorded to the nearest 0.5 ft. Shoulder width was only recorded if edgelines were present, as a distinction between the lane and shoulder could not otherwise be made. For locations without centerlines, including gravel roadways, only the overall surface width was recorded. Gravel or earthen shoulders were not recorded due to difficulties in measurement. To account for the overabundance and relatively low utilization of field access points, driveways were only counted if connected to a structure.

Data Summary and Preliminary Analysis

In total, 6,510 mi of county highways were included in the analysis, of which 55% were paved federal-aid segments, 23% were paved non-federal-aid segments, and 22% were gravel roadways. Examination of the traffic volume distributions showed very little overlap between the various functional classes and pavement surface types, which was not an unexpected results. Consequently, it was decided that separate SPFs would be developed based on functional classification. Furthermore, in order to accurately assess the effects of pavement surface, models that included both paved and gravel roadways were constrained to a low-volume range. For purposes of this study, low-volume was defined as less than 400 vpd. This threshold provided a similar distribution of AADTs between paved and gravel segments and was also consistent with the definition for very low-volume roadways within AASHTO design guidance and the HSM. Considering only those low-volume segments included in the analysis, 29% were paved whereas the remaining 71% were gravel. After preliminary analysis, it was decided that three separate SPFs would be generated, including 1) paved federal-aid county segments, 2) paved non-federal-aid county segments, and 3) non-federal-aid paved and gravel segments with AADTs below 400 vpd. The segment summary statistics associated with each of the three aforementioned county roadway SPFs can be found in Table 1.

SPF Development using Mixed Effects Negative Binomial Models

As crash data are comprised of non-negative integers, traditional regression techniques (e.g., ordinary least-squares) are generally not appropriate. Given the nature of such data, the Poisson distribution has been shown to provide a better fit and has been used widely to model crash frequency data. In the Poisson model, the probability of segment i experiencing y_i crashes during a 1-year period is given by Equation 1:

$$P(y_i) = \frac{EXP(-\lambda_i)\lambda_i^{y_i}}{y_i!} \quad (1)$$

where $P(y_i)$ is probability of segment i experiencing y_i crashes and λ_i is the Poisson parameter for segment i, which is equal to the expected number of crashes per year, $E[y_i]$. Poisson models are estimated by specifying the Poisson parameter λ_i (the expected number of annual crashes) as a function of explanatory variables, the most common functional form being $\lambda_i = \exp(\beta X_i)$, where X_i is a vector of explanatory variables and β is a vector of estimable parameters.

A limitation of this model is the underlying assumption of the Poisson distribution that the variance is equal to the mean. As such, the model cannot handle overdispersion in which the variance is greater than the mean. Overdispersion is common in crash data and may be caused by data clustering, unaccounted temporal correlation, model misspecification, or ultimately by the nature of the crash data, which are the product of Bernoulli trials with unequal probability of events *(25)*.

Table 1. County Road Segment Summary Statistics

Parameter	Level or unit	Min	Max	Mean	Median	SD
Paved federal-aid county segments ($n = 6{,}377$)						
Segment length	mi	0.200	3.79	0.517	0.464	0.290
AADT	vpd	400	12,781	1,911	1317	1,650
Lane width	ft	10.0	14.0	11.1	11.0	0.661
Paved shoulder width	ft	0.0	10.0	1.18	0.5	1.58
Driveway density	driveways/mi	0	69	7.39	6.00	6.99
Curve design speed <55 mph	present along segment	0	1	0.0497	0	0.217
Minor arterial		0	1	0.133	0	0.340
Major collector		0	1	0.865	1	0.341
Minor collector		0	1	0.00125	0	0.0354
All non-deer crashes	annual count per segment	0	11	0.331	0	0.706
Non-deer injury crashes	annual count per segment	0	7	0.092	0	0.325
Non-deer PDO crashes	annual count per segment	0	7	0.239	0	0.571
Deer crashes	annual count per segment	0	14	0.603	0	0.986
Paved non-federal-aid county segments ($n = 2{,}429$)						
Segment length	mi	0.20	2.01	0.576	0.502	0.278
AADT	vpd	3	12,628	572	385	643
Lane width	ft	9.0	15.0	10.6	10.5	0.717
Paved shoulder width	ft	0.0	8.0	0.227	0.0	0.629
Driveway density	driveways/mi	0.00	95.7	17.1	14.4	12.7
Curve design speed <55 mph	present along segment	0	1	0.0618	0	0.241
Minor arterial		0	1	0.0004	0	0.0203
Major collector		0	1	0.0066	0	0.0809
Minor collector		0	1	0.190	0	0.392
Local road		0	1	0.803	1	0.398
All non-deer crashes	annual count per segment	0	4	0.122	0	0.380
Non-deer injury crashes	annual count per segment	0	3	0.0363	0	0.195
Non-deer PDO crashes	annual count per segment	0	4	0.0860	0	0.312
Deer crashes	annual count per segment	0	7	0.212	0	0.545
Low-volume (AADT <400) non-federal-aid county segments ($n = 3{,}514$)						
Segment length	mi	0.20	2.56	0.565	0.501	0.281
AADT	vpd	3	399	155	131	95.0
Surface width	ft	12	36	20.4	20.0	2.75
Driveway density	driveways/mi	0.00	95.7	12.7	10.6	10.2
Curve design speed <55 mph	present along segment	0	1	0.0758	0	0.265
Minor collector		0	1	0.0584	0	0.234
Local road		0	1	0.942	1	0.234
Paved surface	surface type of segment	0	1	0.292	0	0.455
All non-deer crashes	annual count per segment	0	3	0.0653	0	0.269
Non-deer injury crashes	annual count per segment	0	2	0.0197	0	0.142
Non-deer PDO crashes	annual count per segment	0	3	0.0456	0	0.222
Deer crashes	annual count per segment	0	5	0.0623	0	0.265

Note: mi = miles; ft = feet; vpd = vehicles per day; mph = miles per hour; Min = minimum; Max = maximum; SD = standard deviation.

Overdispersion is generally accommodated through the use of negative binomial models (also referred to as Poisson-gamma models).

The negative binomial model is derived by rewriting the Poisson parameter for each segment as $\lambda_i = \exp(\beta X_i + \varepsilon_i)$, where $EXP(\varepsilon_i)$ is a gamma-distributed error term with mean 1 and variance α. The addition of this term allows the variance to differ from the mean as $VAR[y_i] = E[y_i] + \alpha E[y_i]^2$. The negative binomial model is preferred over the Poisson model as the latter cannot handle overdispersion and, as such, may lead to biased parameter estimates (*26*). Consequently, HSM recommends using the negative binomial model for the development of SPFs.

One concern that arises when evaluating the safety performance of the Michigan county road system is the unobserved heterogeneity introduced when collecting data from across various counties and regions of the state. Michigan's county road commissions employ a variety of design and maintenance practices across the state (*24*). Other factors, such as weather, topography, land use, and driver behavior also vary widely across the various regions of the state. To account for these differences, a county-specific random effect was incorporated

Table 2. Mixed Effects Negative Binomial Model Results for Paved Federal-Aid Segments

	Total non-deer crashes				Fatal and injury non-deer crashes			
Parameter	Estimate	SE	z-value	p-value	Estimate	SE	z-value	p-value
Intercept	−5.88	0.151	−38.9	<0.001	−7.31	0.250	−29.2	<0.001
AADT (ln vpd)	0.698	0.0193	36.2	<0.001	0.741	0.0340	21.8	<0.001
Curve with design speed <55 mph not present	baseline				baseline			
Curve with design speed <55 mph present	0.468	0.0456	10.3	<0.001	0.445	0.0783	5.68	<0.001
Lane width <11 ft	baseline				baseline			
Lane width 11 ft	−0.0195	0.0414	−0.470	0.638	−0.0867	0.0728	−1.19	0.233
Lane width 12 ft	0.0299	0.0457	0.650	0.514	−0.0192	0.0793	−0.243	0.808
Lane width >12 ft	0.0343	0.0831	0.410	0.680	−0.305	0.157	−1.94	0.0530
No paved shoulder	baseline				baseline			
1 ft < paved shoulder < 3 ft	0.0450	0.0308	1.46	0.144	0.0354	0.0525	0.673	0.501
Paved shoulder 3 ft or greater	0.0629	0.0333	1.89	0.0588	−0.0208	0.0572	−0.364	0.716
Fewer than 5 driveways per mile	baseline				baseline			
5 to 10 driveways per mile	0.0639	0.0373	1.71	0.0864	0.0521	0.0657	0.793	0.428
10 to 15 driveways per mile	0.0608	0.0371	1.64	0.101	−0.00990	0.0659	−0.150	0.881
15 driveways per mile or greater	0.132	0.0333	3.97	<0.001	0.0540	0.0588	0.918	0.358
Overdispersion parameter	0.463				0.395			
AIC	43,304.1				18,526.0			
Log likelihood	−21,639.1				−9,250.0			

Note: ft = feet; vpd = vehicles per day; mph = miles per hour; ln = natural log; AIC = Akaike information criterian; SE = standard error.

into the analysis, allowing the intercept term to vary across locations and thereby creating a mixed effects model.

It is also noted that the natural log of segment length was included as an offset in each model, with a parameter estimate fixed at 1, thereby forcing the model to treat crashes as a direct one-to-one relationship with segment length. Furthermore, in order to make direct comparisons between low-volume paved and gravel roads, only non-federal-aid roadways were considered and only segments from counties with both gravel and paved roads were included in the low-volume non-federal-aid analysis.

Results and Discussion

The final model results for estimating annual non-deer crashes on rural county two-lane, two-way roads, subdivided into paved federal-aid segments, paved non-federal-aid segments, and low-volume (AADT < 400 vpd) non-federal-aid segments (paved and gravel) are displayed in tables 2, 3, and 4. Coefficient estimates, standard errors, *z*-scores, and *p*-values are provided within each table. Separate analyses were performed for total crashes and FI crashes, which included KABC crash severities.

Paved Federal-Aid County Road Segments

The results of the mixed effects negative binomial models for non-deer crashes on federal-aid county road segments, which are presented in Table 2, yielded several interesting results. First, the presence of a horizontal curve with a design speed below 55 mph was associated with increases in total crashes and injury crashes, with the effect being similar for both the total and FI crashes models. Segments where such curves were present had 60% more total crashes and 56% more FI crashes than segments without such curves. This compares favorably to research from Pennsylvania, in which it was shown that 43% greater total crashes and 48% greater FI crashes occurred on rural road segments that included a 55 mph horizontal curve (*18*).

Increasing driveway density was also associated with increasing crashes, which is to be expected as increasing the number of driveways increases the number of conflict points. Segments with 5 to 10 driveways had approximately 7% more crashes than segments with fewer than five driveways per mile (base condition), whereas segments with 15 or more driveways per mile saw approximately 14% greater crashes. These driveway effects were only observed for total crashes, as no significant effect was found for FI crashes. These results were aligned with prior research that also found total crashes to increase with increasing driveway density on rural two-lane roads (*27*).

Lane width did not have a significant effect on total crashes, but lane widths of greater than 12 ft were found to have fewer FI crashes, which may be due to a reduced likelihood of a lane departure crash. These findings

Table 3. Mixed Effects Negative Binomial Model Results for Paved Non-Federal-Aid Segments

	Total non-deer crashes				Fatal and injury non-deer crashes			
Parameter	Estimate	SE	*z*-value	*p*-value	Estimate	SE	*z*-value	*p*-value
Intercept	−6.229	0.246	−25.3	<0.001	−7.940	0.421	−18.9	<0.001
AADT (ln vpd)	0.730	0.036	20.2	<0.001	0.787	0.0612	12.9	<0.001
Curve with design speed <55 mph not present	baseline				baseline			
Curve with design speed <55 mph present	0.372	0.104	3.58	<0.001	0.564	0.169	3.34	<0.001
Fewer than 5 driveways per mile	baseline				baseline			
5 driveways per mile or greater	0.142	0.095	1.510	0.132	0.105	0.167	0.628	0.530
No paved shoulder	baseline				baseline			
Paved shoulder present	0.0497	0.080	0.623	0.533	0.0237	0.142	0.167	0.867
Lane width <11 ft	baseline				baseline			
Lane width 11 ft	−0.0007	0.062	−0.011	0.991	0.1280	0.110	1.160	0.246
Lane width >11 ft	0.0015	0.103	0.015	0.988	0.2550	0.174	1.470	0.142
Overdispersion parameter	0.546				0.317			
AIC	8,681.1				3,572.4			
Log likelihood	−4,331.5				−1,777.2			

Note: ft = feet; vpd = vehicles per day; mph = miles per hour; ln = natural log; AIC = Akaike information criterian; SE = standard error.

correspond somewhat with previous research associating wider lanes with fewer non-incapacity injury and PDO crashes (*18*) and fewer total crashes (*3, 16*) on state highways. It is worth noting that of the three analyses conducted, lane width was only a significant factor for paved federal-aid county segments, which are most similar to state highways noted in the prior references.

Turing to the shoulder width results, paved shoulders contributed to a marginal increase in total crashes and no substantive impacts on FI crashes. Although the body of research has consistently found that wider shoulders result in fewer crashes (*3, 16–18*), the findings of this analysis are likely confounded by correlation between shoulder width and AADT. Furthermore, as a roadside assessment was not conducted, the impacts of roadside characteristics could not be determined, although functional class and AADT also typically influence such characteristics.

Paved Non-Federal-Aid County Road Segments

The model results for paved non-federal-aid segments, displayed in Table 3, were somewhat similar to those of federal-aid segments. Again, segments that included horizontal curves with design speeds below 55 mph were found to have a greater occurrence of both total crashes and FI crashes. Comparison of the results for federal-aid versus non-federal-aid segments did not show a significant difference between the parameter estimates for curves for either total crashes or FI crashes. However, the larger standard error for non-federal-aid segments indicates greater variability in the effects of such curves on crash occurrence. This may be partially due to smaller sample sizes of non-federal-aid roadways, or to the less consistent signing and delineation standards that are often persistent on non-federal-aid highways, or a combination of both those factors.

Also similar to federal-aid roadways, the effects of lane width and paved shoulder width were not significant for either total crashes or FI crashes. Furthermore, contrary to the results of the paved federal-aid segments, driveway density was not found to be significant. Given that greater than 80% of the included paved non-federal-aid segments were classified as local roads, driver familiarity and travel behavior are likely to be very different than for arterials and major collectors, potentially dampening any effects of these variables.

Low-Volume Non-Federal-Aid County Road Segments (Paved and Gravel)

Non-federal-aid segments with AADTs below 400 vpd showed somewhat different results compared with those for the other segment types, as displayed in Table 4. The presence of horizontal curves with design speeds below 55 mph showed an even stronger effect on crashes compared with the other models. This is not surprising, as such curves were more often present on the low-volume non-federal-aid segments compared with the other county road classes. Furthermore, gravel roads make up greater than 70% of the low-volume dataset, and the reduced surface friction provided by gravel roads is particularly problematic along curves.

Table 4. Mixed Effects Negative Binomial Model Results for Low-Volume (AADT < 400) Non-Federal-Aid Paved and Gravel Segments

	Total non-deer crashes				Fatal and injury non-deer crashes			
Parameter	Estimate	SE	z-value	*p*-value	Estimate	SE	z-value	*p*-value
Intercept	−5.55	0.302	−18.3	<0.001	−6.19	0.492	−12.6	<0.001
AADT (ln vpd)	0.674	0.060	11.2	<0.001	0.584	0.102	5.71	<0.001
Curve with design speed <55 mph not present	baseline				baseline			
Curve with design speed <55 mph present	0.672	0.084	7.97	<0.001	0.708	0.147	4.82	<0.001
Gravel surface	baseline				baseline			
Paved surface	−0.406	0.099	−4.08	<0.001	−0.540	0.149	−3.62	<0.001
Fewer than 5 driveways per mile	baseline				baseline			
5 driveways per mile or greater	0.0663	0.088	0.757	0.449	0.0951	0.151	0.629	0.529
Surface width <20 ft	baseline				baseline			
20 ft < surface width < 22 ft	−0.0390	0.088	−0.442	0.659	−0.0897	0.153	−0.586	0.558
22 ft < surface width < 24 ft	−0.0027	0.097	−0.028	0.977	−0.0951	0.169	−0.563	0.573
Surface width >24 ft	−0.0258	0.110	−0.234	0.815	0.0079	0.187	0.042	0.966
Overdispersion parameter	0.480				0.306			
AIC	8,050.5				3,275.8			
Log likelihood	−4,015.3				−1,627.9			

Note: ft = feet; vpd = vehicles per day; mph = miles per hour; ln = natural log; AIC = Akaike information criterian; SE = standard error.

Paved surfaces were associated with 33% fewer total crashes and a 42% fewer FI crashes compared with gravel segments. This is an important finding, as road agencies often cite speed reductions and subsequent safety improvements as a justification for converting paved roads to gravel roads. Although readers should be aware of the limitations of applying such a factor derived from a cross-sectional study such as this, it remains important to be wary of the potential safety effects associated with converting a paved low-volume road to gravel. Gravel roads inherently provide less friction, are more susceptible to seasonal deterioration during winter and rainy seasons, and are generally more poorly maintained than similar paved roads, leaving them more prone to crashes involving lane departure (*28*). As with the paved non-federal-aid segment models, surface width and driveway density did not significantly affect crashes.

Comparison with SPFs for Michigan State Highways and the HSM

Comparisons were also made between the three county road SPFs, along with SPFs calibrated for non-deer crashes on rural two-lane state highways in Michigan and the uncalibrated HSM base model for total crashes (inclusive of animals). All models were set at the same base conditions, which included 12-ft lanes (or equivalent total surface width), 6-ft shoulders (or widest category), no driveways, and no curves and each SPF was then plotted as a function of AADT within the general range of traffic volumes. Figure 2 displays these comparisons for a) paved county federal-aid and non-federal-aid roads and b) low-volume county non-federal-aid roads (paved and gravel).

Figure 2*a* shows several interesting findings. First, for AADTs below approximately 3,000 vpd, county federal-aid roadways show a higher crash occurrence rate than county non-federal-aid, the HSM, or MDOT highways, which showed the lowest crash occurrence rates. County non-federal-aid paved roadways showed safety performance that was remarkably similar to the HSM base rural two-lane roadway model. At AADTs greater than 3,000 vpd, the federal-aid county crash occurrence rate drops below that for HSM, but remains greater than for MDOT state highways. Figure 2*b* shows similarly interesting findings, particularly that, as expected, gravel roadways show a greater crash occurrence rate than either the paved county non-federal-aid roadways or the HSM models.

Conclusion

The safety performance of county or local roadways is rarely investigated to the same level of detail as that for state highways. However, several states, particularly those in the Midwest and Great Lakes regions, possess a substantial network of rural county highways. Although SPFs exist within HSM and other resources, county-maintained highways typically possess traffic and geometric characteristics that differ considerably from rural highways maintained by the state DOT. Furthermore, nearly all available SPFs are only applicable to paved

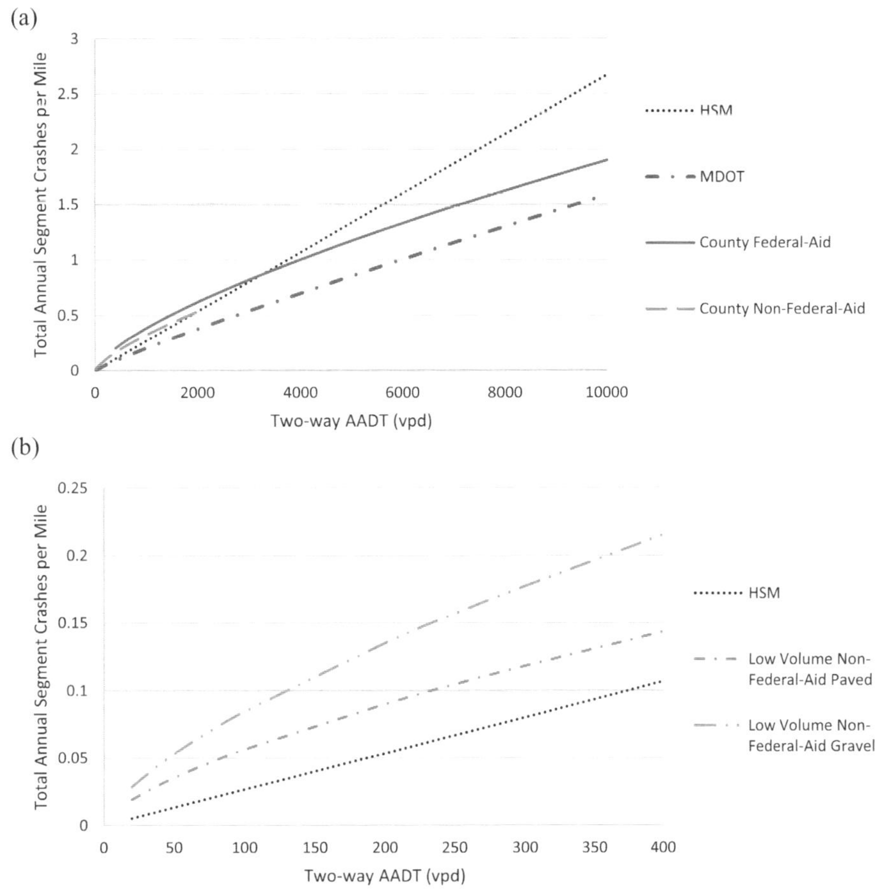

Figure 2. Comparison of total crash SPFs at base conditions: (*a*) paved roadways, by classification and (*b*) low-volume roadways, by pavement surface.
Note: vpd = vehicles per day; HSM = Highway Safety Manual; MDOT = Michigan Department of Transportation.

roads, which further limits the applicability of these models for use by county and local road agencies, which often maintain a substantial network of gravel roads. As greater than one-half of rural crashes in Michigan occur off the state highway system, there is a clear need for additional guidance on how to better design county roadways for safety, given the different design and user characteristics compared with state-maintained rural highways. To that end, research was undertaken to develop SPFs unique to county highways. To accomplish this objective, traffic volumes, roadway inventory, and geometric data were collected for more than 6,500 mi of county highway segments within 29 counties across Michigan and paired with traffic crashes occurring on these segments from 2011 through 2015. Separate SPFs were then generated for paved federal-aid county roadways, paved non-federal-aid county roadways, and paved and gravel non-federal-aid county roadways with traffic volumes below 400 vpd.

Not surprisingly, the county SPF results were generally different than would be expected on state-owned highways, although some similarities were observed. County federal-aid roadways showed a higher crash occurrence rate than county non-federal-aid, HSM, or MDOT highways. Paved non-federal-aid county roadways showed lower crash occurrence rates than their gravel counterparts. Increasing driveway density was found to be associated with increased crash occurrence, although these results were only significant for total crashes on federal-aid county roadways. However, lane width, roadway surface width, and paved shoulder width had little to no impact on total or injury crashes across each of the three county roadway models, although FI crashes showed a lower rate of occurrence on federal-aid county roadways with lane widths greater than 12 ft.

The most consistent geometric-related effects were associated with the presence of a horizontal curve with a design speed less than 55 mph on the segment, which for federal-aid county roadways was associated with approximately 60% greater total crashes and 56% greater FI crashes compared with segments without such curves. Similar curve-related effects were observed for non-federal-aid paved roadways, although the effect was much stronger for FI crashes. The strongest horizontal curve effects were associated with low-volume non-federal-aid roadways, as both paved and gravel segments containing a curve designed below 55 mph experienced more than double the number of crashes (total and FI) compared with segments without such curves.

The results of this research can be utilized toward various safety funding programs within Michigan and beyond. This includes performing network screening and crash prediction estimates to support local agency safety programs, in addition to providing similar support for programs such as the High Risk Rural Roads program.

Acknowledgments

Funding for this research was provided by MDOT.

Author Contributions

The authors confirm contribution to the paper as follows: study conception and design: TG, PS; data collection: SS, AI, MC, TG; analysis and interpretation of results: SS, TG, PS; draft manuscript preparation: SS, TG, PS. All authors reviewed the results and approved the final version of the manuscript.

References

1. *Traffic Safety Facts: Rural/Urban Comparison*. Publication DOT-HS-812-393. NHTSA, U.S. Department of Transportation, 2017.
2. Michigan Department of Transportation. *Crash Reporting Information System Transportation Crash Master*. Michigan Department of Transportation, 2017.
3. *Highway Safety Manual*, 1st ed. American Association of State Highway and Transportation Officials, Washington, D.C., 2010.
4. Garber, N. J., P. R. Haas, and C. Goose. *Development of Safety Performance Functions for Two-Lane Roads Maintained by the Virginia Department of Transportation*. Publication FHWA/VTRC 10-R25. FHWA, Virginia Department of Transportation, 2010.
5. Tegge, R. A., J. Jo, and Y. Ouyang. *Development and Application of Safety Performance Functions for Illinois*. Publication FHWA-ICT-10-066. Illinois Department of Transportation, 2010.
6. Dixon, K., C. Monsere, F. Xie, and K. Gladhill. *Calibrating the Future Highway Safety Manual Predictive Methods for Oregon State Highways*. Publication FHWA-OR-RD-12-07. Oregon Department of Transportation, 2012.
7. Srinivasan, R., and D. Carter. *Development of Safety Performance Functions for North Carolina*. Publication FHWA/NC/2010-09. North Carolina Department of Transportation, 2011.
8. Brimley, B. K., M. Saito, and G. G. Schultz. Calibration of the Highway Safety Manual Safety Performance Function: Development of New Models for Rural Two-Lane Two-Way Highways. *Transportation Research Record: Journal of the Transportation Research Board*, 2012. 2279: 82–89.
9. Bornheimer, C., S. D. Schrock, M. Wang, and H. Lubliner. Developing a Regional Safety Performance Function for Rural Two-Lane Highways. Presented at 91st Annual Meeting of the Transportation Research Board, Washington, D.C., 2012.
10. Lubliner, H., and S. D. Schrock. Calibration of the Highway Safety Manual Prediction Method for Rural Kansas Highways. Presented at 91st Annual Meeting of the Transportation Research Board, Washington D.C., 2012.
11. Srinivasan, S., P. Haas, N. S. Dhakar, R. Hormel, D. Torbic, and D. Harwood. *Development and Calibration of Highway Safety Manual Equations for Florida Conditions*. Publication TRC-FDOT-82013-2011. Florida Department of Transportation, 2011.
12. Alluri, P., and J. Ogle. Effects of State-Specific SPFs, AADT Estimations, and Overdispersion Parameters on Crash Predictions Using SafetyAnalyst. Presented at 91st Annual Meeting of the Transportation Research Board, Washington, D.C., 2012.
13. Savolainen, P. T., T. Gates, D. Lord, S. Geedipally, E. Rista, T. Barrette, B. J. Russo, and R. Hamzeie. *Michigan Urban Trunkline Intersections Safety Performance Functions (SPFs) Development and Support*. Publication RC-1628. Michigan Department of Transportation, 2015.
14. Savolainen, P. T., T. Gates, D. Lord, S. Geedipally, E. Rista, T. Barrette, P. Thompson, and I. Thompson. *Michigan Urban Trunkline Segments Safety Performance Functions (SPFs) Development and Support*. Publication RC-1639. Michigan Department of Transportation, 2016.
15. *Highway Safety Manual – FAQs*. American Association of State Highway and Transportation Officials. http://www.highwaysafetymanual.org/Pages/support_answers.aspx. Accessed October 1, 2017.
16. Wu, H., Z. Han, M. R. Murphy, and Z. Zhang. Empirical Bayes Before-After Study on Safety Effect of Narrow Pavement Widening Projects in Texas. *Transportation Research Record: Journal of the Transportation Research Board*, 2015. 2515: 63–69.
17. Gooch, J. P., V. V. Gayah, and E. T. Donnell. Quantifying the Safety Effects of Horizontal Curves on Two-Way, Two-Lane Rural Roads. *Accident Analysis & Prevention*, Vol. 92, 2016, pp. 71–81. https://doi.org/10.1016/j.aap.2016.03.024.
18. Wang, K., J. N. Ivan, N. Ravishanker, and E. Jackson. Multivariate Poisson Lognormal Modeling of Crashes by Type and Severity on Rural Two Lane Highways. *Accident Analysis and Prevention*, Vol. 99, 2017, pp. 6–19. https://dx.doi.org/10.1016/j.aap.2016.11.006.
19. Harwood, D. W., F. M. Council, E. Hauer, W. E. Hughes, and A. Vogt. *Prediction of the Expected Safety Performance*

of Rural Two-Lane Highways. Publication FHWA-RD-99-207. FHWA, Office of Safety Research and Development, McLean, VA, 2000.
20. Gluck, J., H. S. Levinson, and V. Stover, *NCHRP Report 420: Impact of Access Management Techniques*. Transportation Research Board of the National Academies, Washington, D.C., 1999.
21. Wooldridge, M. D., K. Fitzpatrick, D. W. Harwood, I. B. Potts, L. Elefteriadou, and D. J. Torbic. NCHRP Report 502: *Geometric Design Consistency on High Speed Rural Two-Lane Roadways*. Transportation Research Board of the National Academies, Washington, D.C., 2003.
22. Fitzpatrick, K., L. Elefteriadou, D. W. Harwood, J. Collins, J. McFadden, I. B. Anderson, R. A. Krammes, N. Irizarry, K. Parma, K. M. Bauer, and K. Passetti. *Speed Prediction Models for Two-Lane Rural Highways*. Report FHWA-RD-99-171. FHWA, U.S. Department of Transportation, Washington, D.C., 2000.
23. Zegeer, C. V., R. Stewart, F. Council, and T. R. Neuman. Accident Relationships of Roadway Width on Low-Volume Roads. *Transportation Research Record: Journal of the Transportation Research Board*, 1994. 1445: 160–168.
24. *MI Geographic Framework All Roads Shapefile v16a*. State of Michigan GIS Open Data, 2016.
25. Lord, D. Modeling Motor Vehicle Crashes Using Poisson-gamma Models: Examining the Effects of Low Sample Mean Values and Small Sample Size on the Estimation of the Fixed Dispersion Parameter. *Accident Analysis and Prevention*, Vol. 38, No. 4, 2006, pp 751–766.
26. Lord, D., and P. Y. Park. Investigating the Effects of the Fixed and Varying Dispersion Parameters of Poisson-gamma Models on Empirical Bayes Estimates. *Accident Analysis and Prevention*, Vol. 40, No. 4, 2008, pp. 1441–1457.
27. Ewan, L., A. Al-Kaisy, and F. Hossain. Safety Effects of Road Geometry and Roadside Features on Low-Volume Roads in Oregon. *Transportation Research Record: Journal of the Transportation Research Board*, 2016. 2580: 47–55.
28. Fay, L., A. Kroon, K. Skorseth, R. Reid, and D. Jones. *NCHRP Synthesis 485: Converting Paved Roads to Unpaved*. Transportation Research Board of the National Academies, Washington, D.C., 2016.

The Standing Committee on Low-Volume Roads (AFB30) peer-reviewed this paper (18-06568).

Article

TRR
JOURNAL OF THE TRANSPORTATION RESEARCH BOARD

Transportation Research Record
2018, Vol. 2672(52) 238–250

DOI: 10.1177/0361198118758314
journals.sagepub.com/home/trr
SAGE

Pullout Behavior of Steel Mechanically Stabilized Earth Reinforcements

Timothy A. Wood[1], William D. Lawson[2], Priyantha W. Jayawickrama[2], and James G. Surles[3]

Abstract

Instrumented pullout tests of unprecedented scope and scale explore the pullout behavior for three steel mechanically stabilized earth reinforcement types: ribbed strips, ladder-like strips, and three-wire bar mat grids. These data quantify the distribution of pullout resistance between longitudinal elements and illustrate the nature of certain reinforcement deformations. Consistent with characteristic inextensible pullout behavior and soil-reinforcement interaction, synthesized strain-gage data illustrate linear stress reduction along the embedment length during pullout for all three reinforcement styles. For ladder-like strips, the axial force divides evenly between the two longitudinal elements. For the three-wire bar mat grid, the center bar carries approximately 40% of the axial force, whereas each outside bar carries approximately 30% of the axial force. Observed pullout-induced deformation in the transverse elements of three-wire bar mat grids having widely spaced longitudinal bars is conceptually different from extensible behavior and suggests the need for refinement in current pullout resistance formulations.

Reinforcement extensibility is an important concept for the design of mechanically stabilized earth (MSE) retaining walls with implications for predicted reinforcement loads, failure planes, and pullout resistances. Design practices have assumed that steel reinforcements with straight longitudinal elements are inextensible, that is, the longitudinal stiffness of the reinforcement is much larger than the soil stiffness, and the reinforcement rupture strain is much smaller than the soil failure strain. The implications of inextensibility for pullout resistance are two-fold: the shear stress developed along the soil-reinforcement interface is constant, and the reinforcement stress in the anchorage zone decreases linearly in the longitudinal elements from the failure surface to tail of the reinforcement.

The inextensibility of steel MSE reinforcements may seem obvious. The metallically stabilized earth industry has emphatically claimed inextensibility (*1*), extensive project experience and very limited research support inextensibility (*2,3*) and, most importantly, by default the Simplified Method per AASHTO's *LRFD Bridge Design Specifications* assumes steel reinforcement inextensibility (*4*). However, at least three factors have raised questions about steel MSE reinforcements' inextensibility. First, research has documented deformation in certain ostensibly inextensible steel MSE reinforcements (*5,6*). Second, the Simplified Method acknowledges that "widely spaced" steel MSE grids may deform, and implicitly categorizes these reinforcements as somewhat extensible (*7*). Third, the metallically stabilized earth industry has introduced new semi-extensible steel reinforcements with intentional bends and kinks (*8,9*). Collectively these factors invite inquiry into the assumed inextensibility of steel MSE reinforcements.

Data from an unprecedented program of strain-gage-instrumented pullout tests show the natural variation and trends in longitudinal load distribution and quantify the distribution of pullout resistance between longitudinal elements. In addition, this study provides insight on deformations that may occur in grid-type reinforcements during pullout, and the corresponding design implications. Clarifying inextensible behavior in pullout establishes a baseline for the interpretation of the behavior of semi-extensible and other MSE reinforcement types, and provides a conceptual basis for refined design guidance.

Theory

As introduced in the mid-1960s, reinforced earth theory assumes a reinforced soil mass consisting of linear elements embedded in cohesionless soil. The reinforced soil system effectively creates a coherent mass of composite, cohesive

[1]Department of Civil and Environmental Engineering, The Citadel, Charleston, SC
[2]Department of Civil, Environmental, and Construction Engineering, Texas Tech University, Lubbock, TX
[3]Department of Mathematics and Statistics, Texas Tech University, Lubbock, TX

Corresponding Author:
Address correspondence to Timothy A. Wood: twood3@citadel.edu

material. The resulting composite material gains increased compressive strength by restraining longitudinal soil movement through friction between the soil and tensile (or longitudinal) reinforcing (*10*). The first retaining wall applications of the reinforced earth concept followed these assumptions closely, the use of smooth steel strips to reinforce sandy soil in retaining walls proved largely successful, and the Coherent Gravity Method emerged as a dominant design method (*11*).

Over time, new reinforcement types were introduced including steel ribbed strips, discrete bar mat grids, and continuous welded wire mesh, as well as geosynthetic grids and geotextiles. Not all these reinforcements interact with soil in the same way. A means of characterizing reinforcement types emerged focusing on reinforcement stiffness and the corresponding transfer of load from the soil to the reinforcement; namely, inextensible versus extensible.

In the mid-1970s, McGown and others defined these concepts by comparing tensile strains at reinforcement rupture, typically in the longitudinal direction of the reinforcement, to soil strains at reinforced mass failure:

> Tensile strain inclusions [soil reinforcements] which have in-situ rupture strains less than the maximum tensile strains in the soil without inclusion under the same operational stress conditions should be termed relatively inextensible inclusions…
>
> Tensile strain inclusions which have in-situ rupture strains larger than the maximum tensile strains in the soil without inclusions under the same operational stress conditions should be termed relatively extensible inclusions (*12*).

These definitions of extensibility/inextensibility can guide the evaluation of potential failure planes, demand loads on reinforcement, and pullout resistance.

For properly designed relatively inextensible reinforcement (i.e. pullout does not occur), the rupture capacity of the reinforcement governs the behavior; sufficient strains never mobilize within the soil mass to bring the soil to active condition and failure (*13*). This condition is consistent with Vidal's assumption of static friction between soil and reinforcement and the rigid mass assumption of the Coherent Gravity Method. Furthermore, the relative stiffness of the reinforcement causes the pullout-induced soil strain to be relatively constant along a longitudinal element (*14*). Because very little mobilizing strain dissipates through reinforcement elongation, the soil stress along the length of the reinforcement is assumed constant in the anchorage zone; therefore, the axial force, as modeled, increases linearly from the reinforcement tail to the point of maximum tension.

Early MSE reinforcements only relied on friction between reinforcement and soil to transfer load; however, other reinforcement types such as grids, bar mats, welded wire, and geogrid rely on passive (bearing) resistance. Such reinforcements are subject to all the same mechanistic behaviors based on longitudinal failure strains, but friction and passive resistance on transverse elements work together to transfer load. The combined load transfer mechanism is complex; this paper emphasizes one element of the transfer mechanism, namely, longitudinal inextensibility in common steel reinforcements. For these systems, the soil strain required for passive resistance is larger than that required for frictional resistance (*7*), further widening the gap between the reinforcement rupture strains and the soil failure strains. Implicitly, "the bearing stresses are mobilized simultaneously at all transverse members" for inextensible grids (*15*).

Unified Design Methods

In recent MSE design practice, industry and academia have combined the mechanistic and behavioral distinctions of various MSE reinforcements into unified design methods. Two prominent examples of these unifying approaches are the Structure Stiffness Method (*16*) and the Simplified Stiffness Method (*17*). Both design methods directly engage the reinforcement longitudinal stiffness (i.e. relative extensibility/inextensibility) when predicting wall behavior and reinforcement demand loads. However, this paper illustrates the effect of relative inextensibility on *pullout resistance*. The convenient nomenclature for pullout resistance of the Simplified Method (*4*) suitably illustrates the impact of extensibility/inextensibility on reinforcement pullout.

The Simplified Method predicts pullout resistance using an equivalent friction coefficient, F^*, to capture various reinforcement–soil interface behaviors. Extensibility and its influence on pullout behavior is addressed through a scale effect correction factor, α (*4,7*). These two variables, F^* and α, describe the relationship between reinforcement type and pullout resistance in the Simplified Method. By definition, the scale effect correction factor, α, is intended "to account for a non-linear stress reduction over the embedded length of highly-extensible reinforcements based on lab data" (*7*). Therefore, for inextensible reinforcements such as steel, $\alpha = 1$, and the stress reduction along the reinforcement should be linear.

Previous Research Efforts

For the majority of MSE's history, design methods have assumed steel reinforcements to be inextensible. This assumption has served the steel reinforcement community well by encouraging simpler pullout test methods for inextensible reinforcements and by making clear distinctions between inextensible steel reinforcements and extensible geosynthetic reinforcements (*14*). However, few studies have actually measured relative inextensibility during pullout of steel strips (two instrumented pullout tests) (*2*) and steel grids (12 instrumented pullout tests) (*3*). No such tests exist in gravelly backfill. More recently, *semi-extensible* steel reinforcements have been developed that rely on kinks and bends in the longitudinal elements to increase longitudinal extensibility and reduce reinforcement loads (*8,9*). The development of semi-extensible steel MSE systems would be strengthened by the baseline confirmation of inextensibility for steel MSE reinforcements with straight longitudinal elements.

In addition, research on widely used three-wire bar mat grids has shown normalized pullout resistance to increase more than 200% by reducing the longitudinal bar spacing from 300 mm to 50 mm (*6*). A footnote in the FHWA documentation acknowledges this issue: "For [steel grid] longitudinal bars/wires spacing greater than 6 in. [150 mm] α may be less than 1.0 and pullout tests are required" (*7*). However, this approach implies that wider longitudinal bar spacing actually causes extensibility in the longitudinal bars. These assumptions and approaches indicate a need to quantify the inextensibility of typical steel MSE reinforcements.

The next section describes the test method, instrumentation plan and test program for 43 instrumented pullout tests on three steel MSE reinforcement types. For each reinforcement type, one example test is fully explored, and a mean normalized axial force is defined for each strain-gage location. The synthesis of mean normalized axial force from multiple tests illustrates the generally linear reductions in load along the reinforcement lengths, the load distribution between longitude elements, and the natural variation in the longitudinal load distribution associated with typical MSE installations.

Method

Large-Scale, Instrumented, Laboratory Testing

Texas Tech University's large-scale MSE pullout test box with dimensions of 3.7 m by 3.7 m by 1.2 m deep produced all presented data (*5*). This test box allowed for the testing of MSE reinforcements longer than typical of extensible reinforcement testing (*18*). Reinforcements were placed in three test layers vertically spaced at 0.3 m. A system of nine steel plates, hydraulically connected jacks and reaction beams simulated fill depths up to 12.2 m. Automated systems measured vertical pressure, pullout force, displacement, and strain-gage response. Backfill soil was placed in uniform lifts and compacted using a vibratory plate compactor consistent with Texas Department of Transportation specifications (*19*), and tested using a nuclear density gage (*20*). Reinforcements were tested in two soil types: sandy backfill (SP-SM) and gravelly backfill (GP-GM) (*21*). Sandy backfill achieved 95% relative compaction, C_R, (measured $C_R = 97.3 \pm 2.5\%$). Gravelly backfill was compacted using ordinary compaction until no further consolidation was visible (nominal $C_R = 94.6 \pm 2.0\%$). A hand-operated hydraulic jack applied the pullout force to the reinforcement at a target rate of 2 mm/min. The gripper system enforced uniform displacement of the longitudinal elements.

Electrical resistance strain gages (Micro-Measurement C2A-13-062LW-350, $\pm 1\mu\varepsilon$) were applied to the reinforcement and protected using the manufacturer's procedure (*22*). Opposing strain-gage pairs allowed for accurate estimates of axial force on each longitudinal element. Strain-gage failures occurred during pullout as a result of force on the wiring and abrasion on the gages; therefore, some data are missing from the pullout tests. Initial data at low displacement were ignored because of incidental eccentricities from overburden pressures and alignment of the pullout force system. The average strain between paired strain gages, the longitudinal cross-sectional area, and modulus of elasticity for steel established the axial force at gaged locations along the embedment length. The axial force was normalized by dividing the axial force from the strain gages by the total applied pullout force.

Additional information on the pullout system, backfill gradations, density testing, and un-gaged pullout test results for a much larger test program is published elsewhere (*5,6,23,24*).

Test Specimens

The test program consisted of 43 instrumented pullout tests on three styles of steel MSE reinforcements: 19 tests on ribbed strips, six tests on ladder-like strips, and 18 tests on three-wire bar mat grids. Figure 1*a* shows two instrumented ribbed strips just prior to embedment in the soil backfill. Figure 1*c* shows dimensions for the 50 mm-wide steel ribbed strips. Strain gages pairs were placed on the top and bottom of the strip at each gage position seen in Figure 1,*f–h*. Ribbed strips were tested in sandy backfill under four simulated fill depths and at two embedment lengths as seen in Table 1. In gravelly backfill, ribbed strips 2.4 m long were tested under 1.5 m and 6.1 m of simulated fill (Table 1).

Figure 1*b* shows a combination of instrumented and uninstrumented ladder-like strips prior to embedment. Figure 1*d* shows dimensions for the 100 mm-wide steel ladder-like strips made from 9.1 mm diameter steel bars. These ladder-like strips feature two longitudinal bars spaced 50 mm apart. The transverse bars, 100 mm long, included designed tangs extending 25 mm past the longitudinal bars. Fabrication of these reinforcements includes quality control to ensure the tangs extend beyond the longitudinal bars. These transverse bars were spaced at 150 mm along the longitudinal bars. The ladder-like strips were also strain-gaged in pairs. Figure 1*i* shows each longitudinal wire gaged separately resulting in four strain gages per position along the length of the reinforcement. Ladder-like strips were only tested in sandy soil with embedment lengths of 2.4 m under a range of fill depths (Table 1).

Figure 1*a* shows instrumented three-wire bar mat girds prior to embedment. Figure 1*e* shows three-wire bar mat grid dimensions. The three-wire grid mats did *not* have tangs (as is typical); rather, quality control ensured the transverse bars terminated at the exterior longitudinal bars. Each three-wire grid consisted of transverse bars spaced at 300 mm. The three-wire grids included two longitudinal bar spacings (150 mm and 230 mm) and two longitudinal bar sizes (diameter of 8.8 mm and 12.8 mm). Embedment lengths were determined by the number of transverse bars embedded in the backfill: the 0.9 m three-wire grids had three embedded transverse bars, and the 1.8 m three-wire grids had six embedded transverse bars. For three-wire grids, strain gages measured axial strain in the three longitudinal bars. Additional strain gages were placed on the transverse elements, with data from those gages presented elsewhere (*5*). Again, strain-gage pairs (as

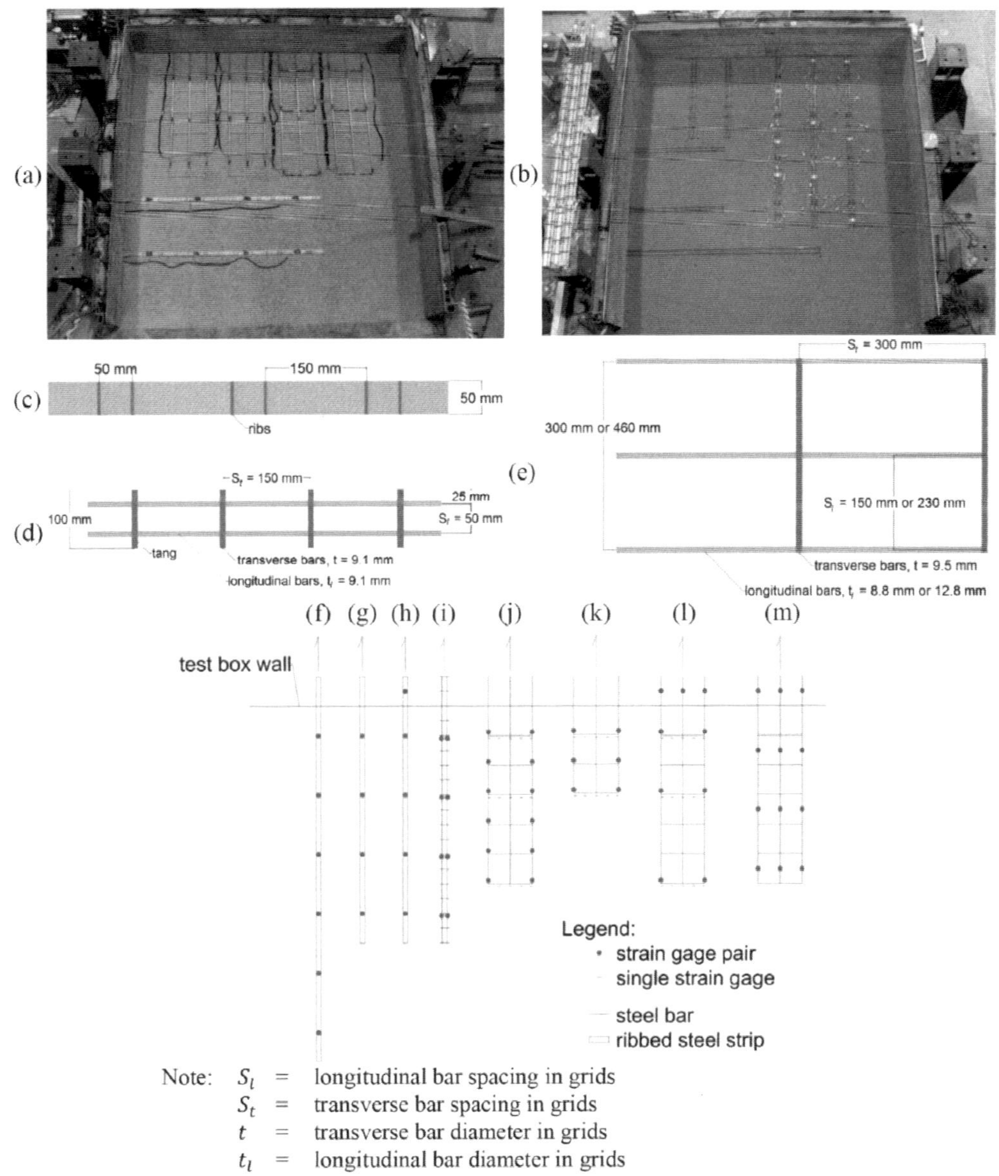

Figure 1. Tested MSE reinforcements: (*a*) ribbed strips oriented left-right, and three-wire steel bar mat grids oriented up-down in 3.7 m × 3.7 m test box, (*b*) ladder-like strips in test box, (*c*) 50 mm-wide ribbed strip sketch, (*d*) 100 mm-wide ladder-like strip sketch, and (e) 300 mm- and 460 mm-wide three-wire steel bar mat grid sketch; strain-gage plans for ribbed strips (*f–h*), ladder-like strips (*i*), three-wire bar mat grids (*j–m*).

seen in Figure 1, *j–m*) allowed bending and axial strains to be differentiated. Table 1 defines testing in sandy and gravelly soils for a range of bar sizes, bar spacings, and fill depths.

Results

Synthesis of instrumented pullout test data for a wide range of reinforcement types, geometries, embedment lengths, and overburden depths requires a novel approach to longitudinal load distribution analysis. Historically, design methods have assumed the inextensibility of steel reinforcements. If so, soil-reinforcement shear stress should develop uniformly along the length of the reinforcement, and the normalized axial force at a particular location in the reinforcement should be very nearly constant throughout a pullout test. Data illustrating a relatively constant normalized axial force support inextensibility. Therefore, the mean normalized axial force can quantify the load carried at each gage location for the whole test. The mean normalized axial forces can illustrate the distribution between the longitudinal elements.

Table 1. Experimental Program for 50 mm-wide, Steel Ribbed Strips in Sandy Backfill and Gravelly Backfill; Steel Ladder-like Strips in Sandy Backfill (S_t = 150mm, t = 9.1mm); and Steel Three-wire Bar Mat Grids (S_t = 300mm, t = 9.1mm) in Sandy Backfill and Gravelly Backfill

L_e (mm)	S_l (mm)	t_l (mm)	Z_p (m)	Replicates	Gage Plan*
Ribbed Strips Embedded in Sandy Backfill					
3.6	–	–	1.5	2	f
3.6	–	–	3.6	2	f
3.6	–	–	6.1	2	f
3.6	–	–	12.2	2	f
2.4	–	–	1.5	2	g
2.4	–	–	3.6	2	g
2.4	–	–	6.1	2	g
2.4	–	–	12.2	2	g
Ribbed Strips Embedded in Gravelly Backfill					
2.4	–	–	1.5	1	h
2.4	–	–	6.1	2	h
Ladder-Like Strips in Sandy Backfill					
2.4	50	9.1	0.9	2	i
2.4	50	9.1	1.8	3	i
2.4	50	9.1	3.6	1	i
Three-Wire Bar Mat Grid Embedded in Sandy Backfill					
1.8	150	8.8	1.5	1	j
0.9	150	8.8	3.6	1	k
1.8	150	12.8	1.5	1	j
1.8	150	12.8	3.6	1	j
1.8	150	12.8	6.1	1	j
0.9	150	12.8	12.2	1	k
1.8	230	8.8	1.5	1	j
1.8	230	12.8	1.5	1	j
1.8	230	12.8	3.6	1	j
1.8	230	12.8	6.1	1	j
0.9	230	12.8	12.2	1	k
Three-Wire Bar Mat Grid Embedded in Gravelly Backfill					
1.8	230	12.8	1.5	1	l
1.8	230	12.8	6.1	2	l
1.8	230	12.8	1.5	2	m
1.8	230	12.8	6.1	2	m

Note: * For gage plan refer to Figure 1, *f–m*.
L_e = total embedment length of the reinforcement.
S_l = longitudinal bar spacing in grids.
S_t = transverse bar spacing in grids.
t = transverse bar diameter in grids.
t_l = longitudinal bar diameter in grids.
Z_p = nominal depth of soil for pullout test (*4*).

Further, the mean normalized axial forces can be used to synthesize data from all tests on a particular reinforcement type and cumulatively evaluate the linear distribution of axial force along the reinforcement length.

Ribbed Strips

Figure 2 presents experimental test results for a representative ribbed strip. Figure 2*a* shows the interaction between displacement and axial force. The load develops similar to hyperbolic relationships suggested by many soil stress–strain models. As expected, strain gages closer to the front of the test box show greater axial force. Figure 2*b* shows axial force plotted against pullout force; each location gains axial force as a nearly linear function of the total pullout force consistent with inextensible behavior. By dividing the axial force by the total pullout force, the normalized axial force is shown versus displacement (Figure 2*c*) and versus total pullout force (Figure 2*d*). For each gage location, the normalized axial force is nearly constant. For both displacement and pullout force, the strains remain below rupture (allowable load of 66 kN) and serviceability limit state (displacement less than 19 mm).

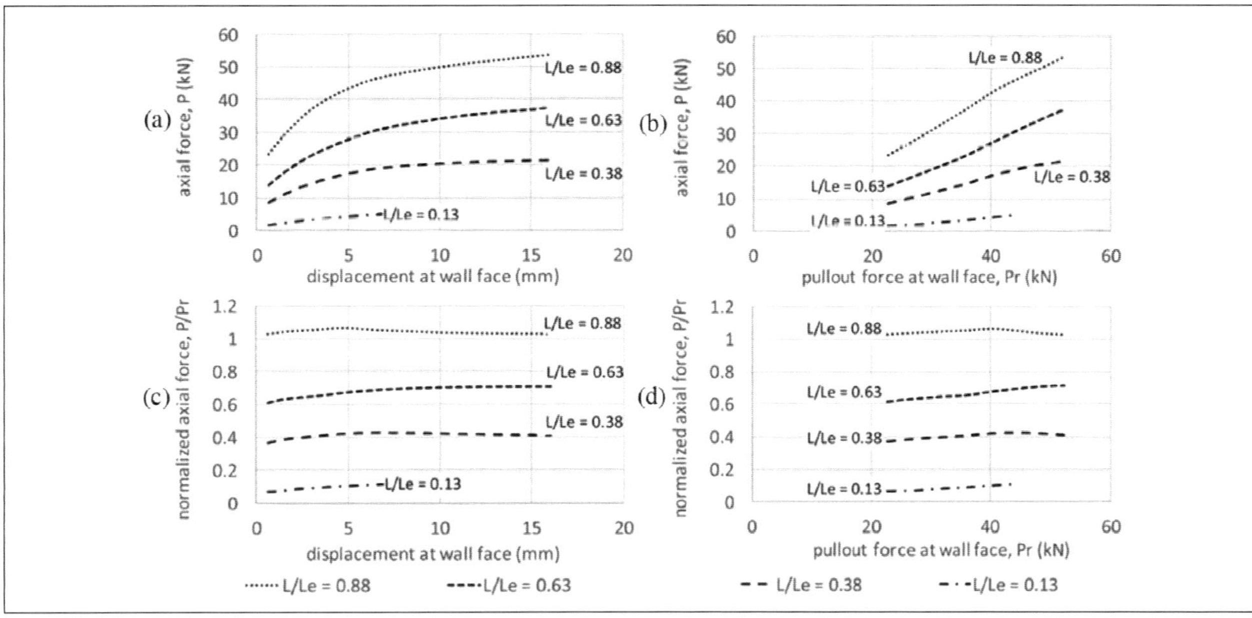

Figure 2. Example steel ribbed strip, embedded 2.4 m under 1.5 m of sandy fill: axial force along reinforcement vs. (*a*) displacement and (*b*) total pullout force; normalized axial force along reinforcement vs. (c) displacement and (*d*) total pullout force.

According to inextensible theory, the axial force in the strip should increase linearly from zero at the tail to the full pullout force at the point of maximum tension (the wall face for a pullout test). Equation 1 defines the theoretical relationship between effective embedment and axial force.

$$\frac{P}{P_r} = \frac{L}{L_e} \tag{1}$$

where: L = effective embedment length of the reinforcement from the tail to point of interest on the reinforcement

L_e = total embedment length of the reinforcement

P = axial force developed by the effective embedment length, L

P_r = total pullout force developed by the total embedment length, L_e (*4*)

Figure 3 shows synthesized data from all 19 instrumented pullout tests on steel ribbed strips. An open circle represents the normalized pullout force from a strain-gage pair, averaged for all pullout forces within service limits (below allowable tensile capacity and pullout displacements less than 19 mm). Overall, the data illustrate the expected relatively inextensible behavior. Each point carries its overburden stress and pullout history, and detailed consideration could explain a portion of the variance. However, by virtue of the large scale and number of tests, this dataset (mean of bias = 0.81, COV = 0.59) is the first to illustrate the natural variance in longitudinal load distribution caused by localized variations in the soil and locked-in strains from reinforcement placement typical of field MSE installations. Nevertheless, the overall trend is clearly linear.

A regression analysis revealed no statistically significant influence from backfill type, total reinforcement length, or backfill depth on the relationship between the normalized axial force and the normalized embedment. Further, statistical hypothesis testing supported the theoretical relationship (Equation 1) as a valid predictor of the axial force during a pullout test. The strain-gage data from this study support the notion of relatively inextensible pullout behavior of steel ribbed strips within service limits. Steel ribbed strips are the simplest reinforcement; strain-gage data confirm the relative inextensibility of steel ribbed strips.

Ladder-Like Strips

Figure 4 shows pullout test results for a representative ladder-like strip. This reinforcement has two longitudinal elements, making it slightly more complex than ribbed strips. Assuming no significant eccentricity, each longitudinal bar should carry half the pullout resistance. Each test location (by embedment length) had a gage pair on the left and right longitudinal bars resulting in coupled axial force data. For this particular example pullout test, the left bar gages at 1 ft from the test box wall did not survive the backfilling process. The other locations show reasonable agreement in axial force between left and right bars. Localized variation in backfill quality and compaction potentially induced some small eccentricity in the reinforcement. Again, the axial force measurements calculated from strain-gage data indicate a linear increase in axial force with total pullout resistance (Figure 4, *a* and *b*). The normalized axial force plots show constant, inextensible behavior (Figure 4, *c* and *d*). The variations in the normalized axial force plots under low displacement are consistent with the removal of incidental eccentricities in the reinforcement pullout system.

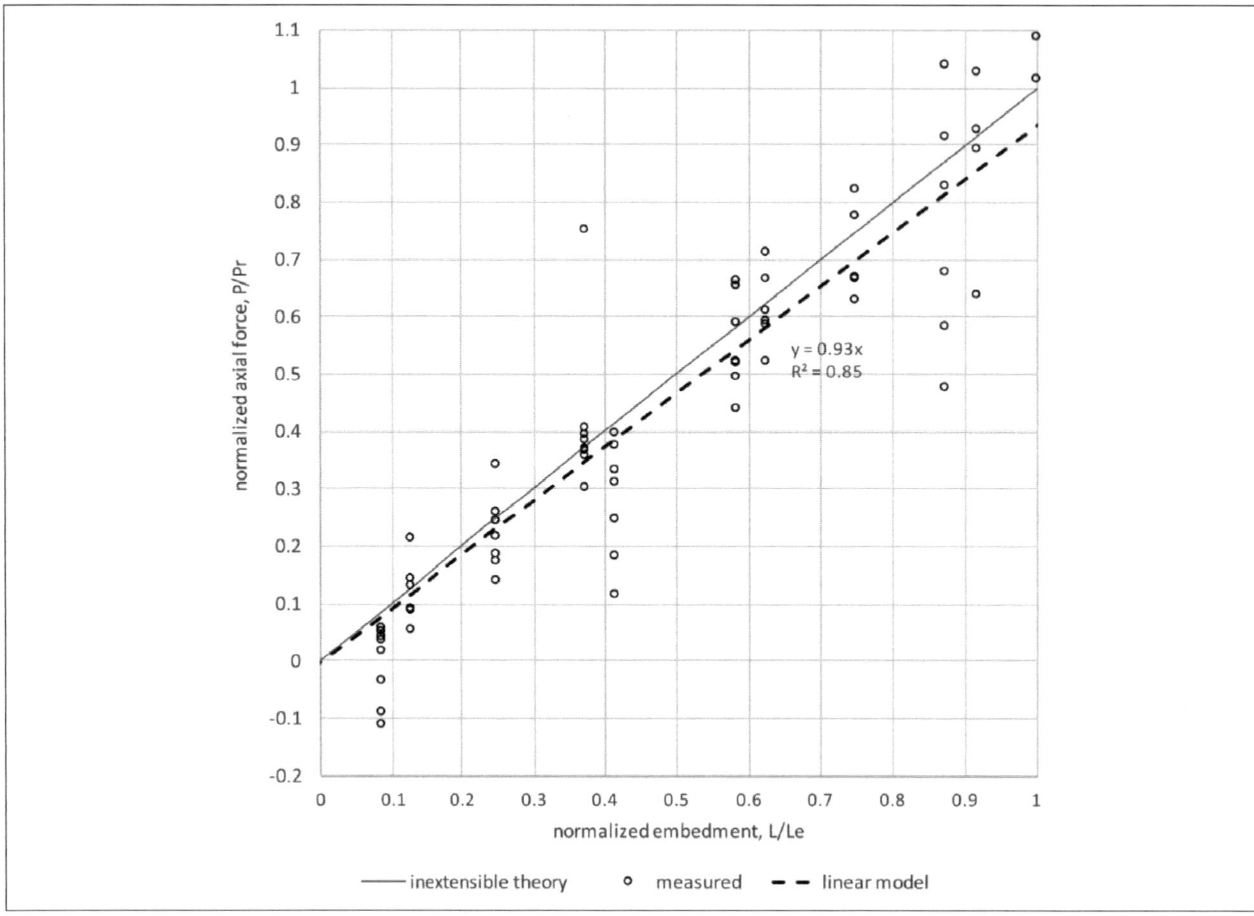

Figure 3. Steel ribbed strip summary; normalized axial force vs normalized length for 19 pullout tests.

Figure 4*e* shows the cross-section distribution of axial force between the longitudinal elements for the example test. As ladder-like strips have two longitudinal wires, each longitudinal element should carry roughly half the load. The relationship between axial force in a single longitudinal element and the effective embedment length is shown in Equation 2.

$$\frac{P}{P_r} = 0.5\frac{L}{L_e} \tag{2}$$

where: L = effective embedment length of the reinforcement from the tail to point of interest on the reinforcement

L_e = total embedment length of the reinforcement

P = axial force developed by the effective embedment length, L.

P_r = total pullout force developed by the total embedment length, L_e (*4*)

Figure 5 shows the theoretical relationship and synthesizes all the measured data for six steel ladder-like strips. Each point summarizes the stress history of a location on a ladder-like strip during pullout. Localized eccentricities caused response variance consistent with typical field installations conditions (mean of bias = 1.01, COV = 0.31); but again, a clear linear trend emerges. These pullout tests were more limited than the tests on ribbed strips; the only manipulated independent variable was depth of fill. These data support the expected stress distribution along the longitudinal elements. Regression analysis shows that depth of fill does not influence the relationship between normalized axial force and normalized embedment. Statistical hypothesis testing supports a linear relationship passing through the origin. The slope of the fitted trend line is statistically different from the slope assumed in Equation 2, but practically little difference exists between the regression slope of 0.53 and the theoretical slope of 0.5. Therefore, the ladder-like strips exhibit pullout behavior consistent with relative inextensibility.

Three-Wire Bar Mat Grid

A previous study observed that relatively stiff steel bar mats experienced plastic deformation of the transverse bars during pullout (*6*). Figure 6*a* shows a bar mat grid as observed after pullout testing and careful exhumation. The transverse bars are clearly deformed, and the current application of the Simplified Method suggests that the impact of this

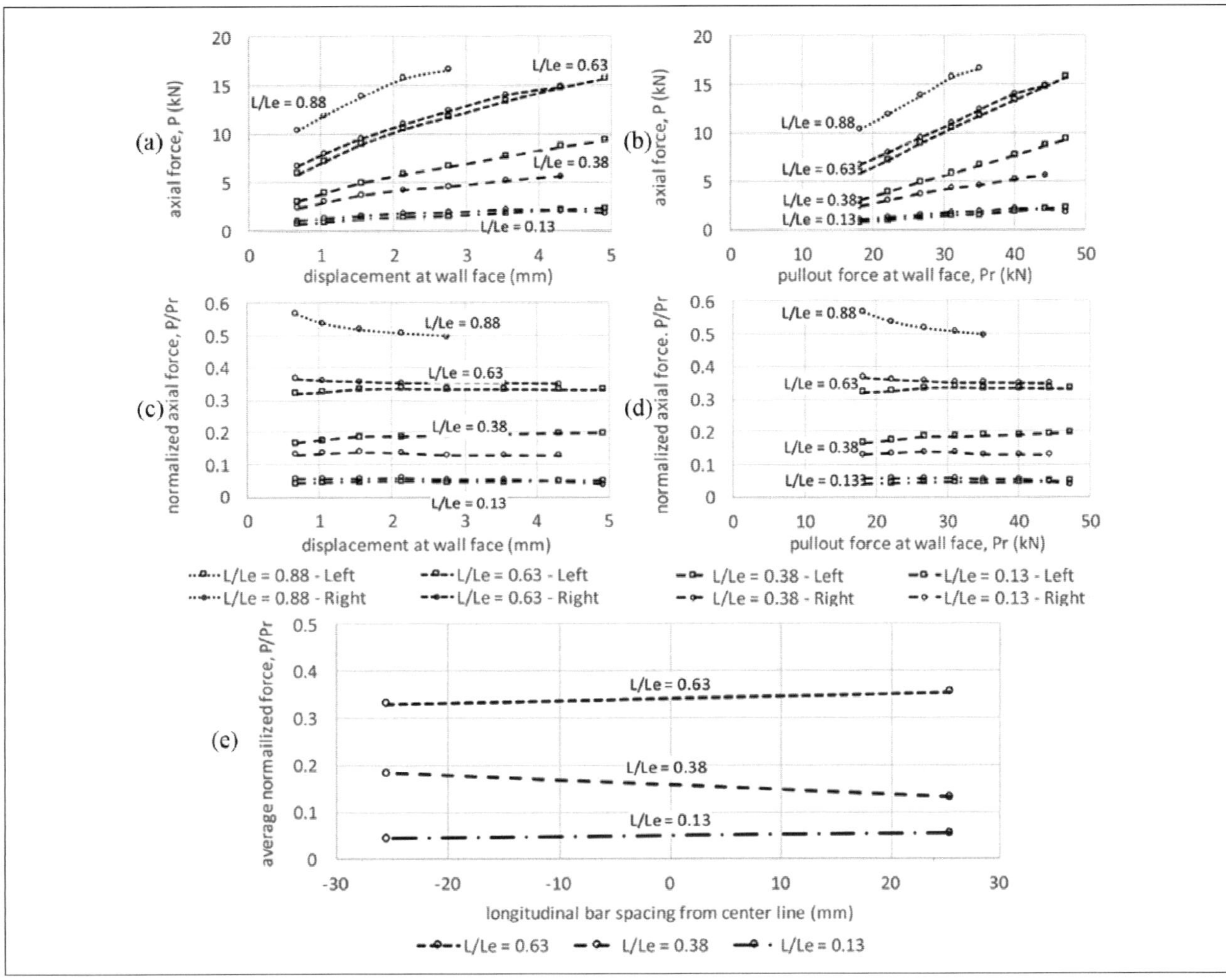

Figure 4. Example steel ladder-like strip, embedded 2.4 m under 1.8 m of sandy fill: axial force along reinforcement vs. (*a*) displacement and (*b*) total pullout force; normalized axial force along reinforcement vs. (*c*) displacement and (*d*) total pullout force, and (*e*) distribution of average normalized axial force between longitudinal bars.

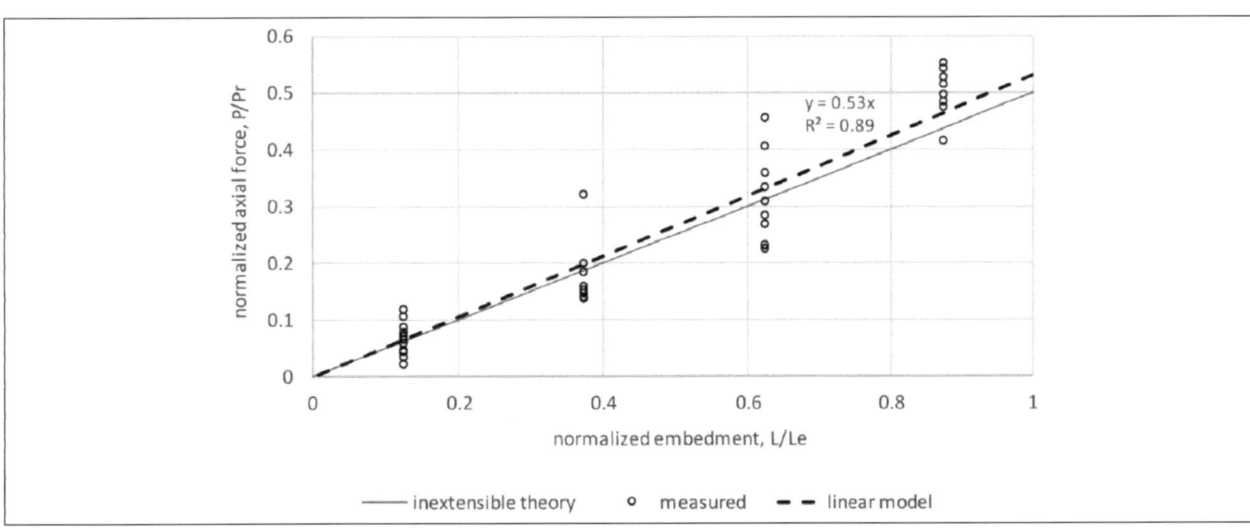

Figure 5. Steel ladder-like strip summary; normalized axial force vs normalized length for six pullout tests.

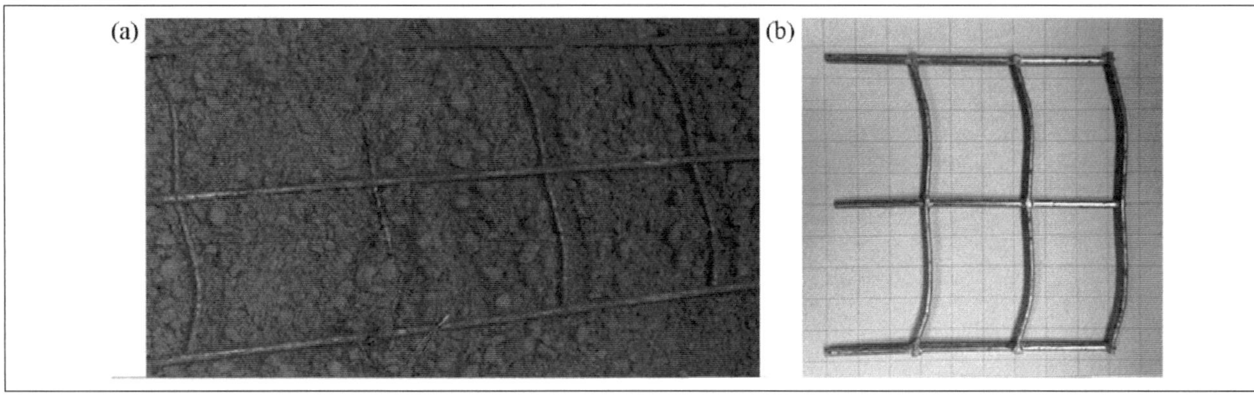

Figure 6. Uniform deformation in transverse bars: (*a*) immediately after pullout testing and (*b*) on a 50 mm by 50 mm grid.

deformation on pullout resistance be quantified through the scale effect correction factor, α, implying extensibility (*7*). However, each transverse bar experienced approximately the same degree of plastic deformation, and similarly sized voids developed behind each transverse bar. Figure 6*b* contrasts uniform deformation of each transverse bar with a square grid. Uniform deformation suggests that all transverse bars experienced the same bending moments and passive resistance. Such uniform passive resistance on each transverse element is qualitatively consistent with inextensible reinforcement behavior. Therefore, the observed deformation is *not* the extensibility the scale effect correction factor, α, is intended to address. Rather, this deformation is better addressed by the pullout resistance factor, F* or some to-be-developed conceptual formulation for describing transverse element deformation (*6*).

Strain behavior during pullout for three-wire bar mat grid reinforcements is more complex than both ribbed strips and ladder-like strips. For three-wire bar mat grid reinforcement, a first-order approximation would suggest each longitudinal bar carries one-third of the load. A more refined tributary area approach would indicate that the outer bars carry one-fourth of the total pullout force whereas the center bar carries one-half of the total pullout force. The differences in behavior between outside and center bars require separate analysis of the axial force.

Figure 7, *a–d*, shows the axial force in the outside bars of a representative three-wire bar mat grid. Much like the ladder-like strips reinforcements, the left and right bars carry similar load magnitudes. Again, the axial force displacement curve (Figure 7*a*) seems appropriately related to soil stress–strain models. Figure 7*b* shows expected linear force development. The normalized axial force varies some under low displacement and large loads, but for the most part the normalized axial force is relatively constant (Figure 7, *c* and *d*).

Figure 7, *e–g*, shows the axial force carried by the center bar for the same representative three-wire bar mat grid. All trends are as expected. However, around 155 kN pullout force and 16 mm displacement, the center bar pullout force appears to maximize at each gaged location. This causes a slight drop in the normalized axial force (Figure 7, *e* and *f*). Further, this corresponds to an increase in the normalized axial force in the outside bars (Figure 7, *c* and *d*). This effect was because of observed yielding, necking, and eventual rupture in the center bar outside the pullout test box. Because the pullout system ensured uniform displacement of all longitudinal bars at the gripper, as the center bar yielded, the center portion of the transverse bars displaced through the soil at a slower rate, generating differential deformation in the transverse bars between the outside and center longitudinal bar welds (Figure 6). Accordingly, the outside bars carried a greater percentage of the load. Therefore, the mean normalized axial force for cumulative synthesis of the load distribution is limited to those loads below yielding.

For steel three-wire bar mat grid, the theoretical relationships are similar to the other reinforcements considered in this paper. Unlike ribbed strips, passive resistance on the transverse bars generates the vast majority of the pullout resistance. If, consistent with relatively inextensible behavior, each transverse bar generates the same amount of passive bearing resistance and the frictional resistance on the longitudinal bars is negligible, the theoretical prediction of axial force at any location along a longitudinal element is shown in Equation 3.

$$\frac{P}{P_r} = m\frac{n}{N} \tag{3}$$

where: m = the distribution of pullout force between longitudinal elements

n = number of transverse elements in a grid reinforcement along the effective embedment length, L

N = total number of transverse elements along the total embedment length, L_e

P = axial force developed by the effective embedment length, L

P_r = total pullout force developed by the total embedment length, L_e (*4*)

The slope, m, is scaled to allocate the total pullout force between longitudinal bars. Figure 7*i* shows a cross-section where, throughout the test, the center bar carries more load

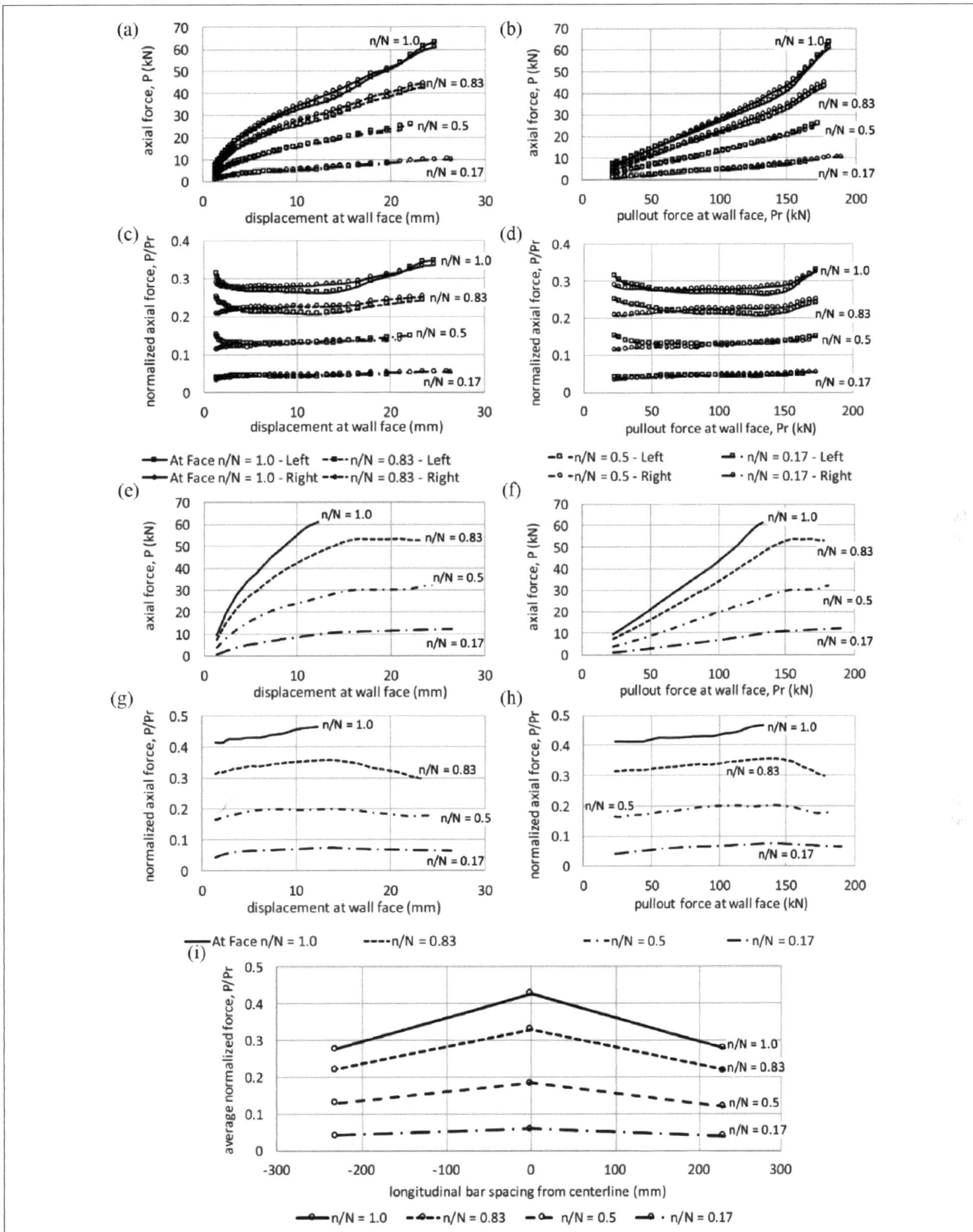

Figure 7. Data from an example steel three-wire bar mat grid, embedded 1.8 m under 1.5 m of gravelly fill: outside bar axial force along reinforcement vs. (*a*) displacement and (*b*) total pullout force; outside bar normalized axial force along reinforcement vs. (*c*) displacement and (*d*) total pullout force; center bar axial force along reinforcement vs. (e) displacement and (*f*) total pullout force; center bar normalized axial force along reinforcement vs. (*g*) displacement and (*h*) total pullout force; (*i*) distribution of average normalized axial force between longitudinal bars.

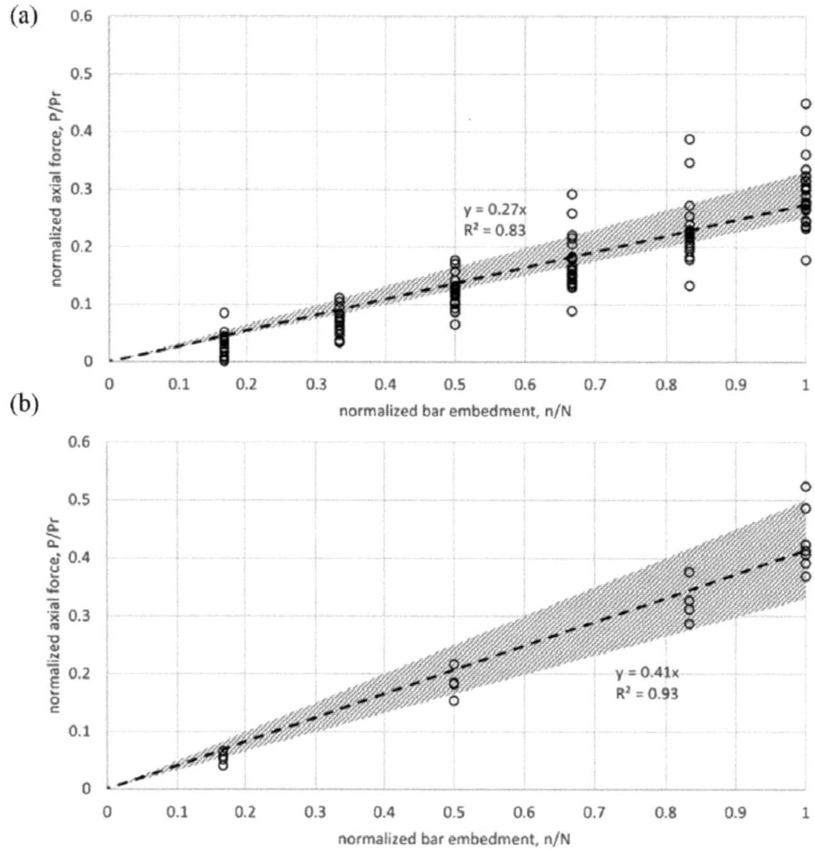

Figure 8. Steel three-wire bar mat grid summary: (*a*) normalized axial force vs. normalized length for outside longitudinal bars for 18 pullout tests and (*b*) normalized axial force vs. normalized length for center longitudinal bars for seven pullout tests.

along the whole length of the reinforcement. The outside bars carry about the same axial force. Therefore, the outside longitudinal bars should have a slope, m, less than the center longitudinal bar.

Figure 8*a* shows a theoretical model for the outside bars as a shaded range from $m = 0.25$ based on tributary area to $m = 0.33$ for an even split between longitudinal bars. In an MSE structure, maximum reinforcement tension occurs at the soil mass's potential failure plane. Uniform displacement of longitudinal elements at the gripper approximates this loading condition. The data for the outside bars in the tested grid show $m = 0.27$ based on a linear regression. Visually, the data seem to follow the linear trend reasonably well (mean of bias = 0.91, COV = 0.34), and quantify typical naturally occurring variance in longitudinal load distribution in grid reinforcements. Statistically, the depth of fill appears to influence the relationship, but practically, the variation attributed to depth of fill is less than 5%. If depth of fill is disregarded, the linear model shows no lack of fit. Backfill type, total transverse bar number, and longitudinal bar spacing have no statistically significant influence on the relationship.

The same trends are seen in Figure 8*b* for the center bar of three-wire bar mat grids. Again, the theoretical range is shown as a shaded region. The slope of the linear regression line from the data is shown as $m = 0.41$. Again, the depth of fill has a statistically significant, but practically inconsequential, influence on the relationship between axial force and the number of embedded bars. A linear relationship showed no lack of fit (mean of bias = 0.94, COV = 0.17). Even in bar mat grids with widely spaced longitudinal bars, pullout behavior is relatively inextensible.

Because steel three-wire bar mats grids are thus shown to be inextensible, the deformation types seen in Figure 6 cannot be evidence of extensibility as defined by failure tensile strains in the longitudinal members. For the Simplified Method, the illustrated inextensibility implies a scale effect correction factor, α, of 1.0, even for widely spaced bar mat grids. This suggests the need for more refined deformation characterizations and design formulations that differentiate longitudinal deformations that are indeed inextensible (as discussed here) from longitudinal deformations resulting from extensible behavior (i.e., geosynthetic reinforcements). For example, the present design formulation might be revised to allow for a mechanistic-empirical factor (not α) that directly accounts for deformation associated with less stiff transverse elements such as widely spaced longitudinal bars or small transverse bar diameters. Other features intended to facilitate longitudinal extensibility, such as kinks or flexible inclusions, could still be modeled using the scale effect correction factor, α. Conceptual clarity on this phenomenon could improve predicted pullout resistance for widely spaced bar mat grids and provide a baseline for interpreting the behavior of semi-extensible and other MSE reinforcement types.

Conclusions

In summary, a significant program of instrumented pullout tests explored pullout behavior for three styles of steel MSE reinforcements. Results support the following conclusions:

- Each tested steel reinforcement type—ribbed strips, ladder-like strips, and three-wire bar mat grids—exhibits relatively inextensible behavior during pullout testing. Therefore, inextensible reinforced earth theory, design assumptions and methods as related to pullout resistance are appropriate for these reinforcements.
- The synthesis of inextensible longitudinal load distribution using the novel approach of mean normalized axial force allows for observation of natural variation in longitudinal load distribution typical of field installations of inextensible soil reinforcements.
- For ladder-like strips, the axial force is distributed evenly between the two longitudinal bars. For three-wire bar mat grids without tangs, the center bar carries approximately 40% of the axial force and each outside longitudinal bar carries approximately 30% of the axial force.
- In three-wire bar mat grids having widely spaced longitudinal bars, pullout-induced deformations exist but are conceptually different from extensibility. A refinement in current policy that directly accounts for deformations in the transverse elements would improve pullout predictions and support a more nuanced formulation of pullout resistance for semi-extensible reinforcements.

The experimental test program for this study confirms that negligible longitudinal elongation in common strip and grid-type steel MSE reinforcements does not practically influence the development of axial force along the length of the reinforcement during pullout. In other words, the tested steel reinforcements are relatively inextensible. With new and innovative products pushing the boundaries of steel MSE reinforcement design outside the original scope of the standard design methods, the MSE community will benefit from the refinement of assumptions concerning inextensible steel reinforcements.

Acknowledgments

Texas Department of Transportation sponsored this research work and provided technical guidance through Marie Fisk, P.E., Marcus Galvan, P.E., and John Delphia, P.E.

References

1. Associate of Mechanically Stabilized Earth. *Why Steel?*. 2016. [Online]. http://www.amsewalls.org/why-steel/. Accessed May 5, 2016.
2. Segesrestin, P., and M. Bastick. Comparative study and measurement of the pullout capacity of extensible and inextensible reinforcements. *International Symposium on Earth Reinforcement*, Fukuoka, Kyushu, Japan, 1996.
3. Bergado, D. T., K. H. Lo, J. C. Chia, R. Shivashankar, M. C. Alfaro, and L. R. Anderson. Pullout Test Using Steel Grid Reinforcements with Low Quality Backfill. *Journal of Geotechnical Engineering*, Vol. 118, No. 7, 1992, pp. 1047–1062.
4. American Association of State Highway and Transportation Officials. *LRFD Bridge Design Specifications, U.S. Customary Units*, 7th ed. AASHTO, Washington, D.C., 2014.
5. Jayawickrama, P. W., J. G. Surles, T. A. Wood, and W. D. Lawson. *Pullout Resistance of Mechanically Stabilized Reinforcements in Backfills Typically Used in Texas: Volume 1*. Texas Department of Transportation, Austin, 2013.
6. Lawson, W. D., P. W. Jayawickrama, T. A. Wood, and J. G. Surles. Evaluation of AASHTO Default Values for Pullout Friction Factor, F* for Steel Grid Mat Reinforcement. *Transportation Research Record: Journal of the Transportation Research Board*, 2014. 2462: 7–17.
7. Berg, R. R., B. R. Christopher, and N. C. Samtani. *Design of Mechanically Stabilized Earth Walls and Reinforced Soil Slopes - Volume I*. Federal Highway Administration, Washington, D.C., 2009.
8. Suncar, O. W. *Pullout and Tensile Behavior of Crimped Steel Reinforcement for Mechanically Stabilized Earth (MSE) Walls*. Utah State University, Logan, Utah, 2010.
9. Tin, N., D. T. Bergado, L. R. Anderson, and P. Voottipruex. Factors Affecting Kinked Steel Grid Reinforcement in MSE Structures. *Geotectiles and Geomembranes*, Vol. 29, 2011, pp. 172–180.
10. Vidal, H. The Principle of Reinforced Earth. *Highway Research Record*, No. 282, 1969, pp. 1–16.
11. Anderson, P. L., R. A. Gladstone, and J. L. Withiam. Coherent Gravity: The Correct Design Method for Steel-Reinforced MSE Walls. *Earth Retention Conference 3*, Bellevue, WA, 2010.
12. McGown, A., K. Z. Andrawes, and M. M. Al Hasani. Effect of Inclusion Properties on the Behavior of Sand. *Geotechnique*, Vol. 28, No. 3, 1978, pp. 327–346.
13. Bathurst, R. J., A. Hernheim, and T. M. Allen. Predicted and Measured Loads Using the Coherent Gravity Method. *Preoceedings of the ICE - Ground Improvement*, Vol. 161, No. 3, 2008, pp. 113–120.
14. Anderson, P. L., R. A. Gladstone, and J. E. Sankey. State of the Practice of MSE Wall Design for Highway Structures. *GeoCongress 2012*, Oakland, CA, 2012.
15. Palmeira, E. M., and G. W. E. Milligan. Scale and Other Factors Affecting the Results of Pull-Out Tests of Grids Buried in Sand. *Geotechnique*, Vol. 39, No. 3, 1989, pp. 511–524.

16. Christopher, B. R., S. A. Gill, J. P. Giroud, I. Juran, J. K. Mitchell, F. Schlosser, and J. Dunnicliff. *Reinforced Soil Structures Volume I. Design and Construction Guidelines*. Federal Highway Administration, Washington, D.C., 1990.
17. Allen, T. M., and R. J. Bathurst. Improved Simplified Method for Prediction of Load in Reinforced Soil Walls. *Journal of Geotechnical and Geoenvironmental Engineering*, Vol. 141, No. 11, 2015, p. 04015049.
18. American Society for Testing and Materials. ASTM D6706-01: *Standard Test Method for Measuring Geosynthetic Pullout Resistance in Soil*. ASTM International, West Conshohocken, PA, 2013.
19. Texas Department of Transportation. *Standard Specifications for Construction and Maintencance of Highways, Streets, and Bridges*. Texas Department of Transportation, Austin, 2014.
20. American Society for Testing and Materials. ASTM D5195: *Standard Test Method for Density of Soil and Rock In-Place at Depths Below Surface by Nuclear Methods*. ASTM International, West Conshohocken, PA, 2008.
21. American Society for Testing and Materials. ASTM D2487: *Standard Practic for Classification of Soils for Engineering Purposes (Unified Soil Classification System)*. ASTM International, West Conshohocken, PA, 2011.
22. Vishay Precision Group. *Instruction Bulletin B-127-14 Strain Gage Installations with M-Bond 200 Adhesive*. Vishay Micro-Measurements, 2010.
23. Lawson, W. D., P. W. Jayawickrama, T. A. Wood, and J. G. Surles. Pullout Resistance Factors for Inextensible Mechanically Stabilized Earth Reinforcements in Sandy Backfill. *Transportation Research Record: The Journal of the Transportation Research Board*, 2013. 2364: 21–29.
24. Jayawickrama, P. W., W. D. Lawson, T. A. Wood, and J. G. Surles. Pullout Resistance Factors for Steel MSE Reinforcements Embedded in Gravelly Backfill. *Journal of Geotechnical and Geoenvironmental Engineering*, Vol. 141, No. 2, 2015, pp. 04014090.

The Standing Committee on Transportation Earthworks (AFS10) peer-reviewed this paper (18-01457).

Article

TRR
JOURNAL OF THE TRANSPORTATION RESEARCH BOARD

Transportation Research Record
2018, Vol. 2672(52) 251–257

DOI: 10.1177/0361198118770166
journals.sagepub.com/home/trr
SAGE

A Process for Optimizing Gradation of Marginal Backfill of Mechanically Stabilized Earth Walls to Achieve Acceptable Resistivity

Jose Luis Arciniega[1], W. Shane Walker[2], Soheil Nazarian[3], and Kenneth L. Fishman[4]

Abstract

The service life of mechanically stabilized earth walls depends on the corrosion rate of the metallic reinforcement used in their construction. The resistivity of the backfill aggregates needs to be measured accurately to estimate realistically the corrosion rate of the reinforcement. Resistivity testing is usually performed using the traditional soil box on the portion of the aggregates that passes a No. 10 or No. 8 sieve to either select or reject the backfill. For a more reasonable characterization of the corrosivity of coarse backfills, it is desirable to use their actual gradations. To that end, several resistivity boxes that were double and quadruple the dimensions of the original box were constructed. In addition to the three standard gradations specified by the Texas Department of Transportation, over 20 backfill materials sampled from sources throughout Texas were fractionated to fines, fine sand, coarse sand, and gravel. Resistivity tests were performed separately on each of these four constituents for each backfill. The results were used to evaluate a relationship that would allow the estimation of the resistivity of any desired backfill gradations from the resistivity values of these four constituents. The proposed model looks promising since the resistivity of the backfill composed of the actual gradation can be estimated with reasonable certainty. The results of this study can potentially help highway agencies and contractors use a number of local quarries that are currently disqualified based on the resistivity values obtained from only testing materials that pass a No. 8 or No. 10 sieve.

Mechanically stabilized earth (MSE) walls have been used widely as an economical alternative to traditional retaining walls. Estimating the rate of corrosion of the metallic reinforcement is crucial for determining the service life of an MSE wall, and as such, an assessment of the corrosion potential of the aggregate used as backfill is required.

The corrosion potential of the backfill is characterized in terms of its resistivity, soluble ion content, and moisture content, and the position of the water table. Resistivity is an intrinsic property of the material that can be used appropriately to estimate its corrosion potential (*1, 2*). As a result, many Departments of Transportation (DOTs) have established limits for the resistivity of material that can be used as backfill. The test is performed within a box with dimensions 153 mm × 102 mm × 44 mm (6 in. × 4 in. × 1 3/4 in.), with two stainless steel plates along the long sides as electrodes. Since the electrical resistance (R) of a material between two electrodes is a function of the electrical resistivity (ρ), the length between the electrodes (l), and the cross-sectional area (A), the resistivity can be approximated, as shown in (*3*), by

$$\rho = RA / l. \tag{1}$$

By passing a 97 Hz square wave current through the two electrodes, the electrical resistance between the electrodes is measured to calculate the resistivity of the material. Elias et al. (*4*) indicated there is little potential for corrosion in MSE walls when the material has a resistivity above 10,000 Ω·cm. A material usually cannot be used as backfill if its resistivity is below a certain level (e.g., 3,000 Ω·cm), as it would be regarded as having a high potential for corrosion.

There is concern that the existing testing procedures are not representative of coarse-aggregate backfills. Procedures such as AASHTO T288 use material that passes the No. 10 sieve, while material used in MSE walls typically contains only a small percentage that passes the No. 10 sieve. Thapalia et al.

[1]Center for Transportation Infrastructure Systems, The University of Texas at El Paso, El Paso, TX
[2]Civil Engineering, The University of Texas at El Paso, El Paso, TX
[3]The University of Texas at El Paso, El Paso, TX
[4]McMahon & Mann Consulting Engineers, P.C., Buffalo, NY

Corresponding Author:
Address correspondence to Jose Luis Arciniega:
jlarciniegaaguilar@miners.utep.edu

Figure 1. Original, doubled dimension, and quadrupled dimension resistivity soil boxes.

(*5*) indicated that the portion that passes the No. 10 sieve was not indicative of the corrosion potential of the material used due to the scarcity of such materials in coarse backfills. This could result in rejection of materials that otherwise would have satisfied the specifications. Several ongoing research studies are attempting to address this concern by increasing the size of the boxes used to measure resistivity in order to accommodate larger aggregates.

The purpose of this study is to explore the possibility of estimating the resistivity of a given backfill knowing the gradation and the resistivity of its major constituents (gravel, coarse sand, fine sand, and fines, as defined by the Unified Soil Classification System, USCS). A model that can approximate the resistivity of an aggregate with a known gradation based on the resistivity values of the four constituents can have practical advantages, with the main advantage being the possibility of adjusting the gradation of a backfill to achieve the specified minimum resistivity. As shown below, this could allow the selection of coarse materials that would traditionally not have been considered.

Methodology

In this research, the resistivity of backfill materials was analyzed with a method based on the Texas Department of Transportation (TxDOT) test procedure Tex-129-E (*6*), which is similar to AASHTO T-288 (*7*). The main difference between the two protocols is that Tex-129-E specifies analyzing material passing a No. 8 (2.4 mm) sieve, while AASHTO T-288 specifies analyzing material passing a No. 10 (2 mm) sieve. For this research, 1.3 kg (2.9 lb) of sample that passed a No. 8 sieve was prepared. The material was compacted to the brim of the resistivity soil box (see Figure 1), and any excess removed from the surface with a straight edge. The compacted specimen was weighed (to calculate the density of the sample) and the resistivity was measured using a soil resistance meter. Deionized (DI) water was then added in increments of 50 mL (0.11 lb) for sand soils and 100 mL (0.22 lb) for clay soils. Each increment of water was a repeat of the testing cycle, taking the material out of the box, mixing in additional water, and compacting. The end point for each test in this study was when either a sheen was observed at the top of the sample or the minimum resistivity was measured (whichever happened first).

Two additional resistivity boxes with dimensions two and four times the original box size were built to accommodate testing of coarse aggregates (see Figure 1). The doubled dimension box was used to test coarse sand, while the quadrupled dimension box was used to test gravel and TxDOT gradations of AS, BS, and DS (see Figure 2). The amount of material used and DI added were proportioned in order to give the same relation by volume as in the original size test procedure. That is, the amounts of material and water used in the doubled dimension box were eight times the original, and the amounts of material and water used in the quadrupled dimension box were 64 times the original. Arciniega (*8*) showed that these three boxes yielded resistivity values that were consistent and independent of the box size. Research at the University of Texas at El Paso (UTEP) is being conducted to implement a modified version of Tex-129-E that accommodates the use of these boxes to test the resistivity of coarse aggregates.

Figure 2*a* shows the grain-size distributions of the samples for Type AS, Type BS, and Type DS. All gradations included only materials finer than the 75 mm (3 in.) sieve. As also shown in Figure 2*a*, Type AS materials contained less than 50% materials passing the 12.7 mm (0.5 in.) sieve, less than 15% passing the No.40 sieve, and less than 5% passing the No. 200 sieve. Type BS materials contained less than 60% passing the No. 40 sieve and less than 15% passing the No. 200 sieve. Type DS backfills contained less than 15% passing the 9.5 mm (3/8 in.) sieve and less than 5% passing the No. 200 sieve. The gradations used for this study were designed to have distributions that passed through the middle of the acceptable ranges prescribed by TxDOT. The weight fractions of these gradations are shown in Figure 2*b* and are listed in Table 1. Type AS contained a higher proportion of intermediate particles and a small fraction of fines (about 1%), while Type DS had a gap-graded distribution with a high fraction of gravel and low fraction of sand and fines with the fines fraction also being low (2%). Type BS material was a well-graded material with a relatively high fraction of fines (8%). In addition to these three gradations, the backfill material was fractionated into gravel, coarse sand, fine sand, and fines as per USCS. Gravel is defined as particles between the 75 mm (3 in.) and No. 4 sieves, coarse sand between the No. 4 and No. 40 sieves, fine sand between the No. 40 and No. 200 sieves, and fines passing the No. 200 sieve. Resistivity tests were also performed on materials passing the No. 8 sieve as required by Tex-129-E.

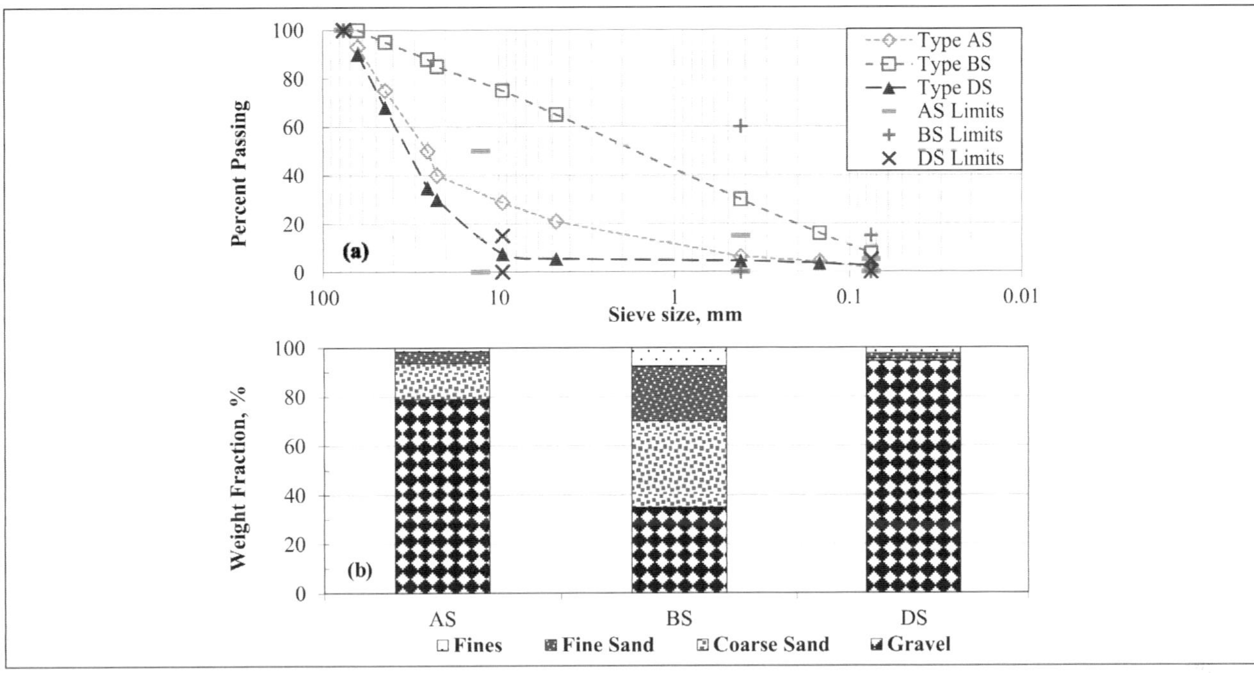

Figure 2. Type AS, BS, and DS gradations: (*a*) cumulative size distributions and (*b*) composition by weight fractions.

Table 1. Composition by Weight Fraction for Type AS, BS, and DS Gradations

Constituents	AS	BS	DS
Gravel	79%	35%	95%
Coarse sand	15%	35%	1%
Fine sand	5%	22%	2%
Fines	1%	8%	2%

Results and Analyses

Table 2 summarizes the observed minimum resistivity values of 18 materials tested using the methodology described. Resistivity values below 3,000 Ω·cm are shown in red, while resistivity values above 10,000 Ω·cm are shown in green. Five of the 18 materials (28%) had a Tex-129-E minimum resistivity of less than 3,000 Ω·cm, but only 9% of the materials with AS, BS, and DS gradations had a minimum resistivity less than 3,000 Ω·cm. Thus, for the 18 materials and three gradations analyzed in this research, approximately 19% of the materials with the AS, BS, and DS gradations would not be allowed for use as backfill based on Tex-129-E but would meet the 3,000 Ω·cm limit if the actual gradation were tested. Half of the resistivity values of the fine sand and two-thirds of the resistivity values of the fines were less than 3,000 Ω·cm, which suggests that most of the soluble ionic content is in the fine sand and fines. All gradations or size fractions for material L had a minimum resistivity of less than 3,000 Ω·cm.

Figure 3 shows the average minimum resistivity of all 18 materials normalized by Tex-129-E per gradation (AS, BS, and DS) and size fraction (gravel, coarse sand, fine sand, and fines). Materials with a Type DS gradation had a higher average resistivity than those with a Type AS gradation, while the materials with Type AS and DS gradations exhibited a higher average resistivity than Type BS. Within the constituent size fractions, the resistivity decreased as the particle size decreased. Again, the fine sand and fines had lower average resistivity values than the gravel and coarse sand (and lower than the fraction tested in Tex-129-E). This pattern indicates that most of the soluble ionic content is in these size fractions, and thus the materials with the Type BS gradation demonstrate a lower average resistivity than the AS and DS gradations due to the higher weight fractions of the fine sand and fines.

From the trends observed within the data, it was hypothesized that the total resistivity of a given material and gradation at saturation could be predicted as a function of the resistivity and weight fraction of the four constituent size fractions (i.e., gravel, coarse sand, fine sand, and fines). This hypothesis assumes that the four constituents are effectively acting as resistors in parallel. An adaptation of the formula to estimate the total resistance of four resistors in parallel was used:

$$\rho_{\text{predicted}} = (W_g / \rho_g + W_{cs} / \rho_{cs} + W_{fs} / \rho_{fs} + W_f / \rho_f)^{-1} \quad (2)$$

where W_i is the weight fraction of constituent i, and ρ_i is the minimum resistivity of the constituent.

Table 2. Minimum Resistivity of Measured Samples by Gradation

	Minimum Resistivity, Ω·cm							
Material	Tex-129-E	AS	BS	DS	Gravel	Coarse sand	Fine sand	Fines
A	5,534	*17,336*	5,334	*22,403*	*29,337*	*12,802*	**2,334**	**1,867**
B	4,001	*17,869*	6,668	*26,670*	*27,737*	6,801	4,467	**2,200**
C	5,334	*14,402*	6,934	*20,003*	*37,338*	*10,668*	3,067	3,400
D	4,867	*20,803*	*10,935*	*20,003*	*28,004*	*33,338*	**2,800**	**2,667**
E	*36,671*	*37,338*	*10,135*	*45,339*	*48,006*	*34,671*	6,934	4,067
F	8,001	*23,203*	*10,935*	*29,337*	*34,671*	*12,268*	6,668	5,134
G	4,201	*20,269*	9,335	*26,670*	*22,936*	*11,201*	5,067	3,600
H	**2,600**	4,001	4,001	4,001	*13,068*	3,867	**2,267**	**2,267**
I	3,134	5,601	**2,667**	5,334	*13,068*	3,200	**1,267**	**867**
J	**2,267**	8,268	3,200	*12,802*	*18,669*	7,468	**1,734**	**1,400**
K	**800**	3,467	**2,160**	7,201	8,001	**2,000**	**1,133**	**1,267**
L	**1,334**	**2,667**	**2,374**	**2,667**	**2,934**	**2,534**	**1,267**	**1,400**
M	*16,002*	*50,673*	*25,337*	*37,338*	*66,675*	*29,337*	9,335	3,267
N	5,334	*16,002*	5,334	*19,202*	*24,536*	*12,002*	5,201	**1,867**
O	9,335	*29,337*	*10,935*	*37,338*	*109,347*	*25,337*	5,467	3,667
P	7,334	*14,669*	7,468	*14,935*	*19,202*	8,134	4,601	**2,534**
Q	3,000	8,001	3,467	9,601	*13,335*	3,734	**1,800**	**1,534**
R	**1,867**	*12,268*	5,067	*16,002*	*17,069*	4,934	**2,534**	**1,667**

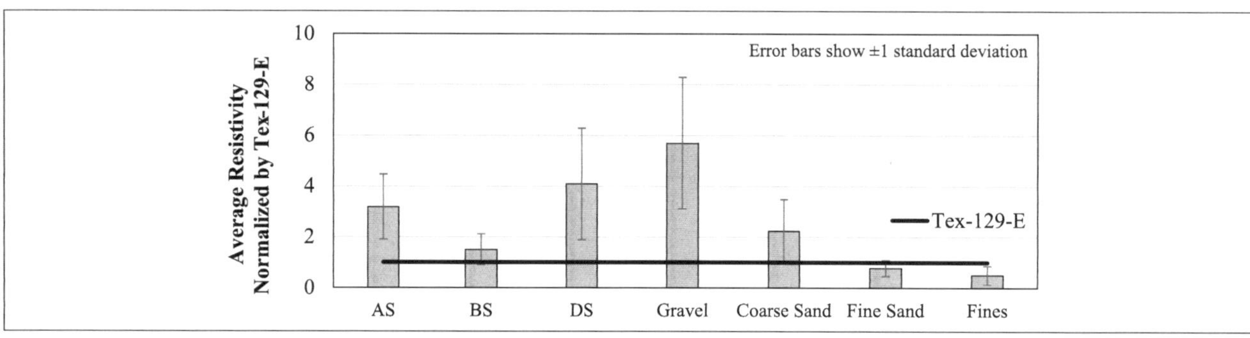

Figure 3. Average resistivity normalized by Tex-129-E.

Table 3. Predicted and Measured Resistivity Values

	AS Gradation resistivity, Ω·cm		BS Gradation resistivity, Ω·cm		DS Gradation Resistivity, Ω·cm	
Material	Measured	Predicted	Measured	Predicted	Measured	Predicted
A	*17,336*	*14,953*	*5,334*	*5,686*	*22,403*	*18,197*
B	*17,869*	*14,807*	*6,668*	*6,732*	*26,670*	*19,463*
C	*14,402*	*18,230*	*6,934*	*7,268*	*20,003*	*24,925*
D	*20,803*	*18,083*	*10,935*	*7,607*	*20,003*	*19,778*
E	*37,338*	*31,973*	*10,135*	*14,647*	*45,339*	*34,477*
F	*23,203*	*22,314*	*10,935*	*11,497*	*29,337*	*27,824*
G	*20,269*	*16,358*	*9,335*	*8,949*	*26,670*	*18,873*
H	*4,001*	*7,937*	*4,001*	*4,006*	*4,001*	*10,554*
I	*5,601*	*6,203*	**2,667**	**2,498**	*5,334*	*8,329*
J	*8,268*	*9,980*	*3,200*	*4,016*	*12,802*	*12,293*
K	*3,467*	*4,427*	**2,160**	**2,099**	*7,201*	*6,231*

(continued)

Table 3. (continued)

	AS Gradation resistivity, Ω·cm		BS Gradation resistivity, Ω·cm		DS Gradation Resistivity, Ω·cm	
Material	Measured	Predicted	Measured	Predicted	Measured	Predicted
L	***2,667***	***2,662***	***2,374***	***2,047***	***2,667***	***2,780***
M	*50,673*	*37,633*	*25,337*	*15,567*	*37,338*	*41,137*
N	*16,002*	*16,216*	*5,334*	*7,882*	*19,202*	*17,673*
O	*29,337*	*38,713*	*10,935*	*17,719*	*37,338*	*51,250*
P	*14,669*	*13,249*	*7,468*	*7,155*	*14,935*	*15,466*
Q	*8,001*	*7,420*	*3,467*	*3,403*	*9,601*	*9,905*
R	*12,268*	*9,632*	*5,067*	*4,440*	*16,002*	*12,456*

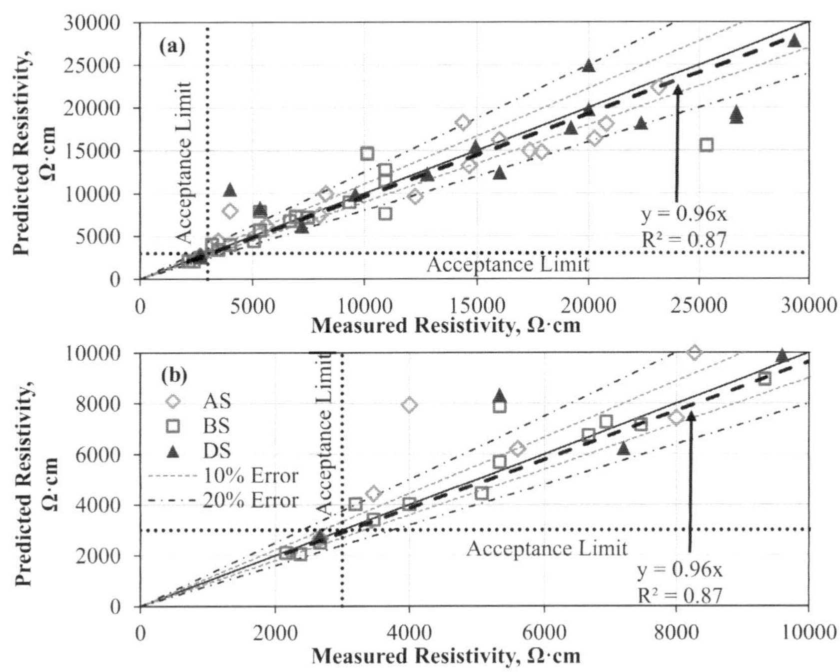

Figure 4. Predicted resistivity versus measured resistivity for (*a*) resistivity less than 30,000 Ω·cm and (*b*) resistivity less than 10,000 Ω·cm.

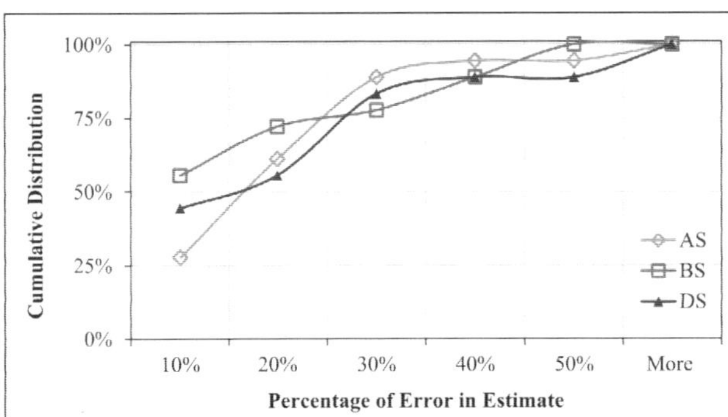

Figure 5. Cumulative distribution of error in the estimate of resistivity per gradation.

The model presented in Equation 2 was tested using the data in Table 2. The values obtained from that model are presented in Table 3. The values in this table are annotated similarly to Table 2. Figure 4 is the cross-plot of the predicted resistivity versus the measured resistivity (shown twice with different boundaries for reference in the interpretation of the data). Only resistivity values under 30,000 Ω·cm were considered in the statistical analysis. With a coefficient of determination (R^2 value) of 0.87 and a standard error of estimate (SEE) of 2,288 Ω·cm, the model can estimate the measured resistivity values reasonably well. The best-fit line produced by the data is very close to showing a one-to-one relationship (slope of 0.96). While there could be concern about the scatter that occurs at a resistivity above 10,000 Ω·cm, these

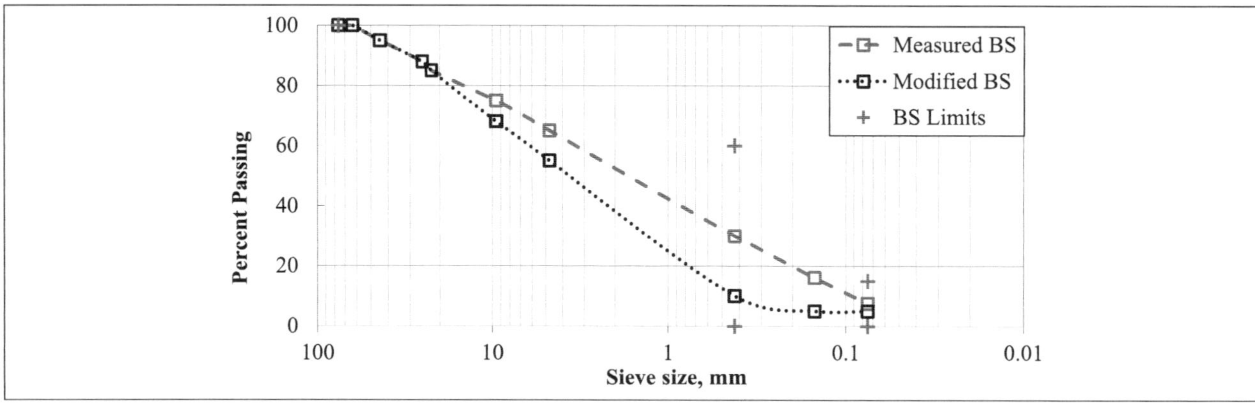

Figure 6. Size distribution of measured BS and modified BS for Material J.

values are beyond the critical range for corrosion mentioned above. When the resistivity is close to the acceptance limit of 3,000 Ω·cm, the prediction errors are less than 20%.

As the cumulative distributions of the percentage of errors for different gradations show (Figure 5), the Equation 2 model appears to be slightly more accurate at predicting the resistivity of the materials with Type BS gradations than of the materials with the Type AS or Type DS gradations. Over 50% of the estimates for the materials with the BS gradation are within 10% error. For all three gradations, over 50% of the estimates are within a 20% error and over 75% are within a 30% error. A very small percentage of the estimates had an error greater than 50%. The distributions of the errors appear to be consistent across the three gradations. This model appears to be useful to predict the resistivity of a given gradation as a function of constituent size fractions which should be backed up by more direct testing. The existence of a tool that reasonably approximates the resistivity of a material can provide a starting point for the empirical adjustment of its gradation in order to reduce the corrosion potential. Since the SEE is 2,288 Ω·cm, one can assume that any material with resistivity greater than 5,500 Ω·cm (a rounded approximation of the acceptance limit of 3,000 Ω·cm plus SEE) obtained with this approximation would most probably pass the 3,000 Ω·cm acceptance limit.

As reflected in Table 2, Material J would fail if analyzed by Tex-129-E or similar procedures (e.g., AASHTO T-288) because the resistivity of 2,267 Ω·cm is less than the minimum 3,000 Ω·cm. However, the material from this quarry can pass the resistivity requirements for Type AS and Type DS. The resistivity of the Type BS gradation for this material is just above the acceptance limit (3,200 Ω·cm) and could be subject to concern about corrosion. If the weight fractional composition of Type BS gradation of Material J could be changed from 35% gravel, 35% coarse sand, 22% fine sand, and 8% fines to 45% gravel, 45% coarse sand, 5% fine sand, and 5% fines (compared in Figure 6), then the model predicts a resistivity value of 6,715 Ω·cm, which is above 3,000 Ω·cm with a 2,288 Ω·cm SEE. One could prepare a sample and measure the resistivity of the proposed gradation using the modified two-electrode boxes to verify that it is in compliance with the specifications.

A concern about this model is the possibility of overestimating the resistivity, as this could result in the selection of a material that would have not passed the 3,000 Ω·cm limit had it been measured directly. However, the error in the estimates seems to be greater when the resistivity is greater than 3,000 Ω·cm, and the model predicts well for resistivity less than 3,000 Ω·cm. Thus, the model is proposed as a viable way to estimate the resistivity of a backfill material through the gradation and the resistivity of the separate particle sizes that constitute it.

The main advantage of this model is that it would allow for the design of backfill aggregate size distribution through the adjustment of the gradation in an empirical manner. This could open up the possibility of using a variety of coarse materials that would not otherwise be considered for use as backfill. By reducing the weight fraction of finer particles, it could be possible to obtain a modified mixture of a material that satisfies the specifications set forth by the DOTs. Providing a method to design backfill for a target resistivity changes significantly the current selection process for backfill in MSE wall. The materials are currently selected or rejected in a pass or fail manner in accordance with AASHTO T-288 or similar procedures. This model could effectively provide an easy medium for characterizing the corrosion potential of a material in terms of its gradation and composition and adjusting these parameters accordingly.

Conclusion

The resistivity of MSE wall backfill materials is an important parameter in the estimation of the corrosion potential, which is in turn a critical factor for the service life of the wall. The concern that current testing procedures do not provide an accurate representation of the resistivity of coarse materials used as backfill has led to numerous studies looking for an alternative. Through the testing of 18 different materials with AS, BS, and DS gradations and their constituent USCS size fractions (gravel, coarse sand, fine sand, and fines), several observations were made. Most importantly, the resistivity of a sample was observed to be correlated with its particle size distribution (i.e., gradations with a greater weight fraction of fine sand and fines had lower resistivity). Based on the hypothesis that the resistivity of a material could be approximated as a function of its composition as a weight fraction of the constituents, a model was developed to predict the aggregate resistivity based on the constituent size fractions acting as resistors in parallel. The proposed model yielded a coefficient of determination (R^2 value) of 0.87 and a SEE of 2,288 Ω·cm. The model can be used as a tool to design the gradation of a material to achieve acceptable resistivity. Use of this model would be valuable as it could provide a starting point for the modification of aggregate size distributions for use in MSE walls, avoiding a large amount of the laboratory trial and error that would otherwise be required. Further testing with a larger number of samples is recommended for further refining the model for wide-scale implementation. In the meantime, it is proposed to verify the results with laboratory testing. Based on the study by Arciniega (*8*), it is recommended to measure the resistivity of the coarse backfills with the actual gradation using resistivity boxes with dimensions that are twice or four times the dimensions of the standard resistivity box.

Acknowledgments

The authors are grateful to the Texas Department of Transportation (TxDOT) and the National Cooperative Highway Research Program (NCHRP) for providing funding and assistance that made this research possible. The authors are also sincerely grateful for support and assistance from the faculty and students at the University of Texas at El Paso (UTEP) Center for Transportation Infrastructure Systems, especially the laboratory manager Jose Garibay, the center director Imad Abdallah, and students Mathew Gonzales, Luisa Morales, and Diana Cabrera.

References

1. Romanoff, M. *Underground Corrosion*. National Bureau of Standards, Circular 579. U.S. Department of Commerce, Washington, DC, 1957.
2. King, R. A. A Review of Soil Corrosiveness with Particular Reference to Reinforced Earth, *TRRL Supplementary Report 316*, Transport and Road Research Laboratory, Crowthorne, Berkshire, England, 1977
3. Serway, R. A., and Beichner, R. J. *Physics for Scientists and Engineers*, 5th ed. Harcourt College Publishers, Orlando, FL, 2000.
4. Elias, V., K. L. Fishman, B. R. Christopher, and R. R. Berg. Corrosion/Degradation of Soil Reinforcements for Mechanically Stabilized Earth Walls and Reinforced Soil Slopes. *Report No. FHWA-NHI-09-087*. Federal Highway Administration, Washington, DC, 2009.
5. Thapalia, A., D. M. Borrock, S. Nazarian, and J. Garibay. Assessment of Corrosion Potential of Coarse Backfill Aggregates for Mechanically Stabilized Earth Walls. *Transportation Research Record: Journal of the Transportation Research Board*, 2011. 2253: 63–72.
6. TxDOT. Tex-129-E Measuring the Resistivity of Soil Materials. Texas Department of Transportation, 2014. ftp://ftp.dot.state.tx.us/pub/txdot-info/cst/TMS/100-E_series/pdfs/soi129.pdf. Accessed July 29, 2017.
7. AASHTO. T-288 Standard Method of Test for Determining Minimum Laboratory Soil Resistivity. American Association of State Highway and Transportation Officials, 2016.
8. Arciniega, J. L. Impact of Grain Size Distribution on Resistivity of Mechanically Stabilized Earth Walls. Presented at 96th Annual Meeting of the Transportation Research Board, Washington D.C., 2017.

The Standing Committee on Transportation Earthworks (AFS10) peer-reviewed this paper (18-03182).

Article

TRR
JOURNAL OF THE TRANSPORTATION RESEARCH BOARD

Transportation Research Record
2018, Vol. 2672(52) 258–267
© National Academy of Sciences:
Transportation Research Board 2018
Article reuse guidelines:
sagepub.com/journals-permissions
DOI: 10.1177/0361198118756901
journals.sagepub.com/home/trr
SAGE

Assessment of Settlement Profile Caused by Underground Box Structure Installation with an Artificial Neural Network Model

Jun Kyung Park[1], Dong Hwan Cho[2], Sharif Hossain[2], and Jeongho Oh[2]

Abstract

Various empirical equations have been proposed to predict the ground settlement profile caused by the excavation of conventional circular tunnels. However, ground movement for the underground box structure has not been fully studied. In this study, ground settlement induced by underground box installations is investigated using two-dimensional finite element analyses. A new formulation to assess the settlement profile applicable to underground box structure is proposed based on parametric analyses of the changes in ground condition, geometric condition of structure, and construction conditions. This paper also presents a method to predict the maximum surface settlement around an underground box structure with artificial neural networks (ANNs), taking into account nine input variables that have direct physical significance. A MATLAB-based multi-layer back propagation neural network model is developed, trained, and tested with parameters obtained from numerical analyses. The maximum settlement from the ANN model, in conjunction with a new formulation to construct the settlement profile, turns out to be promising, by predicting a settlement profile compatible with field measurement data.

In recent years, the rapid growth of urban areas has led to an increase in the number of underground structures that underpass existing infrastructures or facilities (roads, railways, subways, waterworks, etc.) at shallow depths. For safe construction in complex urban environments, trenchless technology is actively undertaken to avoid interference with existing structures, avoiding re-routing of traffic and any sudden and substantial surface displacement of the overlying roadway or railroad. However, the construction of underground box structures using trenchless excavation will inevitably induce varying degrees of ground movement toward the excavation, resulting in ground surface settlement. Thus, an essential element to minimize possible detrimental effects of ground settlement on existing structures depends greatly on the reliability of the ground settlement profile prediction.

To address this important issue, several researchers have attempted to predict maximum surface settlement induced by shallow tunneling for conventional circular tunnels. Because of the complexity and unknown relationships and interactions between the many parameters affecting surface settlement, deriving a closed-form mathematical solution based on analytical methods is very difficult (*1,2*). Many researchers have tried to employ other approaches, including empirical methods (*3–9*), numerical methods (*10,11*), and laboratory studies (including centrifuge tests) based on small model tests (*12*). Although the analytical, empirical, numerical, and laboratory test methods have several advantages in addressing the issue, there are major constraints in application. For instance, in analytical methods there are many simplifications such as a plain strain assumption, elastic behavior, and/or soil isotropy. In laboratory experiments, the effect of some parameters can be well evaluated, though the effect of the problem of scale is somewhat unknown.

Various empirical/analytical equations have been proposed to predict the ground settlement profile caused by excavation of conventional circular tunnels. However, ground movement for the underground box structure has not been fully studied.

Mamaqani developed a surface settlement prediction model using an artificial neural network (ANN) along with finite element analysis for a box jacking method, which is in good agreement with data collected from case studies (*13*).

[1]HBC Inc., Yuseong-gu, Daejeon-si, Republic of Korea
[2]Department of Railroad Infrastructure System Engineering, Korea National University of Transportation, Gyeonggi-do, Republic of Korea

Corresponding Author:
Address correspondence to Jeongho Oh: j-oh@ut.ac.kr

The study also highlighted that the empirical method, suggested by Milligan and Marshal (*14*) and originally developed for tunneling and pipe jacking, tends to overestimate the maximum surface settlement and underestimate the width of the settlement trough channel. A series of large-scale tests conducted by Um et al. showed that maximum heaving and settlement was measured from the center point along with both longitudinal and lateral planes of underground structures (*15*). In addition, when the ratio of cover depth to the diameter of the jacking pipe is larger than 1.5, deformation of the surrounding soil seems to decrease significantly.

ANN is a complex mathematical model, or computational model, inspired by the structural aspects of biological neural networks, and is widely used in the modeling of nonlinear systems and system identification. Many scientists and researchers use and apply this particular biological concept to conduct their studies. Park (*16*) conducted an extensive review on the application of ANN in geotechnical engineering. According to this author, the ANN model has been extensively employed since the early 1990s, for example in constitutive modeling, geo-material characterization, assessment of bearing capacity of pile, slope stability, and evaluation of liquefaction, shallow foundations, and tunnels and underground openings.

In this study, an effort was made to develop an empirical equation to predict settlement profile following installation of underground structures, based on several finite element analysis runs. While previous research conducted by Milligan and Marshall (*14*), Bennett (*17*), Rogers and O'Reilly (*18*), and Liu and Lu (*19*) mainly focused on surface and subsurface ground movements caused by pipe jacking or tunneling, the focus of this paper is ground movement and the underground box structure.

Many box-type structures are being constructed under roadways and railroads at a relatively shallow cover depth in South Korea, as illustrated in Figure 1; therefore, it is necessary to estimate subsurface settlement realistically to mitigate any disasters. In this study, an ANN model is employed to predict the subsurface settlement profile based on finite element analysis results, to compare with a proposed empirical equation and limited field measurements.

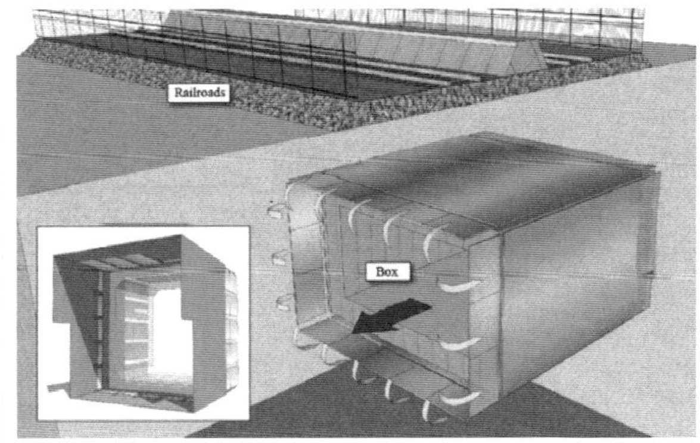

Figure 1. Box structure installation under railroads.

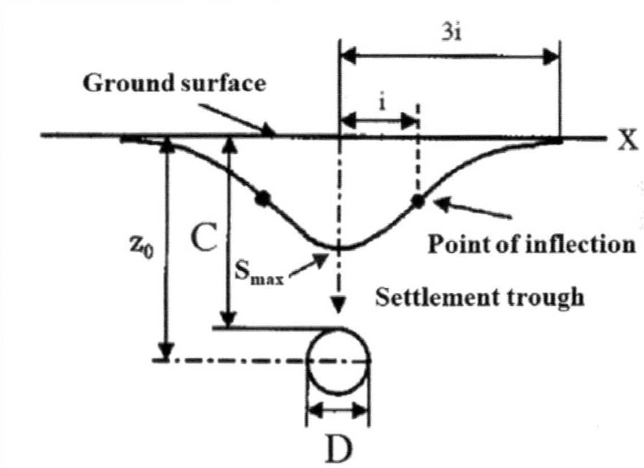

Figure 2. Definition of settlement profiles of Gaussian form [Modified from Peck, 1969 (*3*)].

Settlement Trough for Underground Box Structures

Most of the empirical methods are based on measurements in full-scale circular-type tunnels during boring, or after the completion of construction in the long term. Such results are considered as the most reliable, but they cannot be generalized because of the differences in physical and geometrical properties from one given site to another. These empirical methods need to be verified for application to the construction of underground box structures.

Review of Empirical Methods

There are many research examples clarifying the characteristics of ground motion when a tunnel is excavated. The maximum settlement amount (S_{max}) and the position of the inflection point (i) are depicted by a normal distribution curve on the form of subsidence on the cross-section of the tunnel, as shown in Figure 2.

Peck (*3*) presented the settlement curves of a single-wire tunnel using field measurements, assuming the absence of volume change via

$$S(x) = S_{max} \exp(-\frac{x^2}{2i^2}) \tag{1}$$

where
S = the surface settlement;
x = the transverse distance from tunnel centerline;
S_{max} = the maximum settlement; and
i = the position of inflection point.

Table 1. Different Empirical Equations for the Estimation of Settlement Trough Width, *i*

References	Width of settlement trough, i	Basis for empirical solution
Peck (*3*)	$i/R=(Z_0/2R)^n$; $n=0.8-1.0$	Field observations
Attewell and Farmer (*4*)	$i/R=a(Z_0/2R)^n$; $a=1.0, n=1.0$	Field observations of UK tunnels
Atkinson and Potts (*5*)	$i=0.25(Z_0+R)$; loose sand $i=0.25(1.5Z_0+0.5R)$; dense sand/OC clay	Field observations and model tests
Mair et al. (*6*)	$i=0.5Z_0$	Field observations and centrifuge tests
Clough and Schmidt (*7*)	$i/R=a(Z_0/2R)^n$; $(a=1.0, n=0.8)$	Field observations of UK tunnels
O'Reilly and New (*8*)	$i=0.43Z_0+1.1$; cohesive soil, $3\le Z_0\le 35m$ $i=0.28Z_0-0.1$; granular soil, $6\le Z_0\le 10m$	Field observations of UK tunnels
Rankin (*9*)	$i=kZ_0$; $k=0.5$ for clay	Field observations

Note: Z_0 = the depth of tunnel below ground (at tunnel spring line), R = tunnel radius.

Table 2. Material Properties for the Finite Element Method Analysis

Ground conditions		Underground structure properties	
Material model	Mohr–Coulomb	Material model	Linear elastic
Unit weight (kN/m^3)	18	EA(kN/m)	2.44E+7* / 6.10E+6**
Elastic Modulus (kN/m^2)	200,000	EI (kN/m^2/m)	2.03E+6* / 1.27E+5**
Poisson Ratio (υ)	0.3	Thickness (D)	1* / 0.5**
Cohesion (kN/m^2)	30	W (kN/m/m)	25
Friction angle (deg.)	33	Poisson Ratio (υ)	0.15

Note: *property for box structure and **for circular tunnel.

The inflection point (i) is a parameter determined by the ground condition. From the model test and various field measurement data, several studies have proposed it, as summarized in Table 1.

In addition, Aversin (*20*) and Arato (*21*) proposed a formula through half of the settlement width (L) rather than the inflection point (i):

$$S(x)=S_{\max}(1-x/L)^{4.54}\exp(4.54x/L) \tag{2}$$

$$S(x)=S_{\max}\left(\frac{0.75}{\exp(2x/L)^2}+\frac{0.25}{\exp(2x/L)^4}\right) \tag{3}$$

Evaluation of Surface Settlement by Numerical Analysis Method

Numerical analyses were performed using PLAXIS 2D 8.6 finite element analysis program to estimate the surface settlement profile of underground box structures. To consider the shallow cover depth (C) of underground box structures in the urban environment, three cases (2 m, 3.5 m, and 7.5 m depth) were considered for the numerical analyses. In the case of a cross-sectional shape, it is set as square cross-section (7.5 m × 7.5 m) for box structures for the comparison with the circular cross-section (diameter = 7.5 m).

The Mohr–Coulomb plasticity theory was applied for the plain strain 15-nodes element, and the ground condition was assumed to be weathered soil, properties that account for the majority of the shallow stratum in South Korea, as summarized in Table 2. For the box structure, a linear elastic model was assumed with different properties for box structure and circular tunnel, respectively. For the consideration of real construction conditions in the field, the effects of groundwater and overcut were also included in the numerical analyses. Overcut excavation is required to reduce friction forces and facilitate steering during construction (*13*). In this study, the overcut size was set at 50 and 100 mm.

Maximum Surface Settlement. The maximum surface settlement of the underground box structure was larger than that of the circular tunnel. As the overcut size was increased in the underground box structure, the amount of maximum settlement also increased because of soil collapse into the overcut space, as shown in Figure 3.

The effect of overcut in the circular tunnel became insignificant compared with the underground box structure condition. In addition, the maximum settlement tends to increase with the decrease of the cover depth, particularly in the case in which the groundwater level is assumed to be on the surface.

Settlement Trough. Normalized maximum settlement (S/S_{max}) was introduced to compare the empirical methods summarized

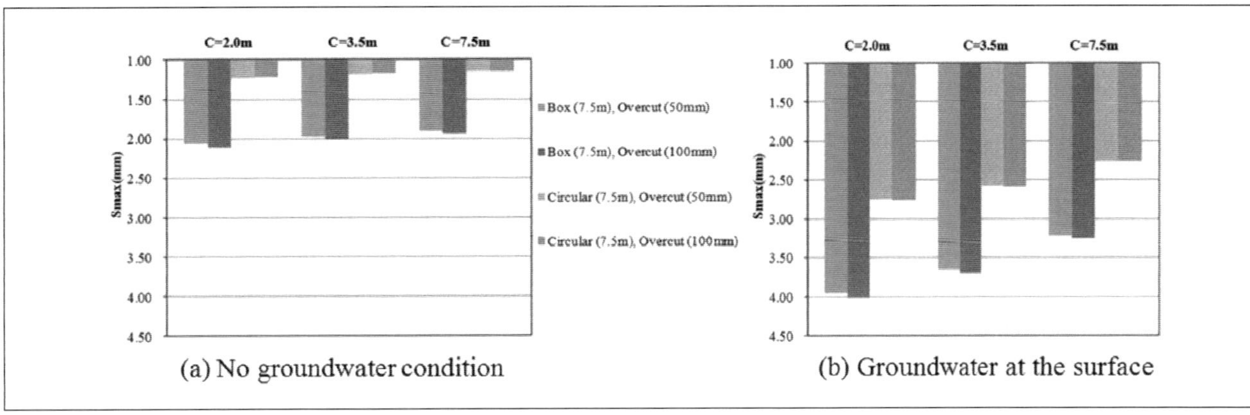

Figure 3. Finite element analysis results: maximum settlement (S_{max}).

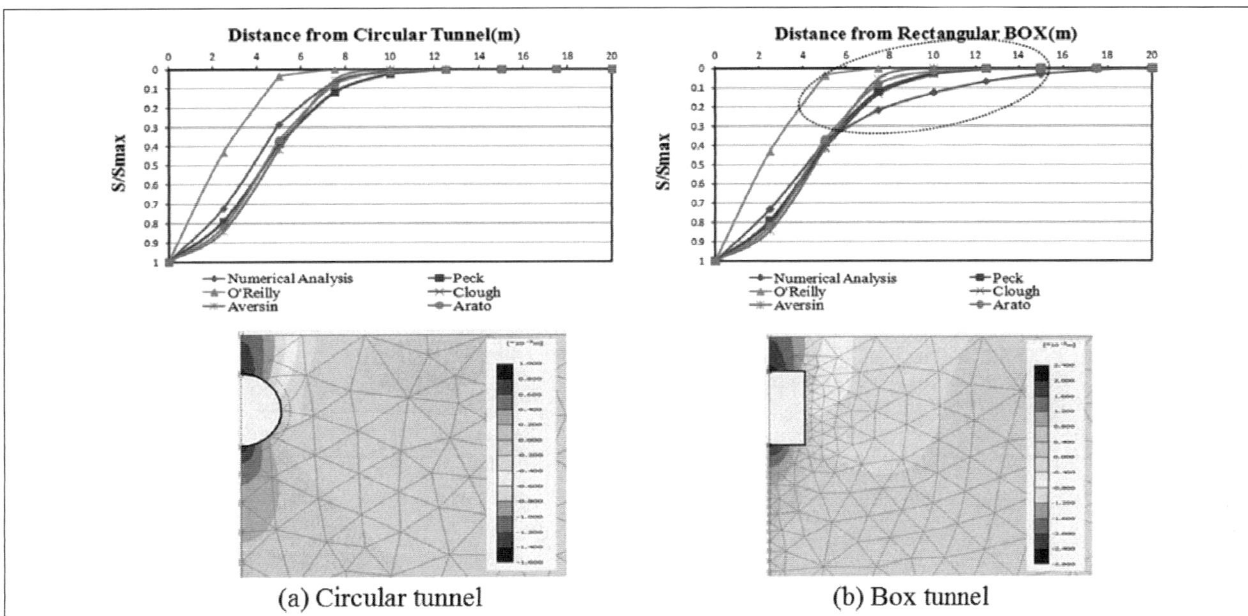

Figure 4. Settlement trough and vertical displacements (C = 3.5m), no groundwater condition.

in the previous section for different shapes of structures. In the case of the circular tunnel not accounting for the groundwater level, the numerical analysis results and the conventional empirical methods generally matched well, as shown in Figure 4*a* except for O'Reilly's method, which might be attributed to the "*i*" value computed. However, in the case of underground box structures, large deviations have been observed after the 1D range from the structure as shown in Figure 4*b*. This is because the stress distribution around the structure in the middle of excavation varies depending on the shape of the structure, and the arching effect of the ground cannot be sufficiently mobilized for the shallow depth of underground structures. Overall, the underground box structure yields a larger maximum settlement than the circular tunnel based on numerical analysis, since the curves, in terms of normalized maximum settlement (S/S_{max}), become smoother with respect to distance from the structure, indicating larger settlement-induced zone.

Figure 5 shows the results obtained when the groundwater level is assumed to be located on the ground surface, along with free drainage condition into the tunnel. It can be seen that there are large deviations between the existing empirical methods and numerical analyses regardless of the sectional shape. This is because the existing empirical methods mainly derive from measurement data for cases where there is no volume change in the undrained condition, in most cases. It is shown that the effect of underground excavation tends to be significant under the conditions in which drainage occurs within granite-weathered soil below groundwater.

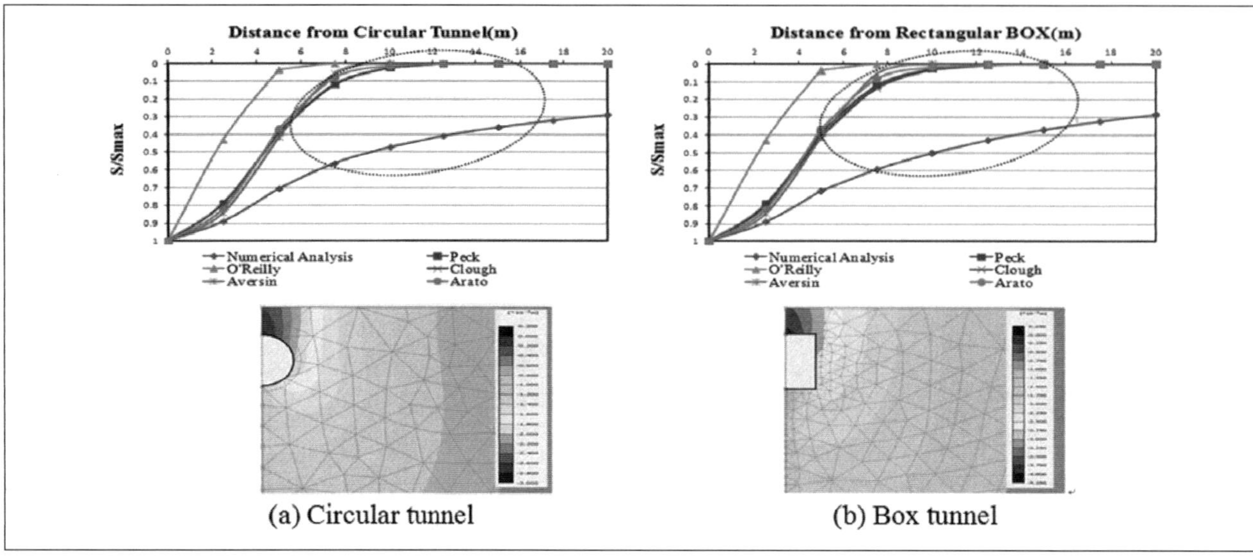

Figure 5. Settlement trough and vertical displacements (C = 3.5m), groundwater at the surface condition.

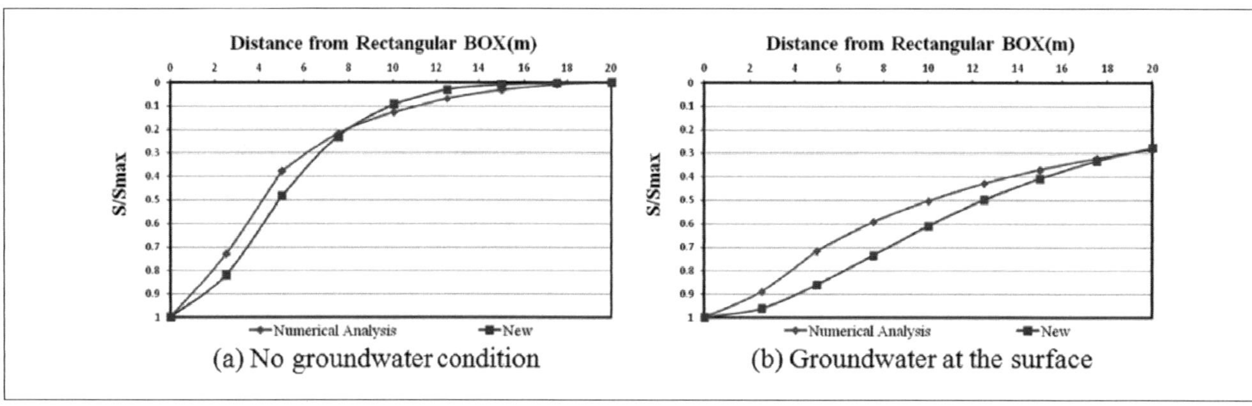

Figure 6. Comparison between numerical analysis and proposed settlement trough for underground box structure (C = 3.5m).

New Suggestion of Settlement Trough for Underground Box Structures

The modified Gaussian curve of the following form is found to be a better fit to the predicted soil settlement profiles from the PLAXIS analyses for underground box structures described earlier:

$$S(x) = S_{\max} \frac{n}{(n-1) + \exp\left[\alpha\left(\frac{x}{i}\right)^2\right]}, n = e^{\alpha}\frac{2\alpha - 1}{2\alpha + 1} + 1 \tag{4}$$

where

n = the shape function parameter controlling the width of the profile; and

α = the correction factor to ensure that i remains at the inflection point.

For the drainage condition, the influence of the tunnel excavation effect was accommodated by changing the inflection point of the settlement shape curve, as shown in Figure 6*b*. It is suggested to use $i = 0.5Z_0$, $\alpha = 0.25$ (for no ground water condition), and $i = Z_0$, and $\alpha = 0.005$ (for ground water at the surface condition) for an underground box structure through the Microsoft Excel Solver analysis. However, these values need to be defined based on further evaluations.

Development of the ANN Model

ANN is a complex mathematical model inspired by biological neurons; it is the emulation of biological neural networks and is widely used in modeling of nonlinear systems and system identification. Generally, its powerful information processing function depends on four factors: input–output properties (activation characteristics) of network

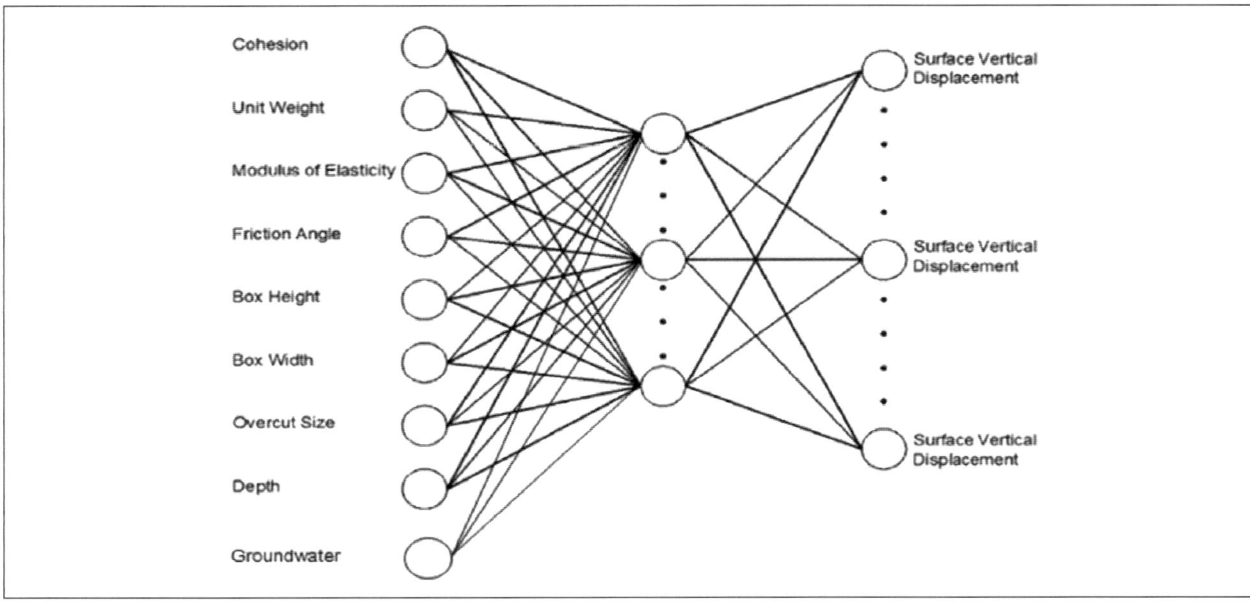

Figure 7. Schematic neural network model in this research [modified from Mamaqani, 2014 (*13*)].

units (neurons), network topologies (connection modes of neurons), connection weights (synaptic strength), and neuron thresholds (special connection weights). An ANN generally consists of an input layer, hidden layers, and an output layer. Every layer includes many neurons, and the numbers of layers and neurons in each layer are determined by the user with respect to the scale of the problem. All neurons in the structure have interconnections between each other, similar to the biological nervous system. The ANN transfers latent knowledge or laws in related inputs to the network's site by processing the inputs, and it includes general laws based on the calculations performed on the numerical inputs or examples. The underground excavation problems based on ANN are generally classified as two types: one is the nonlinear relationship of parameters, which is established by the data provided by finite element analyses data and/or field measurement data; the other is the prediction system for the latter period based on the experience gained and the in-situ measurement data gathered from past projects.

The main objective of this research is to develop an ANN model to estimate maximum surface settlement associated with box structure installation, using the following nine factors: (1) soil cohesion (c), (2) soil unit weight (γ), (3) soil modulus of elasticity (E), (4) soil friction angle (ϕ), (5) box culvert height (h), (6) box culvert width (w), (7) overcut size (s), (8) depth of box culvert from the surface (H), and (9) groundwater condition, as shown in Figure 7. A total of 216 underground box structure models generated from PLAXIS program runs and corresponding maximum surface settlements were recorded.

As described earlier, a neural network consists of a layer of neurons or cells that are connected to each other, receiving data inputs and transferring them to outputs through the hidden layer. A hidden layer is a layer between input and output layers whose neuron is connected to other neurons. There is no exact rule to select the optimal number of neurons in the hidden layer. However, there are some rules of thumb that are available to find the number of hidden neurons. Mamaqani (*13*) recommended that the number of hidden neurons depends on the number of input parameters. He suggested considering 1.5 times the number of parameters in the input layer. Masters (*23*) proposed a three-layer network in which the hidden layer would have sqrt($n \times m$) neurons (where n is input and m is output neurons). In developing an optimal ANN, the data are divided into three subsets, such as training data set, validation data set, and testing data set. The training process is an important step to enhance the ANN model and adjust the weights. The aim of the training process is to reduce errors between the actual output values and target output values by updating the connection weights in the neural network. Weights are adjustable scaler parameters of the neurons, and represent the synapse. A set of input signals x_i ($i = 1,2,\ldots,n$) is summed up with hidden neurons j and computes its outputs y_j as function of f, where, w_{ji}-weight connections from neuron j to neuron i, and f is a sigmoidal function:

$$y_j = f\left(\sum\nolimits_{i=1}^{n} w_{ji} x_i\right) \tag{5}$$

Validation data are used to verify and determine the performance of the trained ANN model. The validation set is for tuning the parameters to show how well the network is performing. In addition, the validation data set is used to calculate the accuracy of the algorithm in terms of mean square error (MSE) using

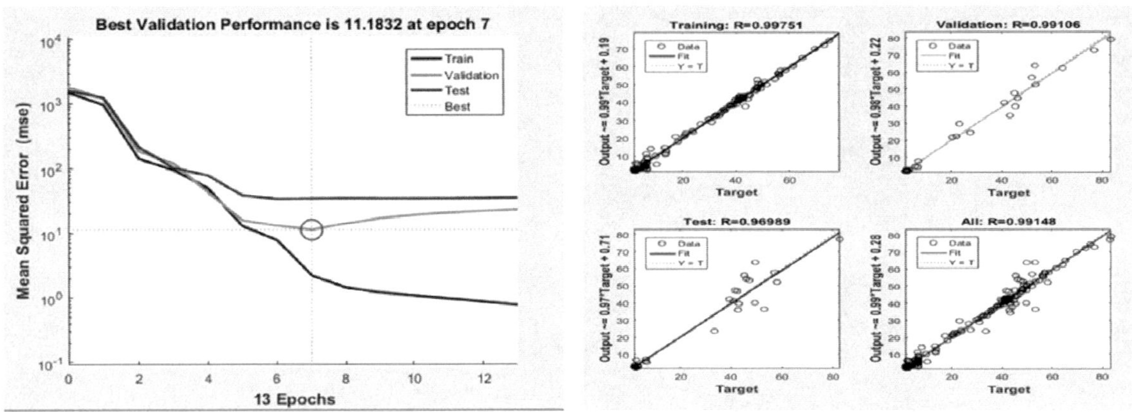

(a) Performance Plot (b) Regression Plot

Figure 8. Performance and regression plot of ANN model for underground box structures.

$$\mathrm{MSE} = \frac{1}{n}\sum_{i=1}^{n}\left(\overline{Y_i} - Y_i\right)^2 \quad (6)$$

where

n = number of data;
$\overline{Y_i}$ = vector of n predictions; and
Y_i = vector of true values.

Once the final model is determined, a testing data set is developed to assess the actual predictive power of the ANN model. The test data set is not used in the model-building process.

Methodology

In a general ANN system, the activation function, also known as the transfer function, is used to convert the result of the weighted sum of a neuron's inputs into a wanted output value. This conversion is achieved by calculating the neuron state by introducing a non-linearity in the functioning of the neuron. Among the transfer functions available, the sigmoidal function has been found most efficient through its better performance (*13,22*).

In this study, feedforward network architectures and the Levenberg–Marquardt training function available in MATLAB, along with a program that fits the purpose of this study, were used to predict the surface vertical displacement based on 216 generated data sets from finite element method analysis. Of the total data set, 70% was used as training data, 15% for validation, and 15% for testing data. The performance of an ANN model can be measured via the MSE of the model, as shown in Figure 8*a*. The MSE varies along with number of iterations of model (called "epoch" in MATLAB), and lower MSE indicates enhanced performance. In this analysis, MSE reached a minimum 11.18 at seven epochs. The training continued for six more iterations before the training stopped. Another step in validating the network is to create a regression plot, which shows the relationship between the outputs of the network and the targets. The performance of the ANN model was also checked in terms of R^2 value obtained from linear regression analysis among variables, as shown in Figure 8*b*. In this analysis, the R^2 value is generally more than 0.95.

Once the ANN model was determined based on the least MSE, shown in Figure 8, the model was applied to the whole dataset to compute the average error percentage using

$$\text{Average Percentage of Error}(\%) = \frac{R_{\mathrm{ANN}} - R_{\mathrm{PLAXIS}}}{R_{\mathrm{PLAXIS}}} \times 100 \quad (7)$$

where

R_{ANN} = the maximum settlement obtained by ANN model; and
R_{PLAXIS} = the maximum settlement obtained by PLAXIS.

Park (*16*) reported that an over-fitting problem or poor generalization capability frequently occurs when an ANN experiences over-learning. Then, such a well-trained model may not perform well on unseen data set because of its lack of generalization capability. Lee and Lee (*24*) utilized neural networks to predict the ultimate bearing capacity of piles. The prediction of the ANN model showed a maximum error of no more than 20% and average summed square error less than 15%. Consequently, the acceptance criterion in terms of average percentage error was set to 15% in this study, based on previous studies. A positive error percentage indicates that the ANN result was overestimated, and vice versa, as shown in Figure 9*a*. The ANN model seemed to be functional because the average percentage of error was 13.8%, less than the criterion.

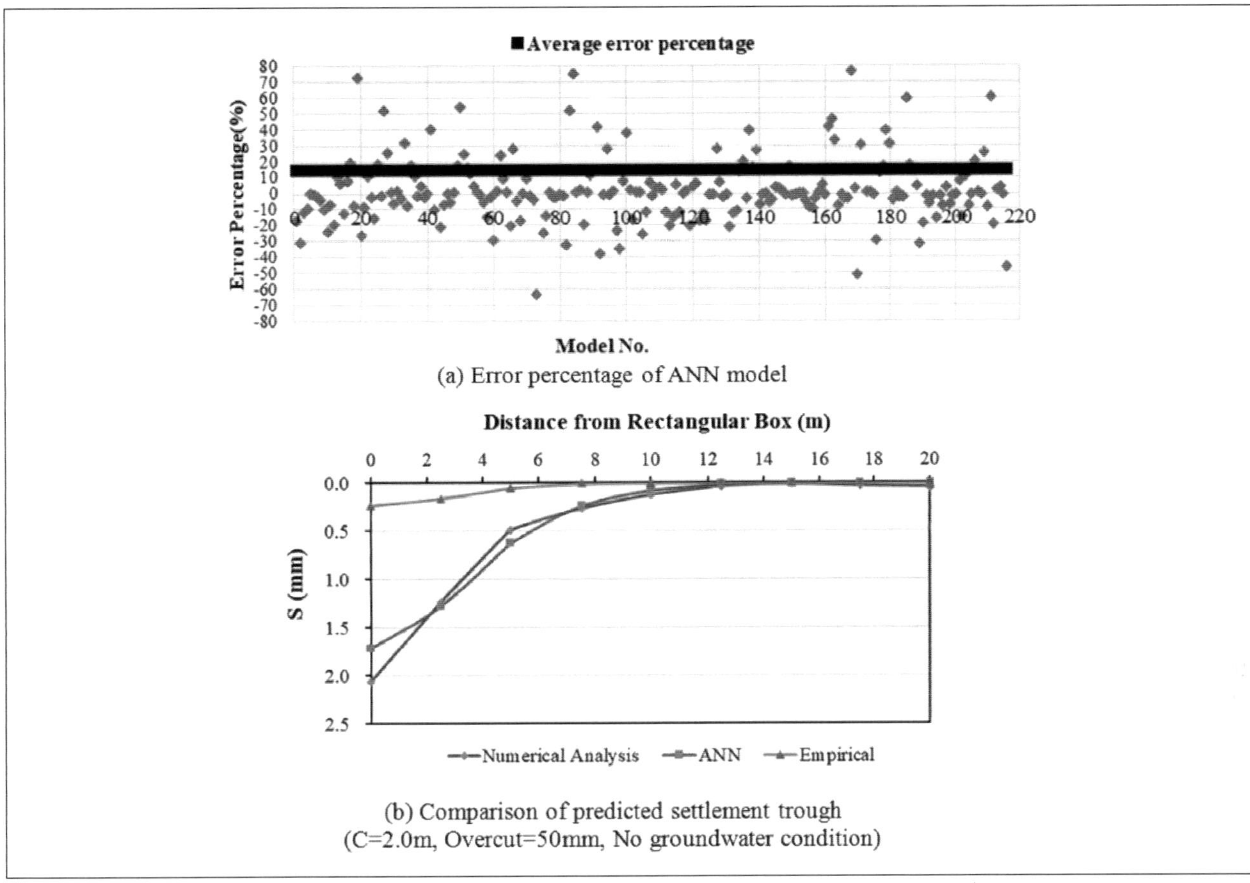

Figure 9. Error percentages of ANN model and comparison of predicted settlement.

The predicted settlement trough for the example case (C = 2.0 m, Overcut = 50 mm, no groundwater condition) results showed that there are good matches between the results obtained from ANN and PLAXIS, as shown in Figure 9*b*. In this comparison, Peck's empirical method was selected. After the estimation of maximum settlement from the ANN method, the remaining part of the settlement trough could be drawn with the newly suggested equation, as described in the previous section for underground box structures.

Application of the ANN Model to a Case Study

The ANN, developed based on the numerical analysis, was used to check its applicability for field measurement data. In this study, the settlement measurement data of the transverse underground box structure located at the 1.88 km branch of the Seoul outer circuit roadway in South Korea were considered (*15*).

The box structure was designed with 26.8 m width and 8.71 m height, and the ground condition comprised weathered granitic soil. The cover depth was 1.6 m and the longitudinal length of the box structure was 60 m, as shown in Figure 10, *a* to *d*. During the construction stage, the maximum surface settlement of 37 mm was measured at the center of the box location.

From the ANN model, the maximum settlement was estimated to be 41.6 mm. With a suggested value in Equation (4) without ground water condition, it was difficult to match with field measurements, so re-calibration was made to find the inflection point and α value ($i = 1.8Z_0$, $\alpha = 0.1$) to construct a transverse settlement trough, as shown in Figure 10*e*. This indicates the feasibility of the suggested equation, which can be adjustable depending on various field conditions. While there is a large difference between the field measurement data and the empirical method suggested by Peck, the difference generated from the suggested equation based on the ANN prediction is relatively insignificant considering the variability of ground condition in the field. This limited validation demonstrated the efficiency of the ANN model in conjunction with the suggested equation in the assessment of the maximum settlement of the underground box structure.

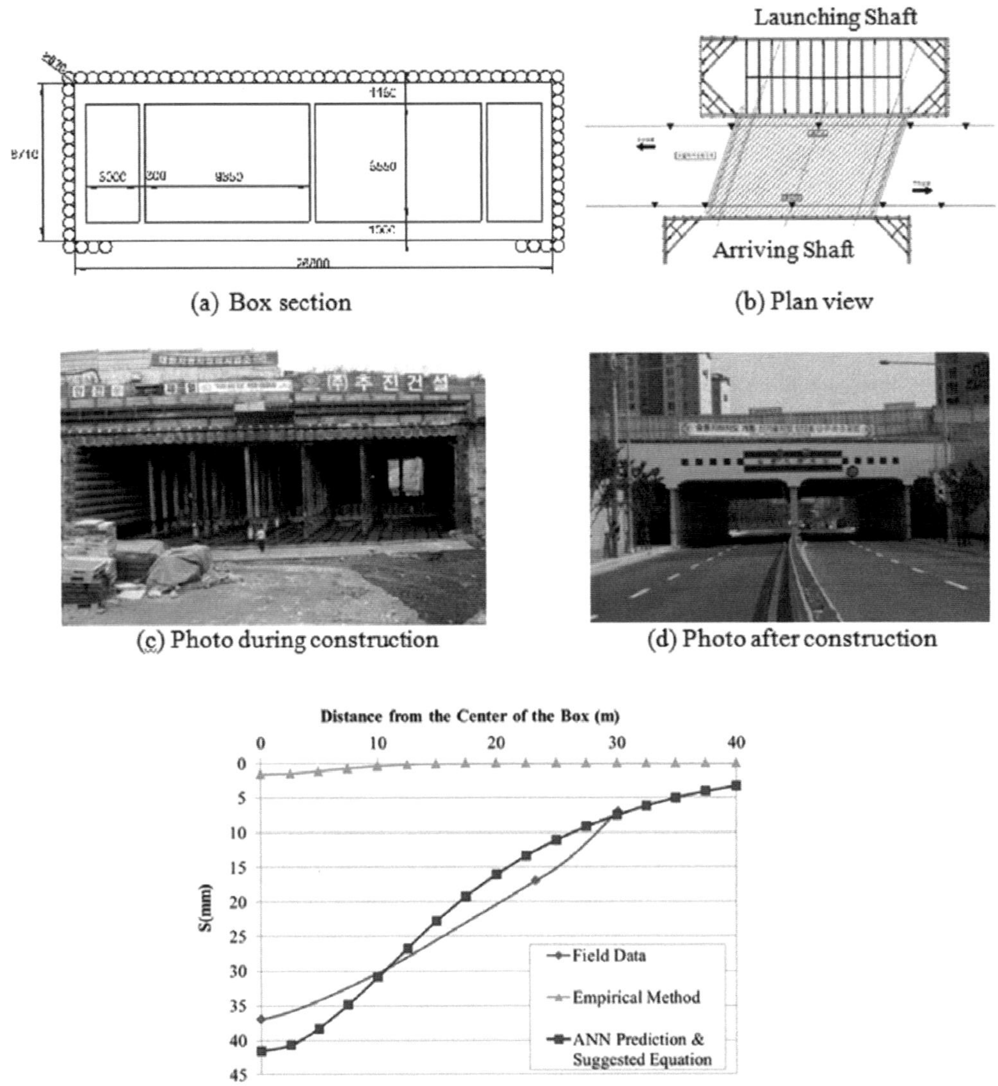

(a) Box section (b) Plan view

(c) Photo during construction (d) Photo after construction

(e) Settlement comparisons between field measurement data, empirical method and ANN model prediction.

Figure 10. Seoul outer circuit roadway project.

Conclusions

A new formulation to assess the settlement profile applicable to an underground box structure is proposed based on parametric analyses of changes in ground condition, geometric condition of structure, and construction conditions. The ANN model is employed to predict the subsurface settlement profile based on finite element analysis results.

- The maximum surface settlement of the underground box structure was generally larger than that of the circular tunnel, based on numerical analysis when identical conditions were given.
- While estimated surface settlements from previous empirical equations were generally found to be compatible with predictions of finite element analysis in the case of a circular tunnel without taking into account ground water level, the estimated surface settlements of box structures based on empirical equations seemed to underestimate beyond 1D location.
- This study proposed a modified equation to incorporate ground water condition and box structure shape, which is more compatible with finite element analysis results.
- To develop an ANN model to predict surface settlement caused by excavation of underground structures,

a total of 216 finite element analysis runs were made taking into account various soil conditions, depth of cover, ground water level, and structure shape. The influence of overcut size on the surface settlement becomes alleviating in case of a circular shape structure, considered as attributed to the arching effect. At a relatively shallow depth of cover, the surface settlement tends to increase, especially in box-shaped structure considered.
- A field case study was employed to limitedly verify the proposed ANN and Equation (4). The i and α values in Equation (4) can be adjusted depending on field conditions, which is deemed feasible.

This study has introduced artificial intelligence for prediction of ground surface settlement based on accumulated numerical analysis data. However, it should be noted that the capabilities of such codes in making accurate predictions is entirely dependent on the quality and the quantity of data used in training the ANN. If the data are deficient or training is inadequate, the proposed neural network-based prediction should be treated with caution. Therefore, the collection and analysis of monitored data should be carefully carried out for guaranteed predictions.

Acknowledgments

This work is part of a research project financially sponsored by the Korea Agency for Infrastructure Technology Advancement. (Grant No. 16SCIP-B108153-02).

References

1. Rowe, R. K. Theoretical Examination of the Settlements Induced by Tunneling: Four Case Histories. *Canadian Geotechnical Journal*, Vol. 20, No. 2, 1983, pp. 299–314.
2. Verruijt, A., and J. R. Booker. Surface Settlements due to Deformation of a Tunnel in an Elastic Half Plane. *Geotechnique*, Vol. 46, No. 4, 1996, pp. 753–756.
3. Peck, R. B. Deep Excavation and Tunneling in Soft Ground. *Proceedings of the International Conference in Soil Mechanics and Foundation Engineering*, 1969, pp. 225–290.
4. Attewell, P. B., and I. W. Farmer. Ground Deformation resulting from Shield Tunneling in London Clay. *Canadian Geotechnical Journal*, Vol. 11, No. 3, 1974, pp. 380–395.
5. Atkinson, J. H., and D. M. Potts. Stability of a Shallow Circular Tunnel in Cohesionless Soil, *Geotechnique*, Vol. 27, No. 2, 1977, pp. 203–215.
6. Mair, R., M. Gunn, and M. O'Reilly. Ground Movements around Shallow Tunnels in Soft Clay, *Proceedings of 10th international conference on soil mechanics and foundation engineering*, Vol. 2, 1981, pp. 323–328.
7. Clough, G. W., and B. Schmidt. Chap 8. Design and Performance of Excavations and Tunnels in Soft Clay. In *Soft Clay Engineering* (E. W. Brand and R. P. Brenner ed.), Elsevier, 1981, pp. 569–631.
8. O'Reilly, M. P., and B. M. New. Settlements above Tunnels in the UK – their Magnitude and Prediction, *Tunneling* Vol. 82, 1982, pp. 173–181.
9. Rankin, W. Ground Movements resulting from Urban Tunneling, *Proceedings of 23rd conference of the engineering group of the geological society*, London Geological Society, 1988, pp. 79–92.
10. Melis, M., L. Medina, and J. Ma. Rodriguez. Prediction and Analysis of Subsidence induced by Shield Tunnelling in the Madrid Metro Extension. *Canadian Geotechnical Journal*, Vol. 39, No. 6, 2002, pp. 1273–1287.
11. Ding, W. Q., Z. Q. Yue, L. G. Tham, H. H. Zhu, C. F. Lee, and T. Hashimoto. Analysis of Shield Tunnel. *Int. J. Numer. Anal. Meth. Geomech.*, Vol. 28, No. 1, 2004, pp. 59–71.
12. Meguid, M. A., O. Saada, M. A. Nunes, and J. Mattar. Physical Modeling of Tunnels in Soft Ground: A Review. *Tunnelling and Underground Space Technology*, Vol. 23, No. 2, 2008, pp.185–198.
13. Mamaqani, B. Numerical Modeling of Ground Movements Associated with Trenchless Box Jacking Technique, Ph.D Dissertation, The University of Texas at Arlington, TX, 2014.
14. Milligan, G. W. E., and M. A. Marshall. The Functions and Effects of Lubrication in Pipe Jacking, In *ITA World Tunnel Congress 1998- Tunnels and Metropolises* (Negro Jr, A. & Ferreira, A. A. ed.), Balkema, Rotterdam, 1998, pp. 739–744.
15. Um, K.Y. et al. *A Study on Construction Method of High-Speed Railway Underpass Structure (in Korean)*, Research Report, Korea Railway Research Institute, 2010.
16. Park, H. I. Study for Application of Artificial Neural Networks in Geotechnical Problems. In *Artificial Neural Networks-Application*. InTech, 2011.
17. Bennett, D. Jacking Force and Ground Deformations Associated with Microtunneling. Ph.D. Dissertation, The University of Illinois, IL, 1998.
18. Rogers, C., and M. O'Reilly. Ground Movements Associated with Pipejacking and Trenching. *Proceedings of the International Conference on Soil Mechanics and Foundation Engineering*, 1991, pp. 907–910.
19. Liu, W. T., and X. Y. Lu. 3D Numerical Analysis of Soil Structure Interaction Behaviors of Pipe Jacking Construction. *Applied Mechanics and Materials*. Vol. 204. Trans Tech Publications, 2012, pp. 534–538.
20. Aversin, S. G. *Rock movements Related to Mining (in Hungarian)*, Nehezipari Kiado, Budapest, 1953.
21. Arato, A. *Deformation due to Cavities, Soil Settlement Effects on Buildings* (ed. E. Dulacska), Elsevier, Amsterdam, 1992, pp. 54–97.
22. Lawrence, J. *Introduction to Neural Network : Design, Theory, and Applications*. California Sceintific Software, Nevada, 1994.
23. Masters, T. *Practical Neural Network Recipes in C++*. Morgan Kaufmann, 1993.
24. Lee, I. M., and J. H. Lee. Prediction of Pile Bearing Capacity using Artificial Neural Networks. *Computers and Geotechnics* Vol. 18, No. 3, 1996, pp. 189–200.

The Standing Committee on Subsurface Soil-Structure Interaction (AFS40) peer-reviewed this paper (18-04174).

Article

TRR
JOURNAL OF THE TRANSPORTATION RESEARCH BOARD

Transportation Research Record
2018, Vol. 2672(52) 268–277

DOI: 10.1177/0361198118773196
journals.sagepub.com/home/trr
SAGE

Evaluation of FHWA Pile Design Method against the FHWA Deep Foundation Load Test Database Version 2.0

Nikolaos Machairas[1], Gregory A. Highley[1], and Magued G. Iskander[1]

Abstract

The efficacy of the FHWA pile design method was explored using data made possible by the 2017 release of FHWA Deep Foundation Load Test Database (DFLTD) v.2. Information contained within DFLTD v.2 was leveraged to evaluate the most common pile design methods against failure loads obtained via in situ static load testing procedures. In the process, the authors developed a custom relational database and software to batch process the available information. The scope was limited to impact-driven, un-tapered, steel and concrete piles, loaded in compression, using a static load test. DFLTD v.2 contains 3,116 unique-combination project/exploration/foundation/test cases of which 213 contained sufficient data to permit batch processing of the results to compute the axial load capacity and interpret the static failure load, according to the study scope. Scatter between measured (interpreted) and predicted capacities is significant; the computed capacity was off by a factor of 2 in many tests. The range in calculated to measured capacities (*Qc/Qm*) was from 0.12 to 8.88, and the mean *Qc/Qm* was 1.48. Preliminary evaluation suggests that the method performs better in clay than in sand, and overpredicts the capacities of long and larger diameter piles. The authors trust that this study will permit engineers and state agencies to better understand the efficacy of the most commonly employed design methods, resulting in a more resilient infrastructure.

The accuracy and precision of empirically produced observations is positively affected by the sample size and reliability of the data used in the study. In sciences in which experimentation for data creation is expensive, time-consuming, or otherwise problematic, the field of study in general relies on sharing of raw data to build on this sample size. Currently accepted design practices for driven piles are mainly empirical or semi-empirical. However, mass data production from pile load tests is expensive and time-consuming. Beyond these inherent limitations in empirical pile design analysis, many of the available load tests have insufficient data from which to gather an interpreted ultimate pile capacity.

Single pile design is typically conducted using ultimate capacity determination from either calculated or interpreted methods (or both in combination). Determination of the ultimate capacity of driven piles using calculated methods is conducted by following a series of steps in a design process, starting with data and properties gathered from a geotechnical investigation such as those recommended by FHWA (*1, 2*), along with pile selection. Interpreted pile capacity calculations occur using load-settlement curve data generated during an axial static load test of a driven pile. Comparison between calculated and interpreted capacities for large data sets provides insight of suitability of use of current design methods under varying pile and soil conditions.

For this study, a program was developed in Python which demands certain soil and pile properties as inputs in order to calculate the ultimate pile capacity using the most commonly used and relied on FHWA recommended design methods (*1*). Using this tool, data available from the FHWA Deep Foundation Load Test Database (DFLTD) Version 2.0 (v.2) (*3*) could be analyzed on a large scale to compare hundreds of interpreted pile capacities with their corresponding calculated values. This produced data is used to compare performance of interpreted vs. calculated (I/C) capacity in sands, clays, and mixed soils. Additionally, the effect of pile type, pile length, and the pile diameter on the I/C ratio is explored.

[1]NYU Tandon School of Engineering, Brooklyn, NY

Corresponding Author:
Address correspondence to Magued G. Iskander: iskander@nyu.edu

The evaluation is made primarily because the DFLTD v.2 has just been released. The scope was limited to impact-driven, un-tapered, steel and concrete piles loaded in compression, using a static load test.

History and Applications of Pile Load Test Databases

Many of the methods in current use for pile design are based on small databases of interpreted load test data. For piles in sand, Nordlund (*4, 5*) developed his method of calculating bearing capacity of piles in cohesionless soils from as few as 41 load tests from eight different test sites. Similarly, for clays, Tomlinson (*6, 7*) employed a small data set of 56 piles to develop his popular Alpha-design method that was based in part on data published by Peck (*8*). These methods were adopted by several design standards, including The Canadian Manual on Foundation Engineering (*9*), the American Petroleum Institute (API) (*10*), and FHWA (*1, 2*). Due to the dependence of soil behavior on geography and site-specific circumstances and the use of limited data employed to launch historical foundation design methods, it is possible that many currently practiced methods of pile design are based on empirical formulas that required gross overgeneralization to develop.

To address this concern, several past geotechnical studies have been undertaken with a focus on the expansion of load test databases in order to improve the methods employed to predict the axial capacity of piles. The first modern effort to organize a database of measured and interpreted pile capacities was performed at the request of the American Petroleum Institute (API) by Professor Olson and his students at the University of Texas (*11, 12*). This effort led to the development of the popular API RP2A method for capacity of piles in sand (*10*). Later efforts were made to improve on both the size of the database and on its interpretation (*13–19*). Later FHWA began a parallel effort to develop a DFLTD under the leadership of Carl Ealy (*20*). Likewise, several highway departments developed state databases of interpreted pile capacities, including Iowa, Illinois, and Louisiana (*21–24*). Similarly, several small databases have been developed at several universities and research centers including the Norwegian Geotechnical Institute and Texas A&M (*25–27*).

The original version of DFLTD includes methods of search for foundations of specific site and pile characteristics from the more than 1,500 load test results available (*20*). Although DFLTD claimed over 1,500 load tests, the authors are not aware of any studies that have been able to employ a substantial portion of these tests in comparing interpreted and computed capacities because most tests lacked crucial information necessary for either interpreting the test or computing the capacity. This is a problem with all piling databases. For example, the Olson database classified tests into five data quality factors, for both soils information and pile load test information. Few tests in the entire Olson database achieved a top tier classification in both categories, and thus nearly all analyses were based on fewer than 100 high quality tests.

In a 2013 memorandum, Abu-Hejleh describes the DFLTD as being "outdated" to the point of impossibility of database expansion (*28*). Abu-Hejleh et al. (*29*) discussed the usefulness of current databases in use in the United States and establishes the suggested minimum data required for load test databases pertaining to Project Data, Subsurface Data at the Load Test Foundation, Test Foundation Data, Load Test Data, and attachment of useful files. The guidelines for developing useful load test databases discussed by Abu-Hejleh et al. were last updated in 2015 (*30*).

FHWA rekindled the effort to gather and distribute load test information on piles as part of its effort to develop a method for predicting the axial capacity of large diameter open ended pipe piles. This effort resulted in the release of the DFLTD v.2 in February 2017 (*3*).

Working with the DFLTD v.2

The release of DFLTD v.2 updated the query process, expanding on the capabilities of the first version (*20*). The graphical user interface within Microsoft Access allows load tests to be filtered based on a predefined set of options to view or export only those containing the desired project, foundation, and soil parameters. There is currently limited functionality to filter test records for data completion in order to locate tests with all necessary parameters to carry out design calculations for the pile foundations included in the database. Furthermore, the process of extracting data, while sufficient for a case-by-case investigation, could not accommodate the need of this research endeavor to analyze cases in batch mode. For these main reasons, the authors worked with the backend of DFLTD v.2 and exported all data to a server-based database system.

Database Migration

One of the most important updates in DFLTD v.2 was organizing data in a relational schema. This was a major limitation of the first version in which data retrieval was programmatic and non-relational. While the delivery platform, Microsoft Access, remained the same, the relational schema allows for complex search queries, expanding the functionality of the provided graphical interface. The industry standard for interaction with Relational Database Management Systems (RDBMS) is the Structured Query Language (SQL). Having a relational backend, DFLTD v.2 can be queried using SQL. Microsoft offers basic functionality to do so within MS Access, therefore DFLTD v.2 was migrated to a RDBMS hosted on Microsoft Windows Server 2012. Microsoft SQL Server Migration Assistant v6.0 for Access (*31*) was used to ensure compatibility across both

Table 1. Empirical Values for ϕ, *qu*, and γ based on SPT Blow Counts Employed in this Analysis (after Bowles, 1977 and CALTRANS, 2004)

Corrected SPT N	0–4			4–10			10–30			30–50			50+		
Ranges	min	avg	max	min	avg	max	min	Avg	max	min	avg	max	min	avg	max
Approximate ϕ (degrees)[‡]	25	27.5	30	27	29.5	32	30	32.5	35	35	37.5	40	38	40.5	43
Approximate γ (lb/ft^3)[§]	70	85	100	90	102.5	115	110	120	130	110	125	140	130	140	150

Uncorrected SPT N	0–2	2–4	4–8	8–16	16–32	32+
Approximate q_u (ksf)	0–0.5	0.5–1.0	1.0–2.0	2.0–4.0	4.0–8.0	8.0+
Approximate γ (lb/ft^3)[§]	100–120	100–120	110–130	120–140	120–140	120–140

[‡]Caltrans guidance for soil friction ϕ: SW: use average ϕ + 1°, SC: use minimum ϕ, ML: use minimum ϕ + 0.5°, GM and SP: use average ϕ, GC: use average ϕ -1°, GW: use maximum ϕ, GP: use maximum ϕ - 1°.
[§]Caltrans guidance for moist unit weight γ: SW, GW: use maximum γ, SP, GP: use average γ, ML, SC, SM: use minimum γ.

platforms. Multiple random checks ensured that data transfer was successful.

This structure allows for data retrieval based on any kind of permissible filtering. Perhaps more important than searching for a specific value based on a set of conditions is filtering for cases that are not missing values.

Load Test Selection

The ability to run complex SQL queries was crucial in filtering out the load test cases within DFLTD v.2 for which it would be impossible to calculate pile capacity due to missing values. In total, DFLTD v.2 in its graphical interface shows 916 projects with 1,798 load tests. What was needed in this study was the unique combination of project/exploration/foundation/test cases of which the total count was 3,116. This count is larger than the load test count because it includes all combinations of explorations per pile for a given project, since a project can have multiple associated piles and explorations. Given that there is limited information on the relative position of explorations to pile foundations, a sensible approach is to analyze for all combinations. Applying a single load test to different soil borings can result in bias in the correlation depending on site variability. For highly variable sites, the method can artificially introduce a large difference between calculated and measured resistances. The opposite is also true for uniform site condition.

The necessary information to calculate the capacity of a pile foundation using the FHWA method and then compare the capacity to the interpreted failure load is contained within multiple RDBMS tables. All crucial properties to carry out the capacity calculations were mapped out and traced back to the corresponding fields of the DFLTD v.2 database. Ideally, load test records are expected to be complete, but it was quickly made obvious that this was not the case. For example, an SQL query on records including depth to water table, a field available in three separate tables, showed that depth to water table was available in only 30% of the total number of unique combinations.

Using reasonable engineering judgment to approximate necessary properties can help prevent the number of applicable test records from dropping too low. Information such as the depth to water table can be approximated, but physical pile properties such as pile depth/diameter and many geotechnical properties cannot. For this project, if the value for water table was recorded in the database, it was always used. About 10% of the load tests in the final selection did not have values for water table stored. For these tests, the water table was set at a depth of 6 ft, which may be too conservative, and possibly contribute to scatter for piles in sand and mixed soil profiles. In the case where other geotechnical properties were missing, the authors only used approximation methods described within the FHWA design manual (*1*). This applies mainly to missing total unit weight, undrained shear strengths, and angles of internal friction, for which Bowles (*32*) has produced empirical values along with the most recent guidance from Caltrans Geotechnical Manual (*33*). Minimum, average, and maximum of the recommended range for soil friction angle for a given standard penetration test (SPT) N-value is determined given the corresponding USCS soil type classification (Table 1). The entire process was algorithmically translated to allow for batch calculations. Undrained shear strength is more straightforward to determine from SPT N-values, albeit less reliable. In order to use the approximation methods, each case must have, at a minimum, SPT N field values. Seven percent of the available records were rejected for not having SPT information. Additionally, pile set-up time, an important parameter that may significantly affect the final ultimate capacity, was not recorded for most tests employed in this study and, as such, could not be included in the analysis. A summary of the filtering process and its results is presented (Figure 1).

Applicable test records must have static test data to compare against the calculated capacities. If static test data were missing, specifically compression test data, the record was rejected. This reduced the pool of available records by 15%. Additionally, the scope was reduced to pile foundations that

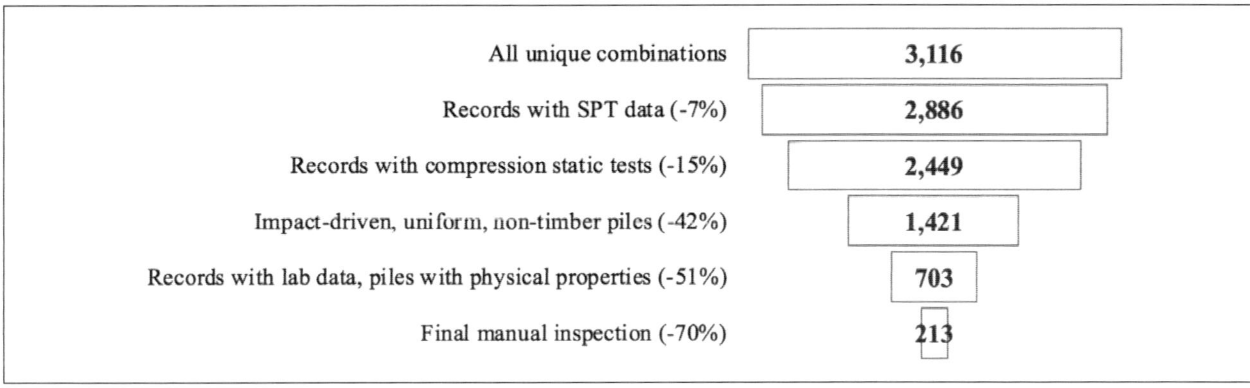

Figure 1. Filtering of available test records.

were impact driven, non-tapered, non-timber, and with values in the length field. With this query the records were reduced by 42%. Then, the search query was expanded with common table expressions (CTE) to check for necessary geotechnical data in several combinations of tables, lab results from different fields, and so forth. Also, with the physical properties for each pile type residing in separate tables, a CTE checked that each pile foundation selected had information on core diameter, side length, strength, and so forth. This step eliminated 51% of the available records.

Using complex SQL queries, the unique test record combinations were reduced from 3,116 to 703. However, this process could not guarantee that capacities would be calculated for these 703 records. Data were gathered from several tables, making manual inspection exhaustive and error-prone. The authors developed a tool that queried the database, retrieved all relevant field values, and organized the information in a format familiar to geotechnical engineers, as shown in Figure 2. This custom tool combined information on unit weight, undrained shear strength, and angle of internal friction from all tables. If any of these properties was missing, the program would attempt to approximate the values from SPT data. Should the process be successful, an effective stress plot was produced on the lower left part. The program also organized all available static test and interpreted failure load data from DFLTD v.2, using Davisson's failure criterion (*34*). The authors are familiar with concerns about use of SPT data, at least solely, in the design calculations of pile foundations. In this study though, this was a last-resort option to maintain a reasonable number of "workable" test records. Should the use of SPT data be rejected, the pool of applicable data would only include a dozen or two.

The form shown in Figure 2, produced for each record in the last selection of 703 records, made it efficient and intuitive to deduce which records would produce results and which would not. By manual inspection of the produced forms, the final number of applicable test records came down to 213 (*35*).

The evaluation is made primarily because the DFLTD v.2 has just been released. Manual manipulation of the data may result in a larger data set. Similarly, inclusion of dynamic load test data will increase the number of records.

FHWA Pile Design Methods

In April 2006, FHWA published a revised report on the Design and Construction of Driven Pile Foundations (*1, 2*). Within this document, FHWA recommends several empirical and semi-empirical design methods for determining the ultimate capacity of single piles. The FHWA report categorizes the design methods as follows: Methods of Static Analysis for Piles in Cohesionless Soils and Methods of Static Analysis for Piles in Cohesive Soils. The report also includes design recommendations for piles driven into mixed soil profiles (i.e., layers of both sand and clay).

Each method recommended within the report calculates ultimate bearing capacity of single piles as being equal to the summation of the side and tip resistances. Side resistance is calculated as the sum of intervals corresponding to segregated soil layers for each of these methods of design. Therefore, a soil profile based on soil type and SPT N-values and other soil properties must be first determined from boring log information and other laboratory data. The report suggests that in some cases an acceptable conservative design decision is to disregard the capacity generated at the pile tip altogether, because of the lack of movement necessary to generate resistance at the tip. However, in this study the capacity was always computed as the sum of the side and tip resistances. The tip resistance in mixed soils is determined based on the end bearing stratum. Calculations require certain pile properties to be known as well, such as embedded length and pile type.

For cohesionless soils, FHWA suggests the following methods for single pile design: the Meyerhof method (empirical) (*36*), the Brown method (empirical), the Nordlund method (semi-empirical) (*4, 5*), the Effective Stress method (semi-empirical) (*37*), and cone penetration test (CPT)

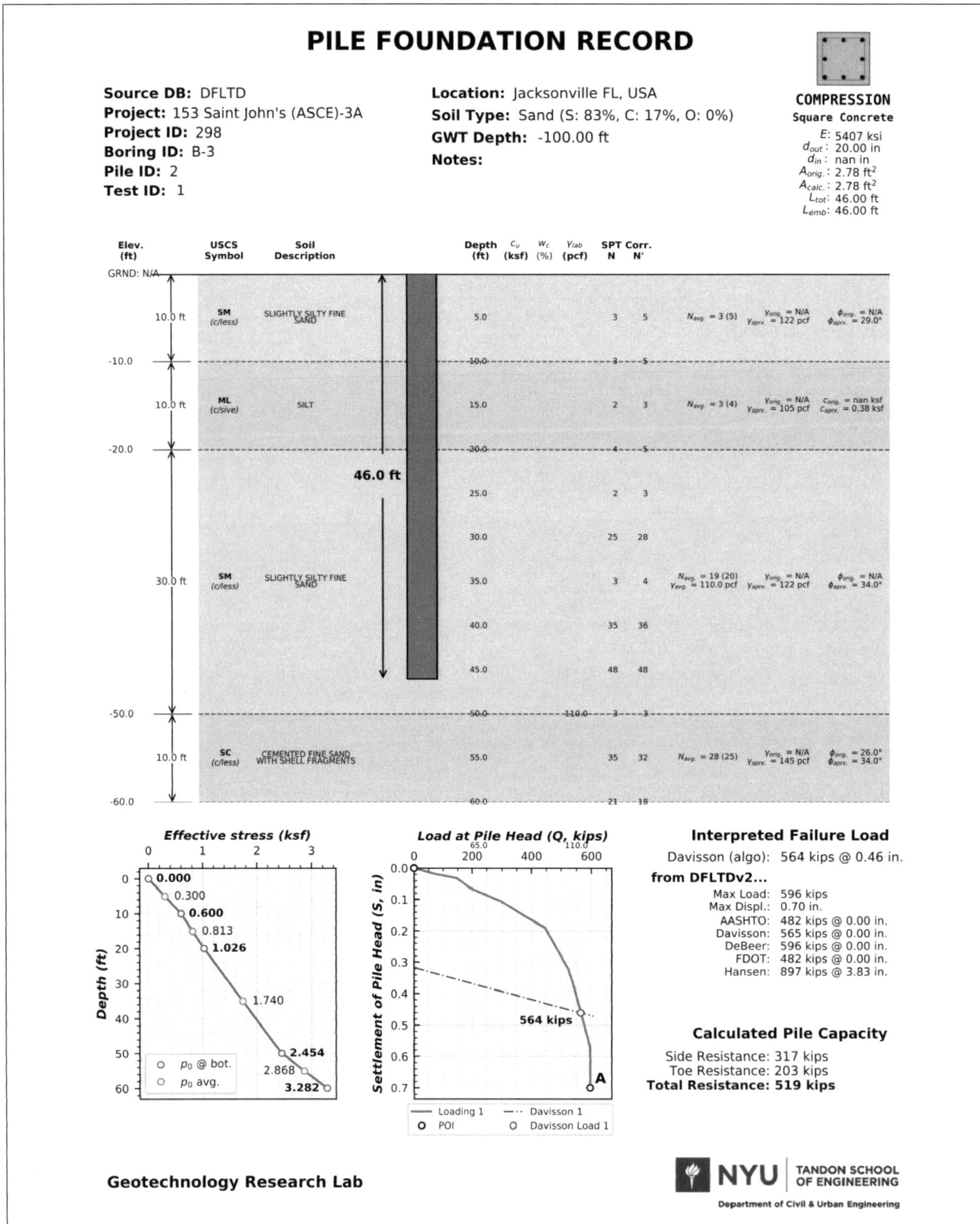

Figure 2. Example of auto-generated pile record form.

methods (empirical) (*38*). The report notes that the methods derived strictly from corrected SPT N-values, Meyerhof and Brown, are less reliable compared with the Nordlund and Effective Stress methods (*1*). The Nordlund method was used in this study for determining capacity in cohesionless soils because of its wide popularity. The method uses corrected SPT N-values (or, preferably, lab-produced strength parameters) to determine the soil friction angle for each observed soil layer and uses a series of published tables and charts to assume correlations for the coefficient of lateral earth pressure and the soil-pile friction angle. These values are used along with the effective overburden pressure to determine the side resistance for each defined layer. Pile tip bearing capacity factors are also correlated from the soil friction angle using charts published by Nordlund. Upper limits are placed on pile tip overburden pressure, p_t, and pile tip resistance, R_t, to limit the magnitude of the computed unit tip resistance and calculate a safer ultimate pile capacity.

For cohesive soils, FHWA suggests the following methods: the Alpha method (or Tomlinson, empirical) (*39*), the Effective Stress method (semi-empirical) (*37*), and CPT methods (empirical) (*38*). The Tomlinson method was selected because of its wide popularity. When dealing with mixed soil profiles, Tomlinson provides for adjustment factors to account for drag-down of weaker soil into stiffer layers, a phenomenon that occurs during pile driving and reduces the side resistance. These factors were accounted for in our calculations.

For steel H and unfilled open end pipe piles, the authors followed FHWA guidance to "use only steel cross section area at pile toe unless there is reasonable assurance and previous experience that a soil plug will form at the pile toe." This information was not available for many of the tests employed, so the authors simply calculated end bearing using the rim area of steel piles. For side resistance, Hannigan (*1*) does not require adding the internal side resistance, for pipe piles and requires the use of the rectangular area for H piles.

Analysis Procedure

In order to batch process all load tests available in DFLTD v.2 it was necessary to develop a suite of algorithms written in Python that can reproduce the Nordlund and Tomlinson design methods exactly as they are outlined within the FHWA design manual. Note that the software application DrivenPiles by Multidimensional Software Creations (MDSC) (*40*), follows a slightly modified design methodology than the one presented in the FHWA design manual (*1*), making it unsuitable for this study. The Python program follows the design procedures laid out in the FHWA design manual to evaluate given soil and pile data and output corresponding ultimate capacity values for each soil type (sand, clay, mixed). This code also serves as a platform from which to compare static design results with interpreted design results.

The FHWA suggests one method for interpreting the pile capacity from a static load test: the Davisson failure criterion (*33*), which has also been programmatically translated to allow for automatic detection of the interpreted failure load. The algorithm detects all loading/unloading cycles of a given static test and can apply the Davisson criterion on all of them, keeping as a final solution the greatest of all interpreted failure loads. For the test records in DFLTD v.2, in which the Davisson failure load was included, the authors found that eight out of ten times the algorithmically produced failure load would match the stored value in the database. Differing results were manually checked and the authors satisfied themselves that their code computes the correct Davisson failure load. This gave more confidence for the reliability of the calculated results.

Several pre-solved cases were used to validate the results produced by the algorithms replicating the Nordlund and Tomlinson pile design methods. Namely, Appendix F of the FHWA design manual (*1*) offers step-by-step example problem calculations covering soil profiles predominantly defined by sand, clay, and mixed layers. The design algorithms were tested against these example problems, returning capacities that were within 5% of the solutions outlined in the manual. Since the example problems included only a single pile type, the authors tested the algorithms with other problems as well as random spot checks of cases retrieved from DFLTD v.2. In all cases the design capacities produced by the algorithms matched hand calculations.

Qc/Qm Comparison in Sand, Clay, and Mixed Soil Profiles

Analyses were performed for piles in the current database, using the FHWA method to compute the calculated capacity (*Qc*). Interpreted failure load, as obtained from the Davisson criterion, is frequently referred to as "measured capacity (*Qm*)." Data is presented for sand, clay, and mixed soil profiles. Sand and clay profiles contained tests in which more than 70% of the capacity was derived from the relevant soil. In order to optimize the visual separation of the points, calculated (*Qc*) and measured (*Qm*) pile capacities are plotted in a log-log scale (Figure 3). As a reference, 1:½, 1:1 and 1:2 (*Qc:Qm*) lines were added on the plots. A histogram of the distribution of *Qc/Qm* is also presented for piles in each profile.

Ideally, *Qc/Qm* should be close to 1. The scatter between measured and predicted capacities is significant; the capacity was off by a factor of 2 in many tests. The range in *Qc/Qm* was from 0.12 to 8.88. The mean *Qc/Qm* was 1.6 in sand, 1.2 in clay, and 1.43 in mixed profiles. The corresponding standard deviation in *Qc/Qm* was 1.4 in sand, 0.56 in clay, and 1.34 in mixed soils. The results for mean and standard deviation for the three predominant soil types

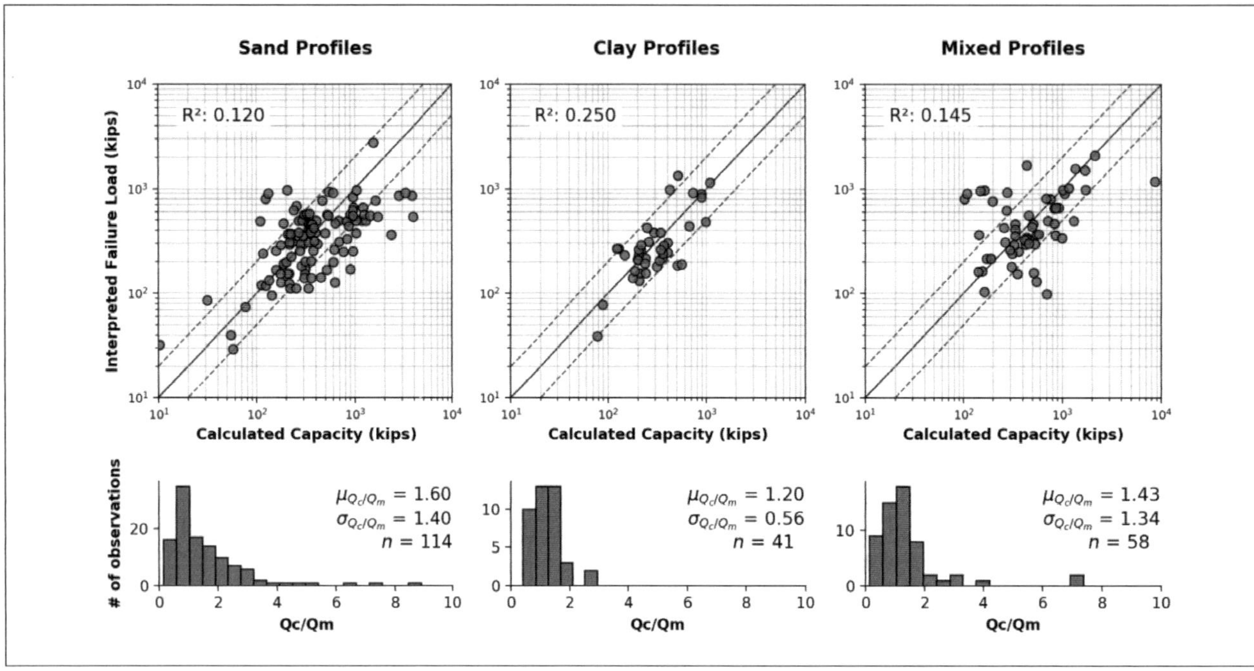

Figure 3. Distribution of calculated (Qc) v. interpreted – also known as (aka) measured (Qm) capacity for all soil profiles.

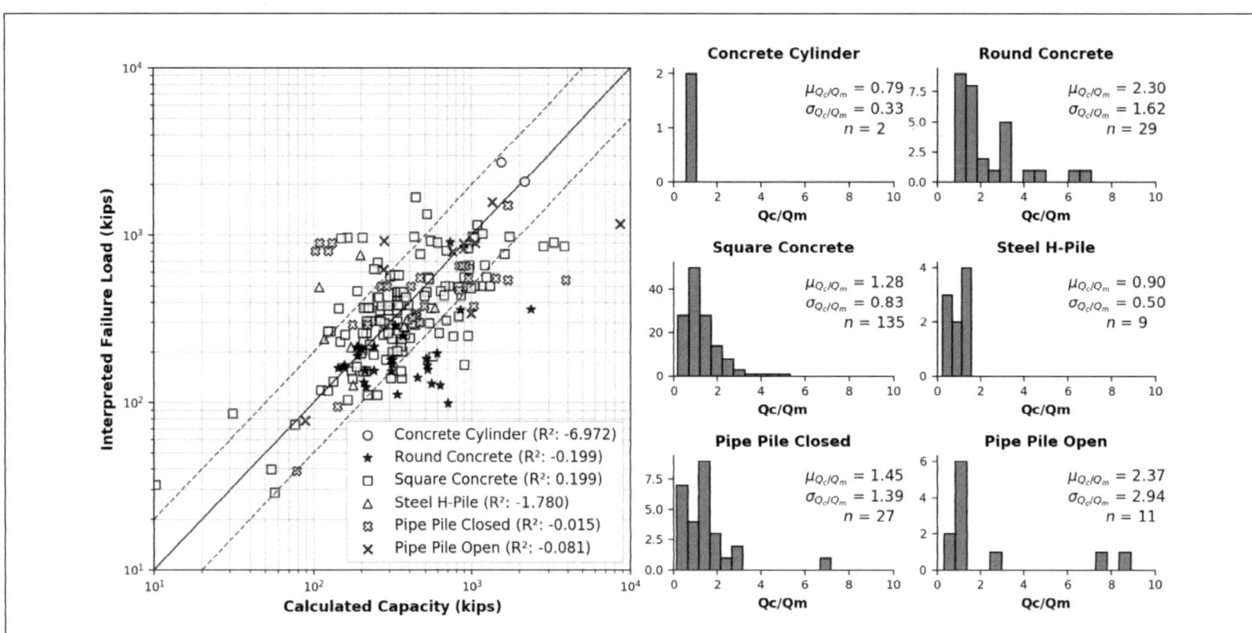

Figure 4. Distribution of calculated (Qc) v. interpreted – aka measured (Qm) capacity for six pile types.

clearly show a better performance in clay than in sand. The variation observed for the sand profiles could be related to the correlation used for soil friction angle; all SPT correlations are problematic. We only employed the correlation recommended by FHWA, to be consistent with its design methodology.

Overall Performance (per Pile Type)

Data are plotted according to pile type as shown in Figure 4. A frequency distribution of Qc/Qm is also presented for piles in each pile type. There are few tests represented for some pile types, so it is difficult to generalize the results. It appears

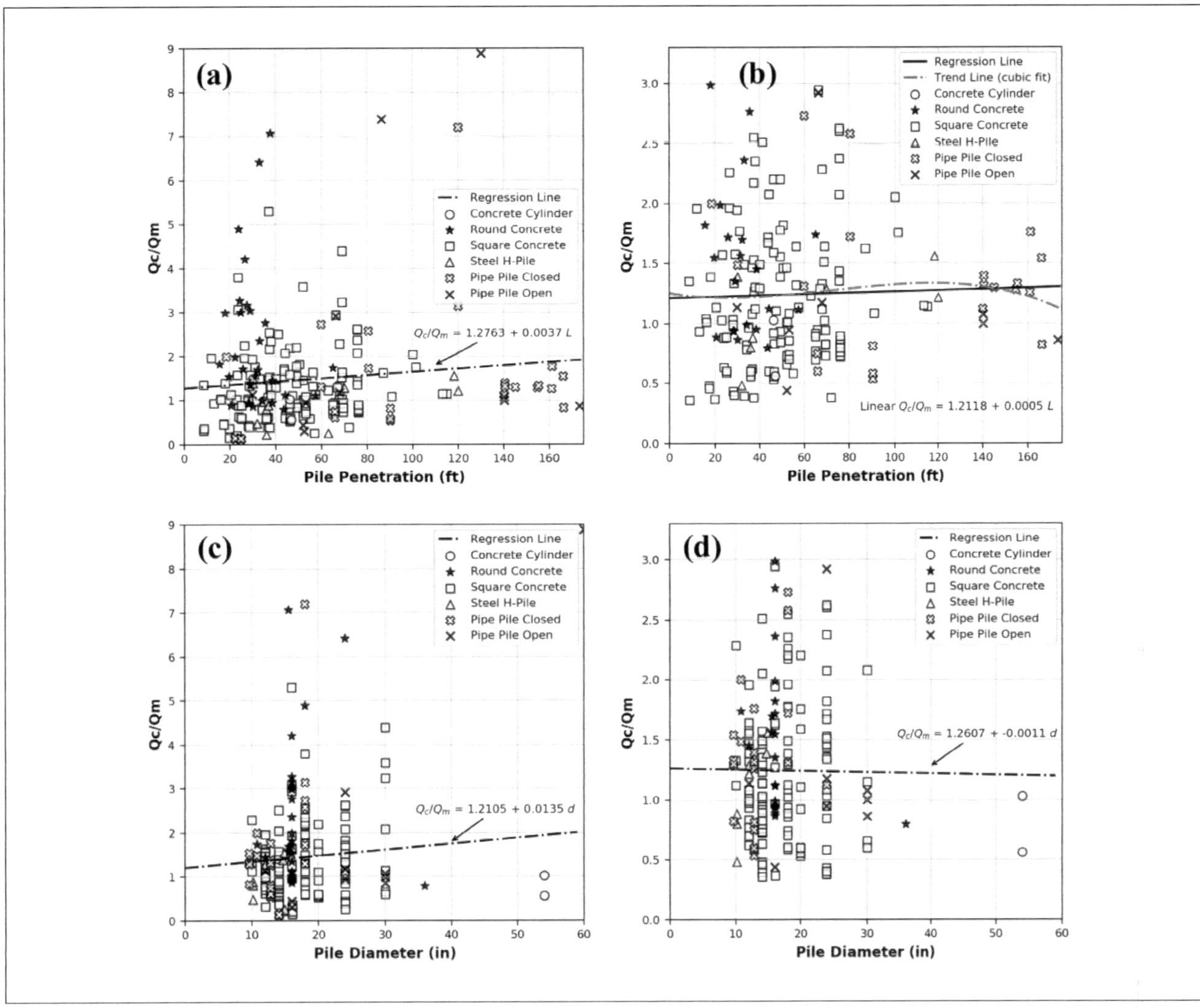

Figure 5. Effect of pile properties on the ratio of calculated (*Qc*) to interpreted – aka measured (*Qm*) capacity for six pile types: (*a*) Effect of penetration length using all tests, (*b*) Effect of penetration length with outliers removed, (*c*) Effect of pile diameter using all tests, and (*d*) Effect of pile diameter with outliers removed.

that the design method performed best for square concrete piles. This is not surprising considering that Tomlinson's database was heavily weighted toward this pile type. On the other hand, round concrete piles and open pipe piles exhibited the highest average (2.30 and 2.37 respectively) and standard deviation values (1.62 and 2.94 respectively) for *Qc/Qm*. The effect of pile shape in calculated capacity is a point for future exploration.

Effect of Pile Penetration Length

Calculated capacity normalized by measured (interpreted) capacity is plotted against pile length (Figure 5). The regression line shows that the ratio is increasing with pile embedded length. A long-standing problem with many design methods for predicting pile capacity is that their use led to underprediction of capacities of short piles and overprediction of the capacity of long piles (*25, 16*). Data may have been skewed by a few long pipe piles. The number of observations for long piles is small and the quality of the data is generally suspect, but the consistency of the data is cause for concern, especially because long piles are often used to support offshore and bridge structures. The linear function produced by the coefficients of the regression line suggests that the effect is large. The regression trend may have been skewed by the outlier points at 120- and 130-ft penetration lengths with *Qc/Qm* values >7, since the remainder of the points for pile penetrations larger than 100 ft have *Qc/Qm* values in the range of 1.2 to 1.4. Nevertheless, all things being equal, regression suggests that capacity can be overestimated by 50% for a 125-ft pile, and by 100% for a 250-ft long pile. Sorting out the length effect depending on pile type resulted in conflicting trends depending on pile shape. The authors believe that the available data may not necessarily produce realistic trend lines for each pile type, and as a result the effort was abandoned.

Overprediction of capacities for long piles does not necessarily indicate problems in the FHWA formulations since the length effect virtually disappears when the analysis is repeated for the 183 piles (85% of total) having $0.33 < Qc/Qm < 3$. In fact, a reverse trend can be observed for piles longer than 100 ft. This suggests that data quality issues, as well as atypical soil properties that would require long piles to be driven (e.g., presence of volcanic sands), may have contributed to the notion of overprediction of the capacity of long piles. In any case, there are fewer than 20 tests with lengths >100 ft, having $0.33 < Qc/Qm < 3$, which is cause for concern.

Effect of Pile Diameter

Calculated capacity normalized by measured (interpreted) capacity is also plotted against pile diameter (Figure 5). The regression line shows that the ratio is increasing with pile diameter, suggesting that capacity can be overestimated by 15% for each additional 12-inch increase in pile diameter. However, when the analysis is repeated for the 183 piles (85% of total) having $0.33 < Qc/Qm < 3$, the trend is reversed. In any case, fewer than a dozen tests having diameters >25 inches are available, which is cause for concern.

Summary and Conclusion

FHWA recommendations have become the accepted industry standard for pile design as demonstrated by its widespread use. In this study the efficacy of the design method was explored using data made possible by the recently released DFLTD v.2. The scope was limited to impact-driven, untapered steel, and concrete piles loaded in compression, using a static load test. DFLTD v.2 was ported to a RDBMS that was queried using SQL. DFLTD contains 3,116 unique-combination project/exploration/foundation/test cases, of which 213 contained sufficient data to permit batch processing of the results in order to compute the axial load capacity and interpret the static failure load.

In general, scatter between measured (interpreted) and predicted capacities is significant; the computed capacity was off by a factor of 2 in many tests. The range in Qc/Qm was from 0.12 to 8.88. The mean Qc/Qm was 1.6 in sand, 1.2 in clay, and 1.43 in mixed profiles. A length and diameter effect were evident when all data were used, but both virtually disappear when the analyses were repeated for the 183 piles (85% of total) having $0.33 < Qc/Qm < 3$. In any case, few long and large diameter piles are available to justify definitive conclusions for large piles. Our preliminary evaluation suggests that the method performs better in clay than in sand, and that the method has a large amount of scatter associated with the predicted capacities. This scatter stems from the fact that the design method does not adequately account for several phenomena that occur during pile driving, such as pile shaking, change in soil fabric due to driving, and drag-down of soils from one layer to the next (*15*, *41*, *42*).

The authors believe that this evaluation permits engineers and state agencies to better understand the efficacy of the most commonly employed design methods, resulting in a more resilient infrastructure. FHWA pile design methods are not perfect, yet they have been successfully employed in practice, in large part because many designers factor in local experience to produce reasonable designs. The bias reflected in this study is more likely attributed to data quality issues, rather than the design methodology. The authors also believe that all design methods need to be continuously updated to reflect current knowledge. This study provides justification to regulating bodies to invest in populating pile databases of measured and predicted capacities in order to re-visit pile design guidelines.

References

1. Hannigan, P. J., G. G. Goble, G. E. Likins, and F. Rausche. *Design and Construction of Driven Pile Foundations Reference Manual - Volume I*. Report No. FHWA-NHI-05-042. National Highway Institute, Federal Highway Administration, U.S. Department of Transportation, Washington D.C., 2006.
2. Hannigan, P. J., G. G. Goble, G. E. Likins, and F. Rausche. *Design and Construction of Driven Pile Foundations Reference Manual - Volume II*. Report No. FHWA-NHI-05-043. National Highway Institute, Federal Highway Administration, U.S. Department of Transportation, Washington D.C., 2006.
3. Petek, K., R. Mitchell, and H. Ellis. *FHWA Deep Foundation Load Test Database Version 2.0 User Manual*. U.S. Department of Transportation Federal Highway Administration, McLean, Virginia, USA, 2016.
4. Nordlund, R. L. Bearing Capacity of Piles in Cohesionless Soils. *Journal of the Soil Mechanics and Foundations Division*, Vol. 89, No. 3, 1963, pp. 1–36.
5. Nordlund, R. L. Point Bearing and Shaft Friction of Piles in Sand. *Proc., 5th Annual Short Course on Fundamentals of Deep Foundations Design*, University of Missouri-Rolla, St. Louis, Missouri, USA, 1979.
6. Tomlinson, M. J. The Adhesion of Piles in Clay Soils. *Proc., 4th International Conference on Soils Mechanics and Foundation Engineering*, Rotterdam, Netherlands, Vol. 2, Thomas Telford Ltd., London, 1957, pp. 66–71.
7. Tomlinson, M. J. Some Effects of Pile Driving on Skin Friction. *Proc., Conference on the Behaviour of Piles*, London, UK, Institution of Civil Engineers, London, 1971, pp. 107–114.
8. Peck, R. B. *A Study of the Comparative Behavior of Friction Piles*. Special Report No. 36. Highway Research Board, Washington, D.C., 1958, p. 72.
9. Associate Committee on the National Building Code. *Canadian Manual on Foundation Engineering*. National Research Council of Canada, Ottawa, 1975.
10. American Petroleum Institute. *Recommended Practice for Planning, Designing and Constructing Fixed Offshore Platforms (RP 2A)*, 18th ed. American Petroleum Institute, Washington, D.C., 1989.

11. Dennis, N. D., and R. E. Olson. Axial Capacity of Steel Pipe Piles in Sand. In *Geotechnical Practice in Offshore*, ASCE, 1983, pp. 389–402.
12. Dennis, N. D., and R. E. Olson. Axial Capacity of Steel Pipe Piles in Clay. In *From Soil Behavior Fundamentals to Innovations in Geotechnical Engineering Honoring Roy E. Olson*, ASCE, 1983, pp. 176–194.
13. Olson, R. E., and K. S. Al-Shafei. Axial Load Capacities of Steel Pipe Piles in Sand. *Proc., 2nd International Conference on Case Histories in Geotechnical Engineering*, University of Missouri, Rolla, Vol. 3, 1988, pp. 1731–1738.
14. Olson, R. E. Axial Load Capacity of Steel Pipe Piles in Sand. *Proc. 22nd Annual Offshore Technology Conference*, Houston, 1990, Paper 6419, pp. 17–24.
15. Iskander, M., and R. E. Olson. Review of API Guidelines for Pipe Piles in Sand. *Proc., Conference on Civil Engineering in the Oceans V (CEO V)*, College Station, ASCE, 1992, pp. 798–812.
16. Olson, R. E., and M. Iskander. Axial Load Capacity of Pipe Piles in Sands. *Proc., 1st Regional Conference and Exhibition on Advanced Technology in Civil Engineering*, Saudi Arabia, ASCE–SAS, Manama, Bahrain, 1994, pp. 383–394.
17. Olson, R. E., and M. Iskander. "Axial Load Capacity of Piles in Sand. *Proc., International Conference on Case Histories in Geotechnical Engineering*, St. Louis, Missouri, Missouri University of Science & Technology, 1998, pp. 293–300.
18. Olson, R. E., and M. G. Iskander. Axial Load Capacity of Un-Tapered Piles in Cohesionless Soils. In *Contemporary Topics in Deep Foundations*, ASCE, 2009, pp. 231–238. https://ascelibrary.org/doi/abs/10.1061/41021(335)29
19. Olson, R. E., and T. J. Shantz. Axial Load Capacity of Piles in California in Cohesionless Soils. In *Current Practices and Future Trends in Deep Foundations*, American Society of Civil Engineers, Reston, Va., 2004, pp. 1–15.
20. Kalavar, S., and C. Ealy. FHWA Deep Foundation Load Test Database. *New Technological and Design Developments in Deep Foundations (GSP 100)—Proc., Sessions of Geo-Denver 2000*, ASCE, Denver, 2000, pp. 192–206.
21. Roling, M. J., S. Sritharan, and M. T. Suleiman. *Development of LRFD Procedures for Bridge Pile Foundations in Iowa Volume I: An Electronic Database for Pile Load Tests (PILOT)*. 2010. Iowa Department of Transportation, Ames, Iowa, United States.
22. Roling, M. J., S. Sritharan, and M. T. Suleiman. Introduction to PILOT Database and Establishment of LRFD Resistance Factors for the Construction Control of Driven Steel H-Piles. *Journal of Bridge Engineering*, Vol. 16, No. 6, 2011, pp. 728–738.
23. Tavera, E., G. Rix, G. Burnworth, and J. Jung. *Calibration of Region-Specific Gates Pile Driving Formula for LRFD.* Louisiana Department of Transportation and Development, 2016.
24. Long, J., and A. Anderson. *Improved Design for Driven Piles on a Pile Load Test Program in Illinois*. Report FHWA-ICT-12-011. Illinois Center for Transportation, University of Illinois at Urbana-Champaign, 2012.
25. Briaud, J.-L., J. Anderson, D. Perdomo, and L. M. Tucker. Evaluation of API Method Using 98 Vertical Pile Load Tests. *Proc., 19th Annual Offshore Technology Conference*, Houston, 1987. pp. 447–451
26. Lacasse, S. *Uncertainties in Offshore Geotechnical Engineering, International Survey of API RP2A Design Parameters for Axial Capacity of Driven Piles in Sand.* Report No: 85307-14. Norwegian Geotechnical Institute, 1988.
27. Paikowsky, S. G. Load and Resistance Factor Design (LRFD) for Dynamic Analysis of Deep Foundations. *Proc., 15th International Conference on Soil Mechanics & Foundation Engineering*, Istanbul, Turkey, 2002, pp. 981–984. Taylor & Francis, London, UK.
28Abu-Hejleh, N., and S. Anderson. *Distribution: Deep Foundation Load Test Database (DFLTD) Memorandum.* U.S. Department of Transportation, Federal Highway Administration, 2013.
29. Abu-Hejleh, N. M., M. Abu-Farsakh, M. T. Suleiman, and C. Tsai. Development and Use of High-Quality Databases of Deep Foundation Load Tests. *Transportation Research Record: Journal of the Transportation Research Board*, 2015. 2511: 27–36.
30. Abu-Hejleh, N., M. Abu-Farsakh, M. T. Suleiman, and C. Tsai. State of Practices in Databases for Deep Foundation Load Tests. *Proc., IFCEE 2015*, American Society of Civil Engineers, Reston, Va., 2015, pp. 237–246. San Antonio, Texas.
31. Microsoft SQL Server Migration Assistant v6.0 for Access. *Microsoft Download Center*, (n.d.). https://www.microsoft.com/en-us/download/details.aspx?id=43690. Accessed April 18, 2017.
32. Bowles, J. E. *Foundation Analysis and Design*. McGraw-Hill, New York, 1977.
33. Caltrans. *Caltrans Geotechnical Manual*. Office of Geotechnical Services, 2014. http://www.dot.ca.gov/hq/esc/geotech/geo_manual/manual.html. Accessed November 5, 2017.
34. Davisson, M. T. High Capacity Piles. *Proc., Soil Mechanics Lecture Series on Innovations in Foundation Construction*, ASCE Illinois Section, Chicago, 1972, pp. 81–112. Chicago, Illinois, US.
35. Machairas, N., G. Highley, and M. Iskander. Evaluation of FHWA Pile Design Method Against the FHWA Deep Foundation Load Test Database Version 2.0. Presented at 97th Annual Meeting of the Transportation Research Board, Washington, D.C., 2018.
36. Meyerhof, G. G. Bearing Capacity and Settlement of Pile Foundations. *Journal of the Geotechnical Engineering Division*, Vol. 102, No. 3, 1976, pp. 195–228.
37. Fellenius, B. Chapter 13 - Pile Foundations. In *Foundation Engineering Handbook* (H.-Y. Fang, ed.), 2nd ed. Van Nostrand Reinhold Publisher, New York, 1991, pp. 511–536.
38. Eslami, A., and B. H. Fellenius. Pile Capacity by Direct CPT and CPTu Methods Applied to 102 Case Histories. *Canadian Geotechnical Journal*, Vol. 34, No. 6, 1997, pp. 886–904.
39. Tomlinson, M. J. *Pile Design and Construction Practice, Fourth Edition*. E & FN Spon, London, 1994.
40. MDSC. *DrivenPiles - User Manual*. MDSC Inc, 2015. North Logan, Utah, US.
41. Randolph, M. Science and Empiricism in Pile Foundation Design. *Geotechnique*, Vol. 53, No. 10, 2003, pp. 847–875.
42. O'Neill, M. W. Side Resistance in Piles and Drilled Shafts. *Journal of Geotechnical and Geoenvironmental Engineering*, Vol. 127, No. 1, 2001, pp. 3–16.

The Standing Committee on Foundations of Bridges and Other Structures (AFS30) peer-reviewed this paper (18-05548), which was selected by the Geotechnical Engineering Section (AFS00) to receive its 2018 Best Paper award.

Article

TRR
JOURNAL OF THE TRANSPORTATION RESEARCH BOARD

Transportation Research Record
2018, Vol. 2672(52) 278–290

DOI: 10.1177/0361198118796733
journals.sagepub.com/home/trr
SAGE

Performance Investigation and Design of Pile-to-Pile Cap Connections Subject to Uplift

Philip P. Iekel[1], Brent Phares[1], and Michael Nop[2]

Abstract

Substructure bridge components are designed to resist gravitational forces such as dead load and vehicular live load, as well as lateral forces including wind, vehicular braking, and centrifugal force effects. Significant lateral forces can create "uplift" conditions on some portions of the foundation. A review of current design techniques regarding uplift in the pile-to-pile cap connection indicates a lack of uniformity in the design process across state agencies stemming from minimal research performed in this area. In addition, approved uplift anchors for use in the field have not been tested. In order to close this gap, twenty-one full-scale steel H-pile specimens were fabricated and tested in Iowa State University's Structural Engineering Laboratory to test the capacity of the pile-to-pile cap connection under static tensile loading. Specimens were cast both with and without anchorage and with 12" and 24" embedment depths in order to understand the behavior of the connection and to determine a suitable anchorage detail and design approach for uplift cases. Findings revealed that: (i) capacity of bare piles is generally underestimated and could be more frequently considered for uplift design; (ii) concrete cracking leads to a loss of bond in these types of connections; and (iii) positive anchorage or embedment that extends above the lower rebar mat of the footing is necessary to develop a high capacity connection.

Although not desirable, it is common for some amount of substructure uplift force to be generated under certain loading conditions. In Iowa, uplift conditions are usually produced from the Strength 1, Strength 3, Strength 5, Extreme Event 2, and Service 1 load combinations found in the American Association of State and Highway Transportation Officials (AASHTO) bridge design code (*1*). Contributing factors include braking, centrifugal force effects, stream forces, temperature, wind, and unbalanced vertical loading. In general, continuous multi-span bridges that incorporate tall piers and long spans have larger uplift forces. Tall piers increase the overturning effect of lateral loads as a result of a larger moment arm. In addition to this, fixed piers on long continuous multi-span bridges typically must resist proportionally more longitudinal loading than expansion piers, resulting in a higher probability of uplift forces. Curved structures can also introduce thermal and friction forces acting transversely, further increasing potential uplift loads transferred to the substructure elements (A. Nelson, personal communication, July 2017).

An investigation of recent Iowa Department of Transportation (DOT) bridge designs show that factored uplift values used in design are in the range of 8–40 kips for one pile. A select few projects on interstate ramps have been designed for uplift values as high as 87 kips (*2*). In the event of such uplift forces the Iowa DOT has a clear design methodology for handling pile-to-soil interaction (*3*). However, once the pile design has been satisfied, attention must be focused on the critical pile-to-pile cap (or pier footing) connection. Very little literature is available on this topic, which has led to a broad range of design approaches that are in effect today. With the larger uplift loads now found in design, it is imperative that details used to resist these forces be based upon sound experimental and analytical research coupled with engineering fundamentals.

Current Practice

In Iowa the overwhelmingly preferred foundation configuration for piers and abutments consists of driven steel

[1]Bridge Engineering Center, Institute for Transportation, Ames, IA
[2]Iowa Department of Transportation, Ames, IA

Corresponding Author:
Address correspondence to Brent Phares: bphares@iastate.edu

piles with a concrete cap (*3*). Figure 1 details a pile foundation which is a representative example of current Iowa projects. Piles are positioned in a rectangular pattern around the base of the column and embedded 12" into the footing. In the case of pier piles, the perimeter piles are usually battered at an angle of 1 horizontal to 4 vertical to resist lateral forces acting on the superstructure. A lower mat of reinforcing steel is placed 1" clear above the tops of the piles and is generally composed of reinforcing bar sizes ranging from #8 to #11. A top mat typically consisting of #5 reinforcing bars is included when uplift forces are present (*3*).

Corner piles are often susceptible to the highest uplift forces. The Iowa DOT currently assumes a nominal bond strength of 0.060 ksi between the pile and pile cap concrete if uplift forces are expected. This nominal bond strength is based on the bond stress recommended in the American Institute of Steel Construction (AISC) design manual for filled steel columns and is thought to be a reasonable lower bound as mentioned in the design manual commentary (*3*, *4*). Two other papers pertaining to bond stress, one by Griffis for encased steel columns and another by Watson related to embedded beam flanges, recommend even higher values (*5*, *6*).

For design, the resistance factor for the strength limit state check is set at 0.45 per the recommendation in AISC for bond stress on filled steel columns (*4*). While the AASHTO code generally specifies a resistance factor

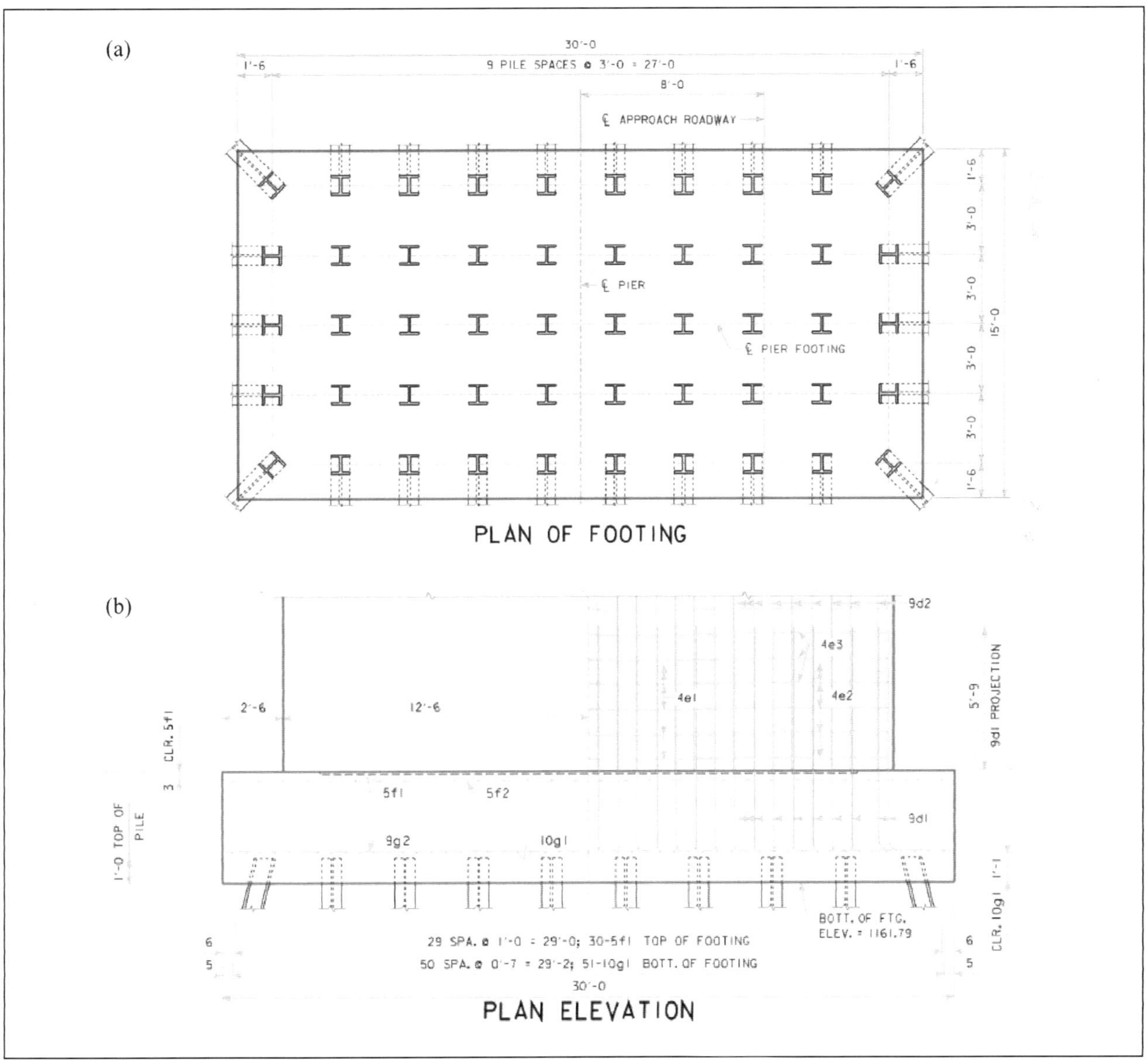

Figure 1. Typical bridge and substructure details: (*a*) plan view of pier footing and (*b*) elevation view of footing.

of 1.0 for Extreme Event, a 0.75 factor used by the Iowa DOT acknowledges the critical nature of the connection (*1*, *7*, *8*).

In the event that the bond strength is not sufficient, positive anchorage must be attached to the pile. The Iowa DOT specifies the use of an F1554 threaded rod through the web of the pile or the use of welded studs to the web. However, a brief survey of recent bridge plans in Iowa showed a range of anchors utilized including, but not limited to, C-channels, L angles, and in one instance V-bars through the web (*2*).

Results from the 2016 North Central States Consortium (NCSC) questionnaire, in which ten state agencies answered questions relating to current design practice for bridges, revealed that methods for uplift design in the pile connection varied greatly (G. Novey, personal communication, October 2016). Pertaining to bare piles subject to uplift forces, Iowa's specification of 0.060 ksi is the largest allowed of all NCSC respondents. Other states either do not consider any uplift capacity, or assume some lesser bond strength (i.e., 1.0 ksf or 10 psi). Relating to anchorage, welded studs to the web of the pile or drilling/torching holes and inserting a bar perpendicular to the pile are the most commonly used. The methodology for setting the ultimate capacity for these anchors is either unknown and assumed or based on the breakout capacity for studs in shear found in ACI 318. North Dakota welds reinforcing steel directly to the pile (longitudinal orientation) using straight bars and sets the yield strength for the bars as the capacity of the connection. The remaining few states do not have uplift details and instead prefer to resize the footing so that no uplift is generated. While resizing the footing is a viable option, designing piles for no uplift is sometimes impractical/impossible and may not be cost efficient. Therefore, appropriate positive anchorage details must be incorporated into design manuals for use, even if the allowable uplift forces are small.

Previous Investigation

Two notable studies have been completed to experimentally evaluate pier piles subjected to uplift forces. In 1972 Horn, Stewart, and Boulware tested six different anchorage details attached to steel and timber piles (*9*). For each detail three specimens were fabricated and tested, twice under static loading and once under 100 cycles of loading equal to 40% of the bearing capacity of the pile. The steel pile anchors consisted of hooked plates bolted to the web and #6 V-bars through the web. They attained average ultimate static capacities of 76–125 kips with 5" of embedment. Current AASHTO code specifies that piles must 'project at least 12" into the pile cap after all damaged material is removed'. This standard is reduced to 6" if there is positive anchorage present (*1*). Horn et al. also tested one bare steel pile which reached an ultimate capacity of 60 kips. Embedment depth for the control specimen is not clearly stated but is assumed to be 6" or less based on other specimens tested. The research concluded that V-bars on steel piles were the most cost-efficient and best-performing anchorage device out of the two that they tested.

GAI Consultants, Inc. recognized the need for further testing with their 1982 publication *The Steel Pile, Pile Cap Connection* (*7*). Twelve HP10 and four PP10-¾ × 0.5" piles with various anchorage details were subjected to static tensile forces to measure the capacity of the connection. Embedment depth ranged from 6–9" in unreinforced blocks and the testing plan included two bare piles. The most proficient performing anchors were straight bars that were welded to the web. The majority of the other anchorage types (cap plates, angles, and bar grids) performed marginally better or worse than the bare pile that was used as a control. They hypothesized that this was due to the fact that these types of anchors only applied load over a limited region near the tip of the pile, thus resulting in lower ultimate capacities. GAI concluded that "the ultimate tension strength of a pile cap-pile anchor system is directly related to the depth of embedment of the tension anchor," rather than the shape of pile. As such, they recommend welded straight or hooked bars be used over anchorage types that do not take advantage of reinforcing bar development length.

Several studies have been conducted on pile-to-pile cap behavior under lateral or simulated seismic loading (*10–12*). These studies largely focus on the moment capacity and load-displacement response with different details in the connection as well as retrofit analysis. Shama investigated the behavior of the pile-to-pier cap under seismic loading as well (*13*, *14*). Pertold recognizes that emphasis has been placed on sections subject to bending moment and highlights the need for more information on the vertical resistance of embedded columns (*8*, *15*). His work isolates the effects of bond and punching resistances in footings. Significant resistance from bond stress is observed and is more than the punching resistance for the given geometry. However, no testing was completed to study the interaction of these two failure mechanisms acting simultaneously.

Heristchian compared the experimental and numerical capacities of three embedded IPE140 sections with and without baseplates. In conjunction with prior research, Heristchian observes that the introduction of the end plate or utilizing tapered sections mobilizes a bearing force that has higher unit stresses than shear and is closely related to the punching resistance of the concrete (*16*, *17*). He concludes that more experimental data and numerical models on pullout behavior are necessary to verify and validate the results.

Preliminary work has been completed on pile-to-pile cap connections under uplift forces. Despite this, questions still remain about the ultimate capacities of specific connections used by various states and the current methodology for designing bare piles subject to uplift. Furthermore, from the NCSC report it is clear that design practice for uplift in the connection is not uniform or very well substantiated. Therefore, the goal of this research is to investigate an often-overlooked portion of substructure design and bring clarity to the process. Design values for current and higher-capacity anchors are recommended, as well as design protocol for bare piles embedded in pile caps and subject to uplift forces.

Experimental Setup and Methods

The focus of this work is on the ultimate capacity of pile-to-pile cap connections and, as such, emulating field conditions was critical. The testing layout and specimen dimensions utilized emphasize this while also seeking to minimize the effect of arbitrary stress introduction at the pile face, which is an important consideration (*16*). The test setup resembles that of Horn 1972 but includes a few modifications/improvements. The pile is cast horizontally between two simulated concrete footings measuring 4' x 4' x 3' and set on rollers. The lower reinforcing bar mat consists of #9 bars both ways spaced at 8.5" and 10" o.c. and the top mat is composed of #5 bars both ways spaced at 12" o.c. Dimensions and reinforcement details were taken from a sampling of representative Iowa projects and conform to the minimum standards given in the Iowa Bridge Design Manual and the AASHTO Bridge Design Manual. Figure 2 shows the layout schematic and detailed dimensions.

A major benefit of the test configuration is that each test produces the lower of two ultimate capacities for the connection being tested (one on each end of the pile)—essentially doubling the data pool. Two 200 kip actuator jacks are set at mid-height between the simulated footings to induce tensile forces on the pile. HP10x57 piles with stiffeners distribute the force as a line load on each side of the pile, resulting in lower unit stresses than the concentrated load forces applied by Horn and GAI (*7, 9*).

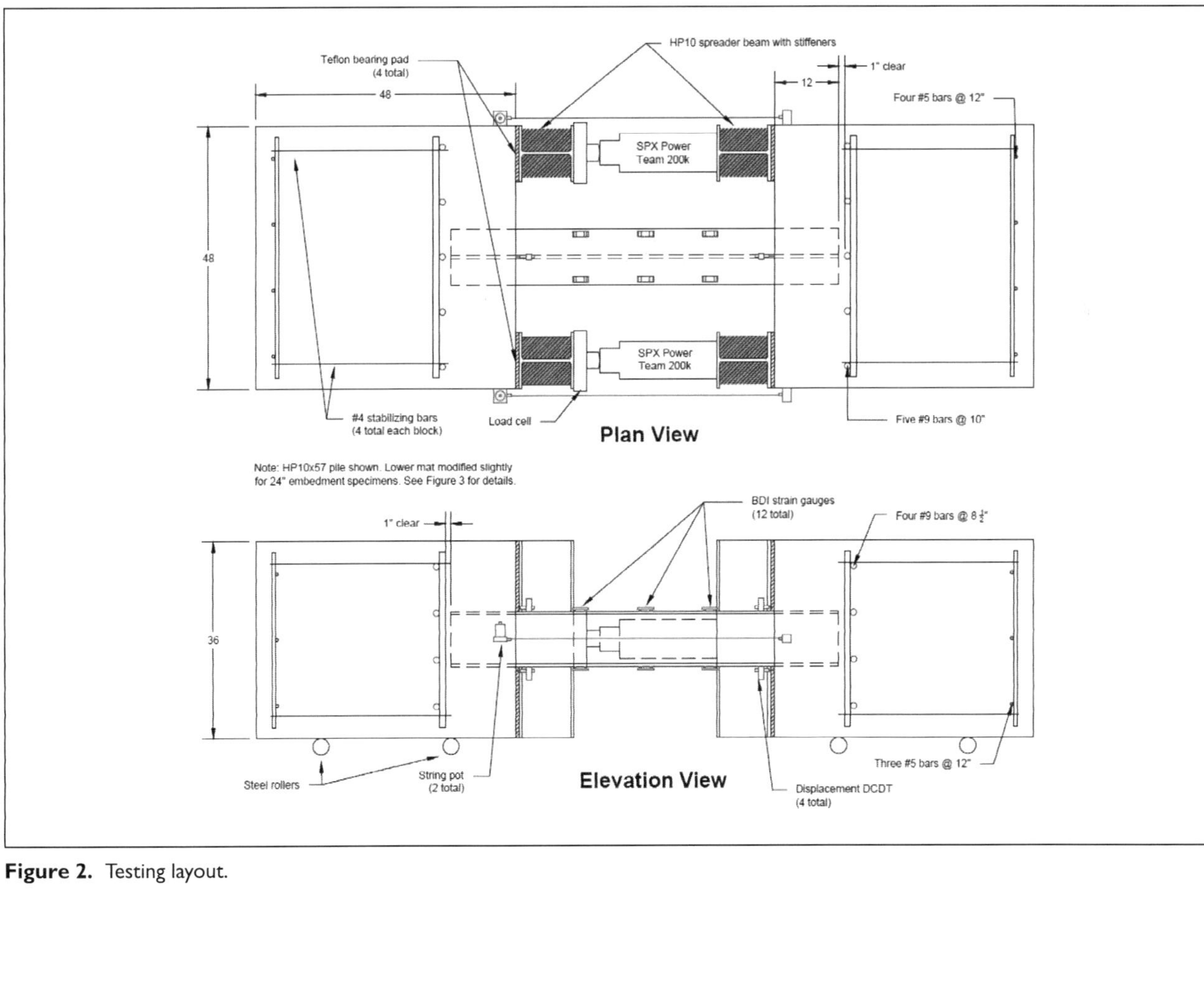

Figure 2. Testing layout.

Table 1. Test Results

Specimen ID	Pile size	Anchor type	f'$_c$ (psi)	f'$_r$ (psi)	Ultimate load (kips)	Slip @ ultimate load (in)
NA1	HP14x73	No anchor	5900	570	107	0.0171
NA2	HP14x73	No anchor	4900	490	114	0.011
NA3	HP14x73	No anchor	4700	510	123	0.0376
NA4	HP10x57	No anchor	6100	560	127	0.0122
NA5	HP10x57	No anchor	4500	490	125	0.0124
NA6	HP10x57	No anchor	4500	490	108	0.0145
TR1	HP14x73	Threaded rod	5900	570	104	0.0131
TR2	HP14x73	Threaded rod	4900	490	112	0.0121
C1	HP10x57	Cage	5900	500	136	0.0206
C2	HP10x57	Cage with bars	5900	500	145	0.0105
E1	HP10x42	24" embedment	5700	560	218	0.0534
E2	HP10x42	24" embedment	6200	610	175	0.017
E3[a]	HP10x57	24" embedment	4600	540	208	0.0146
E4[a]	HP14x73	24" embedment	4300	500	190	0.0090
E5[a]	HP14x73	24" embedment	3600	430	177	0.0088
E6[a]	HP14x73	24" embedment	3600	430	182	0.0052
U1	HP10x42	U-bar (1" drilled holes)	5800	560	277	0.0507
U2	HP10x57	U-bar (1–1 ¼" torched)	7000	680	257	0.0686
V1	HP10x57	V-bar (1 ¼–1 ½" torched)	7000	680	236	0.0385
V2[a]	HP10x57	V-bar (2 ½" torched)	4900	570	203	0.0158
V3[a]	HP10x57	V-bar (2" torched)	4900	570	198	0.0128

[a]Displacement after ultimate load may be suspect.

For all test specimens, a standard Iowa 4C 4,000 psi mix design was used. Twelve 4" × 8" cylinders and three modulus beams were cast for each mix to obtain basic strength properties (results may be found in Table 1). Specimens were cured for at least seven days and obtained a minimum 4,000 psi strength before moving to prevent cracking at the connection (with the exception of E5 and E6, which did not achieve design strength after 28 days).

To understand and quantify the type of loads being induced on the connections, twelve strain gauges were attached to the top and bottom flanges of the pile. From the recorded longitudinal strains, a system of developed equations was used to determine the axial, bending, and torsion stress induced on the pile at each of the three locations. In addition to the strain gauges, displacement was measured at the interface between the pile and footing on the top and bottom flanges with four displacement transducers. Values were averaged for each side and used for subsequent analysis. Two string potentiometers were also utilized to measure the relative movement between pile ends.

Anchorage and Pile Selection

Three different size piles were utilized in this work (HP14x73, HP10x57, and HP10x42) to investigate the effect of size and surface area on the ultimate capacities. Piles were obtained from existing bridges to mimic worst-case material conditions but were visually inspected to ensure straightness and no discontinuities in the embedded portion of the connection. The surfaces of the embedded portions were dry-broomed but were otherwise left unprepared.

Fabrication and testing were completed in groups of two or three specimens, which allowed an iterative approach to be taken in developing the anchorage details. In total, seven different connections were tested with varying degrees of replication. Each connection is discussed briefly below with the correlating detailed sketches given in Figure 3. Design of the anchors was based on published literature, input from Iowa DOT designers, and the research team. Several factors were considered, including effectiveness of connection, ease of installation, no field welds (due to the high variability in quality), and expected ultimate failure. Six piles were embedded 12" with no additional anchorage to serve as control specimens and to evaluate the current bare-pile design methodology. The current approved anchor in Iowa was tested, as well as several piles confined with hoops to increase crack control. Embedment was investigated by increasing the depth to 24" as this is a relatively easy alternative to fabricating positive anchorage. In addition, more involved anchors consisting of U- and V-bars tied to the top mat provided insights into the performance of deeply embedded anchors. In all tests of details that had torched or drilled holes, wood or plastic spacers were installed between the bar and the top of the hole to simulate poor construction quality.

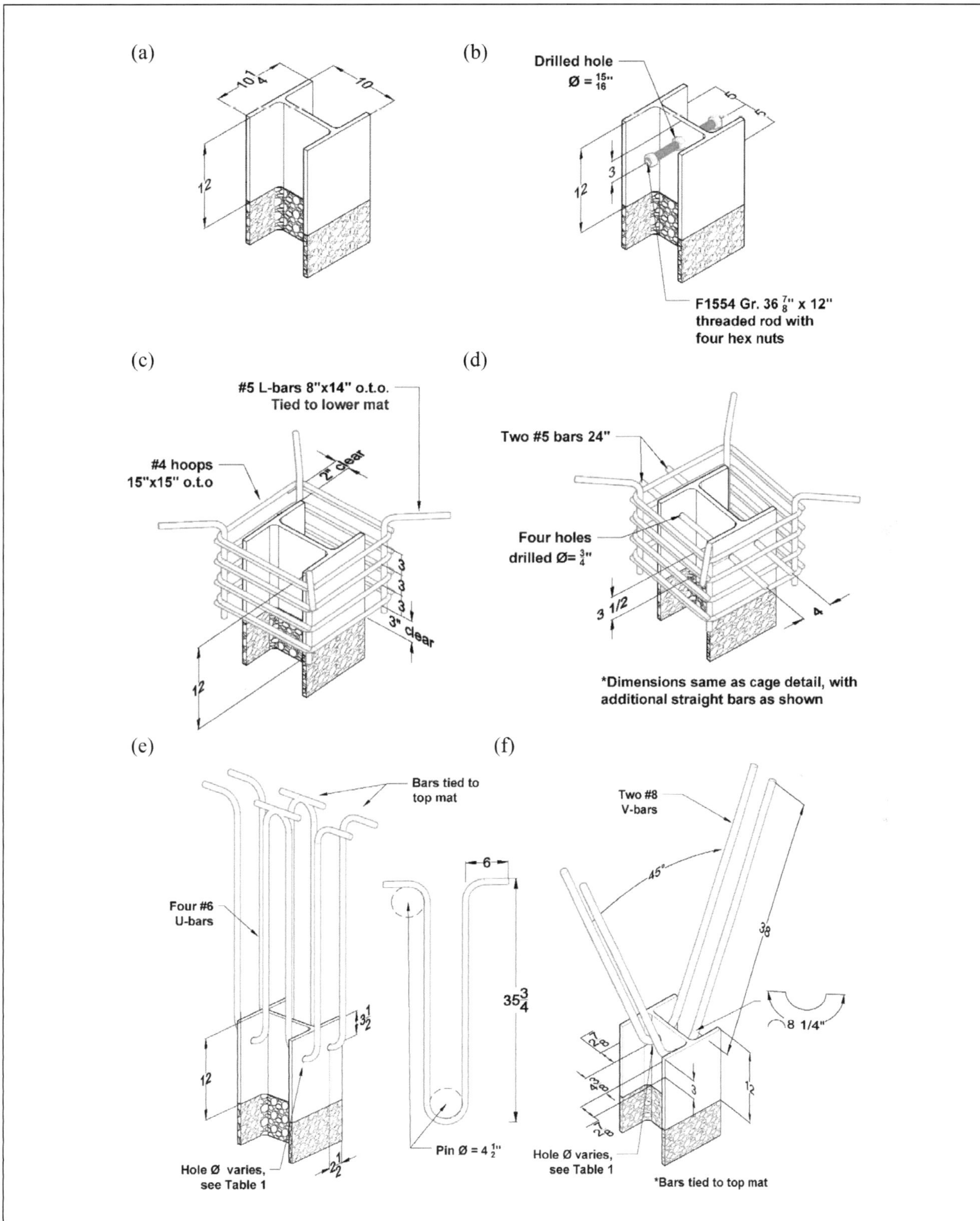

Figure 3. Sketch of anchor details (HP10 shown): (*a*) bare pile; (*b*) threaded rod; (*c*) cage; (*d*) cage with bars; (e) U-bar; (*f*) V-bar; and (*g*) 24" embedment.

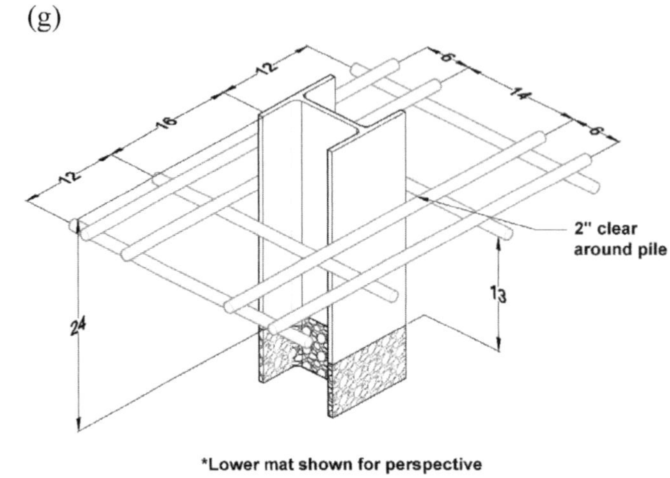

Figure 3. (continued)

Test Results and Discussion

Results from the testing are summarized in Table 1. For completeness, concrete strength properties are included for each specimen, but were not the primary focus of this research. For reference, the relationship between bond and concrete compressive strength for plain bars was considered by Feldman (*17*). Slip data from both ends of the specimen were measured and analyzed; however only data from the block that failed are included in Table 1 and subsequent figures. Owing to the type of cracking observed on some of the specimens, recorded displacements after the ultimate load was attained may not be fully representative of the overall behavior (these specimens are marked with a footnote in Table 1).

Bare-Pile Tests

All six tests with piles embedded 12" into the footing attained substantially higher ultimate capacities than the current design methodology that assumes a 60 psi bond stress (see Figure 4*a*). Size of the pile did not seem to affect the behavior of the connection nor the ultimate capacity for the HP10 and HP14 piles considered here. Initial stiffness values varied and ranged from 50,000–95,000 k/in, and all specimens behaved linearly until 60–70 kips. At that load all specimens had a notable stiffness change. Within each test, displacement between blocks for the most part tracked similarly and then started to deviate between 65 and 90 kips. The overall magnitude of slip at the ultimate load was small and similar for both pile sizes.

The failure was characterized by a sudden formation of diagonal cracks propagating from the pile while load simultaneously dropped to between 0 and 12 kips. Specimen NA3 was able to sustain slightly larger displacements than the other specimens near their ultimate loads, but still displayed a brittle failure. The observed behavior is consistent with previous work (*8*, *9*). A typical failure mode for bare piles is illustrated in Figure 5*a*.

Further investigation revealed that concrete cracking leads to a loss of bond between the pile and concrete, suggesting that design methods based on bond stress may not be accurate. In other words, the concrete is not sufficiently confined, and/or the pile embedded deep enough so as to allow for a pure bond/pullout failure. Fundamental evidence of this fact are the results which show HP14 piles reaching ultimate capacities at or below a HP10 for the same embedment depth, even though HP14s have a 40% larger surface area.

From the testing of the 12" embedded specimens it became apparent that the limiting factor in the connection rests upon crack initiation in the concrete, which in general is impacted by the strength of the concrete, footing geometry, anchorage (if applicable), and the shape and size of section. With respect to strength, Roeder concludes in his study of composite action in circular and square concrete-filled tube (CFT) sections that "bond stress capacity does not appear to be related to concrete strength" (*18*). Based on the apparent lack of a relationship between concrete strength and connection capacity, a similar conclusion might be drawn about the ultimate capacity. For example, specimens NA5 and NA6 were

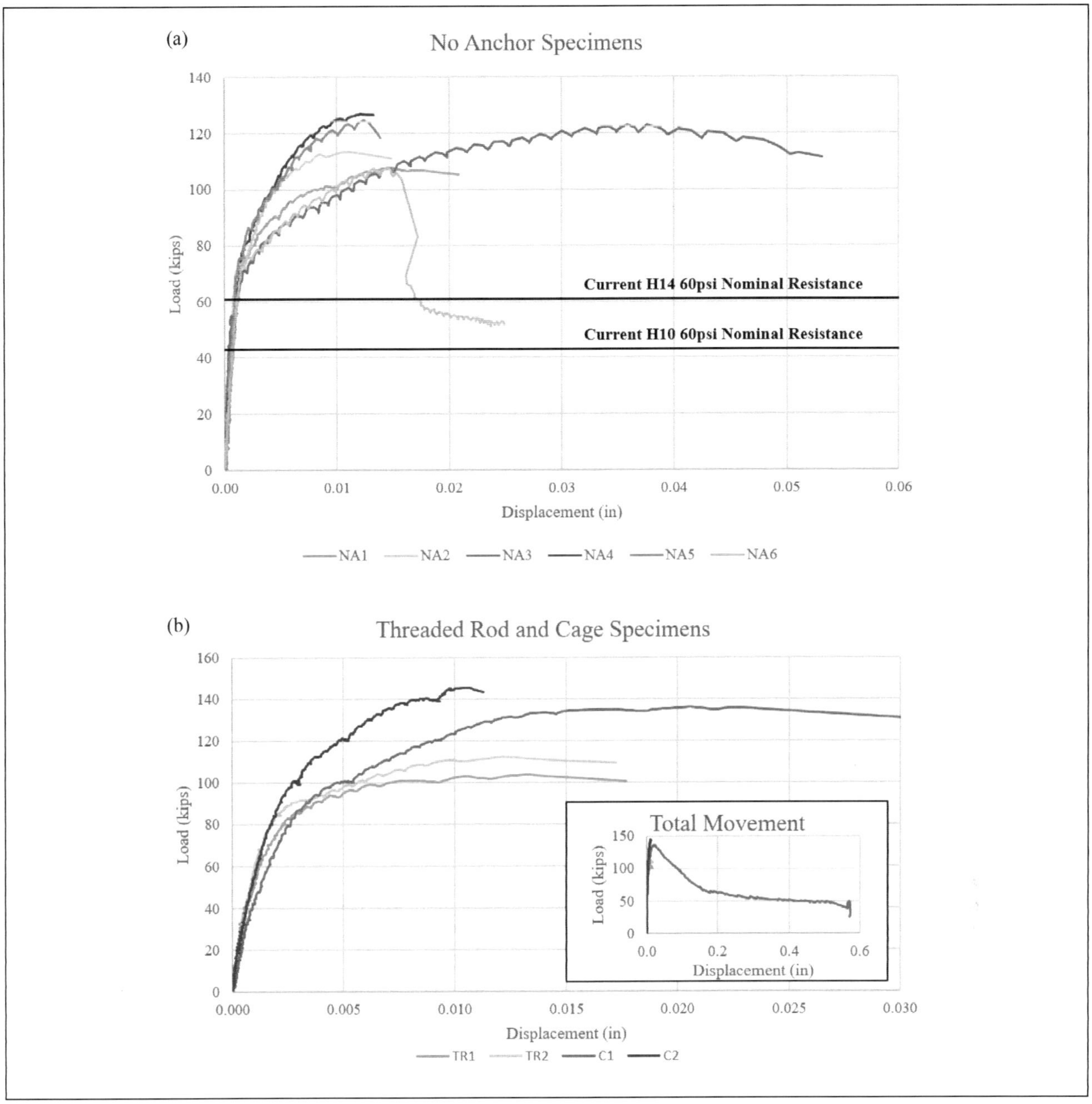

Figure 4. Test results for specimens: (*a*) bare pile and (*b*) threaded rod and cage.

cast from the same ready-mixed concrete truck and have the same strength properties, yet the capacity differs by almost 20 kips. However, because the failure modes more closely resemble punching failure than bond failure, a strong correlation between concrete strength and ultimate capacity would be expected, as Hallgren notes (*19*). The absence of a complementary relationship prompted consideration of the impact of other factors on the connection. Current design equations related to bond, like those put forth by Griffis, only account for the influence of concrete strength (*5*). Likewise, punching-shear-failure design relates to strength, embedment depth, and the critical shear perimeter. A conservative and safe nominal capacity can be drawn from the data presented, but more investigation and research is needed. It is hypothesized that regardless of design protocol, this type of connection carries with it a higher degree of variability which should be reflected during design.

Threaded-Rod and Cage Specimens

Load vs. displacement curves for the threaded-rod and cage specimens are presented in Figure 4*b*. The threaded-rod specimens (TR1 and TR2) behaved similar to the bare-pile tests and achieved marginally lower ultimate capacities than their bare-pile counterparts (NA1 and NA2). The cage and cage-with-bar specimens gave some indication of what impact confining the pile has on connection capacity. No noticeable difference was observed until after the peak load was reached for the cage specimen, highlighted in the lower right-hand portion of Figure 4*b*. The hoops performed adequately, confining the diagonal cracks and allowing the pile to slip out in a more bond-failure type fashion. At ½" displacement the connection still maintained a load of 50 kips. The pile slipped out completely just before 0.6" of slip was reached at about 45 kips. With the additional bars through the flanges, specimen C2 obtained a slightly higher capacity but failed very suddenly. Concrete around the pile was confined, but a large shear plane along the hoops extending to the lower mat caused failure in the connection.

While confining the concrete increased the capacity of the connection to some extent and provided helpful insights, the capacity gain was small for the complexity of the design details. In addition, all specimens including the bare piles failed to engage the lower mat steel, as evidenced by cracking at the depth of the mat/pile. With the exception of specimen C1, anchorage only succeeded in pulling out a larger area of concrete around the pile, and consequently more closely resembles punching failure of a footing. These data are consistent with other anchors tested by GAI Consultants (cap plates, angles, and bar grids) and further supports the hypothesis that anchorage whose primary resisting mechanism is bearing near the tip of the pile is not effective and may actually be detrimental to the connection (*7, 16*). Pertold's findings agree with this as he states that the "punching resistance for a certain geometry of the concrete base could be less efficient at resisting the vertical load from column than bond resistance" (*8*). Based on these findings, it appears that anchorage design should prioritize embedment depth and consider crossing the lower mat in order to engage the steel and the rest of the footing.

Connections with Deeper Embedment Details

Results from the specimens incorporating deeper embedment performed significantly better than the previously discussed specimens, as shown in Figure 6. The 24" embedded specimens demonstrated a wide range of ultimate capacities bookended by two of the three HP10 tests. Specimen E1 had the highest capacity of 218 kips as shown in Table 1. Specimen E2 exhibited behavior most closely associated with pure bond pullout failure, with minimal cracking and very large displacements. At termination of the test, the load was 125 kips with a measured slip of 5" (instrumentation was only accurate to displacements equal to 1 3/8" in magnitude; final measurements recorded by tape), as shown in Figure 6*a*. The ability of the connection to sustain significant load after initial movement can most likely be attributed to the confinement provided by the lower mat of #9 bars. Characteristic of the three tests on the HP14 piles was a tension crack that formed at the center of the face after the peak load was attained which caused the concrete to spall outward away from the pile (Figure 5*c*).

Concrete strength and stiffness of the connection did not correlate with the measured ultimate capacities that resulted. All 24" embedment specimens failed in a ductile manner except one (E1). Performance seems to rest largely on concrete strength and the ability of the lower mat to confine the pile. Additionally, consideration of punching failure through the top of the footing due to compression should be accounted for in piles that extend above the lower mat. Pestotnik recommends at least 1' 6" of clear distance between the tip of the pile and the top of the footing (*20*). Hallgren's research demonstrates that footings without vertical shear reinforcement still have considerable resistance to punching failure (*19*).

Anchorage employing V- and U-bars extending to the top reinforcing steel mat produced the highest strength connections (Figure 6*b*). In Specimen U1 initial diagonal cracking was observed at 180 kips. The connection then resisted a peak load of 277 kips, well exceeding the theoretical yield strength of the bars, while demonstrating resistance against failure due to concrete cracking (Figure 5*d*). Strain hardening was observed in the bars and subsequent failure of the connection occurred when all four bars fractured simultaneously at the flange-hole interface.

Observations can be made concerning hole construction method, size, and placement in the web from specimens U2 and V1-3. U2 holes were torched as specified previously. The overall capacity was about 20 kips less than U1, but as Figure 6*b* shows the steel did not engage uniformly, and as a result bars fractured at four distinct points. Torched holes were incorporated in all V-bar specimens because of constructability concerns with drilling in the web of smaller piles. Specimen V1 bars also did not engage uniformly due to torching, but still attained substantial peak load and failed due to reinforcing steel shear at the hole interface. V2 and V3 were fabricated with "large" torched holes. No failure of the rebar was observed in contrast to the other specimens, although the large jump in displacement near the end of the test could indicate a fracture of one of the bars. After this point the load began to increase, suggesting at least one

Figure 5. Failure mode results: (*a*) typical failure mode for bare and threaded-rod specimens; (*b*) bond failure in specimen E2; (*c*) tension crack typical in HP14 specimens with 24" embedment; and (*d*) U1 and U2 typical cracking.

bar remained attached to the pile to provide resistance. Testing was terminated shortly after because the concrete had sustained excessive cracking which compromised the test setup. All three V-bar specimens still surpassed the yield strength of the bars. As long as there is sufficient bond and development length is ensured, testing demonstrated that these types of connections are perhaps the most efficient and reliable. If hole size is controlled, the failure mechanism is transferred from concrete cracking to steel bar fracture, which may be preferable.

Concerning design strength, even though bars in these types of connections failed in shear at the hole interface rather than in tension in the legs, a good lower bound metric for estimating the capacity of the connection seems to be the total area steel embedded (i.e., number of bar legs times area of one bar times bar yield strength). More experimental and numerical studies with different bar size and quantity included are necessary in order to develop models to more accurately estimate the capacity anchorage details utilizing embedded bars.

Summary and Conclusions

A lack of published literature and codified design provisions pertaining to pile-to-pile cap connections under uplift loading, coupled with survey results from the

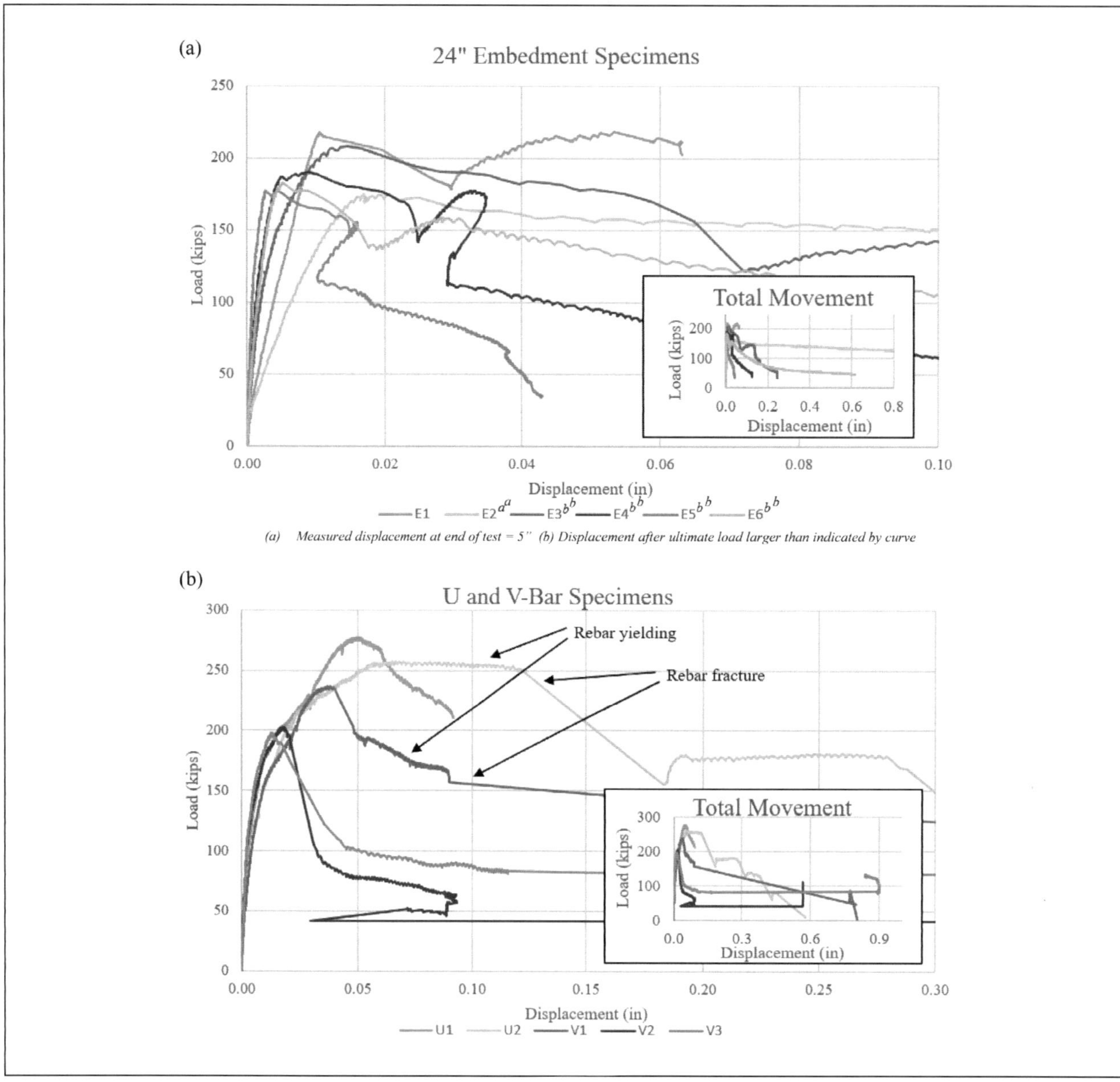

Figure 6. Test results for specimens with deeper embedment: (*a*) 24" Embedment and (*b*) U- and V-bar.

NCSC, and an ever-increasing number of designs facing uplift situations, indicates a need for further investigation of the performance of various uplift connection details. The goal of this work is to provide information on cost-effective and safe alternatives for the pile-to-pile cap connection. In order to do this, twenty-one full-scale steel H-piles were embedded in simulated pile caps and subjected to tensile forces replicating uplift conditions in bridge substructure systems. A variety of anchorage types were tested with the pile embedded 12" into the footing, as well as bare piles embedded both 12" and 24". From the investigation, it can be concluded that:

1) For bare piles embedded 12", the size of the pile (HP10 or HP14) did not have an impact on the ultimate capacity of the connection. Despite several NCSC respondents indicating that their respective agencies do not allow uplift at all, six tests performed with this configuration indicate that nominal ultimate capacities of 104 to 127 kips can be expected. Current methods for estimating bond by the Iowa DOT underestimate the connection capacity by an average of 190% for HP14s and 275% for HP10s. It is believed that the underestimation is the result of an

incompatibility of the design methodology with the actual observed failure mode.

2) Failure modes observed from testing bare piles embedded 12" indicate that concrete cracking leads to loss of bond between the pile and concrete.
3) The results from the threaded rod and cage with bar specimens support published conclusions that C-channels, angles, cap plates, and bar grids seem to produce stress concentrations near the tip of the pile which can lead to lower ultimate capacities than connections without such apparatuses.
4) Utilizing V- and U-bar anchor configurations provides a rational design approach because the capacity of the connection can be calculated from the area of embedded reinforcing steel. The capacity of connections utilizing reinforcing steel extending into the footing have higher capacities than other types of connections.
5) Torching holes (as opposed to drilling) does not appear to have a significant effect on the overall capacity of the connection. However, size of the hole and level placement in the web of the pile is critical to achieving full capacity.

Implementation

It is recognized that across state agencies design methodologies, construction practice, and loading conditions vary, sometimes substantially. Despite this fact, the applicability of this research is clear and the direct implementation of the results, combined with a good working knowledge of individual agency practice, provide safe and cost-effective solutions to an oft overlooked portion of bridge substructure design. Therefore, it is recommended that the following be considered:

1) For bare piles embedded 12" into the footing, design should not follow previous design methodologies that assume a 0.060 ksi bond strength between the pile and concrete. Rather, a nominal resistance of 100 kips should be used for HP10s and larger. Resistance factors should be taken as:

$\varphi = 0.25$ (for strength limit states)
$\varphi = 0.40$ (for extreme limit states)

2) When the nominal resistance of 100 kips provided by 12" of pile embedment is not sufficient, anchorage should consist of V-bars, U-bars, or by increasing the embedment depth to 24". When installing V- or U-bars, details should follow the dimensions and layout provided in Figure 3. Holes can be torched or drilled ¼" larger than the nominal bar diameter. Hole size must be closely controlled during construction. For standard V-bar connections using #8 bars, a nominal resistance of 87.5 kips for each V-bar shall be assumed (totaling 175 kips for two bars), which incorporates the following resistance factors:

$\varphi = 0.25$ (for strength limit states)
$\varphi = 0.40$ (for extreme limit states)

The standard details for U-bars using #6 bars should be designed with a nominal resistance of 50 kips per U-bar (totaling 200 kips for four bars). Development length must be achieved in these connections if the nominal capacities put forth are to be used. If 24" embedment is utilized, a nominal resistance of 175 kips should be used. Designers must also ensure that the bottom mat reinforcement in both directions is detailed so as to clear the piles by 1.5–2" on all sides. Additionally, the bottom mat reinforcement shall not be place more than 13" above the bottom of the footing. Resistance factors for both the U-bar detail and 24" embedment shall follow the ones put forth for V-bars.

3) When the 24" embedment detail is used, designers must ensure the depth of footing concrete above the tops of the piles is deep enough to prevent punching failure of pile through the top of the footing under compressive loads. If the minimum requirement is not met it is recommended that V- or U-bars be used instead.
4) Do not use anchor details that that incorporate C-channels, angles, clips, cap plates, bar grids, and other shallow reinforcement where the primary resisting mechanism is bearing and not development length.

Acknowledgments

The authors of this research would like to thank the Iowa DOT for funding this work and for the expertise that the Office of Bridges and Structures offered over the course of this investigation. Additionally, a large thank you is extended to Doug Wood, laboratory manager for Iowa State University Structures Laboratory, and Owen Steffens for their help in the fabrication process.

Author Contributions

The authors confirm contribution to the paper as follows: study conception and design: Mike Nop, Brent Phares, Philip Iekel; data collection: Brent Phares, Philip Iekel; analysis and interpretation of results: Brent Phares, Philip Iekel; draft manuscript preparation: Mike Nop, Brent Phares, Philip Iekel. All authors reviewed the results and approved the final version of the manuscript.

References

1. American Association of State Highway and Transportation Officials. *AASHTO LRFD Bridge Design Specifications*, 7th ed. Washington, D.C., AASHTO, 2014.
2. Iowa Department of Transportation. *Highway Plans Collection.* Iowa Department of Transportation, Ames, Iowa. www.mydotdocs.iowadot.gov/Home.aspx. Accessed June 3, 2017.
3. Office of Bridges and Structures, Methods Section. *LRFD Bridge Design Manual.* Iowa Department of Transportation, Ames, Iowa. www.iowadot.gov/bridge/design-policies/bridge-design-manual. Accessed June 20, 2017.
4. American Institute of Steel Construction. *Steel Construction Manual*, 14th edition. American Institute of Steel Construction, Chicago, Ill., 2011.
5. Griffis, L. G. *Load and Resistance Factor Design of W-Shapes Encased in Concrete.* American Institute of Steel Construction, Chicago, Ill., 1992.
6. Watson, J., R. O'Neil, R. Barnoff, and E. Mead. Composite Action without Shear Connectors. *Engineering Journal*, Vol. 11, 2nd Quarter, 1974, pp. 29–33.
7. GAI Consultants, Inc. *The Steel Pile, Pile Cap Connection.* American Iron and Steel Institute, Washington, D.C., 1982.
8. Pertold, J., R. Y. Xiao, and F. Wald. Embedded Steel Column Bases: I. Experiments and Numerical Simulation. *Journal of Constructional Research*, Vol. 56, No. 3, 2000, pp. 253–270.
9. Horn, M. W., C. F. Stewart, and R. L. Boulware. *Capacity of Pile Anchors.* Publication CA-HY-BD-4162-1-72-12. California Division of Highways, Sacramento, CA, 1972.
10. Rollins, K. M., and T. E. Stenlund. *Laterally Loaded Pile Cap Connections.* Publication UT-10.16. Utah Department of Transportation Research Division, Salt Lake City, UT, 2010.
11. Xiao, Y., H. Wu, T. T. Yaprak, G. R. Martin, and J. B. Mander. Experimental Studies on Seismic Behavior of Steel Pile-to-Pile-Cap Connections. *Journal of Bridge Engineering*, Vol. 11, No. 2, 2006, pp. 151–159. DOI: 10.1061/(ASCE)1084-0702(2006)11:2(151).
12. Shama, A. A., J. B. Mander, and A. J. Aref. Seismic Performance and Retrofit of Steel Pile to Concrete Cap Connections. *ACI Structural Journal*, Vol. 99, No. 1, 2001, pp. 51–61.
13. Shama, A. A., J. B. Mander, B. A. Blabac, and S. S. Chen. Seismic Investigation of Steel Pile Bents: I. Evaluation of Performance. *Earthquake Spectra*, Vol. 18, No. 1, 2002, pp. 121–142.
14. Shama, A. A., J. B. Mander, and S. S. Chen. Seismic Investigation of Steel Pile Bents: II. Retrofit and Vulnerability Analysis. *Earthquake Spectra*, Vol. 18, No. 1, 2002, pp. 143–160.
15. Pertold, J., R. Y. Xiao, and F. Wald. Embedded Steel Column Bases: II. Design Model Proposal. *Journal of Constructional Research*, Vol. 56, No. 3, 2000, pp. 271–286.
16. Heristchian, M., P. Pourakbar, S. Imeni, M. Reza, and A. Ramezani. Ultimate Tensile Strength of Embedded I-Sections: A Comparison of Experimental and Numerical Results. *International Journal of Advanced Structural Engineering*, Published Online, 2014. DOI: 10.1007/s40091-014-0077-y.
17. Heristchian, M., M. Motamedi, P. Pourakbar, and A. Fadavi. Tensile Behaviour of Tapered Embedded Column Bases. *Asian Journal of Civil Engineering*, Vol. 15, No. 4, 2014, pp. 485–499.
18. Roeder, C. W., B. Cameron, and C. B. Brown. Composite Action in Concrete Filled Tubes. *Journal of Structural Engineering*, Vol. 125, No. 5, 1999, pp. 477–484.
19. Hallgren, M., S. Kinnunen, and B. Nylander. Punching Shear Tests on Column Footings. *Nordic Concrete Research*, Vol. 21, 1998, pp. 1–22.
20. Pestotnik, C. *Design Criteria for Piers.* Iowa Department of Transportation, Ames, Iowa, 1979, pp. 25–26.

The Standing Committee on Foundations of Bridges and Other Structures (AFS30) peer-reviewed this paper (18-04200).

Article

TRR
JOURNAL OF THE TRANSPORTATION RESEARCH BOARD

Transportation Research Record
2018, Vol. 2672(52) 291–306

DOI: 10.1177/0361198118776781
journals.sagepub.com/home/trr
SAGE

A Load-Testing Program on Large-Diameter (66-Inch) Open-Ended and PSC-Instrumented Test Piles to Evaluate Design Parameters and Pile Setup

Md. Nafiul Haque[1], Murad Y. Abu-Farsakh[1], Chris Nickel[2], Ching Tsai[3], Jesse Rauser[2], and Zhongjie Zhang[1]

Abstract

This paper presents the results from a pile load testing program for a bridge construction project at Chalmette, Louisiana. The load testing includes three 66-in. spun-cast post-tensioned open-ended cylinder piles and one 30-in. square prestressed concrete (PSC) pile driven at four different locations along the bridge site in clayey-dominant soil. Both cone penetration tests and soil borings/laboratory testing were used to characterize the subsurface soil conditions. All test piles (TP) were instrumented with strain gauges to measure the load distribution along the length of the TPs and to measure the side and tip resistances, separately. Dynamic load tests (DLT) were performed on all TPs at different waiting periods after pile installations to quantify the amount of setup (i.e., increase in pile resistance with time). Case Pile Wave Analysis Program (CAPWAP®) analyses were performed on the DLT data to calculate the resistance distributions along the TPs. A static load test was performed only on the PSC pile and statnamic load tests (SNLT) were conducted on both pile types. Design parameters such as the total stress adhesion factor, α, and the effective stress coefficient, β, were back-calculated. The α values ranged from 0.41 to 0.86, and the β values ranged from 0.13 to 0.29. The load test results showed that SNLT overestimated the tip resistance as compared with dynamic and static load tests. Moreover, the pile tip resistance was almost constant during the testing period, and setup was mainly attributed to increase in pile side resistance with time.

Open-ended (OE) cylinder piles are widely used for bridge foundations construction, as they provide superior moment resistance and high axial capacity (*1–5*). Small-diameter (14″ or less) OE piles are generally observed to plug early in the driving process and are thought to behave in a similar way to closed-end piles. At larger diameters, however, OE piles usually "cut" through the soil instead of plugging, resulting in a column of soil within the pile whose height approaches the pile embedment depth. This transition to large-diameter OE piles from prestressed concrete (PSC) and small-diameter closed-ended piles have brought to focus some of the gaps or inadequacies in current pile design procedures by American Association of State Highway and Transportation officials and American Petroleum Institute. However, very few full-scale field studies were conducted and reported on large-diameter OE cylinder piles to examine their behavior during driving, load testing, soil setup process, and predicting the axial capacity by total stress and effective stress methods.

During pile driving, the surrounding soil is displaced predominantly radially along the side and vertically and radially beneath the tip, thus generating significant amount of excess pore water pressure (PWP). In addition, the soil within the vicinity of pile face loses its strength as a result of an increase in excess PWP, disturbance of the soil structure, and the soil remolding. As the excess PWP starts to dissipate, the effective stress of

[1]Louisiana Transportation Research Center, Louisiana State University, Baton Rouge, LA
[2]Department of Transportation and Development, Baton Rouge, LA
[3]Ardaman and Associates, Inc., Baton Rouge, LA

Corresponding Author:
Address correspondence to Murad Y. Abu-Farsakh: cefars@lsu.edu

the disturbed soil starts to increase, and consequently setup primarily occurs because of the increase in shear strength and the increase in lateral stresses against the pile (*6–8*). Therefore, pile setup phenomenon is mainly attributed to three main mechanisms: (1) dissipation of excess PWP (consolidation), (2) thixotropic effect, and (3) aging effect (*9*).

Thixotropic effect (or regaining of soil strength of disturbed soil with time) plays a significant role at the early stage of setup, along with consolidation. Any setup that occurs after the completion of excess PWP dissipation (i.e., consolidation) is mainly caused by the "aging" effect (i.e., time-dependent change in soil properties at a constant effective stress). This is caused by the combined effect of secondary compression (creep), particle interference, and clay dispersion. Aging effects increase the soil's shear modulus, stiffness, and dilatancy, and reduce the soil's compressibility (*10*).

Setup was observed in various soil types from cohesive to cohesionless soils (*6*, *8*, *9*). Soft clays usually exhibit more setup than stiff clays. Loose sands and silts have been found to setup somehow similar to soft clays (*9*, *10*). Long et al. determined that although the largest setup occurred in the first 10 days after driving, setup appeared to continue for up to 500 days, and measured side resistances increased two times compared with end of driving (EOD) side resistance (*11*). An increase in pile resistance of up to 100% has been reported in non-cohesive soil. Koutsoftas reported a 125–150% increase of pile resistance in dense sand (*12*).

Test piles (TPs) are usually instrumented with strain gauges to evaluate the load distribution along the pile length. Design parameters, such as the total stress adhesion factor, α, and the effective stress factor, β, that are typically used to estimate the side resistance of the pile can be back-calculated from the measured load distribution of the instrumented pile. Not enough studies have been conducted to recommend the α and β values for large-diameter OE piles, as the instrumentation and performing load testing on them are expensive.

The total stress adhesion factor, α, which is related to the undrained shear strengths (S_u) of clayey soils can be calculated from the method proposed by Tomlinson (*13*) as follows:

$$\alpha = f_s/S_u. \quad (1)$$

where f_s = unit side resistance, and S_u = undrained shear strength. For cohesive soils, the value of α ranges from 0.4 to 1.0 (*14*). The β-method or effective stress method, which relates friction resistance to the effective overburden pressure, was first proposed by Burland (*15*) and is evaluated using the following relationship:

$$\beta = f_s/\sigma'_v \quad (2)$$

where, β is the effective stress coefficient and σ'_v is the effective overburden stress. Many researchers (*16*, *17*) reported the back-calculated values of the β-coefficient, indicating that the value of β ranged from 0.15 to 0.35 for clays and ranged from 0.30 to 0.90 for sands (*17*).

The Louisiana Department of Transportation and Development (LA DOTD) has conducted a load testing program on large-diameter OE cylinder piles at Rigolets Pass Bridge site. This study consisted of driving three 66-in. instrumented OE cylinder piles and an instrumented 30-in. PSC pile at four different locations along the bridge site in Chalmette, Louisiana in clayey-dominant soil to conduct several load tests after driving. This paper provides the results and analyses of this experimental study, which addresses the findings from dynamic load tests (DLTs), static load tests (SLTs), and statnamic load tests (SNLTs).

Objectives

Improving the ability to predict the static resistance during design and to include the resistance from setup of OE cylinder piles is an important task. The current available methods to predict setup and design can be improved further with field load testing data that includes accurate measurements of total resistance, side and tip resistances separately; and load settlement. To achieve these aforementioned concerns, the specific objectives of this study were: (a) instrument the TP with strain gauges to measure the side and tip resistances separately during load testing, (b) perform the laboratory and in-situ testing to characterize the subsurface soil condition, (c) conduct the load tests (i.e., DLT, SLT and SNLT) to measure the amount of setup with time, (d) calculate the rate of increase in pile resistance or setup, and (e) back-calculate the design parameters (α and β) for individual soil layers along the length of the piles.

Project Description and Soil Condition

Test Site

This state project consisted of the construction of the new Rigolets Pass Bridge located in Orleans Parish along route US 90 in Chalmette, Louisiana, which is located northeast of New Orleans (Figure 1*a*). This waterway (pass) connects Lake Pontchartrain with the Gulf of Mexico. The existing bridge was built in 1931 and consisted of two narrow traffic lanes along with a swing span to control major waterway traffic (Figure 1*b*). The bridge was damaged by Hurricane Katrina in August 2005, and required major repairs. In order to safely improve roadway and waterway traffic flow, a new two-lane high-rise structure was built west of the existing structure (Figure 1*c*). The new structure consists of 63

Figure 1. Rigolets Pass Bridge. (*a*) Aerial view of the project, (*b*) Old bridge, (*c*) Statnamic load test, (*d*) New bridge.

bents with a total bridge length of 5489-ft (Figure 1*d*). The bridge reaches 72-ft. at its highest point and allows marine vessels to travel underneath it, unlike the bridge it replaced.

Test Piles

Four TPs were driven along the bridge to perform the load tests and verify the design methodology. Figure 2 shows the photos of the TPs, instrumentation, and the hammer used for driving the piles. The bridge has two span arrangements. TP-1 was performed for the shorter span bridge owing to smaller load demand, whereas the other test piles were performed for the longer span portion and thus greater load demand. TP-1 was a 30-in. square, PSC pile with a 16.5-in. diameter void cast in the center (Figure 2*a*). It had an as-built length of 57-ft. and was driven to a final tip elevation of approximately −41.5-ft. After installation, approximately 10.7 ft. of the pile was cut off to facilitate the load testing. TP-2 was constructed using nine 66-in. open diameter (OD), 16-ft.-long spun sections and one 66-in. OD, 9-ft.-long section (Figure 2*b* and *c*). TP-3 was constructed using eleven 66-in. OD, 16-ft.-long spun sections and one 66-in. OD, 7-ft.-long section. TP-4 was constructed using eleven 66-in. OD, 16-ft.-long spun sections. These sections were post-tensioned using 56 strands, resulting in an as-built length of 153-ft., 183-ft., and 174.9-ft for TP-2, TP-3, and TP-4, respectively. Each OE cylinder TP section has a wall thickness of approximately 6 inches. To accommodate testing, an 8.08-ft. section was cut from TP-3, which resulted in a tested length of 174.9-ft. The width, length, and hammer info of these four TPs are provided in Table 1.

Subsurface Geotechnical Conditions

Laboratory and in-situ testing programs were performed to characterize the subsurface soil conditions at the locations of the TPs. Boreholes were drilled at the four TP locations and high-quality 7.6-cm (3-in.) Shelby tube samples were extracted at different depths for comprehensive laboratory testing. The laboratory testing program included moisture content, Atterberg limits, and unconsolidated undrained (UU) triaxial tests. The in-situ

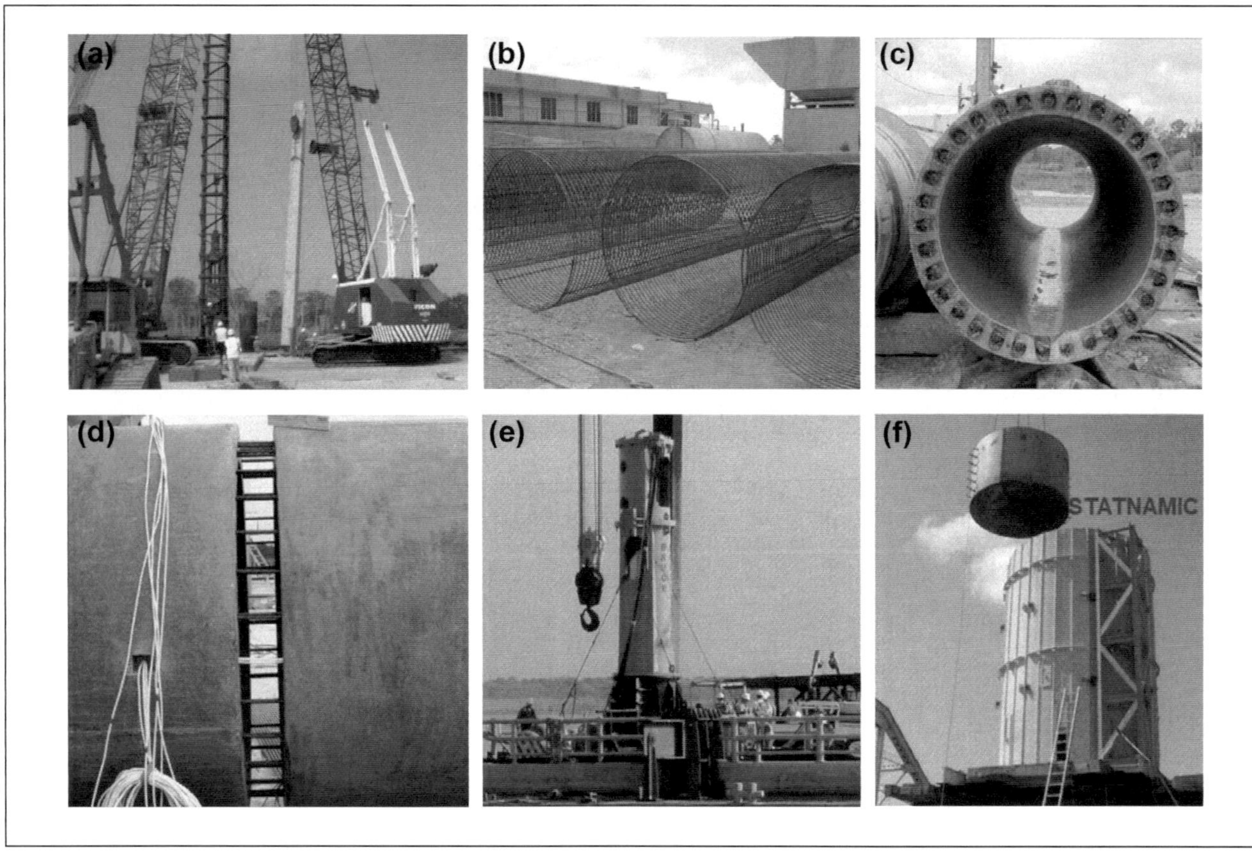

Figure 2. Test piles, instrumentation and statnamic testing. (*a*) TP-1 (30-in. test pile), (*b*) Steel reinforcement, (*c*) 66-in. OE test pile, (*d*) Wires of strain gauges, (e) Bruce hydraulic hammer, (*f*) Statnamic loading system.

Table 1. Test Pile (TP) Information

TP ID	Pile type	Pile width/ diameter inch (mm)	Length ft. (m)		Hammer	Pair of strain gauges	Installation depth of strain gauges (below GL) ft.
			Total	Penetration			
TP-1	PSC	30 (762)	57 (17.4)	41.5 (12.6)	Conmaco 300 E5	5	0, 13.5, 24, 34, 44.5
TP-2	Spun-cast post-tensioned cylinder	66 (1676)	153 (46.6)	119.8 (36.5)	Bruce SGH-3013 hydraulic	3	0, 56, 88.5
TP-3	Spun-cast post-tensioned cylinder	66 (1676)	183 (55.8)	174.9 (53.3)	Bruce SGH-3013 hydraulic	6	0, 51, 82.5, 114, 147, 173.5
TP-4	Spun-cast post-tensioned cylinder	66 (1676)	176 (53.6)	164.3 (50.1)	Bruce SGH-3013 hydraulic	5	0, 3, 51.5, 84, 116, 148

testing program included performing both cone penetration tests (CPT) and standard penetration test near the TP locations. The CPT provided measurements of cone tip resistance (q_c) and sleeve friction (f_s). The profile of CPT tests was used to classify the soil using the Zhang and Tumay probabilistic region estimation method (*18*). Figure 3, *a–d*, presents the soil boring log and CPT profile for TP-1, TP-2, TP-3, and TP-4, respectively. The subsurface soil stratigraphy as revealed from boring and CPT profile showed that the soil at all the TP locations primarily consisted of clay with intermittent sand layers.

Instrumentations

The TPs were instrumented with three to six strain gauge pairs to evaluate the load distribution along the length of

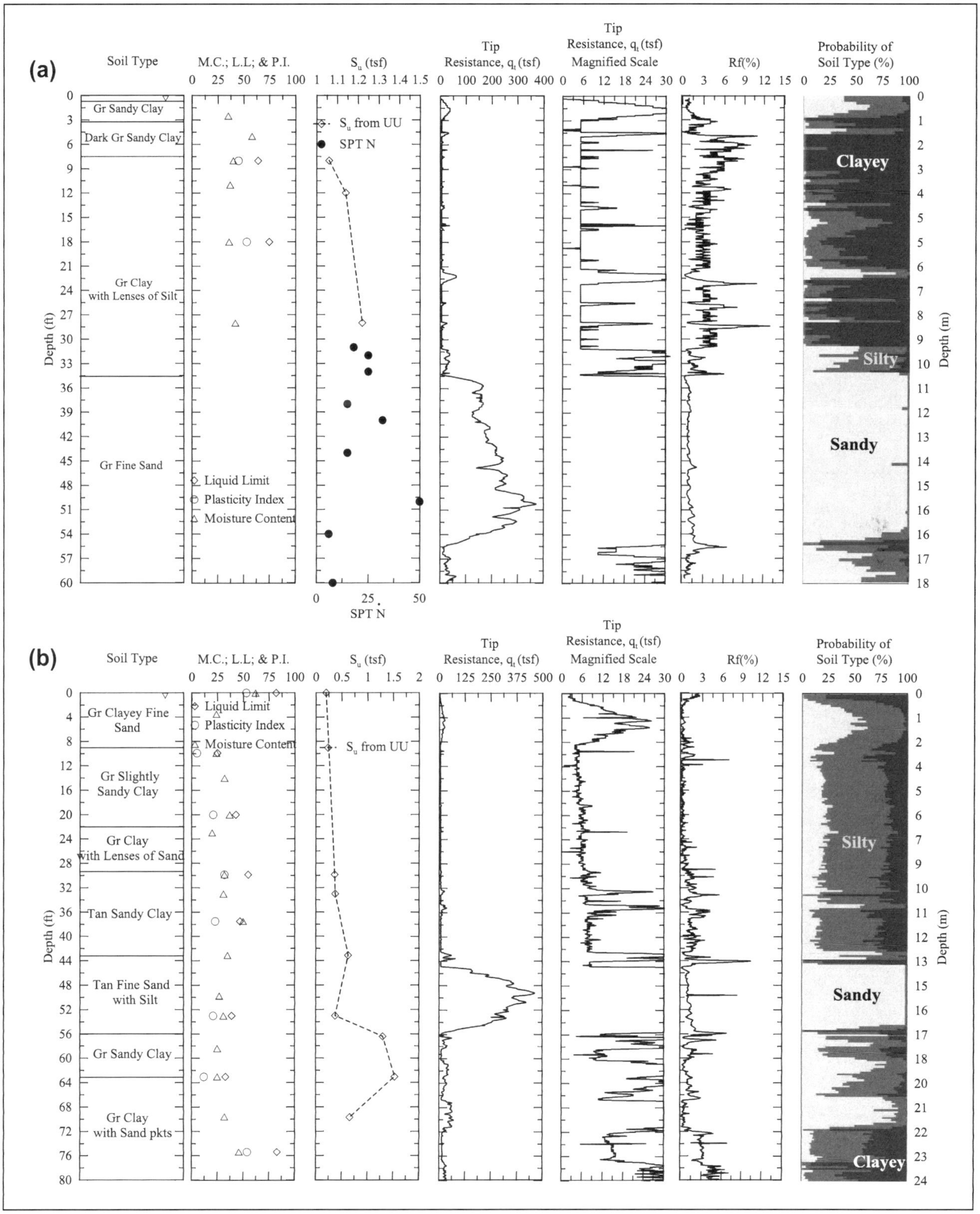

Figure 3. (Continued)

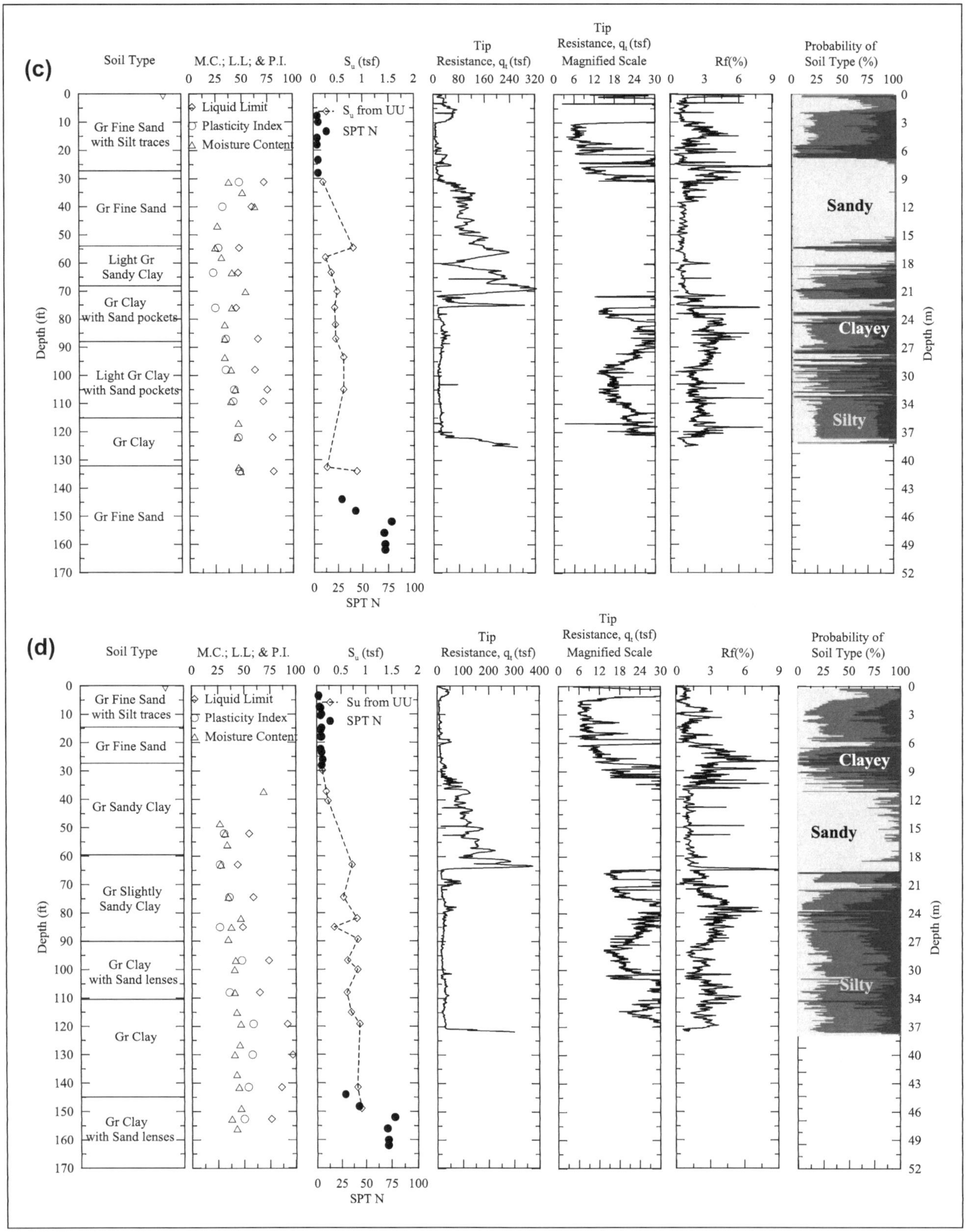

Figure 3. Soil profile, CPT profile and classification at test piles location. (*a*) Laboratory and CPT soil classification of TP-1, (*b*) Laboratory and CPT soil classification of TP-2, (*c*) Laboratory and CPT soil classification of TP-3, (*d*) Laboratory and CPT soil classification of TP-4.

Table 2. Hammer Information

Properties	Hammers	
	Conmaco 300E5	Bruce 8GH 3013
Stroke length (inch)	60	51
Rated striking energy ft.-lbs.	150,000	282,087
Blows per minute (bpm)	40	26
Operating pressure (psi)	150	3917
Weight of hammer (lbs)	58400	125664
Length of hammer (ft.)	20.83	30

the piles and to determine the side and tip resistances, separately. The locations of the strain gauges are presented in Tables 1 and 2. These gauges were manufactured using Micro-Measurements gauge type CEA-06-125UW-350 gauges (Figure 2*d*). All the strain gauges were connected to a MEGADAC data acquisition system manufactured by Optim, Inc. This system monitored the strain gauges, load cell, and accelerometers. Although the system is capable of digitizing and recording at rates of up to 250,000 hertz, for the axial load tests, the data were recorded at the rate of 5,000 hertz for each sensor. Capacitive accelerometers, manufactured by PCB Piezotronics, Inc. were arranged across the top of the pile during SNLT.

Pile Installation

TP-1 was driven with a Conmaco 300 E5 pile driving hammer. The 300E5 has a rated hammer energy of 149.6 ft-kips with a ram weight of 30 kips and a maximum stroke height of 5-ft. The OE cylinder piles (i.e., TP-2, TP-3, and TP-4) were driven using a Bruce SGH-3013 hydraulic hammer (Figure 2*e*). The SGH-3013 has a rated hammer energy of 282.1 kip-ft. with a ram weight of 66-kips and a maximum stroke height of 4.3-ft. Before TP installation, an 8-ft. diameter isolation casing was installed at TP-2 location. The soils inside the casing were removed to an elevation −42.6 ft. for TP-2. The soils in the vicinity of the TP-3 and TP-4 location were removed to an elevation −68.6 ft. and −35.1 ft., respectively. The TPs were then driven to a final pile tip elevation of −119.8 ft., −164.3 ft., and −164.3 ft. for TP-2, TP-3, and TP-4, respectively. The soil inside the OE cylinder piles after driving was encountered at elevation −25 ft., −43 ft., and −22 ft., for TP-2, TP-3, and TP-4, respectively. Hammer cushion reportedly consisted of two layers with a combined thickness of 7 7/8-in. The bottom layer consisted of alternating layers of SUS stainless steel and Klinger material and was overlain by Nylon (PC-MC904). Pile cushion reportedly consisted of 15-in. plywood. Pile capblock assembly weighed 28.6-kips. The pile driving resistances and the results of DLTs for all the TPs are presented in Figure 4.

Load Testing Program

The load testing program was designed to confirm the pile resistance, study the pile setup behavior, and assess drivability. An extensive load testing program was carried out after pile installation using DLTs, SLT, and SNLTs at this construction site. The SLT was performed only on the PSC TP (i.e., TP-1). It was decided that multiple restrikes and one SNLT would be used to study the setup behavior of OE piles (i.e., TP-2, TP-3, and TP-4). The test schedule for conducting DLTs, SLTs, and SNLTs are presented in Table 3. All the side (R_s), tip (R_{tip}), and total (R_t) resistances during each event are presented in Table 3.

Dynamic Load Tests (DLTs)

Restrike events were performed using Pile Driving Analyzer® (PDA®) including two accelerometers and two transducers for PSC piles attached to the pile top. Four accelerometers and four strain transducers were used for the OE piles. PDA® was employed to monitor the driving stresses. The DLTs were performed in accordance with the ASTM-D-4945-89. Four restrikes were conducted on the PSC pile (i.e., TP-1) after EOD to evaluate setup. One DLT was performed on the OE piles to assess the setup. The measured side (R_s), tip (R_{tip}), and total (R_t) resistances during each DLT of all the TPs are presented in Table 3. The setup ratio (i.e., defined as the resistance at a certain restrike or load test over EOD resistance) of side (R_s/R_{so}), tip (R_{tip}/R_{tipo}), and total (R_t/R_{to}) resistances are also calculated and presented in Table 3. The Case Pile Wave Analysis Program (CAPWAP®) analyses of the measured stress and velocity data were performed on the selected EOD blows to determine the initial pile resistance, the side resistance distribution along the length of the pile, and the tip resistance. The first high-energy blow of each restrike event was also analyzed with CAPWAP® to determine the load distributions for restrike resistances. The CAPWAP® analyses of all the TPs along the length of the piles are presented in Figure 4*c–f*. Blow counts and permanent movement (set) of each restrike events are also presented in Table 3. Hannigan et al. (*19*) have shown that a permanent set of less than 0.10-in. may not mobilize the ultimate pile resistances. The permanent set of all TPs during all DLT restrikes was more than 0.10-in. which confirms that full pile resistances most likely had been mobilized.

The CAPWAP® analyses showed that the tip resistance (R_{tip}) was almost constant during the restrikes

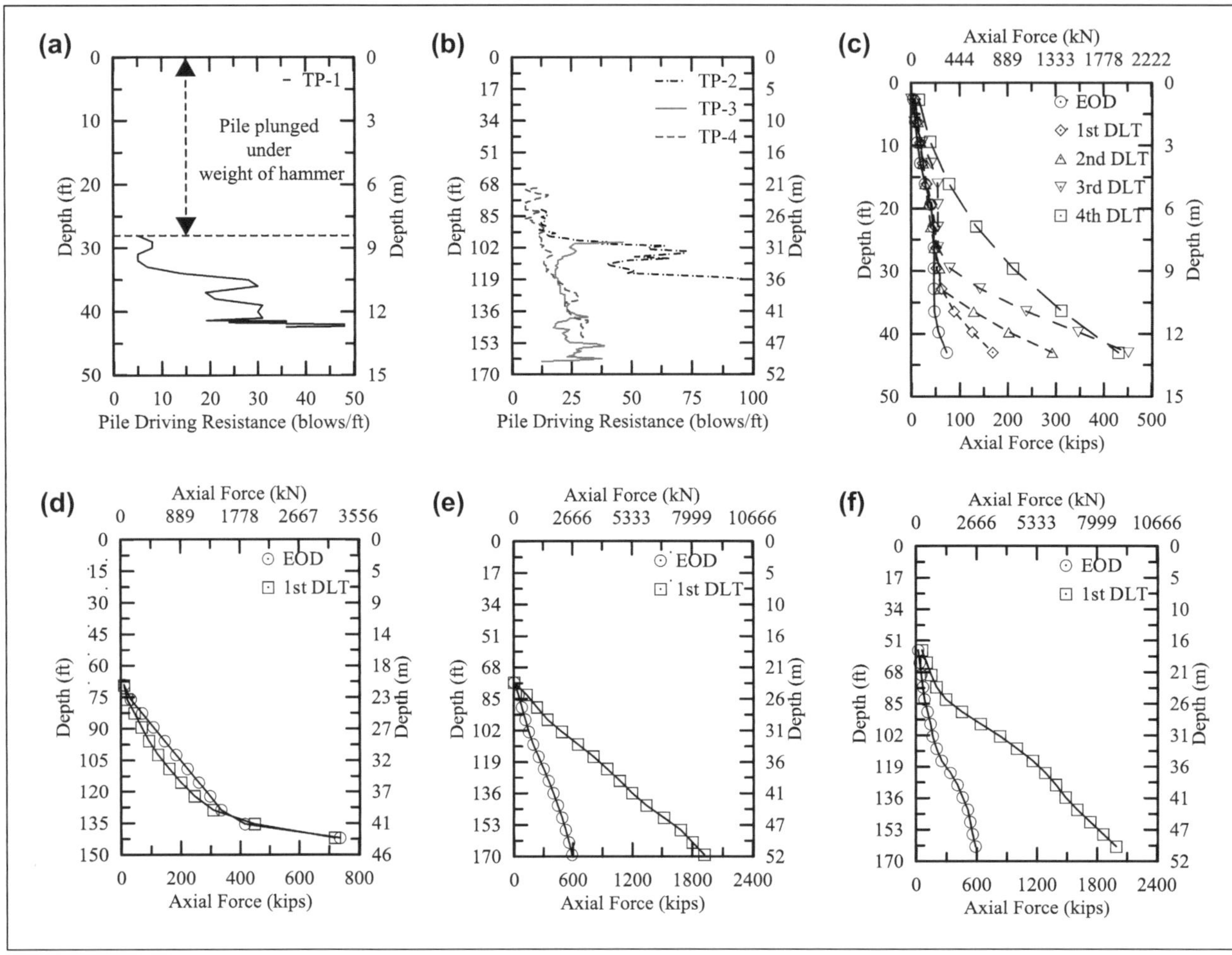

Figure 4. Driving resistance and CAPWAP® analyses of test piles. (*a*) Driving resistance of TP-1, (*b*) Driving resistance of TP-2, TP-3 and TP-4, (*c*) CAPWAP® analyses of TP-1, (*d*) CAPWAP® analyses of TP-2, (*e*) CAPWAP® analyses of TP-3, (*f*) CAPWAP® analyses of TP-4.

Table 3. Load Test Results for Test Piles

			Side resistance, R_s		Tip resistance, R_{tip}		Total resistance, R_t			
Pile ID	Events	Time	kips (kN)	Ratio (R_s/R_{so})	kips (kN)	Ratio (R_{tip}/R_{tipo})	kips (kN)	Ratio (R_t/R_{to})	Blow count (blow/ft.)	Permanent set (in.)
TP-1	EOD	–	52 (231)	1.0	712 (3167)	1.0	764 (3398)	1.0	32	0.38
	1st DLT	23 min	220 (979)	4.2	714 (3176)	1.0	934 (4155)	1.2	27	0.44
	2nd DLT	60 min	274 (1219)	5.3	712 (3167)	1.0	986 (4386)	1.3	27	0.44
	3rd DLT	22 hr	438 (1948)	8.4	714 (3176)	1.0	1152 (5124)	1.5	36	0.33
	4th DLT	15 day	430 (1913)	8.3	630 (2802)	0.9	1060 (4715)	1.4	58	0.21
	SLT	14 day	694 (3087)	13.3	486 (2162)	0.7	1180 (5249)	1.5		
	SNLT	21 day	550 (2446)	10.6	884 (3932)	1.2	1434 (6378)	1.9		
TP-2	EOD	-	737 (3278)	1.0	975 (4337)	1.0	1712 (7615)	1.0	98	0.12
	1st DLT	36 hr	720 (3203)	1.0	525 (2335)	0.5	1245 (5538)	0.7	80	0.15
	SNLT	21 day	1655 (7361)	2.2	1311(5831)	1.3	2966 (13192)	1.7		
TP-3	EOD	-	591 (2629)	1.0	109 (485)	1.0	700 (3114)	1.0	19	0.63
	1st DLT	24 hr	1921 (8545)	3.3	109 (485)	1.0	2030 (9030)	2.9	69	0.17
				4.9						
	SNLT	22 day	2909 (12939)	4.9	168 (747)	1.5	3077 (13686)	4.4		
TP-4	EOD	–	588 (2615)	1.0	118 (525)	1.0	706 (3140)	1.0	31	0.39
	1st DLT	48 hr	1991 (8856)	3.4	125 (556)	1.1	2116 (9412)	3.0	125	0.10
	SNLT	30 day	2760 (12276)	4.7	189 (840)	4.7	2949 (13116)	4.2		

except for TP-2. The side resistance (R_s) increased significantly for all the TPs except for TP-2, which contributed to the setup process that will be discussed later. The situation for TP-2 was complicated by the fact that some damage was noted in the EOD and the following restrike. CAPWAP® modeling suggested a loss of impedance of up to 35% over the bottom 33-ft. of the pile. Obviously, the pile damage could affect the accuracy of calculated results such as pile resistance and maximum tension stresses. However, it appeared that compression forces were transferred through the damage location. Hence, the 36-hour interruption preceding restrike did not exhibit the expected resistance increase owing to pile setup, as compared with the other TPs; instead the soil resistance at the tip apparently relaxed for TP-2.

Statnamic Load Tests (SNLTs)

SNLTs were conducted on all the TPs. The statnamic device consists of a cylinder that supports reaction weights and was placed on top of the pile head. The device and reaction weights were initially supported by a reaction frame constructed around the TP as shown in Figure 2*f*. In this project, SNLT was accomplished with a 4,500 ton Statnamic device equipped with a gravel-catching system. This device used a controlled burn of fuel to generate pressure which was transferred to the test shaft through a load cell. Load, acceleration, and strain were monitored to obtain the foundation response during SNLTs. The pile top acceleration measurements were made with two capacitive accelerometers placed at two points (roughly 180 degrees apart) for TP-1. For the OE piles (i.e., TP-2, TP-3, and TP-4), the pile top acceleration measurements were made with four capacitive accelerometers placed at four points (roughly 90 degrees apart). These measurements for both type of piles showed little to no eccentricity for the loading. From the measured accelerations, pile movements at each accelerometer location were calculated. The segmental unloading point (SUP) method was used to determine SLT equivalent capacities. The SUP discretizes a foundation into segments. The number of segments and their lengths are defined by the locations of the embedded strain gauges. This allows the standard unloading point method to be applied to each segment. Then the total derived static response was calculated as the sum of the derived static responses from the individual segments.

Load-Movement Plots

Figure 5 presents the load versus movement response for the TPs resulting from SLT and SNLTs. A single load cycle was applied using the statnamic device. The equivalent static loads of 1435, 2966, 3077, and 3315 kips were applied for TP-1, TP-2, TP-3, and TP-4, respectively. The total movements during the SNLT at the maximum load were 0.91, 0.82, 1.01, and 1.43 in. for TP-1, TP-2, TP-3, and TP-4, respectively. The SLT was performed only on TP-1 (i.e., PSC pile). The static test was taken to a maximum static load of 1170 kips. The peak movement at the maximum load was 0.62 inches. Although the FHWA failure criterion was not reached for TP-2, the plunging failure occurred at 2,966 kips. It appears that the damage identified by the dynamic testing had an effect on the axial geotechnical resistance of the pile.

Load-Distribution Plots

The main objectives of instrumenting the TPs with strain gauges were to calculate the side and to determine the load distributions along the length of the piles. The strain gauge readings obtained during the load tests were used to estimate the distribution of load transfer along the piles, as presented in Figure 6. To calculate the axial load distribution during load testing, the modulus of concrete was back-calculated for each load test using the strain data measured by the strain gauges that were installed at the ground level. A constant modulus of concrete was used for strain gauge data analysis in each load test. This constant modulus was determined using the strain data measured at the first load increment. The strains recorded at the gauge level near the pile tip comprise both the side and tip resistances of the bottom pile segments. In order to separate the side and tip resistances, the side resistance was extrapolated from the segment immediately above the bottom segment to the tip elevation of the pile, which is then deducted from the measured load to calculate the tip resistance.

Results and Analyses

Side Resistance Mobilization

The interpreted load values against measured movement of strain gauge are presented in Figure 7, *a–d*, respectively. The smooth plots indicate that the strain data are of good quality and are reasonable for further analyses. Figure 8, *a–d* present the calculated unit side resistance (f_s) between gauge levels versus measured strain gauge movement for TP-1, TP-2, TP-3, and TP-4, respectively. The f_s values were determined by subtracting forces between adjacent levels and dividing by the respective surface area (pile circumference × distance between the gauge levels) as the test progressed. The figures ensured that the side resistances were fully mobilized.

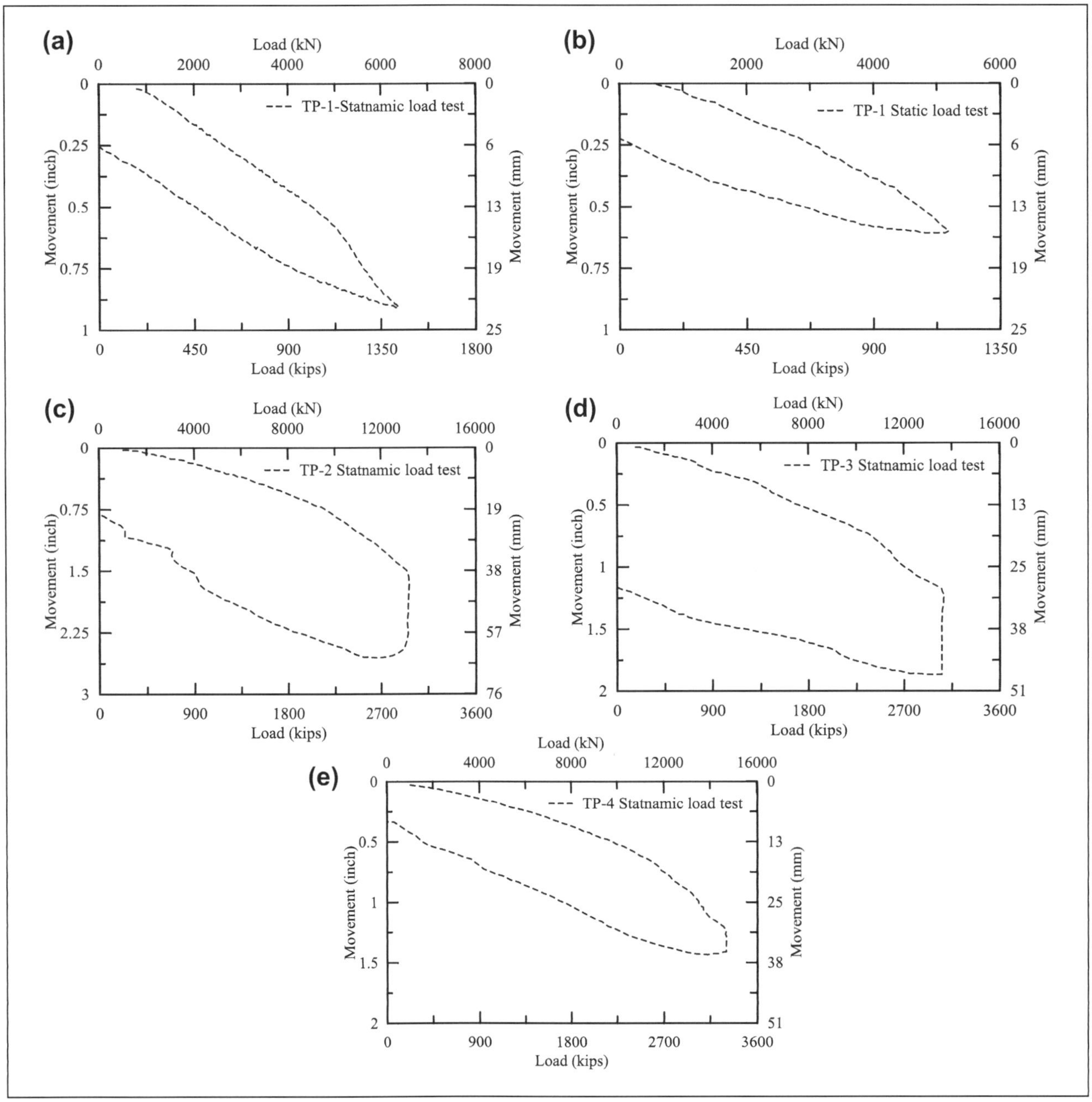

Figure 5. Load-movement plots from static and statnamic load tests. (*a*) TP-1 Statnamic load test, (*b*) TP-1 Static load test, (*c*) TP-2 Statnamic load test, (*d*) TP-3 Statnamic load test, (*e*) TP-4 Statnamic load test.

Back-Calculation of Total and Effective Stress Parameters

Table 4 presents the total stress adhesion factor, α and the effective stress coefficient, β calculated using the average peak unit side resistances obtained from the average strains at each gauge level Equations 1 and 2. The undrained shear strength (S_u) derived from UU was used to calculate the values of α. As observed from the table, the average value of α for all soil layers and all TPs range from 0.41 to 0.86, with an average of about 0.61. Since the unit side resistances (f_s) were calculated from the load tests and the load tests were performed 14 to 30 days after EOD, the α values might be affected by setup process. Both the α and β values are lower for TP-2 compared to the other TPs. The damage at the TP-2 bottom, 33 ft. may contribute to this lower value.

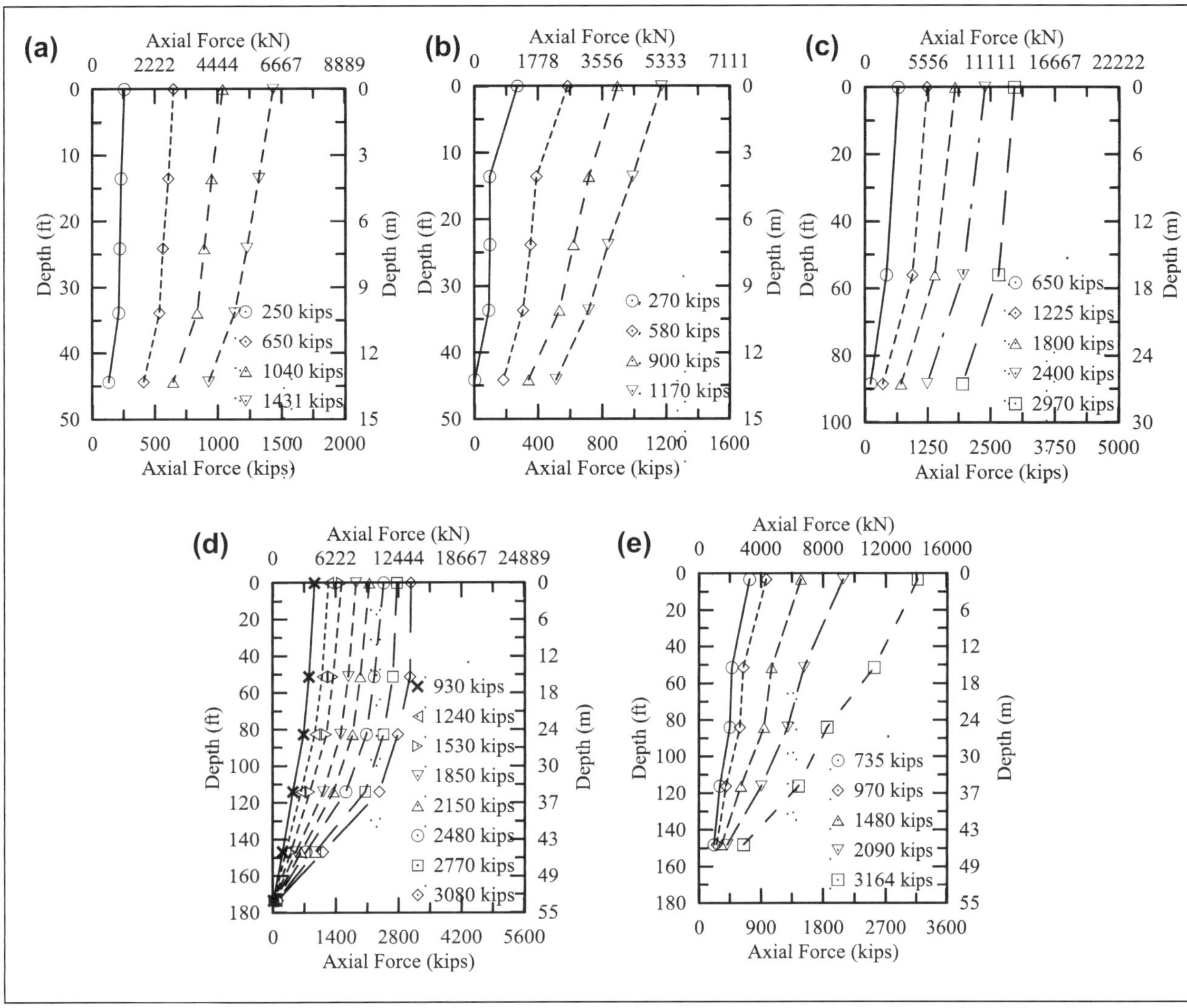

Figure 6. Load distribution plots. (*a*) TP-1 Statnamic load test, (*b*) TP-1 Static load test, (*c*) TP-2 Statnamic load test, (*d*) TP-3 Statnamic load test, (*e*) TP-4 Statnamic load test.

Meanwhile, the calculated β-coefficients ranged from 0.13 to 0.29, which matches very well with reported values in the literature (*16*, *17*). The average value of β-coefficients for all soil layers and all TPs is about 0.22. Figure 9, *a–c* present the distribution of the unit side resistance, α factor, and the β coefficient along the length for all the TPs, respectively. Note that these values were plotted at the midpoint between the strain gauge levels since the f_s, S_u and effective overburden stress (σ'_v) were also calculated in between the two gauge levels. Figure 9*a* indicates that the unit side resistance increased linearly with depth.

Increase in Pile Resistance or Setup

One of the main objectives of this study was to estimate the amount and rate of setup for OE piles, so that it can be incorporated into the design of production piles. All the TPs were tested using DLTs at EOD, restrikes(s) after that, and with at least one SNLT or SLT to quantify the increase of resistance (or setup) with time. The total pile resistance (R_t) estimated using the PDA® and CAPWAP® from the DLTs as well as the resistances measured by the SLT and SNLT are summarized in Table 3. The setup ratio for side (R_s/R_{so}), tip (R_{tip}/R_{tipo}), and total (R_t/R_{to}) resistances are also calculated for each TP and presented in Table 3. The results of testing program showed that all the TPs exhibited significant amount of setup. Figure 10, *a–d* present the measured total (R_t) and side (R_s) resistances versus elapsed time for TP-1, TP-2, TP-3, and TP-4, respectively.

Several methods (*20–24*) were developed to predict pile setup with time after EOD. In this study, the

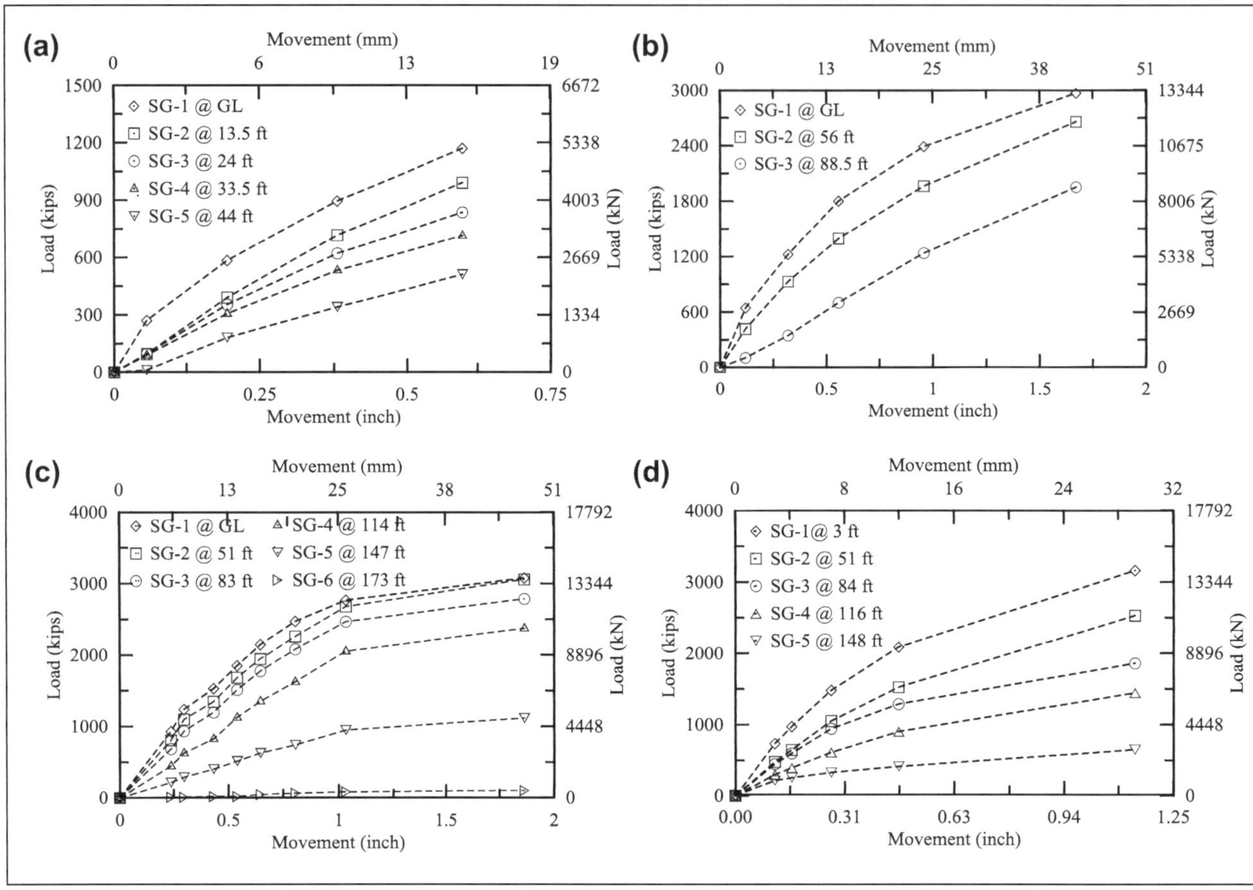

Figure 7. Calculated load versus measured movements. (*a*) TP-1 30 inch PSC test pile, (*b*) TP-2 66 inch cylinder test pile, (*c*) TP-3 66 inch cylinder test pile, (*d*) TP-4 66 inch cylinder test pile.

relationship developed by Skov and Denver is used in estimating pile setup (*20*). They proposed the following equation:

$$\frac{R_t}{R_{to}} - 1 = A \log_{10} \frac{t}{t_o} \quad (3)$$

where R_t = Pile resistance at time, t, R_{to} = Pile resistance at initial time, t_o, t = Time elapsed since end of initial pile driving, t_o = a reference initial time after which the resistance increases with the logarithmic increase of time, and "A" is the rate of pile resistance (R_t/R_o) increase per log cycle of time (t/t_o). In this study, t_o is selected as 1 day. As shown in Figure 10, both the total (R_t) and side (R_s) resistances were best fitted to a linear logarithmic time scale, which suggests that a linear logarithmic time relationship similar to the models proposed by Skov and Denver (*20*) and others (*4, 15*) does exist for this load testing program. The tip resistance (R_{tip}) did not change significantly during all the restrikes for all the TPs with the exception of TP-2. As mentioned earlier, damage was noticed at the bottom of TP-2, 33-ft., which contributed to the lower tip resistance (R_{tip}) during the restrike for TP-2. However, for the other TPs, any change in the total resistance (R_t) was mainly as a result of side resistance (R_s) setup, which is consistent with the findings in the literature (*3, 4*) in which the findings of load tests showed that the side resistance (R_s) mainly contributes to the setup process and the tip resistance (R_{tip}) stays constant over time during the load tests. The load test results also showed that for all the TPs (i.e., TP-1, TP-2, TP-3, and TP-4), the tip resistance (R_{tip}) was always over estimated during the SNLT compared with the tip resistance (R_{tip}) computed by the restrikes, leading to the conclusion that the derived tip resistance (R_{tip}) values from SNLTs are consistently greater than those from DLT.

The calculated setup parameter "A" of total (R_t) and side (R_s) resistances for all TPs are tabulated in Table 4. It is interesting to observe that though both TP-3 and TP-4 were driven at two different locations, both TPs had similar setup ratios (R_t/R_{to} = 4.4˜4.7) and rate of

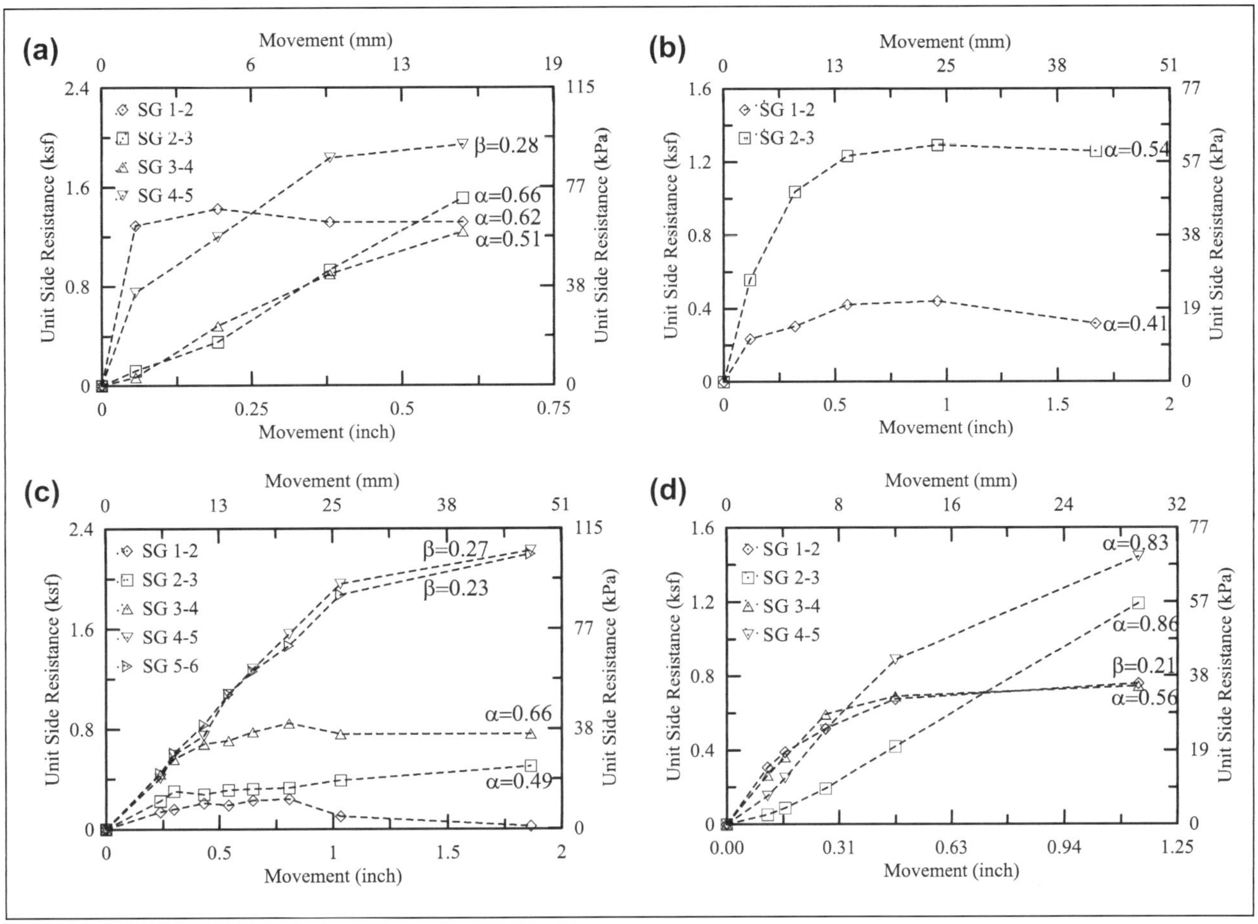

Figure 8. Unit side resistance versus measured pile head movements at each gauge level. (*a*) TP-1 30 inch PSC test pile, (*b*) TP-2 66 inch cylinder test pile, (*c*) TP-3 66 inch cylinder test pile, (*d*) TP-4 66 inch cylinder test pile.

Table 4. Back-Calculated Design (α and β) and Setup Parameters (A) for Test Piles

Pile ID	Segment length (ft.)	Design parameters, α and β		FHWA recommendation (*19*) α	FHWA recommendation (*19*) β	Setup parameter "A" for total resistance, R_t	Setup parameter "A" for side resistance, R_s
TP-1	0–13.5	0.62	0.14	0.2–1.8	0.15–0.35	0.10	0.28
	13.5–24.0	0.66	0.22				
	24.0–33.5	0.51	0.19				
	33.5–44.0	Sand	0.28				
TP-2	0–56.0	0.41	0.19			0.14	0.13
	56.0–88.5	0.54	0.23				
TP-3	0–51.0	Casing				0.23	0.24
	51.0–83.0	0.49	0.24				
	83.0–114.0	0.66	0.29				
	114.0–147.0	Sand	0.27				
	147.0–173.0	Sand	0.23				
TP-4	3.0–51.0	Sand	0.21			0.23	0.23
	51.0–84.0	0.86	0.13				
	84.0–116.0	0.56	0.18				
	116.0–148.0	0.83	0.26				
Average		0.61	0.22			0.18	0.22

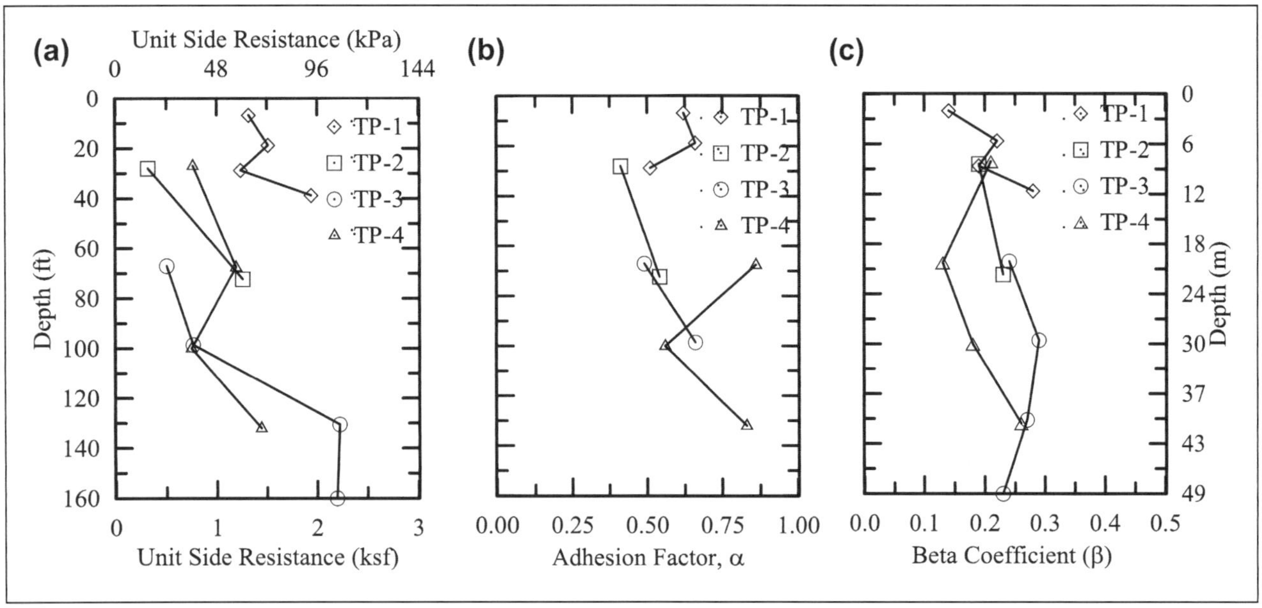

Figure 9. Distribution of (*a*) unit side resistance with depth, (*b*) adhesion factor, α, with depth, (*c*) β-coefficient with depth.

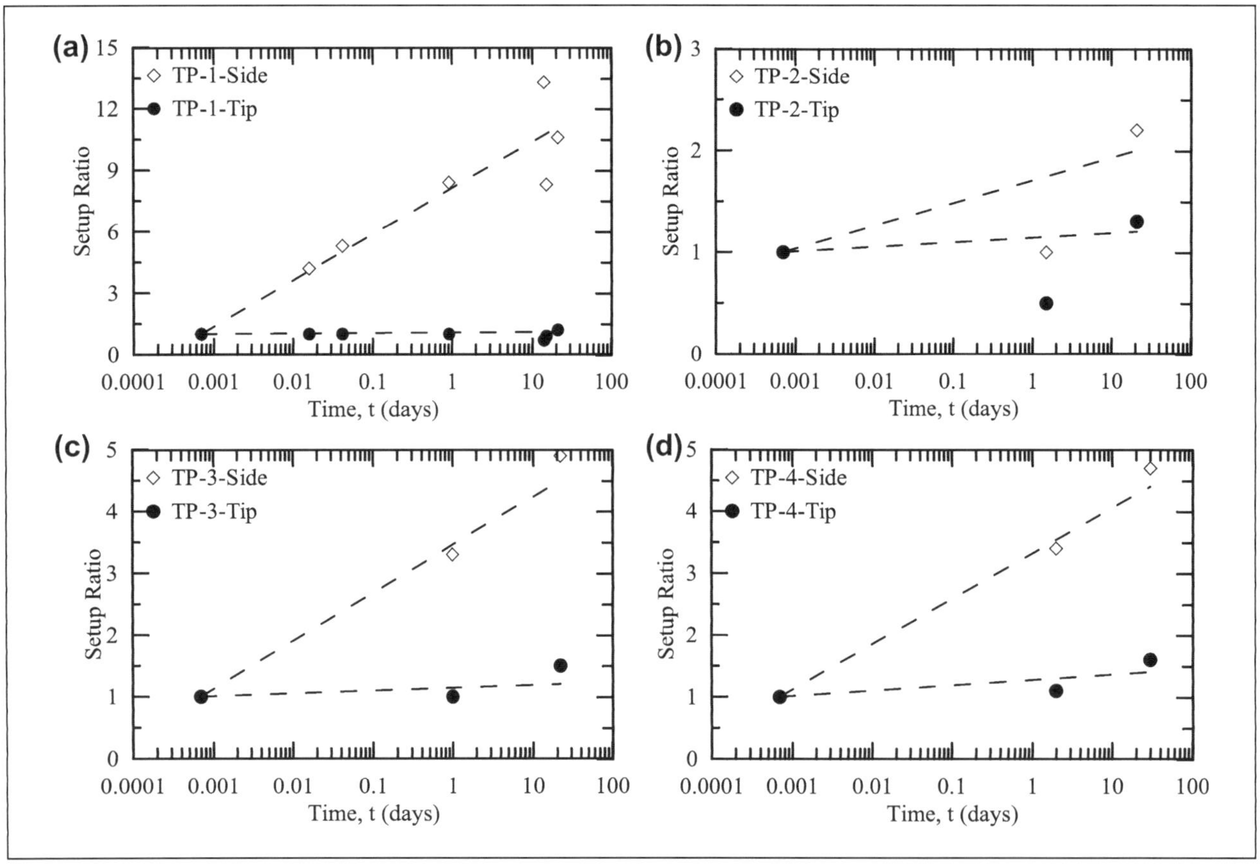

Figure 10. Setup of side and total resistances of all test piles. (*a*) TP-1 30 inch PSC test pile, (*b*) TP-2 66 inch cylinder test pile, (*c*) TP-3 66 inch cylinder test pile, (*d*) TP-4 66 inch cylinder test pile.

setup (A = 0.23˜0.24). TP-2 had a different amount and rate of setup compared with the other two OE piles (i.e., TP-3 and TP-4). The difference is thought to be caused by the damage at the bottom 33-ft. of TP-2. As the setup of the PSC pile (TP-1) is mainly caused by outside friction and the setup rate, "A" for side resistance, of the PSC piles and OE cylinder piles (TP-3 and TP-4) are close (i.e., A_{PSC} = 0.28 and $A_{OPEN\text{-}ENDED\ CYLINDER}$ = 0.23˜0.24), one can postulate that no significant amount of setup was observed as a result of inside friction for the OE cylinder piles.

Conclusions

In this study, a full-scale pile load testing program was performed on four test piles to evaluate the design parameters (i.e., α and β); and to investigate the setup behavior for both PSC and OE cylinder piles. All the test piles were instrumented with strain gauges in order to measure the load distribution along the piles. Subsequent restrikes and one SNLT or SLT were performed at different times after EOD to evaluate the setup behavior with time. Laboratory and in-situ tests were conducted to characterize the subsurface soil properties. Based on this study, the following conclusions can be drawn:

1. The adhesion factor, α, calculated from the SNLTs ranged from 0.41 to 0.86 with an average value of 0.61. The β-coefficient ranged from 0.13 to 0.29, and the average value of β for all soil layers in this study was 0.22. These values are comparable to results previous studies in literature.
2. The unit side resistances of most soil layers (determined using differentiation of loads between strain gauge levels) mostly exhibited strain softening behavior, which ensured that the side resistance was fully mobilized.
3. The distribution of unit side resistance and design parameters (α and β) with depth showed that the unit side resistance and design parameters (α and β) increased almost linearly with depth.
4. All the test piles (i.e., both open and close-ended pile) exhibited significant amount of setup which was mainly owing to increase in side resistance. The tip resistance was almost constant for all test piles during the restrikes except for TP-2. The lower tip resistance for TP-2 can be attributed to the damage caused during driving at the lower 33-ft of this pile.
5. The load test results showed that for all the test piles, the inferred tip resistances (R_{tip}) from SNLT were consistently greater than those determined from DLT.
6. The logarithmic setup parameter "A" for all test piles were back-calculated for both total and side resistances. The results revealed that for the large-diameter OE cylinder piles, the "A" parameter for side resistance was about 0.23 and that for PSC test pile "A" was about 0.28 in this study.

Acknowledgments

This research project is funded by the Louisiana Department of Transportation and Development (State SIO No. 30000134) and Louisiana Transportation and Research Center (LTRC Project No. 11-2GT). The authors acknowledge the GRL engineers for their support and assistance in performing the analyses for the CAPWAP® data.

Author Contributions

The authors confirm contribution to the paper as follows: Md Nafiul Haque and Murad Y. Abu-Farsakh analyzed the data and wrote the article. Md Nafiul Haque, Murad Y. Abu-Farsakh, Ching Tsai and Zhongjie Zhang discussed the results and commented on the manuscript. Chris Nickel carried out the experiment. Jesse Rauser contributed to interpretation of results.

References

1. Szechy, C. H. "The Effect of Vibration and Driving Upon the Voids in Granular Soil Surrounding a Pile." *Proceedings of the 5th International Conference on Soil Mechanics and Foundation Engineering*, Paris, France, July 1961, pp. 161–164.
2. Paik, K. H., and R. Salgado. Determination of the Bearing Capacity of Open-Ended Piles in Sand. *Journal of Geotechnical and Geoenvironmental Engineering*, Vol. 129, No. 1, 2003, pp. 46–57.
3. Brown, Dan A., and W. Robert Thompson, III. *NCHRP Synthesis of Highway Practice 478: Design and Load Testing of Large Diameter Open-Ended Driven Piles*. Transportation Research Board of the National Academies, Washington, D.C., 2015.
4. Dasenbrock, D. "Time Dependent Open-End Pipe Pile Capacity Assessment by Dynamic and Quasi-Static Methods." *Proceedings of the 31st Annual Conference on Deep Foundations*, Washington, DC, January 2006.
5. Paikowsky, S. G., and L. R. Chernauskas. "Dynamic Analysis of Open-Ended Pipe Piles." *Proceedings of the 8th International Conference on the Application of Stress-Wave Theory to Piles*, Lisbon, Portugal, September 2008, pp. 59-76.
6. Abu-Farsakh, M. Y., M. N. Haque, and C. Tsai. A Full-Scale Field Study for Performance Evaluation of Axially Loaded Large-Diameter Cylinder Piles with Pipe Piles and PSC Piles. *Acta Geotechnica*, Vol. 12, No. 4, 2017, pp. 753–772.
7. Chen, Q., Md. N. Haque, M. Abu-Farsakh, and B. A. Fernandez. Field Investigation of Pile Setup in Mixed Soil. *Geotechnical Testing Journal*, Vol. 37, No. 2, 2014, pp. 1–14.

8. Haque, Md. N., M. Y. Abu-Farsakh, C. Tsai, and Z. Zhang. Load Testing Program to Evaluate Pile Set-Up Behavior for Individual Soil Layers and Correlation of Set-Up with Soil Properties. *Journal of Geotechnical and Geoenvironmental Engineering*, Vol. 143, No. 4, 2016, pp. 1–19.
9. Komurka, V. E., A. B. Wagner, and T. B. Edil. *Estimating Soil/Pile Set-Up*. Wisconsin Highway Research Program #0092-00-14. Wisconsin Department of Transportation, 2003.
10. Schmertmann, J. H. The Mechanical Aging of Soils. *Journal of Geotechnical Engineering*, Vol. 117, No. 9, 1991, pp. 1288–1330.
11. Long, J. H., J. A. Kerrigan, and M. H. Wysockey. Measured Time Effects for Axial Capacity of Driven Piling. *Transportation Research Record: Journal of the Transportation Research Board*, 1999. 1663: 57–63.
12. Koutsoftas, Demetrious C. High Capacity Piles in Dense Sands. In *Deep Foundation Congress, Geotechnical Special Publication*, No. 116, Vol. 1, ASCE, Reston, Va., 2002, pp. 632–646.
13. Tomlinson, M. "The Adhesion of Piles Driven in Clay." *Proceedings of the 4th International Conference on Soil Mechanics and Foundation Engineering*, London, UK, Vol. 2, August 1957, pp. 66–71.
14. Tomlinson, M. "Some Effects of Pile Driving on Skin Friction." *Proceedings of the Conference on Behavior of Piles*, London, UK, September 1970, pp. 107–114.
15. Burland, J. B. Shaft Friction of Piles in Clay. *Ground Engineering*, Vol. 6, No. 3, 1973, pp. 30–42.
16. Clausen, C. J. F., P. M. Aas, and K. Karlsrud. Bearing Capacity of Driven Piles in Sand, the NGI Approach. *Proc. of International Symposium on Frontiers in Offshore Geotechnics*, Perth, Taylor & Francis, London, 2005, pp. 677–681.
17. Fellenius, B. H. *Basics of Foundation Design*. Electronic Edition. http://www.fellenius.net/papers/383%20The%20Red%20Book%20-%20Basics%20of%20Foundation%20Design.pdf. Accessed on June 15, 2017.
18. Zhang, Z., and M. T. Tumay. Statistical to Fuzzy Approach toward CPT Soil Classification. *Journal of Geotechnical and Geoenvironmental Engineering*, Vol. 125, No. 3, 1999, pp. 179–186.
19. Hannigan, P. J., F. Rausche, G. E. Likins, B. R. Robinson, and M. L. Becker. *Design and Construction of Driven Pile Foundations – Volume I*. FHWA Report No. NHI16-009. US Department of Transportation, FHWA, Washington, D.C., 2016.
20. Skov, R., and H. Denver. "Time Dependence of Bearing Capacity of Piles." *Proceedings of the 3rd International Conference on the Application of Stress-Wave Theory to Piles* (B. H. Fellenius, ed.), Ottawa, Canada, May 1988, pp. 879–888.
21. Svinkin, M. R. Setup and Relaxation in Glacial Sand-Discussion. *Journal of Geotechnical Engineering*, Vol. 122, No. 4, 1996, pp. 319–321.
22. Svinkin, M. R., and R. Skov. "Set-up Effect of Cohesive Soils in Pile Capacity." *Proceedings of the 6th International Conference on Application of Stress Waves to Piles* (S. Niyama, and J. Beim, ed.), Sao Paulo, Brazil, September 2000, pp. 107–111.
23. Haque, Md. N., M. Abu-Farsakh, Z. Zhang, and A. Okeil. Evaluate Pile Set-up for Individual Soil Layers and Develop a Model to Estimate the Increase in Unit Side Resistance with Time Based on PCPT Data. *Transportation Research Record: Journal of the Transportation Research Board*, 2016. 2579: 17–31.
24. Haque, Md. N., and M. Y. Abu-Farsakh. Development of Empirical Models to Estimate the Increase in Pile Resistance (Set-up) with Time. In *Geotechnical Frontiers*, ASCE, Orlando, Fla., 2017, pp. 193–203.

The Standing Committee on Safety Data, Analysis and Evaluation (ANB20) peer-reviewed this paper (18-01739).

Article

TRR
JOURNAL OF THE TRANSPORTATION RESEARCH BOARD

Transportation Research Record
2018, Vol. 2672(52) 307–315

DOI: 10.1177/0361198118790377
journals.sagepub.com/home/trr
SAGE

Research in using Polyurethane Foam to Mitigate Rock Bolt Installation Difficulties

Benjamin George[1], Lawrence Pierson[1], and Richard Barrows[2]

Abstract

Rock bolts are commonly used on transportation projects to stabilize cut slopes, reinforce undermined or unsupported blocks, and reduce potential for large rock slides. Installing bolts can be challenging. Locations where bolts are useful may not be easily accessible for personnel or large equipment, and targeted blocks generally are marginally stable. In addition, jointed rock with openings and voids, which are typical conditions, complicate hole advancement and full grout encapsulation. These issues can lead to construction delays, claims, reduced corrosion protection, or shorter design life. This paper describes research conducted to address two rock bolting challenges in open-jointed rock: i) advancement of drill holes to design depth, and ii) successful completion of grouting through the application of injected polyurethane foam (PUF). Testing utilized double-stacked concrete blocks aligned in rows to simulate a rock mass with vertical fractures. Steel spacers were placed between the stacks to simulate 2-, 4-, 6-, and 8-inch apertures. A standard rock drill advanced holes using a 3-inch bit. Research showed PUF could: bridge simulated aperture sizes up to 8 inches, survive drilling action during secondary reaming, resist drilling air and grouting pressures, and reduce grout consumption.

This research, conceived and designed by author Pierson, was funded by the Federal Highway Administration, Western Federal Lands Highway Division (WFLHD), in Vancouver, Washington. A discussion of the research need, project goals, and research plan is below. The research was conducted during the summer of 2012 and a report detailing the research was provided to WFLHD in October 2013.

Research Need

Rock bolts are a common rockfall mitigation measure. They are installed to secure rock blocks in place and enhance the overall stability of natural rock outcrops and man-made cut slopes. In many applications, cement grout is used to establish anchorage and/or to encapsulate the bolt. Encapsulation transfers the applied bolt tension across the length of the installation to the surrounding rock and provides long-term corrosion protection. For rock bolts, corrosion is primarily the result of contact of the metal with soluble salts contained in the groundwater (*1*). The presence of grout surrounding the steel can significantly reduce corrosion potential. Cement grout provides a highly alkaline environment in the pH range of 11 to 13, which helps protect steel. At this pH, a passive film forms on the steel that reduces the rate of corrosion. The cement grout provides both chemical and physical protection to the steel, but for this to be effective, it is important the bolt be fully encapsulated.

At some sites, subsurface conditions can lead to excessive grout takes or incomplete grout encapsulation. If excessive grout is required or rock bolt installations are rejected due to incomplete grouting, construction delays and claims can result. Grout loss is typically caused by open joints and voids encountered in a drill hole that allow grout to flow away from the rock bolt and not fill the annular space. Without full encapsulation, corrosion can shorten the bolt design life.

In addition, open joints and voids can disrupt the drilling process. These natural openings provide a path for drilling air pressure to escape, which is needed to clear cuttings from the hole. The loss of drilling air pressure can contribute to failed drilling attempts and loss of drill steel and bits. These problems can affect the project schedule, cause cost overruns and possibly result in less slope reinforcement that does not satisfy design or project goals.

[1]Landslide Technology, Portland, OR
[2]Federal Highway Administration - Western Federal Lands Highway Division, Vancouver, WA

Corresponding Author:
Address correspondence to Benjamin George:
benjaming@landslidetechnology.com

Figure 1. Ecology block testing apparatus and drilling setup.

Mitigating these problems has generally focused on either adjusting the grout design or attempting to block off the openings. Traditional techniques have included multiple stage grouting and pre-grouting, modification of the grout mixture (i.e., addition of fillers, dispersants, accelerators, etc.), using grout socks, caulking of fractures at surface exposures, and various combinations of chemical grouting. In some cases, where the drill hole can be completed but the installations are not properly grouted, the slope may be reinforced with mitigation that will have a shorter design life.

In an attempt to develop a technique to resolve these common rock bolting problems, the use of polyurethane foam (PUF) was investigated. The research approach tested if PUF could be used to fill or bridge across open joints and voids as they are encountered during drilling, or upon completion of drilling, to improve the installation quality and design life. Cornforth Consultants, Inc. (CCI) performed the research in conjunction with a rockfall mitigation contractor, S.C. Industrial Corporation (SCI) of Palm Desert, California.

Research Goals

Research focused on determining the effectiveness of using PUF as a drilling aid and as a grout loss mitigation method for the installation of rock bolts. The following goals were established.

1. Determine the potential for using PUF to help mitigate loss of grout and/or drilling air pressure in simulated discontinuities with wide openings.
2. Identify potential drilling difficulties that may be associated with advancing holes through PUF treated zones.

Figure 2. Steel spacers used to control the simulated joint spacing.

This project represents an initial research effort into the use of PUF in rock bolting applications. Although the research goals were met, as with most research efforts into an untried product and/or methodology, new questions and issues were identified. Further research is needed to address these and to evaluate the use of PUF for other ground reinforcement applications.

Research Procedures

The testing setup utilized concrete ecology blocks placed in aligned stacks to simulate a rock slope with multiple vertical fractures (Figure 1). Spacers were placed between the blocks to form apertures or "open joints" with 2-, 4-, 6-, and 8-inch-wide openings (Figure 2). Although open discontinuities much narrower than this are commonly encountered that cause problems, this research effort attempted to evaluate the limits of PUF in wide joints.

Figure 3. Showing (a) standard 3-inch diameter drill bit, and (b) typical 90-pound hand-drill.

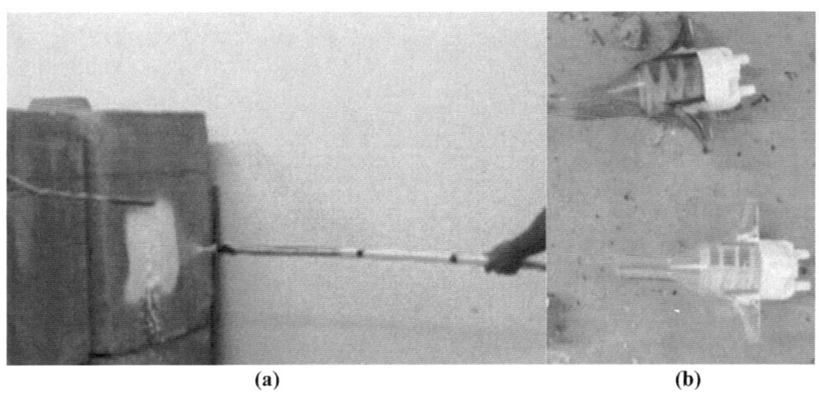

Figure 4. Showing (a) foam injection "wand" with spray pattern from blue nozzle, and (b) Froth-Pak foam mixing nozzles.

In order to simulate a crack with a potentially "infinite" volume, no barriers or siding were placed around the outer sides or across the top of the ecology blocks. The only limiting boundary to the "infinite" volume was the floor. The blocks were lashed together with a cable to form a tight assemblage that allowed the drilling to be performed without shifting the blocks or changing the simulated apertures. Photography equipment was setup next to each simulated joint to document results.

A standard rock drill was used to advance holes through the blocks using a 3-inch diameter bit (Figure 3). The drill was advanced and retracted using an electric winch. The PUF injection was conducted in various phases so that the effects of cured PUF on drilling could be evaluated. Following PUF injection and redrilling, Type II Portland cement was injected to evaluate the resistance of PUF plugs to typical grout pressures and to determine the ability of the PUF seals to retain grout and limit grout takes.

Two foam products were used in the field tests: FOAMPLUG, a product of Foam Concepts, LLC; and Froth-Pak Foam Kit, a product of DOW Chemical Company. Material safety data sheets for each product were collected and can be provided upon request. The FOAMPLUG material did not produce a PUF material with sufficient instantaneous internal strength and adhesion to effectively bridge across the tested apertures. As a result, the Froth-Pak material was used for all subsequent testing. The spray gun mechanism included with the Froth-Pak kit was modified into a wand to facilitate injection of PUF at depths up to 12 feet (Figure 4*a*). Two nozzles were included with each Froth-Pak kit that produced distinctly different spray pattern (Figure 4*b*). The "blue" nozzle provided a narrower pattern and the "clear" nozzle provided a wider pattern. Testing indicated the narrow pattern was the most effective for building a bridge, especially on the wider apertures.

Figure 5. A dry pack of grout placed at the face of ecology blocks during grouting.

Test 1: Mitigating Drilling Advancement Difficulties

This test attempted to simulate the loss of air pressure during drill hole advancement. When this happens in field applications, the cuttings can build up behind the drill bit and the drill can become stuck in the hole. This can result in lost drill steel and bits, abandoned holes and the need to redrill the hole. The test was designed to assess the capability of PUF to resolve these drilling problems by preventing loss of air circulation. The test also evaluated the ability of injected PUF to withstand the air pressure and vibrations caused by drilling operations, and to resist abrasion from passing drill cuttings. In general, the test included:

1) Drilling through the first two stacks of ecology blocks (Stacks 1 and 2) past the 2- and 4-inch apertures.
2) Retracting the drill steel and placing PUF injection equipment in the drill hole.
3) Injecting PUF into the drill hole starting at the "deepest" simulated joint (i.e., 4-inch aperture) and retracting the injection nozzle as filling proceeded.
4) Allowing the PUF to cure for a minimum of 15 minutes.
5) Re-drilling the PUF-filled hole to evaluate effects of cured PUF on drilling action (Figure 5).
6) Evaluating the following performance of the PUF:
   a) Capability of PUF to adequately bridge and fill the open joints,
   b) Effect of the PUF plug on drill performance, and
   c) Ability of the PUF plug to resist drilling air pressure, vibrations and abrasion.

Test 2: Mitigating Grout Injection Difficulties

This test attempted to simulate mitigation of excessive grout injection volumes. The test was designed to assess the capability of PUF to bridge and/or fill wide discontinuities and control grout injection volumes. In general, the test included:

1) Drilling through the next two stacks of ecology blocks (Stacks 3 and 4) past the 6- and 8-inch apertures to the targeted depth of 12 feet within the final stack of the ecology blocks (Stack 5).
2) Injecting PUF from the far end of the hole, filling the remaining 6- and 8-inch joints and allowing the PUF to cure for 15 minutes.
3) Redrilling through the PUF-filled portion of the hole and assessing the drilling characteristics and ability of the PUF to resist drilling action through these wider apertures.
4) Conducting secondary foam injection, as needed, for the 6- and 8-inch apertures to provide sufficient foam placement to allow grouting operations to proceed.
5) Inserting a grouting tremie tube to the far end of the hole and filling the hole with grout.
   a) To assist with tremie tube insertion, it was attached to a small diameter PVC pipe.
   b) To minimize grout loss at the "face" of the test apparatus, the drill hole opening was partially blocked with dry packed grout (Figure 6).
6) Evaluating the following performance of the PUF:
   a) Capability of PUF to adequately bridge and fill the wider 6- and 8- inch apertures,
   b) Serviceability of the PUF plug during redrilling operations, and
   c) Ability of the PUF plug (all aperture openings) to resist nominal grouting pressures and retain grout.

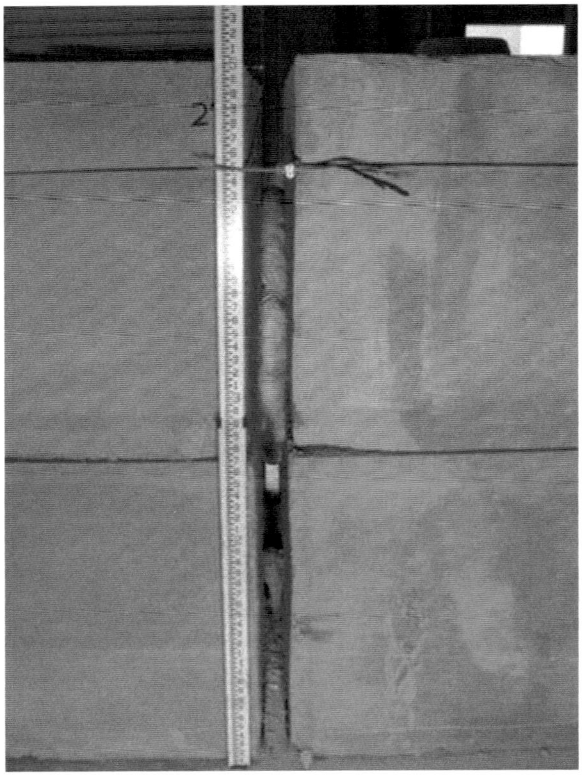

Figure 6. Final condition of the 2-inch aperture after PUF injection and setting.

Research Findings

The research resulted in a proof-of-concept for using PUF as a grout loss mitigation technique when encountering fractured rock during rock bolt installation. Specific results obtained from each of the three tests performed are as follows.

Test 1 Results: Mitigating Drilling Advancement Difficulties

The 2- and 4-inch apertures were effectively bridged with injected PUF within 30 seconds of PUF placement. The final conditions of the 2-inch aperture and the PUF dispersion pattern are illustrated in Figures 7 and 8. A detailed breakdown of PUF injection and the rapid expansion and setting of the PUF in the 4-inch aperture is provided below. The sequence of events has been illustrated in Figure 9.

- 17 seconds into injection: PUF expanded outward on the simulated joint surface.
- 31 seconds into injection: PUF forms a bridge across the aperture.
- 37 seconds into injection: Weight of bridge overcomes PUF internal strength and begins to sag, eventually failing and falling to the floor.
- 48 seconds into injection: New bridge is formed across the aperture along the drill hole alignment.
- From 37 to 48 seconds: PUF accumulating on the floor began to expand upward. However, prior to the PUF on the floor reaching the drill hole the bridge formed along the alignment. Eventually, PUF from the floor and the bridge connected.

Figure 7. Foam dispersion pattern for the 2-inch aperture.

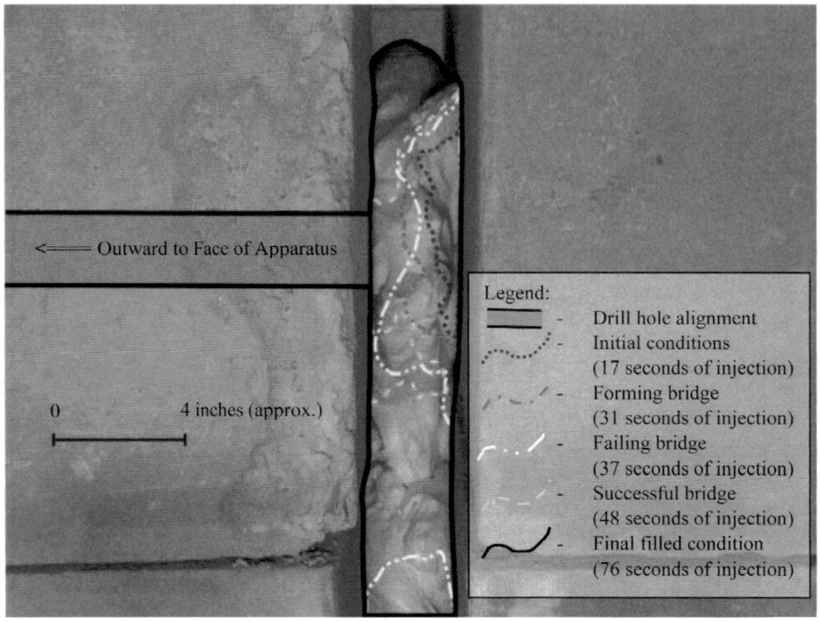

Figure 8. Side view of the PUF bridge that formed in the 4-inch aperture.

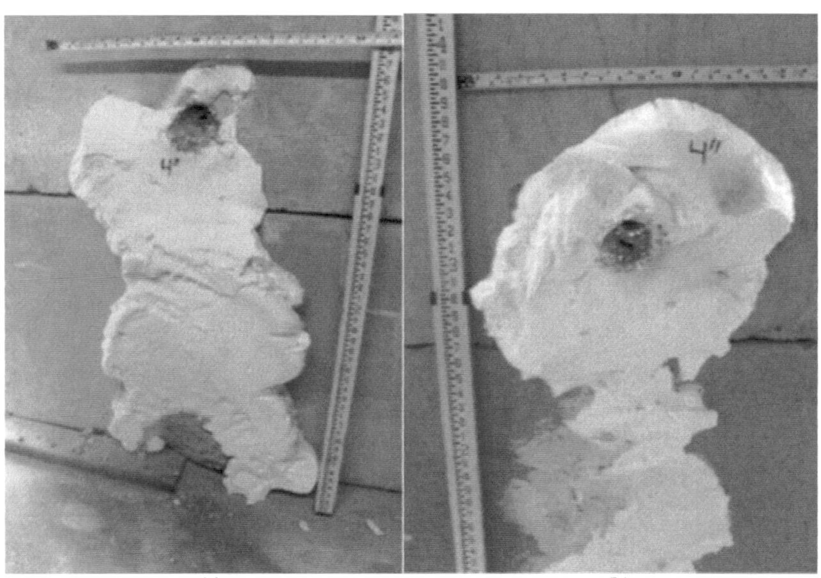

Figure 9. Foam dispersion pattern for the 4-inch aperture: (*a*) outward, and (*b*) inward.

The PUF-filled drill hole and the PUF bridges were redrilled to the original hole depth. The PUF was easily penetrated by the drill. The PUF cuttings were in the order of 0.25-inch in diameter. The PUF bridges within the 2- and 4-inch simulated joints adequately resisted the drilling vibrations and air pressure without loss of air circulation. The PUF in the sealed apertures resisted the abrasion from moving drill cuttings as the hole was advanced through the next two blocks to conduct Test 2.

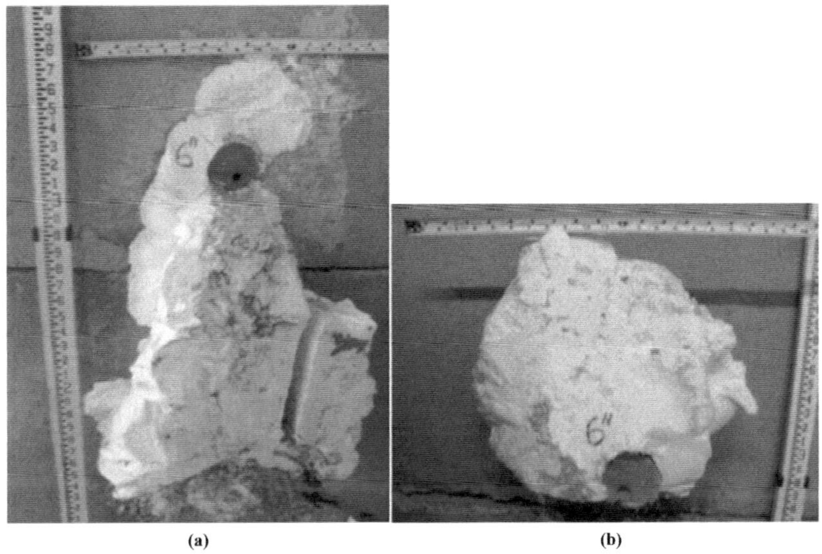

Figure 10. Foam dispersion pattern for the 6-inch aperture: (*a*) outward, and (*b*) inward.

Test 2 Results: Mitigating Grout Injection Difficulties

Foaming the 6- and 8-inch apertures required two stages of injection to effectively form a bridge. The Froth Pak PUF material was not capable of supporting itself long enough to create an unsupported plug between the blocks. It continually collapsed under its own weight until a foam column formed from the floor to help support and span the opening. Although there were no boundaries on the effective width of the simulated fractures, the roughly 3-foot tall support columns formed as very steep-sided triangles with an average width of only 6 inches on either side of the drill hole as shown in Figure 10. The amount of expanded foam required to form the support column across the 6- and 8-inch apertures was approximately 1.9 and 2.5 cu. feet (0.05 and 0.07 cu. meters), respectively, which is a relatively small amount considering the potentially infinite capacity of the discontinuity. In a natural setting, it is likely that the conditions required to form similar support columns would exist and that an equally small volume of expanded foam would be needed to bridge these gaps.

The foam plugs were capable of resisting drilling action, drilling air pressure and grouting pressures. However, in the 8-inch aperture, there were several weaker zones in the plug that allowed some grout to escape. The leaking areas in the foam were mitigated with two stages of grouting, which still reduced the volume of grout required to fill the drill hole.

After grouting, the ecology stacks were separated to measure the grout fillings and document the foam patterns. It was discovered that the 6- and 8-inch apertures were wider than the planned aperture widths due to 2-inches of concrete spalling off the faces of the blocks during drilling. For that reason, the lengths of the grout fillings were actually 8 inches and 10 inches long, respectively, as shown on Figure 11.

Conclusions and Recommendations

This research was performed to determine if PUF could be used in rock bolting applications to:

- Bridge significantly wide, simulated rock joints,
- Aid the drilling process by retaining air pressure used to clear drill cuttings,
- Mitigate loss of grout by effectively bridging/ filling open joints up to 8 inches wide and resist grouting pressures, and
- Resist drilling stresses by not tearing from joint surfaces during drill re-advancement or being damaged beyond use by the abrasion of moving drill cuttings.

Conclusions

This project was an initial phase of testing that provided a proof-of-concept for using PUF to mitigate difficulties for installing rock bolts. PUF was capable of bridging large simulated open joints in excess of 8 inches. Cured PUF material in the simulated joints resisted typical drilling air pressures and abrasion and permitted the effective clearing of cuttings. PUF successfully bridges

Figure 11. Grout cylinder formed in the 6-inch aperture.

that resisted the hydrostatic pressure produced during grouting operations. PUF material was capable of withstanding drilling stresses to allow completion of the drill hole and successful grouting across large simulated apertures. The quick placement and fast setting of the material could allow an effective and efficient deployment of this mitigation tool.

Recommendations

To expand on this study and further advance the understanding of the use of PUF in rock bolting and other ground reinforcement applications, the following research is needed:

PUF products:

a) Research manufacturers' product data to expand the list of alternative and acceptable PUF products that provide a range of properties for different applications and conditions.
b) Work with PUF manufacturers to adjust the attributes of the PUF (i.e. expansion rate, adhesion, internal frictional strength, reaction to water, others) to produce specific mixtures for various subsurface conditions and test these specialized products in similar research or in a field application.
c) Work with manufacturers to refine product ordering and storage requirements and to develop product comparisons and cost data for the different products that may be used in a range of ground reinforcement applications.

Simulated ground conditions:

a) Test the various PUF products under similar simulated ground conditions to develop comparison data and capability/application charts.
b) Test in different simulated materials such as fractured rock, cobles, gravels, and various soils.
c) Test with simulated groundwater conditions or with the presence of water in drill holes.
d) Evaluate the usefulness of PUF to reduce grout loss in other ground reinforcement applications such as soil nailing and tieback installations.

Injection/placement tools and techniques:

a) Develop durable and effective injection systems.
b) Develop injection procedures for specific applications. Injection may vary between rock bolts, soil nails, tiebacks or other ground reinforcement techniques.

Field testing:

a) Install rock bolts or other ground reinforcement elements at locations that have historically required significant grout volumes. Use standard construction techniques to mitigate grout loss and install some elements utilizing PUF injection as a grout loss technique.
b) Compare the grout volumes required for full encapsulation from various mitigation techniques such as grout socks, multiple stage grouting, PUF injection, etc.

Specifications and user manual:

a) Develop construction specifications/guidelines for use of PUF.
b) Develop an "Inspector's Guide" for using PUF in rock bolting and other ground reinforcing applications.

Author Contributions

The authors confirm contribution to the paper as follows: study conception and design: Lawrence Pierson, Benjamin George, Richard Barrows; data collection: Benjamin George, Lawrence Pierson; analysis and interpretation of results: Benjamin George, Lawrence Pierson; draft manuscript preparation: Benjamin George, Lawrence Pierson. All authors reviewed the results and approved the final version of the manuscript.

Reference

1. Samtani, N. C., and E. A. Nowatzki. *Hollow Bar Soil Nails – Review of Corrosion Factors and Mitigation Practice.* FHWA-CFL/TD-10-002. Federal Highway Administration, U.S. Department of Transportation, 2010.

The Standing Committee on Engineering Geology (AFP10) peer-reviewed this paper (18-04809).

Article

TRR
JOURNAL OF THE TRANSPORTATION RESEARCH BOARD

Transportation Research Record
2018, Vol. 2672(52) 316–324

DOI: 10.1177/0361198118777622
journals.sagepub.com/home/trr
SAGE

Estimating Event Likelihood for Rock Slope Assets on Transportation Networks

Aine Mines[1], Darren Beckstrand[1], Paul D. Thompson[2], Bret Boundy[3], Scott Helm[3], and Jeff Jackson[3]

Abstract

Recent federal regulations require states to develop risk-based transportation asset management (TAM) plans for bridge and pavement assets. These requirements are making TAM principles, and their related benefits and tools, familiar to a variety of sections within Departments of Transportation (DOTs) and prompting select DOTs to apply them to select geotechnical assets. The Montana Department of Transportation (MDT) recently completed a multi-year project to update its previous Rockfall Hazard Rating System (RHRS) into a Rock slope Asset Management Program (RAMP), incorporating some of these new tools. Incorporating risk into decision-making is important, but risk posed by deteriorating geotechnical assets is difficult for departments to quantify. Adverse event data are often collected in an ad-hoc fashion, with only the worst events attracting attention. As part of the RAMP project, an MDT-administered adverse event survey was combined with a statewide dataset of detailed rock slope assessments. These data were used to develop initial relationships between rock slope condition and adverse event likelihood per unit area of rock slope face. Annual likelihood of both service disrupting events and accidents was calculated for each rock slope in MDT's database and used to identify high-risk corridors. Using established AASHTO methods, event likelihood was combined with traffic and detour data, generating site-specific estimates of annual safety and mobility risks, expressed in dollars, which can be incorporated into benefit/cost tools at various planning levels and permits dollar-to-dollar comparisons of the risks posed by rock slope assets to other actively managed assets.

Current federal rules require that state transportation departments implement transportation asset management (TAM) programs for pavements and bridges (*1*). The TAM framework incorporates modern best practice to help Departments of Transportation (DOTs) project revenue needs, forecast future asset conditions based on various budget scenarios, and apply smart business practices to the allocation of limited resources. Although TAM plans are currently only required for bridges and pavements, state DOTs are encouraged to apply these modern methods to managing other department assets to plan for and appropriately manage future deterioration. Some DOTs have begun to apply Asset Management Principles to select geotechnical assets (*2, 3*).

An integral component of TAM is the use of risk assessment in project selection. However, accurately assessing risk at this high level depends on estimating event likelihood. Models exist for quantitative risk assessments of highly monitored rock slopes for failures of individual rock masses (*4*), but are not applicable for a statewide highway network within a reasonable budget.

Purpose and Need for Risk-Estimating Tools

In Montana's new Rock slope Asset Management Program (RAMP), rock slopes can be classified into Good, Fair, or Poor sets as determined by category ratings familiar to many geotechnical personnel (*5*). In this system, a Poor condition rock slope is more likely to generate rock on the road than a Good condition slope. How much more likely, and the extent to which this puts user mobility and safety at risk, is more difficult to determine. In the authors' experience, state DOTs do not currently track individual rockfall events and their consequences, complicating correlation of asset condition and event likelihood. However, by

[1]Landslide Technology, Inc., Portland, OR
[2]Consultant, Bellevue, WA
[3]Montana Department of Transportation, Helena, MT

Corresponding Author:
Address correspondence to Aine Mines: ainem@landslidetechnology.com

surveying personnel and applying expert judgement, it is possible to develop initial risk-estimating tools in support of risk assessment at the network level.

The likelihood analysis described in this paper was part of a research project for MDT covering rock slope assets. It describes a method that DOTs or other large asset owners could use to generate preliminary event likelihoods from limited data collected by surveying institutional memory. These estimated likelihoods, which are analogous to annual probabilities, can then be combined with established AASHTO methods for estimating monetary consequences, projecting the annual monetary risk costs to which the DOT and the road user are exposed by failures of a given geotechnical asset.

Incorporating Existing Asset Condition Data

MDT's Existing Rockfall Hazard Rating System (RHRS) Inventory System

Between 2003 and 2005, MDT conducted a research project to collect information on its rock slope assets. The resulting Rockfall Hazard Classification and Mitigation System used the nationally recognized RHRS (*6*) to inventory rock slopes throughout the state that met the following acceptance criteria:

- All rock slopes that were excavated as part of road construction were included for evaluation.
- Natural outcrops within the right-of-way (ROW) were included; natural outcrops outside the ROW were excluded, unless judged highly active with the ability to affect the roadway.
- Rock slopes with no ability (based on expert judgement) or history of producing rocks on the road were excluded from the database.
- Basic site information and photographs were collected for low hazard rock slopes.
- Medium to high hazard rock slopes received a detailed evaluation in 13 RHRS-based rating categories.
- Long slopes were subdivided by either topographic depressions within the slope (e.g., gullies), geologic characteristics (jointed, hard rock versus soft rock subject to rapid differential erosion), or rock slope condition.

In all, 2,653 rock slopes were evaluated between 2003 and 2005, and 1,869 (70%) met the project's acceptance criteria for inclusion in the database. A total of 869 sites (46% of slopes in the database, or 33% of all slopes) received detailed ratings (*7*). In 2016, as part of MDT's improvement of its RAMP, 362 sites were revisited and re-evaluated. This included 126 sites along Montana's Interstate system that had been deemed low hazard in 2003, with only basic information previously collected. Based on maintenance input, one entirely new rock slope was added to the database. The current RAMP contains 1,870 rock slopes, of which 997 have received the detailed ratings required to determine asset condition (*8*). District 1 (D1) in northwestern Montana, which provided the bulk of the data used to develop event likelihoods for this program, contains 421 of the rock slopes that received detailed ratings, or 42% of the total. The location of these rock slopes is shown in Figure 1. The location marker is color-coded to match the rock slope's condition.

Derivation of Asset Condition from Inventory Ratings

Since its development by the Oregon DOT in 1985, RHRS-type programs have been adopted by many transportation agencies (*9*). As discussed in previous papers (*6, 7*), RHRS programs score rock slopes across a variety of rating categories that capture both the likelihood of an adverse event and the likely impacts of that event in relative terms, rather than monetary consequence or probabilistic likelihood terms. Within each rating category, there are four basic exponential scoring options: 3, 9, 27, and 81 points, with judgement dictating when to score between these guidelines. Some rating categories, such as slope height, are calculated mathematically. For all rating categories, the maximum, and worst, possible score is capped at 100. The RHRS score for a site includes components that address both hazard and risk (*6*).

Asset condition, on the other hand, is based primarily on slope characteristics that typically degrade with time. Further, the Condition Index (CI)/Condition State (CS) approaches commonly found in TAM normally use a linear scoring system, where higher values (e.g., 100) correspond with better asset condition. This is the inverse of the RHRS scoring system, so as part of related research for the Alaska DOT's Geotechnical Asset Management (GAM) program, researchers implemented mathematical utility functions to convert RHRS scores to CI scores, which map directly to five asset CSs, further grouped into Good (CS 1), Fair (CS 2 and 3), or Poor (CS 4 and 5) condition. This approach utilizes Good/Fair/Poor terminology codified in legislation while still permitting more nuanced technical assessments. The CSs are determined using the scores in the "ditch effectiveness" and "rockfall activity" RHRS rating categories. This work has already been summarized in previous reports and papers (*5, 10, 11*). The same methods were also applied to MDT's rock slope database, and a site-specific CS was calculated for each assessed rock slope.

Figure 1. Map of Montana's Administrative D1 rock slope assets. Blue asterisks show locations of reported rockfall events. The colored dots mark the location and condition of all assessed rock slopes.

Obtaining Event Data from Available Sources

Survey of Geotechnical Personnel

In order to obtain rockfall event data, the research team developed a simple survey, which MDT sent to the lead geotechnical personnel in each of the state's five administrative districts. The geotechnical managers also contacted maintenance personnel in their districts to get additional information on some events. The event survey was completed in February 2016. It consisted of an Excel spreadsheet-based questionnaire, and the district geotechnical personnel were asked to list all rockfall events they could remember. For each event, respondents were asked to provide the following information:

- Event date (event year at minimum)

- Event location
- If the road was closed, the duration of complete closure
- If a workzone slowdown was required, the duration of this slowdown in days
- Any vehicle or property damage associated with the event
- Any injuries associated with the event
- Other comments

The respondents also included corridors containing multiple rock slopes where events occur regularly. For these corridors, average event occurrence and impacts were recorded (*12*).

In Montana, rock slopes are concentrated in the western part of the state, in Districts 1, 2, and 3. In fact, District 4, which covers the far eastern portion of the state, contains only five of the 997 significant rock slopes inventoried in MDT's updated RAMP. D1 and District 2 (D2) completed the survey questionnaire. The data provided by D1 was detailed and extensive. It included both specific dates and impacts, and estimates of annual events within high-activity corridors. D2 estimated event frequency and impacts for several roadway corridors, but provided no specific events. It is unreasonable to assume that no rockfall events had occurred in districts 3, 4, and 5 in the last 10 to 20 years, even though no specific rockfall events were reported. The research team opted to generate a relationship between event likelihood and rock slope condition for D1 rock slopes and then apply it statewide. This assumes that rockfall events are equally likely statewide, though geologic and climatic conditions vary throughout the state. Improved data collection in all districts would enable the future research necessary to verify that assumption.

Because the survey was in questionnaire format, responses were not standardized, requiring additional work by researchers to normalize the responses. Events with a specific date and RHRS section number were left unchanged. Other survey entries identified corridors that experience regular failures, providing a milepost range and the general adverse event type. To capture event likelihood within these sections, researchers subdivided the individual inventoried rock slopes in the section, estimated the total number of events between 2001 and 2015 (the range of specific dates provided for all events), and divided those estimated events evenly between the rock slope assets. The raw survey data and the edited final dataset used in this paper are available in the documentation for MDT's RAMP project, and can be obtained through MDT's research website (http://www.mdt.mt.gov/research/projects/geotech/rockfall.shtml) (*12*).

Several further assumptions were also required to generate a quantitative event likelihood estimate. In the survey, respondents were asked to provide information on road closures, work zone slowdowns, vehicle/property damage, and injury accidents. Unfortunately, the sample size was too small to correlate closure or slowdown durations with asset condition. Instead, the two categories were combined into a "service disruption" category, and specific durations replaced with a yes/no input. If the respondent replied "possibly" then the researchers modeled that a closure or slowdown followed 50% of adverse events at those sites. The vehicle/property damage was answered on a yes/no basis in the original questionnaire, but again, respondents frequently wrote in "possibly." A smaller frequency was modeled for vehicle and property damage because adverse events requiring payout from the Department are more likely to be remembered by the survey respondents. For this reason, at those sites where respondents wrote "possibly" it was judged that some vehicle or property damage occurred half as often as a service disruption, that is, following 25% of slope failures (*13*).

Allowing indirect responses to the survey encouraged respondents to provide as much data as they could remember, but it also increased the uncertainty in the dataset. It is likely that only particularly significant older events are remembered clearly. Smaller events, even those which still caused temporary road closures, tend to run together in the memory over time, and are likely underreported in the dataset. Further uncertainty was introduced by assuming that if the respondent provided a general frequency and corridor area for an event, such as "once a year between Mile Post 1 and 2.5," the event was equally likely to occur at all sites in that area. Future data collection efforts will refine these initial models. To facilitate data collection, MDT has adopted use of a geotechnical event tracker utilizing their online GIS mapping software for significant events and created rockfall-specific maintenance codes for more routine events.

Other Geotechnical Event Data Sources

This portion of MDT's RAMP program gathered event information by surveying geotechnical personnel in each of the state's administrative districts with some additional input from maintenance personnel. This allowed the research team to include questions about event date, location, and various impacts. However, it relied on respondents' memories, which tend to emphasize serious events at the expense of service disrupting events perceived as "routine." Other DOTs or agencies following the method outlined in this paper may have access to additional data sources, which would strengthen event likelihood correlations. These additional data sources can be from within the DOT or may be obtained from a different agency if records are kept and are consistent.

Table 1. Summary of Edited Survey Responses and RAMP Database Information for District 1

CS	Number of sites in database	Total square footage in database	Annualized number of service disruptions	Annualized square footage involved in service disruptions
1	42	1,891,759	0	0
2	149	8,262,372	0.4	17,093
3	121	5,461,019	2.1	67,666
4	54	3,060,991	3.9	328,644
5	2	282,968	0.6	80,848

For example, the Alaska Department of Transportation and Public Facilities (AKDOT&PF), uses a Maintenance Management System (MMS) to track maintenance costs as part of their daily activities. A specific job code, "Landslide/Rockfall Cleanup," tracks costs associated with unstable slope assets, but can refer to all aspects of unstable slope maintenance, from routine ditch cleaning to emergency repair work. It also does not include work performed by private contractors, who are frequently required to tackle larger emergency events. However, AKDOT&PF's MMS does contain both work date and location information, which, when coupled with surveys or personnel interviews, could help capture more of the routine service disrupting events. In this MMS system, approximately 7,500 geotechnical events are recorded between 2005 and 2016. Similarly, the Idaho State Patrol maintains a call-out log, which includes rock-on-road events reported by roadway users (*14*). Again, coupling this date and location information with a section's own internal information could provide a better data set for correlating asset condition to event likelihoods.

Developing Initial Asset Condition–Event Likelihood Relationship

Once the survey data was processed into a useable format, the CS information for the RAMP sections was appended to the survey. As described in the previous section, each assessed rock slope was assigned a CS using the equations first developed in AKDOT&PF's GAM program (*10*). To generate an annualized event rate that could be used to compare event frequency and asset condition over the date range covered by the survey response, the number of adverse events recorded at each section between 2001 and 2015 was divided by the number of years covered by survey responses (14). There are 421 assessed rock slope assets in D1, but only 48 of them are associated with adverse events in the survey. To more accurately estimate adverse event likelihoods, all rated rock slopes in D1 were also incorporated into the final likelihoods table. For those sites where no adverse events were reported, the annualized rate of adverse events was zero. Due to the small sample size, road closures and traffic slowdowns were combined into a single adverse event type, "service disruption." Some rockfall events, particularly those on Interstate routes, triggered both closures and slowdowns (*13*). For each CS group – 1, 2, 3, 4, and 5 – the annual likelihood of an event occurring somewhere in D1 at a site in that condition was calculated by summing all the annual event likelihoods for individual sites. The raw data compiled from the D1 survey responses are summarized in Table 1, along with additional information from the RAMP database.

Beyond developing district-wide event likelihoods, the research team tasked itself with developing site-specific risk calculations based on slope condition, estimated slope size, traffic volumes, and detour times and distances. Based on experience and expert opinion, a large rock slope in Poor condition should generate more service disrupting events than a small rock slope in Poor condition. Detailed surveying is not intended to be a part of a statewide rockfall condition survey; site visits are intended to be completed in 15–30 minutes. Therefore, only basic slope geometry was collected during site visits. An estimated rock slope area was generated for each site using Equation 1, developed from a review of generalized site geometries and average rock slope dimensions and confirmed with a review of photographic records from approximately 50 sites collected during MDT's 2005 RHRS research project (*11*):

$$\textit{Estimated Rock Slope Area} = 0.65 \times \textit{Slope Height}\ (\textit{ft}) \times \textit{Slope Length}\ (\textit{ft}) \quad (1)$$

Once the rock slope areas were estimated, the calculated annual event likelihood was divided by D1's total inventoried square footage in each CS. This generated the annual proportion of slope area sourcing a rockfall event in each CS. This proportion is taken to be the annual likelihood per square foot based on slope condition. The total square footage combined rock slopes that produced an adverse event and those that did not. For example, assume three slopes in CS 3, each with an area of 10,000 sq. ft., produced recorded rockfall events in a given year, out of a total 5 million sq. ft. of CS 3 slopes. In this year, 30,000 feet of the 5 million sq. ft. in CS 3 produced recorded rockfall events, or 0.6% of the total available CS 3 surface area. Using this example, the likelihood of a single square foot of any CS 3 slope producing a rockfall event is 0.6%/5,000,000, or 1.2E-7%.

Initially, researchers used a linear regression to develop a line of best fit. However, this resulted in impossible negative event likelihoods for CS 1 and 2 slopes. Instead, researchers opted to use the raw calculated likelihoods for CSs 2, 3, 4, and 5 when assessing likelihoods in MDT's updated RAMP. No adverse events were reported for CS 1 slopes, but since even Good condition slopes may have hidden features that could produce rock onto the roadway, these Good slopes still pose a minor risk. To account for this, annual probability per square foot of an event at a CS 1 slope was modeled to 25% of that posed per square foot by a CS 2 rock slope, which equates to about one event every 14 years on a statewide basis. The final likelihoods used in this study are shown graphically in Figure 2.

D1 provided information on vehicle/property damage related to rockfall events. The data were processed using the same method applied to service disruption likelihood. As with the service disruption analysis, slopes in poorer condition are more likely to generate events that cause vehicle/property damage. The survey reported an accident or property damage occurrence of some sort following 49% of service disrupting events. Because the sample size of events causing property damage or injury accidents was so small and included so many assumptions based on indirect survey responses, the likelihood of an accident was modeled at half the likelihood of a service disruption. The term "accident" includes incidents that cause vehicle or property damage, injuries, and/or fatalities. The final event likelihoods for both service disruptions and accidents per square foot based on rock slope condition are presented in Table 2.

Assessing Monetary Risks Based on Event Likelihoods

Part of the attraction of modern asset-management tools is that, when applied properly, they provide decision support to help balance preservation of existing assets against improving users' safety and mobility under limited budgets. A well-established model for this is to monetize safety and service disruption impacts in a likelihood and consequence analysis, which can then be used in benefit/cost analyses. The total event cost is sum of safety, mobility, and recovery costs (*15*).

The annual monetary consequences of a rockfall event are the sum of the estimated safety and mobility impact costs. Both of these costs are based on the likelihood of an adverse event at a given site, and unless the annual likelihood is greater than 100%, the annual monetary consequences of a failure will be a fraction of the hypothetical event cost. Therefore, monetary

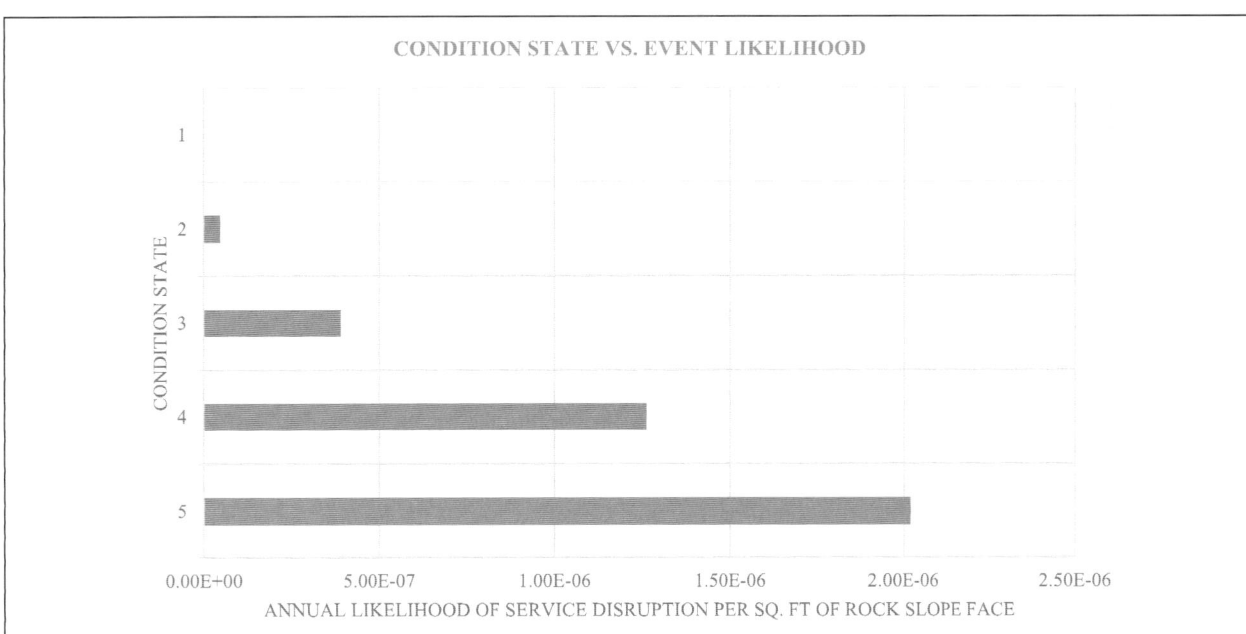

Figure 2. Average annual likelihood of service disruption (defined as a road closure or traffic slowdown) per square foot of rock slope face, based on 2016 MDT survey responses. Note that no events were reported at CS 1 slopes.

Table 2. Relationship between CS and Adverse Event Likelihoods for Montana's Rock Slope Assets, Derived from Condition Assessments and 2016 Adverse Event Survey Responses

CS	Annualized likelihood of service disruption per square foot of rock face (AR_{mob})	Annualized likelihood of accident per square foot of rock face (AR_{acc})
1 – Good	1.19E-08	5.94E-09
2 – Fair	4.75E-08	2.38E-08
3 – Fair	3.91E-07	1.96E-07
4 – Poor	1.26E-06	6.31E-07
5 – Poor	2.02E-06	1.01E-06

consequences should be evaluated over a multi-year period that is equal to the estimated benefit period of any proposed mitigation actions. For MDT's RAMP, annual monetary consequences were estimated using the following pair of equations:

$$AM\$_s = AR_{mob} \times M\$ \quad (2)$$

$$AS\$_s = AR_{acc} \times ACC\$ \quad (3)$$

Where for each slope
$AM\$_s$ = Projected annual cost of service disruption at slope s (\$/year)
AR_{mob} = Annual likelihood of a service disruption at the rock slope
$M\$_s$ = Cost of an individual service disrupting event at the rock slope (\$/event)
$AS\$_s$ = Projected annual safety cost of slope s (\$/year)
AR_{acc} = Annual likelihood of an accident at the rock slope
$ACC\$$ = Average cost per crash in U.S. ($43,525 in 2015$; AASHTO Red Book) (*15*)

For mobility consequences of a service disruption, Equation 4 was used to estimate event costs at each site, based on several assumptions. The average length of time that a road was closed, according to the survey responses, was 6 hours. Therefore, when estimating mobility impacts, the closure time was set to 6 hours for all events. An individual detour length and time was estimated for each site using online mapping tools and assumed start and end points for the average trip through that corridor. The average travel time cost, vehicle occupancy, and vehicle operating costs were all obtained from the AASHTO sources (*15*). Because closure length is greater than 1 hour, travellers would likely use that detour alternate route, the costs of which are described in Equation 4:

$$M\$_s = \frac{ADT_s \times DD_s}{24} \times (DL_s \times VOC\$ + DL_s/DS_s \times TT\$ \times VO) \quad (4)$$

Where
$M\$_s$ = Mobility cost of an individual disruption caused by slope s (\$/event)
ADT_s = Average daily traffic at slope s
DD_s = Duration of the delay caused by an event (in hours, set to 6 in this study)
DL_s = Detour length in miles if the road is blocked at slope s
DS_s = Detour speed in miles per hour if the road is blocked at slope s
$VOC\$$ = Average vehicle operating cost per mile ($ 0.207 in 2015$)
$TT\$$ = Travel time cost, the value per hour of a vehicle occupant's time ($30.50 in 2015$)
VO = Average vehicle occupancy rate (1.3 in 2015)

For safety consequences, all incidents were assumed to be single-vehicle crashes. Only one injury accident, a fatality, was reported in the D1 survey. Due to the (fortunately) small sample size, it was not reasonable to correlate the likelihood of an injury accident with asset condition. Instead, as described previously, the research team developed a CS–vehicle accident relationship and applied the average cost per crash for a single-vehicle accident from the AASHTO Red Book when monetizing consequences, because costs for crashes specifically caused by rockfall events were not available in Montana or the U.S. as a whole. This figure is an average over all vehicle classes and accident types, and takes into account that a small fraction of crashes involve injuries or fatalities. It excludes insurance reimbursement to avoid double counting of costs. It is updated to 2015 dollars using the Consumer Price Index.

Once an annual monetary consequence for each slope has been obtained, the Department can compare them with real or estimated mitigation costs. The resulting benefit/cost ratios can be used in decision support tools.

Incorporating Event Likelihood into Decision Support Tools

Using information from the RAMP database on rock slope CS, estimated square footage, and the per sq. ft. likelihoods in Table 3, researchers generated an event

Table 3. An Example Decision Support Tool for Project Selection within High-Risk Corridors, Developed for MDT's RAMP Program

Roadway functional classification	Target maximum annual likelihood of a service disrupting event in a 1-mile segment
Interstate	1%
Principal arterial, minor arterial, major collector	5%
Minor collector	10%

likelihood for each rock slope in the RAMP database. Researchers then considered how best to integrate the data into a decision support tool. Ultimately, they decided on a corridor segment approach. MDT provided a GIS layer with the department's entire roadway system split into one-mile segments. The annual estimated likelihood of a service disruption for all rock slopes in that one-mile segment was summed using ArcGIS statistical analysis tools, generating per-mile annual event likelihoods.

After considering the results, and comparing them to user expectations, the researchers developed a sample decision support tool, shown in Table 3, which used Montana's existing roadway functional classification system and estimated event likelihood to identify "high--risk" corridors, which could be candidates for safety improvement projects.

Applying this decision support tool across the state of Montana identified 149 "high-risk" one-mile segments on 44 routes, shown in Figure 3. At a meeting with MDT personnel, this tool was used to help select four high-risk corridors for conceptual mitigation design work, most of which contained more than one "high-risk" corridor segment. The final selection among corridors also incorporated the monetary risk to which the department was exposed over a 30-year period, instead of focusing only on event likelihood.

Because monetary risk estimates are based on event likelihood, this work can also be incorporated into other decision support tools, such as those based on benefit/cost analyses. This type of tool is particularly useful when allocating a limited budget between multiple underperforming rock slopes within a corridor. Particularly if the improvement project will span multiple years, prioritization tools incorporating risk can help planners choose the most effective sequence of mitigation work.

Conclusion

The rock slope condition–event likelihood relationship developed for MDT's RAMP can be incorporated into the agency's decision-making processes. It has been used in selecting sites for conceptual mitigation design, and in developing event likelihood and associated risk costs for all inventoried sites. This work can be integrated into the decision support tools that are an integral part of modern asset management.

If MDT continues to collect rockfall event data and impacts over time, as they currently propose to do, these initial relationships can be refined. Eventually, the department could develop free-standing likelihood

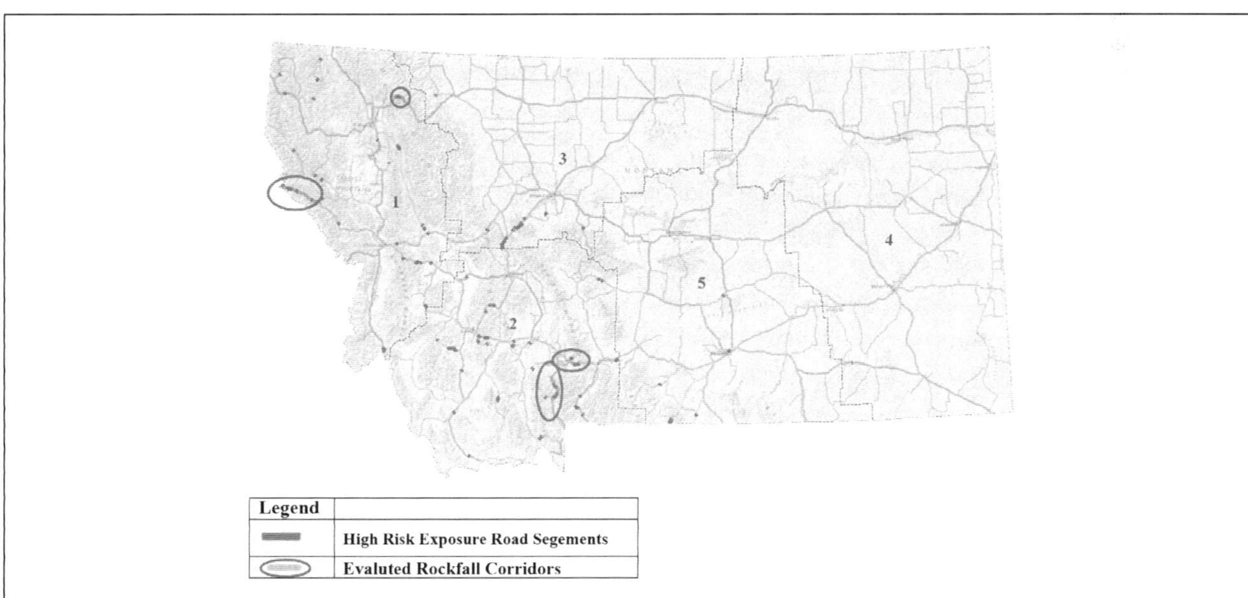

Figure 3. Map of Montana and its administrative districts showing the 1-mile corridor segments that do not meet the sample performance targets in Table 3. The four high-risk corridors selected for conceptual mitigation designs are circled.

correlations for accidents, instead of tying them to the likelihood of a service disruption. MDT could also differentiate between the likelihood of a greater number of adverse event types, such as road closure or traffic slowdowns, instead of grouping adverse events into "service disruption" and "accident" categories. New event data should be obtained as soon as possible after an event happens, and include, at a minimum, event date, specific location, duration of closure, and any injuries or property damage. Additional useful information could include response costs or duration of any traffic slowdowns associated with the event.

Other agencies, particularly those with existing rock slope inventories, could also apply the methods in this paper to estimating event likelihood and risk. Ideally, extensive event data is already compiled for a long time period and easily available. In reality, initial data collection is the most difficult, requiring outreach and expert judgement in interpreting the data and transforming those interpretations into actionable information. However, once an initial asset condition–event likelihood relationship is developed, it will be easier to promote more detailed event tracking both to the maintenance personnel who will be collecting the data and the higher-level planners and managers who will be responsible for funding the risk reduction programs. As the data set improves over time, the condition–likelihood correlation will also become more statistically robust, which will increase its usefulness in decision support tools.

Application of modern methods to managing geotechnical assets is an incremental process, and agencies do not need to be daunted by the initial scarcity of data. Initial work should be seen as a seed that will grow into a risk-forecasting tool that helps the DOT efficiently improve corridor safety for the travelling public. In an age of limited budgets and growing traffic volumes, agencies are faced with a difficult task as they try to reduce risk as efficiently as possible. Developing and utilizing risk assessment tools will help an agency acknowledge, understand, and better manage its risk exposure.

References

1. Department of Transportation, Federal Highway Administration. *23 CFR Parts 515 and 667. Asset Management Plans and Periodic Evaluations of Facilities Repeatedly Requiring Repair and Reconstruction Due to Emergency Events.* Final Rule. Docket No. FHWA-2013-0052. FHWA, 2016.
2. Vessely, M. J., B. Widmann, B. Walters, M. Collins, N. Funk, T. Ortiz, and J. Laipply. Wall and Geotechnical Asset Management Implementation at the Colorado Department of Transportation. *Transportation Research Record: Journal of the Transportation Research Board*, 2015. 2529: 27–36.
3. Beckstrand, D., G. Machan, B. Shaw, and S. Enright. Communicating Geotechnial Risk and Complexity. Presented at 59th Annual Meeting of the Association of Environmental and Engineering Geologists, Kona, Hawaii, 2016.
4. George, M. F., C. M. Christiansen, R. Kromer, and J. Hutchinson. Rockfall Forecasting: A Probabilistic Approach. *Proc., 68th Highway Geology Symposium*, Highway Geology Symposium, Marietta, Ga., 2017, pp. 211–225.
5. Beckstrand, D., A. Mines, B. Black, J. Jackson, B. Boundy, S. Helm, D. Stanley, and P. Thompson. *Montana's Rock Slope Asset Management Program (RAMP).* 68th Highway Geology Symposium, Marietta, Ga., 2017.
6. Pierson, L. A., S. A. Davis, and R. Van Vickle. *Rockfall Hazard Rating System Implementation Manual.* FHWA-OR-EG-90-01. Oregon Department of Transportation, 1990.
7. Pierson, L. A., D. L. Beckstrand, and B. A. Black. *Rockfall Hazard Classification and Mitigation System.* FHWA/MT-05-011/8174. FHWA, U.S. Department of Transportation, 2005.
8. Landslide Technology. *Task 4 Report: Rock Slope Assessments.* State of Montana, Project No. 15-3059V. Montana Department of Transportation, 2016.
9. Turner, K. A., and R. L. Schuster. (eds.). *Rockfall: Characterization and Control.* Transportation Research Board, Washington, D.C., 2012.
10. Beckstrand, D., P. Thompson, R. Kimmerling, M. Vessely, B. Benko, and D. Stanley. Alaska's Geotechnical Asset Management Program. *Proc., 3rd North American Symposium on Landslides*, Association of Environmental and Engineering Geologists, Roanoke, Va., 2017, pp 709–721.
11. Beckstrand, D., A. Mines, P. Thompson, and B. Benko. Development of Mitigation Cost Estimates for Unstable Soil and Rock Slopes based on Slope Condition. Presented at 95th Annual Meeting of Transportation Research Board, Washington, D.C., Washington, D.C., 2016.
12. Landslide Technology. *Task 3 Report: Rock Slope Asset Management Program.* State of Montana, Project No. 15-3059V. Montana Department of Transportation, 2016.
13. Landslide Technology. *Task 5 Report: Determination of Critical Sites.* State of Montana, Project No. 15-3059V. Montana Department of Transportation, 2017.
14. Beckstrand, D., and A. Mines. Jump-Starting a Geotechnical Asset Management Program with Existing Data. *Transportation Research Record: Journal of the Transportation Research Board*, 2017. 2656: 23–30.
15. American Association of Highway and Transportation Officials. *User and Non-User Benefit Analysis for Highways.* American Association of Highway and Transportation Officials, Washington, D.C., 2010.

The Joint Subcommittee on Geotechnical Asset Management of AFP00 and AFS00 (AFP00[1]) peer-reviewed this paper (18-04621), which was selected by the Geological and Geoenvironmental Engineering Section (AFP00) to receive its 2018 Best Paper award.

Article

TRR
JOURNAL OF THE TRANSPORTATION RESEARCH BOARD

Transportation Research Record
2018, Vol. 2672(52) 325–332

DOI: 10.1177/0361198118778925
journals.sagepub.com/home/trr
SAGE

Intelligent Compaction for Quality Control and Acceptance for Soil and Base Compaction through Statistical Analysis

Jimmy Z. Si[1]

Abstract
This paper presents the results of the intelligent compaction data that were collected from various layers including subgrade soil, lime-treated subgrade, cement-treated base and flexible aggregate base layers. Sets of proof-mapping data were collected from each layer upon completion of compaction. The data was then downloaded and analyzed using a computer program. Based on the data analysis and field compaction observation, a new statistical methodology for analyzing intelligent compaction data is proposed. The method is used to assess the uniformity of soil and base compaction quality and this is successfully demonstrated through a case study. A typical normal distribution of an intelligent compaction dataset indicates that a good and uniform compaction is achieved. It is, therefore, possible to assess the compaction quality by evaluating the perfection of normal districtuion of an intelligent compaction (IC) dataset. The compaction uniformity is evaluated by a compaction uniformity index, which is defined as the ratio of the probability within the specified limits in a field compaction data distribution to the probability in a target normal distribution.

Good and long-lasting pavement performance can only be attained by adequate pavement design, appropriate material selection, and proper construction to ensure that the desired density and stiffness of the pavement layers are achieved. The field rolling compaction of earth material to prepare subgrade and build embankment, subbase, and base is one of the most important processes in roadway construction for both flexible and rigid pavements. Flexible pavements are typically designed to meet traffic demands for at least 20 years with regularly scheduled maintenance. Rigid pavements are typically designed for a service life at least 30 years with no major maintenance. One of the common causes of pavements beginning to crack and crumble prematurely is that the base or subgrade layers were not compacted adequately, and oftentimes not uniformly or consistently.

Current standards of most state highway agencies require contractors to build uniform pavement structural layers but with no means to check and quantify the uniformity of compaction during or after construction on each entire layer. Like most state highway agencies, the Texas Department of Transportation (TxDOT) typically uses nuclear density gauges as the primary tools to assess and accept the compaction work by measuring the dry density and moisture of soils and aggregate base materials to ensure that the desired density and moisture are achieved (*1*). Usually, the typical sampling of every 3,000–5,000 cubic yards for dry density and moisture content as quality assurance tests at randomly selected point locations is expected to represent the entire section. On the other hand, measurement of dry density and moisture content does not directly tie the construction quality and pavement performance to mechanistic-empirical design processes where the stiffness and modulus are the key design parameters.

In-situ nondestructive testing (NDT) devices that estimate the stiffness parameters of a constructed pavement structure are now commonly available. Such stiffness parameters provide a direct link to the pavement performance predicted through a mechanistic empirical-based design process. However, similar to density measurements, all NDT tests are point tests. Typically, any point test can only cover less than 1% of the area that is being

[1]Texas Department of Transportation, Austin, TX

Corresponding Author:
Address correspondence to Jimmy Z. Si: Jimmy.Si@txdot.gov

compacted. The rest of the area is assumed to have been compacted at least as well as, if not better than, the area that is actually tested. None of the NDT devices can provide the compaction uniformity information on the entire compacted area. This puts much risk on the owner when accepting the compaction work. After all, it is the uniformity of the compaction that assures the long-lasting pavement performance.

The intelligent compaction (IC) data collected from the TxDOT IC pilot projects have shown that there is no strong correlation bwtween the IC data and density measurements (*2*, *3*). Unlike density gauge or other in-situ stiffness-based device measurements that provide little or no uniform information on assessing the quality of compaction, the IC technology not only makes it possible to collect a large quantity of data that provide uniform information of compaction over the entire area being compacted, but also, for the first time, allows the use of statistical analysis as a tool for the compaction quality control and acceptance (QC/QA). In this paper, a new methodology is proposed based on the statistical analysis of the IC data to assess the compaction quality and is successfully demonstrated through an IC pilot project.

Theoretical Methodology

Intelligent Compaction (IC) technology is a continuous compaction quality control process whose objective is to achieve uniform and consistent compaction during construction. Extensive research has been carried out and extensive literature has been well documented regarding equipment development, data collection, analyasis and interpretation as well applications (*2–12*). Currently, the most popular analysis model is plotting a semivarigram graph. This model involves analyzing three parameters (range, sill, and nugget). It requires extensive knowledge and training for data interpretation and is not, therefore, convenient for implementation on a routine job.

The IC system consists of an onboard stiffness measurement device with a display panel or control box and a Global Positioning System (GPS) device. The display panel mounted on the roller cabinet shows a color-coded map continuously and in real time based on either the stiffness values or the number of passes. Therefore, the quality of compaction can be monitored by the roller operator in real time during the compaction process.

The raw IC data is also automatically and continuously uploaded to the "cloud" or the online data management center. The gridded data can then be downloaded to a computer through a web-based program and further analyzed using a free software program named VETA (*13*). The VETA program is specifically dsigned for IC data analysis. It allows contractors and TxDOT personnel to view and analyze the geospatial IC data for a given compacted area. VETA displays the compaction information such as material stiffness, roller vibration amplitude and frequency, number of rolling passes (pass counts), and roller speed as a color-coded map. The color-coded map can then be used by the inspectors to locate the poor compaction areas (under or over compaction) for further verification testing or corrective actions. This process is referred to as post analysis in this paper to avoid any confusion with the real-time displayed color-coded map that is only available in the in-cab control box during compaction.

It is necessary and helpful to differentiate these two types of color-coded maps. The real-time color-coded maps can be used for quality control purposes mainly for contractors to monitor the compaction process. It tells the roller operator how many passes have been achieved in a given area and when to stop rolling. The color-coded maps as a result of proof-mapping after post analysis using VETA can be used for the compaction acceptance purpose. Figure 1 shows a typical color-coded map of proof-mapping. It can be used as a guide by the state highway agencies for the acceptance of the compaction work. These two types of color-coded maps may use the same color scale but it is not always necessary or practical to do so. This study will focus on the IC data post analysis and its applications.

Currently, at least seven vendors including Bomag, Case/Ammann, Caterpillar, Dynapac, Hamm, Sakai, and Volvo manufacture IC rollers. These vendors offer IC rollers with either a smooth or padfoot single-drum for soil and base compaction. An IC retrofit kit has recently been developed and is commercially available. It is gaining popularity in the U.S. because of its flexibility and cost-effectiveness. This after-market equipment can be retrofitted onto any dynamic vibratory rollers and works the same way as the Original Equipment Manufacturer (OEM) IC rollers (*11*).

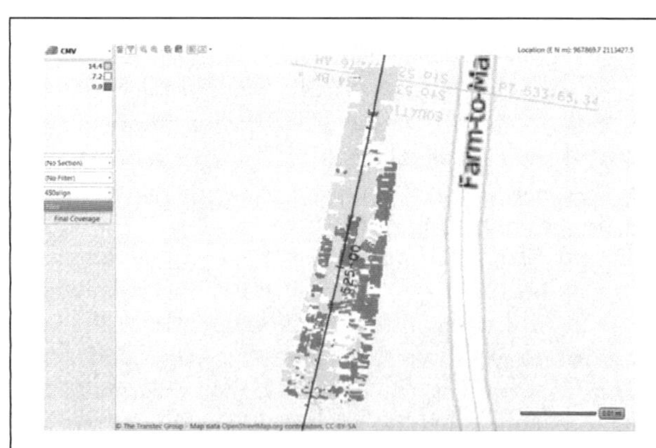

Figure 1. A typical color-coded map of proof-mapping.

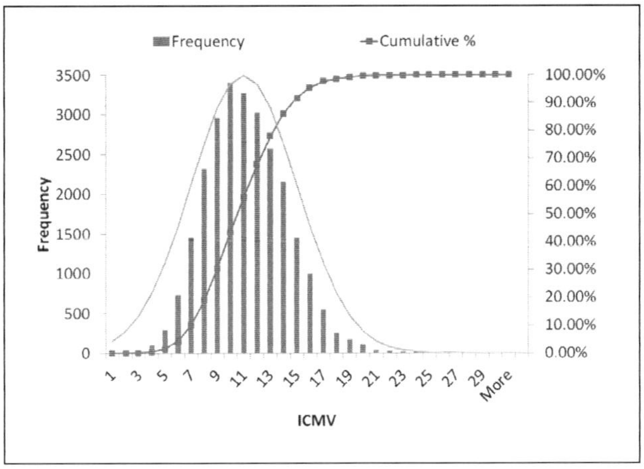

Figure 2. Histogram of IC data normal distribution.

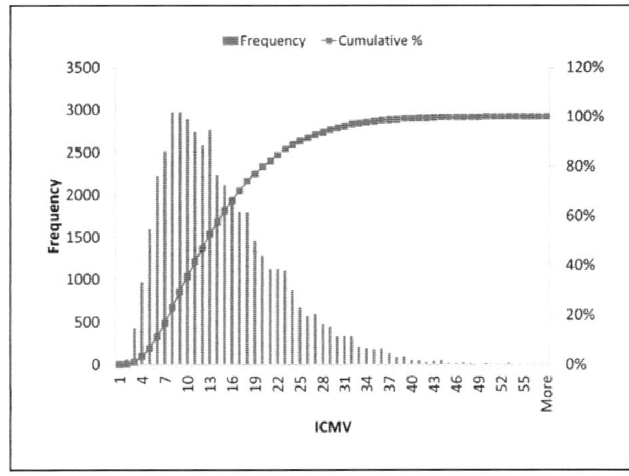

Figure 3. Typical histogram of original IC data distribution.

IC rollers have the capability to record compaction information such as stiffness, roller vibration amplitude and frequency, and roller speed for all passes during the compaction. Regardless of the algorithms that are used by the vendors to calculate the stiffness, all proof-mapping data, which are stiffness-based measurements that are collected upon completion of compaction, can be reasonably assumed in normal distribution. The normal distribution is a symmetric distribution that has been found to adequately describe the IC data populations that are collected during compaction in highway construction.

Figure 2 shows a typical example of a statistical histogram of an IC data set. It shows that the data distribution is close to a normal distribution (continuous line). ICMV is the acronym for the Intelligent Compaction Measurement Values.

Certainly, not all field compaction IC data sets follow a normal distribution curve. Based on TxDOT's database (*11*), most original IC datasets have, in fact, a right-skewed distribution with extremely high values on the right side as outliers, as shown in Figure 3.

Thus, the IC data were transformed by obtaining the square root of the original IC data (*14*). Figure 4 shows a typical transformed IC data distribution, which looks more like a normal distribution. This data transformation is necessary as it is more practical and meaningful for real IC data. In the following discussion, IC data are, therefore, always referred to as the transformed IC data, unless otherwise indicated.

Equation 1 represents a normal distribution in terms of its mean and standard deviation (*15*).

$$f(x) = \frac{1}{\sigma\sqrt{2\pi}} e^{-\frac{1}{2}\left(\frac{x-\mu}{\sigma}\right)^2} \quad (1)$$

where

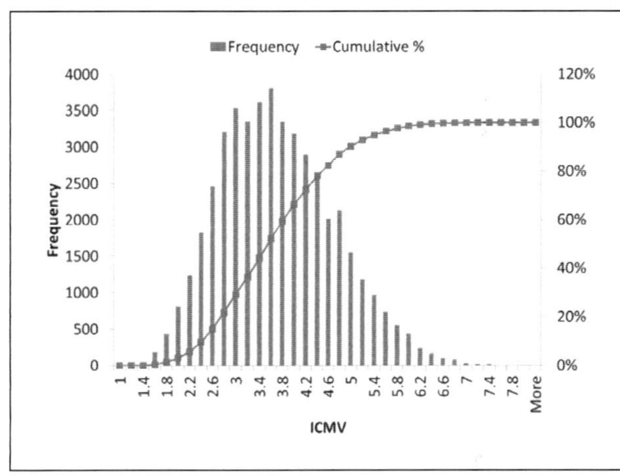

Figure 4. Typical histogram of transformed IC data distribution.

$f(x)$ is the probability density function of the normal distribution;

μ is the mean of the dataset; and

σ is the standard deviation of the dataset.

It can be clearly seen from Equation 1 that a particular normal distribution is uniquely and completely described if the mean and standard deviation associated with the distribution are known. A different normal distribution is created whenever the value of either the mean or standard deviation or a combination of both is changed. Figure 5 shows the conceptual normal distributions for three passes of compaction.

In Figure 5, the ICMV is the Intelligent Compaction Measurement Values; it is an entire set of the IC stiffness or modulus measurements for each pass during compaction. ICTV is the Intelligent Compaction Target Value; this is the average of the ICMV, and is used for the color-coded purpose in later data analysis.

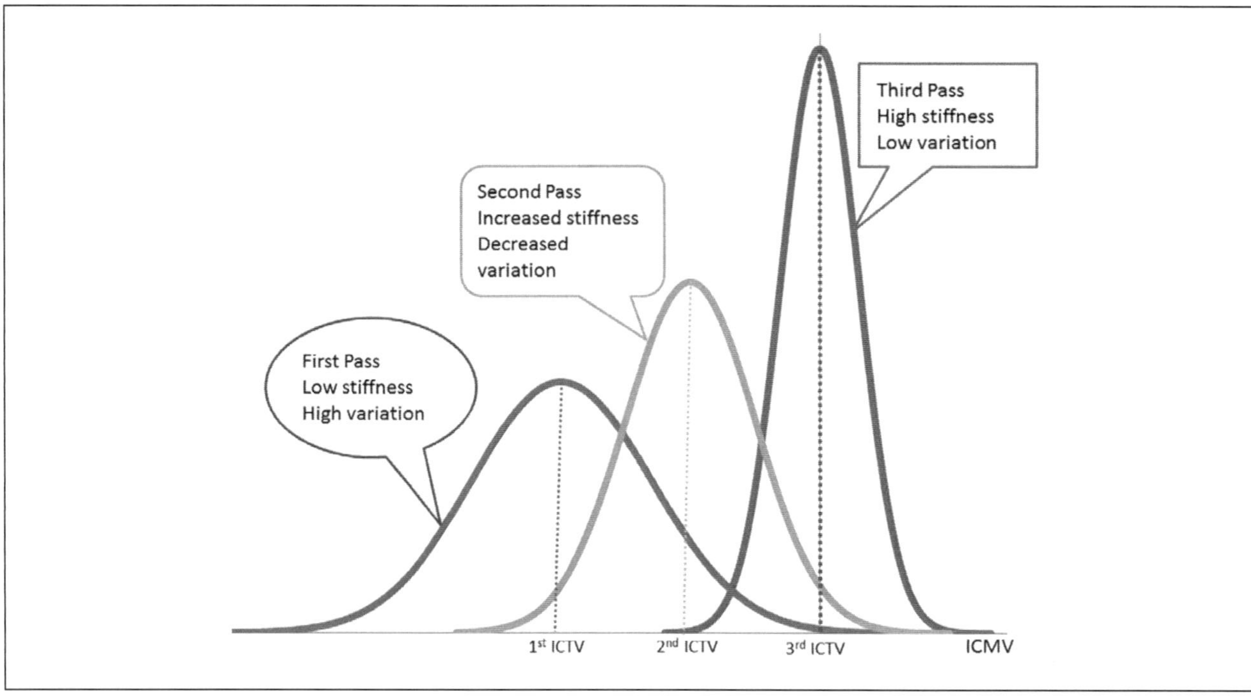

Figure 5. Conceptual compaction progresses in normal distribution.

As shown in Figure 5, there are three normal distributions, each with a different combination of a mean and a standard deviation that define the specific normal distribution. At the beginning of the compaction, the material is typically less stiff, so the average stiffness (1st ICTV) of the first pass is low and the variation of the data set is high so the curve is flatter. As the compaction continues, the material is getting compacted, the average stiffness (2nd and 3rd ICTVs) increases and the variation of the dataset decreases. Obviously, these are the ideal scenarios. In some cases, there may be several normal distributions with the same average stiffness (ICTV) but each with a different standard deviation.

One of the key objectives for compaction is to achieve the highest compaction average stiffness and the lowest variation of the data distribution. In a normal distribution, this means that the distribution curve is taller and thinner as shown in the third pass curve compared with the first and second pass in Figure 5. A taller and thinner shape of the normal distribution curve means that a better and more uniform compaction has been achieved.

In fact, a perfect uniform compaction will never be achieved or even desired in the field for various reasons such as time and cost. The question now is how much uniformity is good enough or adequate for soil and base compaction quality control or acceptance. One way to assess the uniformity of compaction is to calculate and use the probability or the area under the curve in Figure 2 by integrating the equation between the specified limits of interest. Fortunately, it is not necessary to do this because all normal distributions can be transformed into a standard normal distribution for which areas are readily available in the form of a reference table or a computer program.

The equation for the standard normal distribution is shown in Equation 2.

$$f(Z) = \frac{1}{\sqrt{2\pi}} e^{-\frac{Z^2}{2}} \quad (2)$$

where

$f(Z)$ is the probability density function of the standard normal distribution;

$$Z = \frac{(x - \mu)}{\sigma} \quad (3)$$

μ is the mean of the data set in normal distribution; and

σ is the standard deviation of the dataset in normal distribution.

The standard normal distribution is a normal distribution with the mean value of 0 and a standard deviation of 1. The horizontal axis for the standard normal distribution is usually designated as the Z-axis. The Z values are equal to the number of standard deviations above (positive Z values), or below (negative Z values) the mean of 0, as shown in Figure 6.

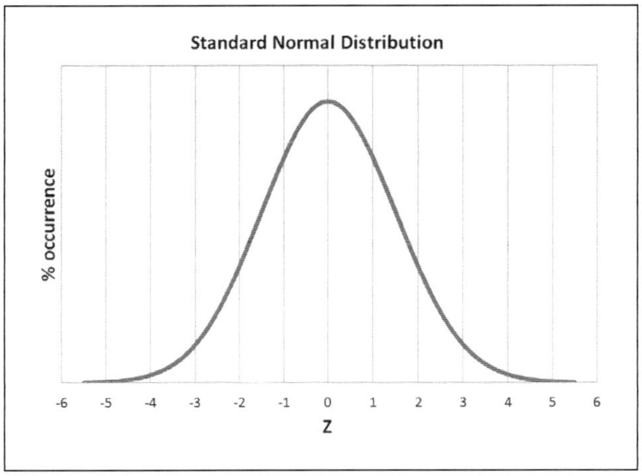

Figure 6. Standard normal distribution.

To transform a normal distribution into the standard normal distribution, the values of Z have to be calculated first using Equation 3. The area under the standard normal distribution curve between the specified limits can then be found using the standard normal distribution Z-table or a computer program.

For the purpose of demonstration, let the lower limit value be 75% of the average (ICTV) of an IC dataset (ICMV). Since the data are assumed in normal distribution, it is a symmetrical or "bell" shaped curve (Figure 6). The upper limit value can, therefore, be set as 125% of the average (ICTV) of the IC data set (ICMV). Oviously it is not always necessary to set up symmetrical limits. In fact, using the standard normal distribution table, the probability or area within any specified limits under any normal distribution curve can be calculated.

Assume the target or desired coefficient of variation (COV) is 10%, therefore,

$$\mu = \frac{\sigma}{0.10} = 10\sigma.$$

The positive Z value is calculated using Equation 3 as

$$Z = \frac{(x-\mu)}{\sigma} = \frac{(1.25\mu - \mu)}{\sigma} = \frac{0.25\mu}{\sigma} = \frac{0.25 \times 10\sigma}{\sigma} = 2.50$$

Thus, by looking at the standard normal distribution table, for $Z = 2.50$, the area under the standard normal distribution curve is 0.4938. This means that 49.38% of the area under the normal distribution curve falls between the average, μ and the upper limit, 1.25 μ. Since the distribution curve is symmetrical, a total of 98.76% of the area under the standard normal distribution curve is between the lower (0.75μ) and upper (1.25 μ limits).

If the COV of a real set of IC data is 20%, then $\mu = \frac{\sigma}{0.20} = 5\sigma$. So, similarly, using Equation 3,

$$Z = \frac{(x-\mu)}{\sigma} = \frac{(1.25\mu - \mu)}{\sigma} = \frac{0.25\mu}{\sigma} = \frac{0.25 \times 5\sigma}{\sigma} = 1.25.$$

Therefore, from the Z-table, for $Z = 1.25$, a value of 0.3944 is determined. This means that 39.44% of the area under the standard normal distribution curve is between the average, μ and the upper limit, 1.25 μ and 78.88% of the area is between the lower (0.75 μ) and upper (1.25 μ) limits.

The analyses and calculations above are based on the data that are in a perfect normal distribution. For soil and base construction, depending on the field conditions when the IC data were collected for a given land or section, the dataset is often not a perfect normal distribution. The coefficients of variation of these datasets are normally greater than those of a normal distribution. It is, therefore, possible to compare a real data distribution with a normal distribution to assess how much the real data distribution deviates from a perfect normal distribution by dividing the area or probability under the real data distribution curve by the area or probability under the normal distribution curve that is set up as a target for specific compaction associated with the specified limits and the coefficient of variation. The result can be used as an index tool for the quality control or acceptance of compaction.

For the same example above, if the dataset is normally distributed with a 10% COV, the probability under the curve within the specified limits is 98.76%. If the data is not normally distributed but assumed normally distributed with a greater COV of 20%, a value of 78.88% of probability within the same specified limits is determined.

Finally, the extent to which the IC data distribution deviates from a desired normal distribution is calculated by dividing the probability of the IC data set that is assumed normally distributed by the probability within the same specified limits in a desired normally distributed population. The result can be defined as the Compaction Uniformity Index (CUI). That is, $\frac{78.88}{98.76} \approx 79.9\%$. This means that the IC data set with 20% COV is about 79.9% of probability close to the desired or target uniform compaction within the same specified limits. CUI indicates how uniform the compaction section has been achieved based on both the specified limits and the coefficient of variation. The benefit of using both specified limits and the coefficient of variation is to allow us to set up different acceptance criteria for different materials and application situations.

For demonstration purposes, as an example in this paper, a dataset with 10% COV and specified limits between 75% and 125% of ICTV (average) of the ICMV was assumed to follow a normal distribution and was used as the target uniformity of compaction which serves as the basis for comparing with any other IC data that are less like normal distribution. The reasonable lower

Table 1. Statistical Analysis Results and CUI for Various Materials

Materials	Min.	Max.	Average	Standard deviation	COV, %	Sample size	Z	Area by Z	CUI, %
Subgrade	1.0	7.5	3.6	1.0	27.8	46,814	0.90	0.3159	64.0
LTS	1.2	8.1	3.6	0.9	25.0	75,025	1.00	0.3413	69.1
CTB	1.4	6.9	3.8	0.7	18.4	69,550	1.36	0.4131	83.7
Flex Base	2.8	9.0	5.3	0.9	17.0	36,257	1.47	0.4292	86.9

Note: LTS = lime-treated subgrade; CTB = cement-treated base.

and upper specification limits and desirable COV of an IC dataset which are expected as the desired or target uniform compaction need to be established based on an extensive series of field trial compaction tests or historical experience before construction for the specific types of materials and applications.

Results and Discussion

TxDOT has been implementing intelligent compaction technology in about 25 projects since 2008. The agency specified the use of the intelligent compaction on soil and base compaction projects to evaluate the effectiveness and efficiency of the technology for potential application in future projects. The following IC data was collected from a TxDOT IC pilot project. The data shown here are only for the purpose of demonstrating the procedure proposed in this paper. TxDOT has not mandated contractors to use any IC information as a tool for either quality control or acceptance of compaction.

It is worth pointing out that the proposed procedure did not include the moisture as a factor simply because currently there are no satisfactory procedures for continuously measuring moisture in the field in real time during compaction. However, by implementing the proposed procedure using CUI, to get the best uniformity of compaction, contractors will need to provide continuous quality construction to keep moisture and chemical additives uniformly distributed and to minimize material segregation. The moisture and material segregation issues can be addressed by contractors through best practice in material management and construction processes. Owners simply need to set up appropriate or reasonable criteria based on a CUI for quality control and acceptance for given types of materials and applications.

The SH24 project is a widening project that adds two new lanes to the existing roadway. A Caterpillar pad-foot roller equipped with the Trimble IC retrofit kit was used for the compaction of subgrade soil, lime-treated soil, cement-treated base, and flexible-base material. The project has been running smoothly and successfully thanks to the understanding of this technology by TxDOT and the contractor as well as good cooperation of both. The IC data have been collected, downloaded, and processed typically on a daily basis. The color-coded maps showing compaction variability were created and used as a guide by district inspectors for selecting the locations for density measurements which are currently required for the acceptance in accordance with the TxDOT IC special specification (*16*).

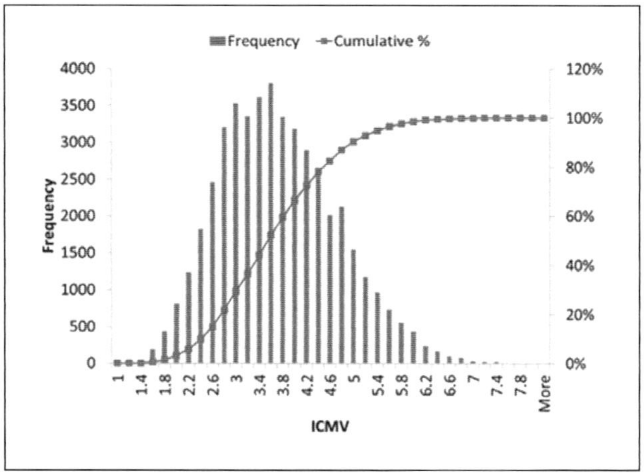

Figure 7. Histogram of IC data distribution for subgrade.

Figures 7 through 10 are examples of IC data distributions for different types of materials from Station 815 + 00 to 826 + 00. The construction sequence is as follows: an 8 in. lime-treated soil layer was constructed on the top of subgrade, followed by an 8 in. cement-treated base (Type D or E, Grade 4, contains 50% RAP) and then a 6 in. flexible (aggregate) base (Type D, Grade 4). The statistical analysis results are summarized in Table 1.

Table 1 shows that, in general, the uniformity of compaction improves after adding subsequent layers based on the CUI values which are from 64.0% for the subgrade to 86.9% for the flexible base. However, the average stiffness of the LTS dataset did not improve compared with that for the subgrade. This could be attributed to the additional moisture content when lime slurry was added and the time when the IC data were collected, since the stiffness is highly sensitive to the moisture content (*11*). In fact, just as the moisture and

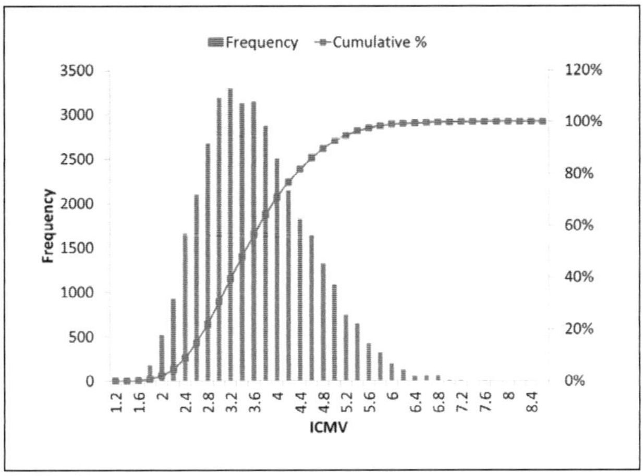

Figure 8. Histogram of IC data distribution for lime-treated subgrade.

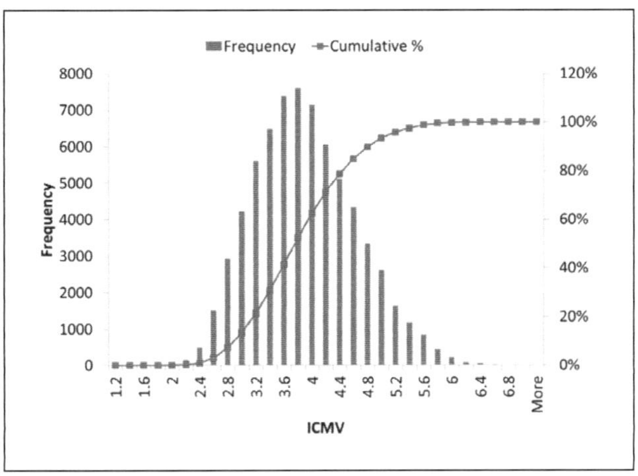

Figure 9. Histogram of IC data distribution for cement-treated base.

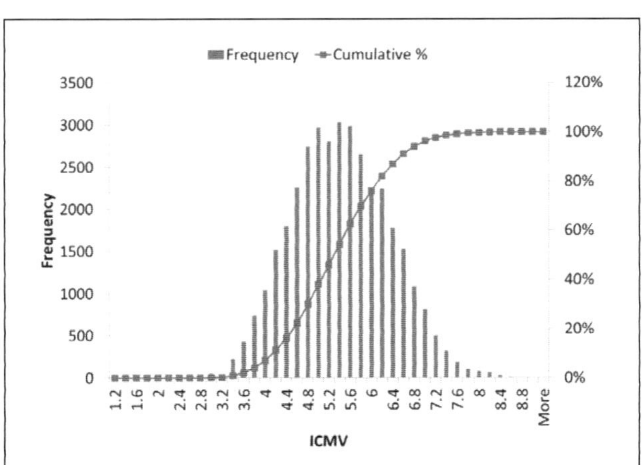

Figure 10. Histogram of IC data distribution for flexible base.

the chemical reactions between the lime and soil keep changing with time, the stiffness of the material also continues to change. This is a good example that explains why it is very difficult, if not impossible, to specify a single target stiffness value for each material without measuring the moisture content and its variation in the field in real time during compaction. It also indicates that to obtain a consistent uniformity value for a fair comparison section by section, it is important to specify a time frame in which the IC data need to be collected especially for any chemical treatment of soils and bases.

Conclusion

The intelligent compaction technology allows for collection of stiffness information directly from a given compacted area in real time during compaction. The analyses of the IC data which have been collected to date demonstrate that for good uniform compaction, the IC data distribution should be in a relatively well defined normal distribution. A new approach that is based on the statistical analysis of the IC data for possible quality control and acceptance was proposed and successfully demonstrated in this paper. By comparing the field IC data distribution with a normal distribution, a new Compaction Uniformity Index (CUI) has been defined based on the statistical analysis of the IC data. Based on these comparisons, the CUI can be used as a good indicator for accessing the uniformity and consistency of compaction.

In general, the more uniform the compaction is, the more likely it is that the IC data distribution will become a normal distribution. Generally, placing subsequent layers of materials or chemical treatment of soils and bases improves the uniformity of the compaction, if designed and constructed properly. Full implementation of this new QC/QA tool based on the statistical analysis of the IC data will result in a uniform and consistently well compacted foundation for the subsequent pavement layers to be built upon. Density or other in-situ NDT point measurements can be used to verify the poorly compacted areas (under or over compaction) that are identified in a color-coded map by the IC proof-mapping. Other construction compaction issues related to the uniformity and consistency due to material segregation and moisture variation can also be assessed by using a Compaction Uniformity Index (CUI) and mitigated through tightening up the acceptance criteria which are pre-defined by specification limits and coefficient of variation.

Author Contributions

The author confirms sole contribution to the paper as follows: study conception and design; data collection; analysis and interpretation of results; and draft manuscript preparation.

The author reviewed the results and approved the final version of the manuscript.

References

1. *Field Method for Determining In-place Density of Soils and Base Materials*. Tex-115-E. TxDOT Test Procedures,Texas Department of Transportation, Austin, Texas, 1999.
2. Anderegg, R., and K. Kaufmann. Intelligent Compaction with Vibratory Rollers Feedback Control Systems in Automatic Compaction and Compaction Control. *Transportation Research Record: Journal of the Transportation Research Board*, 2004. 1868: 124–134.
3. Sandström, A. J., and C. B. Pettersson. Intelligent Systems for QA/QC in Soil Compaction. Presented at 83rd Annual Meeting of the Transportation Research Board, Washington, D.C., 2004.
4. Petersen, D., J. Siekmeier, C. Nelson, and R. Peterson. Intelligent Soil Compaction – Technology, Results and a Roadmap toward Widespread Use. *Transportation Research Record: Journal of the Transportation Research Board*, 2006. 1975: 81–88.
5. Thompson, M., and D. White. Field Calibration and Spatial Analysis of Compaction Monitoring Technology Measurements. *Transportation Research Record: Journal of the Transportation Research Board*, 2007. 2004: 69–79.
6. White, D., and M. Thompson. Relationships between Insitu and Roller-integrated Compaction Measurements for Granular Soils. *Journal of Geotechnical and Geoenvironmental Engineering*, Vol. 134, No. 12, 2008, pp. 1763–1770.
7. Mooney, M. A., and D. J. White. *Intelligent Soil Compaction Systems*. NCHRP Report 676, Project 21-09. Transportation Research Board, Washington, D.C., 2010.
8. Vennapusa, P., D. J. White, and M. Morris. Geostatistical Analysis of Spatial Referenced Roller-integrated Compaction Measurements. *Journal of Geotechnical and Geoenvironmental Engineering*, Vol. 136, No. 6, 2010, pp. 813–822.
9. White, D. J., P. Vennapusa, and H. Gieselman. Field Assessment and Specification Review for Roller-Integrated Compaction Monitoring Technologies. Special Issue: Advances in Instrumentation and Monitoring, *Journal Geotechnical Engineering in Advances in Civil Engineering*, Vol. 2011, 2011.
10. Chang, G., Q. Xu, J. Rutledge, B. Horan, L. Michael, D. White, and P. Vennapusa. *Accelerated Implementation of Intelligent Compaction Technology for Embankment, Subgrade Soils, Aggregate base, and Asphalt Pavement Materials*. FHWA/TPF Research Project, Report No. FHWA-IF-12-002. Transportation Research Board, Washington, D.C., 2011.
11. Nazarian, S., M. Mazari, G. Chang, R. Aldouri, and J. Beltran. *FHWA TIDP 130(096) Research Project Report: Intelligent Compaction Roller Retrofit Kit Validation*. Transportation Research Board of the National Academies, Washington, D.C., 2015.
12. Nazarian, S. M., M. Mazari, I. N. Abdallah, A. J. Puppala, , and L. N. Mohammad. *NCHRP Report 10-84: Modulus Based Construction Specification for Compaction of Earthwork and Unbound Aggregate*. Transportation Research Board of the National Academies, Washington, D.C., 2014.
13. Veta Data Management and Analysis Software. *The Transtec Group, Inc*., Austin, Texas, 2015. http://www.intelligentcompaction.com/veta/. 2015
14. Herrington, P. R., S. Kodippily, and T. F. P. Henning. *Flushing in Chipseals*. New Zealand Transport Agency Research Report 576, 2015.
15. Burati, J. L., and C. S. Hughes. *Material Control and Acceptance – Quality Assurance*. Highway Materials Engineering, Module I, FHWA-NHI Course Number 13123, Report No: FHWA-RD-02-095. U.S. Department of Transportation, Federal Highway Administration, Washington, D.C., 2001.
16. *Intelligent Compaction of Soil and Flexible Base*. TxDOT Special Specification (one-time-use, SS2304), Texas Department of Transportation, Austin, 2015.

The Standing Committee on Geotechnical Instrumentation and Modeling (AFS20) peer-reviewed this paper (18-00113).

Article

TRR
JOURNAL OF THE TRANSPORTATION RESEARCH BOARD

Transportation Research Record
2018, Vol. 2672(52) 333–346

DOI: 10.1177/0361198118782266
journals.sagepub.com/home/trr
SAGE

Spatial Verification of Modulus for Pavement Foundation System

David J. White[1], Pavana Vennapusa[2], Erol Tutumluer[3], William (Bill) Vavrik[4], Maziar Moaveni[3], and Steven Gillen[5]

Abstract

This paper presents the results of a field study recently completed on an Illinois Tollway construction project, in which intelligent compaction (IC) measurements were calibrated on site with in situ elastic modulus (E), resilient modulus (M_r), modulus of subgrade reaction (*k*), and California bearing ratio (CBR) testing. The purpose of the field calibration was to generate geo-referenced spatial modulus maps that can be used for field verification of design input parameters. E values were obtained using light weight deflectometer (LWD), stress-dependent M_r values and static *k* values were obtained using automated plate load testing (APLT), and CBR values were obtained using a dynamic cone penetrometer (DCP). The IC measurements were obtained using a vibratory smooth drum roller outfitted with an index measurement value system, and a roller outfitted with a retrofit kit programmed to produce validated integrated compaction (VIC) measurements through a process of field calibration. Results showed that VIC based calibration can produce correlations with $R^2 > 0.9$ with M_r and *k* values, and the VIC maps can be reliably used for quality assurance/verification testing. Although the index based IC measurement values showed statistically significant relationships, the R^2 values were lower (<0.6). An implementation framework that emphasizes an independent certification approach for field calibration is being developed based on these test results for the Illinois Tollway, to aid in the effective implementation of the modulus-based mapping approach for pavement design field verification.

Intelligent compaction (IC) technologies have been used in the U.S. on over 330 research, demonstration or pilot implementation projects between 2000 and 2016, of which nearly 100 projects were on pavement foundation materials. Several technical articles have been published describing sensor measurements, correlation analysis, data interpretation, and implementation challenges (for examples see references *1–19*). Currently, the Federal Highway Administration (FHWA) has put forth prescriptive specifications that focus on IC equipment and the procedure/format for data reporting. The current specifications lack a detailed framework for calibration (i.e., correlations with independent testing) and validation of results (i.e., accuracy and system quality checks) when considering mechanical soil properties. A version of the current FHWA specifications still requires the IC data to be calibrated with density measurements, which has been proven in numerous correlation studies to be practically impossible (*14, 16*).

The next big leap forward in earthwork construction and pavement foundation construction quality will be realized through developing statistically valid relationships between IC measurement values and mechanical properties of compacted materials that directly link to the design input parameter values. Providing contractors and owners with mechanical property outputs in real-time with nearly 100% spatial coverage of the project will substantially reduce the risk of not meeting the pavement design criteria, thus helping to insure long-term performance. To achieve this goal and advance the current state of the IC technology implementation, the Illinois Tollway Authority sponsored a research project titled "Validation of Intelligent Compaction to Characterize Pavement Foundation Mechanical Properties". One of the primary objectives of the research project is to develop a procedure for field validating the relationship between IC measurements and soil mechanical properties.

[1]Ingios Geotechnics, Inc, Northfield, MN
[2]Ingios Geotechnics, Inc, Little Elm, TX
[3]University of Illinois at Urbana-Champaign, Urbana, IL
[4]Applied Research Associates, Champaign, IL
[5]Illinois Tollway Authority, Downers Grove, IL

Corresponding Author:
Address correspondence to David J. White: David.White@ingios.com

Table 1. Properties of Soil Materials Tested

Parameters	Subgrade	Recycled asphalt pavement (RAP)	Subgrade aggregate or PGE
Classification	Sandy lean clay	Well graded gravel with sand	Poorly graded gravel
USCS	CL	GW	GP
AASHTO	A-7-6 (9)	A-1-a	A-1-a
Particle-sizes			
Max. particle size (in.)	3/8	1.5	5.0
% Gravel	1	68	98
% Sand	32	31	2
% Silt	36	1	0
% Clay	31		
Atterberg limits			
Liquid limit	34	Non-plastic	Non-plastic
Plastic limit	17		
Plasticity index	17		
Standard proctor			
Optimum moisture content (%)	14.0	NA	NA
Max. dry density (pcf)	114		

Note: AASHTO = American Association of State Highway and Transportation Officials; USCS = Unified Soil Classification System; CL = lean clay; GW = well-graded gravel; GP = poorly graded gravel; pcf = pounds per cubic foot.

As part of the research project, field calibration testing and geo-referenced spatial mapping was conducted on pavement foundation materials during construction on the Elgin O'Hare Expressway in Elgin, IL. A Hamm vibratory smooth drum IC roller outfitted with a Hamm measurement value (HMV) system and a caterpillar vibratory smooth drum roller equipped with a retrofit measurement kit that was programmed with a validated intelligent compaction (VIC) system developed by Ingios were evaluated in this study. Field calibration testing involved conducting in situ spot testing using a light weight deflectometer (LWD), dynamic cone penetrometer (DCP), and automated plate load testing (APLT) to develop regression relationships. LWD was used to measure the elastic modulus, dynamic cone penetrometer was used to measure the California bearing ratio (CBR) of the top and bottom layers, and APLT was used to measure in-situ stress-dependent resilient modulus (M_r) from cyclic plate load testing at different stress levels and the modulus of the subgrade reaction (k) value from static plate load testing.

In this paper, results from field testing and mapping conducted on subgrade, subgrade aggregate or porous granular embankment (PGE) material, and recycled asphalt pavement (RAP) base materials are presented. The field calibration testing results were used to assess if statistically significant and reliable regression relationships could be developed so that the IC roller map could be reliably converted to a map that displays the mechanical property of interest, rather than an index-based IC measurement value. Results and approaches presented in this paper are being used to develop the next generation IC specifications for implementation on the IL tollway projects and should be of interest to other state and federal highway agencies.

Materials

Testing was performed in this study on compacted subgrade, 6 in. of PGE placed over the compacted subgrade, and 3 in. of RAP base material placed over the PGE. A summary of the material properties is provided in Table 1. Subgrade consisted of sandy lean clay, PGE material consisted of poorly graded gravel material with 5 in. maximum particle size, and RAP material consisted of well graded gravel with sand.

Methods

Hamm Measurement Value

A Hamm's H11i smooth drum vibratory roller weighing about 24,857 lbs. outfitted with Hamm's IC measurement system was used in this study. The roller was outfitted with a global positioning system (GPS) with corrections from satellite-based augmentation system (SBAS). The IC measurement system reported HMV which is the same as the more commonly referenced compaction meter value (CMV) measurement system. Like CMV, HMV is also an index value and is a dimensionless compaction parameter that depends on the roller dimensions (i.e., drum diameter and weight), and roller operation parameters (e.g., frequency, amplitude, and speed) and is determined using the dynamic roller response. HMV or CMV are calculated as the ratio of the acceleration of the first harmonic component of the vibration ($A_{2\Omega}$) and the acceleration of the fundamental component of the vibration (A<συβ><συβ><συβ>Ω</συβ></συβ></συβ>) multiplied by a constant value (C). The C value used depends on the manufacturer. It must be noted that the HMV or CMV measurements and thereby its correlations with the mechanical property measurements are influenced by drum bouncing (*14, 20*). Based on past studies, HMV or CMV measurements have a measurement influence depth of about 3 to 5 feet (*20*).

In this study, the HMV value reported by Hamm on the on-board display was different than what was exported. After a close review of the data and communication with personnel at Hamm, it was found out that the outputted values are 10x the values displayed on the screen. It was unclear which values (reported on screen or the outputted) were the correct ones to use. For this reason, the HMV values reported in the correlations with the in situ point measurements show 10x higher values than reported in the color-coded maps in this paper.

Validated Integrated Compaction Measurement Values

VIC is a technique that uses advanced data analytics and requires site specific calibration of the roller sensor measurements using in situ plate load test measurements (e.g., in situ resilient modulus). The VIC calibration process utilizes the full spectrum of the drum acceleration signature rather than pre-selected frequencies as in the case of CMV. Recent field calibrations on subgrade and base materials using this approach showed that the coefficient of determination (R^2) > 0.95 are achievable using this technique, compared to R^2 of 0.6 using CMV for the same data (*21*). Because of the high R^2 values and relatively low measurement errors associated with this site calibration process, the calibration relationships can be reliably used to develop the desired mechanical property maps for quality control/acceptance. With VIC, the output maps are validated mechanical property outputs (e.g., M_r or k values for this project).

In this study, a caterpillar smooth drum vibratory roller weighing about 27,450 lbs outfitted with an Ingios VIC retrofit system was used. For comparison purposes, CMV was also simultaneously measured.

Light Weight Deflectometer (LWD) Testing

The LWD was setup with a 11.81 in. diameter plate (Figure 1*a*), and the tests were performed following the manufacturer recommendations (*22*). The elastic modulus values were determined using Equation 1:

$$E_{LWD} = \frac{(1-\eta^2)\sigma_0 r}{D_0} \times F \qquad (1)$$

In which, E_{LWD} is the elastic modulus (psi), D_0 is the measured peak plate deformation (in.), η is the Poisson's ratio (assumed as 0.4), σ_0 is the applied stress (psi) is the 14.5 psi, r is the radius of the plate (in.) which equals 5.9 in., and F is the shape factor depending on stress distribution (assumed as 8/3 for granular materials). The measurement influence depth of LWD testing is about 1 to 2 ft. based on the criteria of 1 to 1.5 times the diameter of the loading plate (*14, 20*).

Dynamic Cone Penetrometer

The DCP tests were performed in accordance with ASTM D6951-03 (*23*). The tests involved dropping a 17.6 lb. hammer from a height of 22.6 in. and measuring the resulting penetration depth (Figure 1*b*). A 30-in. penetrating rod was used. CBR values were determined using Equations 2 and 3, whichever was appropriate, in which the dynamic penetration index (*DPI*) is in units of mm/blow.

$$CBR(\%) = \frac{292}{DPI^{1.12}} \qquad (2)$$

for all materials except CL soils with CBR <10

$$CBR(\%) = 1/(0.017019 \times DPI)^2 \qquad (3)$$

for CL soils with CBR <10

The DCP test results were used to determine an average CBR of a given layer or for a given depth. For tests conducted on 6 in. of PGE over subgrade and 3 in. of RAP over 6 in. of PGE and subgrade, the average CBR of the top and bottom layer are reported. The top layer was either the PGE or PGE + RAP, and the bottom layer was the top 12 in. of the subgrade which represents the top subgrade layer. The average value was reported by calculating the DPI value based on the total number of blows taken to the desired depth.

Automated Plate Load Testing

APLT was developed recently to directly and rapidly measure the in situ M_r and k values through automated cyclic and static plate load testing, respectively (*24*). The major advantages of using in situ cyclic testing to determine M_r using APLT is the ability to perform a conditioning stage such as a laboratory M_r test (*25*) and obtain M_r values at various cyclic and confining stresses (Figure 1*c*).

In Situ Composite Resilient Modulus

Cyclic APLT using a 12-in. diameter loading plate was conducted to determine in situ composite resilient modulus ($M_{r\text{-Comp}}$) at six different stress levels, which involved one conditioning sequence with 500 cycles followed by six loading steps with 100 cycles each. The average of the last 5 cycles from each step were used for calculations.

The $M_{r\text{-}Comp}$ was calculated as the ratio of the cyclic stress divided by the resilient deflection (during unloading) using the Boussinesq's half-space equation:

$$M_{r-Comp} = \frac{(1-\eta^2)\Delta\sigma_{cyclic} r}{\delta_r} \times F \quad (4)$$

In which $M_{r\text{-}Comp}$ is the in situ composite resilient modulus (psi), δ_r is the resilient deflection of plate during the unloading portion of the cycle (determined as the average of three measurements along the plate edge), η is the Poisson's ratio (assumed as 0.4), $\Delta\sigma_{cyclic}$ is the cyclic applied stress (psi), r is the radius of the plate (in.), and F is the shape factor depending on stress distribution (assumed as 8/3 for granular materials). Using the criteria of 1 to 1.5 times the plate diameter for measurement influence depth, the $M_{r\text{-Comp}}$ values have an influence depth of about 1 to 1.5 ft.

The M_r parameter is a highly stress-dependent parameter, and most soils exhibit the effects of increasing stiffness with increasing bulk stress and decreasing stiffness with increasing shear stress. The APLT testing program was designed to assess the in situ composite resilient modulus at six different stress levels. The results were used to model the behavior using the "universal" model (*26*) shown in Equation 5:

$$M_r = k_1^* P_a \left(\frac{\theta}{P_a}\right)^{k_2^*} \left(\frac{\tau_{oct}}{P_a} + 1\right)^{k_3^*} \quad (5)$$

In which M_r is the in situ resilient modulus (psi), P_a is the atmospheric pressure (psi), θ is the bulk stress (psi) equal to $\sigma_1 + \sigma_2 + \sigma_3$, σ_1 is the applied cyclic stress ($\Delta\sigma_{\text{cyclic}}$) used in $M_{r\text{-}comp}$ calculations because there is no confining stress at the surface, $\sigma_2 = K_o\sigma_1$ and $\sigma_3 = \sigma_2$, K_o is the coefficient of lateral earth pressure at rest which is equal to $\eta/(1\text{-}\eta)$, η is the Poisson's ratio assumed as 0.4; τ_{oct} is the octahedral shear stress (psi) $= \sqrt{(\sigma_1-\sigma_2)^2 + (\sigma_2-\sigma_3)^2 + (\sigma_3-\sigma_1)^2} \Big/ 3$; and k_1^*, k_2^*, and k_3^* are the regression coefficients determined from in situ testing (these coefficients are presented here with a * to differentiate with the regression coefficients traditionally developed using laboratory test results).

Modulus of Subgrade Reaction

Static plate load tests were conducted in general accordance with the American Association of State Highway and Transportation Officials (AASHTO) standard for nonrepetitive loading using the static plate load test (*27*) to determine the k value using a 30-in. diameter loading plate setup shown in Figure 1*d*. A thin layer of fine silica sand was used as a bedding material for all tests. Using the criteria of 1–1.5 times the plate diameter for the measurement influence depth, the k values were determined using the 30-in. diameter loading plate to have an influence depth of about 2.5–3.8 ft.

The test standard requires increasing applied stresses up to 30 psi in 5 psi increments. In this study, applied stresses were increased up to a maximum of at least 15 psi in 2.5 psi increments. The test was performed for two loading cycles. Plate deformations were measured at three locations along the edge of the plate. The uncorrected k value was determined using Equation 5.

$$k_u' = \frac{10\,psi}{\delta} \quad (5)$$

In which k'_u is the uncorrected modulus of subgrade reaction (pci), and δ is the deformation corresponding to the 10-psi loading increment (inches). In this study, a plot of applied stress on the x-axis and average plate deflection on the y-axis is prepared for the two loading cycles. Then a second order polynomial curve is fitted separately for both the first and second loading cycles, using the model shown in Equation 6:

$$y = a_1x^2 + a_2x + a_3 \quad (6)$$

In which y is the deflection in inches, x is the applied stress in psi, and a_1, a_2, and a_3 are the regression coefficients. To assess the quality of the regression fit, the coefficient of determination (R^2) value is determined. A minimum R^2 value of 0.98 has been established as required to achieve acceptable results.

Using the second order polynomial fit parameters the average plate deflections corresponding to a target applied stress (σ) are computed using Equations 7 and 8 for the first and second load cycles, respectively:

$$\delta_i = a_1\sigma^2 + a_2\sigma \text{ for } 1^{st} \text{ loading cycle} \quad (7)$$

$$\delta_r = a_1\sigma^2 + a_2\sigma \text{ for } 2^{nd} \text{ loading cycle} \quad (8)$$

In this study a target applied stress of $\sigma = 10$ psi has been used. The k'_u values calculated for the 1st and 2nd loading cycles are reported as $k'_{u(1)}$ and $k'_{u(2)}$, respectively. The k'_u values were then corrected for plate bending to determine k_u following the procedure described in AASHTO T222 and Equation 9 for $k'_u \geq 100$ pci and $\leq$ 1000 pci.

$$k_u = -39.9178 + 5.5076 \times [k'_u]^{0.7019} \quad (9)$$

The k_u values calculated for the 1st and 2nd loading cycles are reported as $k_{u(1)}$ and $k_{u(2)}$, respectively.

Test Results

HMV measurements were obtained in two areas consisting of 6 in. PGE over compacted subgrade, one area with

Figure 1. In situ testing methods used at this site: (*a*) light weight deflectometer, (*b*) dynamic cone penetrometer, (*c*) automated plate load testing setup for measuring in situ resilient modulus using cyclic testing with 12 in. diameter plate, and (*d*) in situ modulus of subgrade reaction with static testing with a 30 in. diameter plate.

3 in. RAP base layer over 6 in. PGE and subgrade and one area with a temporary access road with RAP material over PGE and subgrade. A screen shot of the HMV color-coded map in an area with PGE material is shown in Figure 2, which showed significant variability. The area consisted of about 70 ft. wide of mainline and an on-ramp. Visually, the ramp area showed stiffer conditions than in the mainline. Discussion with field engineers indicated that the ramp area consisted of core-outs involving excavation of soft unstable subgrade soils during construction and replacement with crushed rock material.

A histogram of the HMV measurements from two PGE areas along with LWD and DCP-CBR measurements obtained are summarized in Figure 3. DCP tests conducted at 30 test locations and LWD tests conducted at 55 locations in the PGE areas were selected based on the HMV color-coded map to capture the variability observed. The histogram plots summarize the univariate statistics, that is, mean (μ), standard deviation (σ), and coefficient of variation (COV). HMV measurements showed a COV of 63%. In situ E_{LWD} and DCP-CBR measurements showed COV between 38% and 97%. DCP-CBR of subgrade was more variable (with COV of 97%) than the DCP-CBR of the top PGE layer (with COV of 57%).

Based on the authors' experience, the COV values measured in the PGE test areas are considered high. For reference, ISSMGE reports an acceptable COV to maintain relatively "uniform" conditions of about 20% (*28*).

All in situ test measurements were tagged with SBAS-GPS measurements to spatially pair with the HMV measurements. HMV versus in situ test measurement correlations are shown in Figure 4. The regression relationships showed coefficient of determination (R^2) of about 0.50 with E_{LWD}, about 0.3 with DCP-CBR of top layer, and about 0.63 with DCP-CBR of the top 12 in. of the subgrade layer. Results indicate that the HMV measurements are more strongly correlated with subgrade layer measurements (DCP-CBR of subgrade) than the DCP-CBR of the PGE layer and the E_{LWD} measurements which represent a composite response of both PGE and subgrade within its measurement influence depth of about 1–1.5 ft. below the surface.

Figure 2. Screenshot of intelligent compaction HMV measurements map from HAMM IC roller on 6 in. of compacted PGE over subgrade.

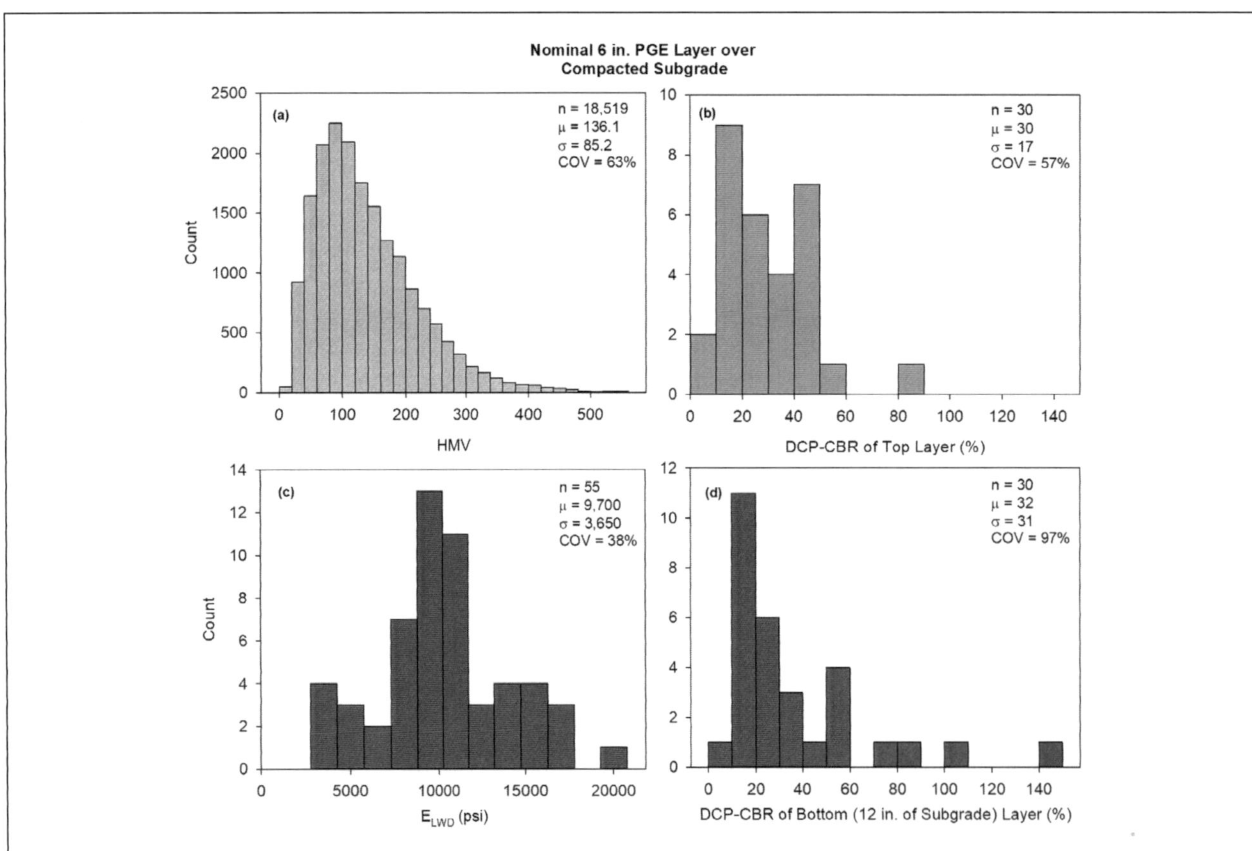

Figure 3. Histograms of different measurements obtained from multiple test sections with PGE over compacted subgrade: (*a*) HMV measurements, (*b*) DCP-CBR of PGE layer, (*c*) E_{LWD}, and (*d*) DCP-CBR of top 12 in. of compacted subgrade layer.

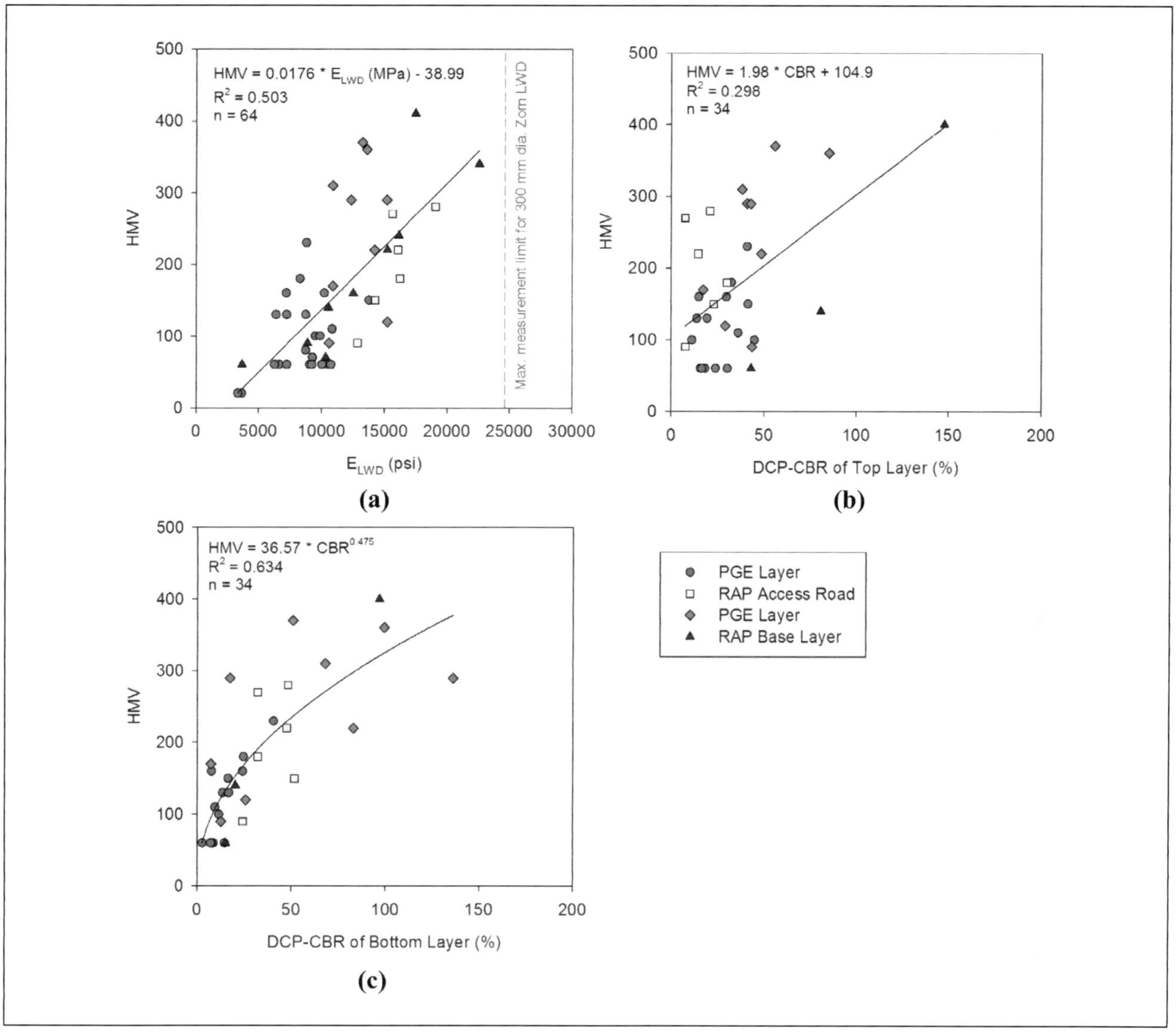

Figure 4. Correlations between HAMM IC roller HMV measurements versus: (*a*) E_{LWD}, (*b*) DCP-CBR of top layer (varied between 6 in. to 9 in.), and (*c*) DCP-CBR of top 12 in. of compacted subgrade layer.

VIC field calibration was performed with cyclic APLTs on 3 in. RAP base material over 6 in. PGE and compacted subgrade material, and directly on compacted subgrade material to capture a wide range of stiffness conditions. Static 30 in. APLTs were performed on 6 in. PGE over compacted subgrade to calibrate with static *k* measurements. CMV measurements were also simultaneously obtained for field calibration.

Results of both the VIC and CMV calibrations with cyclic APLT measurements showing measured versus predicted $M_{r\text{-Comp}}$ values at 15 psi cyclic stress are shown in Figure 5. The 15 psi cyclic stress level is selected here as an example, to match with the stress level applied with the LWD test measurements, but $M_{r\text{-Comp}}$ at all other stress levels measured (5 psi to 40 psi) showed similar trends and R^2 values. E_{LWD} measurements were also obtained at these test locations, and the calibration results are also shown in Figure 5.

VIC – $M_{r\text{-Comp}}$ calibration measurements showed a R^2 of 0.92 and a standard error of about 7.2 ksi for $M_{r\text{-Comp}}$ values that ranged between 4.8 and 95 ksi. The CMV – $M_{r\text{-Comp}}$ calibration measurements showed a R^2 of 0.23 and a standard error of about 21.9 ksi. The two points that fell far away from the best fit line are CMV measurements with significant jumping because of relatively stiff conditions. As noted earlier, CMV measurements are influenced by drum jumping (*20*).

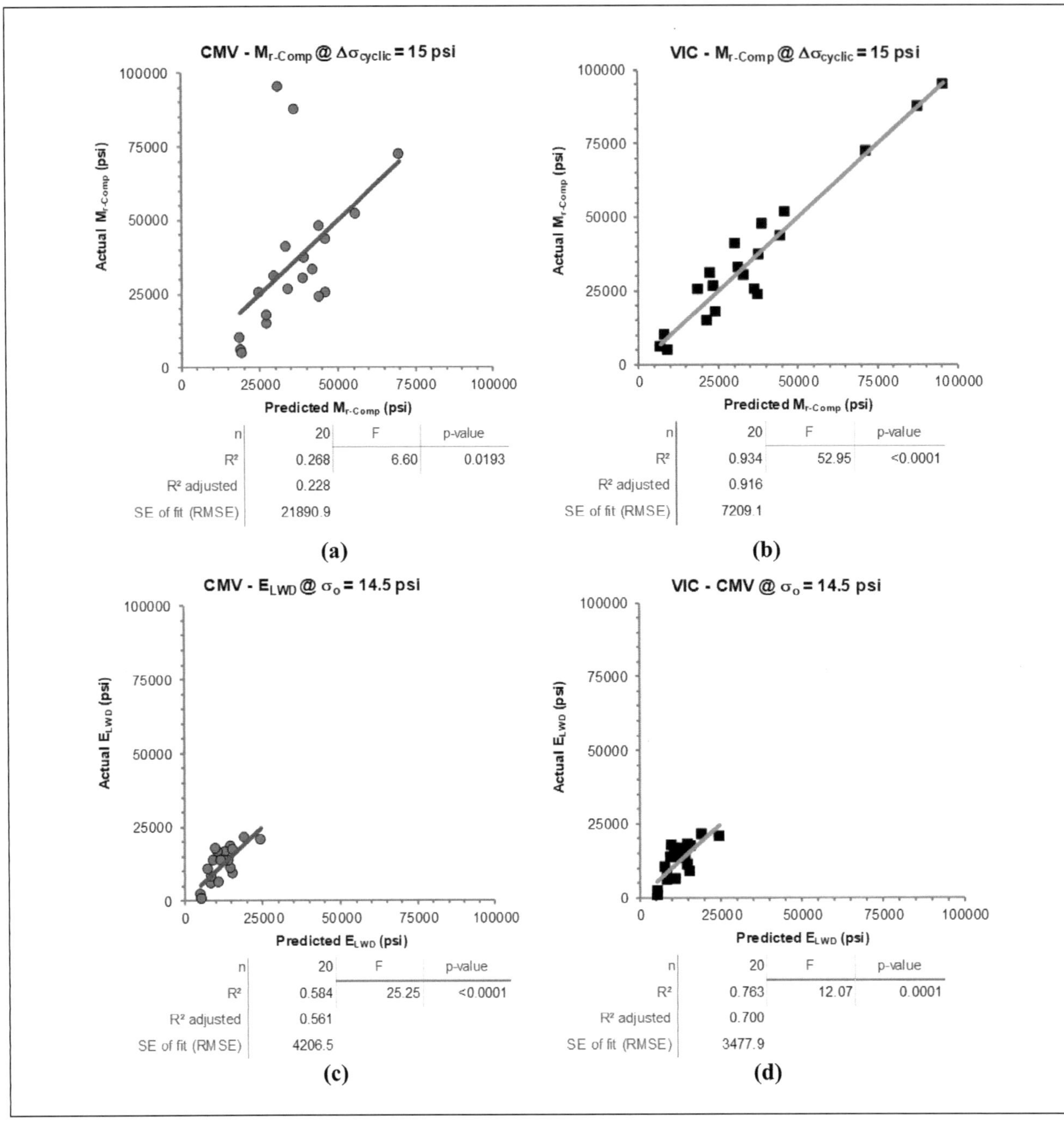

Figure 5. Summary of calibration results showing predicted versus measured values along with a summary of statistics for each calibration relationship: (*a*) CMV – $M_{r\text{-}Comp}$ at $\Delta\sigma_{cyclic}$ = 15 psi, (*b*) VIC – $M_{r\text{-}Comp}$ at $\Delta\sigma_{cyclic}$ =15 psi, (*c*) CMV – E_{LWD} at σ_o = 14.5 psi, and (*d*) VIC – E_{LWD} at σ_o = 14.5 psi.

The VIC – E_{LWD} calibration showed a R^2 of 0.7 with a standard error of 3.5 ksi for E_{LWD} values that ranged between 0.7 ksi to 21.3 ksi. The CMV – E_{LWD} calibration showed a R^2 of 0.56 with a standard error of 4.2 ksi for E_{LWD} values that ranged between 0.7 ksi to 21.3 ksi. E_{LWD} measurements were on average about three times lower than the $M_{r\text{-}Comp}$ values obtained at similar applied stress (~15 psi). The moduli values obtained from LWD are not the same as M_r. This is because LWD measures peak deflections and not rebound deflections, and conditioning cycles (which can take up to several 100 cycles) are not applied with LWD. Although the LWD testing

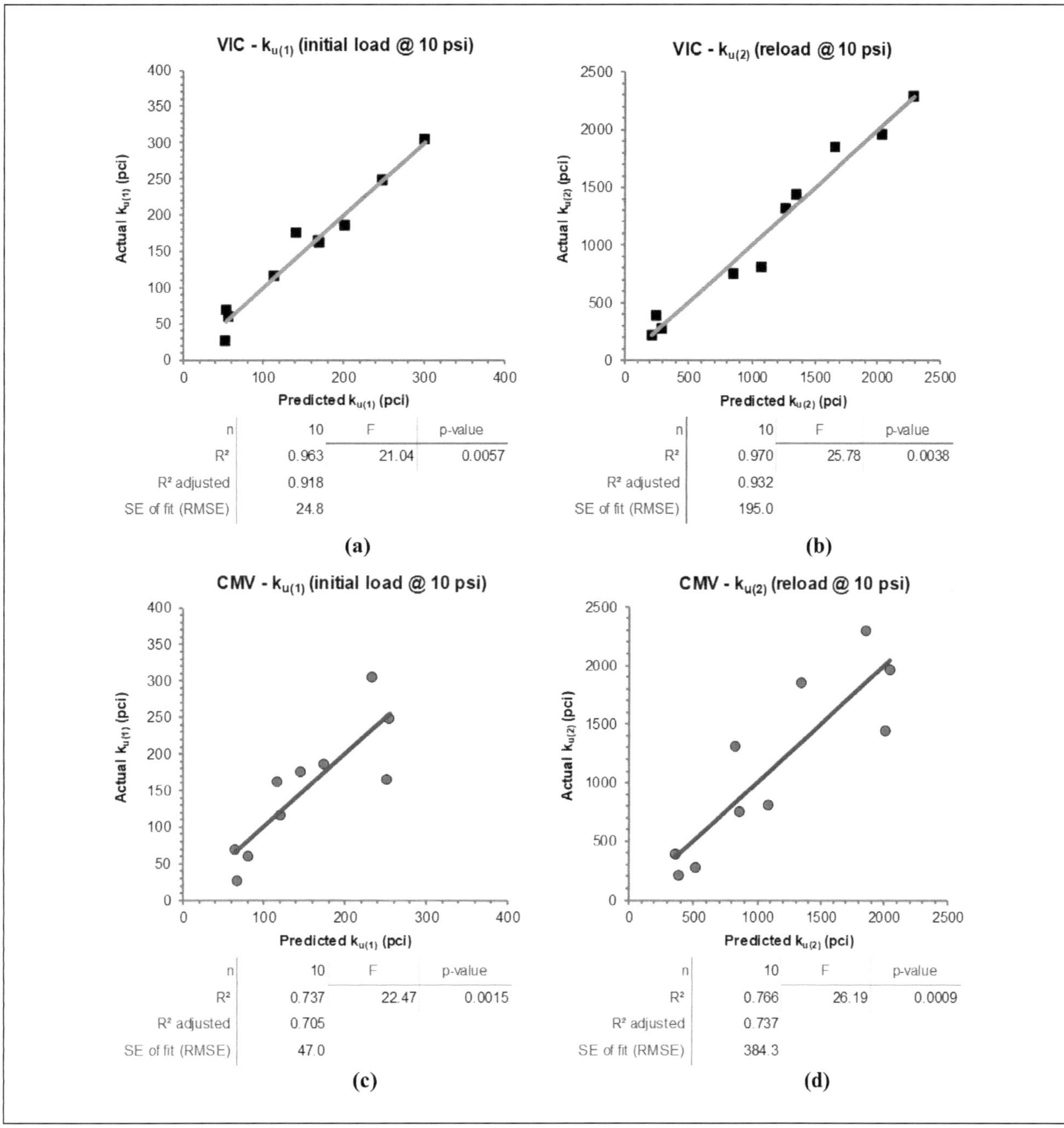

Figure 6. Summary of calibration results showing predicted versus measured values along with a summary of statistics for each calibration relationship: (*a*) CMV – $M_{r\text{-Comp}}$ at $\Delta\sigma_{cyclic}$ = 15 psi, (*b*) VIC – $M_{r\text{-Comp}}$ at $\Delta\sigma_{cyclic}$ =15 psi, (*c*) CMV – E_{LWD} at σ_o = 14.5 psi, and (*d*) VIC – E_{LWD} at σ_o = 14.5 psi.

method provides a rapid measurement, is should not be considered as a direct measure of M_r or compared directly with the design values. It is well-documented that moduli values provided by different LWD manufacturers can be significantly different (in the order of 2–3 times) because of differences in the measured/calculated deflections and applied stresses (*20*).

The results of the VIC and CMV measurement calibration with static APLT measurements showing measured versus predicted $k_{u(1)}$ and $k_{u(2)}$ values are shown in Figure 6. The VIC – $k_{u(1)}$ calibration measurements produced a R^2 of 0.92 and a standard error of about 25 pci for $k_{u(1)}$ values that ranged between 26 and 305 pci. Similarly, the VIC – $k_{u(2)}$ calibration measurements produced a R^2 of 0.93 and a standard error of about 195 pci for $k_{u(2)}$ values that ranged between 212 and 2,291 pci. CMV – $k_{u(1)}$ calibration measurements produced a R^2 of 0.71 and a standard error of about 47 pci for $k_{u(1)}$ values that ranged between 26 and 305 pci. Similarly, the CMV– $k_{u(2)}$ calibration measurements showed a R^2 of 0.74 and a standard error of about 384 pci for $k_{u(2)}$ values that ranged between 212 and 2,291 pci.

Results from the calibration testing and analysis showed that VIC calibration with both $M_{r\text{-comp}}$ and k_u measurements obtained from APLT produced high R^2 values (>0.90). Higher R^2 values and low standard errors suggest a higher reliability in future predictions of the respective measurement values in a production area. Following the VIC calibration, $M_{r\text{-Comp}}$ maps at different stress levels and k_u maps were produced at this site to demonstrate spatial assessment of the compacted layers properties when considering design values (i.e., M_r or k). An example $M_{r\text{-Comp}}$ map at 20 psi cyclic stress on a compacted subgrade area is presented in Figure 7. The average $M_{r\text{-Comp}}$ of the entire area was about 24.3 ksi with a COV of about 78%.

In addition, the $M_{r\text{-comp}}$ versus cyclic stress results from the two selected test locations (labeled as A and B) representing stiff and soft conditions in the area are also included in Figure 7 for reference. Test location A with relatively stiff conditions showed that the $M_{r\text{-Comp}}$ values increased with increasing cyclic stress up to about 19 psi and then decreased with increasing stress. On the other hand, at test location B with relatively soft conditions, the $M_{r\text{-Comp}}$ values generally decreased with increasing cyclic stress.

Subgrade conditions under proof-rolling at test location B showed rutting, which confirmed the relatively low $M_{r\text{-Comp}}$ values (~5 ksi at 15 psi cyclic stress) near that test location. Decreasing $M_{r\text{-Comp}}$ with increasing stress is a characteristic of wet or soft cohesive subgrade soils, and test location B is representative of such a condition. An increase in $M_{r\text{-Comp}}$ with increasing stress is a characteristic of granular material behavior. The decrease in modulus beyond 19 psi stress, which is considered a "break point" (BP) stress, is attributed to underlying wetter subgrade conditions.

The sharp contrast seen in the $M_{r\text{-Comp}}$ map (Figure 7) with a clear boundary of the "red" area, especially near the west half of the test area, is related to how the construction progressed in the area. The final subgrade layer in the area was constructed over nearly a 20 + ft. embankment constructed next to a retaining wall. The south half of the test area was constructed later than the northern half and in a narrow space because of limitations with how the material could be sloped and matched with the progress of the retaining wall construction.

About a 100 ft. long × 30 ft. portion near test location B was recompacted as part of the quality assurance evaluation of the subgrade, as it showed rutting under proof rolling. Immediately after the subgrade repair work, 6 in. PGE layer was placed above the subgrade layer. $M_{r\text{-Comp}}$ map of the PGE layer is presented in Figure 8, which clearly showed reflections of the underlying subgrade layer properties. Average $M_{r\text{-Comp}}$ of the entire area was about 28.8 ksi with a COV of about 43%. On average, the $M_{r\text{-Comp}}$ on the PGE layer was slightly higher than on the subgrade and the COV is reduced from 78% to 43%.

Static 30 in. plate load tests were conducted on the PGE layer, and the results from two select locations (C and D) are presented in Figure 8. Test location C was in a location with relatively stiff subgrade condition and produced a $k_{u(1)}$ of 185 pci and a $k_{u(2)}$ of 1,847 pci, while test location D was in a location with relatively soft conditions and produced a $k_{u(1)}$ of 26 pci and a $k_{u(2)}$ of 212 pci. The predicted $M_{r\text{-Comp}}$ at test locations C and D were 21.6 ksi and 14.1 ksi, respectively.

Summary of Key Findings

This paper presents field calibration test results comparing index-based IC measurement values (HMV) and different mechanical property measurements (i.e., E_{LWD} and CBR). In addition, results from a new VIC approach that uses advanced data analytics and requires site specific calibration of the roller sensor measurements using in situ APLT measurements ($M_{r\text{-Comp}}$, $k_{u(1)}$, and $k_{u(2)}$), are also presented. Based on the onsite calibration results, modulus maps were generated to demonstrate an approach to use the maps for field verification testing. Some key findings from this study are as follows:

- The regression relationships between HMV showed a R^2 of about 0.50 with E_{LWD}, about 0.3 with CBR of top (6 in. PGE) layer, and about 0.6 with CBR of the underlying top 12 in. of the subgrade layer. Results indicated that the HMV measurements are more strongly correlated with the subgrade layer CBR than the CBR of the PGE layer and the E_{LWD} measurements.
- HMV measurements obtained on the PGE layer from two different areas showed a COV of 63%. In situ E_{LWD} and DCP-CBR measurements showed a COV between 38% and 97%. The DCP-CBR of the subgrade was more variable (with a COV of 97%) than the DCP-CBR of the top PGE layer (with a COV of 57%).
- Results from the VIC calibration with both $M_{r\text{-comp}}$ and k_u measurements obtained from APLT produced relatively high R^2 values (>0.90) and with a

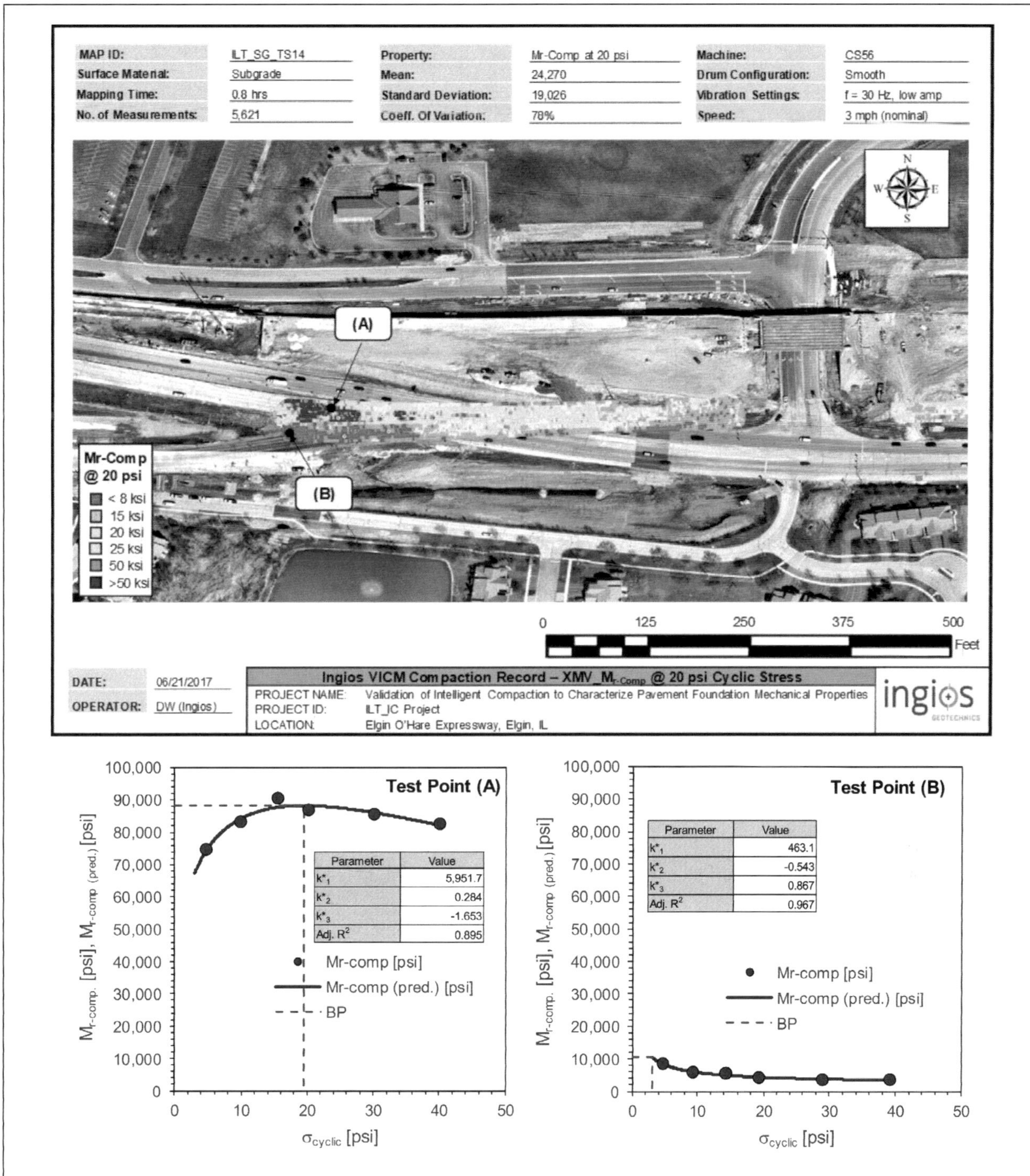

Figure 7. Color-coded map of $M_{r\text{-Comp}}$ values at 20 psi cyclic stress on compacted subgrade along with $M_{r\text{-Comp}}$ versus cyclic stress at two selected test locations (A and B).

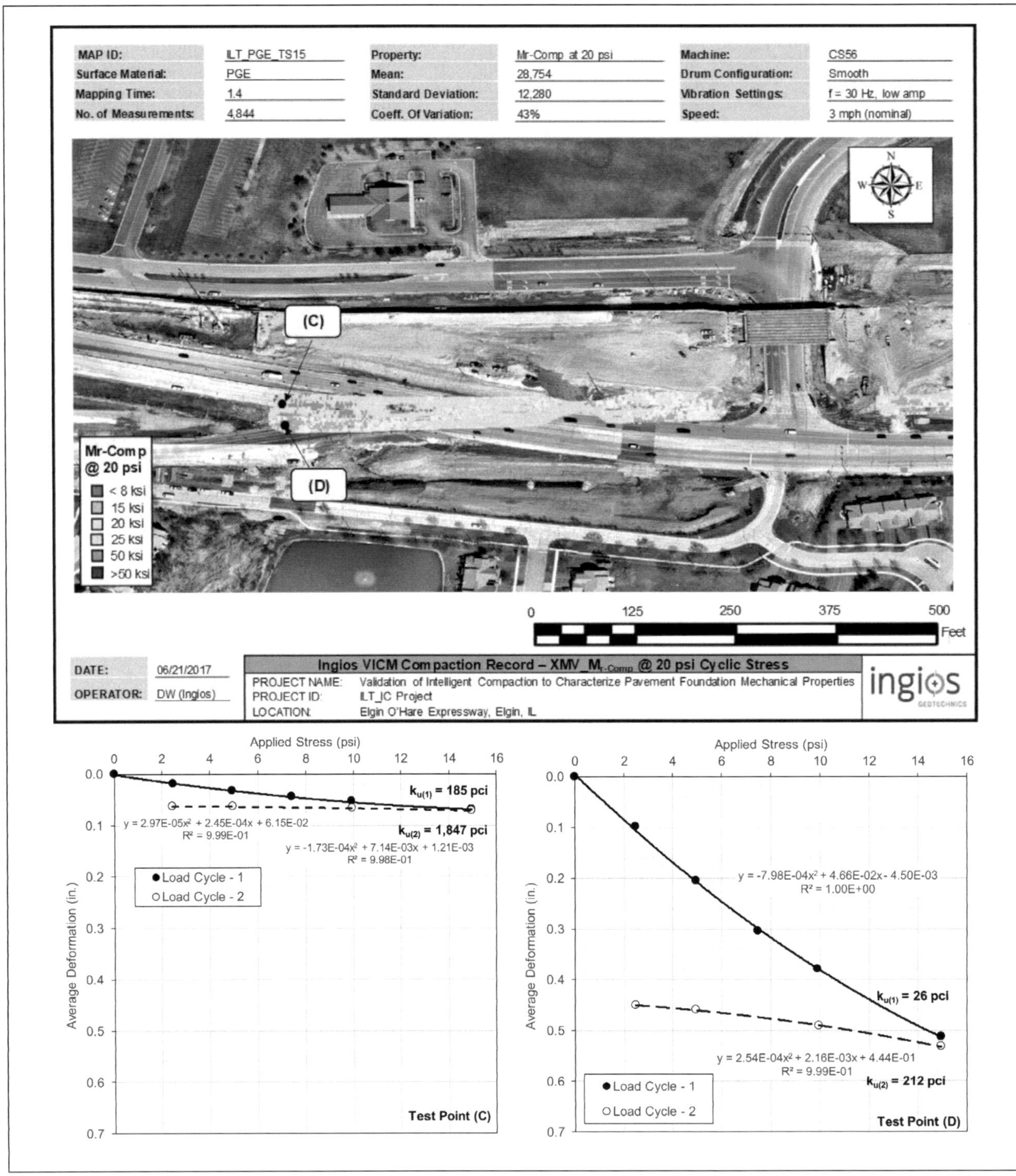

Figure 8. Color-coded map of $M_{r\text{-Comp}}$ values at 20 psi cyclic stress on 6 in. of PGE placed over compacted subgrade along with static plate load test results at two selected test locations (C and D).

relatively low standard error. Higher R^2 values and low standard errors suggest higher reliability in future predictions of the respective measurement values in a production area.

- The VIC – E_{LWD} calibration showed R^2 of 0.7 while the CMV – E_{LWD} calibration showed a R^2 of 0.56.
- The CMV calibration with $M_{r\text{-Comp}}$ produced R^2 of 0.23 and k_u produced R^2 of about 0.71–0.74.
- Results show that the E_{LWD} measurements were on average about three times lower than the $M_{r\text{-Comp}}$ values obtained at a similar applied stress (~15 psi). The moduli values obtained from LWD measurements are not the same as Mr. This is because the LWD measures peak deflections and not rebound deflections, and conditioning cycles (which can take up to several 100 cycles) are not applied with LWD, which limits the usefulness of the E_{LDW} data as a design verification value.
- VIC – $M_{r\text{-Comp}}$ maps in a subgrade area identified a sharp contrast with a clear boundary of relatively low M_r and high M_r values. The sharp contrast and the relatively low M_r values was linked to the construction sequencing followed in the area because of the proximity of a retaining wall.
- A comparison of VIC – $M_{r\text{-Comp}}$ map obtained on the PGE layer with the map obtained on the underlying subgrade layer showed reflections of the soft and stiff areas in the subgrade layer. Although, the average $M_{r\text{-Comp}}$ increased slightly on the PGE layer (from 24.2 ksi on the subgrade to 28.8 ksi on the PGE) while the COV decreased from about 78% on the subgrade to about 43% on the PGE layer.

Based on the results obtained from this project, the authors are developing an implementation framework that emphasizes an independent certification approach for field calibration for use on Illinois Tollway projects, to aid in effective implementation of the modulus-based mapping.

Author Contributions

The authors confirm contribution to the paper as follows: study conception and design: David White, Pavana Vennapusa, Erol Tutumuler, and Maziar Moaveni; data collection: David White, Pavana Vebbaoysa, Erol Tutumuler, and Maziar Moaveni; analysis and interpretation of results: David White, Pavana Vennapusa, Erol Tutumuler, William Vavrik, Maziar Moaveni, and Steven Gillen; draft manuscript preparation: David White, Pavana Vennapusa, Erol Tutumuler, William Vavrik, Maziar Moaveni, and Steven Gillen. All authors reviewed the results and approved the final version of the manuscript.

References

1. Thurner, H., and Å. Sandström. A New Device for Instant Compaction Control. *Proc., International Conference on Compaction*, Vol. II, Assoc. Amicale de Ingénieus, Paris, France, 1980, pp. 611–614.
2. Samaras, A., R. Lamm, and J. Treiterer. Application of Continuous Dynamic Compaction Control for Earthworks in Railroad Construction. *Transportation Research Record: Journal of the Transportation Research Board*, 1991. 1309: 42–46.
3. Thurner, H. Continuous Compaction Control - Specifications and Experience. *Proc., XII IRF World Congress*, International Road Federation (IRF), Madrid, Spain, 1993, pp. 951–956.
4. Brandl, H., and D. Adam. Sophisticated Continuous Compaction Control of Soils and Granular Materials. *Proc., XIVth International Conference on Soil Mechanics & Foundation Engineering*, Vol. 1, International Society of Soil Mechanics and Foundation Engineering (ISSMFE), Hamburg, Germany, 1997, pp. 1–6.
5. Kröber, W., E. Floss, and W. Wallrath. Dynamic Soil Stiffness as Quality Criterion for Soil Compaction. In *Geotechnics for Roads, Rail Tracks and Earth Structures*. A. A. Balkema Publishers, Lisse/Abingdon/Exton(Pa)/Tokyo, 2001, pp. 189–199.
6. Anderegg, R., and K. Kaufmann. Intelligent Compaction with Vibratory Rollers - Feedback Control Systems in Automatic Compaction and Compaction Control. *Transportation Research Record: Journal of the Transportation Research Board*, 2004. 1868: 124–134.
7. Sandström, A. J., and C. B. Pettersson. Intelligent Systems for QA/QC in Soil Compaction. Presented at 83rd Annual Meeting of the Transportation Research Board, Washington, D.C., 2004.
8. White, D., J, E. Jaselskis, V. Schaefer, and E. Cackler. Real-time Compaction Monitoring in Cohesive Soils from Machine Response. *Transportation Research Record: Journal of the Transportation Research Board*, 2005. 1936: 173–180.
9. Camargo, F., B. Larsen, B. Chadbourn, R. Roberson, and J. Siekmeier. Intelligent Compaction: A Minnesota Case History. *Proc., 54th Annual University of Minnesota Geotechnical Conference*, The Minnesota Geotechnical Society, Minneapolis, CD-ROM, 2006.
10. Petersen, D., J. Siekmeier, C. Nelson, and R. Peterson. Intelligent Soil Compaction – Technology, Results and A Roadmap Toward Widespread Use. *Transportation Research Record: Journal of the Transportation Research Board*, 2006. 1975: 81–88.
11. Thompson, M., and D. White. Field Calibration and Spatial Analysis of Compaction Monitoring Technology Measurements. *Transportation Research Record: Journal of the Transportation Research Board*, 2007. 2004: 69–79.
12. Newman, K., and D. White. Rapid Assessment of Cement/Fiber Stabilized Soil Using Roller-Integrated Compaction Monitoring. *Transportation Research Record: Journal of the Transportation Research Board*, 2008, 2059: 95–102.

13. White, D., and M. Thompson. Relationships between In-Situ and Roller-Integrated Compaction Measurements for Granular Soils. *Journal of Geotechnical and Geoenvironmental Engineering*, Vol. 134, No. 12, 2008, pp. 1763–1770.
14. Mooney, M. A., R. Rinehart, D. J. White, P. Vennapusa, F. Facas, and O. Musimbi. *Intelligent Soil Compaction Systems*. NCHRP Report 676. National Cooperative Highway Research Program, Washington, D.C., 2010.
15. Vennapusa, P., D. J. White, and M. Morris. Geostatistical Analysis of Spatial Referenced Roller-Integrated Compaction Measurements. *Journal of Geotechnical and Geoenvironmental Engineering*, Vol. 136, No. 6, 2010, pp. 813–822.
16. White, D. J., P. Vennapusa, and H. Gieselman. Field Assessment and Specification Review for Roller-Integrated Compaction Monitoring Technologies. *Special Issue: Advances in Instrumentation and Monitoring in Geotechnical Engineering in Advances in Civil Engineering Journal*, Vol. 2011, 2011, Article ID 783836.
17. Vennapusa, P., and D. J. White. Interpretation of Dual-Integrated Compaction Measurements on Layered Granular Fill. *Proc., Geo-Congress 2014 – Geo-Characterization and Modeling for Sustainability*, Atlanta, Ga., ASCE, Reston, Va., 2014, 2426–2435.
18. White, D. J., P. Vennapusa, and M. Dunn. A Road Map for Implementation of Intelligent Compaction Technology. *Proc., Geo-Congress 2014 – Geo-Characterization and Modeling for Sustainability*, Atlanta, Ga., ASCE, Reston, Va., 2014, pp. 2010–2018
19. Liu, D., Z. Li, and Z. Lian. Compaction Quality Assessment of Earth-Rock Dam Materials Using Roller-Integrated Compaction Monitoring Technology. *Automation in Construction*, Vol. 44, 2014, pp. 234–246.
20. Vennapusa, P., D. J. White, J. Siekmeier, and R. Embacher. In Situ Mechanistic Characterizations of Granular Pavement Foundation Layers. *International Journal of Pavement Engineering*, Vol. 13, No. 1, 2012, pp. 52–67.
21. White, D. J., P. Vennapusa, and T. Cackler. *ICM Report – Ohio River Bridges East End Crossing I-265, Section 6, Sta. 350 + 00 to 400 + 00*. Report submitted for Walsh Vinci Construction, Inc., by Ingios Geotechnics, Inc. (unpublished report), 2014.
22. Zorn, G. *Operating MANUAL: Light Drop-Weight Tester*. ZFG2000, Zorn Stendal, Germany, 2003.
23. ASTM. *Standard Test Method for Use of the Dynamic Cone Penetrometer in Shallow Pavement Applications*. ASTM D6951-03. American Standard for Testing Methods, West Conshohocken, Pa., 2007.
24. White, D. J., and P. Vennapusa. In Situ Resilient Modulus for Geogrid-Stabilized Aggregate Layer: A Case Study Using Automated Plate Load Testing. *Transportation Geotechnics*, Vol. 11, 2017, pp. 120–132.
25. AASHTO T307-99. Standard Method of Test for Determining the Resilient Modulus of Soils and Aggregate Materials. In *Standard Specifications for Transportation Materials and Methods of Sampling and Testing*, 20th ed. American Association of State Highway and Transportation Officials, Washington, D.C., 2000.
26. AASHTO. *Mechanistic-Empirical Pavement Design Guide: A Manual of Practice*, 2nd ed. American Association of Highway and Transportation Officials, Washington, D.C., 2015.
27. AASHTO T222-81. Standard Method of Test for Nonrepetitive Static Plate Load Test of Soils and Flexible Pavement Components for Use in Evaluation and Design of Airport and Highway Pavements. In *Standard Specifications for Transportation Materials and Methods of Sampling and Testing*, 32nd ed. American Association of State Highway and Transportation Officials, Washington, D.C., 2012.
28. ISSMGE. *Geotechnics for Pavements in Transportation Infrastructure, Roller-Integrated Continuous Compaction Control (CCC), Technical Contractual Provisions – Recommendations*. International Society for Soil Mechanics Geotechnical Engineering, London, 2005.

The Standing Committee on Geotechnical Instrumentation and Modeling (AFS20) peer-reviewed this paper (18-06163).

Article

TRR
JOURNAL OF THE TRANSPORTATION RESEARCH BOARD

Transportation Research Record
2018, Vol. 2672(52) 347–357

DOI: 10.1177/0361198118798721
journals.sagepub.com/home/trr
SAGE

Microstructure-Based Random Finite Element Method Simulation of Frost Heave: Theory and Implementation

Shaoyang Dong[1] and Xiong (Bill) Yu[1]

Abstract

Frost heave can cause serious damage to civil infrastructure. For example, interactions of soil and water pipes under frozen conditions have been found to significantly accelerate pipe fracture. Frost heave may cause the retaining walls along highways to crack and even fail in cold climates. This paper describes a holistic model to simulate the temperature, stress, and deformation in frozen soil and implement a model to simulate frost heave and stress on water pipelines. The frozen soil behaviors are based on a microstructure-based random finite element model, which holistically describes the mechanical behaviors of soils subjected to freezing conditions. The new model is able to simulate bulk behaviors by considering the microstructure of soils. The soil is phase coded and therefore the simulation model only needs the corresponding parameters of individual phases. This significantly simplifies obtaining the necessary parameters for the model. The capability of the model in simulating the temperature distribution and volume change are first validated with laboratory scale experiments. Coupled thermal-mechanical processes are introduced to describe the soil responses subjected to sub-zero temperature on the ground surface. This subsequently changes the interaction modes between ground and water pipes and leads to increase of stresses on the water pipes. The effects of cracks along a water pipe further cause stress concentration, which jeopardizes the pipe's performance and leads to failure. The combined effects of freezing ground and traffic load are further evaluated with the model.

Frost heave is a phenomenon in which soil swells upwards during freezing conditions. It mostly happens in areas with freezing climates and frost-susceptible soils such as silty or clayey soils (*1*, *2*). Frost heave in cold regions can cause pavement upheaval and cracking, failure of retaining walls and pipelines, and other serious damage to transportation infrastructure (*3*–*6*). Previous theories to address frost heave mainly include the segregation potential theory (*4*) and the discrete ice lens theory (*7*). Both theories can be applied to numerical models to predict frost heave. Frost heave of soils is a complex process which involves thermal flow, hydraulic conduction, and stress evolution. It is a coupled multi-physical process which is referred to as the thermo-hydro-mechanical (THM) process (*8*–*13*). The THM process is coupled with an extension of isothermal theory that incorporates the thermal expansion of the solid skeleton as well as the pore fluid (*14*, *15*). Kurashige solved the problem of thermally induced stresses in a poroelastic cylinder by employing the Crank-Nicholson implicit scheme (*16*). The THM process creates a coupling effect that is critical to the behaviors of frozen soils. During the freezing process, the THM coupling effect can lead to decreased water content, increased matric suction, the rise of water pressure head, and the generation of internal stress. A number of researchers have applied the coupled-THM finite element analysis to soil. For example, Nishimura developed a coupled-THM model for frozen soil to study the foundation stability, frost heave, and mass movements in cold regions (*17*). The development of the coupled-THM model to simulate frozen soil requires a sound foundation of theoretical mechanisms, accurate parameters obtained from experimental measurements, and reliable numerical methods to implement in the computational simulations.

The thermal, hydraulic, and mechanical properties of soil are quite sensitive to design parameters. Some of the parameters are randomly distributed throughout the soil sample due to complex geological processes. This random process can be simulated by the random finite

[1]Department of Civil Engineering, Case Western Reserve University, Cleveland, OH

Corresponding Author:
Address correspondence to Xiong (Bill) Yu: xiong.yu@case.edu

element method (RFEM) (*18–20*). These properties may include parameters such as Young's modulus, Poisson's ratio, density, thermal conductivity, heat capacity, coefficient of thermal expansion, and a variety of other parameters (*21–23*). The RFEM is an extension of the traditional FEM which adds randomness to material properties. Fenton et al. combined random field simulation with the non-linear finite element to investigate the active earth pressure and stability of the retaining wall (*24*).

From a microstructural point of view, frozen soil is a four-phase material which includes soil particles, ice, water, and air, arranged randomly due to complex geological processes (*25*). Microstructural models for soil composition have been established (*26, 27*). Helliwell, Tracy, and colleagues obtained the structure of soil by using X-ray computed tomography (CT) images (*28, 29*). The microstructure extracted from the high-resolution images provides a foundation for simulating the physical coupling process, which could improve the understanding of THM-coupled physics at multiple scales, that is, micro-scale or pore-scale of soil samples.

This paper describes the development and implementation of RFEM for the multiphysics THM processes in frozen soils. The spatial randomness of soil parameters is described with this model. The results reveal interesting phenomena associated with freezing/thawing due to soil microstructure.

Theoretical Basis

The theoretical basis for THM processes is well established. Three types of physical fields (thermal, hydraulic, and mechanical) fall into the problems of continuum mechanics and the governing equations of these physical fields abide by the laws of energy conservation and mass conservation. The governing equation for the thermal fields (Equation 1) contains the convection and conduction terms, and is modified from Fourier's equation:

$$\rho_j C_{p_j} \frac{\partial T}{\partial t} + \rho_j C_{p_j} q \cdot \nabla T = \nabla \cdot (k_j \nabla T) \qquad (1)$$

where the subscript j denotes different phases of the poromaterial; ρ_j is the density; C_{pj} is the heat capacity; k_j is the thermal conductivity; T is the temperature; t is time; and q is the flux.

The flow of moisture is described with the Darcy's law (Equation 2):

$$q = -\frac{\kappa_j}{\mu} \nabla p \qquad (2)$$

where p is the pore-water pressure; κ_j is the intrinsic permeability; μ is the dynamic viscosity; and q is the flux.

The governing equation for the mechanical field is Navier's equation, which includes the strain-displacement correlation (Equation 3), the constitutive relationship (Equation 4), and the equation of motion (Equation 5).

The strain-displacement correlation is:

$$\varepsilon_j = \frac{1}{2}[\nabla \mathrm{u}_j + (\nabla \mathrm{u}_j)^T] \qquad (3)$$

where ε_j is the strain tensor and u_j is the displacement vector.

The constitutive relationship for phase j of the porous material is:

$$\sigma_j = \mathrm{D}_j(\varepsilon - \varepsilon_0 - \varepsilon_{th_j} - \varepsilon_{tr} - \varepsilon_{hp}) + \sigma_{0,j} \qquad (4)$$

where Dj is the stiffness matrix of phase j; σ is the stress tensor; σ_0 is the initial stress tensor; ε is the total strain tensor; ε_0 is the initial strain; ε_{thj} is the strain caused by thermal expansion of phase j; ε_{tr} is the strain resulting from the phase change of water, which is $0.09Q$, where Q is the amount of water that changes into ice and 0.09 is the relative change of volume when water changes into ice; ε_{hp} is the strain caused by the change of matric suction, which is calculated by h/H, where h is the water head and H is the modulus related to the matric suction.

The equation of motion is:

$$\nabla \cdot (\mathrm{C}_j \nabla \mathrm{u}_j) + \mathrm{F} = \rho_j \ddot{\mathrm{u}}_j \qquad (5)$$

where C_j is the stiffness tensor of the material; u_j is the displacement vector; ρ_j is the density; and F is the body force.

Experimental Measurements of Frozen Soil

Two identical cylindered silt specimens of 3.3 cm in diameter and 7.2 cm in height were prepared with a Harvard miniature compactor. The specimens had saturated gravimetric water content of 22%. The dry density of the silt specimens was 1.708 g/cm^3 and their initial temperature was 20°C. The specimens were thermally insulated on the sides and at the bottom to create one-dimensional thermal flow. The wrapped specimens were placed in a freezer with a temperature of −20°C. Thermal couples were installed in the silt specimens at 10 mm, 30 mm, and 50 mm from the top of the specimen to measure the temperature distribution within the silt specimens during the freezing process. Two different methods were utilized to measure the vertical deformations of the specimens during the freezing process: manual measurement with a dial gauge indicator and automatic measurement with a laser displacement transducer (Figure 1). The performance of these methods is compared from the measurements of two separate specimens.

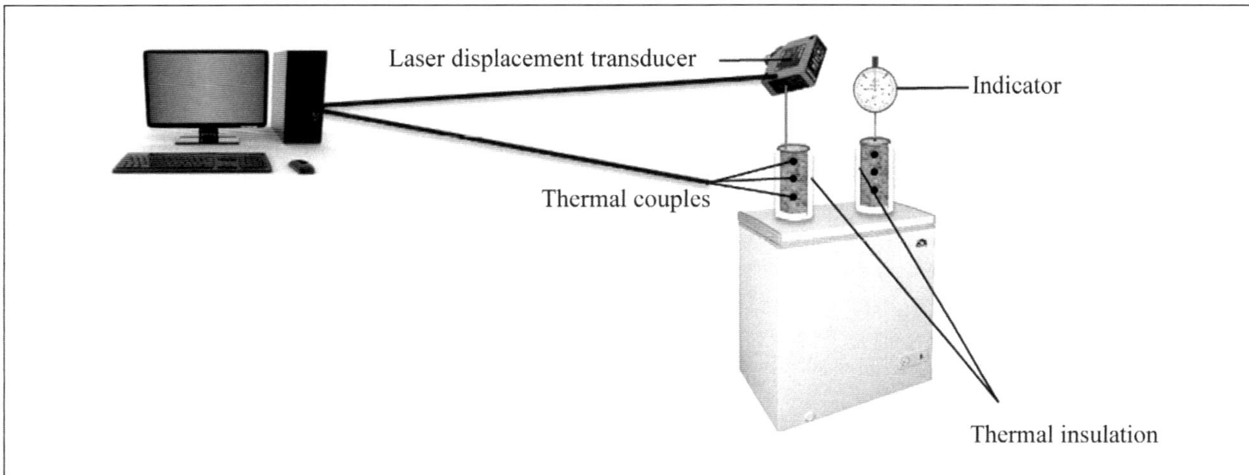

Figure 1. Schematic picture of the experimental system for measuring the frozen soil.

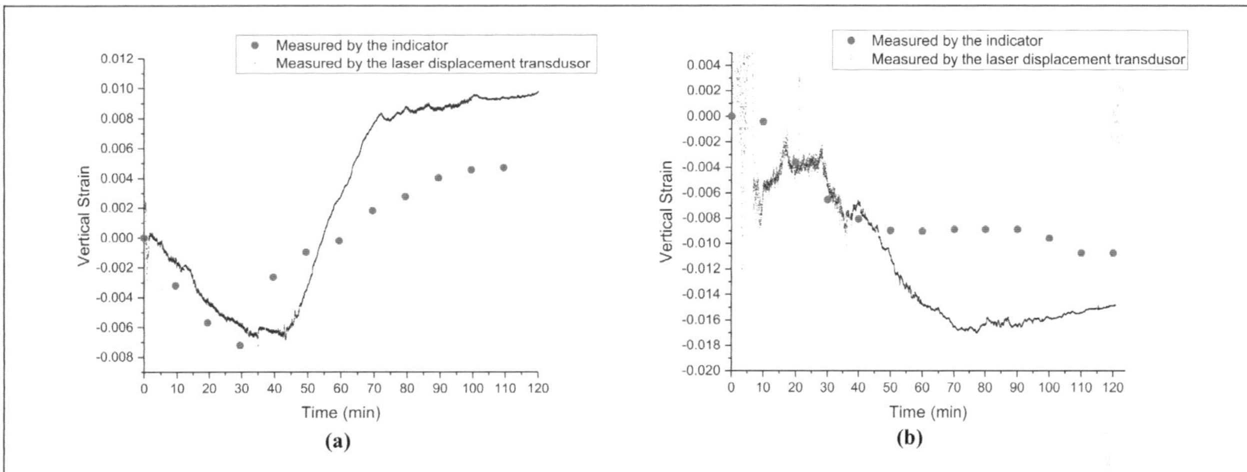

Figure 2. Vertical strains of saturated silt specimen during (*a*) freezing; and (*b*) thawing processes.

The laser displacement transducer and the thermal couples were connected to the computer to monitor and record the vertical deformation and temperature distribution of the silt specimens during the freezing process. The results are plotted in Figures 2 and 3, respectively.

For the silt specimen measured by the indicator, the vertical strain decreased from 0 to −0.007 in the first 30 min due to the compaction of matric suction, then the vertical strain of the specimen increased from −0.007 to 0.005 for the rest of the time due to the frost heave. For the silt specimen measured by the laser displacement transducer, the vertical strain decreased from 0 to −0.0065 in the first 35 min due to the compaction of matric suction, then the vertical strain of the specimen increased from −0.0065 to 0.01 for the rest of the time due to the frost heave. Figure 2 shows that the vertical strain measured by the indicator is smaller than that measured by the laser displacement transducer because the silt specimen was pressed by the indicator so that the vertical deformation was reduced; while the vertical strain measured by the laser was a non-contact measurement. It also shows that the vertical strain measured by the indicator reached the lowest value 5 min earlier than that measured by the laser displacement transducer. Due to contact between the metal stick of the indicator and the silt specimen, the measuring point on the silt specimen froze faster and arrived at freezing point earlier. It can thus be concluded that during the freezing process the vertical strain measured by the indicator is less reliable than that measured by the laser displacment transducer.

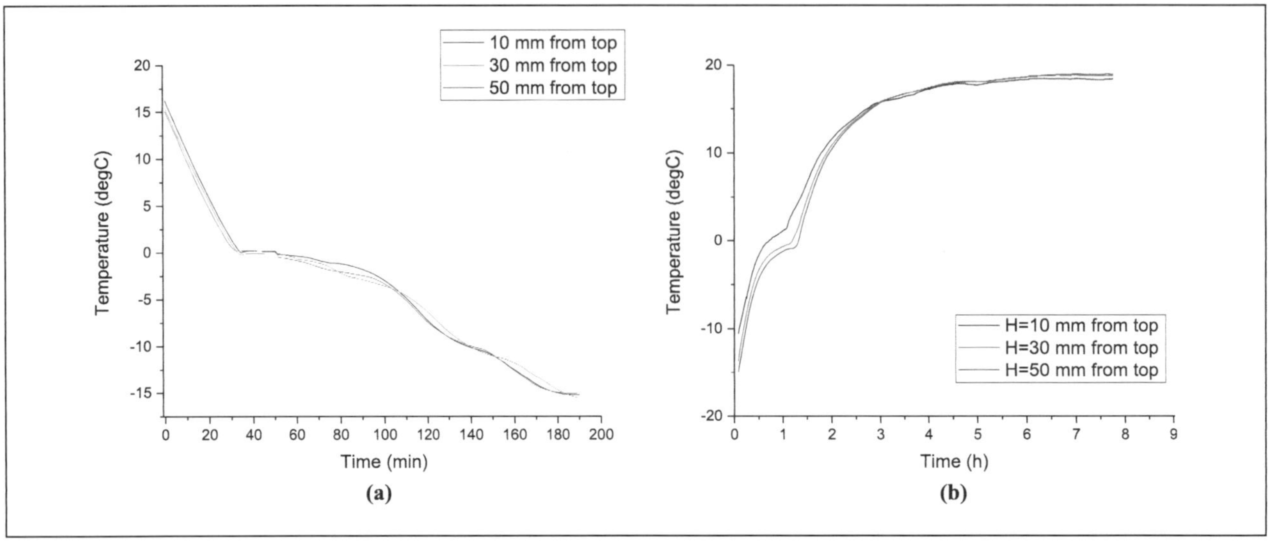

Figure 3. Temperature of saturated silt specimen during (*a*) freezing and (*b*) thawing processes.

As shown in Figure 3, the temperature decreased rapidly above 0°C, and then it decreased at a slower rate at around 0°C due to the latent heat of fusion. The temperature decreased rapidly again below 0°C and gradually approached the temperature in the freezer. Comparing Figure 3 and Figure 2 it can be observed that during the phase change period (35–80 min), the vertical strain of the silt specimen sharply increased due to the expansion of the ice phase.

The frozen silt specimens with thermal couples were taken out of the freezer into the laboratory, where the air temperature was 19°C. At the same time, the thermal couples were connected to the computer to monitor and record the temperature during the thawing process. The laser displacement transducer and indicator were also applied to measure the vertical deformation of the silt specimens. The vertical strains and the temperature distribution of the silt specimens during the thawing process are plotted in Figures 2 and 3, respectively.

For the silt specimen measured by the indicator, the vertical strain decreased from 0 to −0.009 in the first 50 min, then the vertical strain came to a stable level during the period from 50 to 90 min because the matric potential decreased which caused volume expansion counteracting the decreases in the volume due to the phase change of ice. The vertical deformation continued to decrease and gradually became stable at −0.01 for the rest of the time. For the silt specimen measured by the laser displacement sensor, the measured data were not stable in the first half hour. It was observed that the ice crystals on top of the silt specimen melted to water in the first half hour. It is speculated that the ice crystal melting accouts for the unstable measurement of the verticle strain. The vertical strain decreased during 30 to 80 min of thawing due to the phase change of ice. The vertical deformation became stable at −0.015 for the rest of the time. The absolute value of vertical strain measured by the indicator is smaller than that measured by the laser displacement transducer.

As shown in Figure 3, the temperature increased rapidly below 0°C, and then it increased slowly around 0°C due to the latent heat of fusion. The temperature increased rapidly again above 0°C and gradually approached the environmental temperature. It took about six hours for the silt specimen to reach a thermal balance condition (in which the temperature of the silt specimen is the same as the environment). Since the thermal flow traveled from the top of the silt specimen to the bottom, the temperature at a higher location is greater than that of a lower location within the silt specimen at a given time.

Numerical Simulation of Frozen Soil

The physical properties of a silt specimen prepared by Harvard miniature compactor were measured experimentally in the previous part. To simulate the microstructure of frozen soils, firstly, the volumetric content of each phase is calculated from its dry density (1.708 g/cm^3), initial gravimetric water content (22%), and specific gravity (2.65) of the silt specimen. The next step is to generate an m × n matrix in Matlab. In the matrix, m is equal to the height of the silt specimen divided by the diameter of the silt particle, while n is equal to the radius of the silt specimen divided by the diameter of the silt particle. Thus, $m \times n$ equals the total number of elements in the image of two-dimensional silt model. Each element in the matrix represents a particular phase of the silt specimen, and different numbers (0, 1/3, 2/3, 1) are assigned to the

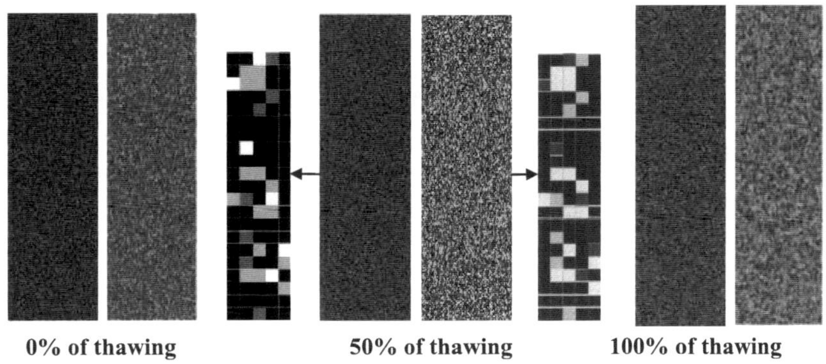

Figure 4. Two-dimensional silt microstructure-based image with different degrees of thawing and enlarged image with 50% of thawing. (Gray-scale images generated from Matlab, color images converted into Comsol.)

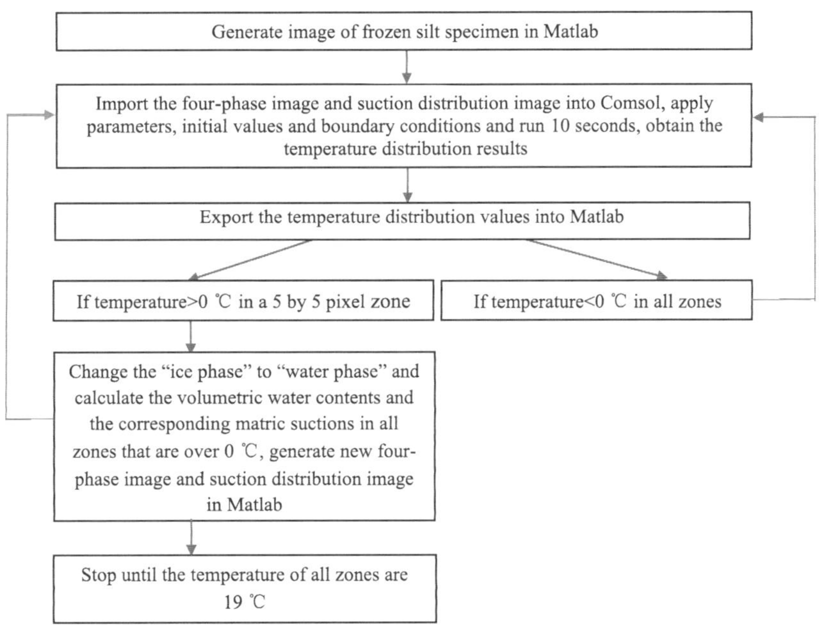

Figure 5. Flow chart of the process of calculating the thermal flow simulation.

elements in the matrix to represent four different phases within the silt specimen. These numbers are randomly assigned, and the probability of their occurrence is equal to the volumetric content of each phase. The final step is to generate an image from the matrix in Matlab. Each pixel of the image represents each element in the matrix, and the color in the pixel corresponds to the number in the element. Four different colors (black, dark gray, light gray, and white) represent the soil particles, ice, water, and air, respectively (Figure 4). The probability of occurrence of each color equals the volumetric content of each phase.

The microstructure image is imported to Comsol and material properties are assigned based on the color scale of the image. Four different colors (dark blue, light blue, yellow, and red) represent the soil particles, ice, water, and air, respectively (Figure 4). Figure 4 shows the two-dimensional silt images at different degrees of thawing after conversion from Matlab to the Comsol model.

With the microstructure-based RFEM model, the thermal parameters can be assigned to each pixel based on the color of the image, which represents an individual phase. The heat capacity, thermal conductivity, and density of the air, water, ice, and silt particles are assigned to

Table 1. Constant Parameters for the Simulation

Constant	Value	Units	Description
R	1.65	cm	Radius of silt specimen
H	7.2	cm	Height of silt specimen
w_0	0.22	1	Initial water content
ρ	1708	kg/m^3	Dry density of silt specimen
d	150	μm	Diameter of silt particle
G_s	2.65	1	Specific gravity of silt specimen
E_s	12.7	GPa	Young's modulus of silt particle
E_i	9	GPa	Young's modulus of ice
E_w	3.9×10^{-5}	Pa	Young's modulus of water
E_a	0	Pa	Young's modulus of air
E_p	205	GPa	Young's modulus of the steel pipeline
μ_s	0.3	1	Poisson's ratio of silt particle
μ_i	0.3	1	Poisson's ratio of ice
μ_w	0.5	1	Poisson's ratio of water
μ_a	0	1	Poisson's ratio of air
μ_p	0.28	1	Poisson's ratio of the steel pipeline
ρ_s	2600	kg/m^3	Density of silt particle
ρ_i	917	kg/m^3	Density of ice
ρ_w	1000	kg/m^3	Density of water
ρ_a	1.29	kg/m^3	Density of air
ρ_p	7850	kg/m^3	Density of the steel pipeline
κ_s	10^{-17}	mm^2	Intrinsic permeability of silt particle
κ_i	10^{-17}	mm^2	Intrinsic permeability of ice
κ_w	3.08×10^{-13}	mm^2	Intrinsic permeability of water
κ_a	1.31×10^{-19}	mm^2	Intrinsic permeability of air
ε	0.3663	1	Porosity of silt specimen
χ_f	4.6×10^{-10}	1/Pa	Compressibility of water
μ	8.9×10^{-4}	Pa·s	Dynamic viscosity of water
k_s	2	W/m·K	Thermal conductivity of silt particle
k_i	2.2	W/m·K	Thermal conductivity of ice
k_w	0.58	W/m·K	Thermal conductivity of water
k_a	0.025	W/m·K	Thermal conductivity of air
k_p	44.5	W/m·K	Thermal conductivity of the steel pipeline
Cp_s	835	J/kg·K	Heat capacity of silt particle
Cp_i	1960	J/kg·K	Heat capacity of ice
Cp_w	4181.3	J/kg·K	Heat capacity of water
Cp_a	1005	J/kg·K	Heat capacity of air
Cp_p	475	J/kg·K	Heat capacity of the steel pipeline
L_f	334	kJ/kg	Latent heat fusion of water
α_s	9×10^{-6}	1/K	Volumetric coefficient of thermal expansion of silt particle
α_i	1.9×10^{-4}	1/K	Volumetric coefficient of thermal expansion of ice
$\alpha_{i\text{-}w}$	-0.05	1/K	Volumetric coefficient of thermal expansion of a mixture of ice-water
α_w	2.07×10^{-4}	1/K	Volumetric coefficient of thermal expansion of water
α_a	3.43×10^{-7}	1/K	Volumetric coefficient of thermal expansion of air
α_p	1.23×10^{-5}	1/K	Volumetric coefficient of thermal expansion of the steel pipeline

each pixel based on different colors corresponding to different phases in the silt specimen. The thermal parameter values of each phase are listed in Table 1.

To simulate the temperature distribution during the thawing process, thermal insulation was applied on the sides and at the bottom of the two-dimensional frozen silt model. The initial temperature of the silt specimen was set to −20°C. The boundary temperature of 19°C was applied at the top of the model during the thawing process.

The calculation procedure of the thermal flow simulation is shown in Figure 5. LiveLink for Comsol was used to link the Matlab process with the Comsol simulation. As indicated in the flow chart (Figure 5), phase changes from ice to water when temperature rises from below 0°C to above 0°C. The thermal parameters are not only a function of spatially distributed position, but also a function of temperature. By setting the thermal parameters as a step function of temperature and introducing the latent heat fusion of water, considering the latent

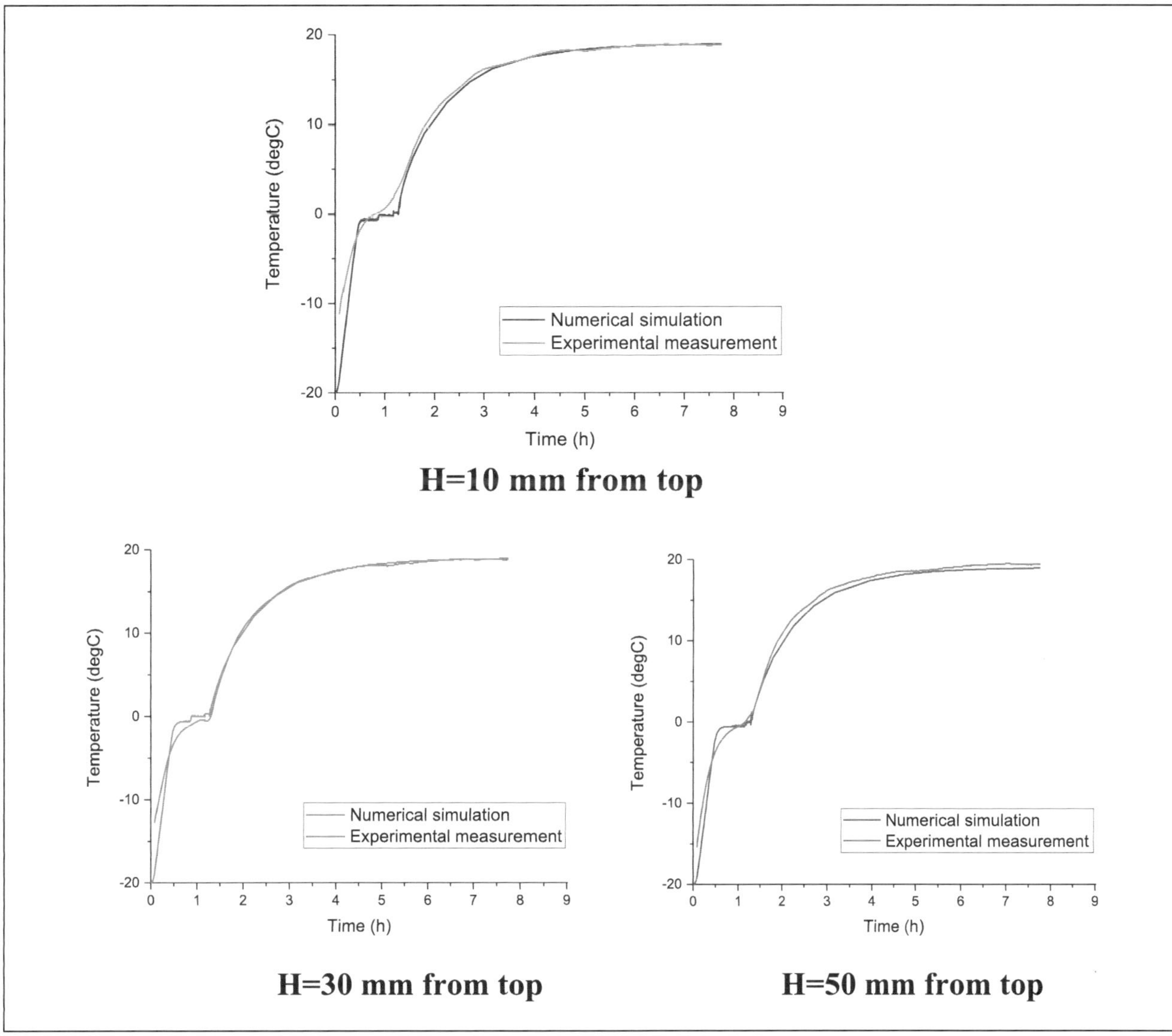

Figure 6. Comparison of the numerical simulation and the experimental measurement at different heights of the silt specimen during the thawing process.

heat during phase transition, the simulation is closer to the reality during phase transition.

Figure 6 compares the numerical simulation and the experimental measurement at different heights of the silt specimen during the thawing process. It takes about 6 h for the silt to reach a thermal balance condition of similar temperature across the specimen. The numerical simulation agrees quite well with the experimental measurement, which validates the reliability of this newly developed microstructure-based RFEM.

The mechanical parameters can also be assigned to each phase based on the microstructure of RFEM. The Young's modulus, Poisson's ratio, density, and volumetric coefficient of thermal expansion of the air, water, ice, and silt particles (Table 1) are assigned to each pixel based on different colors corresponding to different phases in the silt specimen. The thermal properties of water are modeled with a non-constant value to account for the volume change during phase transition (9% of volume expansion when water freezes into ice).

To demonstrate the production of internal stress and vertical strain during the freezing process, a simulation case was designed in Comsol where thermal insulation was applied to the sides and the bottom of the two-dimensional unfrozen silt model. The initial temperature of the silt specimen was set at 20°C. The boundary temperature of −20°C was applied at the top surface of the model to start the freezing process. The heat transfer

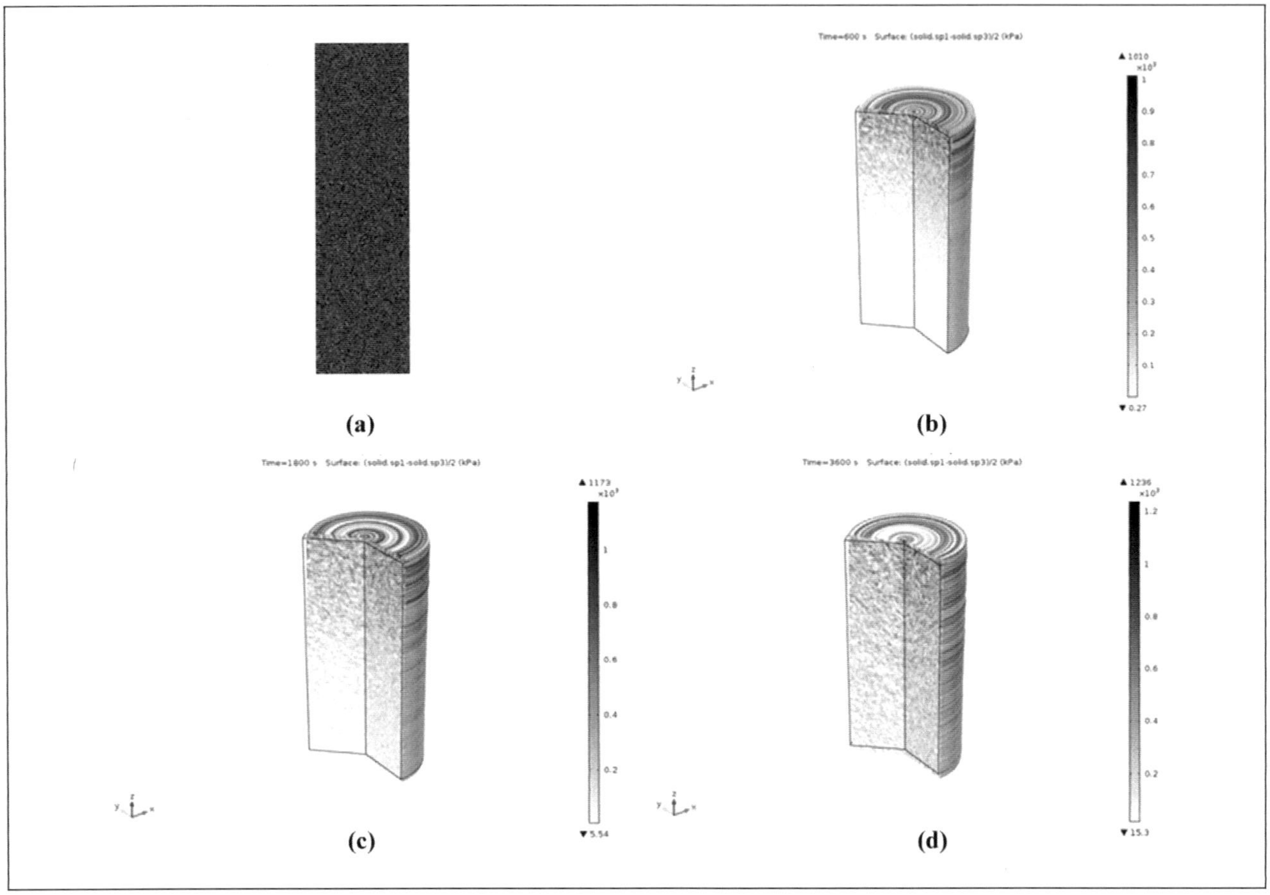

Figure 7. Internal shear stress (Pa) distribution of three-dimensional silt specimen during the freezing process. Microstructure-based image indicating (*a*) the initial conditions before freezing; (*b*) after 10 min; (*c*) after 30 min; (*d*) after one hour.

module, the solid mechanics module, and the poroelasticity module were applied to simulate the temperature distribution and frost heave in the soil specimen. Internal stress was generated by the development of matric suction and thermal expansion due to phase transition. Figure 7 shows the development of internal stress distribution in the silt specimen during the freezing process. As the freezing process starts, there is continuous increase in the internal stress. Overall, the model describes the mechanism of internal stress generation based on physical mechanisms.

Figure 8 plots the maximum vertical strain simulated by the microstructure-based RFEM and compares the results with experimental data. Overall the volume expansion at the end of the freezing process predicted by the RFEM is close to the data measured by the laser displacement transducer. Compared with the experimental data, the simulation results do not show an initial decrease in the soil volume due to increases in the matrix

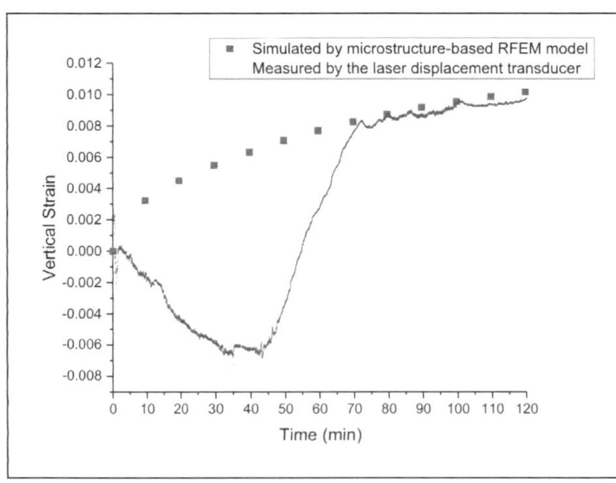

Figure 8. Comparison of the vertical strains of the silt specimen during the freezing process obtained from experimental measurement and numerical simulation.

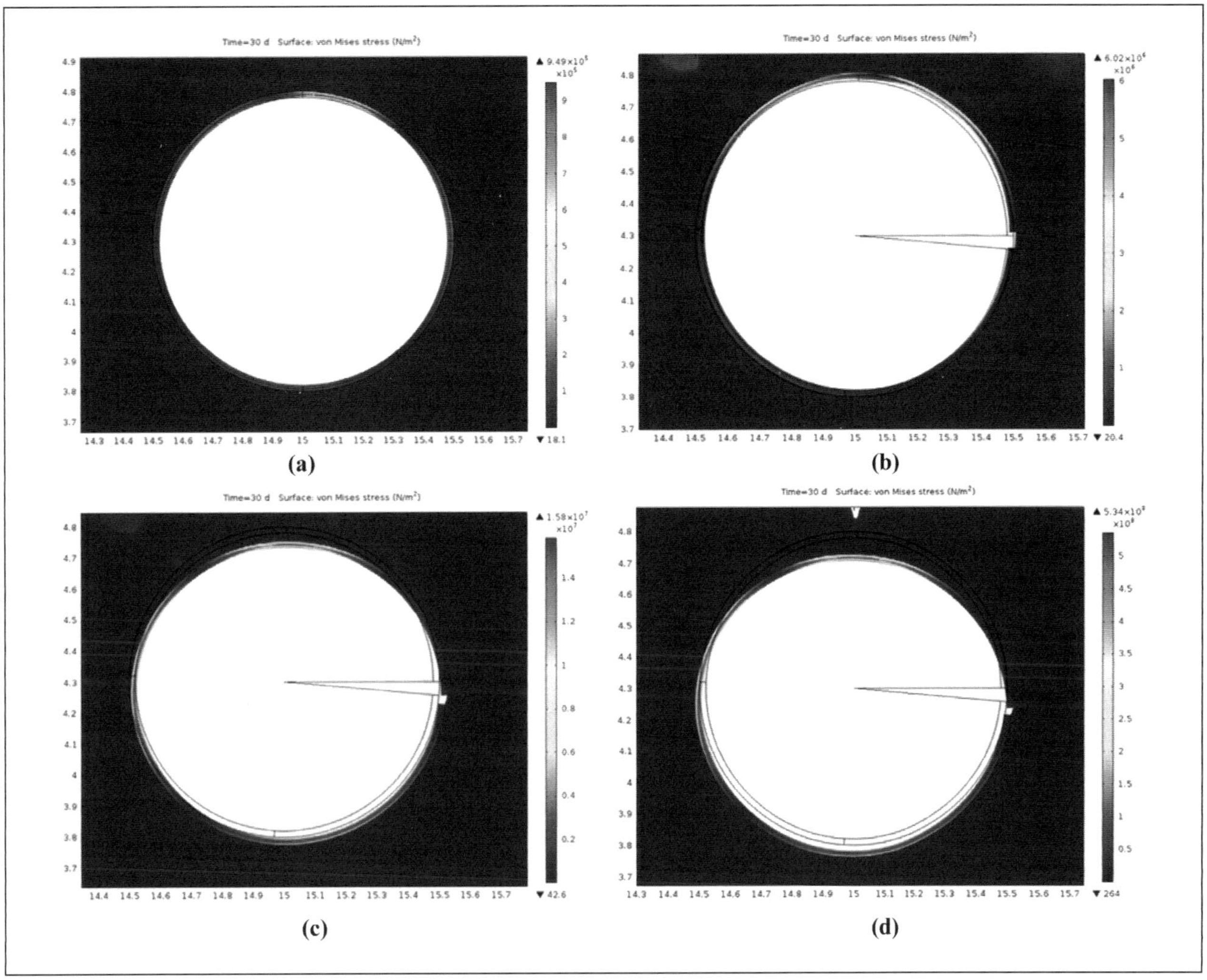

Figure 9. Stress distribution in water pipe buried in a saturated silt layer after 30 days of freezing: (*a*) intact pipe; (*b*) defective pipe; (*c*) pipe with point load of 15 kN on surface; (*d*) pipe with point load of 480 kN on surface.

suction. This may be because the RFEM model has not considered the effects of matric suction plus the variation of the stiffness of frozen soil with different extent of freezing is not accurately accounted for. This is an aspect of the RFEM model that will be further refined in the authors' future work. Overall, the results of simulating experimental freezing/thawing processes indicate that the RFEM model is able to describe accurately the internal stress and volume change due to phase transition associated with freezing.

Simulation Case Study

A water pipe is buried 1.2 m deep in a 30 m × 5 m silt layer. The pipe has a radius of 0.5 m and thickness of 0.02 m. The simulation parameters of the silt layer and the pipe are shown in Table 1.

Thermal insulation is applied to the sides and the bottom of the two-dimensional study region. The initial temperature of the study region is set at 5°C. The boundary temperature of −20°C is applied at the top of the region during the freezing process. Fixed constraint is applied at the bottom of the study region and rollers are applied to the sides of the study region.

After 30 days of freezing, the stress distribution is measured for intact pipe and defective pipe (with a crack). The pipe was also measured with point loads of 15 kN and 480 kN exerted on the silt layer surface above the pipe to simulate traffic load. The results are shown in Figure 9. Frost heave leads to increase of stresses in the pipe. The crack in the defective pipe further leads to stress concentration, and increases the maximum stress in the pipe. Point loads on the surface were found to accelerate internal stress generation, causing deformation and even failure of the pipe.

Conclusion

This research develops a microstructure-based random finite element simulation model (RFEM) to simulate frost heave in saturated soils. This model is able to simulate the bulk behaviors by considering the microstructures. The soil is phase coded and therefore only the parameters of individual phases are needed. This therefore can direct efforts to understand the behaviors of individual phases, which represent more uniform materials. The results are consistent with experimental observations. The model holistically accounts for volume change during the phase transition of water and simulates the production of internal stress due to this process. The modeling approach combines the advantages of a discrete element model in simulating the mechanical behaviors of a particulate system and the computational efficiency of a FEM. With experimental validation, it provides a new and reliable simulation tool to predict frost-induced stress on infrastructures (such as pipelines, retaining walls, etc.). This insight will help guide engineering design in cold regions.

Acknowledgments

This research is partially supported by the Ohio Department of Transportation and U.S. National Science Foundation.

Author Contributions

The authors confirm contribution to the paper as follows: study conception and design: Xiong (Bill) Yu, Shaoyang Dong; data collection: Xiong (Bill) Yu, Shaoyang Dong; analysis and interpretation of results: Xiong (Bill) Yu, Shaoyang Dong; draft manuscript preparation: Xiong (Bill) Yu, Shaoyang Dong. All authors reviewed the results and approved the final version of the manuscript.

References

1. O'Neill, K., and R. D. Miller. Exploration of a Rigid Ice Model of Frost Heave. *Water Resources Research*, Vol. 21, No. 3, 1985, pp. 281–296.
2. Chamberlain, E. J. *A Freeze-Thaw Test To Determine the Frost Susceptibility of Soils*. Special Report 87-1. U.S. Army Corps of Engineers, Cold Regions Research and Engineering Laboratory, Hanover, NH, 1987.
3. Nixon, J. F., N. R. Morgenstern, and S. N. Reesor. Frost Heave-Pipeline Interaction Using Continuum Mechanics. *Canadian Geotechnical Journal*, Vol. 20, No. 2, 1983, pp. 251–261.
4. Konrad, J. M., and N. R. Morgenstern. Frost Heave Prediction of Chilled Pipelines Buried in Unfrozen Soils. *Canadian Geotechnical Journal*, Vol. 21, No. 1, 1984, pp. 100–115.
5. Lai, Y., Z. Wu, S. Liu, and X. Den. Nonlinear Analyses for the Semicoupled Problem of Temperature, Seepage, and Stress Fields in Cold Region Retaining Walls. *Journal of Thermal Stresses*, Vol. 24, No. 12, 2001, pp. 1199–1216.
6. Wu., Q., Y. Liu., J. Zhang., and C. Tong. A Review of Recent Frozen Soil Engineering in Permafrost Regions along Qinghai-Tibet Highway, China. *Permafrost and Periglacial Processes*, Vol. 13, 2002, pp. 199–205.
7. Nixon, J. F. Discrete Ice Lens Theory for Frost Heave in Soils. *Canadian Geotechnical Journal*, Vol. 28, No. 6, 1991, pp. 843–859.
8. Li, N., B. Chen, F. Chen, and X. Xu. The Coupled Heat-Moisture-Mechanic Model of the Frozen Soil. *Cold Regions Science and Technology*, Vol. 31, No. 3, 2000, pp. 199–205.
9. Neaupane, K. M., and T. Yamabe. A Fully Coupled Thermo-Hydro-Mechanical Nonlinear Model for a Frozen Medium. *Computers and Geotechnics*, Vol. 28, No. 8, 2001, pp. 613–637.
10. Liu, Z., and X. Yu. Coupled Thermo-Hydro-Mechanical Model for Porous Materials under Frost Action: Theory and Implementation. *Acta Geotechnica*, Vol. 6, No. 2, 2011, pp. 51–65.
11. Liu, Z., Y. Sun, and X. Yu. Theoretical Basis for Modeling Porous Geomaterials under Frost Actions: A Review. *Soil Science Society of America Journal*, Vol. 76, No. 2, 2012, pp. 313–330.
12. Zhou, M. M., and G. Meschke. A Three-Phase Thermo-Hydro-Mechanical Finite Element Model for Freezing Soils. *International Journal for Numerical and Analytical Methods in Geomechanics*, Vol. 37, No. 18, 2013, pp. 3173–3193.
13. Zhang, Y., and R. L. Michalowski. Thermal-Hydro-Mechanical Analysis of Frost Heave and Thaw Settlement. *Journal of Geotechnical and Geoenvironmental Engineering*, Vol. 141, No. 7, 2015, p. 04015027.
14. Rabin, Y., and P. S. Steif. Thermal Stresses in a Freezing Sphere and its Application to Cryobiology. *Journal of Applied Mechanics*, Vol. 65, 1998, pp. 328–333.
15. Brownell, D. H. Jr, S. K. Garg, and J. W. Pritchett. Governing Equations for Geothermal Reservoirs. *Water Resources Research*, Vol. 13, 1977, pp. 929–934.
16. Kurashige, M. Thermal Stresses of a Fluid-Saturated Poroelastic Hollow Cylinder. *JSME International Journal Series I*, Vol. 35, No. 4, 1992, pp. 386–391.
17. Nishimura, S., A. Gens, S. Olivella, and R. J. Jardine. THM-Coupled Finite Element Analysis of Frozen Soil: Formulation and Application. *Géotechnique*, Vol. 59, No. 3, 2009, pp. 159–171.
18. Fenton, G. A., and D. V. Griffiths. Bearing Capacity Prediction of Spatially Random c-φ Soils. *Canadian Geotechnical Journal*, Vol. 40, No. 1, 2003, pp. 54–65.
19. Griffiths, D. V., G. A. Fenton, and N. Manoharan. Undrained Bearing Capacity of Two-strip Footings on Spatially Random Soil. *International Journal of Geomechanics*, Vol. 6, No. 6, 2006, pp. 421–427.
20. Griffiths, D. V., J. Huang, and G. A. Fenton. Probabilistic Infinite Slope Analysis. *Computers and Geotechnics*, Vol. 38, No. 4, 2010, pp. 577–584.

21. Bharrucha-Reid, A. T. ed. *Probabilistic Methods in Applied Mathematics*. Academic Press, New York, NY, 1968.
22. Sobczyk, K. *Wave Propagation in Random Media*. Elsevier, New York, NY, 1985.
23. Chamis, C. C. Probabilistic Structural Analysis Methods for Space Propulsion System Components. *Probabilistic Engineering Mechanics*, Vol. 2, No. 2, 1987, pp. 100–110.
24. Fenton, G. A., D. V. Griffiths, and M. B. Williams. Reliability of Traditional Retaining Wall Design. *Géotechnique*, Vol. 55, No. 1, 2005, pp. 55–62.
25. Andersland, O. B., and B. Ladanyi. *Frozen Ground Engineering*. Wiley, Hoboken, NJ, 2004.
26. Chang, C. S., and P. Y. Hicher. An Elasto-Plastic Model for Granular Materials with Microstructural Consideration. *International Journal of Solids and Structures*, Vol. 42, 2005, pp. 4258–4277.
27. Ferber, V., J. Auriol, J. Magnan, Y. Cui, E. De Laure, and C. Gerente. A Microstructural Model for the Volume Changes of Unsaturated Silty Soils Due to Wetting. *Proceedings of the 4th International Conference on Unsaturated Soils*, Lisbon, Portugal, 2006, pp. 861–872.
28. Helliwell, J. R., C. J. Sturrock, K. M. Grayling, S. R. Tracy, R. J. Flavel, I. M. Young, W. R. Whalley, and S. J. Mooney. Applications of X-ray Computed Tomography for Examining Biophysical Interactions and Structural Development in Soil Systems: a Review. *European Journal of Soil Science*, Vol. 64, 2013, pp. 279–297.
29. Tracy, S. R., K. R. Daly, C. J. Sturrock, N. M. J. Crout, S. J. Mooney, and T. Roose. Three-Dimensional Quantification of Soil Hydraulic Properties Using X-ray Computed Tomography and Image-Based Modeling. *Water Resources Research*, Vol. 51, No. 2, 2015, pp. 1006–1022.

The Standing Committee on Seasonal Climatic Effects on Transportation Infrastructure (AFP50) peer-reviewed this paper (18-06039).

Sage Publications USA
sagepub.com

CURRAN ASSOCIATES INC.
proceedings.com

ISBN 978-1-5108-8257-7
9 781510 882577